The Emergence of Pressure Blade Making

Pierre M. Desrosiers
Editor

The Emergence of Pressure Blade Making

From Origin to Modern Experimentation

Springer

Editor
Pierre M. Desrosiers
Ste-Catherine 360–4150, H3Z 2Y5
Westmount, Québec, Canada

ISBN 978-1-4614-2002-6 e-ISBN 978-1-4614-2003-3
DOI 10.1007/978-1-4614-2003-3
Springer New York Dordrecht Heidelberg London

Library of Congress Control Number: 2012931928

Printed on acid-free paper

Springer is part of Springer Science+Business Media (www.springer.com)

Foreword

The initial conception of this book occurred during a session at the congress of the International Union of Prehistoric and Protohistoric Sciences held in Lisbon, Portugal, back in September of 2006 and represents an ambitious undertaking covering over 20,000 years of human evolution across nearly all continents. The focal point of the works included herein is pressure blade production,[1] and one will note a plethora of significant archaeological data that corresponds to a diverse range of cultural and environmental contexts spanning various chronological periods.

This session was held in honour of Jacques Tixier and Marie-Louise Inizan, two key pioneering figures in the study of lithic technology, specifically in the application of pressure techniques. While this project was originally to be organized by Noura Rahmani and myself, Noura ended up changing careers which placed me solely at the helm. The impetus for this collaborative undertaking evolved over the subsequent years within the context of a larger project comprised of contributions from both prominent and emerging archaeologists. Three themes are brought forth in the following chapters and include: the history of research in the domain of pressure techniques, syntheses concerning different time periods and geographic areas as well as experimental works. This book presents a collection of the most recent research on the topic and has demanded a great deal of patience from all participants. I sincerely hope the wait was worth it.

The reader will notice that a large number of contributors to this book are not native English speakers, making this a unique opportunity for readers to have access to the data from other countries. This was made possible thanks to the members of an external review committee who invested considerable effort. These members include in alphabetical order: Adrian Burke (Université de Montréal), Tristan Carter (McMaster University), Jenneth Curtis (Parks Canada), Max Friesen (University of Toronto), Lynda Gullason, Raymond Le Blanc (University of Alberta), Susan Lofthouse (Avataq Cultural Institute), David Lubell[2] (University of Waterloo),

[1] Blade production is applied in this instance as a general term and includes bladelets and microblades.

[2] David Lubell was initially selected as a member of the external committee and, later on, ended up contributing to the book.

Michael Shott (University of Akron), Farina Sternke (University of Glasgow), Jonathan Thomas (University of Iowa) and Lucy Wilson (University of New Brunswick). I would like to extend my sincerest gratitude for the collaborative efforts of these great colleagues and friends of mine. Furthermore, review and preparation of the final manuscript was made possible with the help of David Howard and Benjamin Patenaude, to whom I am indebted. However, it has to be said that the final review of the content and English has been the responsibility of each contributor.

Contents

Part III Recent Advances in Experimentation

Contributors

Ciler Altınbilek-Algül Department of Prehistory, University of Istanbul, Istanbul, Turkey

Laurence Astruc Du vilage à l'Etat au Proche et Moyen-Orient, Université Paris Ouest Nanterre, Université Paris I, UMR 7041, ArScan, France

Didier Binder Université Nice Sophia-Antipolis, CNRS, UMR 6130, CEPAM, Nice, France

Frédérique Brunet CNRS-UMR 7041 ArScAn, équipe Asie centrale, Maison René Ginouvès de l'Archéologie et de l'Ethnologie, Nanterre, France

Jacques Chabot CELAT (Laboratoires d'archéologie), Université Laval, Québec, Canada

John E. Clark Department of Anthropology, Brigham Young University, Provo, UT, USA

Carmine Collina Université Nice Sophia-Antipolis, CNRS, UMR 6130, CEPAM, Nice, France

Yan Axel Gómez Coutouly Prehistory and Technology Laboratory, University of Paris 10, Paris, France

Kim Darmark Åland Islands Museum, Mariehamn, Åland Finland

Véronique Darras Archéologie des Amériques, CNRS-Université Paris 1, Nanterre, France

Pierre M. Desrosiers Avataq Cultural Institute, Westmount, Québec, Canada

Oreto Garcia-Puchol Prehistory Department, Universitat de València, València, Spain

Raphaëlle Guilbert Université Nice Sophia-Antipolis, CNRS, UMR 6130, CEPAM, Nice, France

Kenneth G. Hirth Department of Anthropology, Penn State University, University Park, PA, USA

Marie-Louise Inizan Préhistoire et technologie, Université Paris Ouest, CNRS, UMR 7055, Nanterre, France

Peter Kelterborn Rainstrasse 372, 8706 Meilen, Switzerland

David Lubell Department of Anthropology, University of Waterloo, Waterloo, ON, Canada

Antonio Morgado Departamento Prehistoria y Arqueología, Universidad de Granada, Campus universitario de Cartuja s/n; E- 18071 Granada, Spain

Jacques Pelegrin Laboratoire "Préhistoire et Technologie", CNRS et Université Paris Ouest Nanterre, MAE, Nanterre, France

Thomas Perrin Université Le Mirail, Toulouse, France

Noura Rahmani Montréal, Québec, Canada

Mikkel Sørensen Department of Archaeology, University of Copenhagen, Copenhagen, Denmark

Andrei V. Tabarev Division of Foreign Archaeology, Institute of Archaeology and Ethnography, Novosibirsk, Russia

Jun Takakura Archaeological Research Center, Hokkaido University, Sapporo, Hokkaido, Japan

Part I
History of Research

Chapter 1
Introduction: Breaking Stones Without Striking Them

Pierre M. Desrosiers

Human cultural development was a long and steady process in which stone tool manufacture was a fundamental element. With the improvement of lithic technology, humans were able to exploit their environment with greater efficiency as well as colonize and subsist in new territories, innovate, and develop new ways of life. Certain innovations are well documented and include the control of fire, the domestication of plants and animals, as well as the manufacture of ceramics, and the development of writing. Some were invented in different locations and spread by cultural contact and diffusion, while others were invented at different points in time by separate human groups and ultimately became chrono-cultural markers. Furthermore, some of these phenomena have been intensely studied, while others have yet to be thoroughly investigated.

This book seeks to fill part of this information gap. The phenomena in question here are the invention, diffusion, and adoption or reinvention of pressure blade making.[1] The adoption of this production technique corresponds to the exploitation of new environments as well as the appearance of other phenomena such as the Neolithic way of life. Through meticulous experimentation, archaeologists have greatly enhanced their ability to identify pressure blades, and it is now possible to provide a global overview of our current knowledge regarding this technological breakthrough on a global scale.

A brief introduction on the origins of research on pressure blade production will help the reader understand the context in which this book takes place. In doing so, we wish to point out some of the most influential publications on this topic.

---

[1]The term blade is used here without any consideration of size and thus includes bladelets and microblades.

P.M. Desrosiers (✉)
Avataq Cultural Institute, Westmount, QC, Canada
e-mail: servicearch@avataq.qc.ca

P.M. Desrosiers (ed.), *The Emergence of Pressure Blade Making: From Origin to Modern Experimentation*, DOI 10.1007/978-1-4614-2003-3_1,

1.1 The Study of Pressure Techniques

The term "technology" is more commonly interpreted as applied technology, and the first image that comes to mind is anything related to modern technological advances, such as a sophisticated new machine. Nevertheless, for most of the specialists involved in this volume, (either consciously or unconsciously) lithic technology refers not only to a topic of research but to the science of studying stone tool production. This is a social science aimed at understanding the evolution of certain human behaviors by studying the remnants of their production activities. It could be placed within a broader trend in research that connects human behavior to the production of material culture (i.e., Haudricourt 1964; Haudricourt and de Garine 1968; Inizan et al. 1995; Leroi-Gourhan 1943, 1945, 1964, 1965).

Over 99% of what we know about the long history of human evolution is represented by worked stone material; it is no surprise that lithic technology is important to archaeologists (Crabtree 1975). If we consider archaeology's early years when the antiquity of man was being realized, the interest in understanding flintknapping was centrally important. The most accessible examples were those of the gunflint knappers (i.e., Barnes 1937; Clarke 1935; de Mortillet 1908; Schleicher 1927) and a few rare ethnographic cases, often based on secondhand accounts (e.g., Fowke 1891; Holmes 1919; Nelson 1899; Nelson 1916), as well as unique individuals such as Flint Jack, who reproduce antiquities for a living (Vayson de Pradenne 1932). When pressure techniques were first investigated, just how a stone could be broken without being struck and by using softer materials such as antler was not apparent. The most acute observations were probably those made from the stone work of Ishi since it was, after all, a firsthand observation (Heizer and Kroeber 1979; Kroeber 1961; Nelson 1916; Pope 1913).

Holmes (1919: 304–329) had summarized ethnographic data regarding "pressure fracture processes" in a 26-page chapter which probably stands as the first modern research devoted to the topic. His important book probably constitutes the beginning of research on the pressure techniques and other related aspects of lithic technology. If pressure bifacial flaking was readily understood, pressure blade production proved to be a different case however. It seems that the only information about this type of production came from Mesoamerica (Holmes 1919: 322–323).

These accounts had an important impact on the first organized experiments, especially for the use of the T-shaped tool, and the now famous article about Mesoamerican prismatic blades by Crabtree (1968). Crabtree's paper likely constitutes the second most important work to be published on the subject of pressure blade production, and it has inspired a whole generation of archaeologists that investigate this topic. For example, experimentation has since proven that the Mesoamerican pressure tool had a different shape and the core could effectively be held between the feet (Clark 1982).

A third important contribution to the study of pressure blade production is the book *Préhistoire de la pierre taillée 2, économie du débitage laminaire* (CREP 1984). It contains different papers, among which some represent a major breakthrough in our understanding of pressure blade production in the Ancient World and in the development of pressure blade experimentation. It has been a starting point for many of the contributors to this book, and now some of them are presenting their most recent work on this topic in the present volume (Chaps. 2, 7, and 18). Among them, Inizan (Chap. 2) provides insights into the development of research on pressure blades in the Ancient World as well as tracking the most recent evidence for the earliest pressure blade production.

A fourth important reference is *Mesoamerican Lithic Technology, Experimentation and Interpretation* edited by Hirth (2003). This book not only contains the most accurate experimentation results on the subject but places an emphasis on the social importance of craft specialization as well. The social aspects of pressure blade production are now better understood and are the focus of Chaps. 16 and 17 in this volume. A complete history of experimentation applied to the understanding of Mesoamerican prismatic blades can also be found in Chap. 3.

Finally, this volume is intended to compliment the above series of reference works aimed at better documenting the phenomena of pressure blade production, and is the result of a collaborative effort that began in 2006 during a session of the UISPP conference in Lisbon, Portugal. The idea for this session was conceived by Noura Rahmani, who had recently completed a significant research project on the first appearance of pressure blade production in North Africa (Rahmani 2003, 2004; Chap. 4, this volume), by defining a more precise chronological framework for the introduction of pressure in the Maghreb. The notion was put forth that this technique could have either been the result of invention or diffusion. Investigating this hypothesis requires a better understanding of the overall situation regarding pressure blade production in time and space as well as more extensive experiments.

1.2 Contributions in Honor of Tixier and Inizan

Nearly 50 years ago, a meeting between French and American archaeologists working on lithic technology spurred the dawn of a new era (Jelinek 1965; Smith 1966; also see Chap. 2, this volume). If we attribute the identification of pressure blade production to Donald Crabtree and Jacques Tixier in the New and Old Worlds, respectively, we certainly owe our gratitude to Marie-Louise Inizan for having taken this realization to a new level, as will be evident upon reading Chap. 2.

Experimentation has been a key element in the recognition of pressure techniques and has led to the development of modern approaches to lithic technology. Tixier developed a particular approach in France that focused on the development of a terminology permitting the description of this technology in spite of its dynamism (Tixier 1967, 1972, 1978; Tixier et al. 1980). As the cofounder of this new approach, Inizan continuously updated it, thereby making it accessible to a broader audience (Inizan et al. 1999; Inizan et al. 1995). She also tracked pressure technique across the world, including those from North Africa, Middle East, Asia, and North America (e.g., Inizan 1976, 1984; Inizan et al. 1992). This is why the contributions to this book are set up to honor the work of Tixier and Inizan. Both are key researchers who may be considered, along with Crabtree, as the most significant contributors to the development of modern research on pressure blade production. All three have been a major source of inspiration in guiding the careers of many researchers, including many of the authors who contributed to this book.

1.3 Pressure Blade Making: From Origin to Modern Experimentation

As the title dictates, the major research theme in this book concerns the emergence, diffusion, and experimentation relating to pressure blade making. Furthermore, these themes cover a wide range of topics, among them: handling the core, core stabilization, body techniques, craft specialization, diffusion, invention, adoption, product size, know-how, knowledge, identification, and characteristics, as well as the link between pressure blade production and biface pressure flaking.

This book has been divided into three parts, with the chapters arranged according to the following themes: (1) history of research, (2) pressure blade production around the world, and (3) the recent advances in experimentation. Some chapters may have legitimately fit into different sections of the book; however, they were organized according to their primary focus.

The first part of the book is devoted to the history of research and includes two chapters. Chapter 2, by Inizan, is an overview of research developed in the Ancient World and includes the most accurate information about the problem of the origins and invention of pressure blade production. Chapter 3, by Clark, constitutes an extensive overview of the development of research in Mesoamerica and of experiments in pressure blade making. Additionally, in Chap. 18, Pelegrin provides numerous insights with regard to the development of experimentation for the understanding of pressure blade production.

The second part of the book deals with pressure blade production around the world, or more specifically every region where it has been identified. The aim of each chapter is to provide an accurate account of the current knowledge for a given region in terms of the origins, development, and abandonment of pressure blade production. These accounts are inspired by the different traditions of research and the particular interests that have influenced archaeologists working in different areas.

Researchers in Mesoamerica and Maghreb have played an important role in the discovery of pressure blade production. Following Crabtree's contribution, prismatic blade production practically became a topic of research in and of itself in Mesoamerica (Hirth 2003). Due to the intensity of the research and the influence of the American Anthropological approach, social implications linked with the development of pressure techniques were emphasized, particularly with regard to craft specialization, in the evolution of Mesoamerican societies (Chaps. 3, 16, and 17). These elements, along with specific core preparation methods and other aspects of lithic technology, have also been better documented by Darras, as presented in Chap. 17.

In the Maghreb, Tixier and Inizan initiated research on pressure blade production (Inizan 1976, e.g., 1984; Tixier 1967). They developed a specific approach to lithic technology that was influenced by Leroi-Gourhan and psychological approaches focusing on the artisan (Desrosiers 2007). Following this initial activity, research into pressure blade production in the Maghreb has progressed at a slower pace and does not focus on the identification of pressure techniques, with the exception of one study (cf. Sheppard 1987). More recently, Rahmani has brought new life to this topic in North Africa, and the reader will find more of her results in her collaborative work with Lubell (Chap. 4). They present their research from Kef Zoura D (Algeria) and discuss the implications of understanding the circumstances surrounding the adoption of pressure blade production in North Africa.

Following the earliest research in Mesoamerica and in the Maghreb, researchers have progressively identified pressure blade production in many areas of the Ancient World. In the Near East and the Caucasus, the use of pressure techniques for blade production was adopted at the beginning of the Neolithic and progressively evolved until a lever was adopted and employed in their production (Chaps. 5 and 6). In Europe, the adoption of pressure blade production preceded the Neolithic in most regions (Chaps. 7 and 9); however, it became increasingly widespread at the onset of the Neolithic, as demonstrated by the evidence from the Southern Iberian Peninsula (Chap. 8).

Giving particular consideration to Asia, Europe, and the Near East, Darmark addresses the question of "surface pressure flaking" versus pressure blade production in Chap. 10. This work, however, remains inconclusive and brings to light further unanswered questions. Considering the possible relationship between the developments of both aspects is certainly innovative, and it clearly points out a direction for future research.

The first manifestations of pressure blade techniques are found in Asia, as described by Inizan in Chap. 2. More specifically, in the same chapter, Inizan considers the fact that Japan and Korea are now known to contain some of the earliest manifestations of pressure blade manufacture. Detailed information about Japan can be found in Chap. 11 by Takakura, while in Chap. 12 Brunet addresses the question of diffusion or innovation in Central Asia, and Tabarev (Chap. 13) gives an overview of the situation in Russian Far East (Eastern Siberia). With regard to its introduction in North America, Gomez Coutouly presents the context in which pressure blade production emerged in Alaska (Chap. 14), while Desrosiers and Sørensen discuss its diffusion from Alaska to Greenland (Chap. 15).

Part 3 of this book includes two important chapters on improving upon our knowledge acquired through experimentation. In Chap. 18, Pelegrin goes far beyond the simple recognition of the technique of detachment. This new experimental study of pressure blade production greatly improves our understanding of the phenomena of invention and diffusion of pressure techniques. Controlled experiments distinguish the work of Kelterborn (Chap. 19) and provide information toward the systematic study of the mechanics of pressure techniques.

1.4 Toward a Global Understanding of Pressure Techniques

Throughout the course of compiling this book, it had been tempting to propose our conclusions on the invention, diffusion, and reinvention of pressure blade techniques. However, these would not have necessarily been shared by all contributors. Rather, the aim of the book was more to compile a large amount of information on the topic and present the phenomenon as a whole on a global scale. This approach demonstrates the different stepping stones that have emerged in time and space, through carrying out many meticulous experiments and data collection. Over the course of humankind's evolution, many of these stones are still missing which makes stepping between them all the more hazardous. It is to be expected, however, that some of the best architects among us will undertake the task of framing a theoretical bridge that could possibly span these gaps. More importantly, it could be expected that archaeologists will undertake the task of unearthing new stepping stones, the emergence of which will lead us down new paths which follow pressure techniques within the evolutionary framework of humankind.

References

Barnes, A. S. 1937 L'industrie des pierres à fusil par la méthode anglaise et son rapport avec le coup de burin tardenoisien. *Bulletin de la Société préhistorique française* 34(7–8): 328–335.

Clark, J.E. 1982 Manufacture of Mesoamerican Prismatic Blades: An Alternative Technique. *American Antiquity* 47: 355–376.

Clarke, R. 1935 The Flint-Knapping Industry at Brandon. *Antiquity* 9(33): 38–56.

Crabtree, D.E. 1968 Mesoamerican Polyhedral Cores and Prismatic Blades. *American Antiquity* 33:446–478.

Crabtree, D.E. 1975 Comments on Lithic Technology and Experimental Archaeology. In *Lithic Technology. Making and Using Stone Tools*, edited by E. J. Swanson, pp. 105–114. World Series in Anthropology. Mouton Publishers.

CREP (Cercle de recherches et d'études préhistoriques) 1984 *Préhistoire de la pierre taillée 2, économie du débitage laminaire. IIIe table ronde de la technologie lithique, Meudon-Bellevue, octobre 1982*. CNRS, Meudon-Bellevue.

de Mortillet, A. 1908 Les pierre à fusil, leur fabrication en Loir-et-Cher. *Revue de l'école d'Antrhopologie* XVIII.

Desrosiers, P. 2007 Palaeoeskimo Lithic Technology: Constraints and Adaptation. *Lithic Technology* 32(1): 17–38

Fowke, G. 1891 *Stone Art*. Annual Report of the Bureau of Ethnology of the Secretary of the Smithsonian Institution 13. Smithsonian Institution, Washington.

Haudricourt, A.-G. 1964 La technologie, science humaine. *La Pensée* 115: 28–35.

Haudricourt, A.-G., and I. de Garine 1968 La technologie culturelle: essai de méthodologie, edited by J. Poirier, pp. 731–822. Gallimard, Paris.

Heizer, R.F. and T. Kroeber 1979 *Ishi the Last Yahi. A Documenty History*. University of California Press, Berkeley.

Hirth, K.G. (editor) 2003 *Mesoamerican Lithic Technology. Experimentation and Interpretation*. The University of Utah Press, Salt Lake City.

Holmes, W.H. 1919 *Handbook of Aboriginal American Antiquities, Part I, Introductory, The Lithic Industries*. Bureau of American Ethnology Bulletin 60. Smithsonian Institution, Washington.

Inizan, M.-L. 1976 Nouvelle étude des industries lithiques du Capsien d'après les collections de R. Vaufrey conservées à l'IPH à Paris, Université Paris 10, Nanterre

Inizan, M.-L. 1984 Débitage par pression et standardisation des supports : un exemple capsien au Relilaï (Algérie). In *Préhistoire de la Pierre Taillée 2. Économie du débitage laminaire*, edited by CREP, pp. 85–92. CNRS, Meudon-Bellevue.

Inizan, M.-L., M. Lechevallier, and P. Plumet 1992 A technological marker of the penetration into North America: Pressure microblade debitage. Its origin in the Palaeolithic of North Asia and its diffusion. In *Materials issues in art and archaeology III: symposium held 27 April-1 May 1992, San Francisco* edited by P.B. Vandiver, J.R. Druzik, G.S. Wheeler and I.C. Freestone, pp. 661–681 Materials Research Society Symposium Proceedings Vol. 267. Materials Research Society, Pittsburgh

Inizan, M.-L., M Reduron-Balinger, H. Roche, and J. Tixier 1999 *Technolgy and Terminology of Knapped Stone*. Translated by J. Féblot-Augustins. Préhistoire de la Pierre Taillée 5. CNRS, Nanterre.

Inizan, M.-L., M. Reduron, H. Roche, and J. Tixier 1995 *Technologie de la pierre taillée*. Préhistoire de la Pierre Taillée 4. CNRS, Université Paris 10, Meudon-Nanterre.

Jelinek, A.J. 1965 Lithic Technology Conference, Les Eyzies, France. *American Antiquity* 31(2): 277–279.

Kroeber, T. 1961 *Ishi in Two Worlds. A Biography of the Last Wild Indian in North America.* University of California Press, Berkeley.

Leroi-Gourhan, A 1943 *Évolution et techniques. Tome I: L'homme et la matière.* Albin Michel, Paris.

Leroi-Gourhan, A. 1945 *Évolution et techniques. Tome II: Milieu et techniques.* Albin Michel, Paris

Leroi-Gourhan, A. 1964 *Le geste et la parole. Tome I: Technique et language.* Albin Michel, Paris.

Leroi-Gourhan, A 1965 *Le geste et la parole. Tome II: La mémoire et les rythmes.* Albin Michel, Paris.

Nelson, E.W. 1899 *The Eskimo About Bering Strait.* Annural Report of the Bureau of American Ethnology 18. Smithsonian Institution, Washington.

Nelson, N.C. 1916 Flint Working by Ishi. In *Holmes Anniversary Volume: Anthropological Essays Presented to William Henry Holmes in Honor of His Seventieth Birthday, December 1, 1916* edited by F. W. Hodge, pp. 397–402. J. W. Bryan, Washington DC.

Pope, S.T. 1913 Making Indian Arrow Heads. *Forest and Stream* 81: 103–152.

Rahmani, N. 2003 *Le Capsien typique et le Capsien supérieur. Évolution ou contemporanéité, les données technologiques.* Cambridge Monographs in African Archaeology 57. Archaeopress, Oxford.

Rahmani, N. 2004 Technological and Cultural Change Among the Last Hunter-Gatherers of the Magreb: The Capsian (10,000-6000 B.P.). *Journal of World Prehistory* 18(1): 57–105.

Schleicher, C. 1927 Une industrie qui disparaît, la taille des silex modernes (pierres à fusil et à briquet). *Bulletin de la Société préhistorique française* 24(10): 367–369.

Sheppard, P.-J. 1987 *The Capsian of North Africa. Stylistic Variation in Stone Tool Assemblage.* B.A.R. International Series 353. B.A.R., Oxford.

Smith, P.E.L. 1966 Report on the lithic technology conference at Les Eyzies, France, 1964. *Current Anthropology* 7(5): 592–593.

Tixier, J. 1967 Procédés d'analyse et questions de terminologie dans l'étude des ensembles industriels du Paléolithique récent et de l'Épipaléolithique en Afrique du Nord-Ouest.. In *Background to Evolution in Africa*, edited by W. W. Bishop and J. D. Clark, pp. 771–820. Chicago Press, Chicago.

Tixier, J. 1972 Obtention de lames par débitage "sous le pied". *Bulletin de la Société préhistorique française* 69: 134–139.

Tixier, J. 1978 Méthode pour l'étude des outillages lithiques: notice sur les travaux scientifiques de J. Tixier, Université de Paris 10. Nanterre.

Tixier, J., M.-L. Inizan, and H. Roche 1980 *Préhistoire de la pierre taillée - 1. terminologie et technologie.* CNRS, Paris.

Vayson de Pradenne, A. 1932 *Les fraudes en archéologie préhistorique.* Emile Nourry, Paris.

Chapter 2
Pressure *Débitage* in the Old World: Forerunners, Researchers, Geopolitics – Handing on the Baton

Marie-Louise Inizan

2.1 Introduction

We are indebted to Don Crabtree (1968) for the recognition of pressure *débitage*,[1] which endured into the Historic Period among the Aztecs in Mesoamerica. The leading pioneer in the field of replicative experimentation, Crabtree labored for nearly 10 years in the 1950s to obtain systematically regular blades with parallel edges and a constant thickness, similar in character to the Aztec obsidian blades, which were detached from characteristic fluted cores. He immediately saw a parallel between this type of *débitage* and some Paleoindian technologies from North America. Since then, a large amount of research, comprehensively summarized by J. Clark (2003), has been devoted to this subject in the Americas.

This chapter endeavors to recount how pressure *débitage* came to be recognized in the Old World before exploring the implications that ensued. One of the major issues to be addressed is the part this particular blade *débitage* technique played in the identification (or definition) of specific prehistoric cultures ranging from Upper Paleolithic hunter-gatherers to Neolithic agropastoralists in several geographical regions spanning from the Far East to Europe.

[1] The original French meaning of *débitage* is used in this chapter when referring to the production of blanks. On the other hand, the English definition of this French word refers to the waste flakes.

M.-L. Inizan (✉)
Préhistoire et technologie, Université Paris Ouest, CNRS, UMR 7055,
21 Allée de l'Université, 92023 Nanterre, France
e-mail: marie-louise.inizan@mae.u-paris10.fr

P.M. Desrosiers (ed.), *The Emergence of Pressure Blade Making: From Origin to Modern Experimentation*, DOI 10.1007/978-1-4614-2003-3_2,

2.2 The Recognition of Pressure *Débitage* in the Old World

The cornerstone of this research was laid by J. Tixier when he identified pressure *débitage* in the Upper Capsian, an Epipaleolithic culture of the Maghreb dated 9500–5500 B.P. (Rahmani 2004: 89). It is due to J. Tixier's work that research on pressure *débitage* became so prominent for the investigation of prehistoric societies. He favored the use of experimentation as a methodological tool for the purpose of "seeking intentions" (1978: 67), regardless of the technique under investigation.

2.2.1 The Lithic Technology Symposium of Les Eyzies (France)

Although it has never been published, this 1964 symposium held in France (Jelinek 1965) is still viewed as a seminal event in many respects (Fig. 2.1). Not only did it reveal a technique unknown to prehistorians, but it also highlighted the need for

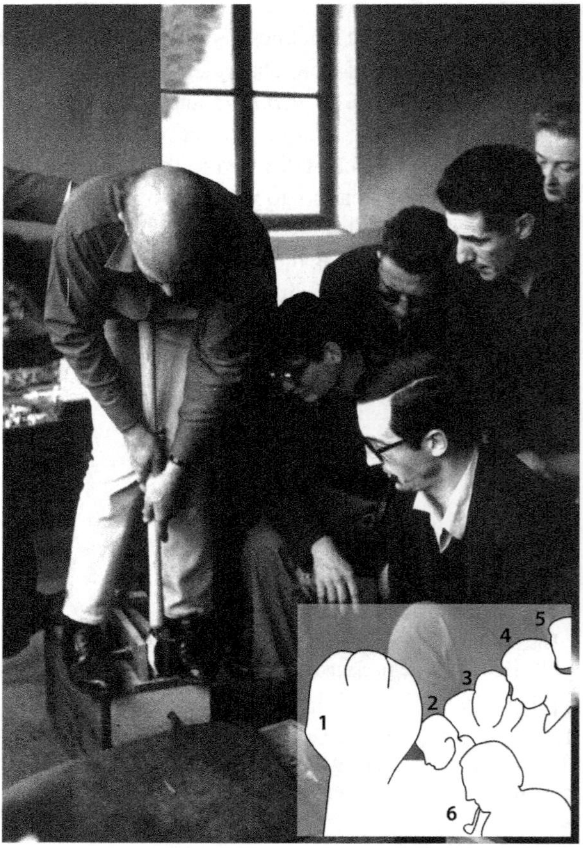

Fig. 2.1 The 1964 Symposium of Les Eyzies: Don Crabtree pressure flakes obsidian in front of H. Irwin (2), J. Epstein (3), J. Tixier (4), M. Wormington (5), J.-Ph. Rigaud (6)

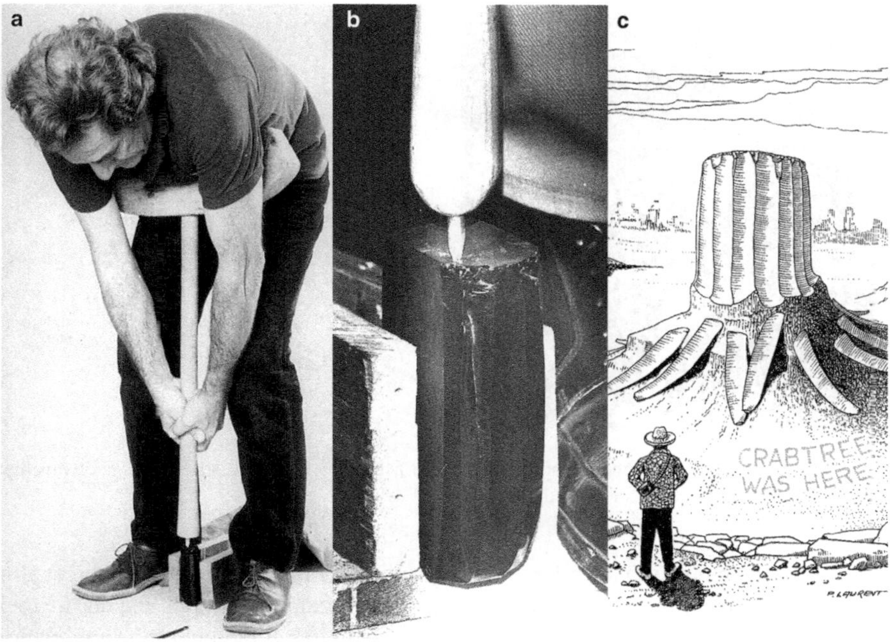

Fig. 2.2 (**a**) J. Tixier pressure flakes obsidian using a copper-tipped pectoral crutch. (**b**) Detail view of the core. (**c**) Drawing by P. Laurent

the use of experimentation to "decipher" prehistoric detachment techniques. It was during this symposium that "we were introduced to the world of pressure *débitage* and retouch, which the French had only just begun to investigate" (Tixier 1978: 25; translated from the original). European researchers were also made aware of the advantages of heat-treating siliceous rocks for the first time. Once again, it was Crabtree who demonstrated that retouch removals detach more smoothly once the material has been subjected to well-controlled thermal treatment. This inspiration was a joint product of his observation of beautifully pressure retouched Paleoindian bifacial points (Folsom, Clovis, etc.), displaying removals "with greasy luster," and of his perusal of ethnographic documents (Crabtree and Butler 1964). Evidence for the use of heat treatment for pressure *débitage* extends as far back as the Upper Paleolithic (Inizan and Tixier 2001).

Crabtree demonstrated some obsidian blade production in front of a few prehistorians, two of whom, F. Bordes and J. Tixier, were experimental knappers themselves. The core was immobilized in a vice and the pressure applied with a copper-tipped pectoral crutch (Fig. 2.2). J. Tixier reflected that this *débitage* technique called to mind that used for bladelet *débitage* in the Upper Capsian of Algeria. As early as 1963 (p. 43), he had drawn a parallel between the "fluted" Mesoamerican cores and those of the Upper Capsian "whose section mimics that of a Doric column," thus defining them before discovering how they were produced. The first publication that mentions pressure *débitage* in the Upper Capsian dates back to 1971 (Tixier 1971: 122).

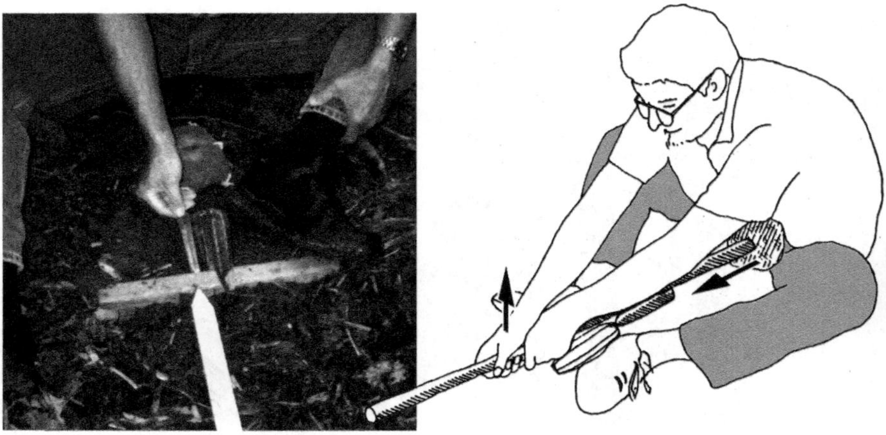

Fig. 2.3 J. E. Clark: experimental body position for achieving blade pressure debitage on obsidian according to the indications left by the Aztecs

It should be noted that Crabtree used obsidian for his experiments, whereas flint was the available raw material in the Capsian. Obsidian is a vitreous and elastic rock and therefore highly suitable for pressure blade making. It facilitates the application of a smaller amount of pressure than flint, and, as a result, the accuracy with which the blades are detached is enhanced. In the course of their trials, Crabtree and Tixier became aware of the varying degree of difficulty induced by the different properties of these two raw materials. In 1969, they successfully pressure-flaked flint for the first time.

2.2.2 Method and Technique

During a symposium in Austria in 1965, J. Tixier emphasized how important it is to distinguish between the terms "technique" and "method": "The technique is the physical means, the method is the intellect that marshals the means," and he added: "We reserve the term 'technique' for the material side of the process" (1967: 807; translated from the original). J. Pelegrin elaborated on this definition: "Techniques are the modes by which detachment is carried out. They always require the use of a least one tool, animated by a gesture made in a particular body position" (Pelegrin 1995: 20; translated from the original). Thus, what Crabtree had discovered was a technique, whereas it was thanks to the analysis of the Florentino codex, one of the major manuscripts dealing with Aztec obsidian productions and a text of which Crabtree had no knowledge, that the Aztec obsidian *débitage* method was reconstructed. The codex is composed of three texts: one which is pictographic, another which is written in Latin, and a third which is written in Spanish. We are indebted to J. Clark (1982) for analyzing the information in this codex and ultimately achieving pressure *débitage* by following the "directions" left by the Aztecs (Fig. 2.3).

2.2.3 Technology and Experimentation

Flowing from the concept of the "operational sequence" [*chaîne opératoire*], technology is a methodological approach to material culture, which was brilliantly developed by A. Leroi-Gourhan (1943, 1964). When applied to prehistoric stone industries, it highlights the importance of the logical study of detachment techniques to address the relationship between different and/or successive techno-complexes. Identifying production techniques is therefore a crucial step in the sequence. In this respect, it should be stressed that sound technical diagnoses rely mainly on experimental replication. However, it is always the observations made on archaeological material that serve as a reference frame.

Replicative experimentation began to develop in earnest in 1980. The themes addressed at the *tables-ronde de technologie lithique* conferences organized in 1980 at Tervuren (Belgium) and in 1982 at Meudon (France) were at variance with the typological traditions that still prevailed at the time in lithic assemblage studies (i.e., to describe, categorize, and compare). At the first conference entitled (*Tailler pour quoi faire?*), it was argued that experimentation, whose usefulness for deciphering production techniques had already been acknowledged, was also necessary for positioning archaeological documents "in a sequence of events extending from acquisition of raw material by prehistoric man to discovery by the prehistorian" (Cahen 1982: 9; translated from the original).

In 1982, the following conference entitled *Economie du débitage laminaire* included a reflection on blade *débitage* and on the need to identify *débitage* strategies as part of a more precise cultural approach. In addition, the issue of pressure *débitage* in the Old World was addressed for the first time (Tixier 1984: 57–70). As a matter of fact, the first identifications of this technique in a variety of chronological contexts across the Old World, e.g., in the Epipaleolithic in the Maghreb (Tixier 1976), in the Mesolithic in Denmark (Callahan 1985), in the Neolithic in Mesopotamia (Inizan 1985, 1986), in the Early Neolithic in Greece (Perlès 1984, 1987), and in the Chasséen in France (Binder 1984), made it clear that this technical phenomenon played a more significant role in prehistoric lithic production than previously thought (Binder and Perlès 1990). It was therefore important to detect its presence in archaeological assemblages through the identification of its associated regular and standardized products. The existence of this blade *débitage* technique opened up new areas of research in the field of lithic technology, particularly for assemblages dating to the Holocene period.

Another peculiarity of pressure *débitage* is that it is exclusively used for the detachment of blade products (blades, bladelets, or microblades), generally during the phase of *plein débitage* (i.e., main blank production phase). Other technical possibilities exist for shaping out the core or, indeed, for executing the initial stage of blade *débitage*. Admittedly, a choice is made based on tradition: "Man has always been granted a wide range of options, he his free to choose and to stand by his choice" (Tixier 1978: 6; translated from the original).

The initial recognition of the technique was but the first step. In fact, the numerous methods used in conjunction with the pressure technique are still being identified in archaeological assemblages and need to be further investigated. While the technique is clearly understood, lithic analysis consistently shows how prolific and complex the methods in the assemblages are, thereby revealing identifiable cultural idiosyncrasies.

Experimentation rapidly proved very useful for the quantification of productivity (Pelegrin, Tixier) and for highlighting both the critical importance of raw material homogeneity and the need for accuracy during the shaping out of the core.

Numerous contributions to the study of several techniques were made by J. Pelegrin during a series of experiments in flintknapping. Pelegrin equipped his pectoral crutch with an antler point and immobilized the core with materials that would have been available to prehistoric knappers (1984: 105–127). He went on to explore the bodily constraints involved in the detachment of "minute to outsize products" before identifying lever-aided *débitage* (Pelegrin 1988). Following the description of the diagnostic criteria for the recognition of the pressure *débitage* produced with a metal point (Pelegrin 1994; Inizan et al. 1994), it was possible to infer the use of such metal points, even in their absence in archaeological material.

Last but not least, Pelegrin's research has had a major bearing on the ability to deduce the technical processes involved in the manufacture of archaeological stone objects in the Old World, resulting in a wealth of cultural interpretations (Pelegrin 2002, 2003, and this volume).

2.3 The Significance of the Identification of Pressure *Débitage* in the Capsian

The diagnostic technological criteria were published together with the study of the Upper Capsian industries of the Aïn Dokkara (Algeria) for the first time in 1976 (Tixier 1976).

Carried out as part of a doctoral thesis (Inizan 1976), a new approach to the Capsian lithic assemblages recovered by R. Vaufrey during excavations conducted in the 1930s opened up hitherto unexplored lines of research pertaining to *débitage* strategies and the interdependence of technical systems. Stored in Paris at the Institut de Paléontologie Humaine, these assemblages enabled R. Vaufrey to define the Capsian as the Epipaleolithic culture of the Maghreb. It is characterized by two cultural traditions: the Typical Capsian and the Upper Capsian. Both were present in the stratigraphy of the site The Relilaï in Algeria (Vaufrey 1933a, b).

At the time of excavation, the recovery of retouched tools was favored, to the detriment of *débitage* products, which were regarded as mere waste and perceived to lack any archaeological value. In spite of the absence of the bulk of the *débitage* products, with the exception of a few blade products and some characteristic waste products such as burin spalls, striking or pressure platform rejuvenation flakes, and thanks to a number of technical characteristics recognized under the supervision of

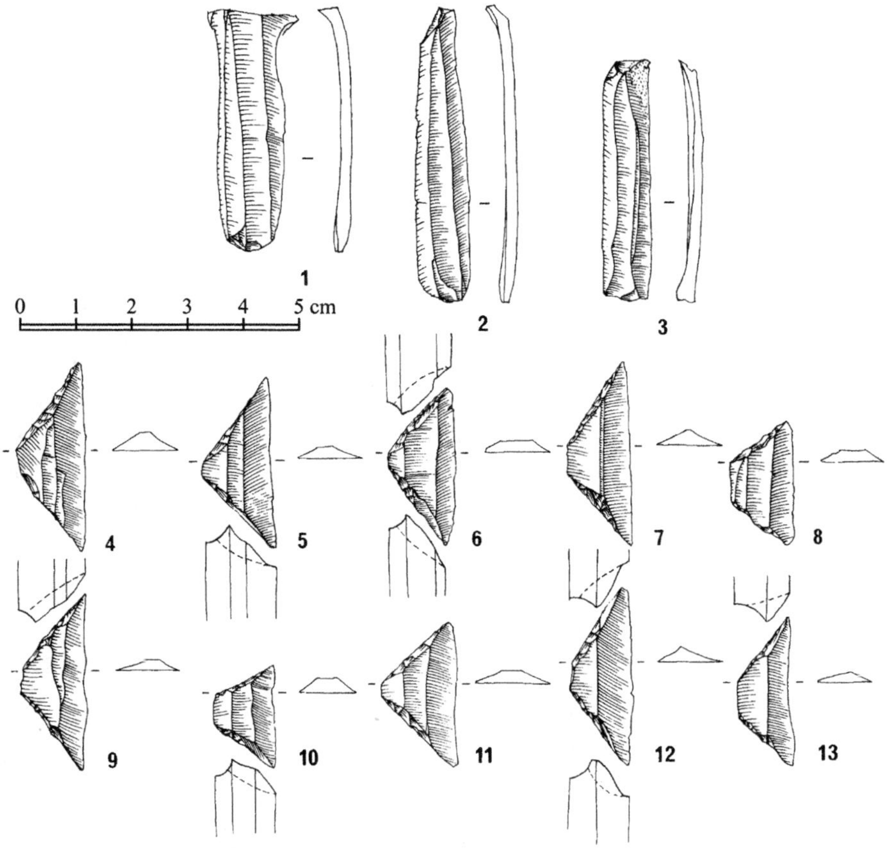

Fig. 2.4 The Relilaï (Algeria): *1, 2, 3* bladelets; *4–13* trapezes obtained from pressure flaked flint blades using the microburin blow technique (Upper Capsian)

J. Tixier, it was possible to postulate the existence of a different *débitage* management (*économie du débitage*) in the Capsian (Inizan 1980: 29). The results of this technological approach demonstrated that pressure *débitage* is present in the Upper Capsian, and established a relationship between this technique and the development of predominantly geometric microliths. Thus, a statistical analysis of 343 trapezes and about 400 unretouched bladelets from The Relilaï showed that the height of the trapezes is a function of the width of the pressure bladelets. Obtained through the microburin blow technique, the geometric elements intended for hafting necessitated the manufacture of regular blanks with a moderate thickness, which could be easily produced using pressure (Inizan 1984: Fig. 2.4).

However, as emphasized by N. Rahmani, the origins of this technical invention in the Upper Capsian still appear to be shrouded in mystery: "Why did Capsians adopt the pressure technique? How did they adopt it, by invention or diffusion?"

(Rahmani 2004: 93). The well-executed character of the pressure *débitage*, attested for by blade dimensions that require elaborate equipment and painstaking core preparation, points toward an adoption of the technique rather than an ex nihilo invention in the Maghreb (Pelegrin 1988).

2.4 From North Africa to Mesopotamia

In 1980, J. Tixier identified pressure *débitage* on obsidian at Oueili, a site of the Ubaid culture in the South of Iraq, which is at least 2,000 km away from the nearest obsidian source (Inizan and Tixier 1983). In the same area, he also identified this technique on flint sickle elements at a younger site, dated to the end of the fourth millennium B.C. Supplemented with bibliographical research, the analysis of several lithic collections uncovered in the early twentieth century in Susiana, Southwestern Iran, subsequently highlighted the importance of this technical phenomenon in Iraq and in Iran (Hole et al. 1969; Inizan 1985, 1986, 1988). In addition, later analyses of lithic material from several sites excavated predominantly in Northern Iraq near Mosul showed that obsidian was always pressure flaked. However, this never took place on site. Pressure *débitage* of flint, carried out on site, was a component of lithic industries as early as 9,000–10,000 years ago in Northwestern Iraq, e.g., at Jarmo, Ml'lefaat, Nemrick, Der Hall, Karim Shahir, etc. (Fig. 2.15) (Ohnuma 1993). This tradition seems to have persisted until the third millennium B.C. (Inizan 1986) and coexisted with another form of blade *débitage* (Inizan and Lechevallier 1994). Indeed, another tradition of blade *débitage*, which is referred to as naviform *débitage* and was standardized through the use of percussion rather than pressure, was present in the Mediterranean and covering the area up to the Euphrates throughout the Neolithic period. It is a bipolar type of blade *débitage*, in which series of removals are alternately detached from opposing striking platforms, resulting in rectilinear end products. These two technological traditions involving high-quality blade *débitage* were identified in the obsidian production workshops at Kaletepe in Cappadocia (Turkey) (Binder 2007; Binder and Balkan-Ali 2001).

As for the Capsian of North Africa, there is evidence of a significant relationship between geometric microliths and a standardized pressure *débitage* of bladelets in the Middle East in the 7th and 6th millennia B.C. At the time when microliths disappeared, another connection emerged between sickle elements and pressure blades. A gradual increase in blade size can be observed in assemblages dating from the 7th to the 3rd millennia B.C., attesting to changes in the knappers' equipment. The small "bullet cores" disappeared at the end of the first half of the sixth millennium B.C. together with the small-sized blades they served to produce (Fig. 2.5). Thick outsized blades (more than 20 cm long), the so-called Canaanite *débitage*, began to appear at the end of the fourth millennium B.C. (Anderson-Gerfaud and Inizan 1994). This corresponds to the development of lever-aided *débitage* (Pelegrin 1988, 1997).

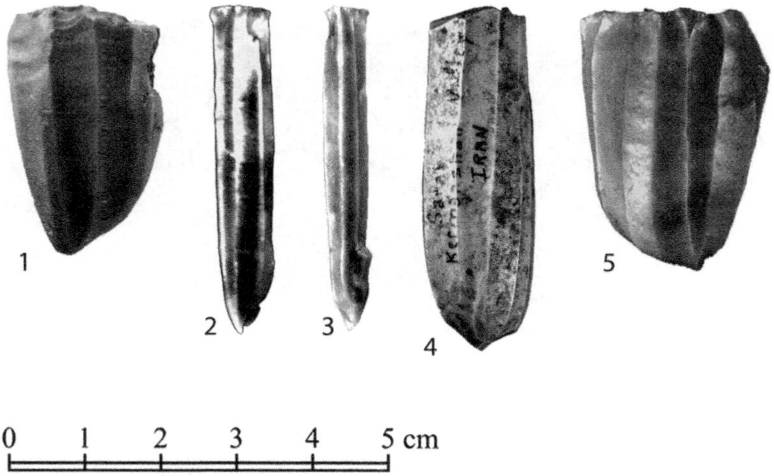

Fig. 2.5 Pressure flaked cores: *1* Mehrgarh (Pakistan); *2, 3* Tepe Guran (Iran); *4* Sarab (Iran); *5* M'lefaat (Iraq)

2.5 Mehrgarh and Central Asia

Archaeological information from Central Asia has dried up since 1979. Collections from Afghanistan and Iran became unavailable for study just when promising work resulting from surveys and excavations carried out in the early 1960s had been published, in which several illustrations suggested that pressure *débitage* was used by settled villagers in Iran (Hole et al. 1969) and Afghanistan (Davis 1978).

This is why from 1984 onward, research on the lithic industries of the Neolithic sites of Mehrgarh and Nausharo in Pakistan was conducted in collaboration with Monique Lechevallier for over a decade. These studies were to serve as a frame of reference for the investigation of the origins of this technique and how it was passed on to the Old World (Inizan and Lechevallier 1985, 1991, 1997).

The first occupations of Mehrgarh in the Kachi Plain, Baluchistan, are dated to the seventh millennium B.C. and occurred in an aceramic context. Pressure *débitage* was carried out almost throughout the entire period of occupation for at least 4,000 years.

M. Lechevallier (2003) has reconstructed the *débitage* strategies for the lithic assemblages from Mehrgarh. They can be summarized as follows. The flint, not available locally, was brought to the site in the form of partially prepared cores, as indicated by the near absence of core preparation flakes and also by the presence of a hoard of nine cores (Lechevallier and Marcon 1998), suggesting that the roughing out took place at the raw material sources. Indirect percussion was used for the initial *débitage* stages, while the removal of bladelets involved exclusively the application of pressure. Heat treatment is documented but was not practiced systematically (Inizan and Lechevallier 1985).

As was previously observed, the geometric microliths and the sickle elements were exclusively produced on pressure blades. This observation holds true for the

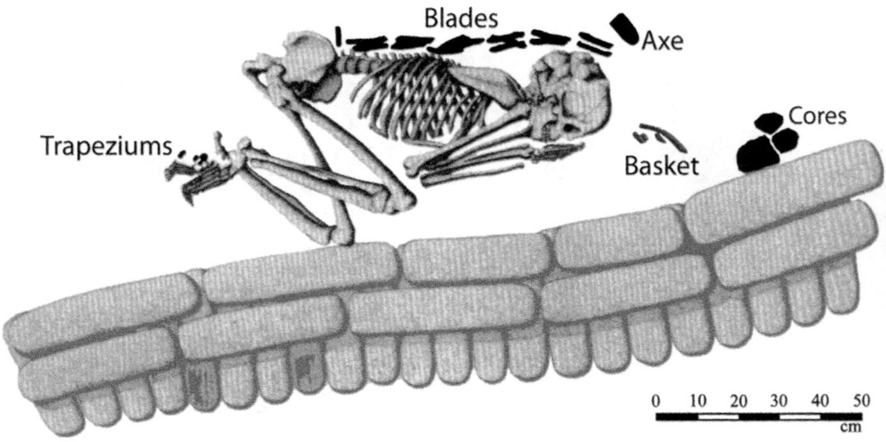

Fig. 2.6 A blade knapper's grave (Neolithic necropolis of Mehrgarh, Pakistan, sixth millennium B.C.)

entire East, something that can be accounted for by an enduring technological tradition. The standardization of very moderately thick, interchangeable elements appears to be associated with hafting. Indeed, thanks to the discovery of several of these elements found obliquely embedded in the bitumen that fixed them to the handle, it has been possible to reconstruct the tools (Lechevallier 1988: 56).

Microwear analysis carried out by P. Vaughan has shown that some of the microliths were intended to be used for hunting and others for cutting plant material (Vaughan 1995).

In 1989, the use of metal for pressure *débitage* in the industries of Mehrgarh and Nausharo was confirmed by J. Pelegrin (Inizan et al. 1994; Pelegrin 1994).

Regarding the funerary rites, the Neolithic necropolis of Mehrgarh offers some insightful data. Dated to the beginning of the sixth millennium B.C., the cemetery contains 150 graves enclosed by low walls. The grave goods associated with the skeletons unambiguously reveal the status of some of the deceased. Thus, the manner in which the grave goods are arranged in grave 114 (three prepared cores by the head of the deceased, 16 blade blanks along his back, which can all be refitted, but the core is missing, and trapezes at his feet) implies that its occupant was a blade producer. He was probably a technically rather than economically specialized craftsman (Perlès 1990: 36–38) (Fig. 2.6).

2.6 Pressure *Débitage* in the Upper Paleolithic of Continental Asia

Pressure retouch was used in the Upper Paleolithic as early as the Solutrean, some 20,000 years ago. However, pressure *débitage* was initially only identified in Neolithic or, at most, in Epipaleolithic contexts. As a result, the use of pressure for *débitage* was regarded as a technical breakthrough, thanks to which it became possible to standardize blade products at the end of prehistory.

We are indebted to J. Flenniken (1987) for recognizing beyond doubt the presence of the Yubetsu-type method of pressure *débitage* (Fig. 2.13) together with the use of heat treatment in the Dyuktaï Paleolithic culture of Siberia, which is dated to 14500 B.P. (Kuzmin and Orlova 1998). The identification of pressure *débitage* in a Siberian Paleolithic culture required a revision of the *débitage* techniques used from the Upper Paleolithic onward in Asian cultural contexts (Seonbok and Clark 1996)

2.6.1 The "Microblade Culture Tradition"

This designation highlights the existence of a culture shared by wide-ranging populations of Final Paleolithic hunters, whose territories spanned Siberia, Mongolia, Northern China, Japan, and extended across the Bering Strait as far as Northern America.

The term "microblade culture" is derived from a morphological recognition of the tiny cores, named "wedge-shaped core," "Gobi core," "true microblade core," etc., rather than from the presence of bladelets and the identification of the technique used to detach these products.

This pressure technique is seldom mentioned, but a correct identification can be arrived at by checking for a series of criteria such as regular arises, curved profiles, small sizes of cores, and bladelets that can even be seen in drawings and would be impossible to obtain by percussion for reasons of inertia (Pelegrin 1988: 49).

Widely used by prehistorians, the term "microblade tradition" is in actual fact a concept referring to a composite tool: a handle or a shaft, fitted with interchangeable lithic elements, obtained by pressure from a handheld core. Thanks to this excellent chronological and typological marker, identified as early as the 1930s (Smith 1974) and referred to as NANAMT (The Northeast Asian-Northwest American Microblade Tradition), one can follow the hunter-gatherers across the wide Siberian steppes and all the way to the American continent. The technology and its bearers easily spread eastward and northward by land during the last glacial maximum (LGM), ca. 18,000–20,000 years ago, when sea levels had dropped by several dozen meters and large land masses became available.

The earliest evidence for microblade pressure *débitage* dates to around 20000 B.P. and can be detected in a large ill-defined area centered in Siberia, Mongolia, and Northern China. Percussion is always used to shape out the cores and to manufacture the heavier tools found in combination with microblade production.

In Korea, an Upper Paleolithic site with evidence of bladelet pressure *débitage* was recently identified (Hong and Kim 2008). The site of Hopyeong-dong was excavated from 2002 to 2004. It is situated within the central part of the Korean peninsula (Fig. 2.15). Hopyeong-dong is one of the earliest sites discovered in this region and bears evidence of a bladelet industry produced by a new technique (pressure *débitage*) that corresponds to the introduction of new raw-material-type obsidian. The use of a pressure technique for the production of obsidian bladelets at this site can be deduced from the refits in conjunction with the numerous, high-quality, illustrations (cf. Hong and Kim 2008). This inference is supported by personal observations of the material during September 2010.

The stratigraphy of Hopyeong-dong presents two successive, culturally distinct, levels. The lower level, dated to 30–27 kyr B.P. (AMS-C14), contains a lithic industry characterized by tanged points, a few quartz and rhyolite blades, and pecked pebbles. Dating to the last glacial maximum (LGM; OIS2 or MIS2), the base of the upper level (24–16 kyr B.P.) corresponds to a radical change in the lithic industry. This departure from the preceding tool kit comprised of end scrapers, burins, and drills introduces the use of high-quality obsidian for the production of bladelets using the pressure technique.

The refits and debris demonstrate that these obsidian bladelets recovered from the uppermost two levels were produced on-site with the exception of the initial core preparation. The Yubetsu method sensu stricto is not present. The *plein débitage* bladelets, which are predominantly fractured, do not exceed 60 mm in length, 5 mm in width, and 2 mm in maximum thickness (Hong and Kim 2008: 356–363), indicating that they can be produced from a core stabilized by hand. A use-wear analysis attests to these bladelet fragments having had multiple functions and indicates hafting in wooden handles (Kononenko 2008). The early appearance of obsidian material at the site of Hopyeong-dong confirms that this part of Korea witnessed the invention and diffusion of pressure bladelet production in the Asian continent (Fig. 2.15).

2.6.2 The Invention and Origin of Pressure Débitage

During the two Franco-Soviet symposiums on the prehistory of Central Asia held in 1982 at Dushanbe in Tajikistan and in 1985 in Paris, some contributions suggested that the origins of pressure *débitage* lie in the Far East and that this technique was first developed in the Upper Paleolithic (Inizan et al. 1992). In addition, a single center of invention was proposed.

The arguments in favor of a single center of invention are based on the enduring presence of the tradition from the Paleolithic onward in Asia and on its absence in Western Eurasia prior to the Holocene. In Asia (Siberia, Mongolia, Japan, etc.), pressure *débitage* is practiced continuously to obtain first microblades and then blades, i.e., "minute to outsize products," from the Paleolithic onward and up to and including the Neolithic period (Pelegrin 1991). In China, Gai Pei (1985: 231) stressed the occurrence of a large number of sites with microblade *débitage*, especially in the Northeast and in Inner Mongolia, and the long chronologies which span from the Upper Paleolithic until the first millennium B.C. In Central Asia, F. Brunet states that "the adoption of pressure *débitage* is a considerable departure … from the ancient Upper Paleolithic traditions" (Brunet 2002: 11; translated from the original).

The process of diffusion probably took place from East to West (Inizan 1991; Inizan and Lechevallier 1994; Inizan et al. 1992).

Fig. 2.7 Refitted core from Bolshoj Yakor (Siberia) (photo M.-L. Inizan)

On the African continent, microblade *débitage* exists in the Capsian of North Africa and in some Neolithic industries of Egypt and of Mauritania (Midant-Reynes 1983; Boeda 1987).

Pressure *débitage*, particularly in the case of bladelets detached from handheld cores, is a technical invention that could be easily transmitted. Larger products, which require devices for the immobilization of the core to facilitate the application of pressure (crutch), suggest the existence of more complex skills and, therefore, were probably more difficult to pass on and to disseminate over a wider area.

In 1990, Prof. Medvedev was kind enough to allow scholars to examine some Siberian material in Irkoutsk for the first time. The key feature was a group of 11 refitted microbladelets (L=12–18 mm; l=2–3 mm) on a narrow-fronted Yubetsu-type core from the site Bolshoi Yakor, located on the bank of a tributary of the Lena River and dated to 11500 B.P. (Fig. 2.7). This refitted group allowed for the recognition of the pressure technique as well as the Yubetsu method and its outstanding efficiency (Ineshin 1993). At Bolshoi Yakor, some of the bifacial preforms were introduced to the site in a prepared state. The cores from the older levels have the narrowest *débitage* surfaces (E. M. Ineshin, personal communication 1991).

Two questions still required an answer: What use were these microbladelets designed for and how were they hafted? Possible explanations were found in assemblages that derived from a number of Siberian sites located on an Arctic island, which had recently been excavated from the thawing permafrost. For instance, at the hunting site of Zhokhovskaja, dated to 8790±90 B.P., several bone and ivory implements were found complete with their hafted microblade insets. In addition, these implements showed that hafting could be achieved with or without the use of adhesives. These 25 objects account for 50% of all the known hafted composite tools from Siberia (Fig. 2.8) (Pituljko 1998). A wooden sledge was also identified. Evidence for a workshop comprising of pressure cores and bladelets is associated with these various pieces of equipment.

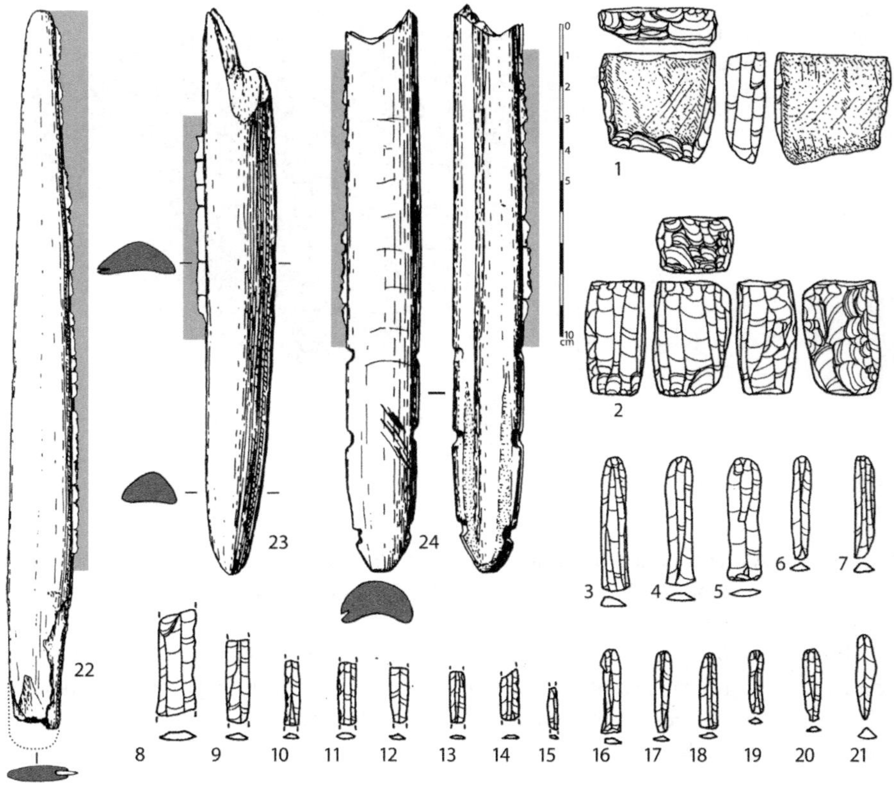

Fig. 2.8 Pressure debitage: core (1–2) and bladelets (3–21); bladelets inserted into bone *baguettes* (22–24) (site of Zhokhovskaja; redrawn by H. Kimura 1999: 10, 12)

2.7 Pressure *Débitage* as Evidence of Mobility

Two examples are developed below to illustrate how the identification of elaborate production techniques helped to outline the migrations of cultural groups. The first example addresses the question for Scandinavia in Northern Europe, and the second highlights the importance of pressure *débitage* methods for understanding the chronology of the colonization of the Japanese Archipelago in the Upper Paleolithic.

2.7.1 Migrating into Northern Europe: Sujala in Lapland

Although mention can be found in the literature of the presence in Northern Europe of the pressure *débitage* technique, this is not backed by a description of the diagnostic criteria. The morphology of the "wedged-shaped cores," for which a Siberian origin is accepted, has also been used as an argument for the presence of this technique (Svoboda 1995).

At higher latitudes, the Finnish site of Sujala in Lapland, which was excavated in 2004, was occupied by reindeer hunters at around 9000 B.P. (Rankama and Kankaanpää 2007: 51). The presence of blade pressure *débitage* (identified by J. Pelegrin) at such an early date was an unexpected discovery, because the technique was supposed to have first reached Scandinavia via Northern Europe at around 7800 B.P., during the Atlantic period. Moreover, it had been generally accepted that Lapland was settled by coastal groups from Norway belonging to the original Ahrensburgian techno-complex. Now, the occupation of Sujala points to the arrival of a group of reindeer hunters who brought with them the knowledge of the blade pressure *débitage* technique, which had been widely used in the Eastern Baltic, and who carried hunting weapons (arrowheads) characteristic of the post-Swiderian culture identified in Russia (and different from those of the Ahrensburgian techno-complex). Migrating from the Northeast and possibly following reindeer herds, these people may well have been the first settlers to arrive in Lapland. They bore witness to a hitherto unsuspected Northern European route for the spread of the pressure *débitage* technique (Fig. 2.15).

2.7.2 Migrating into the Japanese Archipelago

The Japanese Archipelago stretches from latitude 25° to 45° N. The four major islands are Kyushu, Shikoku, Honshu, and Hokkaido. The present-day shorelines gradually took shape after the last glacial maximum (LGM, ca. 18,000–20,000 years B.P.), stabilizing at ca. 10,000–12,000 years B.P. (Ono 1999). Figure 2.9 shows the shoreline displacement and the extent of dry land during the LGM. At that time, the Northernmost Hokkaido Island was part of continental Eurasia, together with Sakhalin Island and the Kamchatka Peninsula in Siberia. It was separated by a strait from the other islands of the Archipelago, which formed a single land mass adjacent to continental Korea (Fig. 2.9). The position of the ancient shorelines had a bearing on human expansion as well as on animal migrations (Fig. 2.10) and on the vegetative cover.

Two fact-finding missions in Hokkaido (1993) and in Kyushu (2000) afforded the opportunity to ascertain that the pressure technique had been systematically used to detach bladelets in Japan for many thousands of years and to study the variety of *débitage* methods. The islands of Japan are of volcanic origin and yield large quantities of high-grade stone raw material that was suitable for pressure blade making, e.g., obsidian, which is primarily found in Northeastern Hokkaido, and various neogene volcanic rocks such as shale and chalcedony. Antler was widely available for use as tools for percussion and pressure techniques.

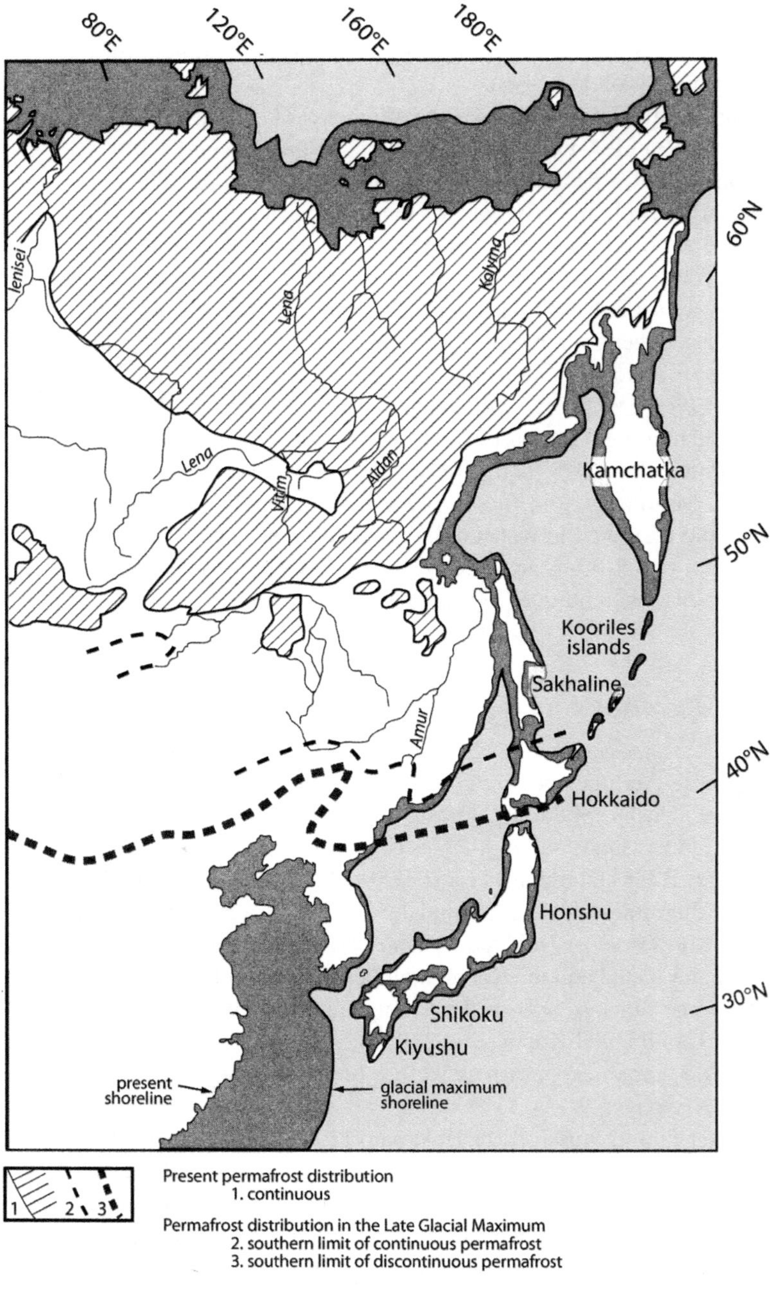

Fig. 2.9 The major post-LGM shorelines of Japan (Modified after Ono 1999)

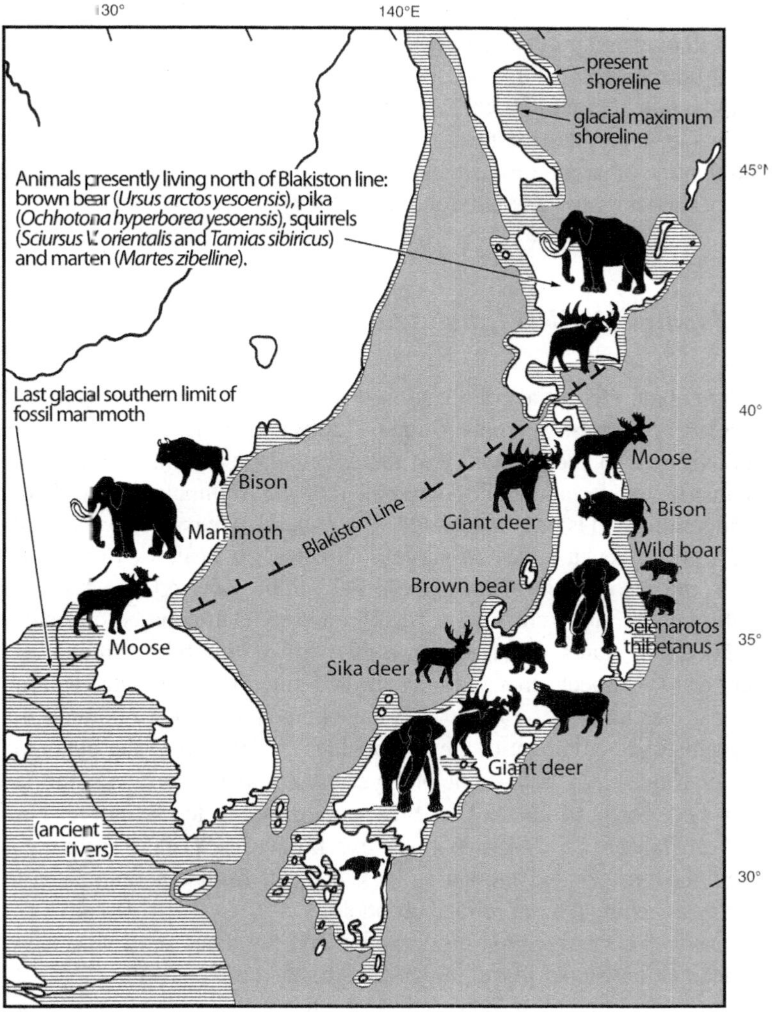

Fig. 2.10 Migration routes of large mammal faunas during the LGM (Ono and Igarashi 1992)

2.8 The "Microblade Tradition" in Japan

This microblade tradition, identified in the Far North of America, in Siberia, Northern China, Mongolia, Korea, and Japan, has been widely published over the last 30 years by a large number of researchers. It was however necessary to date these cultures in order to compare them. Following the 1990 Novosibirsk Symposium, an international meeting dedicated to "The origin and dispersal of microblade industry in Northern Eurasia" was organized by Professor H. Kimura at Sapporo on the Hokkaido Island in 1992 (1993, Japanese/English). This was triggered by the fact that Japan, and more particularly Hokkaido, has the largest number of known and published settlements containing such industries.

The earliest bearers of a microblade tradition appear on what is today the Hokkaido Island at around 20000 B.P. As soon as bladelet industries had been recognized in Japan, a continental origin was attributed to them, and it was generally accepted that the presence of this technical tradition was an expression of a similar subsistence model (Kajiwara and Yokoyama 2003).

2.8.1 Japan's Earliest Inhabitants

Information about the initial colonization of Japan has not been circulated widely, and, therefore, one must welcome Yajima's (2004) recent English publication who gives an overview of the prehistory of the Archipelago, its chronology, and major stone industries. So far, the earliest settlement of the Archipelago dates back to the Upper Paleolithic and is only associated with *Homo sapiens sapiens* populations.

Since 1949, when a series of surveys directed by Prof. C. Serizawa (1971) resulted in the recognition of a Japanese Paleolithic, over 5,000 sites have been located throughout the Archipelago. The chronology is now based on several hundred radiocarbon dates (Ono et al. 2002). Since the breakthrough of AMS (Accelerator Mass Spectrometry) radiocarbon dating, over 400 dates with possible calibrations to as far back as 11850 B.P. have helped to clarify and understand the earliest prehistoric settlement of the Archipelago. They predate the major volcanic eruption in Kagoshima Bay south of Kyushu (AT, Aira-Tanzawa), the ashes of which reached Korea, China, and the Siberian territory of the Primorye. These levels have been dated to 25–24 kyr. With the exception of Hokkaido, ashes are present throughout the territory and alternating loess deposits from the continent contribute further to establishing a more refined chronology.

As a result, the stratigraphic observations and the dates show that the earliest human occupations took place in the South of the Peninsula at ca. 30,000–35,000 years ago. The oldest industries derive from sites located for the most part on the Kyushu Island, in the Kumamoto and Kagoshima prefectures, and on the Tanegashima Island (Fig. 2.11). They do not contain a bladelet component, but they share a "trapezoidal-shaped" tool type (Ono 2004: 29), which was identified at Ishinomoto (31000–33000 B.P.), Nitao (lowest of three cultural levels yielding a total of 150,000 artifacts), Mimikiri (28000 B.P.), Maeyoma (30000 B.P.), and Tachikiri (31000 B.P.) (Fig. 2.12). These small tranchets with a thick cross section were found in association with used cobbles akin to grinding stones, tanged points with a triangular cross section, and *limaces*. These occupations always predate the volcanic eruption (AT1-25–24 kyr).

In the north, on the Hokkaido Island, the earliest occupations recorded so far are located in the south of the island and do not predate 23000–24000 B.P. It is worth repeating here that there was no land bridge linking Hokkaido to the southern part of the Archipelago during the LGM. At Obihiro, Kashiwadai, and Kamioka, bladelet

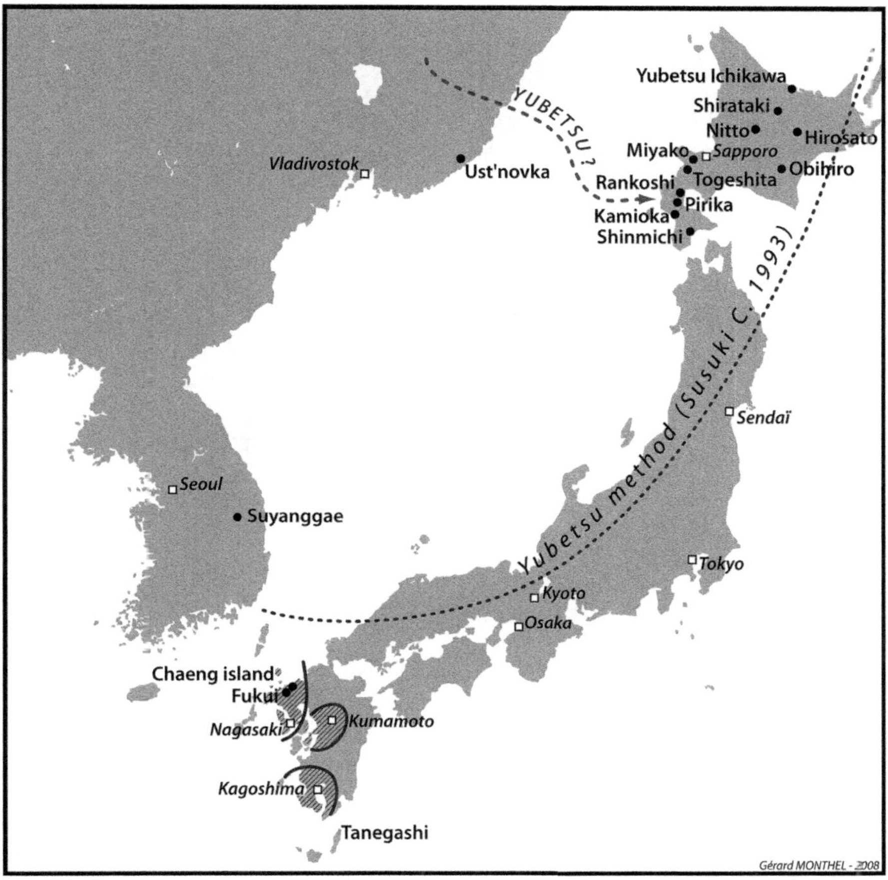

Fig. 2.11 Sites and locations mentioned in the text and the distribution of the Yubetsu method

industries are absent from the oldest occupation levels. This also holds true for the Northeastern Kamishirataki sites (Fig. 2.11). There are similarities between the tool kits of these sites and those of the Honshu sites, e.g., backed blades, big trapezes, etc. The implication is that the earliest populations first settled the south of the Archipelago and then spread northward, almost certainly via the sea.

2.8.2 *Pressure Débitage and the Yubetsu Method*

The Japanese Archipelago plays a significant part in the history of the invention and diffusion of pressure *débitage* in the Upper Paleolithic because this is the area where the Yubetsu method (or technique, depending on the customary use) was first identified. As early as 1959, when prehistoric bladelet industries were discovered in Hokkaido, M. Yoshizaki (1963) provided a description of the Yubetsu bladelet

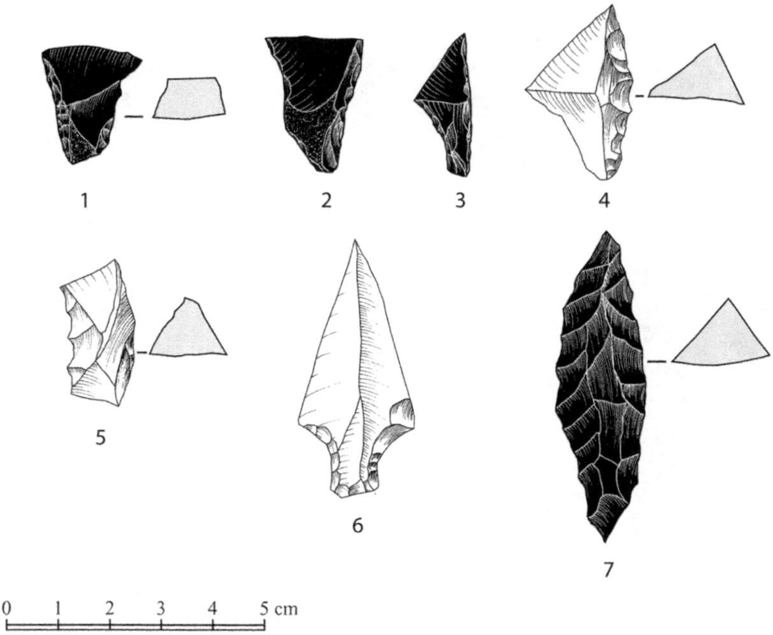

Fig. 2.12 Diagnostic upper paleolithic tools from the island of Kyushu (23000–30000 B.P.): *1, 2, 3, 4* Nitao; *5, 6, 7* Maeyoma

débitage on obsidian, supplemented by drawings. While the method was well understood, the techniques for shaping out the core and detaching the bladelets were not addressed. In spite of the language barrier, knowledge about this original *débitage* method filtered through to the international community of prehistorians thanks to the publication of numerous diagrams showing refitted sequences, particularly from the Hokkaido sites. It was even suggested that pressure *débitage* might have been invented in this obsidian-rich territory (Tixier 1984: 59). However, the use of the term "Yubetsu" to denote a particular method of pressure *débitage* should not be taken as an indication of this invention's geographical origin. The same holds true for the use of the term "Levallois," which also denotes a specific shaping out of the core prior to the detachment of blanks. Besides, there are methods other than the Yubetsu method in Japan.

2.8.3 *The Yubetsu Method*

Schematically, the method entails the preparation of the core to obtain a generally asymmetrical bifacial piece from which one of the ridges is then removed, thus creating a surface that corresponds to a section of the biface. This surface is then prepared by several removals, often supplemented by abrasion. The resulting elongated pressure platform is suitable for the detachment of numerous bladelets or microbladelets of identical length. This effective and productive method was identified on obsidian material from sites located along the Yubetsu River, in Northeastern Hokkaido, not far from the Shirataki obsidian cliffs (Tozawa 1974; Kimura 1992). The lava flows are abundant, and their quality is excellent. Obsidian is highly suitable for pressure *débitage*, but other rocks such as chalcedonies, shales, and schists were also selected for the same purpose throughout the Archipelago (Inada et al. 1993).

In Japan, the term "Yubetsu" is only used for the method involving a strictly bifacial shaping of the core and the detachment of one of the ridges using one or several removals (ski spall) to create an elongated pressure platform, which is no larger than the thickness of the biface. Within the framework of the same concept, there are a number of related methods, e.g., in the *Togeshita* method, the preform is not strictly bifacial, and in the *Rankoshi* method, the bladelets are removed along the length of the biface.

2.8.4 Other Methods

A distinct method involving the reduction of a boat-shaped microcore pertains to another operational scheme of which the *Horoka* method is a good example. The pressure platform, wider than in the previous methods, is obtained by one or several removals. The sides of the cores are then created by means of removals that originate from and join opposite this surface to form a crest resembling the keel of a boat (Fig. 2.13). The platform is semicircular and the *débitage* carried out perpendicular to the platform.

Yet another method, named *Hirosato*, involves a core/burin. This is a *burin busqué* on a truncated blade, where the opposing truncations serve as pressure platforms for the detachment of twisted bladelets.

2.8.5 Some Production Techniques

In Japan, comparative experimental tests involving microblade detachments using direct and indirect percussion as well as pressure were carried out by K. Ohnuma and M. Kubota. These experiments confirm the similarity of the detachment technique, i.e., the use of pressure, between the sites of Shirataki-Hattoridai (Hokkaido) and Karim Shahir and M'lefaat (Iraq). My familiarity with the techniques used in Japan is the result of only two research visits: one to Hokkaido in 1993 and another in 2000, which focused mainly on the Kyushu material. Nevertheless, these short visits offered an insight into the territories located in very different latitudes. In Sapporo, it was possible to examine products from the obsidian quarry of Toma (Shirataki) as well as the material from the sites of Miyako and Yubetsu/Ichikawa. In Imagane, it was possible to access the material from the sites of Pirika and

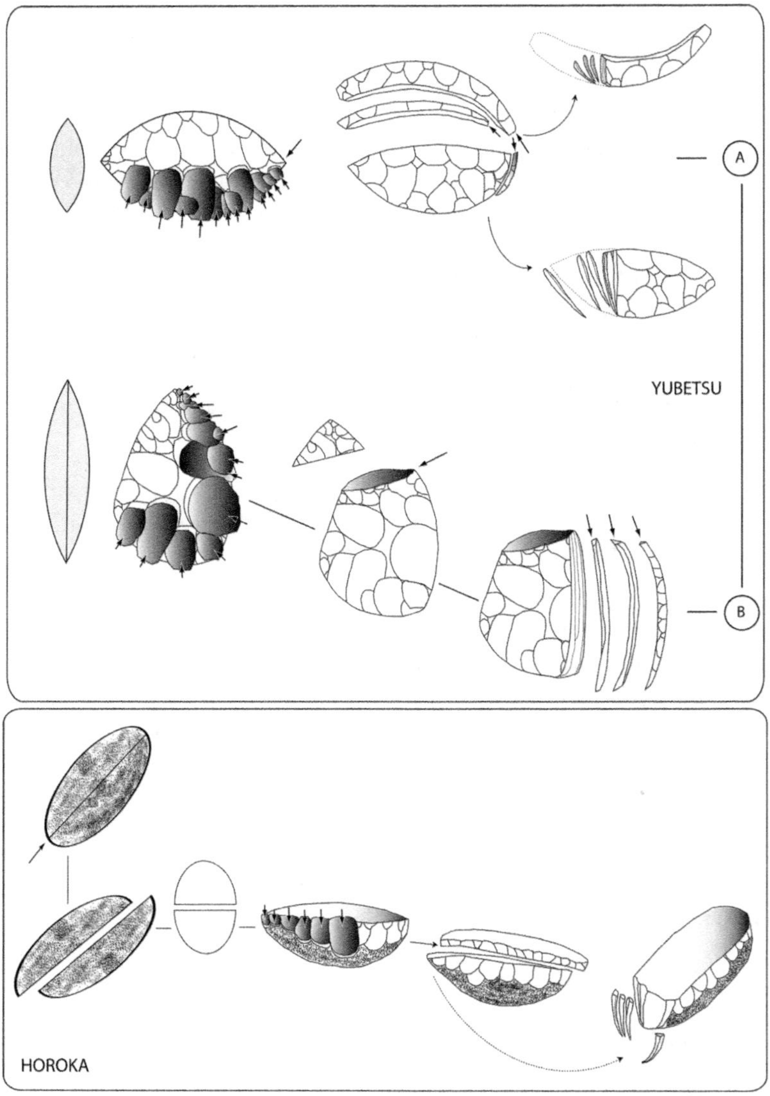

Fig. 2.13 Some pressure debitage methods in Japan: Yubetsu (*A stricto* sensu, *B* Rankoshi); Horoka (G. Monthel *del.*)

Kamioka thanks to H. Terasaki (Fig. 2.11). Percussion with a hard or soft hammer was used on the various raw materials to prepare the core, together with technical procedures such as the abrasion of the overhang, which were systematically carried out when shaping out obsidian cores (Toma), as well as the intentional scratching of pressure platforms.

Numerous bladelets were detached using pressure (Pirika, see below). The many dozens of examined bladelets and cores all displayed the distinctive criteria associated with pressure *débitage* on handheld cores, i.e., regular and parallel arises, small lengths of bladelets, and bladelet removal negatives. Additional criteria on bladelets are small butts and a short bulb below the impact point.

2.8.6 *The Chronology of Bladelet Débitage: Northern Japan (Hokkaido)*

The site of Pirika is a good example to discuss the chronology of Northern Japan. Pirika is composed of three separate locales. At Pirika 1, three archaeological levels were identified and dated to between 20900 and 17500 B.P. (Ono et al. 2003). Numerous refits allowed for the recognition and description of the *débitage* methods and their chronological development. In addition to the main raw material shale, some obsidian is available in the south of the island, where the outcrops are inferior in quality and less abundant than those at Shirataki in the Northeast. None of the archaeological levels yielded any evidence of the presence of the *Yubetsu* method.

Evidence for the use of the *Togeshita* method, i.e., there are no true bifaces with regularized ridges, is already present in the first occupation level. The elongated core is shaped out using percussion and displays a narrow *débitage* surface. The blades are obtained by pressure. Several hundred generally broken bladelets were recovered (there are 721 drawings). Most of them are very regular and thin, with two parallel arises, small butts, and short bulbs.

A trend toward the production of systematically longer bladelets can be observed in level 2 of Pirika 1, owing to the development of the *Rankoshi* method, in which *débitage* is carried out along the length of the core.

It seems that the presence of the *Yubetsu* method stricto sensu in Hokkaido is not present prior to 15000 B.P. Bladelet *débitage* was abandoned at around 12000–11000 B.P. and replaced with blade pressure *débitage* (L=15 cm). The cores are pyramid-shaped with a single pressure platform, as observed for instance at Nitto.

2.8.7 *The Chronology of Bladelet Débitage: Southern Japan (Kyushu)*

In the South, the earliest evidence for the use of microblade *débitage* does not predate 15000 B.P. and is therefore more recent than that in the North. On Chaeng Island (Nagasaki prefecture) just off Kyushu, where three occupation levels were uncovered during excavations, the oldest and undated level contained no evidence for bladelet pressure *débitage*. This technique was used on obsidian in level 2, which dates to 15450 B.P.

High-quality obsidian flows are known to exist in the Nagasaki area. Raw materials from these sources were identified on Chaeng Island and were also transported as far as Korea. Small obsidian and shale nodules were generally flaked along one

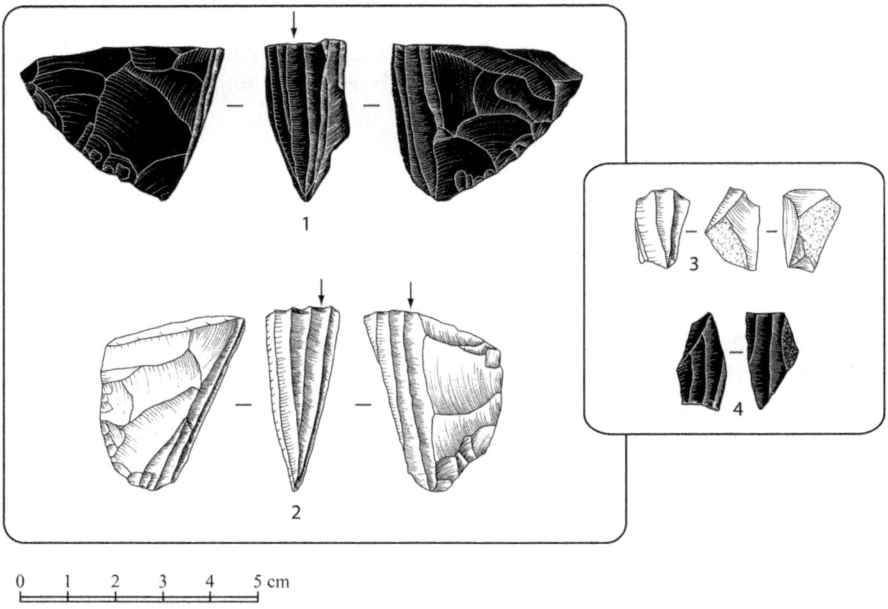

Fig. 2.14 Pressure flaked cores: *1*, *2* (Hokkaido) Yubetsu *stricto* sensu from Miyako, obsidian; *3*, *4* (Kyushu) Nitao, shale; *4* Jyobaba, obsidian

side (Fig. 2.14), while the other side was left cortical. There is no evidence of the Yubetsu tradition on Chaeng Island, but this *débitage* method was identified in Korea, in particular at the site of Suanggae (Lee and Yun 1993).

When the Jomon culture had begun to develop (Incipient Jomon), i.e., when pottery first appeared at around 14000 B.P., bladelet pressure *débitage* was still in use. It disappeared with the onset of the second stage of the Jomon period (Initial Jomon) at around 10000 B.P. Recently, pottery dated by AMS to around 15500 cal B.P. was found at the site of Odai Yamamoto, at the Northernmost tip of Honshu Island (Kobayashi 2007: 92).

Similar observations were made in the Kumamoto area. For instance, at the site of Nitao, microblade *débitage* was present in a horizon dated to 13000 B.P. It is also conducted on small shale and obsidian nodules, flaked along one side.

Around Kagoshima, the obsidian flows are generally of medium-grade quality, owing to their many inclusions. Moreover, the nodules are small, thus imposing a size limit on the products.

Based on the dates and the *débitage* methods used, several stages for the arrival of this technique in Japan can be suggested. First, it seems that a 4,000–5,000-year-long gap existed between the adoption of pressure in the south and in the north of the Archipelago. Second, the analysis of the different production methods and operational schemes showed that Hokkaido and Kyushu did not share a common technological tradition. It is extremely likely that pressure *débitage* was introduced to Kyushu via Korea, whereas the presence of this technique in Hokkaido points to an earlier Siberian tradition.

Similarly, C. Suzuki (1993) suggested the existence of two components of bladelet *débitage* in Japan, which are distinct in terms of their origins and purpose. It is argued that the tradition characterized by the Yubetsu method, which was in use on Hokkaido and in the western part of the Archipelago adjacent to the Sea of Japan, was related to a continental climate, while the much later tradition is thought to have developed in a more temperate climate (Fig. 2.11).

2.9 Discussion and Conclusion

Allowing for the deficiencies of this vast overview in terms of geography, chronology, and technology, there appears to be substantial evidence for the existence of a single center of invention of pressure *débitage* in the Upper Paleolithic of Eurasia. The technique then spread rapidly and afar with its bearers – mobile groups of hunters. Later on in the Holocene, it progressed westward across the Old World, most likely along multiple pathways and through a variety of transmission modes (Perlès 2007). It is also a good marker for the colonization of Northern America (Inizan 2002).

As a prerequisite for the identification of pressure *débitage* on archaeological material, it was necessary to master this technique during experimental replication and to describe its diagnostic characteristics. The discovery that pressure can be used to detach blades by D. Crabtree was certainly a surprise (see above). The subsequent identification of this technique in the Capsian of the Maghreb and in the Upper Paleolithic of Asia was even more unexpected. What J. Clark had reconstructed was the method and technique used by the Aztecs to produce their obsidian pressure blades. It was therefore difficult to imagine that this technique, which was never universal, was developed by Paleolithic hunters during the LGM in Southern Siberia and/or in Northern China.

Thanks to experimental replication, it became obvious that the technical practice of applying pressure to a core, rather than a blow to detach bladelets, was efficient for the production of large numbers of such standardized items. These interchangeable bladelets, intended for hafting in bone, ivory, or wood implements, were found in a vast geographical area spanning Siberia, Mongolia, China, Korea, and extending as far East as Japan. They represent an innovation, which was initially described by prehistorians as "microblade culture" (or "microlithic industry" in China) and considered to be the hallmark of Mongoloid hunters of large mammal faunas, e.g., mammoth.

The dimensions of the bladelets remained stable throughout the Upper and Epipaleolithic, with a mean length of less than 8 cm. Therefore, it can be inferred that there was a change neither in the technology of the implement shafts which accommodate the insets nor in the knappers' skills and equipment.

Unfortunately, due to the fact that detachment techniques are seldom identified and that there are but a few recently excavated, stratified, and dated sites, it is still not absolutely clear when and where this technical innovation first occurred. In China, microblade industries were found in a horizon dated to 21959 ± 100 B.C. at the site of Xiachuan (Gai Pei 1985: 231) and in even older levels at the site of Salawasu.

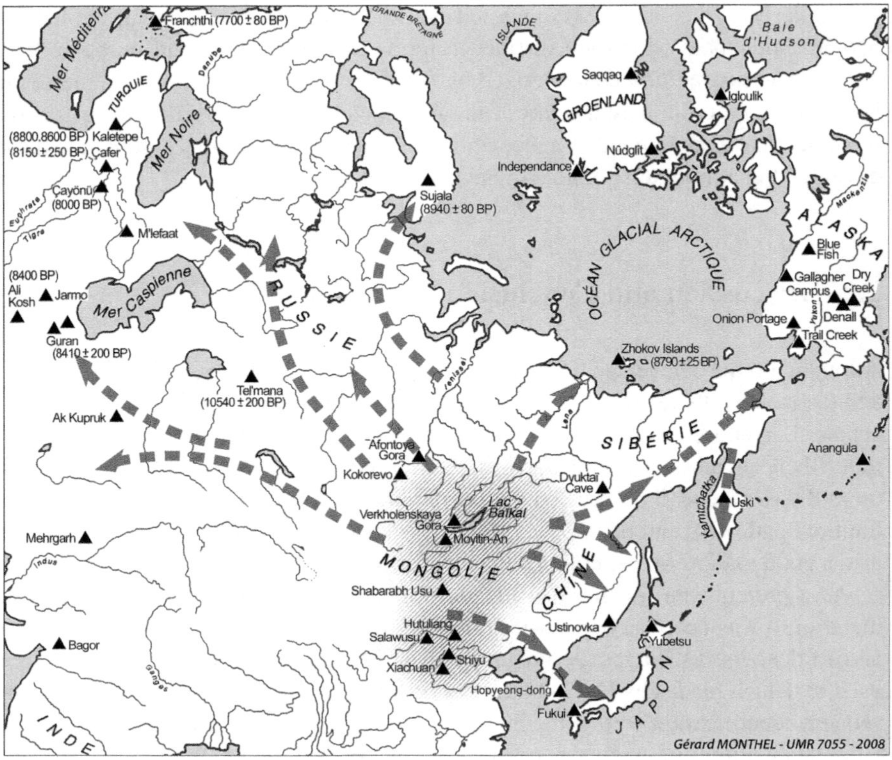

Fig. 2.15 Invention zone of pressure debitage (hatched area) and some hypothetic diffusion routes

Should these dates be confirmed, and should the *débitage* technique be verified, then the tradition of pressure blade making would be more than 20,000 years old.

Compensating for this lack of accuracy, the sites in the Japanese Archipelago are informative regarding the earliest presence of the technique and the different methods implemented. There exists a wealth of excavated, studied, published, and well-illustrated sites. In the north of the Archipelago, the environmental context was identical to that in Siberia, and populations with similar subsistence practices could have easily reached Hokkaido to hunt mammoth, deer, etc. (Fig. 2.10). While the presence of bladelet pressure *débitage* was recorded at sites from the southwestern part of the island which date to ca. 20000 B.P., the first instances of the use of the Yubetsu method stricto sensu only appeared at about 15000 B.P. In the south of the Archipelago (Kyushu), bladelet pressure *débitage* was also practiced from ca. 15000 B.P. onward. However, here evidence for the Yubetsu method is absent, and alternative methods were in use, thus suggesting influences other than those present in the north.

As the Jomon culture developed at the end of the last glaciation, just prior to the Holocene, the use of this technique began to wane but did not disappear entirely.

It continued to be used for the making of obsidian blades on Hokkaido Island, e.g., at Nitto where it was dated to ca. 8000 B.P.

As far as the rest of the Old World is concerned, pressure *débitage* was practiced in various cultures to obtain blades as opposed to bladelets throughout the Epipaleolithic and/or the Mesolithic and predominantly in the Neolithic period. The increase in the dimensions of the blade products signifies that other skills were involved. This invariably would have influenced the way in which the technique was transmitted.

From Central Asia to Iran, the Near East and the North of the Mediterranean, several important sites punctuate the progression of pressure *débitage* (Fig. 2.15).

The Capsian has been repeatedly mentioned in the course of this paper and will also be the subject of some closing remarks. The archaeological evidence can be used to argue against the ex nihilo invention of pressure *débitage* in the Maghreb. On the other hand, the other possibilities for its introduction such as the diffusion by land or by sea cannot be sufficiently supported based on the present state of our knowledge. The technique may have arrived overland via Egypt after having reached the Near East, but this seems unlikely because pressure *débitage* in Egypt is only documented for the final Neolithic. In addition, there are no earlier archaeological sites with assemblages that are characterized by this technique between the Maghreb and the Near East. Diffusion from the Near East by sea could be contemplated, but this in turn poses problems because the Capsian was an inland cultural tradition, i.e., it has never been recorded on the shores of the Mediterranean. Clearly, the origin of pressure *débitage* in the Upper Capsian is a problem that remains to be solved in the future.

Acknowledgments I would like to express my gratitude to Prof. V. Ranov in Dushanbe, who unfortunately went missing in 2006, and Prof. H. Kimura in Sapporo, for their gracious invitations which have allowed me to extend the present research into pressure *débitage*, in particular. I also thank Y. Teraski who provides me access to the Pirika's collections (Imagane, Hokkaido). A special mention should be made for Prof. I. Yamanaka of Kyoto University for his support and the preparation of two long research visits in 1993 and 2000, during which I was provided with access to numerous collections throughout Japan.

I am deeply grateful to Dr. M.Y. Hong and Dr S.J. Park, without whom my research in Korea during September of 2010 would not have been possible. My gratitude also goes out to the director and assistant director of the Gyonggi Cultural Foundation Mr. Y.J. Jo and Mr. S.T. Kim, respectively, as well as Pr. Y.J. Lee and Pr. G.G. Lee of the Chungbuk and Chosun University Museums. Additional thanks are extended to Farina Sternke whose revisions and comments greatly improved this manuscript.

References

Anderson-Gerfaud, P., and M.-L. Inizan 1994 Utilisation du tribulum au début du III° millénaire: des lames « cananéennes » lustrées à Kutan (ninive V) dans la région de Mossoul, Iraq. *Paléorient* 20(2): 85–103.

Binder, D. 1984 Systèmes de débitage laminaire par pression: exemples chasséens provençaux. In *Préhistoire de la Pierre Taillée 2. économie du débitage laminaire*, pp. 71–94. CREP, Meudon, Paris.

Binder, D. 2007 PPN Pressure technology - views from Anatolia. In *Technical systems and Near eastern PPN communities*, edited by L. Astruc, D. Binder and F. Briois, pp. 235–244. Proceedings of the 5th International Workshop, Frejus 2004. APDCA, Antibes.

Binder, D., and N. Balkan-Ali 2001 Obsidian exploitation and blade technology at Kömürcü-Kaletepe, Central Anatolia. In *Beyond Tools. Redefining the PPN Lithic Assemblages of the Levant* (Venice 1998), edited by I. Caneva, C. Lemorini, D. Zampetti and P. Biagi, pp. 1–16. SENEPSE 9, Berlin.

Binder, D., and C. Perlès (avec la collaboration de M.- L. Inizan et M. Lechevallier) 1990 Stratégies de gestion des outillages lithiques au Néolithique. *Paléo* 2: 257–283.

Boeda, E. 1987 Premiers témoins du débitage par pression en Mauritanie. *L'Anthropologie* 91(2): 683–688.

Brunet, F. 2002 Asie centrale: vers une redéfinition des complexes culturels de la fin du Pléistocène et des débuts de l'Holocène. *Paléorient* 28(2): 2–24.

Cahen, D. 1982 Tailler! Pour quoi faire? In *Tailler! pour quoi faire: Préhistoire et technologie II, Recent progress in microwear studies*, edited by D. Cahen, pp. 9–10. Studia Praehistorica Belgica, Tervuren.

Callahan, E. 1985 Experiments with Danish Mesolithic microblade technology. *Journal of Danish Archaeology* 4: 23–39.

Clark, J.E. 1982 Manufacture of Mesoamerican prismatic blades: an alternative technique. *American Antiquity* 47(2): 355–376.

Clark, J.E. 2003 A review of Twentieth-Century Mesoamerican Obsidian Studies. In *Mesoamerican Lithic Technology – Experimentation and Interpretation*, edited by K.G. Hirth, pp. 15–54. University of Utah Press, Salt Lake City.

Crabtree, D.E. 1968 Mesoamerican polyhedral cores and prismatic blades. *American Antiquity* 33(4): 117–132.

Crabtree, D.E. and R. Butler 1964 Notes on experiments in flintknapping: 1, heat treatment silica materials. *Tebiwa* 6: 1–6.

Davis, R.S. 1978 The Palaeolithic. In *The Archaeology of Afghanistan from earliest times to the Timurid period*, edited by F.R. Allchin and N. Hammond, pp. 37–70. Academic Press, London, New York, San Francisco.

Flenniken, J. 1987 The palaeolithic Dyuktai pressure debitage of Siberia. *Arctic Anthropology* 24(2): 117–132.

Gai Pei 1985 Microlithic industries in China. In *Palaeoanthropology and Palaeolithic Archaeology in the Peoples Republic of China*, edited by W.U. Rukang and J.W. Olsen, pp. 225–241. Academic Press, Orlando.

Hole, F., K.V. Flannery, and J.A. Neely 1969 Flint and obsidian artifacts. In *Prehistory and human Ecology of the Deh Luran plain. An early village sequence from Khuzistan, Iran*, edited by F. Hole, K.V. Flannery and J.A. Neely, pp. 74–105. Memoirs of the Museum of Anthropology, University of Michigan, Ann Arbor.

Hong, M-Y., and J-H. Kim (editors) 2008 *Hopyeong-dong Paleolithic site (Namyangju, Gyeonggi Province, Korea).* Korea Land Corporation, Gyeonggi Cultural Foundation Gijeon Institute of Cultural Properties (Excavation Report No. 93), Vol. 2 (Korean/English abstract).

Inada, T., N. Mitsuishi, and K. Kinugawa 1993 The Yubetsu technique in western Japan (Anglais/Japonais). In *The origin and Dispersal of Microblade Industry in Northern Eurasia*, edited by H. Kimura, pp. 38–49. Sapporo University, Sapporo (Anglais/Japonais).

Ineshin, E.M. 1993 Deyatelnostnyi podkhod k izucheniiu planigrafiiarkheologicheskogo pamytnika Bolshoi Yakor (Behavioral approach to the study of the spatial structure of the archaeological site of Bolshoi Yakor). In *Istorichesky opyt osvoeniya vostochnykh raionov Rossii*, edited by A.P. Derevanko, pp. 67–74. Far Eastern branch of Russian Academy of Science, Vladivostok.

Inizan, M.-L. 1976 Nouvelle étude des industries lithiques du Capsien d'après les collections de R. Vaufrey conservées à l'IPH à Paris. Unpublished Ph.D. dissertation, Université Paris X, Nanterre.

Inizan, M.-L. 1980 Séries anciennes et économie du débitage. In *Préhistoire et Technologie lithique* (11–13 Mai 1979 Valbonne), pp.28–30, CNRS, Paris.

Inizan, M.-L. 1984 Débitage par pression et standardisation des supports: un exemple capsien au Relilaï (Algérie). In *Préhistoire de la Pierre Taillée 2: économie du débitage laminaire*, edited by J. Tixier, M.-L. Inizan and H. Roche, pp. 85–92. CREP, Meudon, Paris.

Inizan, M.-L. 1985 Le débitage par pression au Moyen-Orient: premières observations. In *De l'Indus aux Balkans, Recueil Jean Deshayes*, edited by J.-L. Huot, M. Yon and Y. Calvet, pp. 43–45. ERC, Paris.

Inizan, M.-L. 1986 Technologie et Préhistoire récente en Mésopotamie: l'exemple du débitage par pression et de l'économie de l'obsidienne. In *Préhistoire de la Mésopotamie*, pp. 305–315. CNRS, Paris.

Inizan, M.-L. 1988 Technologie lithique: le débitage par pression en Orient. Perspectives de recherche. In *L'Asie centrale et ses rapports avec les civilisations orientales, des origines à l'âge du fer I*, pp. 53–58. Coll. Intern franco-Soviétique. Diffusion de Boccard, Paris.

Inizan, M.-L. 1991 Le débitage par pression: des choix culturels. In *25 ans d'études technologiques en Préhistoire. Bilan et perspectives*, pp. 367–377. XI Rencontres Internationales d'Archéologie et d'Histoire d'Antibes. APDCA, Antibes.

Inizan, M.-L. 2002 Tailler des roches par pression; émergence d'une technique, étapes de sa diffusion dans le monde. In *Matériaux, productions, du Néolithique à l'Âge du Bronze*, Guilaine J. (ed), pp. 33–49. Errance, Paris.

Inizan, M.-L., and M. Lechevallier 1985 La taille du silex par pression à Mehrgarh, Pakistan. La tombe d'un tailleur? *Paléorient* 11(1): 111–118.

Inizan, M.-L., and M. Lechevallier 1991 A techno-economic approach: some examples of blade pressure debitage in the Indo-Pakistani sub-continent. In *9th South Asian Archaeology 1987*, edited by M Taddei and P. Callieri, pp. 43–60. ISMEO, Rome.

Inizan, M.-L., and M. Lechevallier 1994 L'adoption du débitage laminaire par pression au Proche orient. In *Neolithic Chipped Stone industries in the Fertile Crescent*, edited by H.G. Gebel and S.K. Kozlowski, pp. 23–32. Proceedings of the first Workshop on PPN Chipped Lithic Industries, Berlin 1993. ex Oriente, Berlin.

Inizan, M.-L., and M. Lechevallier 1997 A transcultural phenomenon in the Chalcolithic and Bronze Age lithics of the Old World: raw material circulation and production of standardized long blades. The example of the Indus civilization. In *South Asian archaeology, 1995: Proceedings of the 13th Conference of the European Association of South Asian Archaeologists, Cambridge, 5–9 July, 1995*, edited by F.R. Allchin and B. Allchin, pp. 441–449. Oxford and IBH, New Delhi.

Inizan, M.-L., M. Lechevallier, and J. Pelegrin 1994 Lithic technology in Harappan times. In *South Asian Archaeology 1993,* edited by A. Parpola and P. Koskikallio, pp. 245–256. Annales Academiae Scientarum Fennicae, Series B, Vol. 271, Suomalainen, Helsinki, pp. 587–598

Inizan, M.-L., M. Lechevallier, and P. Plumet 1992 A technological marker of the penetration into North America: Pressure microblade debitage. Its origin in the Palaeolithic of North Asia and its diffusion. In *Issues in Art and Archaeology III*, edited by P. Vandiver, J.R. Druzik, G.S. Wheeler and I.C. Freestone, pp. 661–681. Materials Research Society Symposium Proceedings Vol 267, Pittsburgh.

Inizan, M.-L., and J. Tixier 1983 Tell el' Oueili, le matériel lithique. In *Larsa et Oueili, travaux de 1978–1981*, pp. 164–175. ERC, Paris.

Inizan, M.-L., and J. Tixier 2001 L'émergence des arts du feu: le traitement thermique des roches siliceuses. *Paléorient* 26(2): 23–36.

Jelinek, A.J. 1965 Lithic Technology Conference, Les Eyzies, France. *American Antiquity* 31: 277–278.

Kajiwara, H., and Y. Yokoyama 2003 The significance of the Japanese-Russian joint archaeological expedition at the site of Ustinovka complex, primorye region. In *Foraging population of the Sea of Japan during the late Pleistocene-Early Holocene*, edited by A.P. Deverianko and N.A. Konorenko, pp. 18–35. Institute of History, Archeology & Ethnography of the peoples of the Far East, Russian Academie of Science/Tohoku Fukushi University, Novosibirsk.

Kimura, H. (editor) 1992 *Reexamination of the Yubetsu technique and study of the Horokazawa Toma lithic Culture*. Sapporo University, Sapporo (Anglais/Japonais).

Kimura, H. (editor) 1993 *The origin and Dispersal of Microblade Industry in Northerne Eurasia*. Sapporo University, Sapporo (Anglais/Japonais).

Kimura, H. (editor) 1999 *Microblade Industries in Siberia 2*. Archaeology Series 8, 6, 2. Sapporo University, Sapporo (Anglais/Japonais).

Kobayashi, K. 2007 Dates for the first half of the Jomon Period. *Bulletin of the National Museum of the Japonese History* 137: 89–131.

Kononenko, N. 2008 Functional analysis of stone artefacts from the Paleolithic site of Hopyeong-dong, Korea. In *Hopyeong-dong Paleolithic site (Namyangju, Gyeonggi Province, Korea)*, edited by M.-Y. Hong and J.-H. Kim, pp. 746–789. Korea Land Corporation, Gyeonggi Cultural Foundation Gijeon Institute of Cultural Properties (Excavation Report No. 93) Vol. 2 (Korean/English abstract).

Kuzmin, Y.V., and A. Orlova 1998 Radiocarbon Chronology of the Siberian Paleolithic. *Journal of world Prehistory* 12(1): 1–20.

Lechevallier, M. 1988 L'industrie lithique de Mehrgarh. In *Les cités oubliées de l'Indus*, edited by J.F. Jarrige, pp. 56–57. AFAA, Paris.

Lechevallier, M. 2003 *L'industrie lithique de Mehrgarh – Fouilles 1974–1985*. ERC, Paris.

Lechevallier, M., and V. Marcon 1998 Un dépôt de nucléus sur le site néolithique de Mehrgarh, Pakistan. *Paléorient* 24(2): 89–97.

Lee, Y-J., and Y.-H. Yun 1993 Tanged point and Micro-blade Core from Suyanggae, Korea. In *The origin and Dispersal of Microblade Industry in Northerne Eurasia*, edited by H. Kimura, pp. 52–69. Sapporo University, Sapporo.

Leroi-Gourhan, A. 1943 *Evolutions et techniques I: l'Homme et la matière*. Albin Michel, Paris (réédité en 1971).

Leroi-Gourhan, A. 1964 *Le geste et la parole i: technique et langage*. Albin Michel, Paris.

Midant-Reynes, B. 1983 Le débitage par pression: les éléments de faucille de 'Ayn-Asil (Oasis de Dakhla). *Bulletin de l'Institut français d'Archéologie Orientale* 83: 257–262.

Ohnuma, K. 1993 Experimental studies in the determination of manners of micro-blade detachment. *Al-Rafidan* 14: 153–181.

Ono, A. 2004 Recent Studies of the Late Palaeolithic Industries in the Japanese Islands. In *Recent Paleolithic Studies in Japan - Proceedings for tainted Evidence and restoration of Confidence in the Pleistocene Archaeology of the japanese Archipelago*, edited by K. Yajima. The Japanese Archaeological Association, the Ministry of Education, Culture, Sports, Science and Technology of Japan, Tokyo.

Ono, A., H. Sato, T. Tusumi, and Y. Kudo 2002 Radiocarbon dates and archaeology of the late Pleistocene in the Japonese islands. *Radiocarbon* 44: 477–494.

Ono, Y. 1999 Ainu homelands: natural history from Ice Age to modern times. In *Ainu Spirit of a northern people*, edited by W.W. Fitzhugh and C.O. Dubreuil, pp. 32–38. National Museum of Natural History, Smithonian Institution with University of Washington Press, Washington.

Ono, Y., and Y. Igarashi 1992 *Natural History of Hokkaïdo*. Hokkaïdo Daigaku, Sapporo.

Pelegrin, J. 1984 Approche technologique expérimentale de la mise en forme de nucléus pour le débitage systématique par pression. In *Préhistoire de la Pierre Taillée 2: économie du débitage laminaire*, edited by J. Tixier, M.-L. Inizan and H. Roche, pp. 93–103. CREP, Meudon, Paris.

Pelegrin, J. 1988 Débitage expérimental par pression: "Du plus petit au plus grand". In *Technologie préhistorique,* edited by J. Tixier, pp. 37–53. Notes et Monographies Techniques du CRA No. 25. CNRS, Paris.

Pelegrin, J. 1994 Lithic technology in Harappan times. In *South Asian Archaeology 1993,* edited by A. Parpola and P. Koskikallio, pp. 587–598. Annales Academiae Scientarum Fennicae, Series B, Vol. 271. Suomalainen Tiedeakatemia, Helsinki.

Pelegrin, J. 1997 Débitage au Chalcolithique de grandes lames de silex par pression au levier. In *Annuel des sciences et techniques, La science au présent*, Vol 17. Encyclopaedia Universalis, Paris.

Pelegrin, J. 1995 *Technologie lithique - Le Châtelperronien de Roc-de-Combe (Lot) et de la Côte (Dordogne)*. Cahiers du Quaternaire 20. CNRS, Paris.

Pelegrin, J. 2002 Principes de la reconnaissance des méthodes et techniques de taille. In *Tell'Atij, Tell Gudeda, industrie lithique, analyse technologique et fonctionnelle*, edited by J. Chabot, pp. 215–226. Cahiers d'Archéologiqie du CELAT 13, Quebec.

Pelegrin, J. 2003 Blade-making techniques from the Old World. In *Mesoamerican Lithic Technology – Experimentation and interpretation*, edited by K. Hirth, pp. 55–70. University of Utah, Salt Lake City.

Perlès, C. 1984 Débitage laminaire de l'obsidienne dans le Néolithique de Franchti (Grèce). In *Préhistoire de la Pierre Taillée 2: économie du débitage laminaire*, edited by J. Tixier, M.-L. Inizan and H. Roche, pp. 129–138. CREP, Meudon, Paris.

Perlès, C. 1987 *Les industries lithiques taillées de Franchthi (Argolide Grèce) 3: du néolithique ancien au néolithique final.* Excavation at Franchthi Cave, facicule 13, edited by Thomas W. Jacobsen. Indiana University Press, Bloomington and Indianapolis.

Perlès, C. 1990 L'outillage de pierre taillée néolithique en Grèce: appovisionnement et exploitation des matières premières. *Bull de correspondance hellénique* CXIV(I): 1–42.

Perlès, C. 2007 Diffusions, emprunts, refus d'emprunts: les acteurs humains. In *Mobilités, Immobilismes - L'emprunt et son refus (Colloques),* edited by P. Rouillard, C. Perlès and E. Grimaud, pp. 319–333. De Boccard, Paris.

Pituljko, V.V. 1998 *The Zhokhov Site*. Russian Academy of Sciences, Institute for the History and Material Culture, St. Petersburg.

Rahmani, N. 2004. Technological and cultural change among the last Hunter-Gatherers of the Maghreb; the Capsian (10000-6000 B.P.). *Journal of World Prehistory* 8: 57–105.

Rankama, T., and J. Kankaanpää 2007 The earliest postglacial inland settlement of Lapland. In *Kamennyy vek Evropeyskogo Severa,* edited by A.V. Volokitin, V.N. Karmanov and P.J. Pavlov, pp. 44–65. Russian Academy of Science, Ural Branch, Komi Scientific Centre, Institute of language, literature and History, Syktyvkar.

Seonbok, Yi, and G. Clark 1996 Dyuktai culture and New world. *Arctic Anthropology* 26: 1–36.

Serizawa, C. 1971 *Problems of the Early Palaeolithic in Japan*. Vol. 9/3–4. Tohoku University, Dpt. of Archaeology, Faculty of Arts and Letters, Tohoku.

Smith, J.W. 1974 The Northeast Asian-northwest American Microblade Tradition (NANAMT). *Journal of Field Archaeology* 1: 347–364.

Suzuki, C. 1993 Distributional approach to the microlithic Culture of Japanese Archipelago: geography, fauna, flora and population. In *The origin and Dispersal of Microblade Industry in Northern Eurasia*, edited by H. Kimura, pp. 185–190. Sapporo University, Sapporo (Anglais/japonais).

Svoboda, J. 1995 Wedge-shaped core microblade cores from Moravia and Silesia. *Archeologické rozhledy* XLVII: 652–656

Tixier, J. 1963 *Typologie de l'Epipaléolithique de Maghreb*. Mémoire du C.R.A.P.E., Alger, AMG, Paris.

Tixier, J. 1967 Procédés d'analyse et questions de terminologie dans l'étude des ensembles industriels du Paléolithique récent et de l'Epipaléolithique du Nord-Ouest. In *Background to evolution in Africa*, edited by W.W. Bishop and J.D. Clark, pp. 771–820. University of Chicago Press, Chicago and London.

Tixier, J. 1971 Les apports de la stratigraphie et de la typologie au problème de l'homme moderne dans le Maghreb. In *Origine de l'homme moderne*, pp. 121–127. Unesco, Paris.

Tixier, J. 1976 L'industrie lithique capsienne de l'Aïn Dokkara, région de Tebessa. Fouilles L. Balout. *Libyca* 24: 21–54.

Tixier, J. 1978 *Méthode pour l'étude des outillages lithiques.* Doctorat ès lettres, Université Paris X, Nanterre.

Tixier, J. 1984 Le débitage par pression. In *Préhistoire de la Pierre Taillée 2: économie du débitage laminaire*, pp. 57–70. CREP, Meudon, Paris.

Tozawa, M. 1974. *The stone implements of the preceramic Age in Japan.* Archaeological Institute of Meiji University, Tokyo (Japonais: photos sous pochette).

Vaufrey, R. 1933a Notes sur le Capsien. *L'Anthropologie* 43: 457–483.

Vaufrey, R. 1933b Stratigraphie et répartition des faciès capsiens. *L'Anthropologie* 43: 648–649.

Vaughan, P. 1995 Functionnal analysis of samples of Neolithic, Chalcolithic and Bronze Age flints from Mehrgarh (Balochistan, Pakistan). In *Mehrgarh: Field reports 1974–1985. From Neolithic times to Indus Civilisation*, edited by C. Jarrige and J.-F. Jarrige pp. 614–647. Department of Culture and Tourism, Government of Sindh and French Ministry of Foreign Affairs, Karachi.

Yajima, K. (editor) 2004. *Recent Paleolithic Studies in Japan - Proceedings for tainted Evidence and restoration of Confidence in the Pleistocene Archaeology of the Japanese Archipelago*. The Japanese Archaeological Association, The Ministry of Education, Culture, Sports, Science and Technology, Japan.

Yoshizaki, M. 1963 Study of Shirataki Site. In Shirataki Research Group, 8 pl. (English abstract).

Chapter 3
Stoneworkers' Approaches to Replicating Prismatic Blades

John E. Clark

3.1 Introduction

The title of this chapter alludes to articles by Don Crabtree (1966, 1968) on the replication of Folsom points and Mesoamerican prismatic blades. I evaluate Crabtree's contributions in light of subsequent experiments in making pressure and percussion blades, with special attention to Mesoamerican blades. I follow a practitioner's perspective in outlining insights gained from attempts to duplicate blades and cores. Jacques Pelegrin (1990: 118) identifies two kinds of knowledge associated with flintknapping: *connaissances* ("knowledge") and *savoir-faires* ("know-how"). Both are gained through knapping and required for knapping. This distinction corresponds to the difference between "declarative and procedural memory" and includes mental representations, programmed knapping gestures, and intuitive and evaluative operations (1990: 118). What must a knapper be able to think, know, and remember in order to follow a sequence of operations and arrive at a desired end? What did ancient blademakers know? The purpose of this chapter is to track the knowledge and wisdom of blademaking proposed by modern knappers.

Tatsuo Kobayashi (1970) proposed a useful distinction for pressure cores. Blades removed from cylindrical cores, such as characteristic of Mesoamerica, belong to his System B. System A blades are from cores made from bifaces, generally for the production of microblades (cf. Inizan 1991; Parry 1994). Most experiments have been with Type B cores. I follow the historic sequence of blademaking experiments for both types of cores, picking up the narrative a century before serious experimentation began. By all accounts, the pivot point was Crabtree's blademaking experiments (1968). I highlight his work and then consider subsequent experiments from knappers around the world. Where feasible, experiments are described in their order of publication. Experimentation with blades has been much more extensive than the

J.E. Clark (✉)
Department of Anthropology, Brigham Young University, Provo, UT, USA
e-mail: John_Clark@byu.edu

P.M. Desrosiers (ed.), *The Emergence of Pressure Blade Making: From Origin to Modern Experimentation*, DOI 10.1007/978-1-4614-2003-3_3,

published record indicates (e.g., Callahan 1985, 1987b, 1995a: 225; Pelegrin 1984b, c; Kelterborn 2008a), but I am limited to publications and written correspondence.

3.2 A History of Mesoamerican Blade Experiments

Descriptions of the manufacture of Mesoamerican fine obsidian blades start with the Spanish Conquest of the Aztecs (a.k.a., *Mexica* or Mexicans) in the early sixteenth century. Eyewitness accounts provide information that has dictated the questions pursued in experiments. One reason the Mexica technique for blademaking was recorded and reported to the Spanish crown was that the ingenuity involved in making "obsidian razors" was proof of the intellectual abilities and skills of the natives and thus constituted a strong Aristotelian argument for the humanity of Amerindians (Titmus and Clark 2003: 90). Modern knappers lament that these accounts do not contain greater detail, but they provide critical details on the manufacturing process and blademaking tool.

3.2.1 The Speculative Prelude to Replication Experiments

To the great early anthropologist Edward B. Tylor (1861), we owe the first contribution concerning Mesoamerican prismatic blades. He published a translation of the description of blademaking found in Juan de Torquemada's (1615) *Monarquía Indiana,* identified artifacts that corresponded to this old description, and also related these ribbons of obsidian to their cores (called "prisms"). His translation of Torquemada [also republished in John Lubbock's *Pre-Historic Times* (1865: 78)] was widely available in Europe (see Daubreé and Roulin 1868: 20–26). Tylor's account of obsidian razors (Fig. 3.1) begins with his visit to an ancient obsidian quarry. His speculations on their manufacture drew on his knowledge of English gunflint manufacture:

> the workman who makes gun-flints could probably make some of the simpler obsidian implements, which were no doubt chipped off in the same way. The section of a gun-flint, with its one side flat for sharpness and the other side ribbed for strength, is one of the characteristics of obsidian knives. That the flint knives of Scandinavia were made by chipping off strips from a mass is proved by the many-sided prisms occasionally found there, and particularly by that one which was discovered just where it had been worked, with the knives chipped off it lying close by, and fitting accurately into their places upon it … One can often see, on the ends of the Scandinavian flint knives, the bruise made by the blow of the hard stone with which they were knocked off … I have heard on good authority, that somewhere in Peru, the Indians still have a way of working obsidian by laying a bone wedge on the surface of a piece and tapping it till the stone cracks. Such a process may have been used in Mexico.
>
> (Tylor 1861: 95–99)

Later in his book, Tylor presents the Torquemada account which indicates that blades "were cracked off by pressure" rather than indirect percussion (Tylor 1861: 331).

Tylor's remarkable description started with observations of obsidian blades and related them to artifacts from the Old World. He relied on ethnohistoric accounts, ethnographic analogy to gunflint knappers, fracture mechanics, and a close inspection

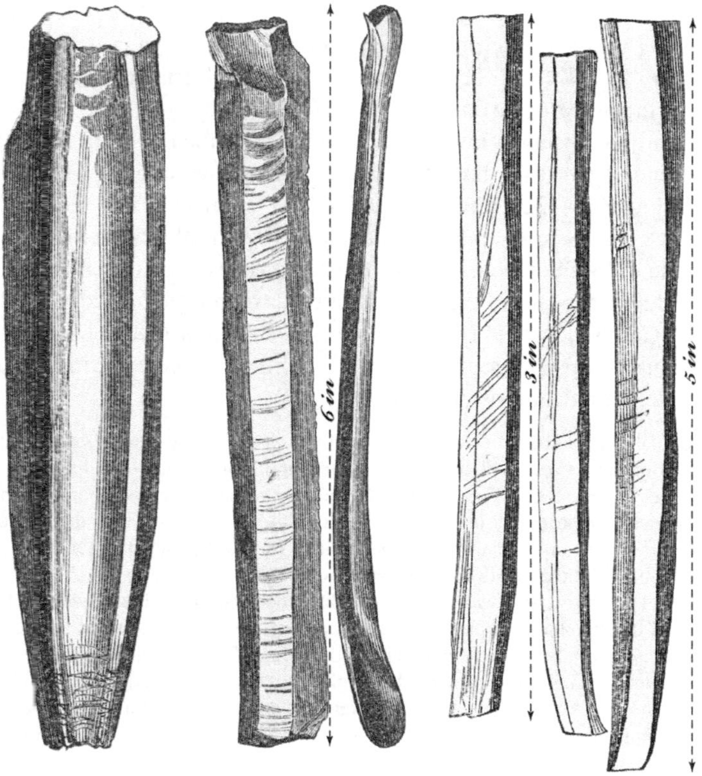

Fig. 3.1 "Fluted Prism of Obsidian," or blade core (*left*), and "Aztec Knives or Razors," or fine blades (*right*), illustrated by Tylor (1861: 96, 98, not to scale)

of cores and blades for the telltale bruises of knapping. It does not seem like much now, but Tylor's identification of prisms as expended cores related to fine blades was a major insight. Four decades later, Adela Breton (1902) still questioned whether these "many-sided objects" were cores, a case made just 2 years earlier by her colleagues William Henry Holmes (1900) and George MacCurdy (1900).

The Torquemada account appears to have been added to Tylor's book as an afterthought. It is not a primary source; Torquemada copied his description of blade-making from Gerónimo de Mendieta's (1971: 406) *Historia Eclesiastica Indiana* (Bk. 4: Chap. 12):

> And they make them [blades], if I can get you to understand, in this manner: they sit on the ground and take a piece of that black stone … That piece which they take is a hand or more long, and thick like the leg or slightly less, and round. They have a staff the thickness of a lance and three cubits long [125 cm] or slightly more, and at the end of this staff they fasten and securely tie a piece of wood one hand long [21 cm], thick like the upper arm or a bit more, and this has its face flat and smoothed, and this serves to make that part weigh more. They put together their unsandaled feet, and with them they press the stone with their chest, and with both hands they take the staff that I said was like a spear shaft, which is also flat and smooth, and they place it against the edge of the face of the stone, which is also smoothed and flat, and then they press towards the chest, and quickly a blade leaps from the stone with its point and two sharp edges …. (my translation)

Tylor also summarized Francisco Hernandez's (1571 [1959]) description of the blademaking tool and interpreted the Latin as describing a crutch-like implement (repeated by Evans 1872: 22; Stevens 1870: 80): "Hernandez ... gives a similar account of the process. He compares the wooden instrument used to a cross-bow. It was evidently a **T**-shaped implement, and the workmen held the cross-piece with his two hands against his breast, while the end of the straight stick rested on the stone" (Tylor 1861: 332). This interpretive leap was in error. By remarkable coincidence, however, there is ethnographic support from other Amerindians for a tool like the one Tylor imagined. This information comes from observations of Indians of the Western territories of the United States, as recounted to George Sellers (1886) by George Catlin (quoted in Holmes 1919: 322–323; reprinted in Moorehead 1910:I: 53–55). Catlin's observations significantly influenced the experiments undertaken to make Mesoamerican prismatic blades:

> ... good flakes could be split from ... [stones by] ... *impulsive pressure*, the tool used being a shaft from 2 and 3 inches diameter, varying in length from 30 inches to 4 feet, according to the manner of using them. These shafts were pointed with bone or buck-horn, inserted in the working end as represented in Figure 2, bound with sinews, or rawhide thongs, to prevent splitting. For some kinds of work the bone or horn tips were scraped to a rather blunt point, others with a slightly rounded end of about one-half inch in diameter .... A water-worn pebble broken transversely was commonly held by being sufficiently imbedded in hard earth to prevent its slipping when held by the foot as the pressure was applied. Large blocks of obsidian or any easily flaked stones were held between the feet of the operator while sitting on the ground, the *impulsive pressure* being given to the tool grasped in both hands, a cross-piece on the upper end resting against his chest, the bone end against the stone in a slight indentation, previously prepared, to give the proper angle and to prevent slipping.
>
> In some cases the stone operated on was secured between two pieces or strips of wood like the jaws of a vise, bound together by cords or thongs of rawhide; on these strips the operator would stand as he applied the pressure of his weight by impulse ...
>
> Figure 2 [b] represents ... the rude sketches made of the flaking tool used to throw off massive flakes, when a sudden percussive pressure was required in addition to the impulsive pressure the man could give. The staffs of these flaking tools were selected from young hard-wood saplings of vigorous growth. A lower branch was utilized, as shown at *a* in Figure 2, to form the crotch in which the blow was struck. Another branch on the opposite side, *a* [prime], was used to secure a heavy stone to give weight and increase the pressure. When the stone to be flaked was firmly held, the point adjusted to give the pressure in the required direction, the staff firmly grasped, the upper end against the chest of the operator, he would throw his weight on it in successive thrusts, and if the flake did not fly off, a man standing opposite would simultaneously with the thrust give a sharp blow with a heavy club represented in the cross section *b* in Figure 2[b], it being so shaped that its force is downward close in the crotch. It has been represented to me that a single blow rarely failed to throw off the flake, frequently the entire depth of the block of stone, sometimes as much as 10 to 12 inches. The tooth or tusk of the walrus was highly prized for tips of the flakers.
>
> (Sellers 1886: 874–875, original emphasis; cf. Holmes 1919: 322–323)

After quoting this passage in his comprehensive work on American stone antiquities, Holmes (1919: 323) noted the similarities of Sellers's description to Torquemada's. Sellers's account is more detailed, and it came with illustrations of the tools (Fig. 3.2). The famous illustration of blademaking published by Holmes (1919: 323, Fig. 182) is his own imaginary reconstruction (Fig. 3.3). Holmes's drawing had more impact than the verbal descriptions themselves. There are enough

Fig. 3.2 Crutch tools for impulsive pressure and assisted pressure illustrated by Sellers (1886: 874, 875; see also, Moorehead 1910: I:19, Figs. 15, 16): (a) "crotch in which the blow was struck," (a') place to secure a heavy stone for weight, and (b) heavy club (shown in cross-section) to strike a blow

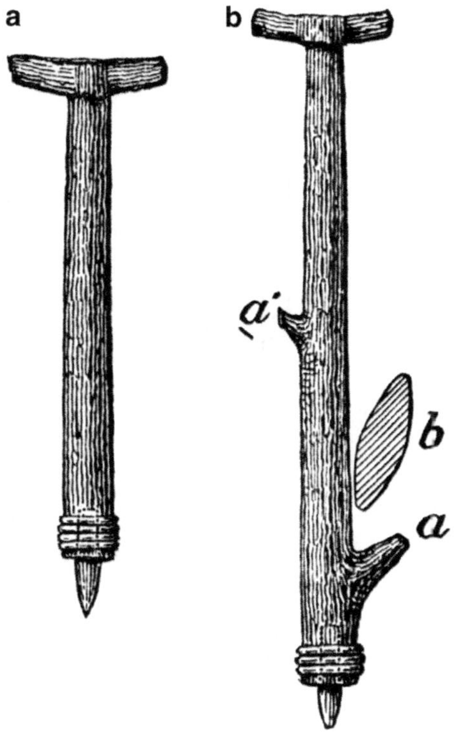

Fig. 3.3 Hypothetical reconstruction proposed by Holmes (1919: 323, Fig. 182) of "Technique of flaking by Mexican Indians as described by Torquemada, and by western United States tribes as described by Catlin"

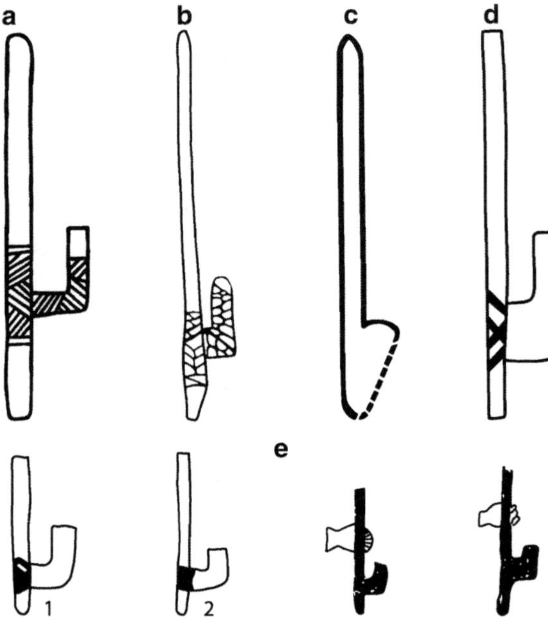

Fig. 3.4 Itzcolotli tools from different sources; note their blunt ends. (**a**) Itzcolotli showing the careful binding of the attached piece of wood to the shaft of a three-cubit-long tool (From Williams and Harvey 1997: 302). (**b**) The tool illustrated by Sahagún (1963: Plate 778) (facsimile, see Thouvenot 1984: Fig. 1 or Clark 1989b: Fig. 1). (**c**) Drawing of the tool from the *Relación de Michoacán* (From Clark 1982: 358, Fig. 1c). (**d**) Ursula Dykerhoff's (1982) drawing of the tool from the *Matrícula de Huexotzinco*. (**e**) Marc Thouvenot's (1984: Figs. 3, 4) depictions of the four tools from the *Matrícula de Huexotzinco* (see also, Prem 1974)

similarities in the descriptions that it appears that ambiguities in Torquemada's summary can be filled in with information from Sellers's account. But combining details from these sources is unwarranted; their amalgamation adversely affected experiments. Different tools and techniques were involved, and these required different working positions, forces, and means of stabilizing cores. Tylor and Holmes assumed the piece of wood tied to the shaft of the Mexican blademaking tool formed a crosspiece of a crutch. It did not. Instead, it formed a hook-like element at one end of a long, staff-like tool called *itzcolotli* (see Clark 1982; Fletcher 1970; Titmus and Clark 2003: 74, Fig. 5.1), as illustrated in Fig. 3.4. The different tools described by Torquemada and Sellers have been the major organizing feature of Mesoamerican blade experiments – a chest crutch versus a hooked lance. Crabtree has long been credited as the first person in modern times to reproduce Mexica razors. An unappreciated irony of his achievement is that he did not make them the Mexican way.

3.2.2 Blade Experiments Before Crabtree

Experiments in lithic technology have been of two types: those designed to understand technology in general and those designed to duplicate specific artifact types. The first starts with experiments and then examines artifacts; the other begins with artifacts and attempts to replicate them. Pelegrin (1991: 120) designates the first approach as "Technical Research." The second approach is known as "Replication Experiments," sensu Jeffrey Flenniken (1981b, 1984), in which an experimenter attempts to duplicate the original conditions and outcomes associated with the manufacture or use of a tool. Historically, the trend has been from Technical to Replication research. Both are valid and useful.

As reported by N. Joly (1883: 212), the first experiments for making Mexican blades were by M. Courtis in 1865:

> M. Courtes [Courtis], member of the French Scientific Commission of Mexico, and M. Chabot, maintain that the Aztecs, in making their obsidian razors, begin by shaping the rock near the quarry whence it was taken. Then after having given to it the form of a prism terminated at one extremity by a blunt point, at the other a flat surface, the workman takes this prism in the left hand, and pressing it against some resisting surface, strikes it at first with light blows, gradually increasing them in force until at last he obtains splinters as sharp as razors, and destined to serve the same purpose.

In his analysis of Aztec blademaking, P. Marcou (1921) provided the 1615 Spanish text of Torquemada and a critique of Tylor's translation of it (see Cabrol and Coutier 1932; Thouvenot 1984). Daubrée and Roulin (1868) had earlier corrected Tylor's translation but did not supply the original text. M. Léon Coutier (in Cabrol and Coutier 1932) tried to follow experimentally the Torquemada description and claimed "it is absolutely impossible for a seated man to effect the abrupt blow that will produce the strong pressure indispensable for detaching a blade" (1932: 580, translation, Olivier de Montmollin). Coutier announced success by indirect percussion instead (see Barnes 1947a: 625). A principal reason for lack of success with the Mexican technique is that Coutier used the wrong tool. Just 2 years before the famous Les Eyzies Lithic Conference in 1964, Coutier demonstrated pressure flaking of flint and glass with a hard wooden tool (Pelegrin 2003: 55), so it was certain that wooden tools could be used on obsidian, as described in Spanish accounts.

A key protagonist in the history of blade experimentation was William Henry Holmes. He described 13 different techniques for chipping stone and many kinds of tools. His coverage defined the limits of the thinkable for most experiments that followed (Holmes 1919). Many tools and holding positions are possible, but fracture was initiated in only four ways: direct percussion, indirect percussion, pressure, and assisted pressure.

The systematic investigation a generation later by William Holmes Ellis at the Ohio Lithic Laboratory, beginning in January of 1938, started with the list of techniques from Holmes's (1919) book and explored them, as well as logical permutations from them. Ellis (1938, 1940, 1965: iv) described and/or tested 19 techniques, including impulsive pressure with a chest crutch, a shoulder crutch, and a "lever and fulcrum" device with a copper tip (Ellis 1965: 36–41). His work was technical research rather than replication, with no attempt to duplicate specific artifacts. Coutier appears to have had some influence on this beginning of the American tradition of formal knapping. Ellis (1965: 9) describes Coutier's visit in 1937, before the founding of the Lithic Laboratory, and his demonstration to Director Henry C. Shetrone of the wooden billet knapping technique.

Ellis quoted Sellers, Torquemada, and Hernandez (the last two from Tylor's translations) and depicted pressure work with a T-shaped chest crutch in a manner similar to Holmes's illustration (Ellis 1965: 37, Fig. XIX). He lists "crutch" in his glossary as a tool with a "bone point inset" (1965: 49). He did not reference blades but implied them under "fluting: channels or grooves in flint caused by the removal of flakes, as from a core or Folsom point" (1965: 49). Ellis apparently did not make any blades by impulsive pressure.

3.2.3 Don Crabtree: Mesoamerican Obsidian Polyhedral Cores and Prismatic Blades

Crabtree started his blade experiments where Ellis left off. He was influenced by Ellis and others, and in his turn, he influenced hundreds directly and thousands indirectly. Crabtree's accomplishments cannot be separated from his biography. Here I focus on critical points of his life circumstances related to the discovery of Aztec blademaking and related matters. In terms of knapping, Crabtree's career spanned 50 years and three stages. In piecing together highlights of his career, I relied on published sources and Crabtree's correspondence, especially that with Jacques Tixier. These letters provide remarkable documentation for Crabtree's activities leading up to the publication of his papers on Folsom points and Mesoamerican cores and blades.

Crabtree traced his interest in stone tools to when he was 5 years old and "received some arrowpoints for running errands for [his] mother to the neighbors" (Crabtree, in Callahan 1979a: 31). What interested him was how the Indians "were able to work stone harder than steel with such perfection" (in Callahan 1979a: 31). Crabtree's first attempts at making arrowheads from the local agate were at the age of 7 (Knudson 1978; Times-News 1970), and he taught himself to use hammerstones and billets by the age of 12 (Times-News 1978; Titmus 1981). Against his father's wishes (Harwood 1999), he kept up this hobby and continued to expand his mastery of different techniques, stone tool types, and raw materials. A trying circumstance arose in 1939 when, at the age of 27, he was diagnosed with cancer. As therapy to recover from cobalt treatments, Crabtree "spent his recuperation period, when his mobility was limited and he was trying to regain muscular strength, flintknapping – making arrowheads, spearpoints, and eccentric forms by the hour" (Knudson 1982: 337). In his convalescence, Crabtree became proficient in making stone tools, and this led to opportunities that brought him to the attention of archaeologists. "In the spring of 1941, fully recovered and with a year of concentrated flintknapping behind him, Crabtree was invited to demonstrate knapping techniques at the American Association of Museums' annual meeting in Columbus, Ohio" (Knudson 1982: 337; see also Crabtree 1966: 6, 22; Titmus 1981; Woods 1981). Contrary to the published account, this followed rather than preceded his 2 months of work in the Lithic Laboratory working with Ellis and Shetrone (letter of R. G. Morgan to Crabtree, March 12, 1941, courtesy of Bradley Lepper). Presumably, Crabtree gained some familiarity with Ellis's (1940) method of experimenting with different techniques and knew of Torquemada's description of blademaking.

Crabtree's budding career was cut short by World War II. During this second stage of his adult life, he helped the war effort, married his wife Evelyn, and worked for a living at a variety of occupations, mostly in the Twin Falls area of southern Idaho. He continued to knap as a serious hobby but did not come to the attention of the scientific community until being rediscovered by Earl Swanson in 1958 (Knudson 1982: 338). A year earlier, Swanson had founded the Anthropology Department at Idaho State College (later changed to Idaho State University) in Pocatello, a city 110 miles on the freeway east of Twin Falls. The Swanson-Crabtree encounter in November of 1958 led to collaboration which lasted until Swanson's untimely death in 1975. Swanson reintroduced Crabtree to the archaeological community by having Crabtree, a craftsman "capable of astonishing excellence in the manufacture of specific kinds of functional stone tools" (Swanson and Butler 1962: 8), give the opening presentation at "The First Conference of Western Archaeologists on Problems of Point Typology" held in Pocatello on March 15, 1962 (Bray et al. 1975: 36; Swanson 1966; Swanson and Butler 1962). "Later that same year, Don suffered a coronary occlusion and was forced to retire from his government position on a disability. However, the participants at the point typology session had become interested in Don's work through Swanson and so with their help [Don] was sent by the National Science Foundation to attend the Lithic Technology Conference in Les Eyzies, France, in November of 1964" (Knudson 1978: 2–3; also Knudson 1982; Smith 1966; Wormington 1985). Retirement allowed Crabtree to dedicate himself almost full-time to flintknapping. Thus began the most productive decade of Crabtree's career. Swanson obtained grant money, mostly from the National Science Foundation, for Crabtree to go to conferences and perform knapping demonstrations, to write articles and a handbook on lithic technology, produce five educational films, and to run a flintknapping fieldschool (ISU Bengal 1965; Knudson 1982; Statham 1978; Swanson 1966). At the time of Swanson's death, he and Crabtree had submitted a sixth NSF grant proposal which requested 2 years of funding for more writing and publication (Swanson and Crabtree 1974). Crabtree intended to explore a block-on-block and a Danish direct percussion technique in making blades; he also intended further studies of cores and blades from Mesoamerica (Swanson and Crabtree 1974: 1–2). Because of Swanson's death, the grant was canceled, and Crabtree never completed his synthesis of blades and blademaking.

The Les Eyzies conference catapulted Crabtree onto the world stage and introduced him to colleagues who became lifelong friends, especially Marie Wormington, François Bordes, Cynthia Irwin Williams, and Jacques Tixier. Each of these scholars influenced and aided Crabtree's career. At the Les Eyzies conference, Crabtree demonstrated the manufacture of obsidian pressure blades and Folsom points (illustrated in Sonneville-Bordes 1967: 50, 51). In turn, he witnessed techniques for making percussion flint blades, a traditional concern of European scholars (see Barnes 1947a, b; Bordes 1947, 1950, 1967). I have not found any concrete record of when Crabtree first made pressure blades, but a 1960 photograph of his display case taken by Gene Titmus shows an Aztec wooden sword edged with obsidian blades (see Times-News 1963, 1978). Crabtree told me it took him 20 years after working with Ellis to figure out how to make blades (Clark 1989a: 131). This would be about the time he met Swanson and Titmus (see below).With encouragement from Tixier, Crabtree started making percussion blades several months after the Les Eyzies conference, and this continued for the next decade (e.g., Bordes and Crabtree 1969b: 1, 7).

Crabtree's blade experiments are best understood within a wider range of activities which included work in fluting Folsom points, a problem that first intrigued him at the age of 16 (Crabtree 1966: 6). His interest in Mesoamerican fluted cores came much later. He described his experimental method more systematically for his Folsom experiments than for his Mesoamerican study. He began trying to make Folsom points in the early 1930s and persisted into the 1970s (Crabtree 1966: 6). Most of these experiments appear to have been casual tinkering rather than formal exercises (see Clark 2002); presumably, many were carried out during his two decades of hobby knapping before 1958. In 1966, he reported 11 broad classes of experiments using different force applications, tools, and holding devices (see the photo-essay in Bray, Swanson, and Farrington 1975). By his own tally, he conducted "hundreds of experiments over a number of years" within these broad classes of techniques (Crabtree 1966: 22). The general classes were (1) direct freehand percussion with hammerstones and billets, (2) direct percussion striking on an anvil, (3) direct percussion with an anvil rest, (4) indirect freehand percussion, (5) indirect percussion with rest, (6) freehand pressure with a short tool, (7) freehand pressure with a longer tool, (8) freehand pressure with a shoulder crutch, (9) pressure with a chest crutch and clamp, (10) pressure with a chest crutch, clamp, and anvil rest, and (11) chest-crutch pressure and indirect percussion with a clamp and anvil (Crabtree 1966: 9). He concluded that Folsom points may have been fluted by direct pressure with rest (10), or by indirect percussion with a clamp and anvil (5), or a combination of the two (11). Much of the work reported for techniques 5 and 11 came from Titmus's collaborative work after 1959, and most of that for technique 10 was solely Crabtree's work (Titmus, 2009, personal communication; cf. Crabtree 1966: 7, 22). Similar techniques were proposed for making Mesoamerican polyhedral cores and prismatic blades.

Crabtree eventually accomplished what Ellis had not, and that was to make fine blades of obsidian using a "chest" crutch, with the core secured in a simple vise. "I suppose one of my greatest thrills in flintworking was the removal of these flakes" (Crabtree, 1979, personal communication). Responding to a question from Errett Callahan (1979b: 11), Crabtree observed: "I think the most challenging [thing] was to understand how a blade was made. That was a great breakthrough – to get a blade with perfectly parallel sides and a trapezoidal cross-section. I think I thought of that for maybe 20 years before I accomplished it. It was far more exciting in a way than the Folsom point." Crabtree mentioned numerous experiments in his landmark study on Mesoamerican cores and blades published in *American Antiquity* (1968), and he demonstrated pressure and direct and indirect percussion techniques in six educational films made from 1968 to 1970 (Bordes and Crabtree 1969a; Crabtree 1969b, 1972a, c–e; Statham 1978; to view, see Lohse 2000).

Crabtree's blade treatise divides into three parts, beginning with an analysis of Torquemada's description, followed by comments on knapping experiments, and ending with results of experiments in high-speed photography for documenting blade manufacture. The high-speed photography began at the end of 1965 (ISU Bengal 1965). Given his purpose to argue for the successful replication of fine blades, it is curious Crabtree started with commentary on Torquemada – nearly equal the space accorded original experiments. Crabtree attempted to reconcile this historic account with his experiments but was ultimately unsuccessful. Observations coming from his experiments made him question Torquemada's account on basic principles. Crabtree claimed to have produced blades by the Mexica technique – *once he modified the technique!* Crabtree observed that if one were to take Torquemada's description of blademaking verbatim, "we have the picture of an Indian sitting flat on the ground, legs straight in front of him, holding a very sharp core between his naked feet, and pressing off blades with a crutch that measures well over 5ft. This simply will not work, and I suggest that the reader convince himself of this by trying this method personally" (Crabtree 1968: 450).

After dealing with Torquemada, Crabtree sketched the parameters of his experiments. Based on years of experience, and through a process of elimination (see Crabtree 1966, 1972a, b: 62; also, Bordes and Crabtree 1969a), he determined that blades and cores made by pressure "have every quality and characteristic of most cores and blades found in Mesoamerica" (Crabtree 1968: 478). In his text, Crabtree alludes to the same variety of experiments described for his Folsom study but does not describe a single one. He made blades with direct handheld percussion using a variety of "percussion tools" (1968: 457). For indirect percussion, he mixed and matched a wide variety of hammers and punches made of hard and soft stone, antler, horn, bone, shell, ivory, and wood; he also reduced different kinds of cores with various holding devices (1968: 459). His experimental percussion cores and blades lacked the regularity and uniformity of archaeological specimens from Mexico, so Crabtree rejected direct and indirect percussion as plausible techniques for making fine blades. He was able to duplicate the final form of Mesoamerican cores and blades, as well as their microcharacteristics, with a technique of applied pressure with a chest crutch. To do so, he found it necessary to immobilize his cores in a wooden vise or clamp, something mentioned by Sellers but not the Spanish friars. For many of his experiments, Crabtree used cores cut from nodules with a diamond saw. Demonstrations of blademaking with such sawn cores became a regular feature of his flintknapping fieldschool, which began in 1969 (Crabtree 1969a).

For his pressure experiments, Crabtree used a simple wooden vise and a chest crutch of semi-flexible wood 32 in. long fitted with bits of copper, antler, ivory, or bone (1968: 452; also, Crabtree 1967: 64–67). He did not mention the number of blades made with different bits or the characteristics of these blades. In demonstrations recorded on film, he uses a copper-tipped chest crutch (see Bordes and Crabtree

1969a; Crabtree 1972e). Most of his blades appear to have been made with the metal-tipped tool, with the focus of his experiments being the reduction of cores of different shapes. Crabtree devoted considerable attention to possible core preforms and the necessity of outfitting them with a straight ridge or two, a bias he learned from Bordes. He described cores of one, two, three, four, and more-than-four starting ridges.

In commenting on Torquemada's description of the blademaking tool, Crabtree alluded to other experiments. One of his most adamant claims was that one cannot hold a core in the feet and push off blades. He experimented with wooden tools and determined that "just a sharpened wooden stick would not be sufficient to make blades" (Crabtree 1968: 449; also Crabtree 1967: 67; Semenov 1964: 59). He speculated that ancient artisans could have used pressure tools with tips of "antler, bone, or jade" – and then he interjected this insightful, parenthetical comment:

> (Since the writing of this paper, I acquired some very hard wood from Mexico ... I made an additional experiment of detaching a prismatic blade with a wooden staff minus a metal or antler tip. Because of the limitation of time, I have, to date only removed three blades in this manner. I seated the rounded distal end of the chest crutch directly over a ridge, applied a thrust of downward and outward pressure, and successfully removed several perfect blades. The blades are true replicas.)
>
> (Each time a blade is removed from the core a new position must be selected on the wooden tip, or the tip must be reworked to expose a new surface. In order to remove a blade from a core, the platform must be isolated so that just the platform area of the blade will contact the wooden pressure tip. The tip is not sharp, but it is very blunt in order to give it strength).
>
> (Crabtree 1968: 449)

Crabtree (1968: Fig. 2) illustrated two extremely large blades (7.5 × 20 cm) made by assisted pressure by Titmus and himself. They were produced by Titmus applying all the pressure he could with his copper-tipped chest crutch and Crabtree lightly striking with an antler billet a protruding piece of bolt stock made to simulate a crotch in the tool (Titmus, May 2009, personal communication; Crabtree 1966: 22, 1967: 64), as illustrated by Sellers (1886: Fig. 2). The size and mass of these assisted pressure blades indicate this technique could easily generate sufficient force to produce the largest blades known archaeologically in Mesoamerica. Whether or not large prismatic blades were made in this manner is a different matter (see Hirth 2003b; cf. Fletcher 1970: 212). Crabtree and I discussed alternative techniques for blademaking, and he provided advice on how to proceed (below). One implication of our discussion was that different techniques of making blades may have been involved in Mesoamerica. This was very much a part of Crabtree's thinking (Crabtree 1975b, 1979: 3).

3.2.4 *Jacques Tixier: Flint and Obsidian Pressure Blades*

Tixier and Crabtree met at the 1964 Les Eyzies conference (Jelinek 1965: 278; Smith 1966) and became lifelong friends. Tixier "learned percussion from François

Bordes, and later, when he attended the lithic technology conference in Les Eyzies, he learned the rudiments of pressure flaking from observing François Bordes and me" (Crabtree 1975a: 114). Crabtree demonstrated his technique for making Mexican prismatic blades at this conference, and in 1969, he and Tixier experimented with making percussion and pressure blades during Tixier's visit to Idaho (Tixier 1984b: 58; Tixier et al. 1980: Figs. 15, 20, 21; Times-News 1969; also, Inizan et al. 1992: 662). [Their long-anticipated collaborative research on Capsian blades (Times-News 1969) has not been published (Pelegrin 1984b: 117, 2003; Texier 1982: 59).] Tixier also gained experience in pressure blademaking and fluting bifaces (see Hirth et al. 2003: 150; Tixier et al. 1980: Fig. 31; Inizan et al. 1992: Figs. 33, 42). Tixier became the pivotal scholar in the French tradition of pressure blademaking and analysis (Inizan 1991; Pelegrin 1981a: 5; Otte 1991). As evident in their exchanged letters, starting just a month after the Les Eyzies conference and until their reunion in Idaho, Tixier urged Crabtree to experiment with methods of replicating Capsian blades and cores. They sent each other raw materials and preformed cores so Tixier could make pressure blades of Oregon obsidian and Crabtree could make percussion blades of French flint; Crabtree also sent antler for tools (Crabtree correspondence). In his package of February 23, 1965, Crabtree sent Tixier a large obsidian blade core prepared "with a chest crutch" and small blades "removed by hand pressure" (Crabtree letter to Tixier, February 23, 1965). Tixier sent Crabtree examples of small artifacts for him to replicate and drawings of cores and blades. On November 21, 1966, Tixier sent Crabtree two obsidian pressure blades produced with a copper-tipped crutch he had made, and he also reported having made pressure blades of glass and un-[heat]treated flint (Crabtree correspondence).

Tixier (2000: 2) summarized their collaboration and interchange by modestly downplaying the tremendous impact he had on Crabtree: "I met Don Crabtree for the first time on November 22, 1964, at the 'lithic technology conference at Les Eyzies'. This meeting went almost unnoticed at the time …. Thanks to Don the world of pressure flaking was revealed to us, the French, to the point where many of us, including me, stayed up two nights amazed by Don's ability and knowledge in the production of obsidian tools, the famous knives of the Aztecs, and PaleoIndian projectile points."

Pelegrin (2003: 55) notes that Crabtree's knapping demonstrations at the Les Eyzies conference:

> impressed the small circle of spectators, including the two Frenchmen who subsequently would reconsider the technique in a fundamentally different way. François Bordes, first and foremost a Paleolithic archaeologist, rethought the issue of core-blade pressure reduction in terms of problems of working Old World flint …. Jacques Tixier, however, continued experimentation with Crabtree and, using the chest crutch pressure technique with flint, identified key attributes of pressure blade reduction in several Epipaleolithic industries … Based on these studies, pressure blade reduction of both flint and obsidian was gradually recognized around the Mediterranean by Tixier and his students ….
>
> (Pelegrin 2003: 55; see also Smith 1966: 592)

Tixier experimented with direct pressure, direct percussion, and indirect percussion in making flint blades (1969, 1972; Tixier et al. 1980; Inizan et al. 1992: 15; see his photo-essay of blademaking in Prideaux 1973: 81–91; cf. Newcomer 1975) and identified blades made by different techniques in the archaeological record (1976, 1984b, 1991b; see also Inizan, Roche, and Tixier 1992). This became possible once Tixier isolated some of the distinguishing features of blades made by different techniques. The critical distinction was between pressure and percussion blades. There is considerable overlap between the two types of blades (Tixier 1984a). Tixier (1984b: 66) argued that there was no "key" to distinguishing pressure from percussion blades, but he did identify two clusters of traits: those that occur on pressure blades and those that frequently occur but are not technical stigmata for them. For the first group of attributes, he observed that pressure blades have uniform, parallel edges and ridges that tend to be straight, of a consistent thinness, especially in the mid-part of the blade, have smooth ventral surfaces (lacking ripple or wave marks, especially in the lower part of the blade), and have narrow butts that very rapidly broaden to the maximum width (Inizan et al. 1992: 65). In the second group of frequently occurring attributes, he included short, fairly pronounced bulbs of force, obtuse platform angles, and cores with well-marked flutes from blade removal (Tixier 1984a, b: 66; Inizan et al. 1992: 63; cf. Inizan et al. 1992: 663; Ohnuma 1993: 159; Texier 1984b). These technical stigmata have been widely used to identify pressure blade technologies in the Old World.

As with Crabtree, Tixier's influence has been fundamental for establishing questions and experimental approaches in lithic studies (1980, 1984a, 1991a, b; Otte 1991), clarifying descriptive terminology (Tixier 1963, 1967, 1974; Tixier et al. 1980; also, Inizan et al. 1992; Inizan et al. 1999), and influencing following generations of analysts and replicators.

3.2.5 Payson Sheets and Guy Muto: Obsidian Pressure Blades and Cutting Edge

At Crabtree's 1971 flintknapping fieldschool, and under his direction, Sheets and Muto (1972) reduced a sawn core and analyzed its products and by-products. This teaching exercise was not to evaluate manufacturing techniques per se, but useful information was generated on the types of errors involved in blade manufacture by beginning blademakers and ways of repairing such damage. The purpose of the experiment was to calculate the overall efficiency of Mesoamerican blade technology in terms of the blade length of acute- and obtuse-angle cutting edge produced per gram, a topic of interest to Crabtree and described in one of his films shot 2 years earlier (Crabtree 1972e; Crabtree letter to Tixer, October 23, 1969; see also Bordaz 1970; Callahan 1979a: 30; Crabtree 1971, 1973, 1977; Eren et al. 2008: 952; Leroi-Gourhan 1957). Because of the analysis of experimental products, however, we learn more from this single core than from the hundreds Crabtree reduced but did not analyze. From an 820 gram core with a rectangular platform measuring 6.0 × 5.8 cm, and with the aid of a copper-tipped chest crutch and a vise, Sheets and Muto made 83 blades plus related debitage products. As to efficiency, they calculated 2.3 cm of acute-angle cutting edge per gram of blades, or 2.1 cm per gram of original core weight (Sheets and Muto 1972: 632). They proposed that these ratios of efficiency might allow archaeologists to monitor obsidian scarcity in prehistoric times. Pressure blades, of course, are much more efficient than small percussion blades.

> For instance, in 1971 Don Crabtree removed a series of 12 percussion blades (cutting edge length 258.4 cm, 385.1 g) from an obsidian core, yielding a CE/M [cutting edge per mass] ratio of 0.74. The CE/M ratio of these blades individually increases markedly with the order of their removal. The earlier blades are relatively thick and irregular, and were removed more for the purpose of shaping the core.
>
> (Sheets 1978: 46–47)

Of interest in this fieldschool exercise, Sheets and Muto removed two blades with a blunt, hardwood tool, a repetition of Crabtree's last-minute experiment in 1967. "The blades removed with the wooden tool had small and diffuse bulbs with accentuated lips, and no 'eraillure' (bulbar) scars. Blades removed with the chest crutch [with a copper bit] had larger bulbs, smaller lips, and more eraillure scars" (Sheets and Muto 1972: 632).

Concern with economic efficiencies brought into question the features of prismatic blades under artisan control. For the analysis of cutting-edge efficiency, these are blade length, width, thickness, and cross section morphology.

> The maximizing blades, those carrying a maximum of length for their weight, are relatively thin and narrow. Thickness and width are controlled by the knapper, within limits. Thicker prismatic blades are produced by seating the tip of the pressure tool farther from the edge of the core .... Manufacturing thicker blades does require more force, so if Crabtree's technique (1968) using a chest crutch and the knapper's weight was used, that weight sets a limit to the thickness and width of blades produced.
>
> (Sheets 1978: 44–45)

Crabtree was inconsistent in his opinion on the role of body weight in making pressure blades. In his principal paper, he observed that all the participants at Les Eyzies were able to detach blades, but the largest blade was made by the smallest person there, Denise de Sonneville-Bordes (Crabtree 1968: 451; cf. Jelinek 1965: 278). The same lesson was repeated at his 1969 fieldschool where Lucy Lewis made some of the best pressure blades (Crabtree 1969a: 5). But Crabtree and his male students kept coming back to body weight as a parameter limiting the size of blades that could be made. Crabtree (1968: 468) records: "I have removed blades 1 in. wide and 8 in. long by the use of the pressure crutch alone, and yet my total weight is only 165 pounds, which makes it impossible for me to exceed this much downward pressure." In contrast, he claimed to have been able to exert a force of 300 pounds through a shoulder crutch (Crabtree 1967: 68).

3.2.6 J. B. Sollberger and Leland W. Patterson: Flint Prismatic Blades and Microblades

Sollberger and Patterson published two series of experiments, with the first focused on the characteristics of flint blades made with different techniques and a follow-up replication experiment for duplicating Iranian microblades. They called their first exercises "replication" experiments, but this is a misnomer because they did not have a replication target in mind (Sollberger and Patterson 1976a).

Sollberger was another American original along the lines of Crabtree, with a special genius for inventing devices to help in flintknapping (Callahan 1995c; Harwood 1987). Sollberger and Patterson were both knappers and part of the knap-in scene of the 1980s, and both dominated the pages of *Flintknappers' Exchange* and *Lithic Technology* with ideas, criticisms, and knapping tips. Patterson (1988) learned from Sollberger, a self-taught knapper. Sollberger's interest in archaeology long preceded his subsequent knapping career. He was involved in Texas archaeology as an amateur and began writing site reports by the late 1940s (Sollberger 1948, 1949). He began knapping in the 1950s but could not get things to work. As Patterson (1988: 20) reports, "Solly once said he got started in flintknapping after watching someone make a simple unifacial scraper, and once he had the 'bug', there was no stopping his drive to become one of the best flintknappers of this era." Sollberger would not claim this status for his own work, but it is an appropriate attribution. His earliest attempt involved a forked stick for a holding device and the use of a copper lever to make a flint dart point (Sollberger 1978: 6; also Harwood 1987). He started serious experimentation about 1967 and was influenced by Crabtree's (1966) publication on Folsom points (Callahan 1978c: 13).

As did Crabtree, Sollberger started by making arrowheads and bifaces (1967, 1968, 1969, 1971, 1976; Sollberger and Patterson 1980), moved on to fluted points (1977, 1978, 1985, 1988, 1989), and then made blades with some of the same techniques used for fluting (Sollberger and Patterson 1976a, 1983). For his part, Patterson began publishing in 1973 and appears to have hooked up with Sollberger about that time; they co-published their first article the next year (Patterson and Sollberger 1974). Their focus on blades came from Patterson's work and interests (1973, 1976, 1981, 1986). Their partnership was a mutually beneficial division of labor in which Sollberger was the knapper, experimenter, and fracture theorist (Callahan 1995b, 1996b; Sollberger 1993; Sollberger and Patterson 1976a, b, 1980, 1983) and Patterson the writer and principal analyst (Patterson 1982, 1990; Patterson and Sollberger 1974, 1978, 1980).

In their major co-paper on flint blades, nine experiments were conducted to discover differences in the attributes of the resultant blades. Two experiments were direct percussion with a hard hammer, another two with a soft hammer, three with indirect percussion (using a wooden vise to hold the core), and two with pressure using an antler. For one experiment, they immobilized the core in a fork-like device (Fig. 3.5a). For the other, the core was handheld (Sollberger and Patterson 1976a: 524). They found that many features on blades were not diagnostic. As with Bordes and Crabtree's (1969b) study, Sollberger and Patterson assessed blade attributes for populations of blades made by different techniques rather than for individual specimens. With his approval, they also corrected Crabtree's speculation concerning the need to create straight ridges on core preforms to make blades. Making a guiding ridge was not necessary (Sollberger and Patterson 1967a: 525).

Sollberger and Patterson's later experiment began with a consideration of microblades (i.e., less than 11 mm wide) and cores from Iran and attempted to duplicate

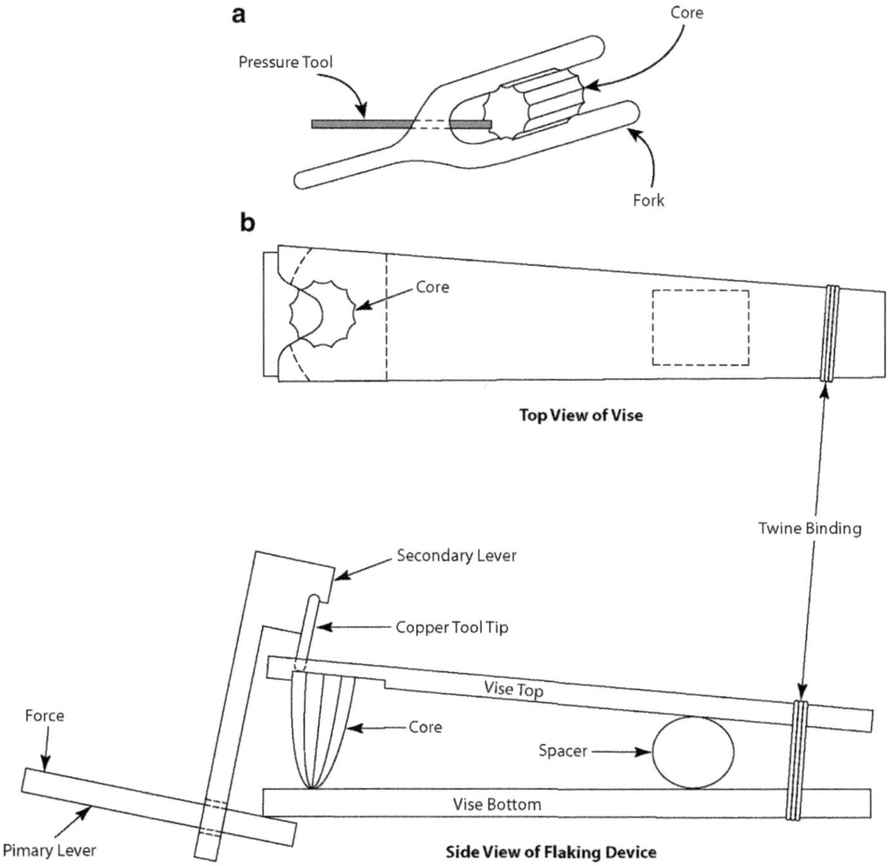

Fig. 3.5 Holding devices used by J. B. Sollberger. (**a**) Forked device to be used with a lever (Redrawn from Sollberger and Patterson 1976a: 524, Fig. 4). (**b**) Vertical clamp for lever pressure (Redrawn from Sollberger and Patterson 1983: 26, Fig. 1)

their features. For these experiments, they used a clever device similar to that employed by Sollberger to flute Folsom points by lever pressure (Fig. 3.5b). It is a combination of a vise with vertical jaws and a lever tool with a copper tip (Sollberger and Patterson 1983: 26, Fig. 1; cf. Sollberger 1978, 1985; for photographs of Solly using this tool, see *Flintknappers' Exchange* [1980] vol. 3: no. 2:16, no. 3:7–8). The tool (Fig. 3.2) is similar in conception to a technique of flaking illustrated by Sellers (Fig. 3.6). Sollberger's device immobilized even small cores without stressing the part of the core platform and face being worked, and it also permitted the slow application of pressure until fracture initiation was achieved. Consequently, stress to cores and blades was minimized, as were manufacturing failures (Sollberger and Patterson 1983: 27). This is really a hybrid technique, half machine, half human. In this experiment, Sollberger and Patterson duplicated their observation about the nonnecessity of preparing an initial ridge on blade cores (1983: 27). They also

Fig. 3.6 Lever pressure used in conjunction with a notched tree (Redrawn from Sellers 1886: 883, Fig. 7; see also, Moorehead 1910: I:21, Fig. 19)

stressed the need to abrade the core platform to reduce tool slippage and of the critical feature of the angle of the platform to the core face. They found it difficult to produce blades if this angle slightly exceeded 90°. They found that blades with their lever-vise were consistently longer than blades made by direct freehand pressure (Sollberger and Patterson 1983: 29).

3.2.7 John Clark: Mesoamerican Obsidian Cores and Blades

My knapping education consisted of watching two old men pressure flake long spearheads from red glass at a scout jamboree when I was about fifteen, studying soon after the photo-essay of Bordes making tools by percussion in the book *Early Man* (Howell 1965: 118–119), years later studying the photo-essays of Tixier making flint blades by direct and indirect percussion in the book *Cro-Magnon Man* (Prideaux 1973: 83–91), reading Crabtree's articles as a college junior in 1975, and viewing several of his films in a class on primitive technology. The film *The Hunter's Edge* impressed me, but I did not internalize details of blademaking beyond the fact that it depended on a vise and a chest crutch. My knapping started about 1967 with making arrowheads from window glass and bottle bottoms with nails, then moving on to obsidian and percussion with a small antler billet.

Necessity motivated me to learn of Mesoamerican blades in 1977 when I was hired to work in southern Mexico on an obsidian assemblage from an early city there. This began my active period of analysis and experimentation, which lasted until I started dissertation research in 1984. I read Sheets's work in 1977 and reread Crabtree's articles, and these guided my course well before I met these scholars in person. I wanted to replicate blades but found I could not get started. I introduced myself via a letter to Sheets and asked for advice. Mail back and forth from Mexico to the United States in those days was glacial, so it was some time before Sheets's helpful reply got me on track. In the meantime, I studied collections of obsidian artifacts.

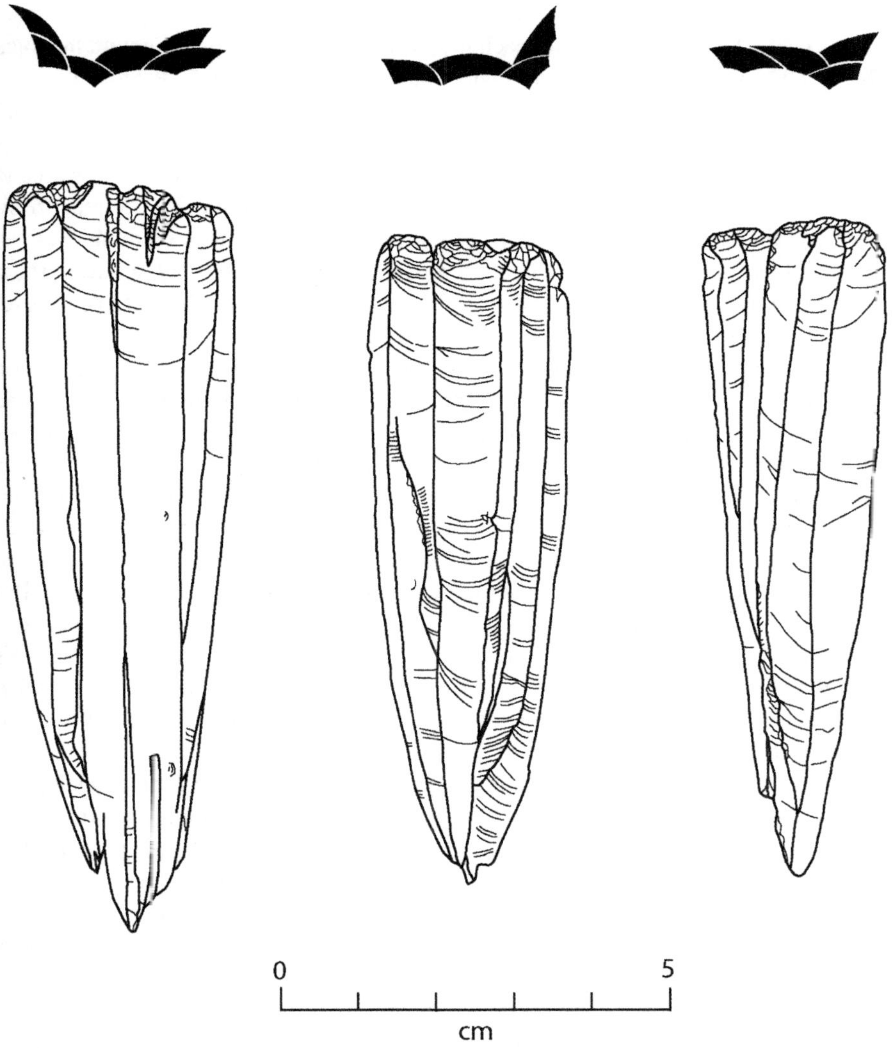

Fig. 3.7 Refitted blades and partial core shells from the Late Classic Maya site of Lagartero, Chiapas. The black silhouettes above each core shell show blade cross-sections and sequences of removal

My first detailed study (unpublished) probably had more impact on me than I have realized. Susanna Ekholm allowed me to study a cache of 73 whole prismatic blades from the Late Classic Maya site of Lagartero (in Matheny 1988; see also Ekholm 1979). I sorted these blades into their original 11+ cores and refitted partial core shells for the interior ring of three different cores (Fig. 3.7). I was fortunate to study a collection of nine percussion cores at a regional museum that corresponded to two stages in Sheets's (1975) behavioral model for blade manufacture (Clark 1977a, b), and these provided a concrete idea of what core preforms looked like. So, over a

4-month period, I became familiar with the ending and beginning stages of pressure blade cores. I had also started to teach myself blademaking, beginning with the thick base of a broken water bottle and finally obtaining obsidian from Guatemala.

I could not follow the knapping tips in Crabtree's (1968) article because I could not hold my percussion cores still. By this time, Sheets had sent me a photograph of the vise used at the 1971 Crabtree fieldschool. With this visual guide, I was able to make a suitable vise, secure my cores, and move into production. Working in primitive conditions, I had no possibility of cutting out cores with a diamond saw, so I made them from scratch by direct hammerstone percussion. With a chest crutch and a vise, I managed to reduce these cores and produce fine blades suitable for use-wear experiments – the main objective of my blademaking.

Because I started from scratch, I learned about the percussion end of the reduction process and of the transition from percussion to pressure. Following Sheets and Muto's (1972) experiment, I classified and analyzed the by-products of reduced nodules all the way down to the pressure blades and expended cores. This was done as part of my Master's thesis. By the time I finished it (Clark 1979), I had met and conversed with Crabtree several times about his technique (Clark 1989a). In writing up the blademaking process for my thesis, I explained how Crabtree had rediscovered the pressure technique for blademaking by following Torquemada's description. But when I completed the comparison, there was scant correspondence between Crabtree's and the Mexica techniques (Clark 1979: 245–253). Critiques of Crabtree's handling of the Spanish sources (Feldman 1971; Fletcher 1970) had already clarified that the Mexica used a different tool than the chest crutch postulated by Tylor, Holmes, Ellis, and Crabtree.

Some of my questioning of Crabtree's reconstruction of the Mexica technique derived from an article in which Sheets (1977) outlined pending questions and opportunities for Mesoamerican lithic studies. He argued that "Coutier's consideration of the possibility of using indirect percussion for blade manufacture deserves more analytic and replicative attention than it has received … replication of core-blade technology.might also consider the possibility of *pulling* the pressure blades off rather than pressing" (Sheets 1977: 143–144, original emphasis). I corresponded with Sheets about his "pulling" technique, and he informed me that it was still a hypothetical possibility rather than a verified procedure: "Let me clarify something. Nowhere have I said that I have pulled prismatic blades off, for I have yet to try it from start to finish. I am pretty sure that it can be done …. If the core were held between the feet, tip toward the worker, and a notched or hooked stick used, then the much greater strength of the body could be used to exert much greater force than the simple weight of the body" (Sheets's letter of September 9, 1978).

The following year, I was hooked on the blade problem. I began my second encounter with obsidian blade technology by making a tool like the one illustrated by Charles Fletcher (1970; see Fig. 3.4b). I took the proportions of his illustrated tool as accurate and converted them to my size by calculating my cubit length as 50 cm and making a tool three cubits long (Clark 1982: 366). I knew nothing about the tool or how it might work, or the adequacy of the descriptions. I put a copper tip in the end of the hook part of this implement because I knew copper bits worked for blademaking. My question was whether someone could sit down, hold a core with naked feet, and press off blades with the tool Fletcher illustrated from Bernardino de Sahagún (1963). I began by testing the tool, core position, and the physics of the procedure as I imagined it. I related my earliest experiments in a letter to Crabtree (October 8, 1979):

> I put the core upside down in the vise, the core facing me, just to see if the mechanics would work … and then I pulled off a blade in the vise, so there was a possibility that the mechanics would work OK, but then my next problem was to figure out how to hold the thing with my feet, and so I started on this and had all kinds of problems. I destroyed my core; I went back in and put it in the vise and used the chest crutch and fixed my core and went out and tried it again, and I started pulling off some blades.

Both Crabtree and Sheets pointed out to me that I had not really "pulled" blades off the core but must have pushed them off if the core was in front of me, with the platform facing me.

These simple experiments were "tinkering," akin to those described by Crabtree. Their focus was to evaluate the reliability of information provided in the Spanish descriptions. I was concerned with the performance characteristics of the manufacturing tool, the material used for the bit, and whether or not one could make blades in a seated position. I verified that blades could be made with this tool while in a seated position (Clark 1982: Figs. 8, 9). My major concern at this point was for the working end of the blademaking tool. I experimented with tropical hardwoods and with oak from a pick handle. I made several blades with a tool of oak wood, but the process of blademaking compacted the wood, so the bit required constant maintenance (see Clark 1982: 371).

The principal conclusion of these first experiments was that the technique described by Torquemada was feasible in all aspects I could check. Coutier (Cabrol and Coutier 1932) and Crabtree (1968) both pronounced Torquemada's description "impossible," but they had used the wrong tool. Sheets proposed that the "pressing towards the chest" mentioned by Mendieta and Torquemada could be a description of a "pulling" motion involved with a lever tool. This suggested to me that the wooden hook of the itzcolotli was the working part of this tool. I am not certain of this (Titmus and Clark 2003: 77, 87, 96), but it is my best guess. The tool certainly functions well in this manner.

By the time I started experimenting with an itzcolotli, I had been making blades the Crabtree/Catlin way – or so I supposed – for 2 years. I measured the working angle used with the chest crutch and tried to design an itzcolotli so I could duplicate the same angles while sitting and holding a core with my feet. Knowing the angle was important for deciding how to position the core on the ground. Most details of my beginning attempts with an itzcolotli are now fuzzy. I could not get the thing to work with a pulling or pressing motion, so I resorted to more of a pushing motion. I remember clearly one occasion when I placed the copper tip of my tool on the core platform and pushed/pulled without much force, and a perfect blade lifted up a few millimeters and then settled down again on the newly created flute on the core. I was astounded. This blade required minimal force, and it ran all the way to the end of the core.

Another learning experience came at the Pachuca Obsidian Conference in January 1981. At this conference, we visited obsidian quarries, examined collections, heard papers on the full range of obsidian topics, and witnessed the Catlin and Mexica techniques side-by-side (Clark et al. 1981, 1982). The memorable experience for me was watching Titmus and Flenniken make blades. In trying to fabricate a core among us, it was obvious to all how difficult percussion preforming of a pressure core could be (see Crabtree 1968: 446, 451; Inizan et al. 1992: 24; Pelegrin 1984b: 126; Titmus and Clark 2003: 91, 93; cf. Bordes and Crabtree 1969b: 3). I made a large conical core by direct percussion, and we finally got it locked tight in Titmus's vise. Both Flenniken and Titmus attempted to remove blades from this core but were unsuccessful. Neither Flenniken nor Titmus was used to making blades from a scratch core, and this was a source of difficulty. I watched Flenniken set up the tool and push and strain to no avail and then give way to Titmus for more of the same. With the core secured in the vise, Titmus stood slightly behind the core and put a chair in front of him to his left. He placed the tip of his chest crutch on the outer edge of the core platform at about a 75° angle, positioned the crosspiece of the crutch on his chest, and held the shaft of the crutch tool with his right hand – while holding onto the back of the chair with his left hand to steady himself (cf. Crabtree 1972e). The reason for holding onto the chair was that he knew from watching Flenniken's attempts that he would need to exert tremendous force to remove a blade, and he was guarding against falling down once the blade broke free.

As it turned out, this precaution proved unnecessary because Titmus did not press off a blade either. Apparently, the first blades off a percussion-preformed core require more force than later ones. I watched these attempts carefully because I had never seen anyone make pressure blades live. It was now my turn to try Titmus's tool. Before attempting a blade removal, I abraded the core platform with my sandstone hammerstone and wiped away the dust. I heard concerned comments from Flenniken and Titmus that the dust would be useful to keep the tool from slipping (supported by Callahan 1985: 36); I removed it for the same reason they advised keeping it. I also set up on the core on the opposite side favored by them. With the core embedded vertically in the lateral jaws of the vise, and with sufficient weight placed on the back end of the vise to prevent the whole contraption from moving, I stood slightly in front of the core, with my back to it, and extended the tip of the chest crutch between my legs and behind me to reach the leading edge of the core platform. The angle of the shaft of the crutch to the plane of the platform was about 105°. Again I heard from Flenniken and Titmus that I was courting disaster.

Standing in front of the core and reaching back to the core platform, I seated the copper bit of the chest crutch between two ridges on the platform to make a trapezoidal blade. Bending over to accommodate myself to Titmus's chest crutch, I locked my elbows over my knees and then bent my knees forward, thereby dropping my weight on the tool at the same time I pushed it forward with both hands. The result was a massive, first series blade that ran the length of the core. I did this several times in short order to take off a series of blades on the portion of the core platform arc exposed in the vise. We then repositioned the core, and Flenniken and Titmus made blades in the same manner. They found that the change in body position and the obtuse tool-to-core angle required less force than they were accustomed to expending. Among us, we took several rings (or series) of blades from this large core, at which point it had become a straight-sided and regularly fluted core – the form Crabtree imagined as the starting core for pressure.

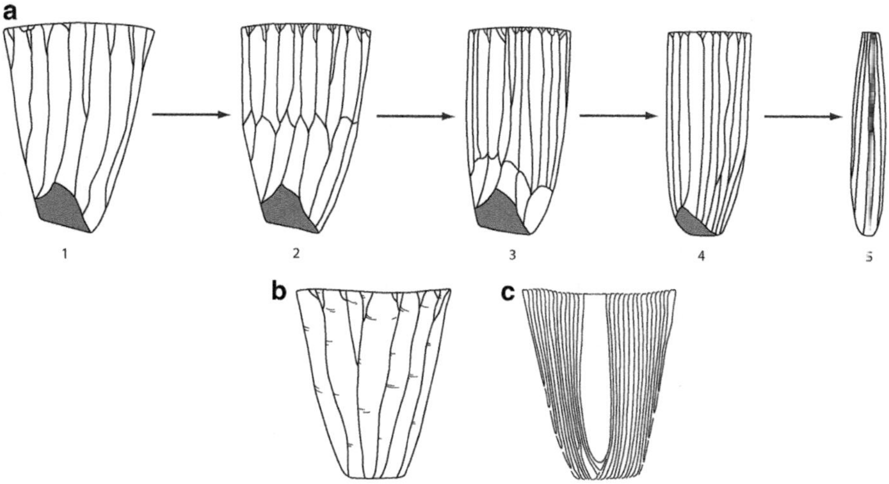

Fig. 3.8 Reduced blade core. (**a**) Sequential changes in the original core. (**b**) The original profile of the core. (**c**) Core and blades from successive rings shown in profile (Redrawn from Clark 1997: 138, 139, Figs. 2, 3)

One of the peripheral benefits of the Pachuca Conference was that Flenniken and Titmus left their chest crutches and vise in Mexico for my use. After returning to southern Mexico, I had the following experience (letter to Titmus, April 23, 1981):

> I did a little biface and blademaking just after I got back .... As I told Jeff ... I took the small chest crutch that he gave me and tried to duplicate the Catlin (in Sellers 1886) technique. [Sellers] actually describes a chest crutch and sitting down and holding the core with the feet. I used pretty much the same angles that I described in my blade paper and that I used in Pachuca with the chest crutch. I had all kinds of success with this technique. It is not even necessary to take off your shoes if you don't want to. In my first reduction of an entire core I produced 290 blades in less than two hours – and that was without hurrying and also stopping to remove hinge fractures, etc [shown in Fig. 3.8]. Needless to say, I was stunned. I still don't know what to think. I don't want to accept it. Perhaps the biggest advantage that I had was the ground platform. It was one of your cores. It is a bit more difficult if there is minor topography on the platform and if it isn't all ground down.

Because of differences in our heights, Flenniken's chest crutch fit me as a stomach crutch, and having a comfortable tool clearly was part of the success of this experiment. I was able to reposition the core in its slight depression with my feet during blading and thereby maintain a rhythm of motion and force in blade removals. This was a new experience for me.

My limited experiences in blading were uncomplicated. As were Crabtree's (1968) experiments, mine were *feasibility exercises* to determine whether or not something could be done – or more precisely, whether I personally could achieve some knapping outcome. The questions I wondered about came from Spanish descriptions of the blademaking technique. One of the last series of verification experiments I was involved in was to check the possibilities of bits for the pressure tool. Some artifacts I was studying at the time suggested to me that some bits may have been made of slate. For the same archaeological collection, I identified stone punches used in indirect percussion that were similar in form, size, and wear patterns to antler punches still being used by the Lacandon Indians of Chiapas, Mexico (Clark 1985, 1988). One punch-like tool, however, lacked the wear on its tip expected had it been used as a punch, so I entertained the notion that it might have been a bit for a pressure tool. I tested a bit made of the same kind of stone as the artifact under study, and I also made the occasion to test bits of shell, bone, chert, obsidian, and tropical hardwood. For experimental control, I employed a chest crutch and vise so I could monitor one variable at a time, namely, the material of the bit. All materials tested served to produce fine blades. I was also able to use these tools to remove blades from foot-held cores (Clark 1985). I observed some differences in the platform characteristics of blades made with wooden bits versus those of stone, but I tabled this observation for later (see Titmus and Clark 2003: 89). After I published these experiments, I next tried to use these same bits in itzcolotli tools deployed while in a seated position and foot-holding a core. I was unsuccessful in these attempts. I do not remember producing anything that would qualify as a fine blade. These negative outcomes suggested to me that bits of these other materials require more force to use than do copper ones.

My intent was to monitor one variable at a time, but this was not possible because the morphology of each tool tip was significantly different for the different raw materials, and some of these differences had significant effects. Using a bit of tropical hardwood (granadillo) in a chest crutch was much more effective than using it for the tip of the itzcolotli had been years earlier. I did not have to resharpen it to remove compacted facets.

The blades made with these wooden bits are the largest blades I have yet made … Some are over 30 mm wide, several millimeters thick, and 17 cm long. To make these large blades, I used a vise to immobilize the core … the wooden bit was attached to a chest crutch for experimental control. I held the chest crutch with my left hand, at the same time pressing with my chest. Just after this pressure was exerted, I hit the bottom of the chest crutch, near the bit, with my free right hand – thus forcing off a large blade …. The blades produced in this manner show pronounced lipping, more so than those produced when only pressure is used (Clark 1985: 1–2; cf. Pelegrin 1984b: 118).

The bit made of chert had a much sharper point. "Most of the blades produced with this tool differ from blades made with other bits; obvious Hertzian cones are visible on the bulbs of pressure, and the eraillure scars are more complex on blades produced with the chert bit" (Clark 1985: 4). This may be because of the raw material and/or its sharper point.

3.2.8 *Pierre-Jean Texier: Pressure Blademaking and Fracture Mechanics*

Judging from publications, the early 1980s witnessed a resurgence in France of experimentation in making pressure blades, with a major event being the blade symposium put together by Tixier in 1982 (*Préhistoire de la Pierre Taillée: 2 Économie du*

Débitage Laminaire, 1984). Two years earlier, Texier reported a simple experiment in making blades from a long, tabular core with a chest crutch, a technique learned from Tixier (see Tixier et al. 1980: Figs. 20, 21). These narrow, tabular cores mimic Type A cores (Kobayashi 1970). The purpose of Texier's experiments appears to have been to understand fracture mechanics rather than techniques for making blades (1984a, b; see Gallet and Texier 1991). Texier described the effect of the transverse arc of the core face in terms of making blades of different thicknesses. Narrower cores produced fewer blades, and narrower blades, per arc than wider cores (1982: Fig. 1).

Many of Texier's observations are apropos to current discussion. He (1982: 58) argued that the characteristics of the butt ends of blades had more to do with the mode of force application (e.g., direct percussion) than the means by which it was delivered (e.g., hammerstone versus billet). His experiments with narrow and wide cores (i.e., Types A and B) led to the observation that narrower blades required less force to remove than wider ones (if one holds thickness equal) and that sharp indentors focus the force application; thus, for fine raw materials such as obsidian, one should work with blunter tools to avoid crushing the platform (1982: 63). As later argued by Pelegrin (below), Texier suggested that large blades implicate a different set of parameters than do the medium-sized blades made in his experiments. The purpose of his experiments was to assess differences between percussion and pressure blademaking. Because of the necessary controls involved in pressure blademaking, such as immobilizing cores in vises, ancient artisans were able to do much more precise work, and with a greatly reduced error rate (1982: 64).

3.2.9 Bo Madsen: Danish Flint Blades

Madsen is the acknowledged master knapper in Denmark and has long been involved in numerous experiments in making flint tools (Madsen 1984, 1989; Hansen and Madsen 1983), with his flint blade experiments going back to at least 1979 and the famous Lejre Seminar (see Callahan 1980a). Madsen became interested in flint tools as a boy and wondered how they were made. He started doing percussion work at the age of 15 when he was a volunteer at the Kulturhistorisk Museum. He made his first percussion blades at 19 (September 1, 2009, personal communication). In 1971, he came across publications by Bordes and Crabtree that got him going. Because of language barriers and technical jargon, it took him several months to work through Crabtree's (1966) paper on Folsom points. As a student, he met and worked with Bordes at a knapping demonstration in 1973 (September 1, 2009, personal communication). In 1975, Madsen met Jacques Pelegrin, and the two have been knapping together ever since. Both worked with Bordes for a month in 1977 and were inspired by him. Madsen started serious work on blades after that time (Callahan 1980a: 20, 1980b: 25, Fig. 1). Madsen met Callahan in 1979 at Lejre and worked with him in arranging the 1979 and 1981 seminars (September 1, 2009, personal communication). He was not satisfied with his blade work at that time (Callahan 1980b: 23) and has continued to work on it (Madsen 1992, 1996).

Results of the 1979 conference are of interest. Twelve knappers of varying ability experimented with different technologies, and also in the creation of an archaeological site (Madsen 1981). Ten to twelve different techniques were demonstrated on how to make blades, and then the participants went to work. The objective of their experiments was to see whether they could define the technical stigmata for direct percussion blades made by stone hammers versus those made with antler billets. In particular, seminar participants wanted to replicate the characteristics of blades from the site of Trollesgave in southern Zealand, Denmark (cf. Fischer 1990).

> With this template in mind, four experimenters each made a minimum of 30 blades with a soft hammerstone; another series of experiments using red deer billets … was also conducted. The experimenters used direct percussion, holding the core either freely in one hand, on the thigh, or in one case on the ground. Hereby a "population" of more than 250 blades was produced in addition to interesting cores, fragments and flakes.
>
> (Madsen 1981: 17)

A total of 601 documented blades were produced and analyzed for over 30 features such as blade curvature and the type of overhang removal (Callahan 1980d: 4; Madsen 1981: 18). Perhaps not surprisingly, there was no clear division among the characteristics on blades that corresponded to the fabricators involved. Rather, the attributes varied along a continuum between the two depending on the elasticity of the fabricator (see Madsen 1981: 18, Fig. 4; below). Madsen has since performed more experiments and used the results as guides for reconstructing the techniques and methods of blademaking at different sites (Madsen 1992, 1996). He considered 28 attributes of blade platforms and provided a useful illustration of them (see Madsen 1996: 69, Fig. 5). The Denmark experiments are valuable as a model of how to conduct experiments and how to analyze the output of experiments.

3.2.10 Katsuhiko Ohnuma: Detaching Microblades

Ohnuma (1993) conducted a significant comparative study of microblades from Iraq and Japan that relied on experimental outcomes to interpret archaeological specimens. His microblade experiments were patterned after earlier experiments carried out with Bergman and Mark Newcomer to distinguish flakes and blades made by hard- and soft-hammer percussion (Ohnuma and Bergman 1983). Ohnuma and Bergman learned basic knapping skills from Newcomer at the Institute of Archaeology, University of London. During that time, they "interacted with many of the people cited" in this chapter, including Tixier, Madsen, and Texier (Bergman, July 27, 2009, personal communication). In their experiments, they followed up on observations Newcomer (1975) made about hard- versus soft-hammer blade production in the European tradition (see Barnes and Cheynier 1935; Barnes and Kidder 1936; Bordes 1947, 1948; Knowles 1953). Ohnuma, Bergman, and Newcomer each made blades and flakes of fine-grained flint with hard and soft percussors and then selected a sample of flakes from each experiment for identification of flaking modes (hard- or soft-hammer percussion) by the other members of the team in a blind test.

Flakes were analyzed individually according to the attributes of their bulbar areas and identified as per knapping mode (see Newcomer 1975). They derived a series of criteria from their blind tests for reliably distinguishing flakes made by hard-hammer percussion versus those made with soft stones or antler percussors. After this early experiment, Ohnuma and Bergman pursued different research interests, with Bergman focusing on analysis and projectile technology (e.g., Bergman 1987, 1993; Bergman and McEwen 1997; Bergman et al. 1988, 1990; Bergman and Newcomer 1983; McEwen et al. 1991; Miller et al. 1986). They met up again for experiments in making microblades.

Ohnuma's microblade experiments followed the same protocol as the flake experiments. These later experiments began in 1991 while Bergman was a research fellow in Japan. Ohnuma and Bergman "had learned some tricks from Madsen while in London, and [they] created bifaces like Kobayashi's System A … [using] a device for immobilizing cores … basically a fork set-up with a notched basal platform" (Bergman, July 27, 2009, personal communication; Ohnuma 1993: 175, note 2). Following the more recent French tradition from Tixier (1984b) to Pelegrin (1984b, 1991, 2003), Ohnuma and Bergman attempted to identify the technical stigmata for blades made by direct percussion, indirect percussion, and pressure. Ohnuma reduced obsidian microcores with these three techniques. He then selected 100 microblades from each experimental series for analysis of metric and nonmetric attributes. Force was applied to some microcores with chest and shoulder pressure; other microcores were held in a natural "graspable vise" (a section of forked tree limb, similar to Sollberger's forked branch [Fig. 3.5a]), and this was held in the hand for a freehand pressure technique (Ohnuma 1993: 162). Another triad of similar experiments was performed by Masaju Kubota, and 50 microblades were selected from each of his experimental series and compared with those from Ohnuma's experiments. Other experiments with these techniques worked cores of siliceous shale. From these experimental blades, diagnostic attributes and metric ratios for distinguishing microblades from the different manufacturing techniques were determined, and these were then used to analyze microcore industries from Iraq and Japan.

A few observations from the study are of particular interest to the general question of pressure blade technologies. Ohnuma's (1993: 172) reaction to the artifacts from two Middle Eastern sites reveals his practitioner perspective:

> The micro-blades from these two sites were so regularly-made that the present author was convinced at first sight that they had been detached by pressure technique; they typically bore such characteristic features as Tixier had proposed for pressure flaked débitage and cores (1984b: 66), i.e. micro-blades with regular thinness/flatness, parallel/straight dorsal ridges/edges, smooth ventral surfaces, narrow butts and short pronounced bulbs, and cores with regular flake scars and pronounced negative bulbs left by uniform and thin/flat micro-blades removed.

This is a common gestalt reaction of flintknappers to collections and a source of irritation between knappers and analysts who don't break rocks. It anticipates Pelegrin's (2003: 57) comment, "one can only recognize what one already knows." The bridge between the two is to make explicit the tacit knowledge that comes from making and handling stone tools of various types. Ohnuma made some of these features explicit. Some attributes were useful for distinguishing pressure blades from percussion blades, others for distinguishing between direct and indirect percussion blades. As with the analysis by Sollberger and Patterson, these distinctions hold for populations of blades compared to each rather than for individual specimens.

3.2.11 Dan Healan and Janet Kerley: Core Immobilization and "Counterflaking"

Getting back to Mesoamerica, Healan and Kerley (1984) published an article based on a knapping experience Healan had at Flenniken's flintknapping fieldschool while learning the technique for reducing cylindrical cores with the use of a Crabtree vise. Healan observed that four blades came off short from his preformed core and exhibited a peculiar kind of edge attenuation and damage. After each failure, Healan placed the stunted blade back on the core to determine what had happened, and he found that the margins of damaged blades had come into contact with the vise. "Areas of contact are always the most salient parts of the body …" (Healan and Kerley 1984: 3). When removing blades from the corners of the core, the future edges of some planned blades contacted the wood of the vise during removal and splintered in a particular way, a pattern Healan christened "counterflaking." This same year, Pelegrin (1984c: 115) independently published the same damage pattern, which he called "inverse retouch" (*retouche inverse*). This has since been presented in English by the awkward term, "reverse scratching" (Pelegrin 2006: 53). The illustration accompanying this last description verifies that it is the same phenomenon described by Healan and Kerley. Reverse or inverse retouch occurred on blades made from cores immobilized in a three-point holding device (see below).

Healan isolated four common features of counterflaked blades, and he and Kerley identified this damage pattern on ten blades found in blade workshop debris from the Postclassic site of Tula, Hidalgo (Fig. 3.9). They correlated the archaeological with the experimental and proposed that the damage to the Tula blades derived from the same cause as the experimental ones and thus were evidence of the use of wooden vises at the Tula blade workshops. As implicated by Pelegrin's observations, this hypothesis was too restrictive.

I became aware of Healan and Kerley's study before it was published and was excited by the possibility of a clear marker for identifying vises, something the Crabtree circle thought about. On July 30, 1983, I spent seven hours trying to duplicate counterflaking. I made cores for pressure reduction in a vise and purposely removed blades in the contact zone where an edge of an intended blade would come off the part of the core in contact with wood. I broke many blades and ruined three cores. Some of these blades exploded coming off their cores. What surprised me more, however, was that some did not. For some, removal of the blade forced the jaws of the vise open for a microsecond and allowed the blade to fly free without damage. After spending a morning traumatizing blades, and not producing any counterflaked ones, it occurred to me to check blades made years earlier with foot-held cores.

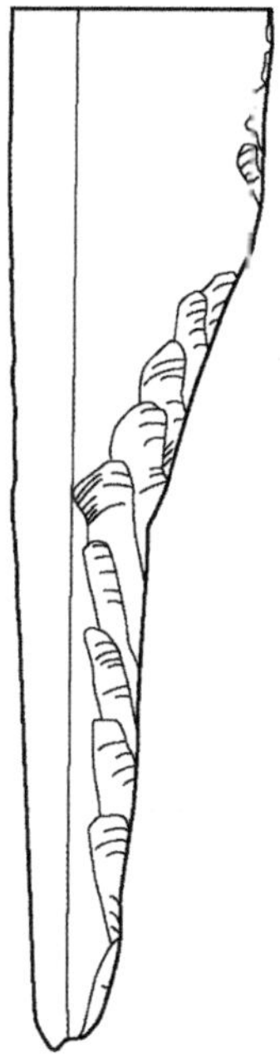

Fig. 3.9 Tip of a counterflaked blade (Drawn from a photograph in Healan and Kerley 1984: 5, Fig. 3e)

I found three counterflaked blades in these experimental collections. I concluded from this that counterflaking is not unique to vise-made blades (Clark 1984). I later found Pelegrin's reference to counterflaked blades associated with fork-held cores, so counterflaking is a general phenomenon of blademaking.

3.2.12 Errett Callahan: Flint Danish Microcores

A compelling case can be made that Callahan has done more to promote experimental archaeology and international dialog about stone tool technology than anyone since Crabtree (see Harwood 1986). Callahan is a master knapper and flint artist who is largely self-taught in the basics, beginning in the late 1950s. Going back to early grade school, Callahan was fascinated with the outdoors and woodlore; he made his first bow and arrow at the age of five and his first (slate) arrowheads in the seventh grade. But he really got started in 1956 while working a summer at the general store at Old Faithful in Yellowstone National Park – chipping out arrowheads from obsidian and glass in his spare time (Callahan and Titmus 1999; Harwood 1986: 1). After about 10 years of self-teaching, he became aware of other knappers.

> I will never forget that day in 1966 when, after plodding along alone for ten years and thinking that I had single-handedly rediscovered the "lost" art of flintknapping, I stumbled upon F. Clark Howell's book, *Early Man* .... In those pages I was confronted with an amazing

> French wizard with such spectacular knapping abilities that my little "arrowheads" suddenly seemed like embarrassing moments of play. I was both deflated and elated. A master far greater than I could ever be; a model for years to come.
>
> (Callahan 1981: 2)

After becoming familiar with Bordes's work, Callahan was profoundly influenced by the writings and work of Crabtree, Bruce Bradley, Sollberger, Titmus, Pelegrin, Madsen, and Richard Warren (Callahan 1995c; Harwood 1986: 2; Watts 1997). As did other American knappers considered here, Callahan started by making arrowheads and bifaces and then moved on to other technologies. Following his magnum opus on Clovis technology (Callahan 1979d, 2000c; see photo-essay in Kopper 1986: 42–43), he progressed to Danish flint technologies, including the production of microblades.

Callahan's (1984, 1985, 2000a) two experimental studies of flint microcores from Denmark were meant to resolve archaeological questions, as well as demonstrate best practices of how replication studies ought to be performed (1995c). He appears to have conducted these experiments in the reverse order of their publication, so I consider them in real time to maintain the experimental progression. His study of Mesolithic Danish microcores from Vedbæk, intended as a tribute to Crabtree in 1982 (1985: 38; also, Pelegrin 1984c: 110), addressed three hypotheses: "(1) Microblades were removed from cores which were secured in a holding device. (2) Microblades were removed from their respective cores by means of a fabricator of an antler-like material. (3) Microblades were removed by hand-held fabricators" (Callahan 1985: 27). To test these ideas, Callahan reduced eight cores. Of these, five were by pressure. He started with his experience with Crabtree's technique and went on from there. By this time, Callahan had his company, Aztecnics, which sells obsidian pressure blades (see Harwood 1986: 2), so he had made "thousands of blades" (Callahan 1995a: 225) by this time. He preferred "a gentle building up of force so that the moment of release comes as a surprise" (Callahan 1995b: 84) – rather than by a "lunging pressure." Experiment 4 combined a copper-tipped tool with a handheld clamp for the core. This was followed by an experiment with the same tool, with the core being held in the hand. Experiment 6 was a handheld core combined with an antler tool. This turned out to be the most difficult exercise and raised issues addressed in two additional experiments.

Callahan's hand clamp was derived from a device first used for fluting Folsom points. This clamp affected how the microcores were designed and prepared.

> The clamp which was used for these experiments [Figure 10] was inspired by a clamp which I witnessed J. B. Sollberger employing for Folsom fluting … (I had previously used a clamp which gripped with two contact points instead of three.) For such a clamp … it was necessary for the core to be quite parallel-sided for a secure seating in the clamp. Otherwise the core would tend to rock back or forth during the stress of pressing …. The cores which were hand-held did not have to be quite so parallel-sided, as the hand could accommodate some degree of irregularity here with no repercussions ….
>
> (Callahan 1985: 34, 36)

In his fourth experiment, 80 microblades were produced with the use of a copper-tipped tool from a clamp-held core. The core in Experiment 5 was handheld,

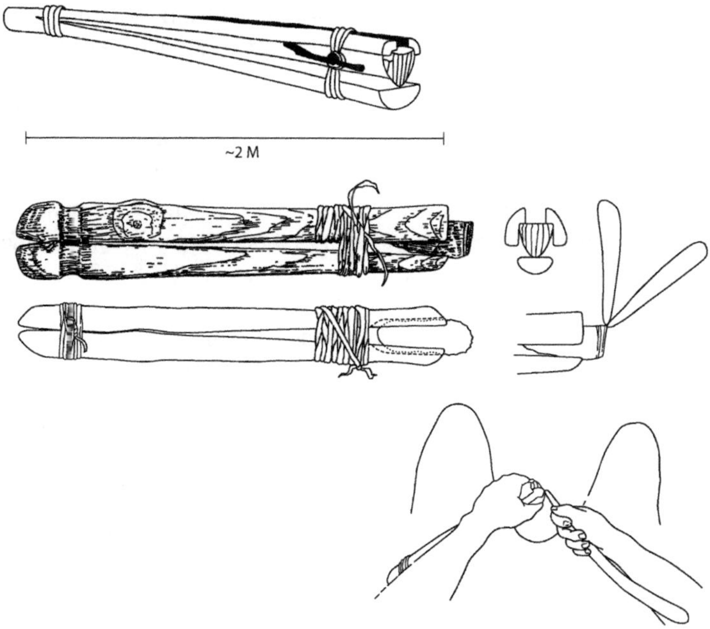

Fig. 3.10 Split vertical vise used by Errett Callahan (Redrawn from Callahan 1985: 33, Fig. 5f–m; see also, Pelegrin 1984c: 110, Fig. 5; Tabarev 1997: 145, Fig. 2)

and 40 microblades were made. "The smallest core which I could grip by hand tight enough to allow the removal of blades was about 5 cm in length" (1985: 34). This parameter relates to hand size, and Callahan claims to have large hands (ibid.).

Experiment 6 was the pivotal one of the series. Callahan reduced the core involved with a short antler pressure flaker while hand-holding the core.

> Despite the use of a shallow core (of minimum depth), it was exceedingly difficult to remove any microblades. A longer antler tine was used for a while, being held in the same manner as the shorter one, but the results were equally poor. In actual fact, 65 microblades were removed; but as detachment required all the strength I could muster, with the antler tip slipping off the platform repeatedly … and as the blades being questionably within the acceptable range of variation, and did not approach the length of the average artifact blade … My inability to remove sizeable microblades surprised me because of the success with which I had removed relatively long blades with antler in the past with much smaller cores …. These cores, however, while being flaked with the same short antler tine I was now using, had been secured in a clamp ….
>
> (Callahan 1985: 31)

Experiment 7 reconfirmed his previous experience. This core was secured in a clamp, and a long antler tine was used as the flaking tool. Callahan (1985: 31) observed "that a primary advantage of the clamp was to increase leverage by having the far end of the clamp resting along the left forearm, thus preventing hand/core movement" (Fig. 3.10). Sixty microblades were made from this core with relative ease, and they ran the full length of the core. Experiment 8, the last, revisited the frustrations of Experiment 6 with the insights gained from Experiment 7. The core was handheld, but a long antler tine was used to remove blades. Thirty blades "were easily removed from the core" (1985: 31).

On the basis of these experiments, and with observations of artifacts in mind, Callahan concluded that "There is a high degree of probability that ... blade removal was made on cores which were hand-held and was effected by pressure with a hand-held fabricator of antler or an antler-like material" (1985: 37, emphasis removed). He made blades with other tools and with the aid of clamps, so some of the deciding logic was based on the "principle of least complexity" (1985: 37) – favoring the simplest way blades could have been produced.

Callahan's second project involved an unusual Danish core with an obtuse platform. Before planning his experiment, he called upon the experience of fellow knappers. He received commentary from eight knappers to help him understand a microcore with platform angles of about 113°, well in excess of the 90° limit mentioned by Sollberger and Patterson (above; Patterson 1981; Sollberger 1986a). It turns out that this unusual Danish core is not unique in its archaeological context. How was such a core produced? Consensus opinion was that it was worked by pressure with an antler tool, with the core immobilized in some sort of holding device, and with an emphasis on outward rather than downward pressure (Callahan 1984: 90). In conjunction with this project, Callahan (1984: 88–89) summarized an important experiment by Anders Fischer (1974) not otherwise reported:

> The best replicas seem to have been made by Fisher. Of all respondents, he, as a Danish archaeologist, has had the most experience in both handling and replicating Maglemose cores. He says, 'Some years ago I did a series of simple experiments ... I found out that nice microblades could easily be made using pressure sticks of red deer antler tine. For the sake of ease, the cores were fixed in a carpenter's bench. The pressure stick was then positioned close to the platform edge, and when quickly pressed downwards and outwards ... regular microblades could be peeled off one after the other – only delayed by the necessary trimming of the platform edge (Fischer 1974: 164). In this process, the platform angle gradually increased from the starting point of around 70 [degrees] to more than 100 [degrees]. The practical limit of flaking was reached at around 110 [degrees] ... *These platform angles do not prohibit controlled flaking patterns*' [Callahan's emphasis].

With these guidelines, Callahan prepared an obsidian core with a sawn platform which he reduced in a vise using a copper-tipped chest crutch (1984: 91). The purpose of this experiment was to determine whether he could produce blades from an obtuse-angled platform rather than to replicate the core under consideration. Once he had preformed a core, he removed a series of blades:

> The downward-outward force relationships were manipulated so that outward force predominated. Thus, downward chest pressure was applied only to the degree which would allow blade removal without having the tool slip from the core. Then, with *no* increase in downward pressure, outward pressure was applied until the blade released. I would estimate that outward pressure was two or three times that of downward pressure. This amount of outward pressure tended to insure blades which terminated at, or shy of, the distal end of the core and to inhibit the tendency for overshooting. Such blades are quite straight and had only the slightest degree of distal curvature.
>
> (Callahan 1984: 91)

This procedure resulted in an exhausted core that was a close match to the archaeological specimen. Particularly interesting was the manipulation of forces in this experiment to resist running blades to the very end of the core. Callahan argued that a different pressure stroke would have given his core a different morphology.

Callahan made 85 blades from his core, and these represented 14 different series of sequential removals. The angle of the platform to the core face increased with each series of blades removed. He terminated his experiment when the angle of his core was 112°, even though another series of blades could have been removed (1984: 92). The increasing platform angle did not adversely affect his removal of blades. Callahan had to take care not to overshoot the distal end of the core. Plunging blades tended to keep his platform-to-core face more stable (1984: 92, 94). Thus, deliberate blade overshots would have been a way to maintain an acute-angled platform and to keep the core in production longer.

3.2.13 *Jacques Pelegrin: Handling and Working Cores of All Shapes and Sizes*

Following Bordes's and Tixier's efforts, Pelegrin has amplified the venerable French traditions of making flint and obsidian blades, emphasizing different techniques for immobilizing and reducing cores of different sizes and shapes (see Pelegrin 1984a, c, 1988, 2003, 2006); he has also relied on his knapping experience to advance theory (1985, 1986, 1990, 1993, 2005). Pelegrin is largely a self-taught knapper who has had significant interaction with the ascending generation of stoneworkers. He began knapping at the age of 12 after seeing the Bordes's photo-essay in the *Early Man* book. At the age of 16, he met Tixier and saw him work briefly but did not receive instruction from him. His most formative influence came from Bordes with whom he worked for six summers, 3 weeks each summer, beginning at 16. Pelegrin made contact with Crabtree just before he died but never saw him knap (Callahan 1982: 63). Pelegrin's early work was with direct percussion techniques, with a later move to pressure techniques and various means of making blades. He started making pressure blades in 1980/1981 (Pelegrin 1984b: 118; Callahan 1982, 1984, 1985: 36; Tixier 1984b: 58). Since this time, Pelegrin has covered much of the same ground charted in Crabtree's (1968) blade study, but in a much more methodical way. Pelegrin has experimented with various kinds of flint and obsidian and made blades from cores of different sizes by direct and indirect percussion (1991, 2003, 2006; Pelegrin et al. 1991; also, Callahan 1982) and by direct pressure and lever pressure (1984a–c, 1988, 2003, 2006). He has also experimented with various holding devices and tools, including flexible and inflexible gut crutches with antler bits. He found crutches with flexible shafts to be superior to rigid crutches for making flint pressure blades, and he suggested that a flexible tool may have been used to make the longest Aztec blades (Pelegrin 1984b: 123, 2003: 59).

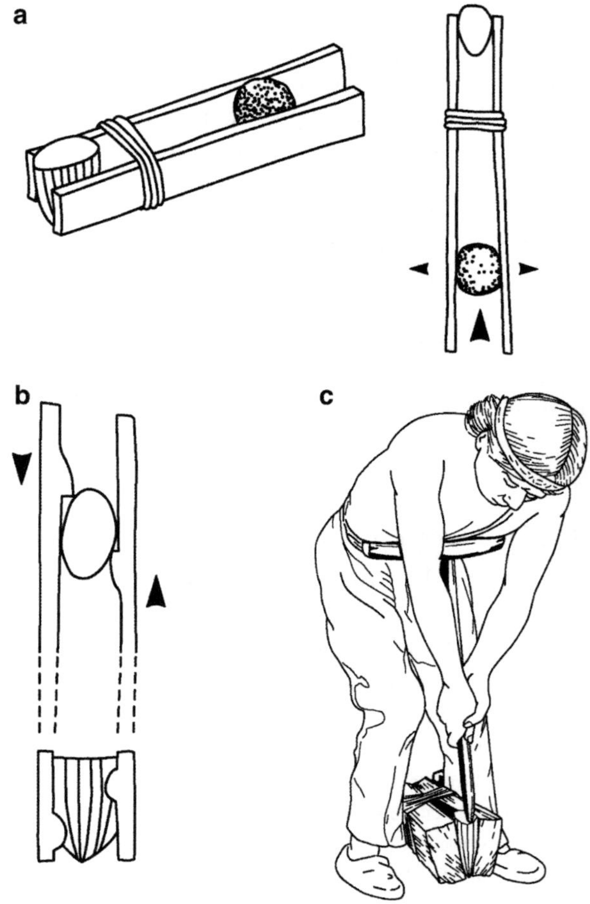

Fig. 3.11 Vises with lateral jaws. (**a**) Side and top views of a Crabtree clamp. (**b**) Variant of a Crabtree clamp with backstop and forestop ((**a**) and (**b**) redrawn from Pelegrin 1984c: 107, Fig. 2). (**c**) Reconstruction of use of lateral clamp (Redrawn from Crabtree and Gould 1970: 187, Fig. 7)

Pelegrin's experimentation with different working positions and ways of immobilizing blade cores is particularly innovative. He investigated differences between Crabtree's technique and my own and experimented with holding cores with clamps, vises, and gravity. Because his guiding interest at the time was flint blades, which by all accounts require more force to make than obsidian blades (1984b: 118), he did not experiment with foot-holding techniques (1984c: 105). Rather, he considered the following variants of vises with lateral jaws, clamps with vertical jaws, and three-point devices for stabilizing cores:

1. The best known option is a wooden vise with flat, lateral jaws (Crabtree 1967: Fig. 2; Sellers 1886: c874) in which two flat boards are tied together, the core inserted at the short end, and the boards spread at the opposite end to secure the core (Fig. 3.11a).
2. Pelegrin (1984c: c107, Fig. 2, translated by Charlotte Laporcherie) proposed a modification of the Crabtree clamp to make it more efficient. He suggested the

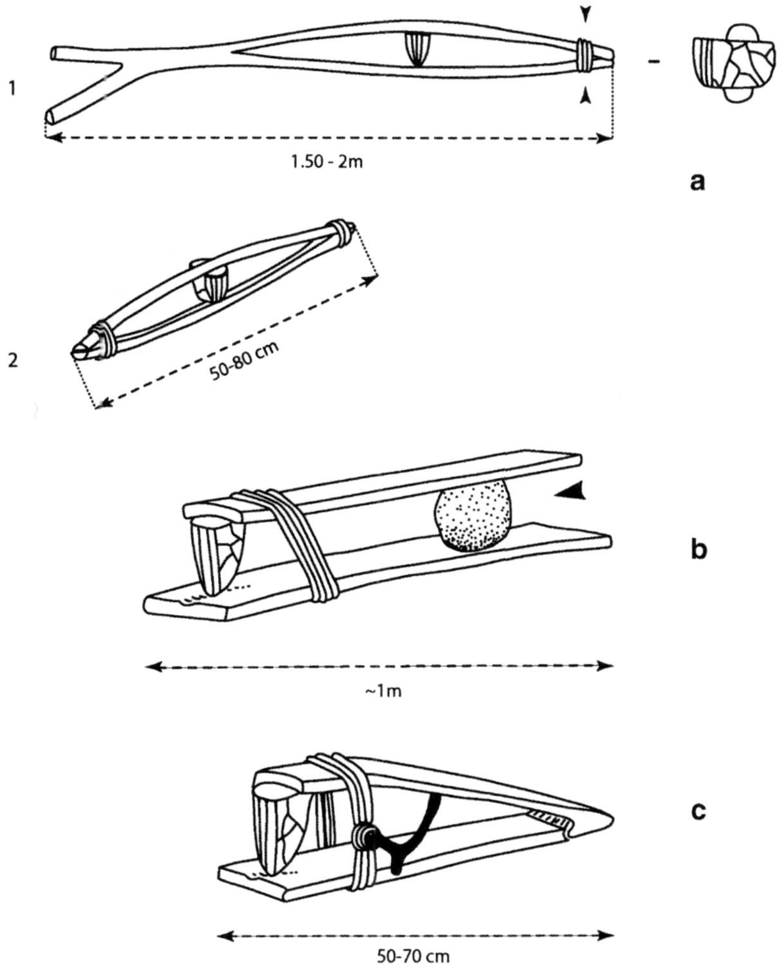

Fig. 3.12 Clamps with vertical jaws. (**a**) Horizontal clamp used by Bo Madsen (Redrawn from Pelegrin 1984c 108, Fig. 3). (**b**) Vertical clamp that is a Crabtree vise turned on its side (cf. Fig. 10a). (**c**) Vertical clamp with back hinge and tightening system ((**b**) and (**c**) redrawn from Pelegrin 1984c: 109, Fig. 4)

addition of two supports to minimize movement, "one forward and at the top of one of the planks, the other at the bottom and more at the back of the second plank" (Fig. 3.11b).

3. Other options for stabilizing cores are to clamp them vertically rather than horizontally. Figure 3.12a illustrates a device used by Madsen. This is a Y-shaped branch, with the two short ends of the forked branch providing stability for the device when placed on the ground, and with the third member split horizontally so a core can be inserted. The core is placed transversely in the split branch and squeezed by tying the two half branches together (cf. Fig. 3.10).

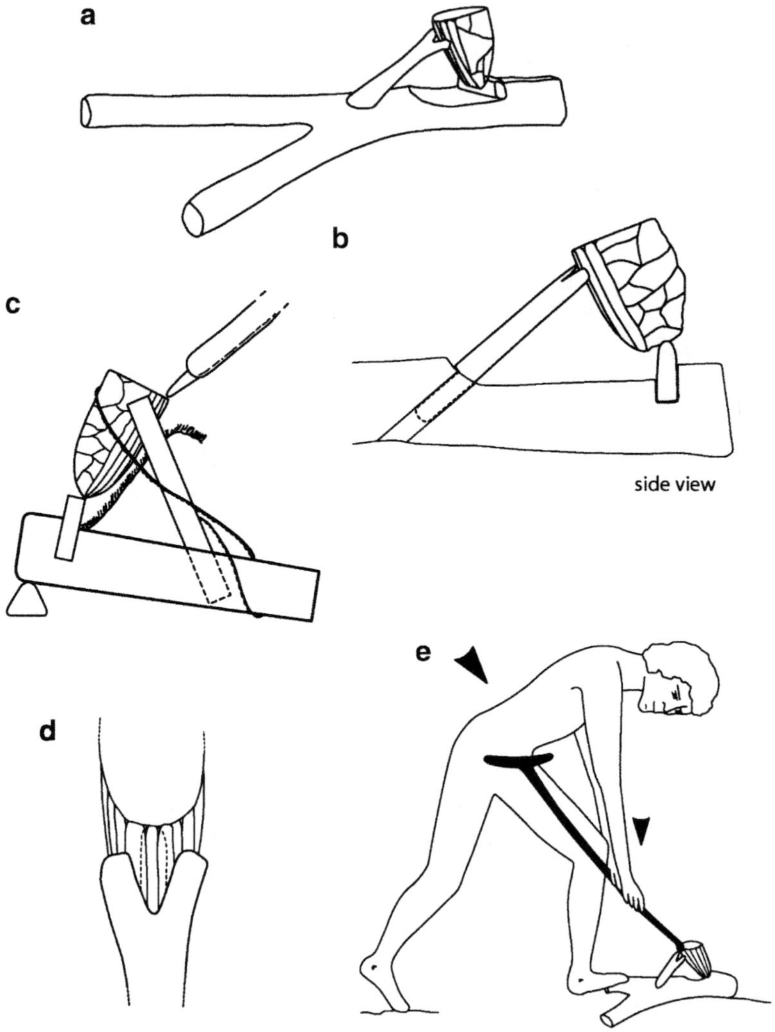

Fig. 3.13 Forked stick device for immobilizing cores. (**a–d**) Positions of the slot and forked stick (Redrawn from Pelegrin 1984c: 111, Fig. 6; see also, Piel-Desruisseaux 1990: 45, Fig. 35). (**e**) Use of devise with a gut crutch (Redrawn from Pelegrin 1984b: 120, Fig. 3)

4. Two versions of a vertical clamp using boards are shown in Fig. 3.12b, c. The first is essentially the Crabtree clamp turned on its side; the second has a hinge at the back and a means of tightening cords at the front of the vise to press on the platform and distal end of the core inserted vertically in the clamp. These are simplified versions of Sollberger's vise (see Fig. 3.5b).
5. Pelegrin's (1984c: 111, Fig. 6) forked device is a modification and extension of the fork apparatus described by Sollberger and Patterson (Fig. 3.5a). Pelegrin's fork requires three points of contact (Fig. 3.13), as with Callahan's device – a distal rest and two points just below the platform. It is a self-tightening device

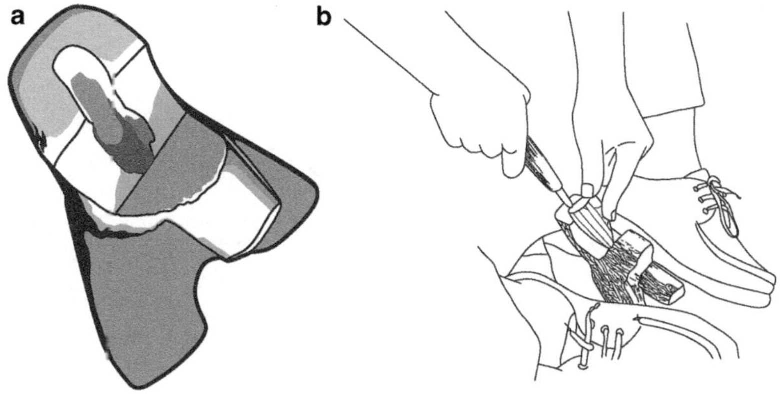

Fig. 3.14 Pelegrin's self-immobilizing core device. (**a**) slotted stick with a stabilizing foot (Drawn from photograph in Inizan et al. 1999: 149, Fig. 73.4; see also, Pelegrin 2003: 62, Fig. 4.8). (**b**) Manner of using the slotted stick (Redrawn from Inizan et al. 1999: 31, Fig. 4.5)

that takes advantage of the forces exerted in working, rather than fighting against them. As illustrated in Fig. 3.13a, this appliance takes advantage of two triangulations. The base support is a Y-shaped branch to be placed on the ground to provide stability (see photo in Inizan et al. 1999: 149, Fig. 73.2). The other points of vertical contact are supplied by a smaller forked branch inserted into the larger one, with a notched place in the main branch to lock in the distal end of the core (Fig. 3.13c). There are two stabilization challenges in making blades; one is to secure a core in a device, and the other is to secure the device itself (cf. Sørensen 2006: 285, Fig. 5).

6. A vertical triangulation setup can be made from a single piece of wood that can provide the horizontal stability and a vertical dimension. This is done by replacing the forked stick with a slotted stick (Fig. 3.14). The two edges of the slotted piece of wood serve the same function as the vertical forked stick (see Pelegrin 2003: 62, Fig. 4.8; also, Inizan et al. 1992b: 96, Fig. 46). Demonstrations of this technique can be seen in two films by François Briois (1992, 1995).
7. The same principles of stabilization through gravity can be built into a permanent system by making the appropriate modifications to a tree root (see Sellers 1886: Fig. 6). As Pelegrin (2003: 58, Fig. 4.2) described it, "one could scoop out a notch on the side of a root at ground surface and set the distal end of the core on a block of wood at the bottom of a hole dug in front of the notch" (Fig. 3.15a).
8. An even simpler system is to "scoop out the center" of a billet-like piece of wood and stand it up "obliquely against a large stone" (2003: 58, Figs. 4.3, 4.4). As shown in Fig. 3.15b, the hollow billet has a pointed end that would be stable – under vertical pressure – in a slight notch in the ground. All of the forces in this stable device are vertical, with the holding device stabilized in the same way and with the selfsame forces that keep the core stable in the slotted device.
9. Slotted devices can be made of different calibers for cores of different sizes. The distance between the opposed edges of the open slot (or the two prongs of

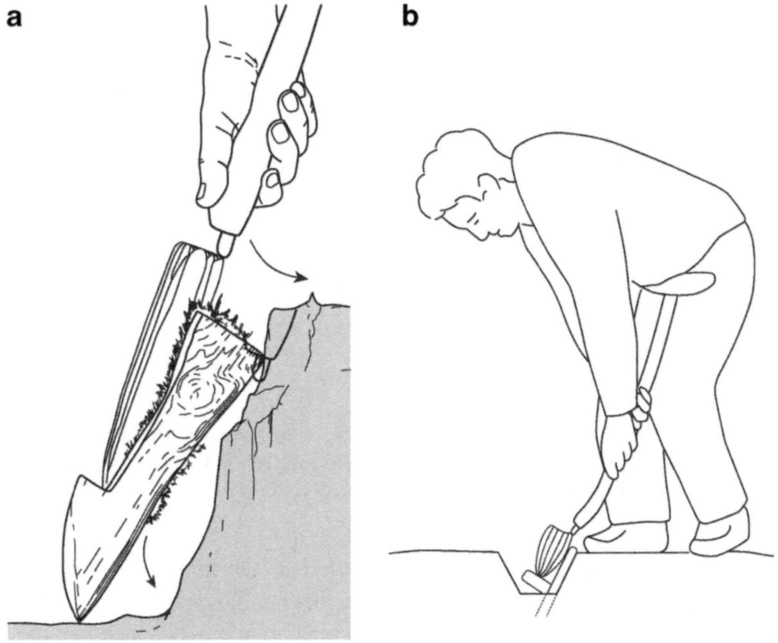

Fig. 3.15 Another of Pelegrin's self-immobilizing devices. (**a**) Using a hole to help secure a core (From Inizan et al. 1992a: 666, Fig. 2). (**b**) Slotted stick holding device without a horizontal foot (Redrawn from Pelegrin 2003: 59, Fig. 4.4)

a forked stick) determines the maximum width of the blades that can be safely removed without damage. Small versions of this device can be held in the hand to remove blades from microcores (Pelegrin 1988: 43, Figs. 2, 3). They can be made of wood or large mammal bone; the "end of a long bone with its medullary cavity can easily be used as a holding device" (Pelegrin 2003: 61, Fig. 4.7; for illustration of experimental microcores, see Inizan et al. 1992a: 664, Fig. 1).

10. Pelegrin employs the same principles for securing cores over 30 cm long as he does for those under 6 cm: a notched rest for the distal end of the core and two points of contact for opposite sides of the proximal end of the core. For his lever pressure technique, he makes a slotted device in a large log of sufficient weight that it is stable under the forces exerted. Because direct pressure is not used, Pelegrin also needs a fulcrum point for the end of his lever tool, as shown in Fig. 3.16.

A crucial insight of Pelegrin's studies is the complementarity between core shapes and means for keeping cores still. Each immobilization device anticipates an optimal core form and shape. In turn, ways of keeping cores still relate to the tools used to make blades of different sizes and the working positions involved. In a rational and efficient world, the size and shape of core preforms would correspond to production techniques and the manners of paralyzing cores for blade production. Morphological transformations of cores during reduction are critical clues for identifying the tools and techniques of ancient blade manufacture. Different techniques

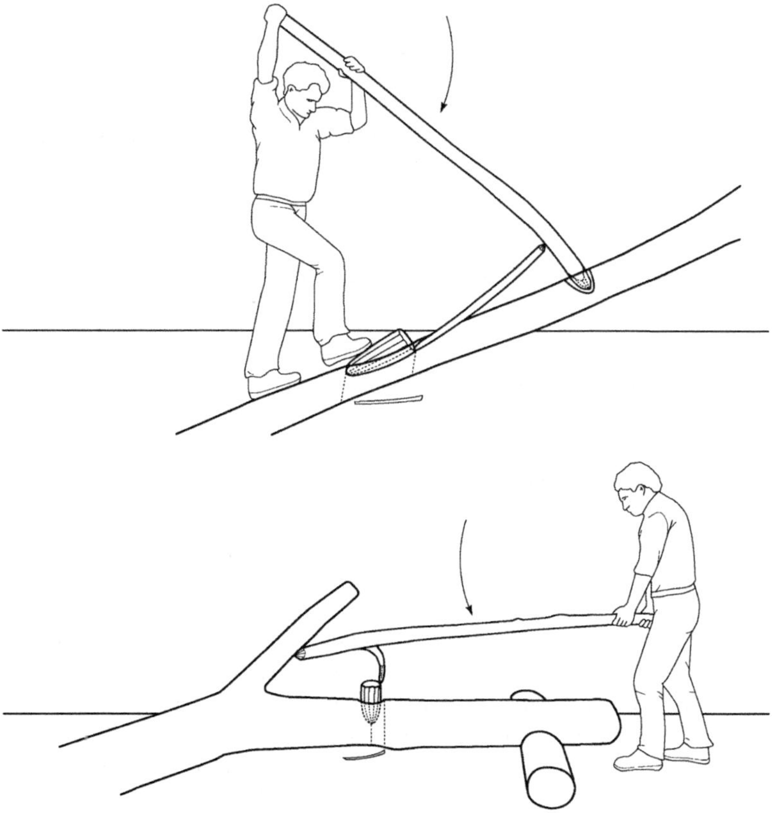

Fig. 3.16 Drawings of slotted logs used with a lever for removing very long blades (Drawn from photographs in Pelegrin 2003: 65, Fig. 4.11)

and core forms also dictate ideal sequences of blade removals, something that can be learned from studying expended cores and from refitting whole blades into their sequence (see Astruc et al. 2007; Kelterborn 2008c; Texier 1984b). The transverse arc of a core face also determines the blade width-to-thickness ratio, the likely number of dorsal ridges, and blade cross section.

In a recent appraisal of Mesoamerican blade experiments, Pelegrin addressed four unresolved issues. These are the techniques needed for making blades from different kinds of cores (large, slender, or small) and the role of indirect percussion in the reduction sequence. Pelegrin (2003: 56) repositioned this last question. Earlier researchers such as Coutier (Cabrol and Coutier 1932) and Crabtree (1968) explored indirect percussion as an option for making fine blades. For Pelegrin, this is no longer a credible hypothesis given the regularity of fine blades and the potentials of indirect percussion (Inizan et al. 1992b: 23), but this does not take the technique completely out of the picture. Indirect percussion was probably used to preform cores in some regions of Mesoamerica (Titmus and Clark 2003: 91); the data for this claim are the punches themselves (Clark 1985, 1988). Pelegrin (1984b, 1988, 2003,

2006) relies on principles of biomechanics and geometry to argue for size limits for blades made from particular raw materials with specific techniques and tools. He has been able to make blades 27 cm long with direct pressure (blade width is more restrictive than length), and blades much longer than that with lever pressure (Pelegrin 2003: 60, 2006). Medium-sized blades can be made in many ways. Small blades or microblades, however, present interesting challenges.

3.2.14 Jeffrey Flenniken: Flint and Obsidian Microcores

Flenniken has replicated a wide range of artifacts (1978, 1981b, 1985; Flenniken and Raymond 1986; Flenniken and Wilke 1989), including microblades (Flenniken 1987; Flenniken and Hirth 2003). He is a self-taught knapper, but he did study with Crabtree in 1973 and learned from him (Flenniken 1981a). At Crabtree's recommendation, Flenniken assumed responsibility for the flintknapping fieldschool when Crabtree retired, a circumstance that led to years of collaboration and interaction with him (Flenniken, in Callahan 1978b: 16, 21). Flenniken learned blademaking from Crabtree and has made hundreds of fine obsidian blades (see Lampe 1991: 659).

Flenniken's (1987) first study of microcores concerned artifacts from eight Paleolithic sites in Siberia related to the Dyuktai blade technique. After studying the cores, blades, and related debitage from these sites, he performed a series of experiments with flint of similar quality and configuration. The Dyuktai technology is a Type A system similar to a Japanese technique demonstrated by Crabtree (1972e) in *The Hunter's Edge*. A thick biface is made, one margin of the biface is removed with a "ski spall" flake to create a platform, and then blades are removed from what is essentially a long narrow core, ideal for holding in a vise (cf. Kelterborn 2003: 127, Fig. 8.9; Texier 1984b: 26, Fig. 2). Much of Flenniken's argument is carried by his drawing of the reduction sequence and his photographs of replicated artifacts rather than his descriptions. He reduced 25 bifacial cores and averaged about 101 "useable" (i.e., trapezoidal cross section) blades per core (Flenniken 1987: 122). Flenniken argued that the Dyuktai blades and cores "could not be produced by any method (i.e., freehand pressure, freehand percussion, hand-vise pressure, indirect percussion, etc.) other than by holding the core tightly in a simple wooden vise ..." (1987: 122). He does not mention, however, any experiments with other techniques, so his confident identification is not supported by argument. He does propose that the blades and cores were made by pressure, and he relied on a list of technical stigmata similar to that proposed by Tixier. Flenniken identified seven attributes:

> (1) blades are small and uniform, (2) blade margins are parallel, (3) all blade arrises (dorsal surface) are parallel, (4) blades are thin and evenly proportioned, (5) blade platforms are small, exhibit minimal platform contact, or show preparation in the form of abrasion, (6) ventral blade surfaces exhibit few, if any, compression rings, [and have] diffuse bulbs of force, and feather terminations, and (7) blade scars present on exhausted cores also exhibit the above attributes .... It is *impossible* to consistently produce percussion blades possessing the above attributes.
>
> (Flenniken 1987: 122)

Of particular interest is an argument Flenniken made for the use of vises. He noted that the exhausted cores are larger than one would expect had they been hand-held and that "usable" stone had been discarded. "The fact that 'usable' stone was discarded as 'exhausted' suggests that the pressure blade cores were held, during blade production, in a wooden or bone vise of some type. The cores were exhausted when they could no longer be held securely in the vise" (1987: 122). Flenniken came to the opposite conclusion for obsidian microcores from the Epiclassic (ca. A.D. 600) site of Xochicalco in highland Mexico.

Flenniken's collaborative project with Ken Hirth and Brad Andrews (Hirth et al. 2000, 2003) with Type B obsidian microblades was inspired by unusual features of Xochicalco obsidian cores. Exhausted blade cores from this site are tiny by any standard and likely indicate a different technique of blademaking than described for the Mexica who lived in the same area five centuries later. Flenniken, Hirth, and Andrews's reconstruction of the blade reduction sequence indicates that "exhausted" blade cores about 18 cm long were imported into Xochicalco and then were broken in half to make microblade cores (Flenniken and Hirth 2003: 99; Hirth 2003a; Hirth et al. 2000: 139, Figs. 7.1, 7.2, 2003). They estimate the size of starting blade cores at 8–9 cm long and 3–4.5 cm in diameter; expended cores were about 5 cm long and 1.1 cm in diameter (Flenniken and Hirth 2003: 99). Given prior history, cores came with very regular, straight ridges. The smallest exhausted cores are about the diameter of a pencil – 0.5–1.0 cm (Flenniken and Hirth 2003: 99; Hirth et al. 2003: 192, Fig. 13.10). How were these cores immobilized during blading? "The problem that the Xochicalco cores present for current technological reconstructions is that they are too small to have been held with the feet" (Flenniken and Hirth 2003: 99). "Experimentation with the foot-held technique by Gene Titmus has established that cores under 8–10 cm in length are difficult to hold with the feet unless the core exceeds 4 cm or more in diameter (see Titmus and Clark [2003])" (Flenniken and Hirth 2003: 99). The main idea is that techniques have inherent limits, and these can be bracketed experimentally.

Flenniken and Hirth explored the possibility that the microcores at Xochicalco were worked with freehand pressure. They pursued two related questions: core size and platform treatment. Most cores at Xochicalco have pecked and ground platforms. In his feasibility exercises, Flenniken made ten small, single-faceted cores by direct freehand percussion; the irregular ridges of these cores were removed by pressure. For each core, Flenniken held it in his left hand in a leather pad and used a deer-antler pressure flaker held in his right hand to make blades 5–7 cm long (Flenniken and Hirth 2003: Figs. 6.2–6.4). The "prismatic blades produced with the handheld technique were very similar to those recovered in the Xochicalco collections" (Flenniken and Hirth 2003: 100–101). The platforms of five other small cores were pecked and ground. Blades were pressed off cores of both types to assess the benefits of the special platform preparation (cf. Adams 2005).

Given the relative benefits of foot-holding and hand-holding techniques, Flenniken and Hirth (2003: 106) suggest that the two were complementary, with one technique picking up were the other left off: it "is precisely at the point where cores become too small and difficult to reduce using a foot-held technique (ca. 8 cm in length and 3–4 cm in diameter) that handheld prismatic pressure blade reduction became feasible." As discussed above, their proposal is not the only option for stabilizing and working small cores. Distinguishing among different ways of securing microcores must depend on features other than size.

3.2.15 *Eugene Gryba: Chert Microblades*

Gryba has been making microblades with handheld pressure since the 1970s (2006: 60, Figs. 5–7) but has only mentioned his experience in relation to fluting Folsom points (1988, 1989). He has made chert blades "in excess of 9.0 cm in length, and channel flakes as large as 2.2 cm wide, 2.3 mm thick, and 12.0 cm long that had an outre passé termination [sic]" (Gryba 2006: 62), and these with antler pressure flakers about 5–7 cm long. Gryba is 1.67 m tall and of slight build (1988: 57) so his success in producing long microblades and flutes in tough stone is not a consequence of his size. As he explains (1989: 66), "The relatively small size of my pressure flakers, and the method I adopted for holding and applying them … evolved from my experiments in microblade production."

Gryba's success derives from avoiding the Crabtree style of pressure blademaking, such as employed by Flenniken. I would characterize Gryba's technique as using a palm "crutch" (without a crosspiece) – pushing his flaker with the heel of his hand. His tools are not hafted, so he relies on thick padding for support (Gryba 2006: 61, Fig. 2). Mechanically, his method is akin to that described for Eskimo curved-handled pressure flakers: the application of pressure from the heel of the hand (see Holmes 1919: 319, Fig. 181). Watching Gryba flute Folsom points is a lesson in humility because it does not seem possible that someone of his lithe frame, with a tool that looks like an unusable slug of antler, can press off such long channel flakes and blades. It is possible because of his way of holding his tool and working.

3.2.16 *P. V. Volkov and E. Iou Guiria: Long Blades with Lever Pressure*

Volkov and Guiria (1991; Guiria 1997) reported a series of blade experiments performed in 1988 which produced more than 2,000 blades and explored the limits of blademaking with different techniques and materials (1991: 385). They illustrated one Type B flint blade from Petrovo-Svistounovo, Ukraine, that is 231 mm long, 26 mm wide, and 3.6 mm thick, having a length to thickness ratio of 1:64 (1991: 386). Their experiments show that such a blade cannot be produced by direct or indirect percussion or unassisted manual pressure. Rather, they calculate that a flint blade this wide and thick requires 300kg of pressure at the point of application (1991: 387). The material a blade is made of makes all the difference. They argued that raw flint blades require two-and-a-half to three times the force to remove as obsidian blades the same size (1991: 389).

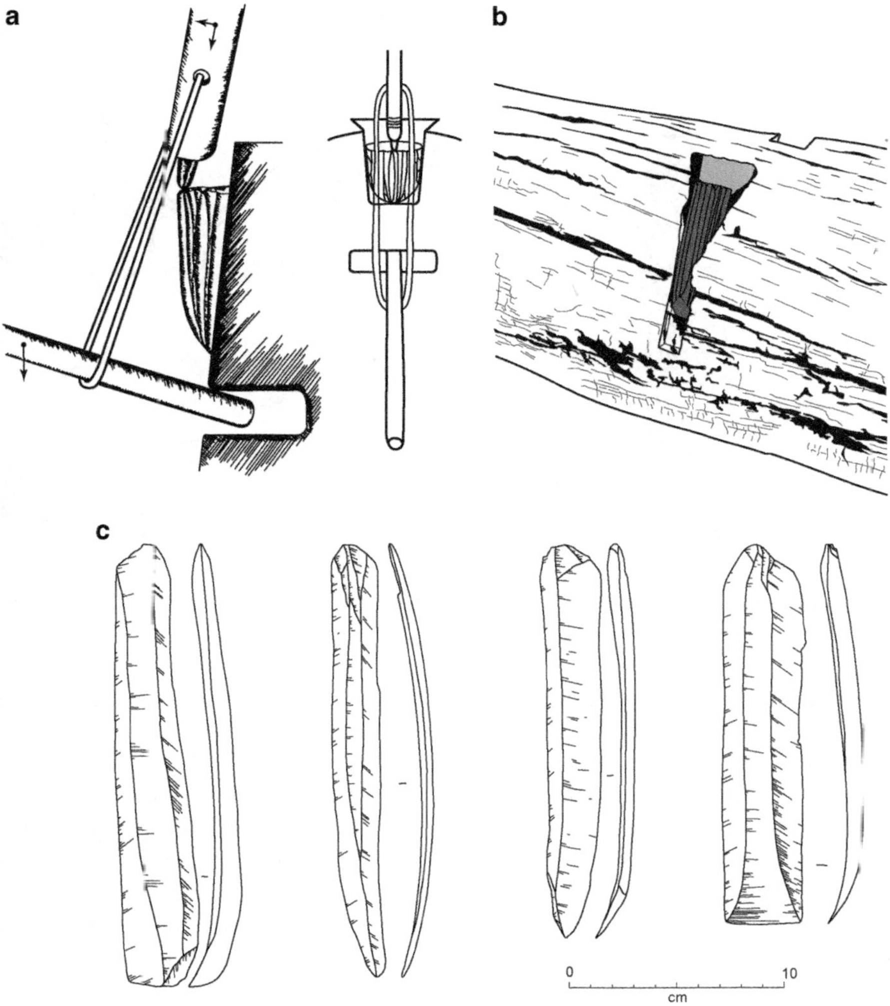

Fig. 3.17 Long blades and the means of making them. (**a**) Combination of a slotted holding device with lever pressure for making very long blades (Redrawn from Volkov and Guiria 1991: 386, Fig. 5). (**b**) Core shown in a slot in a cut tree (Drawn from photograph of Greg Nunn's device). (**c**) Long blades made with lever pressure (Redrawn from Volkov and Guiria 1991: 386, Figs. 7, 8)

One novelty of Volkov and Guiria's experiments is that they compared blades made with unassisted pressure to those produced with assisted, lever pressure. They do not provide details of their manual technique. I presume it was the standard one of Crabtree and Tixier, i.e., use of a copper-tipped chest crutch and a vise. Their version of lever pressure is illustrated in Figs. 3.17 and 3.18. They reported that the handle of their lever was 1.5 m long, but nothing else (1991: 388). As evident in their drawing and cartoons, this device is a modification of Sollberger's lever technique (cf. Fig. 3.5b), but with a rope instead of a wooden crosspiece connecting the

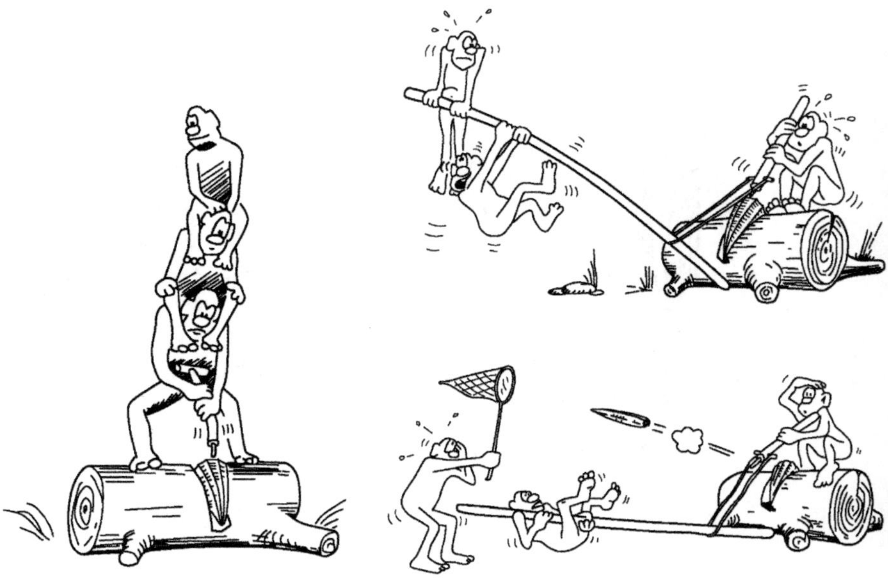

Fig. 3.18 Caricatures of the manufacture of long blades (Redrawn from Guiria 1997: 70, 73, Figs. 12, 17)

lever to the tip of the pressure flaker (see also Kelterborn's double-lever device, this volume). The means of holding the core is not so clear in the original drawing, but it is a self-tightening system. A large slot is cut in a large log or felled tree to hold a core; the slot is narrowest on the outside face of the log and expands or opens up toward the center of the log (Fig. 3.17b). Blades are removed from the exposed face of the core rather than from the back face of the core, as in Pelegrin's systems for long blade removal (Fig. 3.16). The clarifying illustration here is drawn from an experiment conducted by Greg Nunn (see below). Volkov and Guiria's lever technique is a two-person task. The cartoons reproduced in Fig. 3.18 come from Guiria's (1997) synthesis written in Russian. They show that the force exerted by lever pressure exceeds that which could be generated by direct pressure by three knappers in piggyback formation (something actually tried at Les Eyzies; Crabtree, August 27, 1979, personal communication). The cartoons depict the basic form and mechanics of the lever system. One worker slowly applies the pressure through the lever, and the other worker keeps the pressure flaker in correct position on the edge of the core until the blade is released.

Witold Migal (2006: 395) describes a modification of Guiria's and Pelegrin's systems for making long pressure blades found at Neolithic sites in Poland. He used a log for supporting the core:

> the core was made immobile in a carved notch. Blades were detached with an antler-tipped wooden pole put through a hole in the upper part of the log. The point of support for the lever had been constructed as a hole in the wall opposite the cores. This way of working was very compact and connected with the size and shape of cores. The longest blades were 21 cm, and I could exploit the core to about 15 cm of height.

Migal notes that simple machines, such as used to squeeze grapes, could also have been used to make blades (2006: 398, Fig. 8). Long blades and grape seeds appear in the archaeological record at the same time. A significant question addressed by Migal is why long blades had such a restricted modality in length and why some long cores were abandoned before all the possible blades had been removed. He hypothesizes that both phenomena relate to the requirements for immobilizing cores. "Precision in shaping of sides and backs of the cores leads us to the conclusion that a system of immobilization must have existed which excluded the core's further use after a reduction in dimension" (Migal 2006: 392).

A lever pressure technique with cores immobilized in tree trunks is capable of delivering almost unimaginable force to cores and can produce very long, wide, and thick pressure blades, as shown in Fig. 3.17c. Any limitations would be those built into the slots for holding cores and the size of the cores themselves. This system will accommodate much longer levers and significant multiplication of brute human force. As with the smaller, self-immobilizing devices, the same issues of vise stability are in play with a tree as with a small Y-shaped branch vise. "The tree needs to be well braced to eliminate the tree from rolling due to the force of the lever" (Nunn, July 22, 2009, personal communication; see Figs. 3.16, 3.18). Lever pressure is capable of putting more than 500 pounds of pressure on a core, and this is certainly sufficient for rolling even a very large tree trunk.

3.2.17 Philip Wilke and Leslie Quintero: Naviform and Microblade Cores

Wilke and Quintero's (1994; Quintero and Wilke 1995; Wilke 1996) experimental studies of blade cores and microcores provide numerous insights about how to immobilize small cores and of possible telltale traces left on cores and blades secured by different means. Wilke's first knapping experiences and archaeological studies were with projectile points and bifaces, and only later did he turn to blades. He started to knap on his own about 1970 but learned basic skills in a course taught by Jeanne Binning in 1984. He pursued his interest by attending Flenniken's flintknapping fieldschool in 1985, and he assisted in this fieldschool in 1988, 1989, and 1990 (August 3, 2009, personal communciation). At these fieldschools, he witnessed Flenniken make pressure blades with an enlarged version of his Dyuktai technique as well as the Crabtree technique, but it was not until he began teaching pressure blademaking in 1991 that he became proficient with these techniques. He attempted to use the Callahan system but ended up pressing a blade into his leg and gave it up. By 1998, he had also given up vise systems and had begun using jigs inspired by Pelegrin's systems (July 29, 2009, personal communication). For her part, Quintero attended the flintknapping fieldschool in 1990 and is an accomplished knapper and lithic analyst (P.J. Wilke, June 17, 2009, personal communication). Wilke and Flenniken cooperated on several replication studies of projectile points and bifaces (Flenniken and Wilke 1989; Wilke 2002), so it is fair to say that Wilke and Quintero both have been steeped in the Crabtree tradition through Flenniken, Titmus, and others; they are also acquainted with the principal French knappers pursuing blade studies in the Old World.

Quintero and Wilke's first study was of naviform cores and blades (Type A system) from Syria, Israel, and Jordan. Quintero brought some naviform cores to the 1990 fieldschool to solicit opinions on how they were made, and she and the other participants were able to observe attempts to duplicate naviform core blanks, mostly by Titmus. The issue of their production was not resolved, but it did become clear at the 1990 fieldschool that the naviform cores were for "percussion blades, not indirect, as the literature at the time claimed" (P.J. Wilke, July 29, 2009, personal communication). After returning from fieldschool, Quintero and Wilke experimented over the next several years and reduced 300–350 cores. Based on detailed studies of experimental outcomes, they concluded that naviform cores were reduced by direct percussion (Quintero 1998; Wilke and Quintero 1994: 40).

The reduction sequence for making naviform cores bears a formal and logical resemblance to Type A core technologies. The basic concept is to make a blade core by removing the edge of a thick biface or tabular piece of flint. An unusual feature of naviform cores is the manner of setting up platforms. The original biface is shaped like an isosceles triangle with one long side. In contrast to the Japanese and Siberian blade industries, the long edge of the biface or tabular piece of flint becomes the working face of the core from which blades are removed, rather than the platform. Naviform cores have two opposed platforms (the short sides of the triangle), and blades were removed from both. The bidirectionality of these cores served to keep the blades straight, and it also simplified the correction of some kinds of errors. These blades are not as regular or uniform as pressure blades, but they are regular (see Quintero and Wilke 1995: Figs. 1–4; Wilke and Quintero 1994: Plates IV and V). The uniformity of these blades, coupled with their small "punctiform" platforms, had fostered speculation that they must have been made by indirect percussion (Calley 1984, 1986; Cauvin and Coqueugniot 1988; Inizan 1980; Rollefson 1990; Suzuki and Akazawi 1971; Valla 1984). But Wilke and Quintero (1994: 40) have been able to reproduce all features of these blades by direct percussion with a soft hammerstone.

Wilke next studied more regular blades and bullet cores from the east Euphrates and Anatolia region. He went into his experiments of making pressure blades with considerable experience in making percussion blades. Wilke employed a version of Pelegrin's slotted block device, made from an elk metapodial, to extract blades from small cores (Type B), an idea going back to proposed techniques of fluting Folsom points by indirect percussion (see Tunnell 1977: Fig. 9; Warren 1968); see Fig. 3.19. Wilke (1996: 289) replicated small, bullet-shaped cores about 2.5–5.5 cm long and less than 1 cm in diameter. Bladelets from such cores generally qualify as microblades (less than 1.2 cm wide in his scheme). Wilke's experiments were designed to "determine the technological parameters" of bullet-shaped cores from the Near East (1996: 289). He drew attention to the distal shapes of cores and how cores were held, an attribute stressed by Barnes (1947b) in his comparative study of blade cores. Wilke came to the same conclusion as Barnes that "Pointed and Chisel-ended

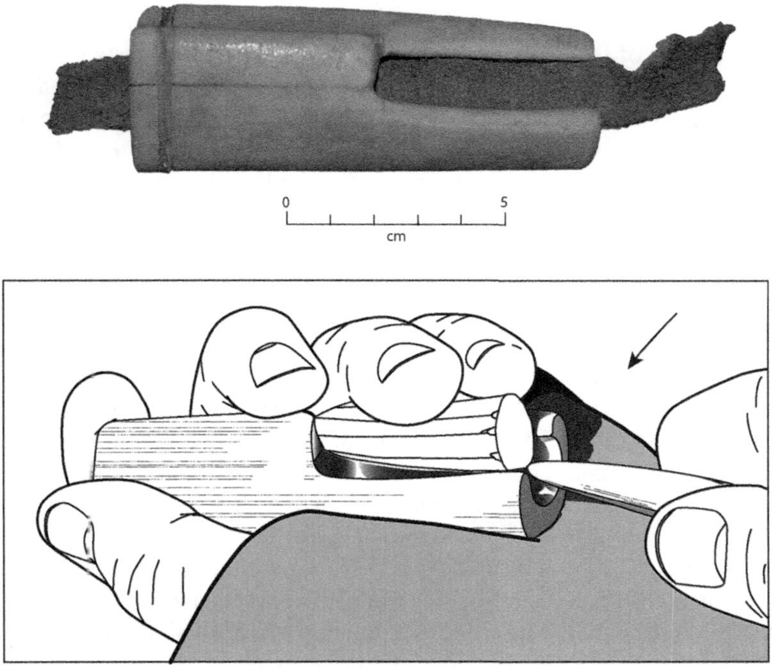

Fig. 3.19 Slotted bone device and its leather strip (to prevent blade breakage) for hand-holding microcores (Photograph courtesy of Phil Wilke, and drawing from a photograph in Wilke 1996: 295, Plate 1)

types of core do not seem suited for use in the palm of the hand. It is probable, therefore, that an anvil was employed and that there were local variations in its character" (Barnes 1947b: 109). In Wilke's technique, the anvil feature was built into his holding device:

> holding the cores in the hand with no distal support … becomes more difficult as core diameter diminishes … experience indicates the technique cannot encourage the consistent production of straight blades … distal core support is essential for detaching straight blades, which in turn maintain the straightness, and the length, of the core face. Distal support causes the detaching force to compress the part of the core where the blade is to be detached, thus directing the fracture to the point of support. Short, straight blades can be produced more consistently using simple but effective hand-held appliances of wood or bone that provide necessary lateral and distal core support, as well as complete core immobilization ….
>
> (Wilke 2007: 222)

Given the regularity of blade margins, ridges, and curvatures, Wilke argued that blades from Near Eastern microcores were made by pressure rather than by direct or indirect percussion (Wilke 1996: 290, footnote 3).

Through trial and error, Wilke identified five critical parameters and/or desiderata of microblade production: (1) "Cores must be immobilized to prevent movement and consequent change in direction of loaded force during blade detachment" (1996:

293). (2) "Cores must be supported at their distal ends, both to direct the detaching force to the end of the core and to stop the detaching fracture at that point. Thus, support encourages the production of straighter blades that run the full length of the core without significant overshot" (1996: 294). (3) "Cores must be supported distally in a manner that leaves room for blades to detach freely from the distal end of the core" (1996: 294). (4) "Cores must be supported in a manner that allows blades to detach freely away from the working face of the core" (1996: 294). (5) "Cores must be supported in a manner that accomplishes all of the above, and that also enables the core to be rotated and maintained easily" (1996: 294). Failure to respect these requirements results in various kinds of suboptimal outcomes. Wilke experimented with microcores made of obsidian and heat-treated chalcedony, flint, and chert. He explored different ways of holding microcores, including holding them on a leather pad in his left hand, various kinds of handheld clamps, and slotted blocks of bone or wood (1996: 293). Over a 3-month period of intensive work, he reduced about 125 experimental cores (June 17, 2009, personal communication). "After much experimentation, it was determined that bullet-shaped microblade cores were effectively reduced using a simple hand-held device that enabled several technological constraints to be accommodated at the same time" (Wilke 1996: 293–300; cf. Pelegrin 1988: 42, 2003: 61). Cores were bullet-shaped by the end of the process because of the way they were processed, but they were not all originally this shape (P.J. Wilke, January 30, 2010, personal communication). As with Sollberger and Patterson's study, Wilke considered blade replication at the level of populations of related products and not of single specimens. It is the consistency of regular blades in predominant frequencies that indicate pressure manufacture.

3.2.18 Andrei Tabarev: Wedge-Shaped Microcores and Holding Devices

Tabarev received flintknapping instruction from Wilke and Quintero at the Lithic Technology Laboratory at the University of California, Riverside (1997: 143), so it is not surprising that his study of wedge-shaped microcores from North-East Asia shares structural features with those just outlined for the Middle East. Tabarev came to these experiments with archaeological background and questions about specific assemblages. In 1995, he was able to view a video which shows Rob Bonnichsen (Bonnichsen et al. 1980) making microblades with a complicated, two-man system for immobilizing cores (Tabarev 1997: 139). Tabarev's experiments attempted to simplify this process. For the most part, he followed the production sequence outlined by Flenniken (1987), but with the suggestion for a different and more efficient hand clamp for holding cores (Fig. 3.20).

Tabarev reduced 70 experimental cores of obsidian, jasper, flint, and flinty tuff (1997: 142) in a three-point holding device modified from a Callahan type clamp (cf. Fig. 3.10). Few details of Tabarev's experiments have been published other than to confirm that cores can be secured for blademaking with simple hand clamps.

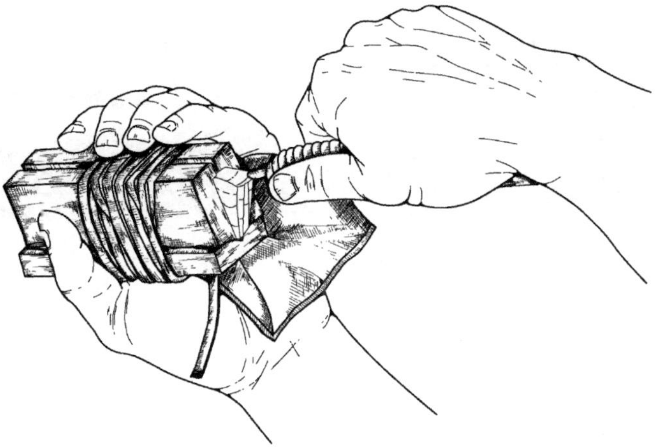

Fig. 3.20 Hand clamp for securing wedge-shaped cores (Redrawn from Tabarev 1997: 146, Fig. 5)

"During the experiments it was apparent that, for the device used [Fig. 3.20], a subtriangular preform would be best" (Tabarev 1997: 143). Of interest was the relationship between Tabarev's holding devices and initial core sizes and shapes, with it "often necessary to adjust the configuration of the microcore in order to fasten it more securely in the clamp device" (Tabarev 1997: 141).

3.2.19 Marc Hintzman: Mesoamerican Microblades

Hintzman (2000) followed the lead of his Master's thesis advisor, Wilke, and conducted experiments to understand obsidian blade production from small cores from a secondary site in the Maya Lowlands of Belize. Hintzman suspected that artisans at the site he investigated got "second-hand cores" – cores worked to near exhaustion elsewhere and imported to this site (Hintzman 2000: 23). These cores were used in the equivalent of a microblade industry.

Minimal details have been provided so far on Hintzman's experiments. He explored three techniques for holding cores that corresponded roughly to those described by Crabtree, me, and Wilke. Wilke designed a larger version of his slotted block device to hold larger cores, and Hintzman used a similar jig in his experiments (Fig. 3.21). Foot-holding did not fare well in these experiments and appears to have been quickly abandoned. Hintzman found he needed a rest for the distal end of the core for foot-holding, but even then the results were unsatisfactory. In contrast, blades made with cores wedged in a self-tightening, slotted device resembled those found at the archaeological site. This device "allowed cores to remain upright, so the blademaker could stand during the reduction process. The result was that essentially no work was required of the blademaker. The weight of the blademaker when applied through a

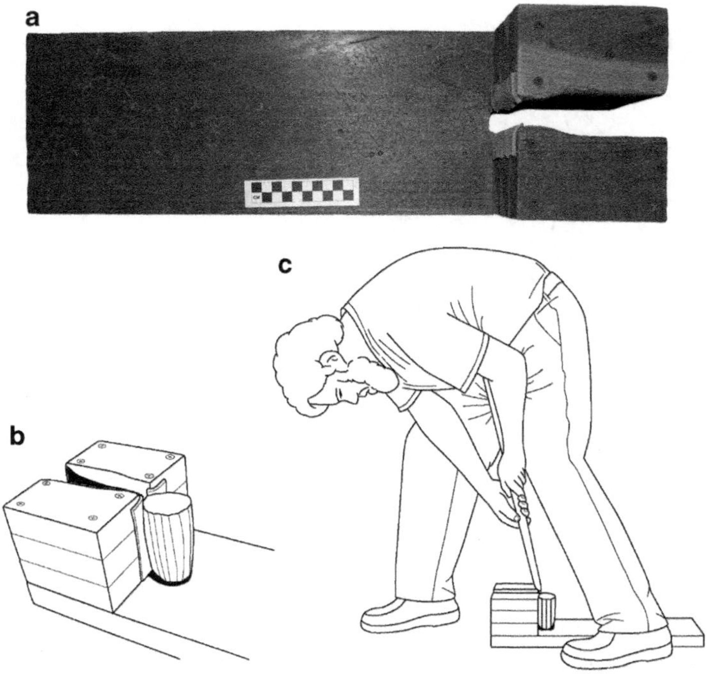

Fig. 3.21 Core self-immobilizing device designed by Phil Wilke for working normal-sized cores. (**a**) Top view. (**b**) Oblique view showing the position of a core. (**c**) Manner of working a core

pressure tool to a core provided all the energy required to detach blades. In this manner, blades frequently were produced for hours at a time" (Hintzman 2000: 39).

Hintzman brings out the issue of the relationship between core geometry and the means for immobilizing cores. His deliberations concern the benefits and disadvantages of pointed versus truncated cores. He observed that many archaeological cores have crushed distal ends, a possible indication of the use of hard supports for cores during blade manufacture (2000: 39). This is a common observation (see Barnes 1947b: 109). Hintzman's speculations would be greatly helped if his experimental cores showed similar kinds of damage; none is mentioned.

3.2.20 Peter Kelterborn: Blading by Double-Lever Machine

Kelterborn has long pursued a rigorous approach to stone-tool replication (1987, 1990a, 1999), beginning with a study of the famous *livre-de-beurre* technique of removing long blades from flint cores from Le Grand Pressigny, France – perhaps by indirect percussion (1980, 1981b) or "lever pressure plus an initiation blow" (1981a: 22). After impressive studies in replicating Gerzean knives (1984) and of debris from an axe workshop from Lake Zürich (1992), he returned to his interest in very long blades, beginning with a detailed description of a flint blade 43.3 cm long from Varna, Bulgaria (1990b). In November of 1992, he was able to spend three hours at the Field Museum of Chicago and examine a large collection of cores (December 12, 1992, personal communication). Kelterborn visited and worked with Crabtree and Titmus on several occasions, beginning in the 1970s and learned much from them (1981b: 15). More recently, he was the driving force behind the Penn State Obsidian Conference in 2000, and he brought out the best in all of us. His chapter in the conference volume (2003) describes meta-experiments. He mentions experiments in making obsidian prismatic blades, but these were to demonstrate the utility of "measurable flintknapping" and "detachment machines" as research tools rather than to elucidate blademaking per se.

Kelterborn hopes to define performance characteristics of various techniques and discover law-like relationships between force application and microattributes on blades, flakes, and cores. His insights derive their credibility from the validity of his meta-experiments, so it is appropriate to accord them some attention. His work provides a reality check on what experimenters have been up to. The cold fact is that few of the "experiments" in blademaking summarized here would qualify as experiments in a scientific venue (see Kelterborn 1987, 1990a; cf. Callahan 1985: 23, 1995c, 1996a, 1999c, 2006b; Saraydar 2008); most have been ad hoc exercises of exploration. Most "experiments" I am aware of were to cure curiosity, with little thought of formality or of eventual needs to describe or illustrate one's efforts in print. Insufficient attention has been accorded to goals, research designs, repeatability, measurable outcomes, and thorough analysis.

As described here, knappers have devised numerous appliances and aids for making blademaking easier. Kelterborn's machines and slab cores are logical extrapolations of Crabtree's use of clamps and sawn cores, as well as the lever jigs created by Sollberger. One problem apparent in old experiments is the difficulty of controlling one variable at a time. Kelterborn does this through machine knapping of standard core forms treated in standard ways. The experimental approach allows him to measure different knapping forces and to replicate consistently his results. He also controls the setup conditions before detachment that may leave no trace on the cores and blades, such as direction, distances, and angles (Kelterborn, this volume). His procedures are "mechanically transparent," repeatable, and predictable (2003: 121).

Before the objective knowledge derived from machine knapping can be of use, however, it must be demonstrated to be analogically appropriate to ancient techniques. Insights from materials science and fracture dynamics have long been associated with the analysis of stone tools, but these insights have yet to take hold – largely because many of them do not make sense of the knapping experience. For example, reading about steel balls dropped on glass prisms has never helped my knapping stroke (cf. Speth 1972, 1974). Crabtree (1975b: 4) often talked of Alaric Faulkner's (1972) dissertation work on fracture mechanics and how these experiments did not approximate the forces involved in making pressure blades. Crabtree wanted to see much more work along these lines.

Kelterborn is a structural engineer as well as a knapper, and he has taken precautions to get things right. He devised a double-lever device to duplicate the human forces involved in pressure blade detachment (Kelterborn, this volume). To test the adequacy of this machine, he made duplicates of obsidian prismatic blades and a Gerzean knife and compared them to the real things. He found no "diagnostic differences" between his possible replicas and the artifacts, so he concluded that "the double-lever machine is a suitable replacement for, and adequately reproduces, what the human body does at the moment of pressure detachment" (2003: 124). Kelterborn compared blades he made with blades made by Titmus with the Mexica tool, and there was no significant difference between them (2003: Figs. 8.5, 8.7).

Kelterborn's next experiment evaluated the adequacy of experimental cores used for deriving replica blades, but this question was entailed in his first experiments. He used cores of thick glass or cut obsidian, as long rectangular slabs ("plank-edge cores") or as blocks that could be worked into prismatic cores. His plank-edge cores relate to a Type A system and are an obviously efficient way to make many blades without having to reset or change conditions. Blades that removed the frosted surface of cut glass or obsidian are not replicas, but those taken from the active keel of these thin cores are.

Kelterborn's third experiment exposed the inadequacy of many previous studies in machine knapping. He compared the manufacture of prismatic blades of trapezoidal cross section to what is (confusingly) termed "blades" in the experimental literature; Kelterborn calls these glass prisms "long rectangular detachments." These are more akin to long burin spalls removed from thin plate glass than to prismatic blades (Kelterborn 2003: Fig. 8.10, footnote 5). Scientific knappers have tried to approximate the forces of blademaking by removing these blade-like (in length and width, but not in cross section) plates of glass (see Faulkner 1972: 98; also, Bonnichsen 1977; Tsirk 2009). Kelterborn established that there are sufficient significant differences between the two products under similar conditions of fracture that one is *not justified* in interpolating the mechanics of blade manufacture based on experiments with forcing off tabular pieces of glass. The differences in "core" geometry are too significant. From this, Kelterborn proposes his Similarity Rule: *"Qualitative fracture laws in flintknapping, and in particular quantitative equations, are only valid within the limits of strictly similar detachment attributes"* (2003: 129, original emphasis). In terms of specifics, "it means that ordinary flakes without dorsal ridges [e.g., from plate glass] cannot be used to describe flakes with one or more ridges or that flake removals with hinge or step terminations cannot be compared to removals with feather terminations" (2003: 129).

Most of us making Mesoamerican obsidian blades have claimed to have "replicated" blades (following Crabtree), but this is an abuse of the concept because no study undertaken to date has even attempted replication, according to the loose guidelines specified by Flenniken (1981b; see also his comments in Callahan 1978b). Kelterborn (1987, 1990a, 2003) provides a higher standard for scientific replication to which we should aspire. What stoneworkers have generated is *savoir-faires*. This is knowledge beyond the capacity of Kelterborn's approach; his experiments favor *connaissances* (see Kelterborn 2003: 123).

Kelterborn (2008a) is compiling a virtual assemblage of expended Mesoamerican obsidian blade cores to guide parameter research. His longest documented archaeological cores (macrocore class, 18–24 cm long) appear to have come from Michoacan (see Hester 1978) and to be the sorts of exhausted cores imported into Xochicalco for recycling into microcores (above). Some of these exhausted cores are not worked all the way around their perimeter, and they evince little evidence of plunging blades. They average 15–16 blade removals in their circumference. These cores show evidence that the "detachment technique by pressure clearly aimed at a consecutive blading order" rather than an "occasional blading order" (2008a). The sequence of blade removals may be telling about the techniques used for immobilizing cores, whether by hand, three-point rests, feet, lateral clamps, or end blocks (2003: 127, Fig. 8.9, 2008b).

3.2.21 Gene Titmus: Mexica Blading with Wooden Tools

Although he began blading soon after Crabtree did, only recently has Titmus committed some of his insights to the written record (Titmus and Clark 2003). Titmus started making arrowheads on his own when he got out of the U.S. Air Force and moved to Shoshone Falls to work for Idaho Power Company. He found many arrowheads as a youth and in his eventual work, and he was curious as to how they were made, so he set about learning how it could be done (Callahan 1980c: 21; Crabtree 1980; Hall 2002; *South Idaho Press* 1981). He met Crabtree, who lived nearby, in 1959, as a person who shared his interests. Titmus had been knapping for about a year and a half before this acquaintanceship (Callahan and Titmus 1999: 66). Crabtree encouraged Titmus to develop his own knapping technique and refrained from giving him instruction (Titmus 1981). By the time they met, Titmus was adept at pressure flaking bifaces; he learned about percussion from watching Crabtree work (May 2009, personal communication).

Titmus's blademaking efforts are of particular interest in evaluating practitioner knowledge because he has the most varied and extensive experience. Of all knappers, he knew Crabtree the longest and the best, and he helped Crabtree with his Folsom point and blade experiments (Fig. 3.22). Titmus has made thousands of blades the chest-crutch way (Callahan 1980c: 24; Crabtree 1980), fluted scores of Folsom points with either indirect percussion or pressure (Callahan 1980c: 23; Titmus 1979, 2002), and produced over 5,000 blades the Mexica way (Titmus and Clark 2003: 86). Crabtree (1980: 21) was particularly impressed with Titmus's "skill in detaching Meso-American blades by pressure." Crabtree introduced me to Titmus in 1979. Titmus observed Crabtree make blades on occasion but had moved away from southern Idaho in 1961, so he did not have much contact with Crabtree for 10 years. Titmus saw me work with an itzcolotli at the Pachuca Conference, and I demonstrated the Catlin technique on his lawn a few years later. Titmus pursued the Mexica technique out of basic curiosity: "One day I just decided I had to see how this damn wooden thing [itzcolotli] worked" (May 2009, personal communication).

Fig. 3.22 Don Crabtree (*left*) and Gene Titmus (*right*) in Kimberly, Idaho, 1979 (Author photo)

As evident in his commentary on my analysis of blade workshop debris from Ojo de Agua, Chiapas, Titmus had experimented with the Mexica technique by 1986:

> I have limited experience making blades Aztec style … I was more than a little amazed at how little inward pressure is needed to remove a blade – it's almost all upward (outward) that is needed. I could make blades travel through irregular surfaces that using the Crabtree method would have stepped the blade off. The blades also follow a very curved surface and still remain the same thickness the length of the blade, not so with Crabtree's method. The blades also look like aboriginal blades – Crabtree's method always produced blades that did not look like aboriginal blades that I have and have seen. I still can't visualize how I can hold the core immobile with my feet.
>
> Back to plunging blades. Using the Crabtree method you can produce plunging blades at will anytime your core diameter gets down to about 2 inches and under. All that is necessary is to set the tip of the crutch quite a ways back from the margin, have the angle of the crutch in line with the face of the core (or the top of the crutch tipped slightly away from the face of the core), apply almost all downward pressure, and you've got it. Thinking about this in terms of the Aztec method and in terms of a skilled craftsman, it is hard for me to believe that a plunging blade would be a common mistake, even when core dimensions are very small.
>
> (Titmus letter, February 24, 1987)

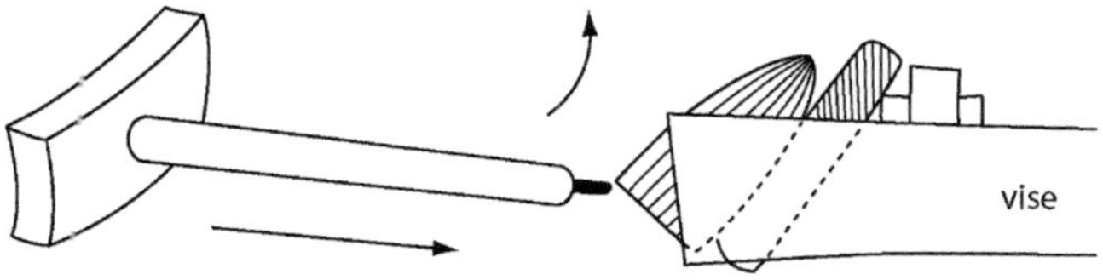

Fig. 3.23 Titmus's experimental method of holding a core for use with an itzcolotli (Redrawn from sketch)

This was wonderful and insightful commentary that I eventually answered (letter to Titmus, June 18, 1987): "You have hit upon the major weakness of my experiments. In my limited experience with the Aztec technique I have yet to produce a plunging blade. I am presuming, however, that this is a function of limited experiments [and experience]."

Just before Halloween of 1987, Titmus sent me a note and a sketch: "I made some blades last night – just wedged the core in the end of the vise and supported it in the back with a piece of wood [Fig. 3.23] and boy did the blades fly just with arm pressure. Will practice your technique come spring." Soon after, Titmus had resolved the problem of foot-holding and was making wonderful blades. As he describes it, his first attempts at blademaking with foot support for his core were "like trying to thread a needle with boxing gloves" (G.L. Titmus, May 2009, personal communication). Eventually, he mastered one way to use an itzcolotli tool.

Titmus made blades with an itzcolotli for more than 10 years before the Penn State Conference, where he demonstrated and explained his technique to colleagues (Titmus and Clark 2003). In a real sense, his experiments are of a different sort than the rest reported here. His was a long-term effort to gain proficiency rather than to check feasibility. Titmus was following a dictum we often heard from Crabtree that understanding comes from mastery, and mastery from work: "From one core what you need is 500 cores, and you start knowing a lot more about it. And after you've done 500 cores then you start admiring what the aboriginals were able to do" (in Clark 1989a: 133). Titmus is still well short of 500 cores, but over the years, he has experimented with the size, composition, and configuration of the itzcolotli tool and the flexibility of its handle (cf. Pelegrin 1984b) and with the materials and configuration of the working bit of this tool. He has experimented with copper, antler, various kinds of hardwood, and plain and notched wooden bits (Titmus and Clark 2003: 89). He has also experimented with working positions and force applications (knapping gestures and techniques in French terms [Inizan et al. 1992a; Pelegrin 2003: 57; Tixier 1967; Tixier et al. 1980]), and with the sizes and shapes of cores, especially platform and distal preparation. Along the way, he also paid attention to what he produced and why. He noticed signal features of different techniques with different tool bits. Cores used in his experiments were from three kinds of Oregon obsidian (Titmus and Clark 2003: 89). For most of them, he sawed the platform, but he also reduced scratch cores with plain facet platforms; he also conducted one experiment with a core with a pecked and ground platform (Titmus and Clark 2003: 91–93).

Titmus developed his own way of using an itzcolotli that requires a special way of immobilizing cores (Fig. 3.24). He uses a backstop and a fore-stop and wedges

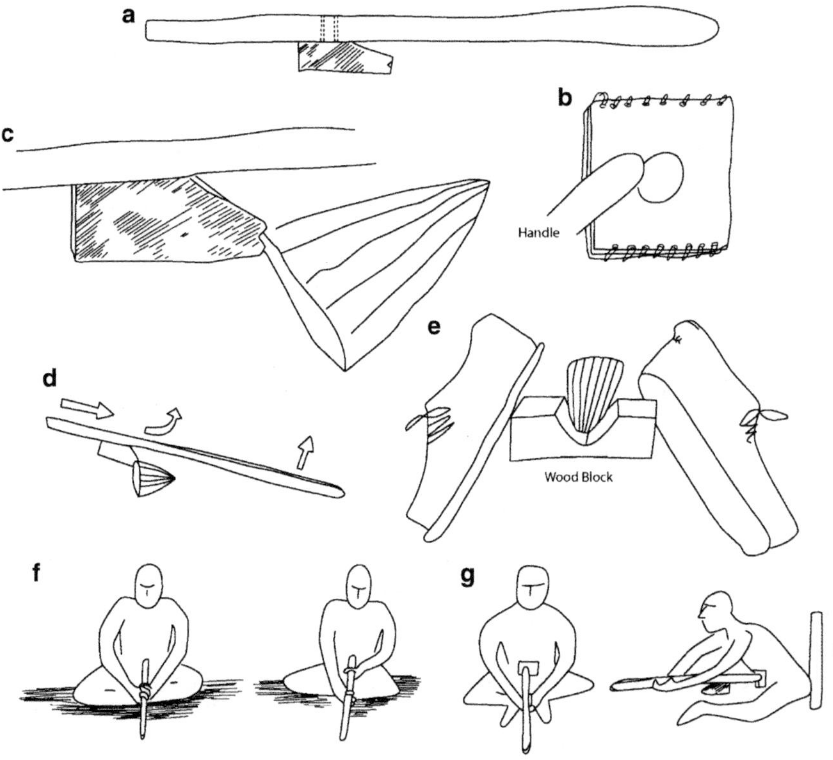

Fig. 3.24 The method used by Gene Titmus to immobilize cores with his feet (Sketches by Rob Fergus, see Titmus and Clark 2003: 81, Fig. 5.5)

himself between them (Titmus and Clark 2003: 81–85, Figs. 5.5–5.11). This compresses his body like a wound-up spring. With a strong, pole fence bracing his back and a small, notched board partially buried in front of him, Titmus places the distal end of his core in the notch of the fronting board and wedges the core in place. With his legs and feet, he applies pressure to the core from the top down, sufficient to keep the proximal and platform end of the core from moving too much during blade removal. Blades are lifted from the upper face of the core. Titmus places the butt end of the pressure tool against a thick leather pad on his abdomen, the distal end of the wooden hook of the tool contacts the upper edge of the core between his feet, and he places both hands on the shaft of the tool. He pushes and lifts/pulls at the same time, and if he does it correctly, a blade comes off up and does a lazy pirouette and lands softly on the ground just beyond his feet. If he forces things, blades fly off the core, and in front of him, and many break as they come off the core (cf. Clark 1982: 372).

The shapes of Titmus's exhausted cores make me think there is more to learn for the Mexica technique. Years ago, Titmus gave me four reduced cores and their blades. I had a graduate student glue the blades back on their cores so we could study the blades in sequence (Figs. 3.25, 3.26). One thing apparent in the exercise so far is that Titmus's cores are bullet-shaped (widest at the platform and tapering to a pointed tip), an infrequent form for expended cores in Mesoamerica. As evident in

Fig. 3.25 Top and side views of a reconstructed core reduced by Gene Titmus with an itzcolotli tool. Note the regularity in the size of the individual blade platforms

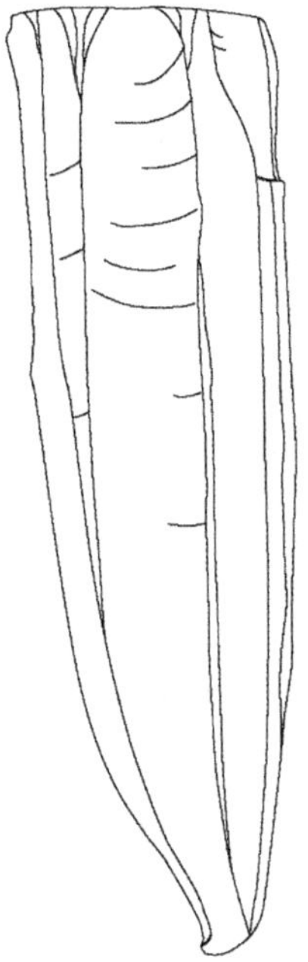

Fig. 3.26 Exhausted obsidian core reduced with itzcolotli pressure by Gene Titmus. Note the hooked distal end where the core was supported on the wooden stop block

one of my experiments, blade platform angles and the shape of a pressure core can change with ring position (Fig. 3.8). The blades Titmus makes from his cores do not exhibit this regular transformation in platform angles. Rather, his cores maintain about the same platform-to-face angle from beginning to end. I have seen a few archaeological cores like these (Crabtree 1972e; Holmes 1919: Fig. 98, center and lower right), but most are widest near the distal end (see Holmes 1919: Fig. 98, left). Titmus generates more force with his backstop and end block than I do without them. He also uses a 75° angle that requires such force rather than my 105° approach. I suspect these inputs account for the differences in the geometry of the expended cores we both produce with the same tool. What I surmise from these differences between Titmus's experimental specimens and Mesoamerican artifacts is that his technique, for all its marvelous successes, still lacks something. My suspicion is that it has too much extra, namely, the backstop, end block, and surplus energy.

3.2.22 Greg Nunn: The Blade Spectrum

Nunn is a highly skilled knapper best known for his replicas of Type IC Danish daggers (2005, 2006). He has been making and learning about percussion, indirect percussion, and pressure blades of flint and obsidian for about 17 years (2007a, b) but has not published on his blade experiments, presuming that his observations must have been made by predecessors (July 2009,[1] personal communication). His knapping history shares features with most American stories presented here, with a few unusual twists. He started serious knapping in 1986 at the age of 30 and began replicating Type IC Danish daggers in 1992, but his interest in knapping and making things goes back to early grade school when he watched his father try to make arrowheads by heating flint chips in an oven and then using a feather to drop cold water on these heated flakes. A knowing neighbor informed them that Indians used antler to flake arrowheads rather than heat and cold, and his father then tried this technique with some success. Nunn remembers studying intensely Bordes's photo-essay in *Early Man* while in fourth grade and being impressed with Bordes's percussion techniques – but not attempting them at that time. His family moved to a ranch on Wilson Mesa in the La Sal mountains east of Moab, Utah, in 1971. There, he found arrowheads and other artifacts during routine work, and he experimented in trying to make duplicates of them from the tough local agates and chalcedonies available. In 1986, he became serious with understanding how these tools were made. He discovered Crabtree's published works and studied them carefully and was disappointed to find that Crabtree was no longer living. Nunn's self-education concerned pressure flaking. He started learning basic percussion skills in 1988 at Callahan's Cliffside workshop in Lynchburg, Virginia. In 1992, he was Callahan's assistant at the dagger-making workshop held at Glass Buttes, Oregon. Titmus visited this workshop and demonstrated Aztec blademaking with an itzcolotli tool and foot-holding technique.

[1]All dated comments in this section refer to information communicated to me by Greg Nunn.

Nunn's learning experience with blading is particularly informative because he was able to observe and work with some of the best knappers in the world, but at the end of the day, he still had to figure out critical details of techniques he observed. His learning cycle has been to observe, try the witnessed technique, and then work out problems not appreciated during the primary observing. Nunn began by exploring pressure blades and then later moved to percussion blades. What he read and saw had to be translated into personal, embodied knowledge. Nunn was and is motivated by curiosity and the joy of learning through doing. In the space of a few years during the early 1990s, he was able to watch Callahan, Bradley, and Titmus make pressure blades of different sizes and materials. Nunn relates his beginning experiences as follows:

> My first exposure to blade making was watching Errett Callahan doing a short demonstration removing obsidian microblades using a forked stick and a hand-held pressure-flaker (1988). My next exposure was Bruce Bradley (1989) who demonstrated to me at his house [in Cortez, Colorado] how to make hand-held pressure microblades without a holding device. He and I were both making nice pressure blades from flint and obsidian that day. I really like this approach because a holding device was not necessary, and I can make extremely nice blades this way.
>
> (July 23, 2009)

Nunn met with Guiria and Bradley in 1990 at Bradley's house in Cortez, and he watched these two make giant pressure blades with lever pressure, with the longest blade measuring 28 cm in length (July 23, 2009; Migal 2006: 395 lists his longest blade as 27 cm). Nunn did not participate in the experiment but did provide some tools for the project. He subsequently tried this technique. For the Cortez experiments, large cores were self-immobilized in a slot cut into the side of a tree; there was no distal support for these cores. In his own experiment, Nunn started by having a friend haul the trunk of an old cottonwood tree to his property so he could make the slotted tree system needed to immobilize large cores. With the help of another friend, Bruce Mace, Nunn was able to perform the two-person lever pressure technique described by Volkov and Guiria. The pressure tool used was a pick handle of hickory 90 cm long, with a fashioned bit of moose antler "off to one face of the handle. This offset keeps the wood handle edge from contacting the core while the blade is being removed. If the bit is centered, the wood will contact the core too much, often damaging the tool" (July 25, 2009). For a lever, they used a log pole 2.6 m long. A rope connected the pressure bit to the lever. The hole in the pick handle pressure tool was next to the bit. The locations of the rope on the bit and on the lever are both significant, and the whole system has to be "tuned" (July 25, 2009). Of special interest, his friend operating the lever "did not know a thing about flintknapping. Under [Nunn's] direction he was able to operate the lever and, at the same time … create a dialog of communication to perform the process. Throughout that year [they] made blades that would be the equivalent to the total [work] of two weeks or so" (July 24, 2009). They were able to make obsidian blades about 25 cm long and 5 cm wide.

Nunn's subsequent experiences centered on making blades in the middle of the size spectrum and were inspired by Callahan's and Titmus's techniques.

> I went to Errett Callahan's Cliffside workshop again (1991). He demonstrated his technique of removing nice pressure blades with his holding device, using a gut crutch and pushing blades away. He seemed to be using his chest more … He did mention that a slightly curved-stick gut crutch may help in removing blades. I went home and made the same device, exactly like Errett's, and a curved gut crutch with a copper tip. I made some small cores with percussion, then switched over to pressure using my replica vise. I was able to remove very narrow blades away from me about 5 cm long, but I could not build enough pressure to make the blades expand any wider, and it hurt my chest. While I was attempting to make blades, I was fortunate to have a very large neighbor kid there acting as a spectator. Out of my frustration, I had him step on the vise. I then walked to the front of the vise/core, placed the gut crutch on the core, and using my chest removed a very nice blade towards me! I was impressed with the instant results. I then readjusted the core and removed another blade … then, I slid the crutch down towards my belly and removed another blade with even more ease and less pain than previously successful blades. That was pretty cool and a major jump forward. What I didn't like about my success was I had to have someone stand on the vise.
>
> (July 24, 2009)

These simple accommodations made during practice and knapping drive home two important points: (1) the need to secure the vise a core is held in, and not just the core, and (2) the different energy requirements of using a crutch tool in different working positions.

Nunn's next major experiment was inspired by Titmus's demonstration of Aztec blademaking, but he also continued to work with the chest crutch system and modified his fixation system, his tool, and his bit. Rather than rely on a Callahan or Crabtree type clamp, Nunn took his inspiration from the slotted tree system of holding cores. To accommodate cores of different sizes, in 1994 or 1995, he built a "small version of the slotted log to accept small cores about 5 to 8 cm long" (July 24, 2009).

Later, he refined and multiplied his log device into a more comprehensive system by making a graduated series of sockets in three stationary logs. Nunn's current workshop has three logs of different diameters. His small log with the two smallest sockets is 11.5 cm in diameter, and the other two logs with larger sockets are 15 cm in diameter (July 25, 2009). These logs are positioned to form three sides of a rectangle 4 × 5 ft (Fig. 3.27). Each log is anchored with long bolt stock on each end to keep it from rolling, but each can easily be repositioned to change the angle of the different sockets if need be. The holes or sockets in the logs are the shape of truncated cones, being widest at the top and tapering to a flat base. On one side, they have a narrow, vertical slot to allow a removed blade to fly free, similar to the device used and independently made by Wilke (see Fig. 3.21). The truncated cones cut into the logs differ in diameter and thus allow one to work the same core in the same workshop by progressively moving to smaller sockets as the core becomes smaller (Fig. 3.27). A core rests on the flat bottom of the tapered socket and is pressed against the side of the socket with the vertical slot. Given the size and shape of the sockets, cores are not as stationary in them as in Pelegrin's three-point devices, and it is necessary to secure them by placing wooden wedges at the top to squeeze them into their sockets. Blades are then removed by standing in front of the core and using a flexible gut crutch to remove them through the vertical slot cut in the log. A piece of cloth is placed in a depression just outside

Fig. 3.27 Greg Nunn's slotted log system for immobilizing cores (Photographs courtesy of Greg Nunn)

the slotted log to keep blades from breaking when driven downward to the ground. Nunn also improved his blading by modifying his crutch tool:

> Sometime before the turn of the century I saw a photograph of Pelegrin's curved-stick gut crutch with the antler bit set into the tip at a particular angle [see Inizan et al. 1992a: 96, Figure 46 Inizan et al. 1999: 149, Figure 73; Pelegrin 1984b: 119, Figure 2, 2003: 60,

> Figure 4.5]. I noticed the stick also had some whittling – kind of like a bow – and no bark. My curved cherry wood gut crutch had bark. I then picked up my gut crutch and modified it by scraping the bark off and scraping some wood grain off with a draw knife and flint blade. This accentuated the curve. I also reset the antler tip like Pelegrin's crutch. I went outside and gave it a try. It had noticeably better performance, and I could use the flex of the stick to my advantage – much more than before.
>
> (July 24, 2009)

With his refined fixation system and crutch, Nunn is able to remove obsidian blades 15 cm long without undue force. This accomplishment is another testament of the correct application of energy because Nunn weighs 130 pounds, so the size of his blades and their ease of manufacture are not related to body mass.

Getting back to small cores, Nunn's ongoing experiments with microblades ironically owe more to watching Titmus make Aztec blades than having witnessed Callahan and Bradley make microblades with hand clamps or freehand pressure. The Titmus system centers around a different tool, gestures, and holding system, with perhaps the most important element for Nunn's current system being a notched board for holding the core with the feet (Fig. 3.24). "After Gene did his demo at Glass Buttes, I went home and built the Aztec tool. I had instant success. I made about 15 blades from a percussion core. Then everything went to pieces after that. Most likely, I wore my tool out" (July 24, 2009). The working bit of Nunn's itzcolotli tool is of ironwood and is 28.3 cm long and 4.2 cm wide, tapering to 8 mm at the very tip. His handle is semi-flexible. In response to my questions about his blading experience, he gave this technique another try, with some success. He observed, "I keep crushing my platform, and if I do not have the correct angle [the core platform] chews up the tip of my tool" (July 23, 2009). Tool maintenance is clearly an important concern with wooden-tipped tools. A few days later, he tried again and removed five 8-cm-long blades in sequence. "The core was small in diameter, and I couldn't hold it any more. I removed blades barefooted! The wood tool is so large, and the core so small in diameter that I am afraid I could run a blade, or a fragment, through my foot" (July 27, 2009). Recently, he removed about 40 fine blades in a row (August 12, 2009). Other than Titmus and myself, Nunn is the only knapper I know who has had this much success with making blades with foot-held cores.

After his earlier, declining success rate with his itzcolotli tool, Nunn blended the two techniques he was familiar with and began making pressure blades with a gut crutch in a seated position. His short crutch is modeled after Pelegrin's shoulder crutch (Inizan et al. 1992a: 96, Fig. 46c; Pelegrin 2003: 62, Fig. 4.8). For small cores and microcores, he prefers a buried, notched board to his socketed logs. Nunn holds microcores with his feet (shoes on or off, depending on the size of the core) in the manner described by Titmus. Nunn sits with his back to a large slab of rock which leans against the wall of his house. He uses a longer crutch for medium-size cores. Both crutches have curved, flexible handles and moose-antler tips.

The published record of Nunn's blading system presents a paradox to the avid reader because the two-page photo-essay of his work (Nunn 2007a) shows a different way of holding cores than the one just described. In principle, the illustrated procedure is patterned after his slotted log system and Pelegrin's tree-root system (No. 7, above). The system in the photo-essay is much more complicated than

Nunn's slotted log system. When he uses his padded crutch, Nunn stands over a core and first sets it by pressing the tip of his tool into the center of its platform. Once the core is set, he positions his moose-antler tip on the core's outer edge in the area facing the open vertical slot in the log. His working angle is nearly vertical or slightly acute. Pressure from his torso forces the tip of his tool into the core – while keeping the core immobile and also setting the tip of his tool in the crutch. Slight or moderate outward force serves to release a blade.

I visited Nunn on July 15, 2009, and he showed me the basics of his core-holding devices, tools, and blademaking techniques. I had never used a self-immobilizing means for holding cores before, so I was looking forward to seeing – and feeling – how Nunn's system worked. His tools were too short for me, but I made some adjustments in using them. In making microblades with Nunn's short gut crutch, I had to remove his stone back support to give myself sufficient room behind his notched wooden backstop to hold the core with my feet. This left my torso half a foot away from the crossbar of his crutch, so I placed it on my inner thigh and applied pressure by squeezing with my legs at the same time I pushed with my arms. I removed a few blades in this manner.

I was curious to see what properties Nunn's system might possess, so I suggested we take one of his discarded cores and take off a major hinge termination by reversing the core. The sockets in his logs taper in the same way conical cores do, so standing a core on its head works against the built-in angles of his holding devices. I was curious to see how an upside-down core could be worked. We secured the top of this core by pounding in three wooden wedges between the core rim and the open space in the back of the socket. We had previously created a small platform on the distal end of this core by removing the tip by percussion, but the end was still strongly curved, so removing a reversed blade from the core was a challenge and required extra force because the platform was significantly offset inward from the core's face. We were successful in removing a hinge termination with a reversed blade. Later we rejuvenated this same core by removing a tablet from the proximal end and creating a new platform. A few good blades and some more miscues later, we had another hinged mass to remove. For this, I kneeled in front of the core and used a 125° angle to drive a thick blade directly under the hinge termination. In my eagerness to remove this problem, I forgot about stressing the tool, and it cracked with the removal of a plunging blade. Nunn commented that I had put about three times more stress on his tool than it was meant to take.

In an earlier attempt, I tried to remove a blade without wedging the top of the core in place. With the release of the pressure, the core literally jumped out of its hole and did a back flip. Fortunately, no major damage was done to the core. Once we reset the core and battened it down with wedges, a blade came off with normal pressure. An obvious lesson from these exercises is that a practical system of making blades must be an effective combination of designed cores, holding devices, tools, and energy delivery. My inept use of Nunn's system demonstrates that it is possible to use tools and holding devices incorrectly. It is also possible to deliver force improperly or ineptly/inefficiently, and it is clearly possible to shape a core inappropriately for any given system. This is good news because it should be possible analytically to infer aspects of a total system from parts of it, such as the size and shape of beginning and finished cores or the maximum or minimum size of blades.

3.2.23 *James Winn: Obsidian Pressure Blades and Holding Blocks*

Winn is another self-taught knapper who has developed great skill in a range of technologies, including the manufacture of blades. He first became interested in stone tools in 1979 while residing in Oregon. A neighbor took him arrowhead hunting – a few artifacts later he was hooked. He became interested in how some of the finer artifacts were made, and he subsequently discovered D. C. Waldorf's *The Art of Flintknapping* and started learning the art. Beginning in the 1980s, Winn has progressed through different technologies. He prefers to focus on a particular technique until he reaches satisfactory levels of competence, and then he moves on to a different technique (July 31, 2009, personal communication).

Given the emphasis here on stoneworkers' perspectives, and the process of learning from various resources (artifacts, written descriptions, illustrations, movies, teachers), the pivotal points of Winn's story are instructive, as related in his own words:

> I decided to try making a simple percussion core sometime around 2002 .... I had never observed anyone making a core before. Having become quite proficient at bifacial percussion I expected that it would be fairly easy to make a fairly nice core .... I quickly discovered that it was not a simple process. The mindset is totally different than [for] bifacing, and it quickly became evident that I did not have a clue as to how to proceed ....
>
> The following year I decided to give it a go again, this time attempting to remove pressure blades off a sawed cube of obsidian. I was armed with sketches showing Crabtree's core-holding device. I can't remember where I saw them, but it showed a pair of 2 × 4's holding a core at the end, with nothing more than a rock wedged between the wood to apply pressure to secure the core [see Crabtree 1967: 72, Figure 2a]. I built a similar device. I used a copper-tipped crutch and attempted to remove blades outwardly [from the core face] as Crabtree described doing. No matter how hard I pushed the rock wedge in, I could not get the core stabilized enough, and the core would rotate in the vise and result in a failed or short blade removal. I had a few successful blade removals, but the failures won out ....
>
> Around 2005, my wife and I ... visited a museum in Athens ... and they had a nice display of blade cores ... I spent a long time studying them and decided I had to give [blademaking] another try. Shortly after that we were traveling through Utah and stopped at a visitor's center. Inside I found the book *Mesoamerican Lithic Technology* [Hirth, editor 2003] .... This piqued my interest even more. Soon after I discovered that Don Crabtree had been videoed making blades, and I purchased the old video from Idaho State University [Lohse 2000]. This was the first video I had seen of anyone making blades, and so I decided to try using Crabtree's method again. I had the same problem as before, with the core rotating, so I decided to add a C-clamp to the sides of the 2 × 4's to further squeeze the core. This worked OK, but the time spent on wedging and tightening the C-clamp was excessive, and I thought there must be a better way. I re-watched the Crabtree video several times and then noticed something I had missed. There appeared to be a bolt rod between the 2 × 4's squeezing the core! That was left out of all the descriptions I had read of this method ... So I discarded the rock wedge and C-clamp and drilled a hole through the 2 × 4's and installed a large diameter bolt. It worked well enough, but I found I had to reposition the core after each blade removal and it still took an excessive amount of time. I decided to abandon this method entirely and go back to the book, *Mesoamerican Lithic Technology*. My ideas for using a block of wood with a hole in it, as well as my crutch types, were primarily inspired by Jacques Pelegrin's [2003] article and pictures shown in that book.
>
> (July 31, 2009, personal communication; cf. Winn 2008a)

Winn's experience of trying to learn blading through reading Crabtree's (1968) article and scrutinizing his videos (Bordes and Crabtree 1969a; Crabtree 1972e) parallels my own. Things appear simple until one attempts to duplicate them from descriptions, and then descriptive deficiencies become apparent that the author did not anticipate. One lesson from the autodidactic exercises reviewed here is that all articles, descriptions, illustrations, and videos underdescribe what takes place in blade production. This is why learning from a proficient knapper whom one can question during the process is so important. Many critical features of the process are not part of discursive knowledge. Lacking live instruction, most of us have had to work out ways that give acceptable results. So far, Winn has progressed through three stages in his blading experience. He went from Crabtree vises to handheld methods *a là* Flenniken (Winn 2008a) and then switched to a Pelegrin system (Winn 2008b).

Winn's video (2008a) on cores and blades was inspired from Flenniken and Hirth's (2003) experimentally aided study of Xochicalco blades and covers direct and indirect percussion blades of flint and obsidian and handheld pressure blading. Two novelties of Winn's indirect percussion blade technique are a tall wooden anvil that he positions between his knees while in a seated position and his practice of scoring core platforms with a chert flake so blades will break at the scored mark. Both practices led to increased regularity of his indirect percussion blades and cores. For pressure blades, Winn compared the relative efficiencies of multifaceted cores with those with pecked and ground platforms. He followed the Flenniken and Hirth method for grinding platforms on a slab of rock on which about a pint of stiff, sharp slurry composed of pulverized chert flakes, dirt, and water had been placed (see Adams 2005). Winn first grinds away the salient ridges of a multifaceted platform and then uses a large chert flake to peck it flatter; he then finishes the platform with more grinding. A small core about 4 cm in diameter takes just under 15 min to prepare. The largest cores take twice as long.

In pressure reduction of these small cores, Winn uses stiff rubber pads with central grooves that allow for the removal of blades. Pads of different sizes are made of sections of old conveyor belt glued together. His flakers are Ishi sticks with blunt copper or antler tips. In contrast to Flenniken's preferred gesture of holding the core horizontally like a biface (Flenniken and Hirth 2003: 101, Figs. 6.2–6.4), Winn holds his cores vertically in his left hand and between his legs and uses his tool to muscle off blades into the vertical groove of his pad. This pad is open-ended and lacks distal support, so he protects the palm of his hand with leather. An experience of pushing a blade through a wad of soft leather made him opt for a palm protector of stiff buffalo leather, but he still has to be careful not to push blades through several layers of this leather. A built-in distal support connected to his slot would resolve this problem, such as used by Pelegrin and Wilke (above). At the time of his video, Winn (2008a) had less than a week's experience with his handheld core technique, but he found it superior to the Crabtree method. When cores get too small for his holding pad, Winn shifts to a smaller one with a narrower slot; thus, he is able to work cores down to exhaustion.

Winn's next blading experiences involved larger cores and holding devices similar to those used by Nunn (Fig. 3.28). Winn immobilizes his cores by placing them in socketed log blocks, with the size of the socket and its block varying according to the size of the core to be worked. Winn's and Nunn's systems both were independently

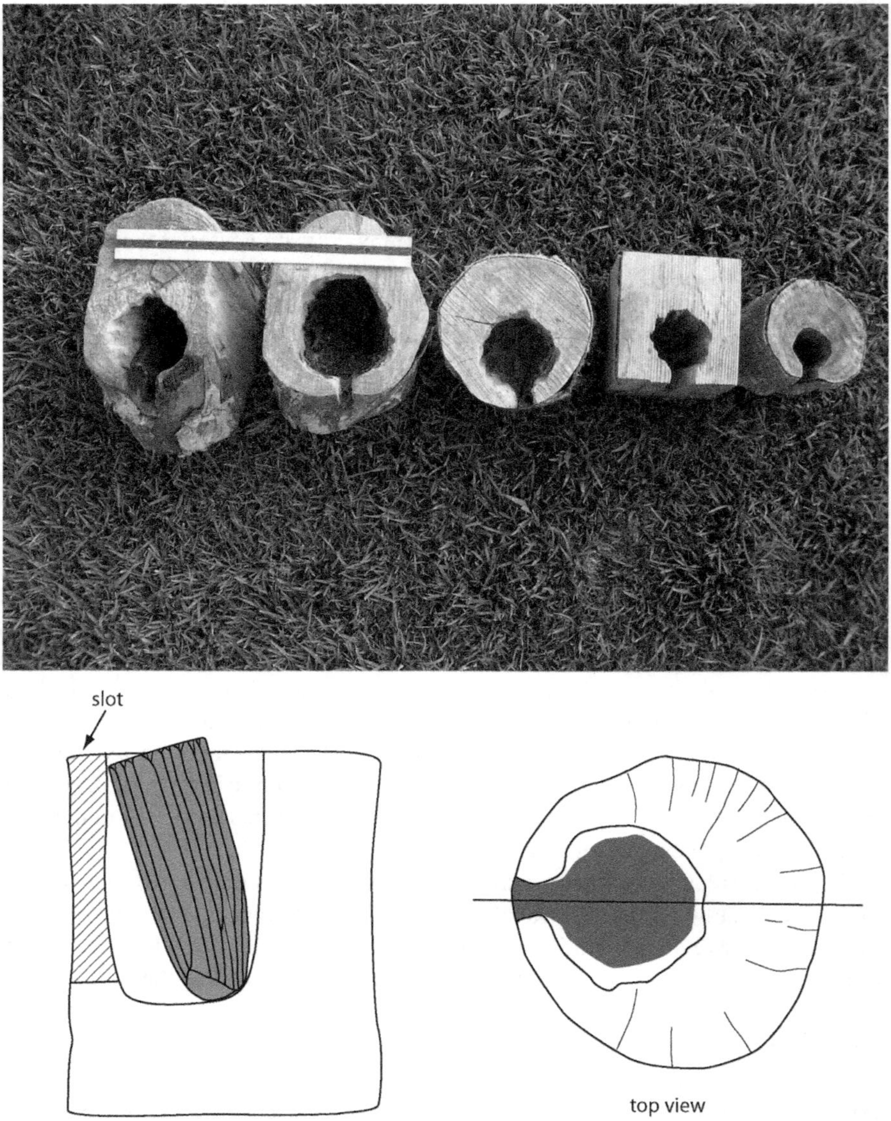

Fig. 3.28 James Winn's socketed blocks for immobilizing cores (Photograph courtesy of James Winn) a drawing showing how cores are secured in them

inspired by Pelegrin's devices (above). One significant difference is that Nunn carves his sockets into horizontal logs, so his sockets cut through the grain of the wood. Their depth is limited by the diameter of the log. In contrast, Winn uses small sections of logs cut off square at both ends, much like a section of log to be split into firewood. He then carves a socket toward one edge of the log block and cuts a vertical slot. His sockets cut into the grain of the wood rather than across it (Fig. 3.28). Some of Winn's sockets are roughly the same truncated-cone shape as Nunn's but not as carefully carved. Some of the irregular ridges of wood left from drilling within his sockets may keep his cores more stable. For other sockets, the hole flares out from top to bottom (on the side away from the open vertical slot). This allows Winn to work cores that are in an inclined rather than vertical position. Winn uses copper-tipped and blunt antler-tipped tools patterned after Pelegrin's curved, flexible tools. He does not anchor his socketed blocks; rather, their flat bases are sufficiently stable that they do not slide on a concrete surface. He can remove blades 22 cm long that require significant pressure.

Winn rarely uses wooden wedges to pin the upper part of his cores in their sockets (August 4, 2009). Cores become naturally wedged in their sockets with the application of downward force on the outside rim of the core – with the distal end of the core forced against the inner (opposite) edge at the bottom of the socket (Fig. 3.28). Once a blade comes free, the core recoils and springs back to a more upright position. Sometimes the removed blades are too wide to exit cleanly through the slot. Some blades remain in the socket with their core, with the backward recoil of the core apparently giving them sufficient space that they do not break. Some blades roll against their cores and acquire spontaneous retouch on their distal ends. The overall impression is that this is a rather simple technology and that obsidian blades are stronger than they look.

Winn demonstrates the reduction of cores of different lengths, all of them with rather small diameters. Given his graduated series of socketed blocks, he is able to work cores down to pencil size. He uses nearly a vertical angle in applying force to a core and relies on impulsive pressure, with a noticeable thrust of pressure to break blades free. For most cores, the platform is positioned level with the top of the socket, or just below it. One core, however, was about 5 cm longer than the socket depth and thus stuck out. For this core, the three points of contact were its distal end and two flanking points on the core's face about 5 cm down from the platform. This brings up the question of core length. Winn's and Nunn's systems of socketing cores attend to core diameter more than length. For short cores, one can put pieces of wood in the bottom of a socket to provide greater elevation; for long cores, one has fewer options for accommodating lengths not anticipated by socket size. How far above the rim of a socket can a core protrude and still be worked safely? Nunn can work cores that rise 7 cm above their sockets and still be safe, but he speculates that beyond this the core may be too unstable (July 28, 2009, personal communication). Winn informed me (July 31, 2009) that if cores stick up too far above their sockets, "the whole block can pivot from leverage." This is sufficient reason to stabilize the socketed block. That such stabilization is necessary only in exceptional circumstances is intriguing, given the difficulties of keeping cores stable in a Crabtree vise. Sockets cut into a tree stump would provide the best stability. It is interesting that Sellers (1886: 882, Fig. 6; also in Moorehead 1910:I: 21, Fig. 18) illustrates an analogous system. The next best thing would be to bury partially the block, or anchor it in the ground.

3.3 Blade and Core Connaissances

The major controversies raised by experiments in making Mesoamerican blades concern the distinction between *Replication* and *Technical Research*. Arguments for replication are premature because the three steps required for it have *not* been attempted in any of the knapping exercises reported here, namely: (1) careful description and specification of the assemblage to be replicated, (2) designed experiments that bracket the assemblage variability by different technical means, and (3) detailed comparisons of specimens from different experiments to artifacts from target assemblages to assess their goodness of fit. Admitting that past exercises fall short of true replication experiments does not lessen their value as technical research. As Pelegrin (1991: 120; see also Tixier 1984a) explains, technical studies skirt the uncertainty problem inherent in replication (i.e., equifinality). Technical research can proceed with certainty and establish diagnostic features of various knapping techniques, tools, and gestures. All experiments help specify the connection of specific techniques to particular outcomes. In concluding this chapter, I focus on such technical knowledge (*connaissance*). The ability to infer knapping techniques, tools, and holding devices from blades is based on understanding what happens to cores and blades during manufacture under different conditions.

3.3.1 Controlling Blade Parameters

A primary concern for technical research is the effect of human input on blademaking. What features of blades were controlled by knappers and which were not? To make this determination, one must factor out the effects of raw material and environmental conditions. Many knappers pursuing a variety of techniques have concluded that they can partially control the length, width, thickness, straightness, regularity, ridge and lateral margin character, transverse cross section, platform size, and surface features of blades. These insights started to spill forth from Bordes and Crabtree's (1969b: 6–7) efforts to duplicate Corbiac blades:

> In our experiments, generally the widths and lengths of the blades were variable and were controlled by the form of the working face of the core. The more attenuated the ridge and the narrower the core, the narrower the blade. The thickness of the blade is also controlled by the position of the punch and the design of the platform in relation to the core. The nearer the punch is placed to the leading edge of the core, the thinner will be the transverse section of the blade.

These insights concern geometry and force application and are true for all blades. Crabtree (1968: 464) records that blade "types are governed by the manner in which the pressure tool is placed on the edge of the core. The triangular blade is made by directly following one ridge, and the trapezoidal type is made by positioning the tip of the pressure tool in line with, but between, two ridges." Blades follow ridges. "Prismatic blades will be no straighter than the ridge left on one face of the core" (Crabtree 1968: 464).

Blade length is more difficult to control than width or thickness, and it also is more restrained by an absolute limit. Pelegrin (2003: 60) observed that blade length relates to the stability of the core and the flexibility of the pressure tool that removes it. Potential maximum blade length is determined by the overall length of the core from which it comes. The main obstacle in making long pressure blades is not the difficulty of pressing them off but in fashioning long cores from which such blades can be taken (Kelterborn 2008a, this volume; Titmus and Clark 2003: 93; Volkov and Guiria 1991: 382).

Blade thickness and width are related variables, and both depend on the arc or diameter of the core face, the placement of the tool in relation to this arc, and the force applied in detachment (Pelegrin 1984a, 2003: 60; Titmus and Clark 2003: 82). Some features of blades also relate to the size, material, and morphology of the tool tip used in detachment. It should become possible to identify the technical stigmata resulting from different kinds of hammers, punches, and pressure bits (see Titmus and Clark 2003: 84–85).

The most striking difference between pressure blades and those made by direct or indirect percussion is the regularity of the straight, parallel edges and dorsal ridges of pressure blades (Crabtree 1968: 457, 459; Pelegrin 1984b, 2003). For Crabtree (1968: 462), the principal problem in transitioning from a percussion to a pressure core was how to straighten ridges and begin making blades with straight, parallel sides: "to make a perfect blade you've got to have a perfect surface to work with. Any irregularity you have could affect the next blade, unless you intersect and take off a thick blade" (Crabtree, in Clark 1989a: 131). The key is to remove double-ridged blades rather than single-ridged ones because removing two widely separated ridges at a time naturally straightens the ridges on a core's face (see Bordes and Crabtree 1969b: 5; Clark 1989a: 131).

Core morphology is not the only factor behind straight blades; the actual alignment of force application is also critical (Crabtree 1968: 474; Pelegrin 1984b: 121; Titmus and Clark 2003: 84; Whittaker 1994: 225; Wilke 1996: 300). The outward force for blade removal has to be aligned with the ridges of the core or the blade removed will be "malformed" (Crabtree 1968: 476), "twisted" (Pelegrin 2003: 61; Titmus and Clark 2003: 85; G.R. Nunn, July 2009, personal communication), or "slightly … helical" (Wilke 1996: 300) due to "axial torsion" (Kelterborn 2008a: 3). Movement of the core during blading can cause a broken or malformed blade, and either outcome hurts the core. Viewed analytically, the kinds and frequencies of twisted blades in an archaeological assemblage could be evidence of the type of core stabilization involved (Kelterborn 2008a, this volume; Titmus and Clark 2003: 83–84).

The relative curvature or flatness of a blade in profile appears to be another characteristic influenced by the means of force application and core immobilization. Crabtree (in Bordes and Crabtree 1969b: 8; also, Collins 1999: 30; Crabtree 1972b: 12; Flenniken 1987: 122; Kelterborn 1980, 1981b: 12; Patten 2009: 68; Pelegrin 1991, 2006) argued that "little curved" or "straight" blades made with indirect percussion resulted from "using a rest, for it prevents movement of the core as the blades are detached and simultaneously causes force to be exerted at the base of the core when the blow is delivered on the upper end. Cores not supported by a rest will produce strongly curved blades." Wilke (1996: 294) makes a similar point for the reduction of microcores: distal core support "encourages the production of straighter blades that run the full length of the core without significant overshot." Another factor with blade curvature is that some transverse and longitudinal curvature is necessary for blade removal, especially for very long blades (Kelterborn 2008a: 3; Pelegrin 2003: 63, 2006); "if you have a little curvature, the blades come off much easier; you can give it that outward force and it will follow" (D.E. Crabtree, August 1978, personal communication).

A distal support or anvil may be apparent in blade profiles and from manufacturing marks on blades and cores (Semenov 1964: 53). For his experiments with Siberian microcores, Flenniken (1987: 121) noticed that the distal end of the "ski spall" removed to create the long platform "exhibited slight undulations, usually terminating in a small hinge as a result of the anvil use," and in other cases it showed "crushing" as well. "Frequently, a small flake was also accidently removed from the distal end of the ski spall and/or ski spall scar as a result of rebound from the anvil" (Flenniken 1987: 121). This last flake sounds analogous to impact flakes broken from the tips of projectile points (cf. Fischer 1989; Flenniken 1985; Flenniken and Raymond 1986; Kelterborn 2001). Similar flakes could result on the distal or proximal ends of a core from the use of a vertical clamp. An important caution in interpreting crushing on the margins of cores is to separate preforming features from damage incurred during blademaking. Gryba (2006: 59) argues that much "crushing" on microblade cores relates to preform preparation and not "from use of a hard anvil during blade removal."

Features of blade platforms or butts have received more attention from knappers and analysts than have distal ends of blades, and deservedly so. As remnants of the core platform from which the blades were detached, butts preserve information on the preparation of the platform surface (plain, abraded, pecked/ground, multifaceted), the angle of the core platform to its face, and the tactics for building precision into blademaking. The size and shape of the platform on a detached blade, and characteristics of both the ventral and dorsal surfaces at its proximal end, are clues to the type of force application, the tool used to detach the blade, and even the morphology of the tool bit and surface area of contact. For the dorsal surface, one notes features of the removal, or not, of overhang left by previous blade removals. For the ventral surface, of particular interest are the bulb of force and the presence or absence of eraillure flakes, lipping, Hertzian cones, cracks, concentric rings, ripples, and fissures.

As to tactics, one can build precision into the preparation and isolation of individual platforms to minimize the chance of improper contact and insufficient force application. Pelegrin describes experiments with the Levallois technique and setting up individual platforms so "you can't miss" (in Callahan 1982: 68). On the other hand, a general platform with predictable qualities can be constructed and precision built into the blademaking tools and/or their placement. Titmus induced precision by notching his pressure tool; in a similar manner, and independently, Pelegrin built precision into manufacture of indirect percussion blades by notching his punch (in Callahan 1982: 63). Another option is to have pointed tools and to be precise in their placement. Given a sufficient population of blades made by such means, it should be possible to reconstruct the platforming concept or tactic for a particular assemblage.

Platform preparation and overhang removal were under a knapper's control, and the platform angle was also under partial control. Platform sizes are a natural consequence of knapping techniques and gestures as applied through various kinds and sizes of tools (see Fig. 3.25). It is worth emphasizing that there is no necessary correlation between the size and angle of platforms and the width, thickness, length, or longitudinal curvature of blades. Rather, platform sizes relate to the type of pressure tool used and the total area of contact. Blunt wooden bits contact more surface than do metal or hard stone bits, and they result in larger platforms, all other things being equal. Pointed versus diffuse contact may relate to features on bulbs of force. Blades made with wooden and antler tools evince a lower frequency of eraillure scars and have more lipping than do blades made with harder, more pointed bits (Clark 1985; Pelegrin 2006; Sheets and Muto 1972). The locations and shapes of eraillure flakes also vary according to the type of platform preparation, with pecked and ground platforms affecting these features the most. Cracked platforms or partial Hertzian cones on the ventral faces of blades are evidence of a hard and pointed tool, such as a copper bit (Pelegrin 2006), or perhaps even flint-tipped tools (Semenov 1964: 50–54, Fig. 11). These features of a blade's bulbar area can help distinguish among the kinds of force application (e.g., direct pressure, lever pressure, and direct or indirect percussion) and the material and morphology of the tool used to remove them. Pelegrin (2006) argues that pronounced lipping is a mark of indirect percussion (see also Kelterborn 1980).

In his foundational paper, Crabtree (1968: 449, 451, 469) claimed that the small platform size of Mesoamerican fine blades was evidence that they were made by pressure rather than by indirect or direct percussion. This is not necessarily the case, as Crabtree discovered soon after in experiments with direct percussion with "edge-ground cobble" hammerstones (Crabtree and Swanson 1968). It is possible to make percussion blades with very small platforms (see Newcomer 1975: 100; Quintero and Wilke 1995; Wilke and Quintero 1994). The analytical literature related to Mesoamerican blades describes changes through time in platform size, shape, and preparation, all of which may indicate an evolution in methods and techniques of blade production. Early blades have isolated, individual platforms; later blades were from cores with platforms prepared for blade removals in groups; and the latest blades had pecked and ground platforms, with little attention accorded removal of platform overhang. Experiments conducted so far have produced credible replicas of early and middle blades, but not Postclassic blades that correspond to the descriptions of the Mexica technique. More experimentation is needed to understand platform treatments, their relationship to blademaking tools, and knapping gestures.

Platforms represent interesting attributes because they were under knapper control, but knappers were also constrained by the laws of force propagation. Platform angles vary as a consequence of blademaking itself, becoming progressively less-acute with the removal of each ring of blades (Callahan 1984: 92, Fig. 15; Titmus and Clark 2003: 92; see Fig. 3.8). For cores with unpecked platforms, removal of pressure blades becomes difficult if the platform angle exceeds 90° (see Callahan 1984; Patterson 1986; Sollberger 1986a). Pecked and ground platforms allow a knapper to remove blades with obtuse angles more easily. Maintenance of an acceptable platform angle would have been important for knappers wishing to maximize production of regular blades.

The distal shapes of blades are governed by core forms and ridge patterns, the amount of force applied during blade detachment, and whether or not a blade runs the full length of its core before termination. Cylindrical cores with truncated ends favor the production of square-end blades, and pointed cores favor the production of pointed and plunging blades. It is possible to produce a pointed blade from a truncated core by stopping it before the end of the core, but this adversely affects a core's ridge pattern. Blunt cores help prevent blades from overshooting; blade overshots are more frequent on pointed cores.

3.3.2 Blade Cores and Their Metamorphoses

Scholars tend to view blade technology as a means of obtaining a large volume of standardized items through controlled and skilled knapping. The hidden irony in this generalization is that blade production has to deal with a morphing core during the process. As Titmus pointed out, a major skill is to produce uniform blades from a core that is constantly changing:

> Pressure core diameter decreases as blades are removed, and average blade width must also decrease correspondingly …. Perhaps the most difficult part of pressure blade removal is maintaining the correct blade width relative to the diminishing diameter of the core in order to maintain side-by-side blade scar ridges that allow for continuous removal of two-ridged (trapezoidal) blades. This requires a consistent amount of force for each removed blade …. Standardization of blade width and thickness relative to core circumference helps maintain the core in a cylindrical shape.
>
> (Titmus and Clark 2003: 91)

Another transformation for most Type B cores was changing platform angles with successive rings of blades. This transformation occurs in the ideal circumstance in which a flat platform has been prepared. For most cores, there are two platform angles a knapper needs to monitor: the platform-to-face angle and the angle of the platform to the main axis of the core – a hypothetical pivot point running through the center of the core. Ideally the core platform should be perpendicular to its central axis. For cores with pecked and ground platforms, it was. For cores with single-faceted or multifaceted platforms, it was not. Cores may have concave or "dished" platforms (Barnes 1947b: 103) from the original percussion blow. Thus, changing platform angles due to core reduction had to be monitored in light of any tilt (or convexity or concavity) in the core platform during rotation around its axis (see Fig. 3.7). "When removing blades, one needs to control the angle between the pressure tool and the core platform, as well as the amount of pressure exerted on the core through the tool. Even when pressure is held constant, a minor change in the working angle can result in an error. If a knapper fails to compensate for minor differences in platform curvature [tilt], he can unknowingly alter the angle between his tool and the core platform" (Clark 1985: 9).

The most obvious changes in a core during manufacture are a reduction in its diameter and mass. Barring knapping mishaps or tactics that remove the end of a core, there is no necessary reason for a core to become shorter. The top of the core is removed faster than its middle or end because of the greater thickness of bulbs of force compared to the body of blades, so cores tend to become more pear-shaped as blademaking proceeds. This means that the longitudinal curvature of blades also changes by ring position. Most curvature is confined to the distal portion of late series blades. Platform angles increase, and the geometry of their midsections changes as a function of decreasing core diameter (Fig. 3.8; also, Pelegrin 1984a: Fig. 2; Texier 1982: Fig. 1). If one views blade cross sections as chords cutting off arcs, these arcs progressively represent a greater proportion of a core's circumference as it shrinks in diameter (see Callahan 1995a: 235; Hay 1978: 174–203, 208–214; Hay and Rogers 1978).

Pressure core preforms come with a set number of ridges, and a knapper removes these and establishes more regular ridges. Kelterborn (2008a) calculates from 14 to 16 ridges for pressure cores. With the production of trapezoidal blades and a diminishing core diameter, something eventually has to give; the number of ridges on the core and the distance between them are both reduced. This results in the production of narrower blades, blades with dorsal ridges that are closer together, some thicker blades, and occasional blades that remove an extra ridge from a core. Blades with more than two parallel ridges are expected from the final stages of the reduction sequence. During blade manufacture, cores get smaller, lighter, of reduced diameter, more parallel-sided, and even pear-shaped. These shifts lead to the concern for small cores expressed by many experimenters. Not only do knappers have to shift techniques during the reduction of a core, they sometimes have to change tools, working stances, force applications, and the manner of securing the core.

3.3.3 *Core Stability*

The issue motivating most experimentation with Mesoamerican blades has been core stability. Arguments about tools, bits, working positions, and force applications boil down to the problem of keeping cores still during blading. Preceding descriptions provide a range of options for core immobilization. Near the end of his life, Crabtree recognized that he had simplified the problem by thinking of cores as ideal forms. Pelegrin (2003, 2006) and Kelterborn (2008a) suggest that a better way to approach cores and blades is to group them by size class, the presumption being that cores of different classes have to be stabilized in different ways and reduced by different techniques. Kelterborn (2008b) identifies parameters for each size class and the limits of different holding techniques and tools.

One can accept the following facts as established: (1) Exhausted blade cores represent a range of sizes, platform angles, platform preparations, face curvatures, ridge numbers, and ridge regularities. (2) Pressure core preforms necessarily represented a similar range of forms and sizes. (3) The process of making pressure blades transformed cores in geometrically predictable ways, ceteris paribus. (4) Cores of different shapes and sizes presented different challenges and opportunities for keeping them stable during blademaking. (5) There are many ways to secure blade cores, and they have different benefits and inefficiencies. (6) Some holding techniques occasionally leave marks on the cores and blades so held. These facts are generally appreciated by experimenters. As pointed out by Pelegrin, Flenniken, and Wilke, one logical deduction from these facts is the likelihood that very small cores were held in the hand and larger cores were not. For analysts, the last two points are central because it should eventually be possible to identify technical stigmata for different holding devices and reconstruct from artifacts the distribution of techniques in time and space. Potentially diagnostic stigmata of holding devices include marks left on cores and blades and special kinds of knapping errors.

Experiments suggest that some damage to blades may be associated with certain holding devices. Healan (above) identified counterflaking as manufacturing marks, and he opened the prospect of looking for fine-grained evidence for contact points between cores and their supports. Most of the devices described by Pelegrin can be characterized as "three-point rests" (Kelterborn 2008b). Titmus's technique also relies on three contact points. In these techniques, a rest is needed for the distal end of a core, and additional support is required for opposed sides of its proximal end. Pelegrin's devices are self-tightening and take advantage of the force exerted through the pressure tool to lock a core in place without additional effort. Blades are pushed off the lower margins of their cores. In contrast, foot-holding counters the force used in blademaking. The feet push a core down, and impulsive pressure in the opposite direction removes a blade from the core's upper surface. The Mexica sources clearly state that blades were lifted from their cores (Clark 1982: 361).

As proposed by Healan, counterflaking results when a blade touches a vise board as it comes off its core, so this can occur on either of the two sides of the core near the jaws of the vise. Distal support of a core on an anvil rest can occasionally damage it. The hardness of this anvil might be apparent in the type of damage. For his foot-holding technique, Titmus noticed that sometimes a blade was removed that ran into the wooden support, and this encounter left a small hinge at the end of the core (Fig. 3.26). Pelegrin (2003: 63) reported lipping on the ends of indirect percussion blades made with anvil support. Kelterborn (2002) observed subtle lipping on the distal ends of some blade cores in the collections of the Field Museum. These are promising observations. There is a reasonable expectation that any part of a core touching a resistant surface has the potential to be damaged during its reduction.

Wilke's (1996) experiments stressed the need to truncate the microcores used in his slotted device to provide them support and, at the same time, to allow blades to release freely. His appliance sets the parameters for acceptable pressure core preforms, and it also determines the sizes and forms of cores that cannot be constrained (cf. Flenniken 1987). Once cores are too narrow to be pressed against the two supports flanking the slot of his device, Wilke cannot reduce them further. An obvious solution to this problem is to have more than one appliance for immobilizing cores, such as Nunn's socketed-logs system. Pelegrin (2003: 62, Fig. 4.8) illustrates a graduated series of devices he uses to process blade cores. In this case, any standardization evident in exhausted cores will conform to the size of the narrowest holding device. Cores that are too small for the device could be channeled into another system, as described by Flenniken and Hirth (2003). Wilke (1996: 300) observed that with a technique of handheld cores, one can expect blades with greater curvatures and a higher frequency of overshot blades. His observations signal the possibility that the types and frequencies of errors might be diagnostic of hand-holding techniques. Exploration is needed of the efficiencies of different core forms vis-à-vis different ways of stabilizing them (cf. Bonnichsen et al. 1980). I suspect that cores were shaped to fit specific fixation systems.

3.3.4 *Counterflaking and Core Immobilization*

Core geometry, manufacturing damage, and patterns of damage all come into play in evaluating counterflaking as an indicator of core immobilization. The critical junctures in core reduction sequences differ for cores held in clamps, forked devices, or with the feet. One implication of the formal transformations of Type B cores is that potential contact points on these cores (as per holding device) change during reduction. It follows that the potential for contact damage varies among techniques according to shifting contact points. As mentioned, counterflaking has been observed for blades made with a Crabtree clamp, Pelegrin's forked stick device, and for foot-held cores. This distribution disqualifies counterflaking as a marker of a single holding technique. But the overall occurrence of counterflaked blades in a reduction sequence may be diagnostic. For Pelegrin's devices (Figs. 3.13, 3.14, 3.15, 3.16), one would expect counterflaking to occur on the proximal sectors of blades where they touch the end of his forked stick or the margins of a slot. Since probable points of contact remain the same (because of his graduated series of devices), counterflaked blades can be expected from beginning to end in the reduction process. Blades from the same core might also evince distal lipping from anvil support. In contrast, the overall pattern of counterflaking for lateral clamp-made and foot-held blades should differ because the contact points were not the same.

The variety of clamps and vises provides many possibilities (Figs. 3.10, 3.11, 3.12, 3.13, 3.14, 3.15, 3.16, 3.17, 3.18, 3.19, 3.20). A Crabtree clamp with flat, lateral jaws, such as employed by Healan, can be used with or without an anvil rest for a secured core. The main difference from a three-point rig is that opposed sides of a secured core are "slightly embedded into the wooden jaws of the clamp" (Crabtree 1968: 453), so the parts of the core in contact with the vise are more extensive than for Pelegrin's forked system. For conical cores, most of the contact that keeps them stable is at their platforms, their widest part. As a core becomes more cylindrical, a greater portion of its sides contact the vise, so the core also becomes more stable. As it gets whittled down and more straight-sided, the core becomes easier to hold with less side pressure. It also has more regular ridges. These coordinated transformations lead to the practical consequence that regular blades are easier to predict and make, thereby lessening the probability of removing a blade that would brush against a vise board and acquire counterflaking damage upon release. The scenario for foot-held cores is the virtual opposite of that for a Crabtree vise. For conical cores, the larger they are the less chance of lateral contact of blades. Cores become increasingly harder to hold as they get smaller and more straight-sided because they provide less purchase for the feet. Consequently, the likelihood of removing a blade that contacts the feet increases as a function of diminishing core diameter and length. Thus, counterflaking can be expected to occur on late series blades.

In sum, based on points of potential contact of cores with holding devices, I expect counterflaking to occur on the proximal ends of fork-made and lateral clamp-made blades. This microdamage may occur throughout the sequence for fork-made blades but only in the first few series for lateral clamp-made blades. For foot-held cores, counterflaking should occur on late series blades and in the proximal, medial, or distal portions of these blades. Distal lipping on cores and blades is not expected with my variant of the Mexica technique, but it should occur with Titmus's version. Distal lipping should be more frequent with fork-made and socket-made blades because cores are forced into a support. Considered together, these markers of contact should allow reconstruction of types of core fixation systems.

3.4 Coda

The purpose of this chapter has been to summarize contributions of past blade experiments and to establish a foundation for further research. The information presented supports a variety of conclusions, depending on a reader's background and interests. When viewed as technical research, all past experiments in blademaking have contributed to the pool of useful facts and knowledge. None of the exercises, however, rises to the level of a "replication experiment," and claims for such analytical rigor are unfounded. Replication studies are a worthy goal but are still a far way off. In the meantime, experimenters would be well-served to continue technical research with the care advocated by Callahan, Pelegrin, and Kelterborn.

In Mesoamerican studies, the missing piece has been detailed analyses of archaeological collections. Those individuals with the talent, time, and inclination to conduct experiments generally lack access to adequate collections – in reality, or virtually. In contrast, those studying collections lack the time and/or ability to conduct experiments. Treatments of chipped stone artifacts from archaeological sites rarely provide the thick descriptions, photographs, or drawings needed to design or guide a replication program. Familiarity with actual collections is necessary. Hence, prevailing conditions favor collaboration between knappers and analysts, such as exemplified by the cooperation of Crabtree and Swanson.

At the Lisbon conference that inspired this chapter, I asked Pelegrin and Kelterborn about the next generation of scientific knappers and was disheartened to learn that more are not in training. The same is true of the Americas, as described by Hirth and Kelterborn (2000: 73): "A point of serious concern emerging from the [Penn State] conference is that there is a real shortage of young, outstanding flintknappers with technical abilities and an analytical interest in indirect percussion (punching) and pressure blade research." The enthusiasm generated by Bordes, Crabtree, Tixier, Titmus, Callahan, Flenniken, Bradley, and Patten for scientific knapping has not been passed on to another generation – at least not obviously so. Our graying masters lack apprentices, and time is short (see Kelterborn 2005; Patten 2005a). There are understandable reasons for the current state of affairs having to do with the incentive structures of academia.

I think it accurate to claim that the perceived need (and/or prestige) of scientific flintknapping has declined substantially during the past 20 years. Of the reasons for this, two deserve comment. One reason may be that major questions appear to have been resolved, with lithic studies settling down and becoming a traditional specialty with prescribed methods. At the moment, descriptive methods appear to be driving the questions addressed in most lithic studies – an unhealthy state of affairs. As the preceding review of blademaking demonstrates, most major questions remain unresolved. Questions should drive methods. Crabtree (1975b: 4) observed that "knowledge of lithic technology is still in its infancy" and one could spend "several lifetimes" attempting to understand the technology of a single people. Mesoamerican studies are barely to the point that experiments can be designed on the basis of past accomplishments.

Another potential reason for the current state of affairs is more troubling, and it is the notion that flintknapping promotes analytical "conceit" and "myopia" (Thomas 1986a, b) or even a "holier-than-thou" attitude (K.G. Hirth, July 28, 2009, personal communication). Crabtree's message was that if one understands how stone tools are made, one can better analyze them. Crabtree's demonstrations were revelatory to most scholars, but the euphoria of his fieldschools has evaporated. The profession is now at a point where some analysts abstain from knapping because they believe it adversely affects their objectivity (Jim Woods, June 2009, personal communication). David Hurst Thomas's (1986a, b) criticisms of Flenniken's pronouncements on "anthropological" knapping (1984, 1985; Flenniken and Raymond 1986) appear to have been the tipping point (Callahan 1995c, 1999c: 4). Thomas chided claims from replicators that knapping experience was an absolute necessity. He argued that the most interesting approaches to stone tools came from nonknappers. Neither Thomas's nor Flenniken's conclusions follow from proposed facts. Knapping experience does not necessarily make one perspicacious or biased.

These are philosophical issues meriting serious discussion. Both entrenched positions distract and disappoint. Some stone-breakers believe bleeding over one's own chippage ought to be a rite of passage. In contrast, some untouchables promote knapping virginity as an analytical virtue. The acceptable truth behind both exaggerated views is that knapping experience changes how one sees past worlds. Thomas surely is correct on this point. One cannot understand the Crabtree phenomenon on any other basis. Epistemologically, however, neither camp has all truth or virtue on its side. Knapping experiences allow one to see things not appreciated before, but sight gained is innocence lost. Some of the most interesting questions I have been asked about stone tools have come from students lacking knapping experience. Lithic studies need variety and open dialog. This was Crabtree's main message, and I believe it to be Thomas's as well. More experiments rather than fewer are needed, but they should be theoretically appropriate, methodologically grounded, and applied to a broader range of questions than in the past (see Binford 1979). Analysts of all sorts can improve the field and help devise more and better experiments for addressing relevant issues of scholarly merit.

Crabtree and Tixier were motivated by their interest in technology and remaking things. This was my initial passion, but I was soon drawn into issues of production, trade, and political economy, and I designed simple experiments to extract metrics for reconstructing commodities, products, the flow of goods, and their consumption (e.g., Clark 1988). Pelegrin (2002) calls these "quantitative experiments." Recent theory has piqued my interest in cognition and meaning, and I think knapping experiments are apropos for studying such phenomena. Most knappers have a notion that they are in some way recreating knowledge and experience of the past (see discussion in *Primitive Technology Newsletter*, Nos. 1 and 2, 1995, 1996). My original intent in this chapter was to view the utility of replication from a perspective of phenomenology and embodiment theory (see Hodder and Hutson 2003). Of all crafts, the chances of understanding what ancient artisans knew and felt in their bones – their savoir faires – are excellent for flintknapping. The opportunity to learn how the ancients may have constructed meaning through the manufacture and use of artifacts is also good because sufficient progress has been made in mapping out some of the technical knowledge and parameters involved. In my preliminary attempt to assess blademaking know-how, I realized I had to deal with *connaissance* first, and that entangled me in the present effort to reconstruct a time line of experiments, questions, and knapper training and experiences. The potential for linking techniques and gestures to mind is a current growth opportunity of lithic experimentation (see Dobres 2000, 2010; Roux and Bril 2005). That said, we still need better understanding of basic techniques and methods, their distribution in time and space, and their costs and benefits under different cultural conditions.

Past practitioners have provided useful guidelines to follow. Avocational knapping is on the rise in the United States (Harwood 2001; Whittaker 2004), so there are more opportunities to learn knapping skills than ever before, even though opportunities within the academy continue to dwindle. There are also many more knapping guides, self-help books, films, and videos available for self-instruction. Knapping skills can be focused for the good of science, in or out of the academy (Kelterborn 2005; Patten 2005a), as epitomized by the work of Crabtree, Titmus, Sollberger and Patterson, Callahan, Kelterborn, and Patten. I have included in the references the major publications of these extramural scholars to aid any who might be interested in following their lead in pursuing questions of chipped stone technologies in innovative ways[2].

---

[2]Callahan (1978a), Callahan (1979c), Callahan (1987), Callahan (1999a), Callahan (1999b), Callahan (2000b), Callahan (2000d), Callahan (2001a), Callahan (2001b), Callahan (2001c), Callahan (2006a), Madsen (1988), Madsen (1993), Patten (1999), Patten (2005b), Pelegrin (1981b), Sollberger (1986b), Titmus (1980), Titmus (1985), Titmus and Woods (1986), Titmus and Woods (1991a), Titmus and Woods (1991b), Titmus and Woods (1992), Titmus and Woods (2003), Trachman and Titmus (2003), Waldorf (1993).

Acknowledgments I received constructive comments for which I am grateful on previous drafts of this chapter from Christopher Bergman, Sandy Clark, Kenneth Hirth, Peter Kelterborn, Bo Madsen, Fred Nelson, Greg Nunn, Bob Patten, Leslie Quintero, Gene Titmus, Philip Wilke, James Winn, and James Woods. Illustrations were provided by Kisslan Chan, Arlene Colman, and Greg Nunn; I appreciate their contributions. I am particularly grateful to Phil for bringing much of the French literature to my attention and loaning me his books. I thank Bradley T. Lepper for copies of Crabtree's correspondence with the Ohio Lithic Laboratory. I especially am grateful to Charlotte Laporcherie and Olivier de Montmollin for translating articles in French for me and to Dr. Inga Tlatova for her summary of a Russian text.

References

Adams, Steve 2005 Experiments with Mesoamerican Obsidian Blade Production and Usage. *Bulletin of Primitive Technology* 29: 48–50.

Astruc, Laurence, Bernard Gratuze, Jacques Pelegrin, and Peter Akkermans 2007 From Production to Use: A Parcel of Obsidian Bladelets at Sabi Abyad II. In *Technical Systems and Near Eastern PPN Communities*, edited by L. Astruc, D. Binder and F. Briois, pp. 327–341. Éditiones APDCA, Antibes.

Barnes, Alfred S. 1947a The Production of Long Blades in Neolithic Times. *American Anthropologist* 49: 625–630.

Barnes, Alfred S. 1947b The Technique of Blade Production in Mesolithic and Neolithic Times. *Proceedings of the Prehistoric Society* 13: 101–113.

Barnes, Alfred S., and A. Cheynier 1935 Étude sur les Techniques de Débitage du Silex et en Particulier des Nuclei Prismatiques. *Bulletin de la Societé Préhistorique Française* 32: 288–299.

Barnes, Alfred S., and H. H. Kidder 1936 Difféntes Techniques de Débitage à La Ferrasie. *Bulletin de la Societé Préhistorique Française* 33: 2672–288.

Bergman, Christopher A. 1987 Hafting and Use of Bone and Antler Points from Ksar Akil, Lebanon. In *La Main et L'outil: Manches et Emmanchements Préhistoriques*, edited by D. Stordeur, pp. 117–126. Travaux de la Maison de l'Orient, Lyon.

Bergman, Christopher A. 1993 The Development of the Bow in Western Europe: A Technological and Functional Perspective. In *Hunting and Animal Exploitation in the Later Paleolithic and Mesolithic of Eurasia*, edited by G. L. Peterkin, H. M. Bricker, and P. Mellars, pp. 95–105. Archaeological Papers of the American Anthropological Association, No. 4. Washington, D. C.

Bergman, Christopher A., and Edward McEwen 1997 Sinew-Reinforced and Composite Bows: Technology, Function, and Social Implications. In *Projectile Technology*, edited by H. Knecht, pp. 143–160. Plenum Press, New York.

Bergman, Christopher A., E. McEwen, and R. Miller 1988 Experimental Archery: Projectile Velocities and Comparison of Bow Performances. *Antiquity* 62: 658–70.

Bergman, Christopher A., and Mark H. Newcomer 1983 Flint Arrowhead Breakage: Examples from Ksar Akil, Lebanon. *Journal of Field Archaeology* 10: 238–243.

Bergman, Christopher A., Mark B. Roberts, Simon Collcutt, and Penny Barlow 1990 Refitting and Spatial Analysis of Artefacts from Quarry 2 at the Middle Pleistocene Acheulean Site of Boxgrove, West Sussex, England. In *The Big Puzzle: International Symposium on Refitting Stone Artifacts*, edited by E. Cziesla, S. Eickhoff, N. Arts and D. Winter, pp. 264–281. Studies in Modern Archaeology, Vol. 1. Holos, Bonn, Germany.

Binford, Lewis R. 1979 Interview. *Flintknappers' Exchange* 2(1): 19–25.

Bonnichsen, Robson 1977 *Models for Deriving Cultural Information from Stone Tools*. National Museum of Man, Mercury Series, No. 60., Ottawa, Canada.

Bonnichsen, Robson, D. L. Keenlyside, and R. Morlan 1980 *Experiments in Lithic Technology*. Educational Film. National Museum of Man, National Museum of Canada, Ottawa.

Bordaz, Jacques 1970 *Tools of the Old and New Stone Age*. David and Charles Newton Abbot, Great Britain.

Bordes, François 1947 Étude Comparative des Différentes Techniques de Taille du Silex et des Roches Dures. *L'Anthropologie* 51: 1–29.

Bordes, François 1948 Les Couches Moustériennes du Gisement du Moustier (Dordogne): Typologie et Techniques de Taille. *Bulletin de la Societé Préhistorique Française* 45: 113–125.

Bordes, François 1950 Principes d'une Méthode d'Étude Comparative des Techniques de Débitage et de la Typologie du Paléolithique Ancien et Moyen. *L'Anthropologie* 54: 19–34.

Bordes, François 1967 Considerations sur la Typologie et les Techniques dans le Paléolithique. *Quartär* 18: 25–55.

Bordes, François, and Don Crabtree 1969a *Blades and Pressure Flaking*. [Film] J. Desmond Clark and Glynn Isaacs, advisors; written and directed by Clyde B. Smith. Department of Anthropology, University of California, Berkeley.

Bordes, François, and Don Crabtree 1969b The Corbiac Blade Technique and Other Experiments. *Tebiwa* 12(2): 1–21.

Bray, Warwick M., Earl H. Swanson, and Ian S. Farrington 1975 The New World. Elsevier Phaidon, New York.

Breton, Adela 1902 Some Obsidian Workings in Mexico. *Proceedings of the 13th International Congress of Americanists*, pp. 265—268. American Museum of Natural History, New York.

Briois, François 1992 *Préhistorie de la Pierre Taillée 1: De l'Eclat à Lame*. Film Produced by Université Toulouse le Mirail, SCPAM.

Briois, François 1995 *Préhistorie de la Pierre Taillée 2: Les Outils de Pierre*. Film Produced by Université Toulouse le Mirail, SCPAM.

Cabrol, Alexis, and Leon Coutier 1932 Contribution a l'etude de la taille de l'obsidienne au Méxique. *Bulletin, Societé Préhistorique Française* 29(12): 579–582.

Callahan, Errett 1978a Craftsman: Robson Bonnichsen. *Flintknappers' Exchange* 1(2): 16–24.

Callahan, Errett 1978b Craftsman: Jeff Flenniken. *Flintknappers' Exchange* 1(3): 16–24.

Callahan, Errett 1978c Craftsman: J. B. Sollberger. *Flintknappers' Exchange* 1(1): 12–17.

Callahan, Errett 1979a Craftsman: Don Crabtree. *Flintknappers' Exchange* 2(1): 29–34. [Interview, 28-12-78]

Callahan, Errett 1979b Craftsman: Don Crabtree. *Flintknappers' Exchange* 2(2): 8–13.[28-12-78, cont.]

Callahan, Errett 1979c Craftsman: Don Crabtree. *Flintknappers' Exchange* 2(3): 22–26. [28-12-78, cont.]

Callahan, Errett 1979d The Basics of Biface Knapping in the Eastern Fluted Point Tradition: A Manual for Flintknappers and Lithic Analysts. *Archaeology of Eastern North America* 7: 1–179.

Callahan, Errett 1980a Craftsman: Bo Madsen. *Flintknappers' Exchange* 3(2): 20–24.

Callahan, Errett 1980b Craftsman Bo Madsen. *Flintknappers' Exchange* 3(3): 23–25.

Callahan, Errett 1980c Craftsman: Gene Titmus. *Flintknappers' Exchange* 3(1): 19–25.

Callahan, Errett 1980d International Work Seminar I: A Pilot Project in Flintknapping Workshops. *Flintknappers' Exchange* 3(1): 4–5.

Callahan, Errett 1981 A Few Moments with Francois Bordes. *Flintknappers' Exchange* 4(3): 2.

Callahan, Errett 1982 An Interview with Flintknapper Jacques Pelegrin. *Contract Abstracts* 3(1): 62–70.

Callahan, Errett 1984 I Hate to Bicker, But …: A Study of Microblade Cores with Obtuse Platform Angles. *Lithic Technology* 13(3): 84–97.

Callahan, Errett 1985 Experiments with Danish Mesolithic Microblade Technology. *Journal of Danish Archaeology* 4: 23–39.

Callahan, Errett 1987 Anatomy of a Decision. In *Flintknapping: An Emic Perspective*, edited by R. H. Harwood, J. E. Atwood and R. Bailey, pp. 18–23. Harwood Archaeology, Palmdale, California.

Callahan, Errett 1987b Flintknapping in Scandinavia. In *Flintknapping: An Emic Perspective*, edited by R. H. Harwood, J. E. Atwood and R. Bailey, pp. 6–17. Harwood Archaeology, Palmdale, California.

Callahan, Errett 1991 How to Make Science (Heaven Forbid). *Bulletin of Primitive Technology* 1(2): 10–11.
Callahan, Errett 1995a Blades from Middle Neolithic Battle Axe Culture Graves in Sweden: A Preliminary Technological Study. Appendix 2, in *Slutvandrat?Aspecter på Övergången från Rörlig till Bofast Tillvaro*, by Helene Knutson, *Aun* 20, pp. 225–237. Societas Archaeologica Upsaliensis, Uppsala.
Callahan, Errett 1995b Sollberger Memorial. *Bulletin of Primitive Technology* 10: 76–77.
Callahan, Errett 1995c What is Experimental Archaeology? *Primitive Technology Newsletter* 1: 3–5.
Callahan, Errett 1996a In Defense of Level II. *Primitive Technology Newsletter* 2: 8.
Callahan, Errett 1996b Book Review: "Flintknapping: Making and Understanding Stone Tools, by John C. Whittaker. *Bulletin of Primitive Technology* 11: 82–85.
Callahan, Errett 1999a Ishi Sticks, Iceman Picks, and Good-for-Nothing Things: A Search for Authenticity in Pressure Flaking Tools. *Bulletin of Primitive Technology* 18: 60–68.
Callahan, Errett 1999b Methodology and Documentation. *Bulletin of Primitive Technology* 18: 43–48.
Callahan, Errett 1999c What is Experimental Archaeology? In *Primitive Technology: A Book of Earth Skills* (From the Society of Primitive Technology), edited by D. Wescott, pp. 4–6, Gibbs Smith Publishers, Salt Lake City.
Callahan, Errett 2000a Experiments with Danish Mesolithic Microblade Technology. *Bulletin of Primitive Technology* 20: 62–68.
Callahan, Errett 2000b Mastery: A Reconsideration of Standards. *Bulletin of Primitive Technology* 20: 9–10.
Callahan, Errett 2000c *The Basics of Biface Knapping in the Eastern Fluted Point Tradition: A Manual for Flintknappers and Lithic Analysts*, fourth edition. Piltdown Productions, Lynchburg, Virginia.
Callahan, Errett 2000d What is Traditional Flintknapping? *Bulletin of Primitive Technology* 20: 11.
Callahan, Errett 2001a Archaeological Evidence of Rotator Cuff Injury? *Bulletin of Primitive Technology* 21: 44–47.
Callahan, Errett 2001b Craft of Art? *Bulletin of Primitive Technology* 22: 85–86.
Callahan, Errett 2001c Why Traditional? *Bulletin of Primitive Technology* 24: 84–85.
Callahan, Errett 2006a Neolithic Danish Daggers: An Experimental Peek. In *Skilled Production and Social Reproduction*, edited by J. Apel and K. Knutsson, pp. 115–137. Societas Archaeologica Upsaliensis, Stone Studies 2, Uppsala.
Callahan, Errett 2006b Three Levels of Investment in Reconstruction: Therapy, Experience, and Experiment. *Bulletin of Primitive Technology* 32: 11–12.
Callahan, Errett, and Gene Titmus 1999 Flintknapper's Syndrome: A Caution to Flintknappers. *Bulletin of Primitive Technology* 17: 66–70.
Calley, S. 1984 Le Débitage Natoufien de Mureybet: Étude Préliminaire. *Paléorient* 10(2): 35–48.
Calley, S. 1986 L'Atelier de Qdeir 1 en Syrie: Exploitation des Nucléus Naviformes a la Fin du PPNB, 6e Millénaire: Première Approache. *Paléorient* 12(2): 49–67.
Cauvin, M.-C., and E. Coqueugniot 1988 L'Oasis d'el Kowm et le Kébarien Géométrique. *Paléorient* 14(2): 270–282.
Clark, John E. 1977a A Macrocore in the Regional Museum in Tuxtla Gutierrez, Chiapas, Mexico. *Lithic Technology* 6: 30–32.
Clark, John E. 1977b Mesoamerican Large Polyhedral Cores. *Katunob* 10(4): 79–93.
Clark, John E. 1979 *A Method for the Analysis of Mesoamerican Lithic Industries: An Application to the Obsidian Industry of La Libertad, Chiapas, Mexico*. MA, BYU.
Clark, John E. 1982 Manufacture of Mesoamerican Prismatic Blades: An Alternative Technique. *American Antiquity* 47: 355–376.
Clark, John E. 1984 Counterflaking and the Manufacture of Prismatic Blades. *Lithic Technology* 13(2): 52–61.
Clark, John E. 1985 Platforms, Bits, Punches and Vises: A Potpourri of Mesoamerican Blade Technology. *Lithic Technology* 14(1): 1–15.

Clark, John E. 1988 *The Lithic Artifacts of La Libertad, Chiapas, Mexico*. Papers of the New World Archaeological Foundation, No. 52. Brigham Young University, Provo, Utah.

Clark, John E. 1989a A Conversation with Don Crabtree. In *La Obsidiana en Mesoamérica*, edited by M. Gaxiola G. and J. E. Clark, pp. 131–135. INAH, Colección Científica 176. Mexico.

Clark, John E. 1989b Obsidian: The Primary Mesoamerican Sources. In *La Obsidiana en Mesoamérica*, edited by M. Gaxiola and J. E. Clark, pp. 299–319. Instituto Nacional de Antropología e Historia, Mexico City.

Clark, John E. 1997 Prismatic Blade-making, Craftsmanship, and Production: An Analysis of Obsidian Refuse from Ojo de Agua, Chiapas, Mexico. *Ancient Mesoamerica* 8: 137–159.

Clark, John E. 2002 Failure as Truth: An Autopsy of Crabtree's Folsom Experiments. In *Folsom Technology and Lifeways*, edited by J. E. Clark and M. B. Collins, pp. 191–208. Special Publication 4, *Lithic Technology*.

Clark, John E., Margarita Gaxiola, and Dolores Soto de Arechavaleta 1981 Conferences on Mesoamerican Chipped Stone. *Lithic Technology* 10(1): 1–11.

Clark, John E., Margarita Gaxiola, and Dolores Soto de Arechavaleta 1982 Lithic Studies in Mesoamerica: Two Symposia. *Lithic Technology* 11(2): 35–38.

Collins, Michael B. 1999 *Clovis Blade Technology*. University of Texas Press, Austin.

Crabtree, Don E. 1966 A Stoneworker's Approach to Analyzing and Replicating the Lindenmeier Folsom. *Tebiwa* 9: 3–39.

Crabtree, Don E. 1967 Notes on Experiments in Flintknapping. *Tebiwa* 10: 60–73.

Crabtree, Don E. 1968 Mesoamerican Polyhedral Cores and Prismatic Blades. *American Antiquity* 33: 446–478.

Crabtree, Don E. 1969a Report to National Science Foundation on Lithic Technology Field School. Manuscript in the possession of Gene Titmus.

Crabtree, Don E. 1969b *The Shadow of Man*. Idaho State University Film, Earl H. Swanson, editor. Distributed by Information Materials Incorporated (IMI), Burbank, California.

Crabtree, Don E. 1971 Review: Tools of the Old and New Stone Age by Jacques Bordaz. *American Anthropologist* 73(6): 1390–91.

Crabtree, Don E. 1972a *Ancient Projectile Points: A Study of Pressure Flaking*. Idaho State University Film, Earl H. Swanson, editor. Distributed by Information Materials Incorporated (IMI), Burbank, California.

Crabtree, Don E. 1972b *An Introduction to Flint Working*. Idaho State University, Pocatello.

Crabtree, Don E. 1972c *The Alchemy of Time*. Idaho State University Film, Earl H. Swanson, editor. Distributed by Information Materials Incorporated (IMI), Burbank, California.

Crabtree, Don E. 1972d *The Flintworker: Principles of Flaking Stone*. Idaho State University Film, Earl H. Swanson, editor. Distributed by Information Materials Incorporated (IMI), Burbank, California.

Crabtree, Don E. 1972e *The Hunter's Edge: Prehistoric Blade Making*. Idaho State University Film, Earl H. Swanson, editor. Distributed by Information Materials Incorporated (IMI), Burbank, California.

Crabtree, Don E. 1973 The Obtuse Angle as a Functional Edge. *Tebiwa* 16: 46–53.

Crabtree, Don E. 1975a Comments on Lithic Technology and Experimental Archaeology. In *Lithic Technology: Making and Using Stone Tools*, edited by E. Swanson, pp. 105–114. Mouton Publishers, Paris.

Crabtree, Don E. 1975b The Potential of Lithic Technology. In *Primitive Art and Technology*, edited by J. S. Raymond, B. Loveseth, C. Arnold and G. Reardon, pp. 1–6. The University of Calgary, Calgary.

Crabtree, Don E. 1977 The Obtuse Angle as a Functional Edge. In *Experimental Archaeology*, edited by D. Ingersoll, J. E. Yellen and W. Macdonald, pp. 38–51. Columbia University Press, New York.

Crabtree, Don E. 1979 Letters. *Flintknappers' Exchange* 2(3): 2–3.

Crabtree, Don E. 1980 Introduction [to Gene Titmus] *Flintknappers' Exchange* 3(1): 19

Crabtree, Don E., and Richard A. Gould 1970 Man's Oldest Craft Re-Created. *Curator* 13: 179–188.

Crabtree, Don E., and Earl H. Swanson, 1968 Edge-Ground Cobbles and Blade-Making in the Northwest. *Tebiwa* 11: 50–58.

Daubreé, and Roulin 1868 Rapport sur une Collecion d'Instruments en Pierre Découverts dans l'Ile de Java, et Remontant a une Époque Antérieure a Celle ou Commence, pour ce Pays, l'Histoire Proprement dite. In *Comptes Rendus des Séances de l'Académie des Sciences*, t. LXVII. Institut Impérial de France, Académie des Sciences. Imprimerie de Gauthier-Villars, Paris.

Dobres, Marcia-Anne 2000 *Technology and Social Agency: Outlining a Practice Framework for Archaeology*. Blackwell, Oxford.

Dobres, Marcia-Anne 2010 Archaeologies of Technology. *Cambridge Journal of Economics* 34: 103–114.

Dyckerhoff, Ursula 1982 Itzcolotli, der Druckstab der Aztekishchen Obsidian-Hanwerker. *Mexicon* 4(2): 21–22.

Ekholm, Susanna 1979 The Lagartero Figurines. In *Maya Archaeology and Ethnohistory*, edited by N. Hammond and G. R. Willey, pp. 172–185. University of Texas Press, Austin.

Ellis, Henry Holmes 1938 Lithic Problems. *American Antiquity* 4(1): 63–64.

Ellis, Henry Holmes 1940 *Flint-working Techniques of the American Indians: An Experimental Study*. Ohio State Museum, Columbus.

Ellis, Henry Holmes 1965 *Flint-working Techniques of the American Indians: An Experimental Study*. Reprint, Ohio State Museum, Columbus.

Eren, Metin I., Aaron Greenspan, and C. Garth Sampson 2008 Are Upper Paleolithic Blade Cores More Productive than Middle Paleolithic Discoidal Cores? A Replication Experiment. *Journal of Human Evolution* 55: 952–961.

Evans, John 1872 *The Ancient Stone Implements, Weapons, and Ornaments of Great Britain*. D. Appleton and Company, New York.

Faulkner, Alaric 1972 *Mechanical Principles of Flintworking*. Unpublished Ph.D. dissertation, Department of Anthropology, Washington State University.

Feldman, Lawrence H. 1971 Of the Stone Called Iztli. *American Antiquity* 36: 213–214.

Fischer, Anders 1974 An Ornamental Flint Core from Holmegard V, Zealand, Denmark: Notes on Mesolithic Ornamentation and Flintknapping. *Acta Archaeologica* 45. Copenhagen.

Fischer, Anders 1989 Hunting with Flint-tipped Arrows: Results and Experiences from Practical Experiment. In *The Mesolithic in Europe: Papers Presented at the Third International Symposium. Edinburgh 1985*, edited by C. Bonsall, pp. 29–39. Edinburgh.

Fischer, Anders 1990 On Being a Pupil of a Flintknapper of 11,000 Years Ago. In *The Big Puzzle: International Symposium on Refitting Stone Artifacts*, edited by E. Cziesla, S. Eickhoff, N. Arts, and D. Winter, pp. 447–464. Studies in Modern Archaeology, Vol. 1. Holos, Bonn, Germany.

Flenniken, J. Jeffrey 1978 Reevaluation of the Lindenmeier Folsom: A Replication Experiment in Lithic Technology. *American Antiquity* 43: 473–480.

Flenniken, J. Jeffrey 1981a Don Crabtree. *Flintknappers' Exchange* 4(1): 2.

Flenniken, J. Jeffrey 1981b *Replicative Systems Analysis: A Model Applied to the Vein Quartz Artifacts from the Hoko River Site*. Washington State University, Laboratory of Anthropology, Reports of Investigations 59. Pullman.

Flenniken, J. Jeffrey 1984 The Past, Present, and Future of Flintknapping: An Anthropological Perspective. *Annual Review of Anthropology* 13: 187–203.

Flenniken, J. Jeffrey 1985 Stone Tool Reduction Techniques as Cultural Markers. In *Stone Tool Analysis: Essays in Honor of Don E. Crabtree*, edited by M.G. Plew, J. C. Woods and M. G. Pavesic, pp. 265–276. University of New Mexico Press, Albuquerque.

Flenniken, J. Jeffrey 1987 The Paleolithic Dyuktai Pressure Blade Technique of Siberia. *Arctic Anthropology* 24(2): 117–132.

Flenniken, J. Jeffrey, and Kenneth G. Hirth 2003 Handheld Prismatic Blade Manufacture in Mesoamerica. In *Mesoamerican Lithic Technology: Experimentation and Interpretation*, edited by K. Hirth, pp. 98–107. University of Utah Press, Salt Lake City.

Flenniken, J. Jeffrey, and Anan W. Raymond 1986 Morphological Projectile Point Typology: Replication Experimentation and Technological Analysis. *American Antiquity* 51(3): 619–627.

Flenniken, J. Jeffrey, and Philip J. Wilke 1989 Typology, Technology, and Chronology of Great Basin Dart Points. *American Anthropologist* 91: 149–158.
Fletcher, Charles S. 1970 Escapable Errors in Employing Ethnohistory in Archaeology. *American Antiquity* 35: 209–213.
Gallet, Michel, and Pierre-Jean Texier 1991 Un Appareil de Mesures pour la Caracterisation et l'Evaluation des Contraintes Exercees sur un Nucleus los d'un Debitage Experimental de Lames par Pression: Perspectives Experimentales, Implications Archeologiques. In *Archeologie Experimentale: Tome 2 – La Terre: L'os et la Pierre, la Maison et les Champs*, pp. 102–109. Acte du Colloque International, Paris.
Guiria, E. U. 1997 *Tehnologiceskij Analiz Kamennych Industrij: Metodika Mikro-makroanaliza Drevnih Orudij Truda*. Saint Petersburg.
Gryba, Eugene M. 1988 A Stone Age Pressure Method of Folsom Fluting. *Plains Anthropologist* 33: 53–66.
Gryba, Eugene M. 1989 A Mousetrap 10,000 Years Too Late. *Plains Anthropologist* 34: 65–68.
Gryba, Eugene M. 2006 An Assessment of the Free-hand Pressure Flaking Technique of Precontact North America. *Lithic Technology* 31(1): 57–77.
Hall, Dan Alan 2002 A Passion for Ancient Technology: Gene Titmus. *Mammoth Trumpet* 17(2): 4–9
Hansen, Peter V., and Bo Madsen 1983 Flint Axe Manufacture in the Neolithic: An Experimental Investigation of a Flint Axe Manufacture Site at Hastrup Vænget, East Zealand. *Journal of Danish Archaeology* 2: 43–59.
Harwood, Ray 1986 A Thinking Man's Flintknapper: The Errett Callahan Story. *Flintknapping Digest* 3(2).
Harwood, Ray 1987 J. B. Sollberger: Master of Folsom. *Flintknapping Digest* 4(2).
Harwood, Ray 1999 *History of Modern Flintknapping*. World Flintknapping Society.
Harwood, Ray 2001 Points of Light, Dreams of Glass: An Introduction into Vitrum Technology. *Bulletin of Primitive Technology* 21: 24–36.
Hay, Conran A. 1978 *Kaminaljuyu Obsidian: Lithic Analysis and the Economic Organization of a Prehistoric Mayan Chiefdom*. Unpublished Ph.D. dissertation, Department of Anthropology, The Pennsylvania State University, University Park.
Hay, Conran A., and Robert Rogers 1978 A Method for Determining Polyhedral Core Size from Prismatic Blade Debitage. *Lithic Technology* 7: 46–52.
Healan, Dan M., and Janet Kerley 1984 Edge Damage Induced by Core Immobilization in Prismatic Blade Manufacture. *Lithic Technology* 13: 1–10.
Hernández, Francisco 1959 *Historia Natural de Nueva España* (Vol. 2). Universidad Nacional Autónoma de México, Mexico.
Hester, Thomas R. 1978 Preliminary Notes on the Technological Analysis of Obsidian Artifacts from Villa Morelos, Michoacan, Mexico. In *Archaeological Studies of Mesoamerican Obsidian*, assembled and edited by T. R. Hester, pp. 131–158. Ballena Press, Socorro.
Hintzman, Marc W. 2000 *Scarce-Resource Procurement and Use: The Technological Analysis of an Obsidian Blade Workshop in the Lowlands of Belize*. Unpublished Master's Thesis, Department of Anthropology, University of Californian, Riverside.
Hirth, Kenneth G. (editor) 2003 *Mesoamerican Lithic Technology: Experimentation and Interpretation*. University of Utah Press, Salt Lake City.
Hirth, Kenneth G. 2003a Experimentation and Interpretation in Mesoamerican Lithic Technology. In *Mesoamerican Lithic Technology: Experimentation and Interpretation*, edited by K. Hirth, pp. 3–9. University of Utah Press, Salt Lake City.
Hirth, Kenneth G. 2003b The Kaminaljuyu Production Sequence for Obsidian Prismatic Blades: Technological Analysis and Experimental Inferences. In *Mesoamerican Lithic Technology: Experimentation and Interpretation*, edited by K. Hirth, pp. 170–181. University of Utah Press, Salt Lake City.
Hirth, Kenneth G., Bradford Andrews, and J. Jeffrey Flenniken 2003 The Xochicalco Production Sequence for Obsidian Prismatic Blades. In *Mesoamerican Lithic Technology: Experimentation and Interpretation*, edited by K. Hirth, pp. 182–196. University of Utah Press, Salt Lake City.

Hirth, Kenneth G., J. Jeffrey Flenniken, and Bradford Andrews 2000 Flaked-Stone Obsidian Tools and Their Behavioral Implications. In *The Xochicalco Mapping Project*, edited by K. Hirth, pp 136–150. University of Utah Press, Salt Lake City.

Hirth, Kenneth, and Peter Kelterborn 2000 A Mesoamerican Pressure Blade Conference at Penn State University. *Lithic Technology* 25(2): 72–73.

Hirth, Kenneth G., Gene L. Titmus, J. Jeffrey Flenniken, and Jacques Tixier 2003 Alternative Techniques for Producing Mesoamerican-Style Pressure Flaking on Obsidian Bifaces. In *Mesoamerican Lithic Technology: Experimentation and Interpretation*, edited by K. Hirth, pp. 147–152. University of Utah Press, Salt Lake City.

Hodder, Ian, and Scott Hutson 2003 *Reading the Past: Current Approaches to Interpretation in Archaeology*, third edition. Cambridge University Press, Cambridge.

Holmes, William H. 1900 The Ancient Aztec Obsidian Mines of the State of Hidalgo, Mexico. *American Association for the Advancement of Science, Proceedings* 50: 313.

Holmes, William H. 1919 *Handbook of Aboriginal American Antiquities, Part I: Introductory, the Lithic Industries*, Bulletin 60. The Smithsonian Institution Bureau of American Ethnology, Washington D.C.

Howell, F. Clark 1965 *Early Man*. Life Nature Library, Time-Life Books, New York.

Inizan, Marie-Louise 1980 Sur les Industries à Lames de Qatar. *Paléorient* 6: 233–236.

Inizan, Marie-Louise 1991 Le Débitage par Pression: des Choix Culturels. In *25 Ans d'Études Technologiques en Préhistorie*. XIe Recontres Internationales d'Archéologie et d'Historie d'Antibes, pp. 367–377. Éditiones APDCA, Juan-les-pins.

Inizan, Marie-Louise, Monique Lechevallier, and Patrick Plumet 1992a A Technological Marker of the Penetration into North America: Pressure Microblade Debitage, Its Origin in the Paleolithic of North Asia and Its Diffusion. *Material Issues in Art and Archaeology III*, edited by P.B. Vandiver, J.R. Druzik, G.S. Wheeler and I.C. Freestone, pp. 661–181. Material Research Society, Pittsburgh.

Inizan, Marie-Louise, Michèle Reduron-Ballinger, Hélène Roche, and Jacques Tixier 1999 *Technology and Terminology of Knapped Stone*. Cercle de Recherches et d'Etudes Préhistoriques, Paris.

Inizan, Marie-Louise, Hélène Roche, and Jacques Tixier 1992b *Technology of Knapped Stone*. Préhistorie de la Pierre Taillée 3. Cercle de Recherches et d'Etudes Préhistoriques, Meudon.

ISU Bengal 1965 *Archaeologists Get Grant. The ISU Bengal* December 8, 1965, Pocatello, Idaho.

Jelinek, Arthur J. 1965 Lithic Technology Conference, Les Eyzies, France. *American Antiquity* 31(2): 277–79.

Joly, N. 1883 *Man Before Metals*. Kegan Paul, Trench and Company, London.

Kelterborn, Peter 1980 Zur Frage des Livre de Beurre. *Jahrbuch der Schweizerischen Gesellschaft für Ur- und Frühgeschiche* 63: 5–24.

Kelterborn, Peter 1981a The Livre de Beurre Defends Itself. *Flintknappers' Exchange* 4(3): 22.

Kelterborn, Peter 1981b The Livre de Beurre Method. *Flintknappers' Exchange* 4(3): 12–20.

Kelterborn, Peter 1984 Towards Replicating Egyptian Predynastic Knives. *Journal of Archaeological Science* 11(6): 433–453.

Kelterborn, Peter 1987 Principles of Experimental Research in Archaeology. In *Bulletin of Experimental Archaeology*, No. 8:11–12. Department of Adult Education, University of Southhampton.

Kelterborn, Peter 1990a Preconditions and Strategies for Experimental Archaeology. In *Le Silex de Sa Genèse À l'Outil, Tome 2, Cahiers du Quaternaire No. 17*, pp. 599–602. Editions du CNRS, Paris.

Kelterborn, Peter 1990b Varna Blade Analysis. Manuscript in possession of the author.

Kelterborn, Peter 1992 Eine Beilwerstatt im Seegubel, Jona SG. *Jahrbuch der Schweizerischen Gesellschaft für Ur- und Frühgeschiche* 75: 133–138.

Kelterborn, Peter 1999 Analysen und Experimente zu Herstellung und Gebrauch von Horgener Pfeilspitzen. *Jahrbuch der Schweizerischen Gesellschaft für Ur- und Frühgeschiche* 83: 37–64.

Kelterborn, Peter 2001 Replication, Use and Repair of an Arrowhead. *Bulletin of Primitive Technology* 21: 48–57.

Kelterborn, Peter 2002 Report 1: The Field Museum in Chicago. Manuscript in possession of the author.

Kelterborn, Peter 2003 Measurable Flintknapping. In *Mesoamerican Lithic Technology: Experimentation and Interpretation*. K. Hirth, ed., 120–131. University of Utah Press, Salt Lake City.

Kelterborn, Peter 2005 What Widening Gap? *Lithic Technology* 30(2): 86–88.

Kelterborn, Peter 2008a C. Condensed Synthesis: HOW DID THEY DO IT? Manuscript in possession of the author.

Kelterborn, Peter 2008b *The Flintknapper's Alphabet*. Manuscript in possession of the author.

Knowles, F. H. S. 1953 *Stone-Workers Progress*. Occasional Paper on Technology, No. 6. Pitt Rivers Museum, University of Oxford.

Knudson, Ruthann 1978 Don E. Crabtree: Research Associate in Prehistoric Technology, University of Idaho. Manuscript in possession of James Woods.

Knudson, Ruthann 1982 Don E. Crabtree, 1912-1980. *American Antiquity* 47(2): 336–343.

Kobayashi, Tatsuo 1970 Microblade Industries in the Japanese Archipelago. *Arctic Anthropology* 7(2): 38–58.

Kopper, Philip 1986 *The Smithsonian Book of North American Indians, Before the Coming of the Europeans*. Smithsonian Books, Washington, D.C.

Lampe, David 1991 The Knappers. *The World & I*, February: 652–659.

Leroi-Gourhan, André 1957 *Prehistoric Man*. Philosophical Library, New York (translation from French by Wade Baskin: *Les Hommes de la Préhistoire – Les Chasseurs*).

Lohse, E. S. 2000 *Don Crabtree: A Reader in Stone*. (CD with all of Don Crabtree's articles and Educational Films, with articles by E. S. Lohse) Idaho Museum of Natural History. Idaho State University, Pocotello. [for information, see crabtree@iri.isu.edu]

Lubbock, John 1865 *Pre-Historic Times, as Illustrated by Ancient Remains, and the Manners and Customs of Modern Savages*. Williams and Norgate, London.

MacCurdy, George G. 1900 The Obsidian Razor of the Aztecs. *American Anthropologist* 2: 417–421.

McEwen, Edward, R. L. Miller, and Christopher A. Bergman 1991 Early Bow Design and Construction. *Scientific American*, June: 76–82.

Madsen, Bo 1981 An Experimental Site in Lejre. *Flintknappers' Exchange* 4(1): 16–20.

Madsen, Bo 1984 Flint Axe Manufacture in the Neolithic: Experiments with Grinding and Polishing of Thin-butted Flint Axes. *Journal of Danish Archaeology* 3: 47–62.

Madsen, Bo 1988 Le Centre Expérimentale de Lejre au Danemark. *Cossiers, Histoire et Archaélogie* 126: 64–69.

Madsen, Bo 1989 Some Perspectives in Experimental Archaeology. *Arcaeologica*.

Madsen, Bo 1992 Hamburgkulturens Flintteknologi i Jels (The Hamburgian Flint Technology at Jels). *Istidsjægere ved Jelssøerne*. Skrifter fra Museumsrådet for Sønderjyllands Amt 5. Haderslev.

Madsen, Bo 1993 Flint – Extraction, Manufacture and Distribution. In *Digging into the Past: 25Years of Archaeology in Denmark*, edited by S. Hvass and B. Storgaard, pp. 126–130. The Royal Society of Northern Antiquaries, Copenhagen, Aarhus.

Madsen, Bo 1996 Late Palaeolithic Cultures of South Scandinavia – Tools, Traditions and Technology. In *The Earliest Settlement of Scandinavia and its Relationship with Neighboring Areas*, edited by L. Larsson, pp. 61–73. Acta Archaeologica Lundensia, Series In 8th, No. 24, Almquist and Wiksell International, Stockholm.

Marcou, Ph. 1921 Procédé des Aztèques pour la Taille par Éclatenment des Couteaux ou Rasoirs d'Obsidienne. *Journal de la Societé des Americanistas de Paris*, 13(n.s.): 17–24.

Matheny, Deanne G. 1988 *The Northwest Plaza Burials at Lagartero, Chiapas, Mexico*. Unpublished Ph.D. dissertation, Department of Anthropology, University of Utah, Salt Lake City.

Mendieta, friar Gerónimo de 1971 *Historia Eclesiastica Indiana: Obra Escrita a Fines del Siglo XVI*. (Second Edition), Biblioteca Porrúa 46. Editorial Porrúa, Mexico.

Migal, Witold 2006 The Macrolithic Flint Blades of the Neolithic Times in Poland. In *Skilled Production and Social Reproduction*, edited by J. Apel and K. Knutsson, pp. 387–398. Societas Archaeologica Upsaliensis, Stone Studies 2, Uppsala.

Miller, R., E. McEwen, and C. Bergman 1986 Experimental Approaches to Ancient Near Eastern Archery. *World Archaeology* 18(2): 178–195.

Moorehead, Warren K. 1910 *The Stone Age in North America*. Houghton Mifflin Company, Boston and New York.

Newcomer, M. H. 1975 "Punch Technique" and Upper Paleolithic Blades. In *Lithic Technology: Making and Using Stone Tools*, edited by E. Swanson, pp. 97–102. Mouton Publishers, Paris.

Nunn, Greg R. 2005 *Replicating the Type IC Neolithic Danish Dagger*. DVD, Greg Nunn/Paleo Technologies, Castle Valley, Utah.

Nunn, Greg R. 2006 Using the Jutland Type IC Neolithic Danish Dagger as a Model to Replicate Parallel, Edge-to-Edge Pressure Flaking. In *Skilled Production and Social Reproduction*, edited by J. Apel and K. Knutsson, pp. 81–113. Societas Archaeologica Upsaliensis, Stone Studies 2, Uppsala.

Nunn, Greg R. 2007a A Blade-Making Workshop: A Demonstration. *Bulletin of Primitive Technology* 33: 46–47.

Nunn, Greg R. 2007b Using the Soft Hammerstone, the Tool of the West. *Bulletin of Primitive Technology* 34: 57–62.

Ohnuma, Katsuhiko 1993 Experimental Studies in the Determination of Manners of Micro-blade Detachment. *Al-Rāfidān* 14: 153–181. Kokushikan University, Institute for Cultural Studies of Ancient Iraq, Tokyo.

Ohnuma, Katsuhiko, and Christopher Bergman 1983 Experimental Studies in the Determination of Flaking Mode. *Bulletin of the Institute of Archaeology* 19: 161–170, London.

Otte, Marcel 1991 Jacque Tixier, La Préhistorie et l'Ethnologie. In *25 Ans d'Études Technologiques en Préhistorie*. XIe Recontres Internationales d'Archéologie et d'Historie d'Antibes, pp. 11–14. Éditiones APDCA, Juan-les-pins.

Parry, William J. 1994 Prismatic Blade Technologies in North America. In *The Organization of North American Prehistoric Chipped Stone Tool Technologies*, edited by P. J. Carr, pp. 87–98. International Monographs in Prehistory, Archaeological Series, No. 7, Ann Arbor, Michigan.

Patten, Bob 1999 *Old Tools – New Eyes: A Primal Primer of Flintknapping*. Stone Dagger Publications, Denver.

Patten, Bob 2005a Comments to Odell. *Lithic Technology* 30(2): 85–86.

Patten, Bob 2005b *Peoples of the Flute: A Study in Anthropolithic Forensics*. Stone Dagger Publications, Denver.

Patten, Bob 2009 *Old Tools – New Eyes: A Primal Primer of Flintknapping*, second edition. Stone Dagger Publications, Denver.

Patterson, Leland W. 1973 Some Texas Blade Technology. *Bulletin of the Texas Archaeological Society* 44: 89–111.

Patterson, Leland W. 1976 Blade Technology in the Texas Archaic. In *The Texas Archaic*, edited by T. R. Hester. Special Report 2, pp. 14–22. Center for Archaeological Research, University of Texas, San Antonio.

Patterson, Leland W. 1981 The Analysis of Striking Platform Geometry. *Flintknappers' Exchange* 4(2): 18–20.

Patterson, Leland W. 1982 Replication and Classification of Large Size Lithic Debitage. Lithic Technology 11(3): 50–58.

Patterson, Leland W. 1986 Additional Comments on Obtuse Platform Angles. *Lithic Technology* 15(1): 43–45.

Patterson, Leland W. 1988 J. B. Sollberger, Archaeologist and Flintknapper. *Bulletin of the Texas Archaeological Society* 59: 19–21.

Patterson, Leland W. 1990 Characteristics of Bifacial Reduction Flake-Size Distribution. *American Antiquity* 55(3): 550–558.

Patterson, Leland W., and J. B. Sollberger 1974 Lithic Shaping Retouch. *Texas Archaeology* 18(3): 13–16.

Patterson, Leland W., and J. B. Sollberger 1978 Replication and Classification of Small Size Lithic Debitage. *Plains Anthropologist* 23(40): 103–112.
Patterson, Leland W., and J. B. Sollberger 1980 Mayan Blade Manufacture and the Bow and Arrow. *Louisiana Archaeology* 7: 83–100.
Pelegrin, Jacques 1981a Experiments in Bifacial Work. *Flintknappers' Exchange* 4(1): 4–7.
Pelegrin, Jacques 1981b Letter. *Flintknappers' Exchange* 4(3): 2.
Pelegrin, Jacques 1984a Approche Technologique Expérimentale de la Mise en Forme de Nucléus pour le Débitage Systématique par Pression. *Préhistoire de la Pierre Taillée: 2 Économie du Débitage Laminaire*, pp. 93–103. Cercle de Recherches et d'Études Préhistoriques, Paris.
Pelegrin, Jacques 1984b Débitage par Pression sur Silex: Nouvelles Expérimentations. *Préhistoire de la Pierre Taillée: 2 Économie du Débitage Laminaire*, pp. 105–116. Cercle de Recherches et d'Études Préhistoriques, Paris.
Pelegrin, Jacques 1984c Systèmes Expérimentaux d'Immobilisation du Nucléus pour le Débitage par Pression. *Préhistoire de la Pierre Taillée: 2 Économie du Débitage Laminaire*, pp. 105–116. Cercle de Recherches et d'Études Préhistoriques, Paris.
Pelegrin, Jacques 1985 Réflexions sur le Comportement Technique. In *La Signification Culturelle des Industries Lithiques: Actes du Colloque de Liège du 3 au 7 Octobre 1984*, edited by M. Otte, pp. 72–91. BAR International Series 239, Oxford.
Pelegrin, Jacques 1986 *Technologie Lithique: Une Méthode Appliquée à l'Étude de deux Séries du Périgordien Ancien, Roc de Combe couche 8, La Côte niveau III.* Unpublished Ph.D. thesis, University of Paris X-Nanterre.
Pelegrin, Jacques 1988 Débitage Expérimental Par Pression du Petit du Plus Petit au Plus Grand. *Technologie Prehistorique* 25: 37–53.
Pelegrin, Jacques 1990 Prehistoric Lithic Technology: Some Aspects of Research. *Archaeological Review from Cambridge* 9: 116–125.
Pelegrin, Jacques 1991 Sur une Recherche Tecnique Experimentale des Techniques de Debitage Laminaire. In *Archeologie Experimentale: Tome 2 – La Terre: L'os et la Pierre, la Maison et les Champs*, pp. 118–128. Acte du Colloque International, Paris.
Pelegrin, Jacques 1993 A Framework for Analyzing Prehistoric Stone Tool Manufacture and a Tentative Application to some Early Stone Industries. In *The Use of Tools by Human and Non-human Primates*, edited by A. Berthelet and J. Chavaillon, pp. 302–314. Clarendon Press, Oxford.
Pelegrin, Jacques 2002 Conducting Lithic Experiments. *Bulletin of Primitive Technology* 23: 30.
Pelegrin, Jacques 2003 Blade-Making Techniques from the Old World: Insights and Applications to Mesoamerican Obsidian Lithic Technology. In *Mesoamerican Lithic Technology: Experimentation and Interpretation.* K. Hirth, ed., pp. 55–71. University of Utah Press, Salt Lake City.
Pelegrin, Jacques 2005 Remarks about Archaeological Techniques and Methods of Knapping: Elements of a Cognitive Approach to Stone Knapping. In *Stoneknapping: The Necessary Conditions for a Uniquely Hominin Behaviour*, edited by V. Roux and B. Bril, pp. 23–33. McDonald Institute for Archaeological Research, Cambridge.
Pelegrin, Jacques 2006 Long Blade Technology in the Old World: An Experimental Approach and Some Archaeological Results. In *Skilled Production and Social Reproduction*, edited by J. Apel and K. Knutsson, pp. 37–68. Societas Archaeologica Upsaliensis, Stone Studies 2, Uppsala.
Pelegrin, Jacques, Aurélia Peltier, Isabelle Sidéra, Danielle Sordeur, Anne Vincent, and Gérard Deraprahamian 1991 Chasse-Lame en Os? Une Etude Experimentale. In *Archeologie Experimentale: Tome 2 – La Terre, L'os et la Pierre, la Maison et les Champs*, pp. 63–73. Acte du Colloque International, Paris.
Piel-Desruisseaux, Jean-Luc 1990 *Outils Préhistoriques: Forme, Fabrication, Utilisation*, Second Edition. Masson, Paris.
Prem, Hanns J. 1974 *Matrícula de Huexotzinco.* Akademische Druck. Verlagsanslalt, Grax, Austria.
Prideaux, Tom 1973 *Cro-Magnon Man.* The Emergence of Man, Time-Life Books, New York.

Quintero, Leslie A. 1998 *Evolution of Lithic Economies in the Levantine Neolithic: Development and Demise of Naviform Core Technology*. Unpublished Ph.D. dissertation, Department of Anthropology, University of California, Riverside.
Quintero, Leslie A., and Philip J. Wilke 1995 Evolution and Economic Significance of Naviform Core-and-Blade Technology in the Southern Levant. *Paléorient* 21(1): 7–33.
Rollefson, G. O. 1990 Neolithic Chipped Stone Technology at 'Ain Ghazal, Jordan: The Status of the PPNC Phase. *Paléorient* 15(1): 119–124.
Roux, Valentine, and Blandine Bril 2005 *Stoneknapping: The Necessary Conditions for a Uniquely Hominin Behaviour*. McDonald Institute for Archaeological Research, Cambridge.
Sahagún, Fr. Bernadino de 1963 Florentine Codex: general history of things of New Spain, Book 11–Earthy things. Translated by Charles E Dibble and Arthur J. O. Anderson. *Monographs of the School of American Research and the Museum of New Mexico, No. 14, Part XII*. University of Utah Press. Salt Lake City.
Saraydar, Stephen C. 2008 *Replicating the Past: The Art and Science of the Archaeological Experiment*. Waveland Press, Inc., Long Grove, Illinois.
Sellers, George Ercol 1886 Observations on Stone Chipping. *Smithsonian Report for 1885*, Part 1: 871–891.
Semenov, S. A. 1964 *Prehistoric Technology: An Experimental Study of the Oldest Tools and Artefacts from Traces of Manufacture and Wear*, Translated by M. W. Thompson. Harper and Row Publishers, Inc. Great Britain.
Sheets, Payson D. 1975 Behavioral Analysis and the Structure of a Prehistoric Industry. *Current Anthropology* 16: 369–391.
Sheets, Payson D. 1977 The Analysis of Chipped Stone Artifacts in Southern Mesoamerica: An Assessment. *Latin American Research Review* 12: 139–158.
Sheets, Payson D. 1978 From Craftsman to Cog: Quantitative Views of Mesoamerican Lithic Technology. In *Papers on the Economy and Architecture of the Ancient Maya*, edited by R. Sidrys, pp. 40–71. Monograph VIII, Institute of Archaeology, University of California, Los Angeles.
Sheets, Payson D., and Guy Muto 1972 Pressure Blades and Total Cutting Edge: An Experiment in Lithic Technology. *Science* 175: 623–634.
Smith, Philip E. L. 1966 Lithic Technology. Current Anthropology 7(5): 592–593.
Sollberger, J. B. 1948 Lamb's Creek Mound. *The Record*, Dallas Archaeological Society 8(1).
Sollberger, J. B. 1949 A Prehistoric Rockshelter in Kerr County, Texas. *The Record* 8(3).
Sollberger, J. B. 1967 A Partial Report on Research Work Concerning Lithic Typology and Technology. *Bulletin of the Texas Archaeological Society* 39: 95–109.
Sollberger, J. B. 1968 The Paleo Type Flake Knife. *Bulletin of the Texas Archaeological Society* 38: 45–46.
Sollberger, J. B. 1969 The Basic Tool Kit Required to Make and Notch Arrow Shafts for Stone Points. *Bulletin of the Texas Archaeological Society* 40: 231–240.
Sollberger, J. B. 1971 A Technological Study of Beveled Knives. *Plains Anthropologist* 16(53): 209–218.
Sollberger, J. B. 1976 Bifacing Patterns on Prismatic Flakes. *Bulletin of the Texas Archaeological Society* 47: 261–267.
Sollberger, J. B. 1977 On Fluting Folsom: Notes on Recent Experiments. *Bulletin of the Texas Archaeological Society* 48: 47–52.
Sollberger, J. B. 1978 Lever Flaking as a Credible Alternative to Hand-held Pressure Flaking. *Flintknappers' Exchange* 1(1): 6–7, 1(2): 25.
Sollberger, J. B. 1985 A Technique for Folsom Fluting. *Lithic Technology* 14: 41–50.
Sollberger, J. B. 1986a Comments On: I Hate to Bicker … But. *Lithic Technology* 15(1): 41–43.
Sollberger, J. B. 1986b Lithic Fracture Analysis: A Better Way. *Lithic Technology* 15(3): 101–105.
Sollberger, J. B. 1988 On Replicating Fluted Projectile Points. *Bulletin of the Texas Archaeological Society* 59: 1–17.
Sollberger, J. B. 1989 Comment On: A Stone Age Pressure Method of Folsom Fluting by Eugene M. Gryba. *Plains Anthropologist* 34: 63–64.

Sollberger, J. B. 1993 Updating Fracture Mechanics: From Hertz to Archaeology to Cones. *Bulletin of Primitive Society* 1(6): 65–66.

Sollberger, J. B., and L. W. Patterson 1976a Prismatic Blade Replication. *American Antiquity* 441: 517–531.

Sollberger, J. B., and L. W. Patterson 1976b The Myth of Bipolar Flaking. *Lithic Technology* 5: 40–42.

Sollberger, J. B., and L. W. Patterson 1980 Attributes of Experimental Folsom Points and Channel Flakes. *Bulletin of the Texas Archaeological Society* 51: 289–299.

Sollberger, J. B., and L. W. Patterson 1983 A Pressure Method for Microblade Manufacture. *Lithic Technology* 12(2): 25–31.

Sonneville-Bordes, Denise de 1967 *La Préhistoire Moderne*. Pierre Fanlac, Périgueux.

Sørensen, Mikkel 2006 Rethinking the Lithic Blade Definition: Towards a Dynamic Understanding. In *Skilled Production and Social Reproduction*, edited by J. Apel and K. Knutsson, pp. 277–296. Societas Archaeologica Upsaliensis, Stone Studies 2, Uppsala.

South Idaho Press 1981 *Ancient Tool Technology Shown*. South Idaho Press, January 25, Burley, Idaho.

Speth, John E. 1972 Mechanical Basis of Percussion Flaking. *American Antiquity* 37: 34–60.

Speth, John E. 1974 Experimental Investigations of Hard-Hammer Percussion Flaking. *Tebiwa* 17: 7–36.

Statham, William P. 1978 *Experimental Lithic Technology: Observations and Comments on its Conduct and Documentation, as Exemplified by the Crabtree Collection*. Unpublished M.A. Thesis, Department of Anthropology, Idaho State University, Pocatello.

Stevens, Edward T. 1870 *Flint Chips: A Guide to Pre-Historic Archaeology*. Bell and Daldy, London.

Suzuki, C., and T. Akazawa 1971 Manufacturing Technique of the Stone Artifacts from Palmyra, Syria. *Journal of the Anthropological Society of Nippon* 79(2): 105–127.

Swanson, Earl H. 1966 An Introduction to Crabtree's Experiments in Flint Knapping. *Tebiwa* 9(1): 1–2.

Swanson, Earl H., and B. Robert Butler 1962 The First Conference of Western Archaeologists on Problems of Point Typology. *Occasional Papers of the Idaho State College Museum*, No. 10. Pocatello.

Swanson, Earl H., and Don Crabtree 1974 Application for Research Grant (Renewal), National Science Foundation. Manuscript on file with the Museum of Natural History, Idaho State University, Pocatello.

Tabarev, Andrei V. 1997 Paleolithic Wedge-shaped Microcores and Experiments with Pocket Devices. *Lithic Technology* 22(2): 139–149.

Texier, Pierre-Jean 1982 Le Débitage par Pression Pectorale à la Béquille: Une Approche Expérimental au plus prés des Paramètres à l'Origine de la Rupture Fragile des Roches Dures. *Studia Praehistorica Belgica* 2: 57–64.

Texier, Pierre-Jean 1984a Le Débitage par Pression et la Mécanique de la Rupture Fragile Initiation et Propagation des Fractures. *Préhistoire de la Pierre Taillée: 2 Économie du Débitage Laminaire*, pp. 139–147. Cercle de Recherches et d'Études Préhistoriques, Paris.

Texier, Pierre-Jean 1984b Un Débitage Expérimental de Silex par Pression Pectorale à la Béquille. *Bulletin de la Societé Préhistorique Française* 81(1): 25–27.

Thomas, David H. 1986a Contemporary Hunter-Gatherer Archaeology in America. In *American Archaeology Past and Future*, edited by D. J. Meltzer, D. D. Fowler and J. A Sabloff, pp. 237–276. Smithsonian Institution Press, Washington.

Thomas, David H. 1986b Points on Points: A Reply to Flenniken and Raymond. *American Antiquity* 51(3): 619–627.

Thouvenot, Marc 1984 Le Débitage des Lames d'Obsidienne par les Aztèques selon les Textes du XVIe Sièle. *Préhistoire de la Pierre Taillée: 2 Économie du Débitage Laminaire*, pp. 149–157. Cercle de Recherches et d'Études Préhistoriques, Paris.

Times-News 1963 *T.F. Resident Spends Years Copying Ancient Man's Attempt at Stonework.* Times-News April 28, Twin Falls, Idaho.

Times-News 1970 *Area Man Displays Flint-knapping in New York Show*. Times-News March 1. Twin Falls, Idaho.

Times-News 1978 *Crabtree Expert in Flintknapping*. Times-News February 21, Twin Falls, Idaho.

Times-News 1969 *Noted French Archaeologist Visits Friends in Kimberly*. Times-News April 21, Twin Falls, Idaho.

Titmus, Gene 1979 Letters. *Flintknappers' Exchange* 2(3): 2.

Titmus, Gene 1980 Large Obsidian Boulder Reduction. *Flintknappers' Exchange* 3(3): 21–22.

Titmus, Gene 1981 Don Crabtree [obituary]. *Flintknapper's Exchange* 4(1): 2–3.

Titmus, Gene 1985 Some Aspects of Stone Tool Notching. In *Stone Tool Analysis: Essays in Honor of Don E. Crabtree*, edited by M.G. Plew, J. C. Woods and M. G. Pavesic, pp. 243–263. University of New Mexico Press, Albuquerque.

Titmus, Gene 2002 An Analysis of the Folsom Preform. In *Folsom Technology and Lifeways*, edited by J. E. Clark and M. B. Collins, pp. 227–247. Special Publication No. 4. *Lithic Technology*, Tulsa, Oklahoma.

Titmus, Gene L., and John E. Clark 2003 *Mexica* Blade-Making with Wooden Tools: Recent Expérimental Insights. In *Mesoamerican Lithic Technology: Experimentation and Interpretation*. K. Hirth, ed., 72–97. University of Utah Press, Salt Lake City.

Titmus, Gene L., and James C. Woods 1986 An Experimental Study of Projectile Point Fracture Patterns. *Journal of California and Great Basin Archaeology* 8(1): 37–49.

Titmus, Gene L., and James C. Woods 1991a A Closer Look at Margin "Grinding" on Folsom and Clovis Points. *Journal of California and Great Basin Anthropology* 13(2): 194–203.

Titmus, Gene L., and James C. Woods 1991b Fluted Points from the Snake River Plain. In *Clovis: Origins and Adaptations*, edited by R. Bonnichsen and K. L. Turnmire, pp. 119–131, Center for the Study of the First Americans, Oregon State University, Corvallis.

Titmus, Gene L., and James C. Woods 1992 Operación Q206: Una Cantera Caliza. In *El Informe de las Investigaciones Realizadas en la Temporada de 1991 en el Sitio Arqueológico, Nakbe, Petén*, edited by R. D. Hansen. Report Submitted to the Guatemalan Government.

Titmus, Gene L., and James C. Woods 2003 The Maya Eccentric: Evidence for the Use of the Indirect Percussion Technique in Mesoamerica from Preliminary Experiments Concerning their Manufacture. In *Mesoamerican Lithic Technology: Experimentation and Interpretation*. K. Hirth, ed., 132–146. University of Utah Press, Salt Lake City.

Tixier, Jacques 1963 *Typologie de l'Epipaléolithique de Maghreb*. Memoire du C.R.A.P.E. Alger, 2. Paris.

Tixier, Jacques 1967 Procédés d'Analyse et Questions de Terminologie Dans l'Étude des Ensembles Industriels du Paléolithique Récent et de l'Epipaléolithique en Afrique du Nord-Ouest. In *Background to Evolution in Africa*, edited by W. W. Bishop and J. D. Clark, pp. 771–820. University of Chicago Press, Chicago.

Tixier, Jacques 1969 *Outils Prehistoriques: Á la Recherche des Techniques de Taille*. Film produced by B. Mercier for the Institute Pedagogique National, Nationial de la Recherche Scientifique, Paris.

Tixier, Jacques 1972 Obtention de Lames par Débitage "sous le pied." *Bulletin de la Société Préhistorique Française* 69: 134–39.

Tixier, Jacques 1974 Glossary for the Description of Stone Tools, with Special Reference to the Epipaleolithic of the Maghreb. Translated by M. H. Newcomer. *Newsletter of Lithic Technology, Special Publication.*

Tixier, Jacques 1976 L'Industrie Lithique Capsienne de l'Ain Dokkara, Région de Tebessa, Algérie. *Lybica*, Vol. XXIV:21–53, Alger.

Tixier, Jacques 1980 Expériences de Taille. In *Prehistoire et Technologie Lithique*, 11–13 mai 1979. Publications de l'URA 28: cahier 1: 47–49. Éditions du C.N.R.S.

Tixier, Jacques 1984a Lames. In *Préhistoire de la Pierre Taillée, 2: Économie du Débitage Laminaire; Technologie et Expérimentation*, pp. 13–19. Cercle de Recherches et d'Etudes Préhistoriques, Paris.

Tixier, Jacques 1984b Le Débitage par Pression. *Préhistoire de la Pierre Taillée: 2 Économie du Débitage Laminaire*, pp. 57–70. Cercle de Recherches et d'Études Préhistoriques, Paris.

Tixier, Jacques 1991a Cogitations non Conclusives. In *25 Ans d'Études Technologiques en Préhistorie.* XIe Recontres Internationales d'Archéologie et d'Historie d'Antibes, pp. 391–394. Éditiones APDCA, Juan-les-pins.

Tixier, Jacques 1991b Et Passez au Pays des Silex: Rapportez-nous des Lames! (Drawings by Michèle Reduron). In *25 Ans d'Études Technologiques en Préhistorie.* XIe Recontres Internationales d'Archéologie et d'Historie d'Antibes, pp. 235–243. Éditiones APDCA, Juan-les-pins.

Tixier, Jacques 2000 Tribute to Don Crabtree. In *Don Crabtree: Idaho's Own Expert Stone Tool Maker*, IMNH, pp. 2. Special Publication 00-1, Idaho Museum of Natural History, Pocatello.

Tixier, Jacques, Marie-Louise Inizan, and Hélène Roche 1980 *Préhistorie de la Pierre Taillée 1: Terminologie et Technologie.* Cercle de Recherches et d'Etudes Préhistoriques, Valbonne Cedex.

Torquemada, Juan de 1975 *Monarquía Indiana* (fifth edition). Original 1615. Editorial Porrúa, Mexico.

Trachman, Rissa M., and Gene Titmus 2003 Pecked and Scored Initiations: Early Classic Core-Blade Production in the Central Maya Lowlands. In *Mesoamerican Lithic Technology: Experimentation and Interpretation*, edited by K. Hirth, pp. 108–119. University of Utah Press, Salt Lake City.

Tsirk, Are 2009 A Knapping Dilemma with Transverse Blade Breakage. *Lithic Technology* 34(2): 111–118.

Tunnell, Curtis 1977 Fluted Projectile Point Production as Revealed by Lithic Specimens from the Adair-Steadman Site in Northwest Texas. *The Museum Journal* 17: 140–168. West Texas Museum Association, Texas Tech University, Lubbock.

Tylor, Edward B. 1861 The Manufacture of Obsidian Knives, etc. *Anahuac: Mexico and the Mexicans, Ancient and Modern.* London: Longman, Green, Longman and Roberts.

Valla, François 1984 *Les Industries de Silex de Mallaha (Eynan) et du Natoufien dans le Levant.* Mémories et Travaux du Centre de Recherche Française de Jérusalem 3. Paris, Association Paléorient.

Volkov, P. V., and E. Iou. Guiria 1991 Recherche Expérimentale sur une Technique de Débitage. In *25 Ans d'Études Technologiques en Préhistorie.* XIe Recontres Internationales d'Archéologie et d'Historie d'Antibes, pp. 379–390. Éditiones APDCA, Juan-les-pins.

Waldorf, D. C. 1993 *The Art of Flint Knapping*, fourth edition. Mound Builder Books, Branson, Missouri.

Warren, R. 1968 A Brief Survey of the Possible Techniques used in Fluting of Folsom Points. Manuscript on file at the University of Colorado Museum, Boulder, Colorado.

Watts, Steve 1997 A Tribute to Errett Callahan. *Bulletin of Primitive Technology* 14: 8–10.

Whittaker, John C. 1994 *Flintknapping: Making & Understanding Stone Tools.* University of Texas Press, Austin.

Whittaker, John C. 2004 *American Flintknappers.* University of Texas Press, Austin.

Wilke, Philip J. 1996 Bullet-shaped Microblade Cores of the Near Eastern Neolithic: Experimental Replicative Studies. In *Neolithic Chipped Stone Industries of the Fertile Crescent and Their Contemporaries in Adjacent Regions*, edited by S. K. Kozłowskī and H. G. K. Gebel, pp. 289–310. Studies in Early Near Eastern Production, Subsistence and Environment, No. 3. *Ex Oriente e. V.*, Berlin.

Wilke, Philip J. 2002 Bifacial Flake-Core Reduction Strategies and Related Aspects of Early Paleolithic Lithic Technology. In *Folsom Technology and Lifeways*, edited by J. E. Clark and M. B. Collins, pp. 345–370. Special publication 4, *Lithic Technology.*

Wilke, Philip J. 2007 Book Review: *Mesoamerican Lithic Technology: Experimentation and Interpretation. Lithic Technology* 32(2): 219–232.

Wilke, Philip J., and Leslie A. Quintero 1994 Naviform Core-and-Blade Technology: Assemblage Character as Determined by Replication Experiments. In *Neolithic Chipped Stone Industries of the Fertile Crescent.* Studies in Early Near Eastern Production, Subsistence, and Environment, 1, edited by H. G. Gebel and S. K. Kozłowskī pp. 33–60. Berlin, Ex Oriente.

Williams, Barbara J., and H. R. Harvey 1997 *The Codice de Santa María Asunción: Facsimile and Commentary: Households and Lands in Sixteenth-Century Tepetlaoztoc.* University of Utah Press, Salt Lake City.

Winn, James 2008a Disc No. 5. *Blades and Cores*. Video available from the author.
Winn, James 2008b Disc No. 6. *Novaculite Laurel Leaf, Blade Cores, and More*. Video available from the author.
Woods, James C. 1981 Don E. Crabtree (1912–1980). *Idaho Archaeologist* 4(4): 1–3.
Wormington, H. Marie 1985 Foreword. In *Stone Tool Analysis: Essays in Honor of Don E. Crabtree*, edited by M.G. Plew, J. C. Woods, and M. G. Pavesic, pp. vii–viii. University of New Mexico Press, Albuquerque.

Part II
Pressure Blade Production in the World

Chapter 4
Early Holocene Climate Change and the Adoption of Pressure Technique in the Maghreb: The Capsian Sequence at Kef Zoura D (Eastern Algeria)

Noura Rahmani and David Lubell

4.1 Introduction

The Late Pleistocene in the Maghreb is known for the long primacy and stability of the Iberomaurusian, which lasted for more than 10,000 years. However, the Early Holocene saw the development of a mosaic of industries, often considered as cultures (following the typological approach traditionally employed). These can be distinguished geographically as well as by the variability of their respective material culture, but the relationships amongst the makers of these diverse industries are still unexplained despite nearly a century of research (Lubell et al. 1984). Do these industries represent different responses of the same group, or sub-groups, to the varying geographic and environmental characteristics of the Maghreb, or are they simply different manifestations of distinct groups employing diverse adaptations to changing contexts and environments? Site function, as well as availability of game and raw materials, certainly played a key role, and their importance in the definition and shaping of these industries should not be underestimated.

In this contribution, we present these different industries but focus on the Capsian which we examine further through a preliminary study of the ongoing analysis of the lithic material from the different chrono-stratigraphic units at Kef Zoura D, a site in which both major varieties of the Capsian occur (Jackes and Lubell 2008). We briefly review the Early to mid-Holocene paleoenvironmental data for the Maghreb to underline the most likely relationship between climatic changes and technological adaptations. In addition, the well-defined chrono-stratigraphic units at Kef Zoura D allow us to describe the lithic technology and to explore the nature of the changes that occurred in the eastern Maghreb during this period.

N. Rahmani (✉)
Montréal, QC, H2E 2C8 Canada

D. Lubell
Department of Anthropology, University of Waterloo, Waterloo, ON, N2L 3G1, Canada
e-mail: dlubell@uwaterloo.ca

P.M. Desrosiers (ed.), *The Emergence of Pressure Blade Making: From Origin to Modern Experimentation*, DOI 10.1007/978-1-4614-2003-3_4,

4.2 Early Holocene: A Mosaic of Lithic Industries

Besides Late Iberomaurusian which occurs in only a few inland sites, several techno-complexes discovered in Tunisia and Algeria display different typological characters and are thus not ascribable to the Iberomaurusian (Fig. 4.1). Amongst these is the group called Southern Tunisian Bladelet Industries (STBI), which appears in coastal and inland southern Tunisia and is characterized by a high predominance of backed bladelets with trihedral points. Because of this, Gobert and Howe (1952) initially argued it was close to the Iberomaurusian, but later, it was clearly distinguished because of many typological differences (Gobert 1962). The stratigraphic position and the chronological attribution of the STBI are still inadequately defined except for the assemblage of Sidi Mansour (Horizon Collignon) which would belong to the alluvium of Glacis 2 of Coque (1962) and thus be earlier than the Capsian, as previously suggested by Castany and Gobert (1954). Although no mention is made of them in recent publications (e.g. Linstädter 2008; Zielhofer et al. 2008), the STBI may be contemporary with the Typical Capsian. Page (1972) obtained four radiocarbon dates for occurrences in Wadi El Akarit which point to an Early Holocene age. The dates are 8635±260 and 9185±210 B.P. on land snail shell; 8235±180 and 8415±80 B.P. on marine shell. If we subtract 800 years from each of the land snail shell dates prior to calibration (see Jackes and Lubell 2008 and Lubell et al. 2009 for explanation) and use the pan-Mediterranean ΔR of 58±85 for the marine shell dates (Reimer and McCormac 2002), the four dates are statistically

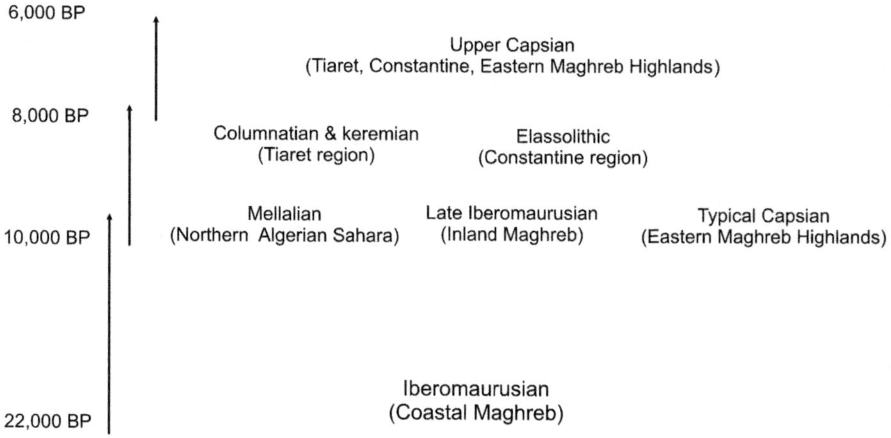

Fig. 4.1 Late Pleistocene/Early Holocene in the Maghreb: a mosaic of industries. Dates used here are uncalibrated

identical, with a pooled mean of 8314±86 B.P. or ~8970–9250 cal B.P. at 1σ (see Lubell et al. 1992 for original data on these and other dates discussed here).[1]

At Columnata and Cubitus in the region of Tiaret (Western Algeria), Cadenat (1948, 1963, 1966, 1970) recognized a hypermicrolithic Epipaleolithic industry rich in tiny segments and microburins. At Columnata, it has a transitional position since it occurs between Late Iberomaurusian and Upper Capsian layers. Called Columnatian, it is characterized by a high frequency of microbladelets (Roubet 1968) and can be dated between 9000 and 9300 cal B.P. on the basis of two charcoal samples.

The same hypermicrolithism was also recorded at Koudiat Kifène Lahda, in an industry which precedes the Capsian and is called Elassolithic (Roubet 1968). This microlithic aspect of the lithics was also recognized by Tixier (1954) at El Hamel in layer A which follows a Late Iberomaurusian layer. It seems that elassolithism, or accentuated microlithism, which is associated with some Epipaleolithic assemblages that precede the Upper Capsian in the western Maghreb, is characteristic of the ninth millennium B.P. (Camps 1974: 111).

In addition to these industries, another rather unique variant should be mentioned, the Keremian from Kef El Kerem in the Tiaret region (Cadenat and Vuillemont 1944). The original character of the industry was highlighted by Tixier when studying the Bois des Pins and Jumenterie assemblages in the same area (Tixier 1967; de Bayle des Hermens and Tixier 1972). The Keremian is primarily defined by the high percentage of scrapers (>40%; Tixier 1967: 807). Almost 250 km from Tiaret, in the Bou Saâda region, the assemblage at Zaccar I seems to present this same aspect (Ferhat 1977). These industries remain imprecisely defined chronologically but are thought to date to the Early Holocene.

Finally, in the Ouargla and Wadi Mya areas of the Algerian Sahara, research by Aumassip, Trécolle, Marmier and Tixier at eight sites identified an Epipaleolithic industry which was stratigraphically immediately below a Neolithic layer at three of them – Hassi Mouillah, Deux Oeufs and El Hadjar – called 'Neolithicized Capsian' (Marmier and Trécolle 1968; Marmier et al. 1978–1979), which is characterized by a very high frequency of straight backed bladelets and darts (*aiguillon droit*) and low frequencies of just about everything else – scrapers, borers, backed flakes and blades, truncations, geometric microliths, pieces with continuous retouch (Aumassip et al. 1983: 52). This well-defined assemblage, called Ouarglian or Mellalian, may date from the Early Holocene to around 7700 cal B.P. (average of two bone samples from Ouargla) but could be earlier, given the charcoal samples from El Hadjar (7970–8310 cal B.P.) and Hassi Mouilah (9450–9800 cal B.P.). Generally, these industries are contemporary with Typical Capsian but occur in different regions of the Maghreb. Following the widespread dispersal of Capsian groups at around 8000 cal B.P., these industries are replaced, or more precisely superseded, by Upper Capsian as is seen stratigraphically in many cases.

---

[1]All calibrations in this paper use CALIB 6.0.1 and Intcal09.14c or Marine09 as appropriate (Reimer et al. 2009). Calibrated dates are given as cal B.P. (at 1σ) and uncalibrated dates as B.P.

4.3 The Capsian

Of all the Maghreb techno-complexes, the Capsian is the best known. It covers the widest geographic area, is found at thousands of sites and is characterized by a rich material culture. Capsian sites, called either *escargotières* or *rammadiya* (Gobert 1937, and see discussions in Lubell et al. 2009; Mulazzani et al. 2009a: 32), are accumulations of land snail shells, ash, burnt rocks, knapped flint, worked bones (both human and non-human) and mammalian faunal remains (for further details, see Lubell et al. 1984; Sheppard 1987; Rahmani 2003). Typological studies (e.g. Camps 1974; Vaufrey 1936; Tixier 1963, 1967) highlighted the existence of two phases (sometimes seen as two facies), *Capsien typique* and *Capsien supérieur*. The first, which we will call Typical Capsian, is now known to have lasted from at least 9500 to around 8000 cal B.P. and is characterized by large tools – mainly burins, scrapers made on blades. The second, which we will call Upper Capsian, followed the first, lasted until at least 6000 cal B.P. and is characterized by smaller tools, especially microliths, made mostly on bladelets.

Capsian sites occur primarily on the high plateaux of eastern Algeria and southern Tunisia, although recent work suggests there may be variants outside this region that can be labelled as 'Capsian' (e.g. Barker et al. 2009; Barich et al. 2010; Mulazzani (ed.) 2011; Mulazzani et al. 2008, 2009a, b).

Detailed studies of the Capsian have shown that ca. 8200 cal B.P., there was a change in lithic production (Sheppard 1987; Jackes and Lubell 2008; Rahmani 2003, 2004). Typical Capsian assemblages before this date are dominated by blade production involving knapping schemes derived from simple or complex core preparation. These indicate use of both soft and hard hammer percussion for blades and flakes, which are then retouched to produce a variety of tools. Succeeding Upper Capsian assemblages are dominated by bladelet production, and, as shown in Fig. 4.2, these were normally produced by pressure technique that required preparation of sophisticated mitred cores. Morphometric analysis shows the production of consistent range of bladelet blanks, which in turn allowed the production of standardized tools.

This change, formerly recognized by typological studies as reflecting different Capsian industries, is now explained as a result of the adoption of pressure technique which characterizes all Upper Capsian assemblages and which defines a major technological change from the Typical Capsian (Rahmani 2003, 2004). The date of 8200 cal B.P. defines the time frame for the adoption of the pressure technique amongst Capsian groups and allows us to consider it as a techno-chronological marker with strong cultural meaning. Interestingly, this technological change is contemporaneous with an environmental shift documented by both global environment studies (e.g. Alley and Ágústdóttir 2005) and sediment and faunal analyses of Capsian sites (Jackes and Lubell 2008; Lubell et al. 1984; Sheppard 1987; Rahmani 2003).

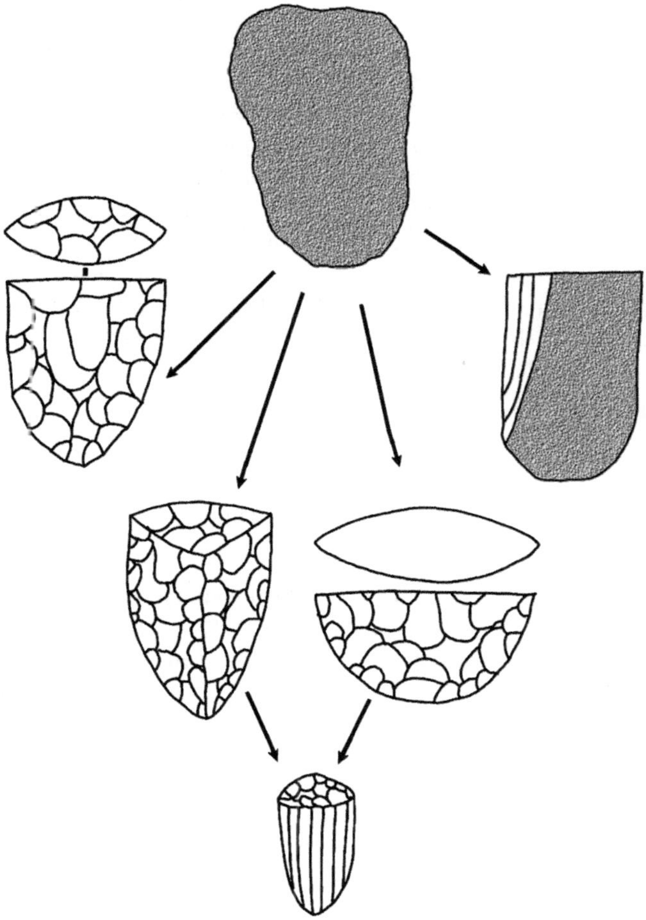

Fig. 4.2 Pressure flintknapping during Upper Capsian: one technique, various methods

4.4 Holocene Environmental Changes

The reconstruction of past environments shows that during the Holocene, North Africa was subject to many fluctuations (e.g. Hassan 2002; Linstädter 2008). After the aridity of the Late Pleistocene, the Early Holocene saw a return of moister conditions and the appearance of vegetated zones in the Sahara (Cremaschi 2002; Vernet 2002). These conditions lasted mostly until around 8000 B.P. when aridity started to increase and reached its peak at 7000 B.P. Alternations of arid and moist intervals characterize Holocene climate, and again around 6500–5500 B.P., another phase of aridity took place but was still moister than today.

In sum, the Early to mid-Holocene shows variable climatic conditions that would have had significant implications for human occupation and adaptation in the region. The changes affected past vegetation and the distribution and availability of resources. According to Hassan (1996: 85), 'the most influential climatic events are the abrupt, severe droughts that punctuated the moist intervals of the Holocene', and this is particularly the case for the change around 8000–7000 cal B.P. in some, but not all regions of North Africa (Hassan 2002 and various essays in the same volume). This climatic shift is today well documented (Alley and Ágústdóttir 2005) and was one of the major environmental factors that affected Capsian foragers (Couvert 1972; Cremaschi 1998; Jäkel 1979; Linstädter 2008; Lubell et al. 1981–1982; Rognon 1976). In fact, this change is clearly evidenced in the archaeological records of Capsian sites with well-dated deposits covering the period of change, which brings us to the case study of Kef Zoura D.

4.5 Kef Zoura D and the Télidjène Basin

Southwest of Tébessa, there is a large antisynclinal depression, the Télidjène Basin (on maps produced in the past 40 years, this is often called Bahiret Tlidjen). In the escarpment surrounding this depression, there are many rock shelters that offered strategic positions for prehistoric settlements. Kef Zoura D is one of these, located in the southwestern part of the basin at the eastern end of Djebel Arhour el Kifène and 3 km in a straight line from Relilaï (Vaufrey 1936), at an elevation of about 60 m above the basin floor (Fig. 1 in Jackes and Lubell 2008). Discovered by Grébénart (1976), it was later excavated by Lubell (Lubell et al. 1981–1982; Jackes and Lubell 2008). Use of carefully controlled excavation procedures allowed us to distinguish and follow the stratigraphy, which when coupled with faunal, sedimentological and lithic analyses have made it possible to identify a chrono-stratigraphic succession of two major archaeological levels, attributed to an earlier Typical Capsian and a later Upper Capsian.

Besides its strategic position, Kef Zoura D is located in a geological context very rich in raw material that has excellent knapping properties and where several flint sources were identified in the immediate surroundings (Vaufrey 1955; Lubell et al. 1981–1982; Jackes and Lubell 2008). One of these, a Senonian brown-grey flint of high quality, was commonly used by the knappers of Relilaï and Kef Zoura D to produce Typical and Upper Capsian assemblages.

4.6 Stratigraphic Units and Chronology

Even though excavation methods were appropriate to the nature of the deposits and attempts were made to follow the very complicated natural stratigraphy, real levels were not established but rather five archaeological units. These units were to some extent defined during excavation – first in a 1976 test and then by more extensive work in 1978 – by following visible stratigraphy but primarily post-excavation using the faunal, sedimentological and lithic data recorded and then confirmed by radiocarbon dating. We present them here based on the more complete discussion in Jackes and Lubell (2008). All depths are expressed as centimetres below datum which was set arbitrarily at 50 cm above the surface of the deposits at the rear of the shelter in 1976. Figures 2 and 4 in Jackes and Lubell (2008) provide further details, showing that Units I, II and III are inclined from the front to the rear of the shelter. Not enough of Unit IV was exposed to know if it is also inclined.

Unit I occurs at or near the surface from 70 to 90 cm in the 1976 test, in the surface deposits of squares C20 and C21 and in a thin lens as far as squares E20 and E21. It is characterized by ash and crushed shells and has the highest frequency of *Otala* sp.

Unit II underlies Unit I in squares D20 and C20 to a depth of 120 cm and appears from the surface to 100 cm in squares E20, E21 and partially in square F20. It includes a depression filled with loose whole snail shell that is mainly *Helix melanostoma*.

Unit III underlies Unit II in squares D20, D21, E20, E21, F20 and F21 at a depth between 80 and 120 cm and appears at the surface partially in F21 and completely in G20 and G21. It is characterized by an increase in the smaller land snails, especially *Helicella setifensis*.

Unit IV corresponds to the deposits in G20 below 110 cm. Compared to Unit III, it has darker deposits and very different assemblages of both land snails and lithics which cluster more closely with those of Unit V.

Unit V represents the oldest deposits in the site and is found in the T20-5 test pit in deposits in front of the shelter which begin at 150 cm below datum and extend to a depth of 300 cm. Despite being on the talus slope in front of the shelter, it is an in situ assemblage of Typical Capsian material (including two cores that can be largely refitted).

Units I, II and III are Upper Capsian and cover three or more periods of occupation during a time span of about 1,500 years. Units IV and V are Typical Capsian and lasted for about 1,000 years. In its broad lines, this stratigraphy is very similar to the neighbouring site of Relilaï where Upper Capsian was recorded in the interior and to the rear of the shelter, whereas Typical Capsian occurred in the slope deposits (Grébénart 1976; Jackes and Lubell 2008).

Kef Zoura D has a set of 19 coherent radiocarbon dates, of which 17 are based on charcoal samples and two on land snail shell. The Typical Capsian assemblage found in Units IV and V is dated ca. 10420–9470 cal B.P. The Upper Capsian assemblages in Units II and III are dated ca. 8365–7500 cal B.P., while the one in Unit I should date younger than 6780 cal B.P., based on a single charcoal sample (2σ ranges; for data, see Jackes and Lubell 2008: Table 1 and Fig. 3).

4.7 Typical Capsian Lithic Technology

Found mostly in the slope deposit in front of the shelter which begins below the level of the modern surface (Jackes and Lubell 2008: Fig. 2), the Typical Capsian assemblage from Unit V represents the oldest dated Capsian occupation in North

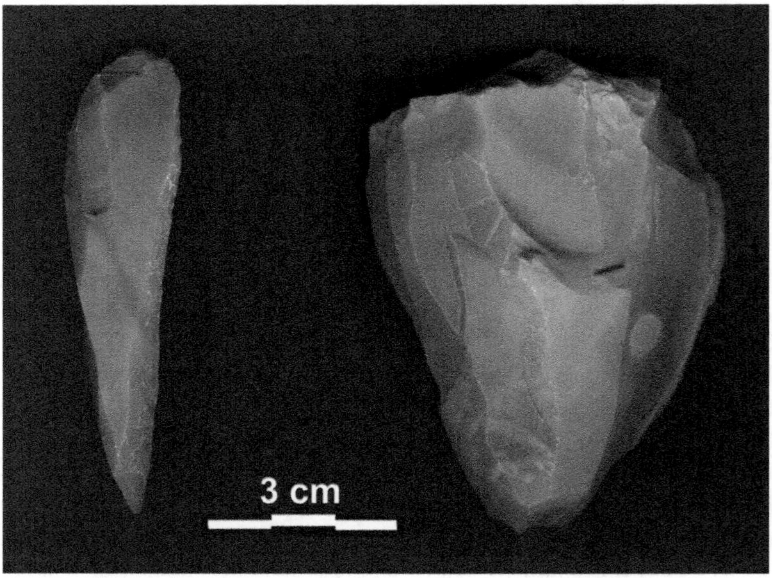

Fig. 4.3 Examples of typical Capsian debitage: a backed blade and a core

Africa. In typological terms, it is a very classic Typical Capsian. Compared to Relilaï (Grébénart 1976; Rahmani 2003), it shows the same characteristics with high frequencies of both burins and backed bladelets. This contradicts the hypothesis that Typical Capsian rich in burins is a chronological facies and thus recent (Camps 1974). On the contrary, we think that these typological characteristics can be explained in technological and functional terms.

Firstly, the entire lithic assemblage of Unit V is made using the high-quality Senonian grey-brown flint which is readily available locally. It was thus mainly acquired with a low-cost procurement process, and the knappers did not conserve the material which was found in such profusion. In Unit IV, the same behaviour is documented, but occasional introduction of black flint as blanks and finished tools is also recorded.

Secondly, the lithic assemblages from Units IV and V show a dominant scheme of blade production from less-prepared single platform cores (Fig. 4.3). Hard or soft hammer percussion was used to produce blades with different dimensions and characteristics. Blades produced by hard hammer are thick, with wide sections associated frequently with a curved profile and an irregular shape. Blades produced by soft hammer percussion are thinner, displaying a narrower section and a straight profile as well as a regular shape. Most blades produced by soft hammer were transformed into backed blades (Fig. 4.3), while those produced by hard hammer were often transformed into burins.

Thirdly, there is no distinct scheme of bladelet production from prepared pyramidal cores, and most backed bladelets are in fact made on transformed burin spalls.

The production of triangular section bladelets on blades originally produced with the burin technique is very significant. This characteristic is better underlined by the relative frequencies of burins and backed bladelets in both Typical Capsian assemblages, emphasizing the relationship between the percentages of bladelets and burins in Units IV and V and the fact that they make up more than 50% of the retouched tools (the differences between them are not statistically significant: $\chi^2 = 1.65$, d.f: 1, $p \leq 0.20$). The burins are here considered as bladelet cores, and the blanks produced were primarily transformed into backed bladelets and occasional perforators. Both assemblages contain end-scrapers, notches and denticulates that were manufactured on less carefully prepared blanks, especially blades and flakes that often retained extensive cortex.

4.8 *Chaînes Opératoires* of Tool Production in Upper Capsian Units

For the Upper Capsian in Units I, II and III, raw material procurement was essentially local, although the assemblages also display an important feature: the introduction of exogenous black flint brought from tertiary deposits in the northern Tellian region (cf. Rahmani 2004: 90).

Core preparation followed the same steps and volumetric dimension as the Relilaï mitred cores, underlining the important similarity (Rahmani 2003, 2004) between the two assemblages. Prepared cores were made from medium-sized nodules selected from the available local material. The flint knappers started first by creating the striking platform and then preparing two or three crests that were used to shape the core. Great care was devoted to the preparation of the core, and the use of indirect percussion is clearly attested by characteristic scars and bifacial flakes (Fig. 4.4). This attention paid to core preparation is actually rewarding because pressure bladelet production required shaped geometric cores but could progress on core with less maintenance and only minimal edge and platform trimming. Experimental replications have shown that the best-prepared cores are the most productive ones (Pelegrin 1984), and this pattern is clearly attained in Upper Capsian units.

On the mitred cores, the production starts by the removal of crested bladelets often by indirect percussion, followed by under-crested bladelets and serial bladelets mainly detached by pressure technique. The technique is precise and efficient and guaranteed the production of generally identical bladelets. After a series of bladelets was produced, the mitred core became a fluted core exhibiting regular successive scars. At the end of production, the fluted cores were frequently exploited by hard hammer percussion to produce expedient tools, but they still display past sequences of bladelet production by pressure. This is very common in Upper Capsian industries and results in a scarcity of fluted cores in archaeological assemblages.

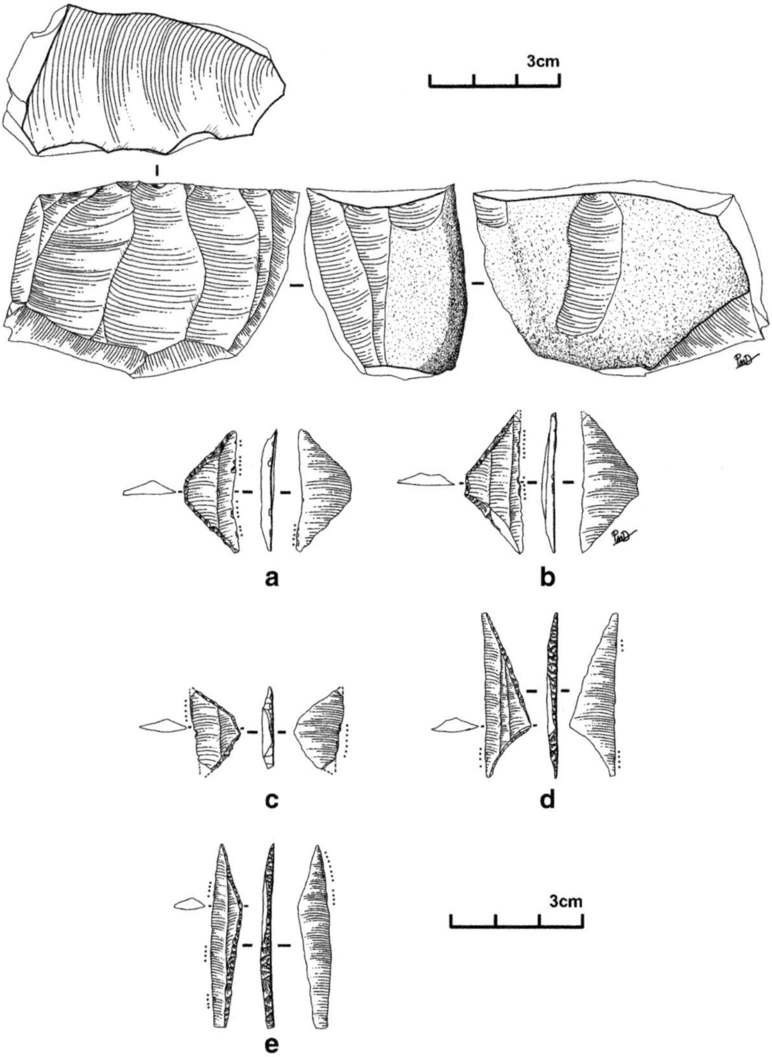

Fig. 4.4 Upper Capsian debitage: prepared mitred core (top) and geometric microliths (a–e)

The bladelet blanks produced are very regular, straight, thin and with consistent dimensions. They were selected to produce tools such as the trapezes and triangles common in Unit III and the notches and denticulates which characterize Units I and II.

Particularly in Unit III, emphasis was on the production of geometric microliths using microburin technique, which resulted in many microburins as waste. The microliths obtained have a narrow range of variation for thickness and width and were likely the result of serial production. Their standardization points to use as interchangeable elements in a complex hafting system, a constraint that was overcome by the new production technique (Fig. 4.4).

Compared to the underlying Unit IV, Unit III attests to the introduction of pressure technique and its orientation towards microlith manufacture. The disparity is striking since the technique, the *chaîne opératoire* and the final tools all changed, but most importantly, the desired end products were completely different. This corresponds to a change in lithic technology, previously underlined by Sheppard (1987), and suggests a technological response to changing conditions.

The Unit I and II lithic assemblages show that bladelet production by pressure was oriented to the production of notches and denticulates. This again attests to a change, not in the technique of production but in the transformation of the blanks and the toolkit. This final Upper Capsian is known as the Aïn Aachena facies (Tixier 1976) and dates ca. 7500 cal B.P. for the upper levels at Aïn Dokkara. The increase in notched tools might be correlated with an intensification of plant processing (cf. Clarke 1976).

4.9 Introduction of Tertiary Black Flint: Clue for North–South Movement?

The Upper Capsian of Kef Zoura D also shows the introduction of an exogenous raw material, black flint from the Northern Tellian region more than 100 km from the Télidjène Basin (cf. Rahmani 2004). Finished tools made on this material are found occasionally in Unit IV, but become far more common in Unit III following the technological change, where it occurs especially as blanks and finished tools and occasionally prepared and exploited cores. The amount of black flint decreases in Unit II and increases again in Unit I where it is represented mostly by notched pieces (Fig. 4.5).

The presence of this flint suggests regular North–South connections, and its increase after the technological change, when conditions seem to have been more arid, could be explained by movement between the Tébessa and Constantine regions, either following wild game during cyclic migration or for prolonged hunting trips which may have introduced a new settlement dynamic. This was the case especially for Unit III, the oldest Upper Capsian at Kef Zoura D with a high frequency of microliths.

4.10 Discussion

The five units identified at Kef Zoura D through the use of tight stratigraphic control during excavation and subsequent analyses illustrate changes in the lithic technology and the tools produced which can be directly dated due to the consistent radiometric data. Between Unit IV and Unit III, a change in lithic technology took place,

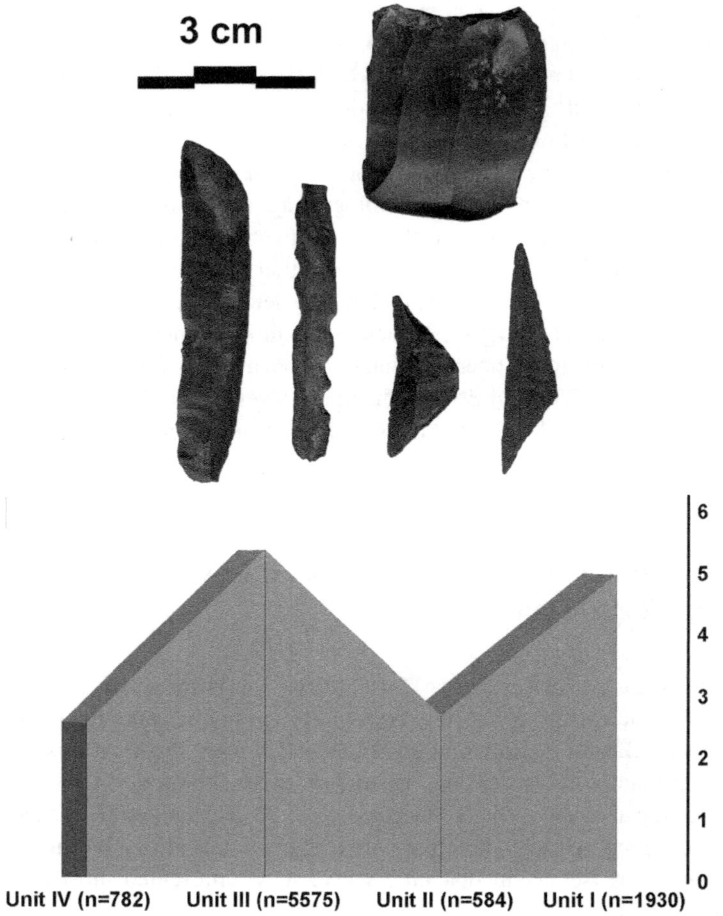

Fig. 4.5 Introduction over time of exogenous tertiary black flint in the Kef Zoura D assemblages. The upper part of the figure shows examples of artefacts made using black flint, the lower part shows the frequency in each unit. The differences are statistically different: $\chi^2=16.40$, d.f: 3, $p \leq 0.001$

initiated by the adoption of pressure technique which is dated to around 8200 cal B.P. This introduction is significant culturally since it explains the typological differences highlighted for almost a century between Typical Capsian and Upper Capsian. Most importantly, this change characterizes a shift and underlines a chronological distinction attesting further to the chrono-stratigraphic succession of Typical Capsian to Upper Capsian.

In the Maghreb, an episode of aridity took place around 8000 cal B.P., and this would have had significant implications at the archaeological level. At Aïn Misteheyia, Kef Zoura D and Medjez II, there is evidence for a change in the consumed fauna which corresponds to the onset of this increase in aridity (Lubell et al. 1984; Lubell 2005; Jackes and Lubell 2008). Before the arid period, faunal remains are primarily made up of large herbivores (*Bos*, *Equus*, *Alcelaphus*) and larger land snails (*Helix melanostoma*), while after the episode of aridity, smaller herbivores are more abundant (*Gazella*, lagomorphs) as well as smaller xerophilous land snails (*Helicella sitifensis*, *Leucochroa candissima*). Sheppard associates the appearance of pressure technique with the climatic change and suggests, 'If the technological change followed the environmental change then increased environmental stress may either have assisted in the spread of the new technology (e.g. through enlargement of area exploited and an increase in inter-band contact), or have provided the selective forces which promoted its development' (Sheppard 1987: 233). The 8200 cal B.P. event is considered to have been abrupt and short. Thus, any changes in adaptation by Capsian foragers would have been more likely a reaction to these conditions rather than a long-term prepared response involving thorough decision, especially as archaeological data do not indicate transitional stages.

By ca. 7000 B.P., the Maghreb climate was more arid than previously, but still moister than today, with steppe on the highlands, Mediterranean forest in some areas of higher elevation and broad expanses of grassland in what is today desert. The change in the toolkit in Units I and II at Kef Zoura D shows a clear preference for notches and denticulates obtained on first choice blanks struck by pressure. This change points to a variation in subsistence and other needs and stresses a mode of adaptation oriented towards the exploitation of different resources than before.

Another element of consideration is the fact that pressure technique permits the possibility of producing more bladelets with fewer raw materials and most likely facilitates the movement of Capsian groups who were becoming less dependent upon the proximity of raw material sources (cf. Rahmani 2004). This idea is well supported by the increased frequency of black flint at Kef Zoura D after the adoption of pressure technique and the general expansion of Capsian territory.

4.11 Conclusion

Kef Zoura D provides an exceptional opportunity to follow Holocene forager adaptations and to understand the changes that occurred over time within Capsian societies. Capsian industries show evidence of increasing complexity during three millennia, as seen in the changing lithic technology and in the adjustment of the toolkit. The microlithic technology associated with the high standardization of the retouched tools reflects a shift from a focus on large game hunting to more varied and specialized activities and, as a result, reveals the high degree of flexibility in the Capsian foraging system.

In fact, besides lithic technology studies, Kef Zoura D documents a change in food-gathering behaviour also found at the contemporary site of Aïn Misteheyia in the Télidjène Basin (Jackes and Lubell 2008). Study of land snails, and faunal remains, coupled with the limited evidence available from charcoal (D'Andrea et al.

1995) and phytoliths (Shipp et al., n.d.) at both sites shows that there were changes in the subsistence regime that correspond to the technological changes.

The significance of these changes is related to changes in past environments, subsistence, settlement patterns and land use. Early to mid-Holocene Capsian foragers undertook many readjustments of their subsistence practices which can be seen in the lithic data by the development of a special technology that reveals both significant flexibility and increasing complexity. Overall, the patterns show that this long-term change points to an increased emphasis on a wide variety of smaller food resources that reproduce more rapidly. Interestingly, these can be seen as expressions of processes involving the broad-spectrum revolution (Lubell 2004).

Acknowledgements The excavations at Kef Zoura D were funded by a grant from the Social Sciences and Humanities Research Council of Canada to DL. Subsequent laboratory research benefited from this grant and additional financial help from the University of Alberta. Permission to conduct the field research came from the Ministère de l'Education Supérieur et de la Recherche Scientifique and the Ministère de l'Information et de la Culture in Algiers. The Centre National de Recherches Préhistoriques, Anthropologiques et Historiques, Alger (then known as the Centre de Recherches Anthropologiques, Préhistoriques et Ethnographiques) provided essential logistical assistance. NR's work was funded by a post-doctoral grant from les Fonds québécois de la recherche sur la société et la culture (FQRSC) which also supplied funding for her participation at the Pan-African Congress, Botswana 2005, where an earlier version of this paper was first presented. We thank Pierre M. Desrosiers for producing the figures and especially the drawings in Fig. 4.4. DL continues to be grateful to both Mary Jackes and Peter Sheppard for their ongoing collaboration in the analysis of Kef Zoura D.

References

Alley, R.B. and A.M. Ágústdóttir 2005 The 8k event: cause and consequences of a major Holocene abrupt climate change. *Quaternary Science Reviews* 24: 1123–1149.

Aumassip, G., F. Marmier, J. Tixier, and G. Trécolle 1983 *L'Epipaléolithique Nord-Saharien; Ouarglien ou Mellalien*. Quick Copy, Bordeaux.

Bayle des Hermens, R. (de) and J. Tixier 1972 Le gisement Kérémien de la Jumentérie de Tiaret (Algérie). *Actes du 6ème Congrès panafricain de Préhistoire*, Dakar 1967, 288–293.

Barich, B.E., E.A.A. Garcea, C. Giraudi, G. Lucarini, and G. Mutri 2010 The latest research in the Jebel Gharbi (Northern Libya): environment and cultures from MSA to LSA and the first Neolithic findings. *Libyca Antiqua,* Nuova Serie, 5: 237–252.

Barker, G., A. Antoniadou, H. Barton, I. Brooks, I. Candy, N. Drake, L. Farr, C. Hunt, A.A. Ibrahim, R. Inglis, S. Jones, J. Morales, I. Morley, G. Mutri, R. Rabett, T. Reynolds, D. Simpson, M. Twati, and K. White 2009 The Cyrenaican Prehistory Project 2009: the third season of investigations of the Haua Fteah cave and its landscape, and further results from the 2007–2008 fieldwork. *Libyan Studies* 40: 1–41.

Cadenat, P. 1948 La station préhistorique de Columnata (commune mixte de Tiaret, Département d'Oran). *Bulletin de la société de géographie et d'archéologie* 70–71: 3–66.

Cadenat, P. 1963 Sondage au « Cubitus », station épipaléolithique des environs de Tiaret. *Libyca* 11: 75–122.

Cadenat, P. 1966 Vues nouvelles sur les industries de Columnata. *Libyca* 14: 189–207.

Cadenat, P. 1970 Le Columnatien, industrie épipaléolithique d'Algérie. *Bulletin de la Société d'Études et de Recherches Préhistoriques, Les Eyzies* 20: 40–50.

Cadenat, P and G. Vuillemot 1944 La station préhistorique de Kef el-Kerem (Djebel Nador). *Bulletin de la Société de Géographie et d'Archéologie d 'Oran* 65: 52–65.

Camps, G. 1974 *Les Civilisations préhistoriques de l'Afrique du Nord et du Sahara*. Doin, Paris.

Castany, G. and E.G. Gobert 1954 Morphologie quaternaire, paléontologie et leurs relations à Gafsa. *Libyca* 2: 9–37.

Clarke, D. 1976 Mesolithic Europe: the economic basis. In *Problems in Economic and Social Archaeology*, edited by G. de G. Sieveking, I.H. Longworth and K.E. Wilson, pp. 449–481. Duckworth, London.

Coque, R. 1962 *La Tunisie Présaharienne: Études géomorphologiques*. A. Colin, Paris.

Couvert, M. 1972 Variations paléoclimatiques en Algérie : traduction climatique des informations paléobotaniques fournies par les charbons des gisements préhistoriques (note préliminaire). *Libyca* 20: 45–48.

Cremaschi, M. 1998 Geological Evidence for Late Pleistocene and Holocene Environmental Changes in South-Western Fezzan (Central Sahara, Libya). *Proceedings of the 8th U.I.S.P.P. Congress, Forli*, 8–14 September 1996. ABACO, 1037–1051. Italy.

Cremaschi, M. 2002 Late Pleistocene and Holocene climatic changes in the central Sahara: the case study of the southwestern Fezzan, Libya. In *Droughts, Food and Culture: Ecological Change and Food Security in Africa's Later Prehistory*, edited by F.A. Hassan, pp. 65–81. Kluwer Academic/Plenum Publishers, New York.

D'Andrea, A.C., C.T. Shay and D. Lubell 1995 Wood charcoal and palaeoecological reconstructions: an example from Kef Zoura D, Algeria. Paper presented at the 28th Annual Meeting of the Canadian Archaeological Association, Kelowna, B.C.

Ferhat, N. 1977 Le gisement Zaccar I près de Bou-Saâda (Algérie). *Libyca* 25: 85–99.

Gobert, E.G. 1937 Les escargotières: le mot et la chose. *Revue Africaine* 81: 631–645.

Gobert, E.G. 1962 La préhistoire dans la zone littorale de la Tunisie. *Quaternaria* 6: 271–307.

Gobert, E.G. and B. Howe 1952 L'Ibéromaurusien de l'Oued el-Akarit (Tunisie). In *Actes du 2ème Congrès Panafricain de Préhistoire*, edited by L. Balout, pp. 575–594. Arts et Métiers Graphiques, Paris.

Grébénart, D. 1976 *Le Capsien des Régions de Tébessa et d'Ouled-Djellal, Algérie: contribution à son étude. Études Méditerranéennes*. Éditions de l'Université de Provence, Aix-en-Provence.

Hassan, F.A. 1996 Abrupt Holocene Climatic Events in Africa. In *Aspects of African Archaeology, Papers from the 10th Congress of the PanAfrican Association for Prehistory and Related Studies, Harare*, edited by G. Pwiti and R. Soper, pp. 83–89. University of Zimbabwe Publications, Harare.

Hassan, F.A. 2002 Palaeoclimate, food and culture change in Africa: an overview. In *Droughts, Food and Culture: Ecological Change and Food Security in Africa's Later Prehistory*, edited by F.A. Hassan, pp. 11–26. Kluwer Academic/Plenum Publishers, New York.

Jackes, M. and D. Lubell 2008 Environmental and cultural change in the Early and Mid Holocene: evidence from the Télidjène Basin, Algeria. *African Archaeological Review* 25(1–2): 41–55.

Jäkel, A. 1979 Run-off and Fluvial Formation Processes in the Tibesti Mountains as Indicators of Climatic History in the Central Sahara during the Late Pleistocene and Holocene. *Palaeoecology of Africa* 11: 13–33.

Linstädter, J. 2008 The Epipalaeolithic-Neolithic-transition in the Mediterranean region of northwest Africa. *Quartär* 55: 41–62.

Lubell, D. 2004 Are land snails a signature for the Mesolithic-Neolithic transition in the circum-Mediterranean? In *The Neolithization of Eurasia – paradigms, models and concepts involved,* edited by M. Budja. *Neolithic Studies 11, Documenta Praehistorica* XXXI: 1–24.

Lubell, D. 2005 Continuité et changement dans l'Épipaléolithique du Maghreb, In *Le Paléolithique en Afrique, l'histoire la plus longue*, edited by M Sahnouni, pp. 205–226. Errance & Artcom, Paris.

Lubell, D. with contributions by J. Feathers & J.-L. Schwenninger 2009 Post-Capsian settlement in the eastern Maghreb: implications of a revised chronological assessment for the adult burial at Aïn Misteheyia. *Journal of African Archaeology* 7(2): 175–189.

Lubell, D., A. Gautier, E.T. Leventhal, M. Thompson, H.-P. Schwarcz, and M. Skinner 1981–82 The Prehistoric Cultural Ecology of Capsian Escargotières, Part II: Report on investigations conducted during 1976 in the Bahiret Télidjène, Tebessa Wilaya, Algeria. *Libyca* 30–31: 59–142.

Lubell, D., P.J. Sheppard, and M. Jackes 1984 Continuities in the Epipaleolithic of northern Africa with emphasis on the Maghreb. In *Advances in World Archaeology* 3, edited by F. Wendorf and A.E. Close, pp. 143–191. Academic Press, New York.

Lubell, D., P. Sheppard, and A. Gilman 1992 The Maghreb. In *Chronologies in Old World Archaeology, 3rd Edition*, edited by R.W. Ehrich, pp. I: 305–308 and II: 257–267. University of Chicago Press, Chicago.

Marmier, F. and G. Trécolle 1968 Stratigraphie du gisement d'Hassi Mouillah. Règion de Ouargla (Algérie). *Bulletin de la Société Préhistorique Française*, CSRM, n° 4: 121–127.

Marmier, F., J. Tixier, and G. Trécolle 1978–79 Etude de l'industrie du gisement d'Hassi Mouillah III: l'industrie lithique néolithique. *Libyca* 26–27: 81–115.

Mulazzani, S. (ed.) 2011 L'habitat capsien de SHM-1 et les occupations dans le sahel tunisien. Résultats des recherches préhistoriques à Hergla (2002–2007). Antiquités Africaines, CNRS, Aix-en-Provence.

Mulazzani, S., R. Boussoffara, V. Azzarà, L. Belhouchet, M. Boldrin, A. Boujelben, E.R. Crema, M.A. Mannino, C. Del Grande, C. Mazzanti, and S. Scaruffi 2008 Between the Sahara and the Mediterranean : a study of the Middle Holocene communities of the Sebkhet Halk el Menjel (Tunisia). The first three campaigns 2001–2004. In *SOMA 2005 : Proceedings of the IX Symposium on Mediterranean Archaeology, Chieti (Italy), 24–26 Februrary 2005*, edited by O. Menozzi, M.L. Di Marzio and D. Fossataro, pp. 383–390. BAR International Series 1739, Archaeopress, Oxford.

Mulazzani, S., F. Cavulli, V. Azzarà, S. Scaruffi and R. Boussoffara 2009a Structures d'habitat Nord-Africaines: la fouille de la rammadiya cotière Holocène de Shm-1 (Hergla, Tunisie). In *Defining a Methodological Approach to Interpret Structural Evidence*, edited by F. Cavulli, pp. 31–42. BAR International Series 2045, Archaeopress, Oxford.

Mulazzani, S., L. Belhouchet, and Z. Jeddi 2009b Aspects techno-typologiques de l'industrie lithique des chasseurs-pécheurs-cueilleurs de Shm-1 (Hergla - Tunisie). Le premier niveau d'occupation durant le VIIème millénaire cal B.C.. *Africa* LXIV (3–4): 458–473.

Page, W.D. 1972 The Geological Setting of the Archeological Site at Oued el Akarit and the Paleoclimate Significance of gypsum soils, Southern Tunisia. Unpublished Ph.D. dissertation, University of Colorado, Boulder.

Pelegrin, J. 1984 Approche technologique expérimentale de la mise en forme de nucléus pour le débitage systématique par pression. In: *Préhistoire de la pierre taillée 2 : Économie du débitage laminaire. Technologie et Expérimentation.* Paris: Cercle de Recherches et d'Études préhistoriques, 93–104.

Rahmani, N. 2003 *Le Capsien typique et le Capsien supérieur, évolution ou contemporanéité? Les données technologiques*. Cambridge Monographs in African Archaeology 57, B.A.R. International Series 1187, Archaeopress, Oxford.

Rahmani, N. 2004 Technological and Cultural Change Among the Last Hunter-Gatherers of The Maghreb: the Capsian (10000–6000 B.P.). *Journal of World Prehistory* 18(1): 57–105.

Reimer P.J. and F.G. McCormac 2002 Marine radiocarbon reservoir corrections for the Mediterranean and Aegean seas. *Radiocarbon* 44: 159–166.

Reimer, P.J., M.G.L. Baillie, E. Bard, A. Bayliss, J.W. Beck, P.G. Blackwell, C. Bronk Ramsey, C.E. Buck, G.S. Burr, R.L. Edwards, M. Friedrich, P.M. Grootes, T.P. Guilderson, I. Hajdas, T.J. Heaton, A.G. Hogg, K.A. Hughen, K.F. Kaiser, B. Kromer, F.G. McCormac, S.W. Manning, R.W. Reimer, D.A. Richards, J.R. Southon, S. Talamo, C.S.M. Turney, J. van der Plicht, and C.E. Weyhenmeyer 2009 IntCal09 and Marine09 Radiocarbon Age Calibration Curves, 0–50,000 Years cal B.P.. *Radiocarbon* 51: 1111–1150.

Rognon, P. 1976 Essai d'interprétation des variations climatiques au Sahara depuis 40.000 ans, *Recueils de Géographie Physique et Géologie Dynamique* 18: 252–282.

Roubet, C. 1968 Nouvelles observations sur l'Épipaléolithique de l'Algérie orientale. Le gisement de Koudiat Kifen Lahda. *Libyca* 16: 55–101.

Sheppard, P.-J. 1987 *The Capsian of North Africa. Stylistic Variation in Stone Tool Assemblages.* B.A.R. International Series 353, Archaeopress, Oxford.

Shipp, J., A. Rosen, and D. Lubell In press, Phytolith Evidence of Capsian Subsistence Economies.

Tixier, J. 1954 Le gisement préhistorique d'El-Hamel. *Libyca* 2: 78–120.

Tixier, J. 1963 *Typologie de l'Epipaléolithique du Maghreb.* Mémoires du C.R.A.P.E., No. 2. Arts et Métiers Graphiques, Paris.

Tixier, J. 1967 Procédés d'analyse et questions de terminologie concernant l'étude des ensembles industriels du Paléolithique récent et de l'Épipaléolithique dans l'Afrique du Nord-Ouest. In *Background to Evolution in Africa. Proceedings of a symposium held at Burg Wartenstein Austria,* July-August 1965, edited by W. Bishop and J.D. Clark, pp. 771–820. University of Chicago Press, Chicago.

Tixier, J. 1976 L'industrie lithique de l'Aïn Dokkara, Région de Tébessa, Algérie. *Libyca* 24: 21–54.

Vaufrey, R. 1936 Stratigraphie capsienne. *wiatowit* 16: 15–34.

Vaufrey, R. 1955 *Préhistoire de l'Afrique. Tome premier. Maghreb.* Paris: Masson.

Vernet, R. 2002 Climate during the Late Holocene in the Sahara and the Sahel : evolution and consequences of human settlement. In *Droughts, Food and Culture: Ecological Change and Food Security in Africa's Later Prehistory*, edited by F.A. Hassan, pp. 47–63. Kluwer Academic/ Plenum Publishers, New York.

Zielhofer, C., D Faust, and J. Linstädter 2008 Late Pleistocene and Holocene alluvial archives in the Southwestern Mediterranean: changes in fluvial dynamics and past human response. *Quaternary International* 181: 39–54.

Chapter 5
Pressure Blade Production with a Lever in the Early and Late Neolithic of the Near East

Ciler Altınbilek-Algül, Laurence Astruc, Didier Binder, and Jacques Pelegrin

5.1 Introduction

Methods associated with the pressure technique in Near East evolved significantly during the Pre-Pottery Neolithic. In this chapter, we present the evolution of this technique used for the detachment of obsidian and flint blade(let)s in the Tigris and Euphrates High Valleys and on the Anatolian plateau since the Early Pre-Pottery Neolithic B (EPPNB), during the middle of the ninth millennium cal B.C.

In the High Valleys, the methods associated with the pressure technique evolved significantly during the Pre-Pottery Neolithic, leading to more regularized and standardized products. In this context, the appearance of large obsidian blades produced by pressure with the use of a lever provides interesting insight to understand the social aspects of this production including the technological experimentation, the innovation and the exchanges that took place during that period. This provides explanation models that can be compared with the prehistoric context of the Balikh Valley. Such a comparison will permit us to understand if the chronological, technological and social contexts of introduction are similar or not in both regions.

C. Altınbilek-Algül (✉)
Department of Prehistory, University of Istanbul, Istanbul, Turkey
e-mail: cileraltinbilek@gmail.com

L. Astruc
Du vilage à l'Etat au Proche et Moyen-Orient, Université Paris Ouest Nanterre, Université Paris I, UMR 7041, ArScan, France
e-mail: laurence.astruc@gmail.com

D. Binder
Université Nice Sophia-Antipolis, CNRS, UMR 6130, CEPAM, Nice, France
e-mail: didier.binder@cepam.cnrs.fr

J. Pelegrin
Laboratoire "Préhistoire et Technologie", CNRS et Université Paris Ouest Nanterre, MAE, Nanterre, France
e-mail: jacques.pelegrin@mae.u-paris10.fr

P.M. Desrosiers (ed.), *The Emergence of Pressure Blade Making: From Origin to Modern Experimentation*, DOI 10.1007/978-1-4614-2003-3_5,

More specifically, we focus on the first evidence for the early production of large obsidian blades using the pressure technique with a lever in Late Pre-Pottery Neolithic B (LPPNB) contexts from the site of Çayönü Tepesi and in the beginning of the Pottery Neolithic (PN) context from the site of Sabi Abyad I. This comparative study permits us to discuss different aspects of pressure technique including the existence of specialists.

5.2 The Tigris and Euphrates High Valleys and the Anatolian Plateau

5.2.1 The Spread of Pressure Blade Production in the Region

The archaeological sequence at Çayönü Tepesi (Fig. 5.1), one of the major Pre-Pottery Neolithic (PPN) settlements in the High Valleys of the Tigris and Euphrates Rivers, provides evidence of the introduction of pressure techniques for bladelet production during the Early Pre-Pottery Neolithic B (EPPNB), in the second half of the ninth millennium cal B.C. This massive introduction overrode, but did not suppress, the technical traditions that were popular during the previous stages. Thus, during the Pre-Pottery Neolithic A (PPNA) and the beginnings of the Early Pre-Pottery Neolithic B (EPPNB), most bladelets were produced locally using soft percussion; the lithic assemblage is completed by importations of blades obtained

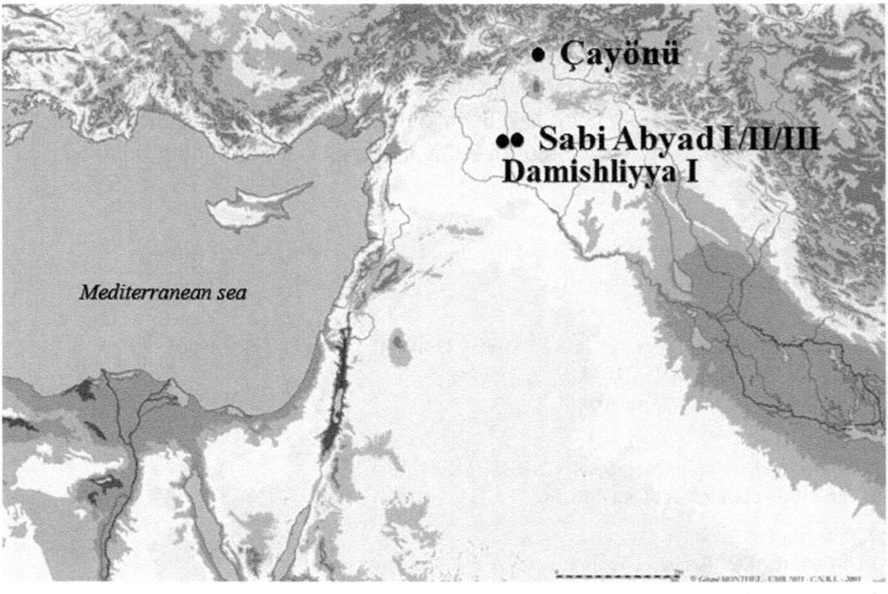

Fig. 5.1 Location of Çayönü Tepesi and Sabi Abyad settlements

from bidirectional or naviform cores, produced by direct percussion too. There is no evidence of the use of pressure to produce blades in the PPNA or PPNA-EPPNB transition occupations at Çayönü Tepesi which suggests that at this time, the High Valleys were linked, instead, to the lithic tradition of the Levantine Corridor (Binder 2008).

It was also during the ninth millennium cal B.C. (EPPNB) that pressure techniques appeared on the Anatolian plateau. The obsidian prismatic bladelet production from the Cappadocian workshops, in particular from the Göllüdağ outcrops, notably the well-known Kömürcü-Kaletepe workshop, spread throughout the whole Near East (Schillourokambos early phase A, Dja'dé, Mureybet, Tell Aïn El Kerkh) between 8700 and 8200 cal B.C. (Binder 2002, 2005; Binder and Balkan-Atlı 2001).

As previously suggested (Cauvin 1994; Inizan and Lechevallier 1994), the appearance of pressure blade production at Çayönü indicates links with Zagros, Caspian and Central Asia, as well as with Yubetsu-Gobi. As Inizan and Lechevallier argue, there is a geographical boundary between pressure and naviform use areas. However, the situation remains unclear for northern Anatolia and the Caucasus, where the early phases of the Neolithic or corresponding occupations are unknown. Despite some export of Cappadocian obsidian bladelets throughout the Levant at the beginning of the EPPNB (e.g. Dja'dé 3), several centuries before the appearance of blade production by pressure in Çayönü, the links between the pressure methods in use in Cappadocia and eastern Turkey are not clear. Radiocarbon dates are rare and often imprecise, and blade series are also uncommon or have been insufficiently studied.

5.2.2 Trends in Pressure Blade Production at Çayönü Tepesi

Çayönü displays a significant evolution of pressure methods from the middle of the ninth millennium cal B.C. until the adoption of ceramics in the beginning of the seventh millennium (Binder 2007). Pressure was used to produce blades from three types of raw material: (1) obsidian from Bingöl and Nemrut Dağ outcrops (100–150 km northeast), (2) local grainy flints that were exploited during the entire sequence by percussion and later by pressure, and (3) fine-grained flints imported as cores-on-flakes or as finished tools.

The Çayönü pressure technique is represented by the following evidence:

1. 'Channeled building sub-phase', dating from 8600 to 8200 cal B.C. (end of the Early Pre-Pottery Neolithic B [EPPNB] to beginning of the Middle Pre-Pottery Neolithic B [MPPNB])

   Excavations at building DI revealed (1) local microblade pressure production using imported cores-on-flakes and (2) wider central bladelets from obsidian or local flint. Obsidian accounts for about half of the blanks removed by pressure. The types of debitage produced by working flint and obsidian are similar: semi-conical core shapes, with high or very high transversal convexity, removed in small sequential series. Pressure platforms are orthogonal to the surface and systematically facetted; microblade butts are generally overhanging. Exhausted cores are bullet-shaped and often exhibit a residue of the inferior face on the core-flake. Blade widths are bimodal: bladelets produced from local flint are more than 8 mm wide; microblades produced from imported core-flakes are between 4 and 8 mm wide. Obsidian bladelets are between 4 and 15 mm wide and follow the same distribution pattern as the flints. These features indicate that the pressure detachment was done partly by hand for the microblades and partly with the use of a short crutch while sitting for the bladelets (Pelegrin 1988, 2003, this volume). Some of the characteristics observed on the proximal parts of these blanks, such as the overhang, the marked bulbs and the small platforms, could suggest the use of native copper pressure flakers (Binder 2007, 2008).

2. 'Cobble-paved building sub-phase', dating from 8250 to 7650 cal B.C. (MPPNB)

   Blades produced by pressure represent about half of the blades and bladelets recovered from the building CM series. The flint debitage resembles that from the obsidian from the Channeled building sub-phase and represents about one-third of the pressure blanks. Two-thirds of the flint pressure bladelets are wider than 8 mm. Most of the obsidian bladelets are wider than 8 mm; they were probably flaked in situ from cores with flat platforms, similar to Kaletepe P and Cafer (lower deposits) items, respectively dated to approximately 8300–8200 and 8250–7850 cal B.C. (Binder 2007, 2008).

3. 'Cell building sub-phase', dating from ca. 7500 to 7250 cal B.C. (Late PPNB)

   The CF and DS series show a major reduction in the quantity of flint blades produced by the pressure technique. Heat treatment of flint is evident but seems to be very marginal. Obsidian cores are shaped in situ; they have a low transversal convexity; there are few microblades; and the blade production is primarily represented by parallel and regular pieces. The platforms are flat and inclined or steeply inclined. The blades were probably produced with the use of a short crutch (Pelegrin 1988, 2003, this volume).

4. 'Large room building sub-phase', dating from ca. 7300 to 6750 cal B.C. (Final PPNB)

   The BF building assemblages provide abundant evidence of the production of obsidian bladelets produced by pressure (80% of the blade total) which are very regular (75% with extraction designs 212'; cf. Binder 1984; Binder and Collina, this volume). Compared to the cell building sub-phases, the widths of the bladelets have significantly increased (ca. 12 mm, based on proximal fragments and whole blades). Microblades disappear during this phase. A single, wide obsidian blade was identified in the sample studied by Binder. Flint was still produced by pressure during this sub-phase but in low proportions. The presence of a flint conical core with a faceted platform may illustrate the beginning of diversification in pressure production techniques.

5. Pottery Neolithic phase(s)

   Early Pottery Neolithic contexts in Çayönü are difficult to assign to a chronological period, and radiocarbon dates are not available. The analysis of an assemblage collected from an architectural complex in Trench P25G allows us to identify components for these phases representing obsidian tool production that are similar to one from the large room building sub-phase. During this phase, the production of obsidian blades by pressure became common (Algül 2008).

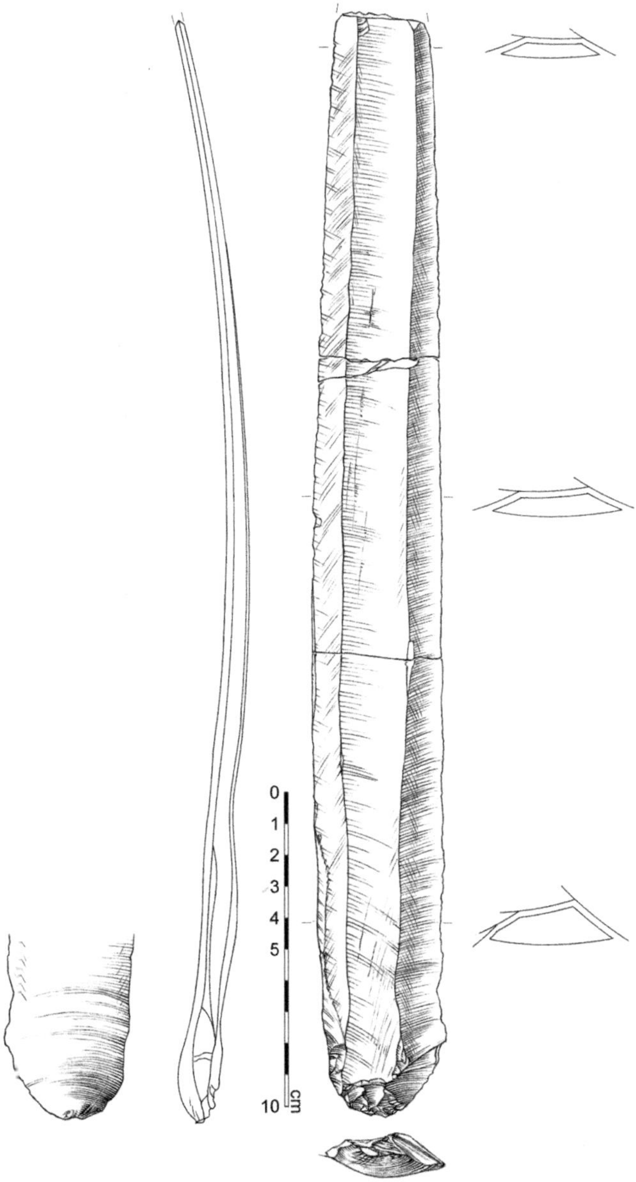

Fig. 5.2 Large blade from Çayönü Tepesi, CV building, Cell Building sub-phase 3 (Late PPNB)

In summary, Çayönü pressure debitage exhibits three trends: (1) an increasingly greater reliance on obsidian compared to flint through time, (2) a constant increase in the width of central bladelets and (3) a transition from a semi-conical type of removal sequence with faceted orthogonal platforms to a more frontal type with flat inclined platforms, which resulted in a more standardized product.

At the end of the PPN, large blades began to be produced (Figs. 5.2, 5.3, 5.4). Fourteen of these large blades have been identified from the cell building sub-phase

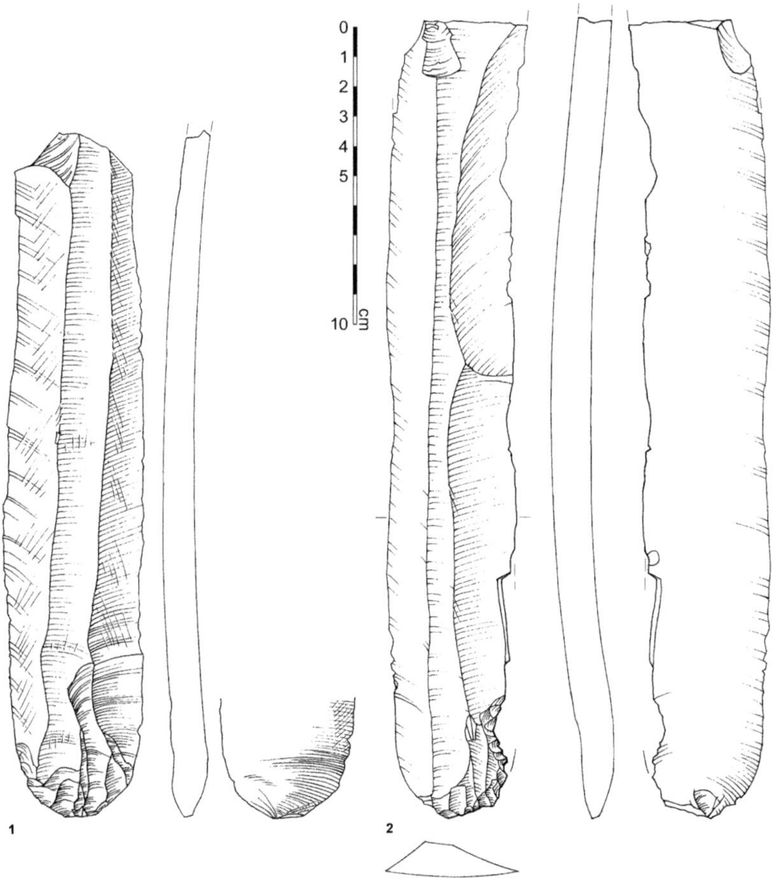

Fig. 5.3 Large blades from Çayönü Tepesi. *1* CV Building, Cell Building sub-phase 3 (Late PPNB), *2* Cell or Large Room Building sub-phase

to the Pottery Neolithic phase. This is a preliminary count based on a sample of the whole assemblage collected from Çayönü: eight blades in the cell building sub-phase (Buildings DE, CV, CE and CY, stages c2 and c3 / Late PPNB); four from the large room (Building BF, stage lr1 / Final PPNB); one from either the cell or large room building sub-phase (18 M, open area); and one fragment from the Pottery Neolithic phase. Among the 14 large blades, four are Çayönü tools, four are Çayönü tools which have been recycled as end scrapers, four are end scrapers and five are unretouched blades with traces of wear. Currently, the large pressure blades are well situated within the Çayönü sequence, dating from the second part of the eighth millennium cal B.C. and at the beginning of the seventh millennium, with a maximum date range of between 7340 and 7080 cal B.C.

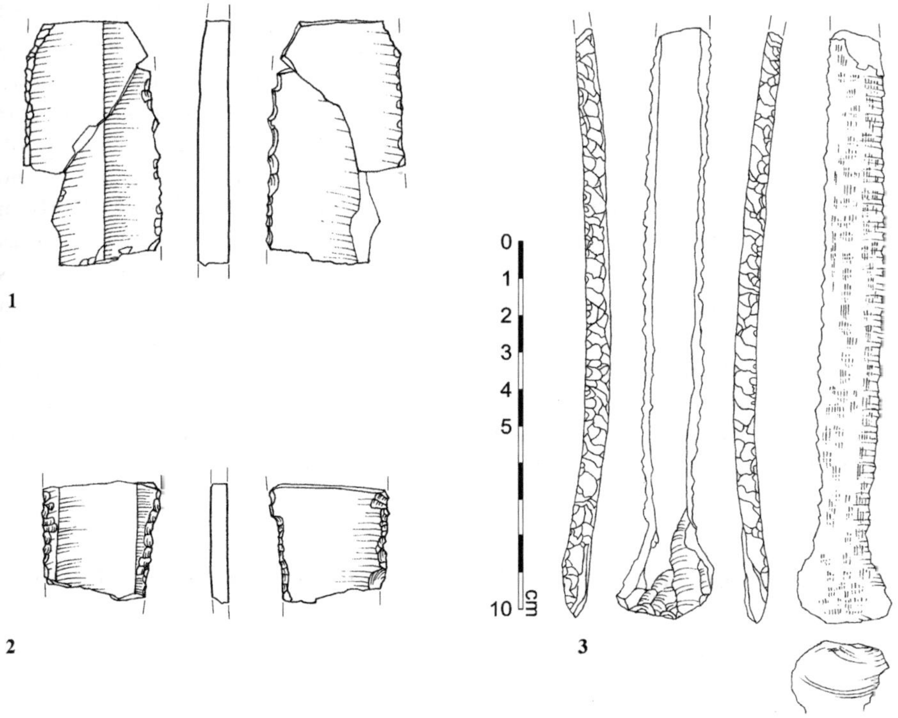

Fig. 5.4 1. Medial fragment of a light large blade with edges damaged, DE building, Cell Building sub-phase 2 (Late PPNB), 2. Blade with bi-lateral retouch, Pottery Neolithic, 3. Çayönü tool, BF building, Large Room Building subphase (Final PPNB) (After Caneva 1994)

5.3 The Balikh Valley and Northern Mesopotamia

5.3.1 Pressure Blade Production in the Region

The introduction and development of pressure blade production in the Balikh Valley from 8500 to 6200 cal B.C. is well documented at four neighbouring sites (Fig. 5.1): Sabi Abyad II (mainly Middle and Late PPNB, and PN) (Verhoeven and Akkermans 2000), Sabi Abyad I (operations 1-2-3, from Early PN to Early Halaf), and to a lesser extent at Sabi Abyad III (Late PPNB/Early PN levels) and Damishliyya I (Late PPNB and PN) (Akkermans 1988). Both excavations and technological studies are in progress for Sabi Abyad I (operation 3) and Sabi Abyad III, and detailed data are not presently available. The situation is rather different from that of Çayönü Tepesi, as pressure technique is clearly evident in the Balikh Valley only from the Middle PPNB onwards and only for obsidian. The obsidian originates in eastern Anatolia, specifically the Bingöl and Nemrut Dağ areas (Astruc et al. 2007; Cauvin et al. 1998). The homogeneous nature of the pressure-flaked materials is striking: they include rectilinear blades or bladelets from cores bearing plane or facetted orthogonal platforms and truncated bases, although variations in the size of the blanks do occur over time.

This specific tradition of obsidian blades produced by pressure is different from the pressure techniques associated with Cappadocian obsidians (as evidenced at Kömürcü-Kaletepe on the Gollü Dağ, EPPNB). It is the main tradition for this period (Late PPN and PN) in northern Mesopotamia, notably east in the Khabur Valley at Tell Sekher el Aheimar and at Kashkashok II (Nishiaki 2000), in the Sinjar at Tell Maghzaliah (Bader 1989), and possibly further south, at Bouqras (Roodenberg 1986). At these sites, pressure-flaked eastern Anatolian obsidian blades were introduced as blanks and, in some occupation levels, represent more than 60% of the assemblages. Bipolar blade production is absent or rare for obsidian and limited for flint. Future studies of these northern Mesopotamian assemblages will focus on the diachronic variations between these three microregions and will allow us to better relate them to the High Valleys, where both pressure technique and large blade production predate the northern Mesopotamian tradition.

5.3.2 Trends in Pressure Blade Production in the Balikh Valley

5.3.2.1 Sabi Abyad II and Damishliyya I (End of Middle PPNB to Early PN)

Pressure bladelets made of obsidian are well represented in Sabi Abyad II and Damishliyya I (Copeland 2000; Nishiaki 2000), making up, respectively, 55% and from 6% to 16% of the two assemblages. The blades are very regular and standardized in form, with widths ranging from 5 to 15 mm and thicknesses of 1–3 mm. They are made from semi-cylindrical or cylindrical cores with plane or facetted orthogonal pressure platforms. The bases of the preforms are truncated, as demonstrated by the quadrangular morphology of most of the distal parts of the bladelets. Careful study of a cluster of 21 blades in occupation levels attributed to the end of Middle PPNB from Sabi Abyad II allows us to reconstruct the method of production and to argue for the introduction into the settlement of a parcel of bladelets originally produced at, or close, to the obsidian sources (chemically identified as located in the regions of Bingöl and Nemrut Dağ in eastern Anatolia, 250–300 km to the northeast; Astruc et al. 2007).

At the present stage of our research, there is no evidence that the pressure technique was used locally by the inhabitants of the Balikh Valley to produce blanks but rather that obsidian blades detached by pressure were introduced into the village as finished products through regional exchange networks (Astruc et al. op. cit.). The parcels of blades introduced to the Balikh communities were stored in domestic spaces, exchanged with neighbours and used locally. This is a key difference with Çayönü, where we argue that the inhabitants, themselves, produced blades using the pressure technique.

5.3.2.2 Sabi Abyad I (Early PN to Early Halaf)

The end of the Late PPNB and the Early PN are currently under study at Sabi Abyad III and Sabi Abyad I (operation 3). The pressure technique and the method of production associated with it are well represented in the Early PN layers at these two sites and at Sabi Abyad II and Damishliyya I. This method of production persists throughout the sequence until Early Halaf and is the dominant technique used to produce obsidian tools. Chronological variation in the amount of obsidian introduced in these living spaces and in the types of obsidian tools produced are apparent, especially based on the results of the 2007 and 2008 excavation seasons.

In 2005, a nearly complete large blade (Fig. 5.5) was recovered from the courtyard of a storage building (Sabi Abyad I, operation 2, Astruc 2011 approximately 6200 cal B.C.). In the neighbouring open space, six fragments of large blades were also discovered (Fig. 5.6). These fragments belong to the typological group of side-blow blade-flakes (SBBF) or side-blow blade-flake cores (Braidwood 1960). They are, in fact, related to a very specific technique of breaking blanks by using percussion on an anvil (Nishiaki 1996). SBBF were recognized in Kashkashok II, Sekher-el-Aheimar and Sabi Abyad I as by-products of this technique. Wear patterns representing different activities occur on every specimen anterior to the intentional breakage or truncation of the blank. The SBBF technique is therefore a technique of rejuvenation and/or a technique of calibration of the blanks in the longitudinal axis, like a truncation for instance. Complementary to this, from Sabi Abyad I, operation 3, three fragments of large blades were recovered (Fig. 5.7).

The large blades from the Sabi Abyad I site are mainly produced from obsidian of a high quality, which is green and translucent except for one specimen, which is made from an opaque and bedded raw material that has a slightly rougher 'touch' and grain. Similar high-quality obsidian served as raw material for the large blades found at Çayönü. Despite the degree of fragmentation of the blades and the absence of proximal fragments, the identification of the pressure technique of production using a lever is obvious. Our objective is to describe the evidence for this type of blade production which has not been sufficiently recognized in previously studied collections largely because of the fragmented state of the specimens.

In the Balikh Valley, there is a striking continuity in raw material procurement and in the mode of preparation for the detachment and production of obsidian blade(let)s from 7500 to 6200 B.C. (cal). Very large blades are also produced by the pressure technique, with the use of a lever. This particular technique was clearly in use by 6100 B.C., based on the find of seven fragments, as well as several specimens from Sabi Abyad I, operation 1, dating to 6200 B.C. (Copeland 1989), and three fragments from Sabi Abyad I, operation 3, ca. 6100–6500 cal B.C. Although the initial introduction of large blades within the Sabi Abyad sequence is still in question, excavations at Sabi Abyad III bring more evidence of the way that the Balikh communities became integrated into the obsidian trade networks. Current hypotheses on obsidian production, exchange and use will be evaluated in order to understand the nature of the specializations and the structure of the networks. Variation in the relative use of flint and obsidian, for example, does not seem to

Fig. 5.5 Large blade from Sabi Abyad I, operation 2, proximal end truncated with the SBBF technique

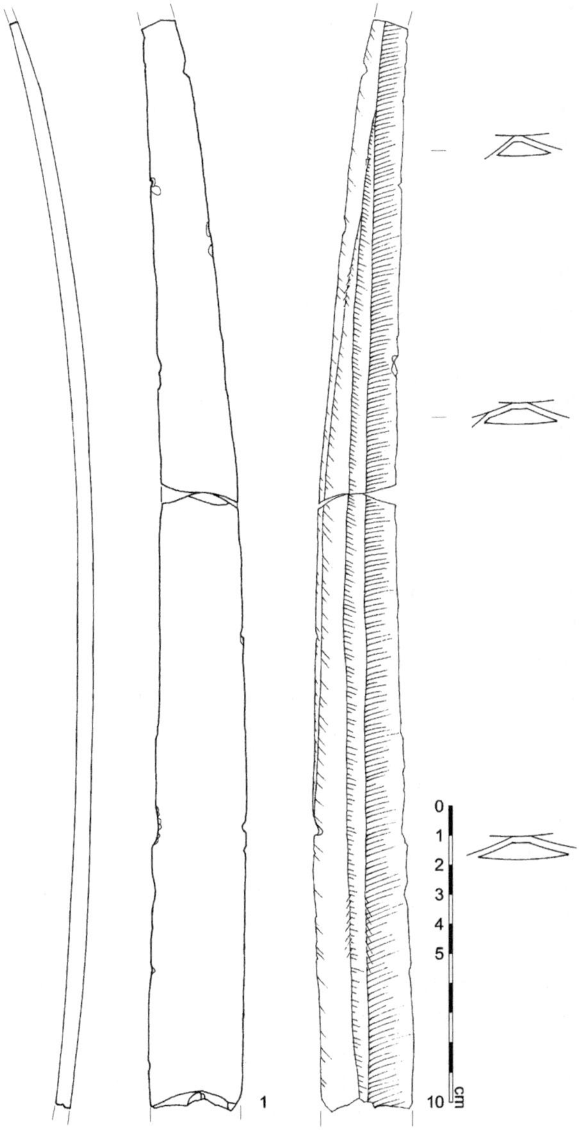

follow a linear evolution. Similarly, the presence of very large blades is not necessarily linked to a constant increase in blade widths through time. Finally, the main changes in the sequence occur not at the end of PPN, as at Çayönü Tepesi, but during the subsequent 'Initial PN' (following Nieuwenhuyse's terminology) or Early PN period, with a diversification in the size and the nature of the products.

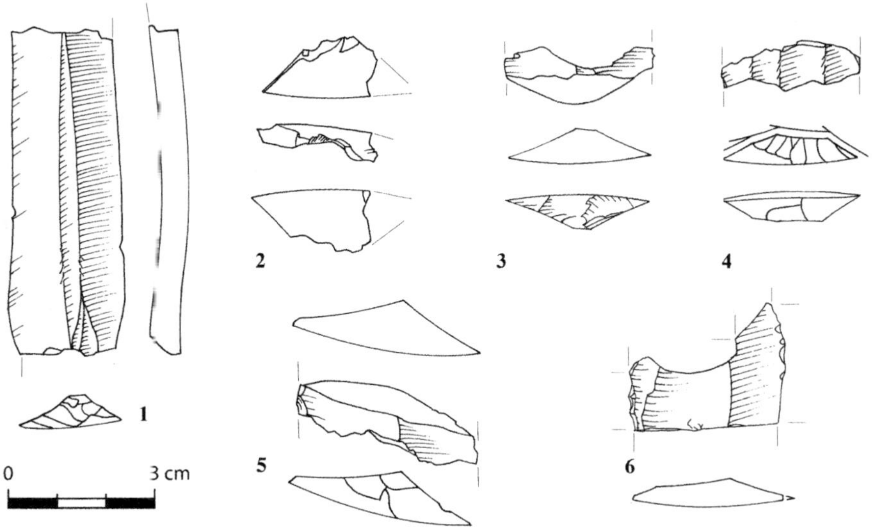

Fig. 5.6 Fragments of large blades from Sabi Abyad I operation 2. *1, 4* Truncated large blades. *2, 3, 5* Side-blow-blade flakes. *6* Mesial fragment of a large blade

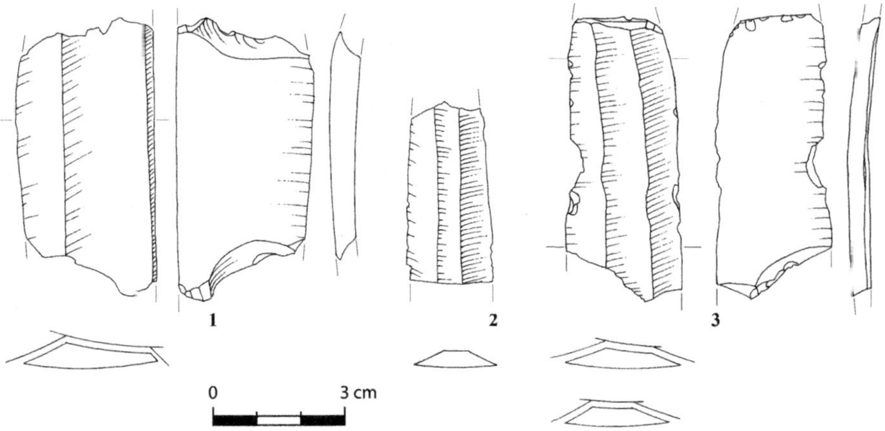

Fig. 5.7 Fragments of large blades from Sabi Abyad I, operation 3. *1* SBBF core. *2, 3* Mesial fragments of large blades

5.4 Technological Analysis of the Large Blades

5.4.1 Description of the Archaeological Specimens

Six blade fragments were recovered from Çayönü Tepesi: one nearly complete blade, three fragments of large and regular obsidian blades, and two mesial fragments (Figs. 5.2, 5.3, 5.4, 5.8):

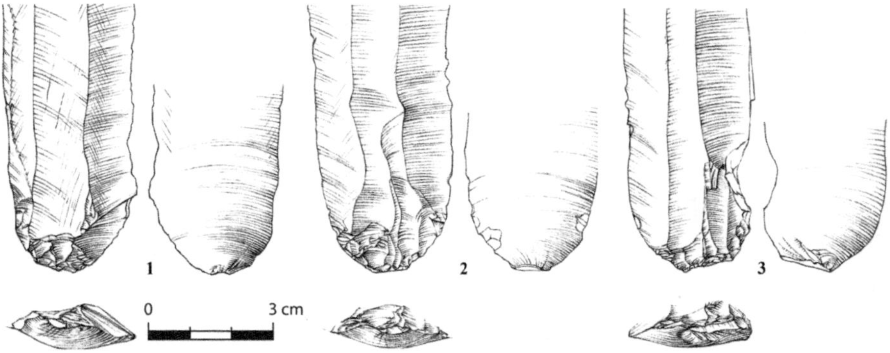

Fig. 5.8 Detailed views of the proximal portions of three large blades from Çayönü Tepesi (Figs. 5.2, 5.3)

- A nearly complete blade, 28.5 cm long, was found broken in three main fragments (Fig. 5.2, ÇT S3-1 CV building/cell 3). The mesial and distal fragments of the blade have been previously described (Binder 2005: Fig. 5), but the proximal part was only recently identified and refitted. The original length of the blade may have been as much as 33 cm, if we estimate that 5 cm are missing from the present distal end, which is uncurved and measures 24 mm in width and 3.8 mm in thickness. The proximal section is 31.9 mm wide and 8.4 mm thick, and the mesial section is 29.5 by 5 mm. The regularity of the blade's edges and arrises (Inizan et al. 1995) is impressive, and its thinness slightly decreases towards the distal end. The profile is moderately curved without inflexion or undulation. These characteristics are consistent with pressure blade production using a lever. The blade is four-sided in the proximal section, the code of *débitage* being 4321 (Binder 1984), but it becomes trapezoidal and asymmetric (321) in the medial section. A long scar measuring more than 68 mm long, together with rather hinged short scars, is evidence of core preparation to detach this large blade (Fig. 5.8(1)). The butt is small, ovoid and dihedral (7.8 mm wide and 2.1 mm thick). The butt shows a tiny inclination to the left edge, and its edge angle is greater than 90° (approximately 95°). The pressure point is located on the dihedral, defined by two tiny flake scars. The lip is clearly developed, and the absence of any cracking or damage suggests the use of a pressure stick armed with an antler point to detach the blade.
- A large proximal fragment, 17.2 cm long, comes from a blade that was probably 20–25 cm in length (Fig. 5.3(1), ÇT70 R2-10/4 CV building/cell 3). It is as wide as the previously described specimen (32 mm) and somewhat thicker (7.8 mm under the bulb, decreasing regularly to 6.4 mm at its mesial break). Its ventral face is perfectly regular, without any undulation, and the profile is almost straight (Fig. 5.8(2)). The butt is small (8.8 mm wide and 2.8 mm thick), with an oval and slightly concave surface that bears two tiny flake scars, probably produced by pressure, giving a platform angle of 80°. The detachment of the blade was

prepared by tiny bladelet-like removals from the core front towards the face of the core, which reduced the overhang and isolated the point of compression. Under a clear lip, the bulb is thick and high with a little concavity under the bulb. A clear ripple is visible just under the bulb, 22 mm under the lip, as well as in the bulb negative of two of the three blade scars on the dorsal face (right and middle). From our experimental reproduction of pressure blade production using a lever, this kind of ripple, frequent but not constant, is due to micromovements of the core in its wooden device when building up the full pressure that is detaching the blade (Pelegrin, this volume). Based on its width and thickness, this blade was detached by pressure using a lever. The clear lip and absence of cracking on the butt indicates the use of an organic pressure point (Fig. 5.8).

- Another long (20.5 cm) proximal fragment comes from a large obsidian blade that was probably at least 25 cm long before it broke (Fig. 5.3(2), ÇT 18 M 1–20/ cell or large room building sub-phases). This blade appears to have been detached after a previous and unsuccessful attempt at its right side leaving a hinge at 11.5 cm from the top; hinge that was prolonged by a rippling splinter. However, the previous blade scars to the central and left side were regular and helped to correctly guide the blade, which has a uniform shape 30 mm+/−1 mm wide and 10 mm thick, and very discrete undulations of the profile. Prepared in the same way than the preceding described blade, the butt slants 10° laterally with an 80° platform angle (Fig. 5.8(3)). It is 13.5 mm wide and 3.2 mm thick and asymmetrical, the fracture initiating at its higher, left corner with no visible crack: a crack would indicate the use of a hard, metallic material for the pressure flaker. The bulb is rather prominent but without any concavity under the bulb and bears a ripple 16 mm beneath the lip. These features indicate a lever pressure detachment, probably with an organic point.
- The fourth piece recovered from the cell period (ÇT 84 18 M 3–6, related to DE building/c2) is a short mesial fragment of a light, large blade (Fig. 5.4(1)). The edges are damaged, but the initial width can be estimated as 32 mm, and the thickness is 6.2 mm. The regularity of the scars is very high with a straight profile, showing that the original blade was detached by lever pressure.
- A large 'Çayönü tool' previously described by Caneva et al. (1994) appears to have been made from a light, large blade similar to the one just described (Fig. 5.3, ÇT 70 U 3–0, related to BF building/Lr1). It comes from the subsequent 'large room' phase but helps to reconsider the blade blanks from earlier 'Çayönü tools' from the 'cell' phase and to understand one of the functions of large blades in this archaeological context. This tool, 12.3 cm long with a missing distal portion, was made from a large blade that might have reached about 20 cm in length, based on the existing profile. The original width of at least 25 mm has been significantly reduced by steep retouch (the initial arris to the right is totally removed), except at the proximal end which is less modified and 22 mm wide. The regularity of the blade blank is very high, with sides lacking any undulations and a regular thickness (5.2 mm under the bulb, 5.8 mm halfway, 4.3 mm at the distal break). The profile is slightly curved, a little more in the proximal portion. The butt was prepared with tiny axial removals and is thin

(8 mm wide and 1.5 mm thick), with an 80° platform angle. A marked ripple lies on the bulb 12 mm below the butt. The original blade was detached by pressure, and the remaining section is just wide enough to suggest that it was detached using a lever.

- A short mesial fragment (2.5 cm long) (Fig. 5.4(2), ÇT'89 P24I 5-28/5-29) comes from the 'Pottery Neolithic' layer, a much later occupation than the previous levels. The extreme regularity of the two upper arrises and the blade's thinness (3.2 mm) testify to the detachment of the blade by pressure. Presently 25 mm wide, the blade was originally 27 or 28 mm in width before it was retouched, an indication that it was made using the pressure lever technique.

From Sabi Abyad, ten fragments of long blades were recovered. These include seven fragments found in the courtyard of a burned building (operation 2, V6 sector, around 6100 cal B.C.[1]) and three specimens found in Sabi Abyad I, operation 3 (sectors I03, E03, E04), which provide evidence that this technique was in use by 6500 cal. BC. (Figs. 5.5, 5.6):

- Two long fragments were refitted to reconstitute a nearly complete blade (Fig. 5.5), which has its proximal end truncated just under the bulb and its distal point missing (possibly lost during production). The present length is 28.6 cm but was probably 2 cm longer at the proximal end and 1.5 cm longer at the distal end for an original length of approximately 32 cm. The blade was detached from a core that may, itself, have been 34 or 35 cm long, considering that two of the previous blade scars were a little longer than the blade itself, and that the core platform was probably somewhat reduced during earlier blade removals. Considering the profile and the arrises of the blade, the core front was an elongated and slightly convex bullet-shape. At that stage of the core reduction, the slightly convex profile and the regularity and thinness of the blade (from 4.8 mm thick at the proximal end to 4.3 mm at a few cm from the distal end) demonstrate a well-mastered pressure blade technique. The use of a lever is probable because this 24-mm-wide blade was necessarily preceded by the removal of wider blades in order to 'open' and regularize the production surface of the core (this opening included at least one or two crested blades and several lateral, under-crested blade). The section of the blade is initially trapezoidal and symmetric (212′) and then becomes slightly asymmetrical with an adjacent fourth lateral blade scar. The blade was probably a central blade belonging from at least the third series of blades detached from the core.
- One fragment represents the proximal portion of a large blade truncated at the bulb, the medial break resulting from a snap with a ventral tongue (Fig. 5.6(1)). The piece is 6.7 cm long, 24–21 mm wide, and 6.3 mm thick. The profile is rather

[1] The detailed study of the stratigraphy and its correlation to a new set of radiocarbon dates from tell Sabi Abyad I, operation 3, is in progress. The approximate dates provided here are therefore preliminary.

curved but very regular, as are the edges and converging arrises. The blade appears to have been detached by pressure using a lever. The blade material is a green but slightly grainy and bedded variety of obsidian, similar in appearance to one of the blades from Çayönü (Fig. 5.3(1)).

- Five fragments of large blades (Fig. 5.6(2–6)) are made from the same variety of green translucent obsidian as that of the blade illustrated in Fig. 5.5, and they originate from the same V6 sector of Sabi Abyad. Four of the pieces are 'side-blow blade-flakes'. The fifth piece is a damaged fragment of a blade. The fragments provide an estimation of the size of the section of the original blade they come from:

Figure 5.6(2) A 37.5-mm-wide and 6-mm-thick triangular section that may indicate an 'early' blade

Figure 5.6(3) An estimated width of 32 mm and a thickness of 6 mm, probably from a triangular section blade (different from the preceding Fig. 5.6(2))

Figure 5.6(4) A 31-mm-wide and 7-mm-thick symmetrical trapezoidal section

Figure 5.6(5) A 28-mm-wide and 6.6-mm-thick prismatic section with three arrises

Figure 5.6(6) An asymmetrical trapezoidal section with an estimated width of 34 mm and a thickness of 6 mm

Although the detachment technique of the original blades cannot be ascertained from these fragments, each of them lies in the range of blades produced by lever pressure.

From Sabi Abyad I, operation 3, three fragments of large blades were recovered (Fig. 5.7):

- Found in an open area of sector I03 (6250/6200–6050 cal B.C., Fig. 5.7(1)), one fragment comes from a large blade and is truncated by an inverse notch and snapped at its distal end. From the mesial to the distal end, the section decreases in size from 26 mm by 5.8 mm to 23 mm by 4 mm, with an increasing curvature, indicating that it comes from the distal half of the blade blank. A slight undulation of the dorsal side and arrises is mirrored on the ventral side. The overall regularity and slight curvature testify to a pressure technique, very probably with a lever, given that the mesial section of the blade is larger, about 28–30 mm. From an open area in sector E04 (6550–6500 cal B.C., Fig. 5.7(2)), a mesial fragment 4 cm long with a distal inverse notch comes from the distal half of a large blade (the width decreases from 20 to 17 mm, and it is 4 mm thick). The regularity and symmetry of the section suggest that the fragment is that of a central blade detached from a very well-treated pressure core, possibly using a lever.
- In sector E03 (6750–6600 cal B.C., Fig. 5.7(3)), a mesial fragment of a very regular obsidian blade was recovered. Truncated by an inverse notch at both ends, it is 6.3 cm long, 31.4 mm wide, and 6 mm thick. The remarkable regularity of the edges, the arrises and the ventral side, and the wide width of the blank suggest that the blade blank was detached by pressure using a lever.

5.4.2 *Blade Production Using the Pressure Technique with a Lever*

The detachment of blades by the pressure technique is characterized by regularity, reduced curvature and thinness (Pelegrin 1988: 48; 2003: 63; Tixier 1984: 66). Indeed, the mechanical conditions of a pressure technique, immobilization of the core, permanence of the compression along the fracture propagation and absence of shock, which would generate vibrations and therefore undulations, are the only means of detaching such a regular and fragile column of volcanic glass. The blades presented here bear the scars of two to four previous removals that are also highly regular, implying a very controlled and repeatable mechanism of detachment.

We made careful experiments on obsidian both using indirect percussion and pressure (standing pressure and pressure with a lever), and this after years of experience of these techniques with flint as a raw material (Pelegrin 2002a). Obsidian blades can be detached in series using indirect percussion, but they are far to be as regular as pressure blades (Pelegrin 2000, 2003, 2006, this volume). In this respect, we fully share Crabtree's opinion (1968: 459) that 'the impact from the percussor causes excessive undulations and waves on both the core and blade; the dimensions of the blade cannot be controlled with regularity; the bulbs of force are much too large, and the curve of the blades and termination of the ends cannot be controlled' (see also Figs. 5.4, 5.8). In addition, the fragility of obsidian leads to a high rate of proximal breaks when trying to produce relatively thin blades. These proximal breaks, which are rarely produced by pressure detachment, occur even more frequently with the use of indirect percussion than with direct percussion. They clearly occur during the detachment itself (and not after, as do simple medial breaks) because they produce distal ripples and hinged termination of the blade, thus spoiling the regularity of the distal end of the core. The extreme sensitivity of obsidian to breakage explains why, beyond 12–15 cm in length, irregularity of curvature and termination seems inevitable, even when using an elastic support for the core (which has a regulating effect on the detachment of flint blades) (Pelegrin 2000, 2002b, 2003).

There are two practical ways to produce large blades by pressure: using the full weight of the body transmitted by a crutch in a standing position and using a lever. During a recent colloquium held at Pennsylvania State University (Hirth 2003), some of the most experienced specialists agreed that more than length, it is the width of a blade that is dependent on the force of the pressure, as Crabtree (1968: 468) stated: 'the wider the blade, the greater the amount of pressure that is required'. In working flint, for example, the maximum width of pressure blades detached using a relatively long crutch placed at belt level by a person in a standing position can reach about 20 mm when using an organic (antler) pressure point and even 21 or 22 mm when using a copper pressure point (harder than antler, copper helps to detach thicker butts). Blades with these maximum widths have been observed in different archaeological contexts (Pelegrin, this volume). With obsidian, our attempts at using the standing pressure crutch technique produced blades with widths of up to 26 mm, confirming an earlier observation of ours that obsidian could yield blades which were 30% wider than flint, using an identical technique and level of effort (Pelegrin 1988, see also Kelterborn, this volume). Crabtree (1968: 468) concluded that the maximum size of the obsidian blades that he could produce using his standing technique was '1 in. wide and 8 in. long', while the 'Mexica' technique reconstructed by Clark and replicated by Titmus could be used to detach blades up to 24 mm wide (Titmus and Clark 2003). The width of the almost complete blade from Çayonü (Fig. 5.2; 31.9 mm) is clearly larger than that which can be achieved with the standing pressure technique; a more powerful device had to be used to detach the blade, one which involved the use of a lever, such as the one we used in our experiments (Pelegrin, this volume).

Our analysis of the proximal portions of four large obsidian blades found at Çayonü Tepesi (Figs. 5.4(3), 5.8) indicates that the point used for their detachment was probably made of an organic material. Three of the detachment butts are ovoid and plane, the fourth one is ovoid and dihedral; the clear lips and the absence of cracks on the butts favour an organic point, probably antler (experiments from Pelegrin in Astruc et al. 2007). This is even more apparent for the fourth butt: the point of pressure is located on the dihedral which did not suffer of any damage that would be caused by a copper point. At Sabi Abyad I, a proximal fragment of a large blade has been found at the surface of the Tell, in the operation 3 area. Its ovoid, plane butt is similar to those of Çayonü Tepesi's large blades.

5.5 Discussion

The analysis of the large blades of Çayönü Tepesi and Sabi Abyad I brings a new perspective to lithic specialization within Neolithic communities in the Near East. The production of large blades using a lever occurred as early as the second half of the eighth millennium cal B.C. at Çayonü Tepesi, likely between 7340 and 7080 cal B.C. This is the earliest evidence of this remarkable technique. It was thus testified in the Balikh Valley a thousand years later, between 6100 and 6500 cal. B.C.

5.5.1 The Degree of Production Specialization

The production of large obsidian blades demonstrates a remarkable level of technical specialization for these early periods. Pressure detachment with a lever was a technique likely practised by a few highly qualified specialists, who were possibly already fully trained in the standing pressure technique. To carry out this type of blade production, successive choices had to be made in order to reach the optimal exploitation of both raw material and technical investment and to avoid accidents that would lead to the waste of several blades or of the entire core. Risk levels associated with the various techniques would have been under constant evaluation, and substantial experience in pressure blade production would have been necessary to develop and control the whole production system, to manufacture the tools and to control the numerous practical details or adjustments.

Experimental research (Pelegrin 1988) has shown that the technical knowledge needed to produce medium-sized blades by standing pressure is considerable. However, the necessary expertise is much greater when the goal is to produce a standardized series of long blades. At both sites, the lengths of the nearly complete blades – 27.2 cm at Çayönü Tepesi and 28.6 cm at Sabi Abyad – allow us to estimate the length of the original cores as 32 cm or more. A very high level of understanding of the mechanical properties of obsidian is necessary to shape such huge cores and to produce these large, wide blades.

The initial core preparation has to be of very high quality, as any irregularity on the production surface will have a direct effect on the regularity of the ventral surfaces and edges of the blades. Once the critical roughing out by stone percussion is finished (no deep or hinged scars are allowed), the next stage is a patient shaping using direct stone percussion or indirect percussion for the detachment of transversal flakes, alternating from three to four axial crests; then the detachment of several large covering flakes by direct percussion, using a hard wood hammer, and, finally, shaping the crests by a subtle direct or indirect percussion or even by pressure flaking. The goal is to correct the volume that will be transformed into blades by defining the convenient convexities and avoiding any deviation – bumps or hollows – from an ideal of ±2 mm. Experimental reproduction by J. Pelegrin has shown that crested or under-crested blades (the first series of blades which serve to remove the pre-shaped surface of the core) can tolerate such irregularities if they are broad and thick enough, without reproducing these irregularities on their scar or without becoming hinged.

Difficult choices also have to be made when conducting the subsequent blade removal. The repartition of arrises on the core has to be strictly controlled, leading to different possible rhythms of *débitage* (convergent, divergent, inserted and adjacent unidirectional or alternating) (Astruc et al. 2007). In this respect, it is crucial to realize that each blade detachment is anticipated not only to visualize the final product but to control the effect of its removal on the geometry of the core. This requires meticulous attention to the preparation of each detachment not only to avoid accidents such as edge crushing, hinging and excessive plunging but to actually detach the expected blade with the most precision.

5.5.2 From Producers to Users

At Çayönü Tepesi and Sabi Abyad I, these large blades represent the highest recognized degree of specialization in lithic technology, attesting to a production technique that remained constant from the second part of the eighth millennium to the seventh millennium cal B.C. The large blades from Çayönü Tepesi and Sabi Abyad are so similar that they could have been made by the same craftsmen and remind us that the specialists involved in this type of blade production were part of a common technical tradition, which was transmitted through time by way of apprenticeships in the acquisition of the raw material and in the technical knowledge of production.

At Çayönü, it is difficult to estimate the relative importance of the large blades until further excavations are completed. At Sabi Abyad I, operation 2, the assemblage recovered from the burned building and its adjacent open areas reveals that these products represented a small proportion of the obsidian blades collected. For both Çayönü Tepesi and Sabi Abyad I, no evidence of the in situ production of large blades has been identified. Instead, these blades appear to have been introduced to the settlements as finished products. With the aid of experiments providing quantitative data, Pelegrin (Astruc 2007) determined that a core that is 12–15 cm in width and shaped with three axial crests can potentially produce up to 70–80 blades, of which 50 would be first choice blades (among which 80% are with a symmetric trapezoidal section, code 212′). The time input, according to Pelegrin, can be estimated as 2–3 h for shaping the core and 3–4 h for reducing it into blades. For larger blades produced by a pressure technique with a lever, these figures can be reduced to 20–30 blades per core produced within a full day of work. That means that a few specialists having easy access to obsidian and/or working seasonally on the outcrops could each produce several hundreds of large blades per year. One or few little groups of such knappers could therefore be at the origin of a direct or indirect diffusion on a large geographical scale. These large blades were exchanged within the obsidian trade networks from eastern Anatolian sources to Upper Mesopotamia: located in the High Valleys, Çayönü Tepesi lies 80 km from Bingöl and 250 km from the Nemrut Dağ area, while Sabi Abyad I is located in the Balikh Valley some 300 km from both sources.

Large obsidian blades are rare in both assemblages. Although both the functional patterns and the tool curation are different at Çayonü Tepesi and Sabi Abyad I (in the former site, the typology of the large blades includes notably Çayönü tools and scrapers, in the latter, the typological range is limited to SBBF and truncated blades (Algül 2008)), these tools do not appear to be related to specific activities or technical operations. Instead, their use seems embedded in everyday life with no special attention or treatment accorded to them. They are not found in caches, in funerary or symbolic contexts or in any other specific situations. While the size and quality of the products may reflect technological experimentation by the producers, these remarkable blades were most probably manufactured to be used in social contexts, including inter-community exchanges. They represent a great deal in terms of values, emulation and social image, but they do not seem to be related to rituals that could be the basis of long-distance diffusion of socially valorised objects (Pétrequin et al. 2006).

5.5.3 Concerning the Historical Aspects

The evolution of the pressure technique to detach blades within the High Valleys of the Near East between 8500 and 6000 cal B.C. has been interpreted by Binder (2007) as representing a long-standing tradition of craftsmen progressively exploring all the technical possibilities offered by the pressure technique, a behaviour that was directed by social demands, including the use of these products as status and/or identity markers.

For 1,500 years, eastern Turkey was a centre of highly specialized lithic production and the head of a trade network which covered a large part of the Near East. During this period, the Cappadocian workshops, very active during the ninth millennium, saw their influence wane considerably from the beginning of the eighth millennium cal B.C. to the middle of the seventh, a probable consequence of the autonomy taken by the Aşıklı-Musular-Çatalhöyük culture confronted to the cultures from the Levantine Corridor and the Mesopotamian High Valleys. During this time, pressure detachment does not seem to be in use in Central Anatolia. It is then re-introduced at Çatalhöyük VIB during the second part of the seventh millennium and spreads towards the Lakes District, the Marmara and the Aegean, perhaps as a consequence of the reactivation of eastern influence (Binder 2005).

Unfortunately, technological studies of the Çatalhöyük assemblage are currently not precise enough to discuss the evidence for pressure blade production with a lever at this site. However, the blade analysis conducted by Connolly (1999) indicates that a significant proportion of the blades are wide, and it seems a possibility that some were detached by pressure. On the other hand, conical pressure cores with orthogonal faceted platforms from phase VI could be similar to the shapes known or supposed at Sabi Abyad and Bouqras (Bialor 1962), indicating a common tradition. A re-examination of these studies could help us to appreciate the role played by Central Anatolia in the diffusion of the lever pressure technique between eastern Turkey and the Aegean, where lever pressure is in evidence during the very first stages of the Neolithic, approximately 6200 cal B.C. (Pelegrin in Perlès 2004: 28–29; Perlès 2004).

Acknowledgements The first technological study of the Çayönü Tepesi assemblage, including the initial recognition of the lever technique was conducted by D. Binder during a collaboration between the 'Mission de Préhistoire Anatolienne' (Ministère des Affaires étrangères et européennes, Direction générale de la Coopération internationale et du Développement) and the Department of Prehistory at the University of Istanbul in 2004. Further documentation and study were made by C. Altınbilek-Algül with the collaboration of L. Astruc. Drawings are by M. Grenet, C. Altınbilek-Algül, and L. Astruc. We are grateful to I. Caneva and the members of the Çayönü Tepesi team for their support and help. The analysis of the Sabi Abyad lithic assemblage was conducted by L. Astruc with the constant support of P.M.M.G. Akkermans, the Leiden University and the Rijksmuseum van Oudheden of Leiden and the team 'Du village à l'Etat au Proche- et Moyen-Orient', ArScAn-UMR7041 du CNRS, Maison de l'Archéologie et de l'Ethnologie. Drawings are by L. Astruc. J. Pelegrin is responsible for the technical descriptions and interpretations of the large blades. We are grateful to G. Monthel (Préhistoire et Technologie, UMR 7055, Maison de l'Archéologie et de l'Ethnologie) for his work on the preparation of the oral presentation for the Lisbon Symposium. A complement to the corpus of drawings from the Çayönü Tepesi assemblage and the rework of all the drawings for this publication by M. Grenet were undertaken as part of the programme ANR-08-BLAN-0318-01 'ObsidiennesUs: Obsidiennes, Pratiques techniques et Usages en Anatolie' held by CNRS-IFEA in collaboration with the Department of Prehistory at the University of Istanbul and the Ecole Centrale de Lyon.

References

Akkermans, Peter M.M.G. 1988 The Soundings at Tell Damishliyya. In *Hammam et Turkman*, edited by Mauritz N. van Loon, pp. 9–67. NHAI, Istanbul.

Algül, Çiler G. 2008 Obsidian Technology During the Transitional Stage between the Pre-Pottery and Pottery Neolithic Periods. Unpublished Ph.D. dissertation, Prehistory Department, Institute of Social Sciences, University of Istanbul, Istanbul.

Astruc, Laurence, Bernard Gratuze, Jacques Pelegrin, and Peter M.M.G.Akkermans 2007 From production to use: a parcel of obsidian bladelets at Sabi Abyad II. In *Technical Systems and Near Eastern PPN Communities. Proceedings of the 5th International Workshop. Fréjus, 2004*, edited by Laurence Astruc, Didier Binder qnd François Briois, pp. 327–342. APDCA, Antibes.

Astruc, L. 2011 A note on the complexity of lithic assemblages, in E. Healey, S. Campbell, O. Maeda (Ed.), *The State of the Stone Terminologies, Continuities and Contexts in Near Eastern Lithics*. Proceedings of the Sixth PPN Conference on Chipped and Ground Stone Artefacts in the Near East, Manchester, 3rd–5th March 2008, Includes papers of the Fourth PPN Workshop on Chipped Lithic Industries, Niğde/Cappadocia, 4th–8th June 2001 (coordinated by Nur Balkan-Atlı) Studies in Early Near Eastern Production, Subsistence, and Environment 13. Berlin, *ex oriente* : 257–263.

Bader, Nikolai O. 1989 *Earliest Cultivators in Northern Mesopotamia: the investigations of the Soviet archaeological expedition in Iraq at the settlements of tell Magzalia, tell Sotto and Kültepe*. Editions Nauka, Moscow.

Bialor, Perry A.1962 The chipped stone industry of Çatal hüyük. *Anatolian Studies* 12: 67–110.

Binder, Didier 1984 Système de débitage laminaire par pression: exemples chasséens provençaux. In *Economie du débitage laminaire: technologie et expérimentation (IIIe table ronde de technologie lithique, Meudon-Bellevue, oct. 1982)*, edited by Jacques Tixier, pp. 71–84. CREP (*Préhistoire de la pierre taillée*, 2), Paris.

Binder, Didier 2002 Stones making sense: what obsidian could tell about the origins of Central Anatolian Neolithic. In *The Neolithic of Central Anatolia,* edited by Frédéric Gérard and Laurens Thissen, pp. 79–90. Ege Yayınları, Istanbul.

Binder, Didier 2005 Autour de l'obsidienne: un aspect des processus d'interaction entre agro-pasteurs et chasseurs-cueilleurs en Anatolie centrale. In *Populations, économies et environnements, du Néolithique et de l'Âge du Bronze,* edited by Jean Guilaine, pp. 117–134. Errance, Paris.

Binder, Didier 2007 PPN Pressure Technology: views from Anatolia. In *Technical Systems and Near Eastern PPN Communities. Proceedings of the 5th International Workshop. Fréjus, 2004,* edited by Laurence Astruc, Didier Binder and François Briois, pp. 235–243. APDCA, Antibes.

Binder, Didier 2008 Technologie lithique et comportement social dans le PPN de Çayönü tepesi: un aperçu à travers l'analyse des matières premières. *Paléorient* 34(1): 5–21.

Binder, Didier, and Nur Balkan-Atlı 2001 Obsidian and blade technology at Kömürcü-Kaletepe (Central Anatolia). In *Beyond tools: redefining the PPN lithic assemblages of the Levant,* edited by Isabella Caneva, Cristina Lemorini, Daniella Zampetti and Paolo Biagi, pp. 1–16. Ex oriente (SENEPSE), Berlin.

Braidwood, Robert J., and Bruce Howe 1960 *Prehistoric investigations in Iraqi Kurdistan*. The University of Chicago Press (Studies of Ancient Oriental Civilization, 31). Chicago.

Caneva, Isabella, Anna-Maria Conti, Cristina Lemorini, and Daniella Zampetti 1994 The Lithic Production at Çayönü: A Preliminary Overview of the Aceramic Sequence. In *Neolithic Chipped Stone Industries of the Fertile Crescent 1,* edited by Hans G. Gebel and Stephen K. Kozlowski, pp. 253–266. Ex-oriente (SENEPSE), Berlin.

Cauvin, Jacques 1994 *Naissance des divinités - Naissance de l'agriculture: La révolution des symboles au Néolithique*, CNRS (Empreintes), Paris.

Cauvin, Marie-Claire, Alain Gourgaud, Bernard Gratuze, Nicolas Arnaud, Gérard Poupeau., Jean-Louis Poidevin, and Christine Chataigner 1998 *L'obsidienne au Proche et Moyen Orient : du volcan à l'outil.* Maison de l'Orient Méditerranéen et Archaeopress (British Archaeological Reports, International Series, 738), Lyon et Oxford.

Conolly, John 1999 *The Catalhöyük flint and obsidian industry: Technology and typology in context.* Archaeopress (British Archaeological Reports, International Series, 787), Oxford.

Copeland, Lorraine 1989 The Flint and Obsidian Artifacts of Tell Sabi Abyad. In *Excavations at Tell Sabi Abyad,* edited by Peter M.M.G. Akkermans, pp. 237–284. Archaeopress (British Archaeological Reports, International Series, 468), Oxford.

Copeland, Lorraine 2000 The flint and obsidian industries. In *Tell Sabi Abyad II. The Pre-Pottery Neolithic B settlement. Report on the Excavations of the National Museum of Antiquities Leiden in the Balikh Valley, Syria,* edited by Marc Verhoeven and Peter M.M.G. Akkermans, pp. 51–89. NINO, Leiden.

Crabtree, Don E. 1968 Meoamerican Polyhedral Cores and Prismatic Blades. *American Antiquity* 33(4): 446–478.

Hirth, Kenn 2003 *Mesoamerican Lithic Technology – Experimentation and Interpretation.* The University of Utah Press, Salt Lake City.

Inizan, Marie-Louise, and Monique Lechevallier 1994 L'adoption du débitage laminaire par pression au Proche-Orient. In *Neolithic chipped stone industries of the Fertile Crescent, and their Contemporaries in Adjacent Regions,* edited by Hans G. Gebel and Stephen Kozlowski, pp. 23–32. Ex oriente (SENEPSE), Berlin

Inizan, Marie-Louise, Michèle Reduron-Ballinger, Hélène Roche, and Jacques Tixier 1995 *Technology and Terminology of Knapped Stone.* CREP (Préhistoire de la Pierre Taillée, 5; online website ARKEOTEK), Nanterre.

Nishiaki, Yoshiro 1996 Side-Blow Blade-Flakes from Tell Kashkashok II, Syria: a Technological Study. In *Neolithic Chipped Stone Industry of the Fertile Crescent, and their Contemporaries in Adjacent Regions,* edited by Hans G. Gebel and Stephen K. Kozlowski, pp. 311–325. Ex-oriente (SENEPSE), Berlin.

Nishiaki, Yoshiro 2000 *Lithic Technology of the Neolithic.* Archaeopress (British Archaeological Reports, International Series, 840), Oxford.

Nishiaki, Yoshihiro 2011 Preliminary notes on the Pre-Pottery and Pottery Neolithic lithics from Tell Seker al-Aheimar, the upper Khabur: the 2000–2001 seasons. In *The State of the Stone: Terminologies, Continuities and Contexts in Near Eastern Lithics,* edited by Elizabeth Healey, Stuart Campbell and Osamu Maeda, pp. 457–464. Ex oriente Berlin.

Pelegrin, Jacques 1988 Débitage expérimental par pression: du plus petit au plus grand. In *Technologie préhistorique,* edited by Jacques Tixier, pp. 37–53. CNRS (Notes et monographies techniques, 25*)*, Paris.

Pelegrin, Jacques 2000 Les techniques de débitage laminaire au Tardiglaciaire : critères de diagnose et quelques réflexions. In *L'Europe centrale et septentrionale au Tardiglaciaire. Confrontation des modèles régionaux de peuplement,* edited by Boris Valentin, Pierre Bodu and Marianne Christensen, pp. 73–86. APRAIF (Mémoires du Musée de Préhistoire d'Ile-de-France, 7), Nemours.

Pelegrin, Jacques 2002a Principes de la reconnaissance des méthodes et techniques de taille. In *Tell'Atij, Tell Gudeda: Industrie lithique; Analyse technologique et fonctionnelle,* edited by Jacques Chabot, pp.215-226. CELAT, Université Laval (Cahiers d'archéologie du CELAT, 13; Série archéométrie, 3), Québec.

Pelegrin, Jacques 2002b La production des grandes lames de silex du Grand-Pressigny. In *Matériaux, productions, circulations, du Néolithique à l'Âge du bronze,* edited by Jean Guilaine, pp. 125–141. Errance, Paris.

Pelegrin, Jacques 2003 Blade-Making Techniques from the Old World: Insights and Applications to Mesoamerican Obsidian Lithic Technology. In *Mesoamerican lithic technology. Experimentation and Interpretation,* edited by Ken G. Hirth, pp. 55–71. The University of Utah Press, Salt Lake City.

Pelegrin, Jacques 2006 Long blade technology in the Old World: an experimental approach and some archaeological results. In *Skilled production and social reproduction. Aspects on traditional stone-tool technology,* edited by Jan Apel and Kjel Knutsson, pp. 37–68. Societas Archeologica Upsaliensis, Upsalla.

Perlès, Catherine 2004 *Les industries lithiques taillées de Franchthi (Argolide, Grèce): tome III: Du Néolithique ancien au Néolithique final.* Indiana University (Excavations at Franchthi cave, Greece, 13), Bloomington and Indianapolis.

Pétrequin, Pierre, Anne-Marie Pétrequin, Michel Errera, Serge Cassen and Claude Croutsch 2006 Complexité technique et valorisation sociale: haches polies de Nouvelle-Guinée et du Néolithique alpin. In *Normes techniques et pratiques sociales: de la simplicité des outillages pré- et, protohistoriques. XXVIe Rencontres internationales d'Archéologie et d'Histoire d'Antibes,* edited by Laurence Astruc, François Bon, Vanessa Léa, Pierre-Yves Milcent and Sylvie Philibert, pp. 419–433. APDCA, Antibes.

Roodenberg, Jacob J. 1986 *Le Mobilier en pierre de Bouqras: Utilisation de la pierre dans un site néolithique sur le Moyen Euphrate (Syrie).* NINO, Leiden.

Titmus, Gene L., and John E. Clark 2003 Mexica Blade Making with Wooden Tools. Recent Experimental Insights. In *Mesoamerican Lithic Technology – Experimentation and Interpretation,* edited by Ken G. Hirth, pp. 72–97. The University of Utah Press, Salt Lake City.

Tixier, Jacques 1984 Le débitage par pression. In *Economie du débitage laminaire: technologie et expérimentation (IIIe table ronde de technologie lithique, Meudon-Bellevue, oct. 1982)* edited by Jacques Tixier, pp. 57–70. CREP (*Préhistoire de la pierre taillée,* 2), Paris.

Verhoeven, Marc, and Peter M.M.G. Akkermans 2000 *Tell Sabi Abyad II: The Pre-Pottery Neolithic B Settlement.* Report on the Excavations of the National Museum of Antiquities Leiden in the Balikh Valley, Syria. NINO, Leiden.

Chapter 6
Two Examples of Pressure Blade Production with a Lever: Recent Research from the Southern Caucasus (Armenia) and Northern Mesopotamia (Syria, Iraq)

Jacques Chabot and Jacques Pelegrin

6.1 Introduction

Techniques of blade production by pressure, especially with the aid of levers, are subjects of great potential in archaeology since they represent a highly effective and refined means of producing blades. These techniques are, however, optional (as one could do without them and they are therefore not ubiquitous) in contrast with direct percussion techniques which are simpler to perform and very widely practised.

Blade production through application of pressure implies a specific knowledge regarding the use of special tools and knapping methods and a high degree of know-how, especially in the preparatory shaping of cores. As an elaborate evolution of tool-making techniques, pressure-related blade production can be usefully regarded as a marker of particular cultural traditions and of the diffusion of technical innovation (Inizan 1991).

Accordingly, it is important to identify these techniques in each archaeological tradition or period where they were practised and to determine if they were locally invented, acquired by imitation or transferred by craftsmen. In doing so, archaeologists studying blade technology can contribute to the building of a phylogeny or cultural genealogy of the specialized tool-making techniques, a project which is still at its beginning. Although this goal is beyond the scope of the present study, these particular production techniques permit archaeologists to demonstrate the

J. Chabot (✉)
CELAT (Laboratoires d'archéologie), Université Laval,
1030 av. des Sciences humaines, Québec G1V 0A6, Canada
e-mail: jacques.chabot@hst.ulaval.ca

J. Pelegrin
Laboratoire "Préhistoire et Technologie", CNRS et Université Paris
Ouest Nanterre, MAE, Nanterre, France
e-mail: jacques.pelegrin@mae.u-paris10.fr

P.M. Desrosiers (ed.), *The Emergence of Pressure Blade Making: From Origin to Modern Experimentation*, DOI 10.1007/978-1-4614-2003-3_6,

presence of specialized craftsmen (Pelegrin 1988) and craft specialization, essential for socio-economic studies in prehistory.

In this chapter, we detail the results of two recent studies from the Near East and the Southern Caucasus that identified the production of blades by a pressure technique with the help of a lever.

The first part of our chapter concerns the Southern Caucasus during the sixth millennium B.C., a time where we view the origins of a Neolithic economy, socio-economic practices that probably came from Northern Mesopotamia. This is an obsidian-rich region, but one that is poorly known archaeologically, and our identification of a pressure technique with a lever represents new information alongside the recognition of contemporaneous use of indirect percussion and an ordinary standing pressure technique.

The second part of our chapter deals with so-called Canaanean flint blades from Northern Mesopotamia at the turn of the fourth to third millennium B.C. These distinctive, large blades are long known, but as they are known to have been produced in different types of flint, and throughout various regions, for about one and a half millennia, it cannot be assumed that they were all made using the same techniques. From the study presented here, two variants of a pressure technique with a lever and a copper point are identified, together with other blades detached by indirect percussion.

6.2 Obsidian Blades from the Late Neolithic Southern Caucasus (5900–5300 cal. B.C.)

From 1999 to 2004, different archaeological investigations were conducted on and around the site of Aratashen in the Ararat plain by an international team under the direction of Christine Chataigner from the 'Maison de l'Orient Méditerranéen' (Mission Caucase/Université de Lyon 2), including French, Armenian and Canadian specialists.

The site of Aratashen lies 25 km from Yerevan, the capital of Armenia, and 5 km southwest of Vagharshapat (Echmiadzin). It is situated near a bend of the Kasakh river a few kilometres ahead of its confluence with the Araxe river (Fig. 6.1). The site's stratigraphy includes two Neolithic layers (between 5900 and 5300 B.C.) and one Chalcolithic layer (4800–4500 B.C.), which are typical of the cultural phases of the Ararat plain region (Fig. 6.2). The site was excavated by R. Badalyan (University of Erevan) and P. Lombard (Université de Lyon 2), with J. Chabot being in charge of the lithic material.

The study of the lithic material, including more than 20,000 obsidian artefacts, demonstrated that agricultural tools made from large blades comprised a significant part of the assemblage in keeping with what one views in other regions of the Near East. From 2005, new excavations were initiated on the neighbouring site of Aknashen, providing a similar lithic assemblage. Preliminary chemical analysis of the obsidian artefacts demonstrated that they were made of raw material from eight different regional volcanic sources (Fig. 6.1).

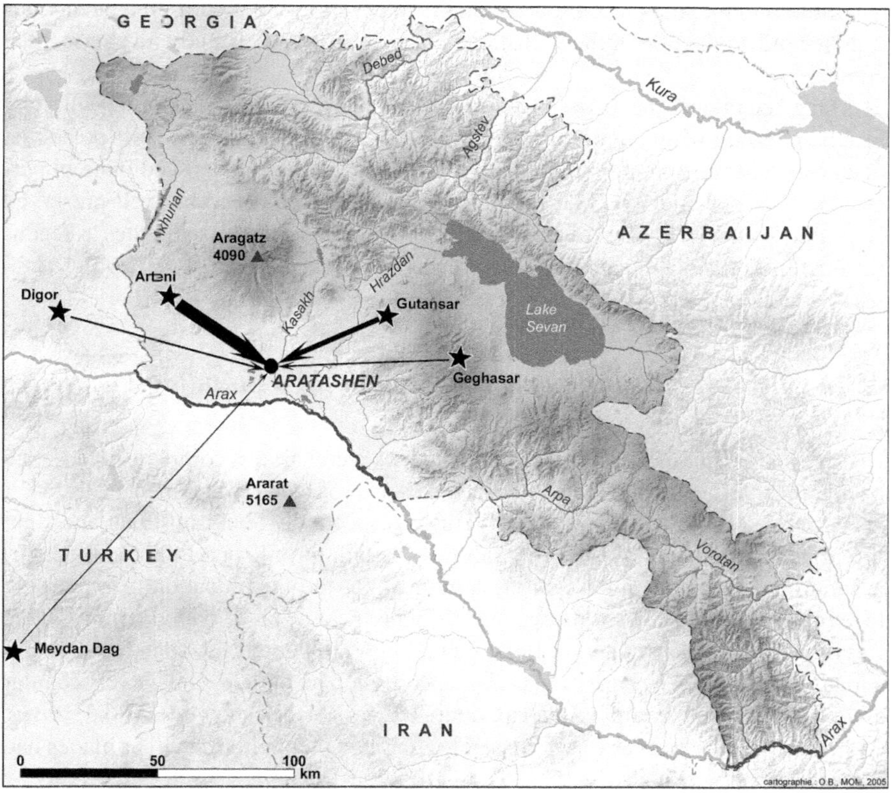

Fig. 6.1 Map of Southern Armenia showing the location of Aratashen and the many sources of obsidian exploited at the site

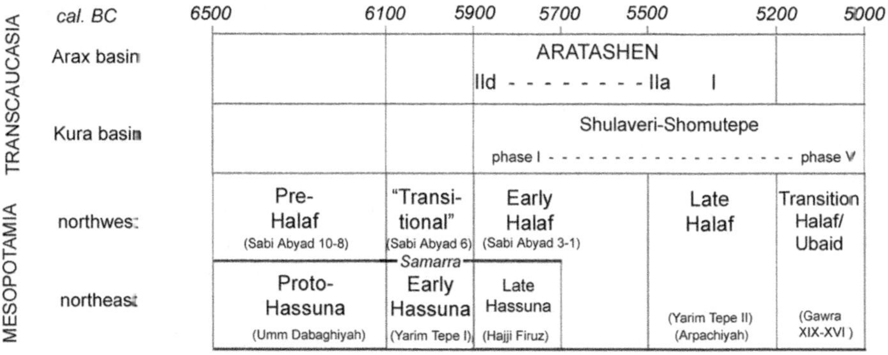

Fig. 6.2 Location of Aratashen within the regional chrono-cultural scheme

Our aim in the first part of this chapter is to describe and identify the techniques used for the production of these obsidian blades, the first such study in an armenian context.

Three techniques have been identified at Aratashen for the *débitage* of long blades: pressure with a crutch (standing position represented by 64 well-preserved diagnostic specimens), pressure with a lever (24 clear specimens) and indirect percussion (29 specimens). All the diagnostic blades came from well-preserved Neolithic levels. Here we provide a description of each of these techniques as reconstructed from the diagnostic blades (for terminology, see Inizan et al. 1999).

6.2.1 Pressure Technique with a Crutch (in a Standing Position)

Clear evidence of this technique is provided by several almost complete blades, and a few cores, left in a good state after the last series of blade detachments (Fig. 6.3). As Crabtree (1968) and Tixier (1984) stated respectively about obsidian and flint, the 'triad' of arguments: regularity, (almost) straight profile and thinness (or lightness) all together constitute a strong argument for the identification of a pressure technique. With obsidian, a mechanical argument can even be added regarding the fragility of this glassy material. It would be impossible to detach such regular, thin and almost straight columns of glass by a percussion technique. Only a very regular and prolonged time compression can do it. The few blades shown (Fig. 6.3) are no exception because there are several cores with scars that indicate similar blades had been produced repeatedly.

The core shown in Fig. 6.3 was probably abandoned because it was considered to be exhausted. This indicates that the minimal length of the expected blades was about 12 cm. Because of its pointed tip, the next series of blades would almost certainly overpass and rapidly shorten the core. Scars from larger removals are preserved on one side of the core, which indicate a shaping of the pre-core with axial removals, using direct stone percussion or indirect percussion from a large, flat platform opened by a large flake. Another core (Fig. 6.4) is even longer, with an obtuse base allowing for the detachment of pressure blades up to 17 cm long. This long, barrel-shaped core is certainly not exhausted, although it bears some irregularities at the back of the last productive side (see profile Fig. 6.4). By its shape, it resembles some Mesoamerican obsidian cores (Hirth 2003; Hester 1978) and similarly involves the same question: how could such a long pre-core be shaped? Here we see a few opposite and oblique scars from the obtuse tip, which could complete a possible axial shaping with heavy percussion blades (see infra). The width of these pressure blades and of the last blade scars on the cores seems to be about 15–20 mm (Fig. 6.4), data that fits quite well with Pelegrin's mode 4 of pressure blade production (see Pelegrin, Chap. 18, this volume), that is using a long crutch in a standing position.

The platform preparation consists of removing the overhang left by previous detachment by scratching with a fine-grained stone so that the remnant platform or

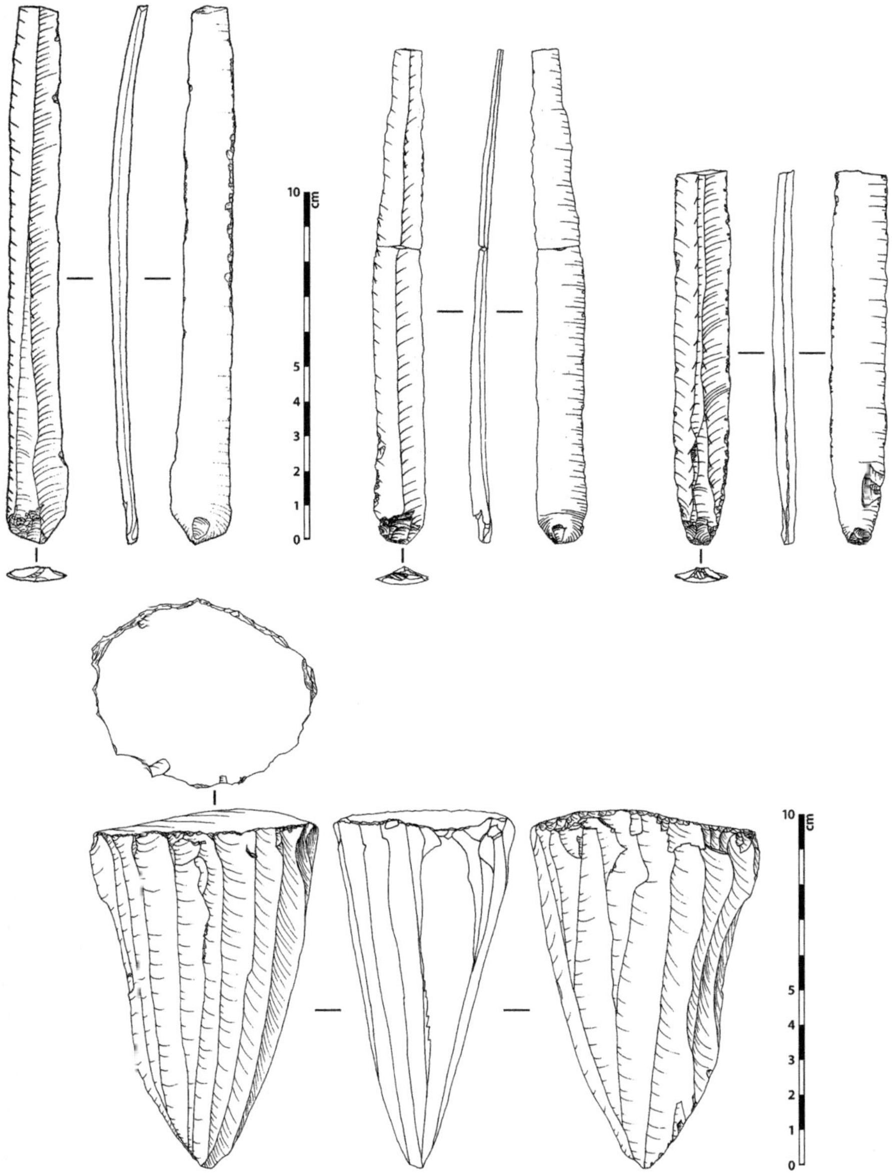

Fig. 6.3 Three slender blades – a possible reason for their discard – and a corresponding conical blade core bearing evidence of having been reduced by a sitting or standing pressure technique

butt of the blades is rather small, with an edge angle ca. 80°–90°. A small lip is visible at the back of the best preserved butts, and no concentrated impact mark. This is indicative of the use of an organic material, presumably antler for the pressure point.

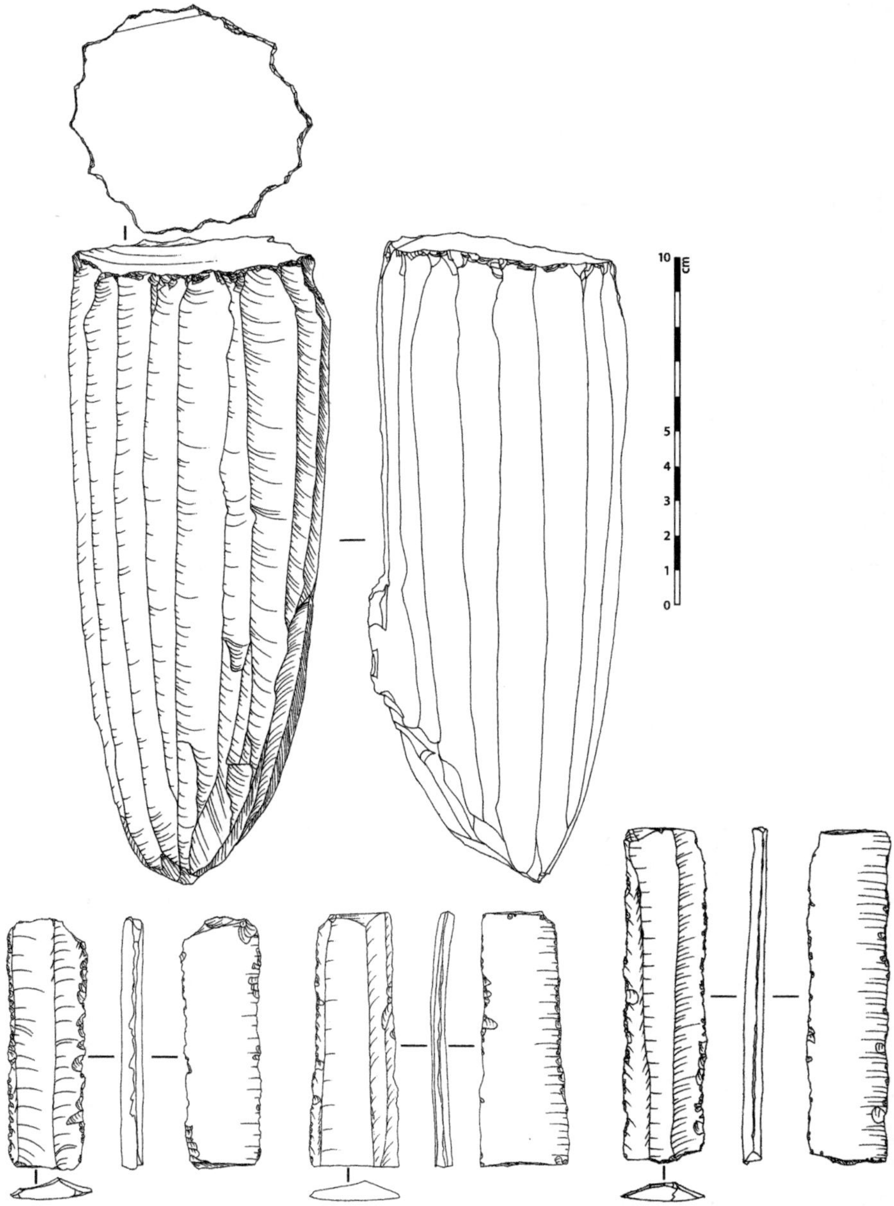

Fig. 6.4 A long barrel shape core and three corresponding blade fragments. The width of the blade fragments and that of the blade scars on the core indicate a standing pressure technique. But the core, considering its diameter, was possibly reduced earlier by lever pressure

6.2.2 Pressure Technique with a Lever

Alongside these medium width blades are blades which are significantly wider (well over 30 mm) and much longer, yet are also quite regular and thin (Fig. 6.5). They are without doubt detached by pressure, but the width of these blades cannot be achieved by the standing pressure technique described previously. As Crabtree stated, referring to his great experience with pressure obsidian blade production, the maximum size of the obsidian blades he could produce using his standing technique was 'one inch wide and 8 in. long', that is about 25 mm wide and 20 cm long (Crabtree 1968: 468, see also Sheets 1978 about archaeological prismatic blades with a maximum width of 2.6 cm and thickness of 0.4). Using a different standing technique, Pelegrin was able to manufacture blades up to 27 mm wide and 28 cm long from a crested core with a plain platform (neither pecked and ground nor sawn), but except for an extraordinarily heavy craftsman, it seems difficult to exceed a blade width of 3 cm when working with a plain platform and using the standing pressure technique.

The conclusion is that the wide and regular obsidian blades excavated at Aratashen were detached by a technique that involved some reinforced pressure, that is with a lever (see Pelegrin, Chap. 18, this volume for experimental reproduction). The nature of the pressure tool tip, however, cannot be ascertained.

Armenia, which is so rich in obsidian, was a consistent source of large and regular obsidian blades produced as early as the sixth millennium and possibly earlier, depending on expected discoveries from new sites under excavation, or via the technical analysis of existing collections. Altinbilek et al. (this volume) report pressure blades produced in Anatolia as early as the eighth millennium (probably using obsidian from Nemrut Dag and/or Bingöl, that is about 250 km southwest of the Armenian sites under discussion).

From the blades detached by lever pressure, it is possible to define the main characteristics of the cores they came from, with a plain flat or flat-facetted platform forming an orthogonal (about 90°) edge angle, which leads us to imagine a core shape similar to that prepared for the production of medium length and width pressure blades. It is therefore possible to imagine that some of the cores may have been initially shaped and/or reduced by lever pressure until they could be further exploited by standing pressure, which could be the case for the core from Fig. 6.4.

At the moment, it is difficult to elucidate the specific raw blank type chosen (prismatic/tabular or nodular shape) and the subsequent means by which these large cores were pre-shaped. The dimensions of the blade fragments indicate an average shape of 10–15 cm for the width of the core at the platform and up to a length of 25 cm or possibly even 30 cm. As for the very large 25–30 cm Mesoamerican pressure cores from the 'preclassic' period initially presented by Hester (1978), two possibilities can be evoked. If the raw material consists of regular prismatic fragments, it might be possible to correct the pre-core with limited shaping and to start directly on the corner(s) with the detachment of very large cortical and semi-cortical blades by lever pressure. If not, the whole volume has to be shaped by percussion removals, which is certainly not possible with axial flaking only, once a large platform has been opened, but has to be done by transversal flaking, that is by the way of several crests.

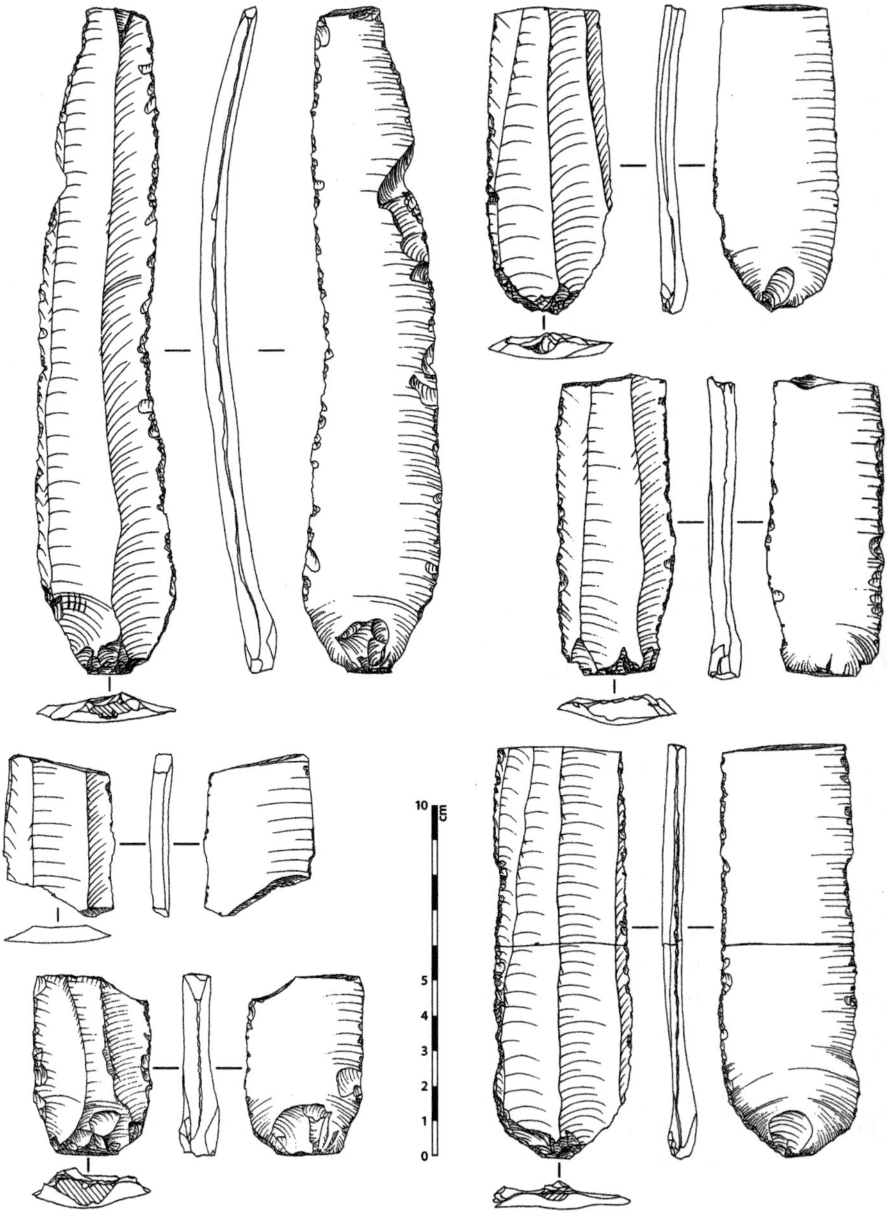

Fig. 6.5 One near-complete blade and five blade fragments, their regularity and width (32–38 mm) indicate a lever pressure technique. On three of the plain orthogonal butts a crack is visible which indicates the use of a copper tipped pressure tool

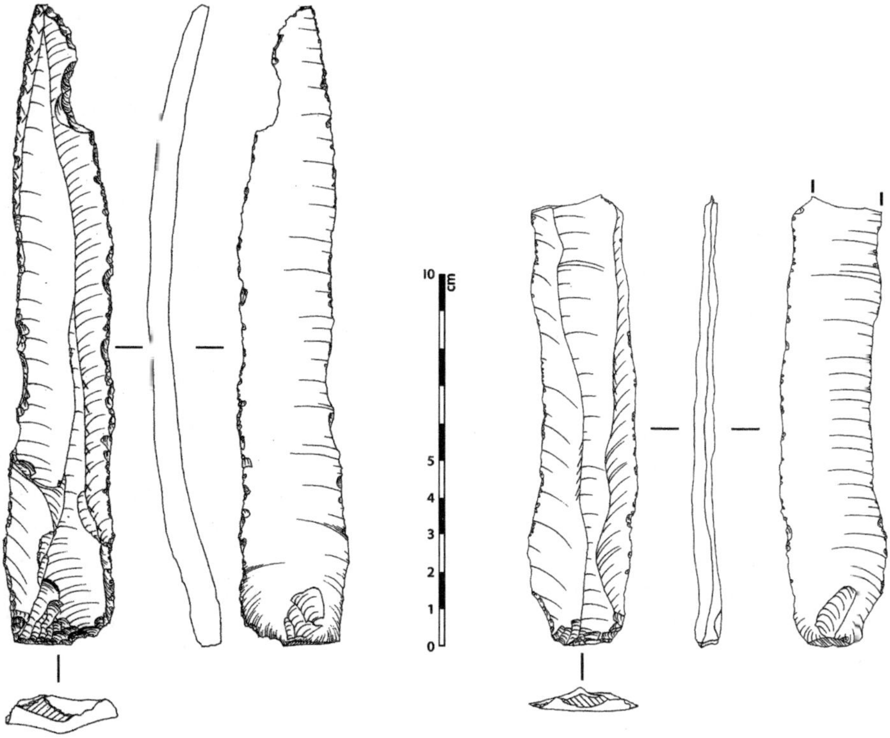

Fig. 6.6 Two blades probably detached by indirect percussion looking at the mediocre regularity of the profile and dorsal ridges

6.2.3 Indirect Percussion

Blades of a respectable size, but lacking some of the pressure features, were also recovered from the excavations at Aratashen. Two examples are presented here for description and diagnosis (Fig. 6.6). They are less than regular, with undulating scars (dorsal arises) and a wavy ventral side. Their butt is quite thick (clearly thicker than pressure blades which usually have a more careful overhang reduction) and again flat and orthogonal that is detached from a large plain platform. We do not believe that such blades can be detached in a series considering their previous dorsal scars by direct stone percussion, neither by organic nor by soft stone percussion (which would deserve an acute platform angle and a more refined edge rounding). These two blades in particular, on the left side of Fig. 6.6, are much more typical of indirect percussion. They could originate from a workshop where the pressure technique would not be known or used, which is doubtful, or they could probably be some end-selected by-products from the shaping of medium size pressure cores. It has indeed been shown, as a technical possibility, that medium size conical cores can be shaped by indirect percussion (Pelegrin 2003); however, more evidence of this 'punch technique' would be welcome to prove this hypothesis.

A lot of information is still missing in order for us to reconstruct a general view of the chrono-cultural sequence related to the blade production of the Armenian Late Neolithic. We are also hindered by the fact that so little is known of the Neolithic in the Southern Caucasus and that we know practically nothing of the detachment techniques known from the same period (or earlier) in regions to the north and to the east.

We do at least know that the culture represented in the Araxe basin by the sites of Aratashen and Aknashen-Khatunarkh shows very close links with the Shulaveri-Shomutepe culture, which developed throughout the sixth millennium to the North in the Kura basin. Architecture, as well as ceramics, bone and lithic industries, testifies to the fact that these two cultures share a common root, and it could be amongst this common stock that the technique of pressure with a lever developed (Arimura et al. 2010).

However, it is currently considered that the introduction of domestication into the Southern Caucasus (in particular the culture of the hexaploïd naked wheat, which was largely widespread in both the Araxe and Kura basins in the sixth millennium) could in fact have been brought into the region by populations coming from Northern Mesopotamia (Zohary and Hopf 2004). This could constitute an interesting route to explain the transfer to the Caucasus of the lever pressure technology, the technique possibly coming from Eastern Anatolia, where it is attested during the second half of the eighth millennium and into the seventh millennium as well (Altınbilek et al., Chap. 5, this volume).

The production of similar large blades seems to continue in Armenia into the Chalcolithic period during the fifth millennium and the first half of the fourth millennium (Badalyan et al. 2009). It is noteworthy that the Araxe basin is located only 250 km from the southeast bank of the Black Sea and the Kura basin is only 100 km away. It is then possible that this tradition might be the source of some of the Eneolithic technological traditions of North-Eastern Bulgaria, where pressure with a lever technique on flat pressure platforms appears as early as 4900 B.C. (Manolakakis 1996, 2005).

6.3 Canaanean Blades from Early Third Millennium Northern Mesopotamia (Northeast Syria, North Iraq, Southeast Turkey)

The so-called Canaanean flint blades were first distinguished on the basis of their large size and regularity in the Levant by R. Neuville (1930) and were subsequently described by J. Cauvin at Byblos, Lebanon, where they were imported from the Early Eneolithic before increasing in size in the Early Bronze Age.

More recently, similar flint blades have been described in Northern Mesopotamia by I. Caneva (1993), C. Edens (2000), J. Chabot (2002), J. Chabot and P. Eid (2003), Y. Nishiaki (2003) and A. van Gijn (2003) (Fig. 6.7).

These large and regular blades are usually found as 5–7 cm long fragments with one or two glossy edges and sometimes bitumen residue from hafting, following

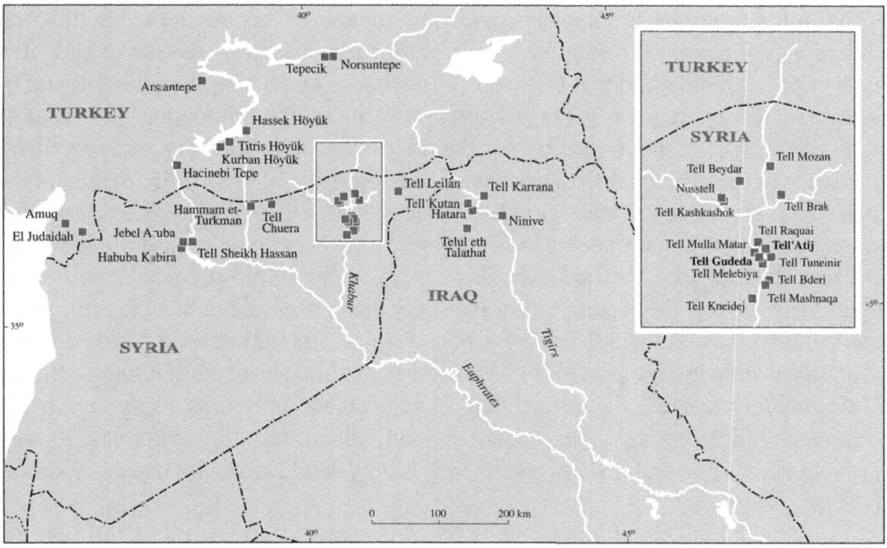

Fig. 6.7 Sites in Northern Mesopotamia where Canaanean blades have been identified

their use as sickles or threshing sledge elements (Collin 1992; Anderson and Inizan 1994; Anderson et al. 2004; Chabot 2002). From here on, we will call such fragments 'Canaanean elements'.

There is evidence for the production of such long blades amongst a few south-eastern Anatolian settlements of the Upper Euphrates, including Hassek Höyük (Behm-Blancke 1992), Hacinebi (Edens 2000; Stein 2000) and Titris Höyük (Hartenberger et al. 1999; Matney et al. 1999). Other areas of production are now known in Israel and Palestine, albeit with some differences in the butts of these blades that probably relate to local traditions of platform preparation (Shimelmitz 2009).

In general, these Canaanean blades have an exceedingly wide geographic distribution, having been recovered from sites throughout Northern Mesopotamia and the Levant down to the northern Neguev. Temporally, their production extends from about 4000 to 2500 cal. B.C. in the North (from Late Chalcolithic to Uruk and Nineveh V – Early Bronze II) and 3600 to 1950 cal. B.C. in the South (from the Early Bronze to Middle Bronze I; Rosen 1997), though a recent reference suggests a somewhat earlier appearance before the Early Bronze Age (Bar and Winter 2010).

6.3.1 *Lever Pressure Technique with a Copper Point*

Judging from the different collections that we could examine, a significant number of these large Canaanean blades seem to have been detached by pressure, which, given their width (usually over 2.5 cm), had to have involved the use of a lever (see Pelegrin, Chap. 18, this volume).

With the proximal fragments of wide and regular blades having a dihedral butt that slopes forward (forming an obtuse platform and flaking surface angle), it is further possible to argue that the lever pressure tool had a copper point. The use of a copper tip can frequently, but not always, result in the production of a discrete transversal crack or micro-crushing on the ridge of the butt (see below, and Fig. 6.8b). A decade of experimental work by one of us (J.P.) from 1986 to 1995 (mainly at the Lejre Research Centre) has demonstrated that both indirect percussion and lever pressure using organic materials fail to detach blades with an obtuse dihedral butt because the tool slips forward, whereas copper as a soft, plastic metal offers sufficient 'grip' on a grainy enough flint. However, lever pressure with an organic point can work well on a dihedral *non-obtuse* platform, while indirect percussion on such a platform preparation provides no advantage (cf. Pelegrin, forthcoming article). Unfortunately convex, or flat-facetted, or plain blade butts that form an average orthogonal angle are much less technologically diagnostic; they can relate to both indirect percussion and lever pressure, unless they display a distinctive crack on the butt (see below).

Fortunately, the recent excavations by M. Fortin at Tell 'Atij and Tell Gudeda (North Syria, Early Bronze Age/Nineveh V) have generated several hundred Canaanean elements, including long and/or well-preserved proximal fragments, many of them with the aforementioned crack in their butts (Fig. 6.8). The recurrent combination of regular form and almost straight profile, plus a wide but not so thick section (the critical associations for the recognition of pressure blades [cf. Tixier 1984]), together with dihedral butts (some obtuse [Fig. 6.8]), demonstrates that many of the Canaanean elements from Tell 'Atij and Tell Gudeda were detached by lever pressure using a copper point which was used for the detachment of many of the Canaanean elements from Tell 'Atij and Tell Gudeda (Chabot 2002; Pelegrin 2006). Incidentally, further evidence for the use of a copper-tipped punch could be observed on some crested blades bearing very small and clear circular cracks, obviously produced by a very pointed tool that is necessarily metallic.

Another technologically indicative feature on one of the Tell Gudeda blade fragments is a specific form of edge damage (called 'counterflaking' by Healan and Kerley 1984 [also 'spontaneous retouch' by Mark Newcomer]) that results from the edge of the blade being detached by coming into contact with a hard or firm material, likely the device that had been used to immobilize the core (Pelegrin 2006: 52–53). This data provides additional evidence for the use of a pressure technique (specifically a lever, given the blade width), as the immobilization of the core for blade detachment is something that one does not associate with percussion techniques. Finally, our experiments have proven that lever pressure can produce some ripples on the bulb, or shortly thereafter (due to micro-movements of the core in its immobilizing device during compression and early detachment). Such ripple marks are visible on some of the archaeological blades or on some of their previous blade scars on the upper side.

It is worth noting that a substantial fraction of similar 'Canaanean' elements from Tell 'Atij and Tell Gudeda are fragments of blades that had been detached by indirect percussion (or 'punch' technique). This is clearly the case with some of the proximal

Fig. 6.8 Seven proximal Canaanean elements from Tell 'Atij, with dihedral butt and frequent ripple marks under the bulb

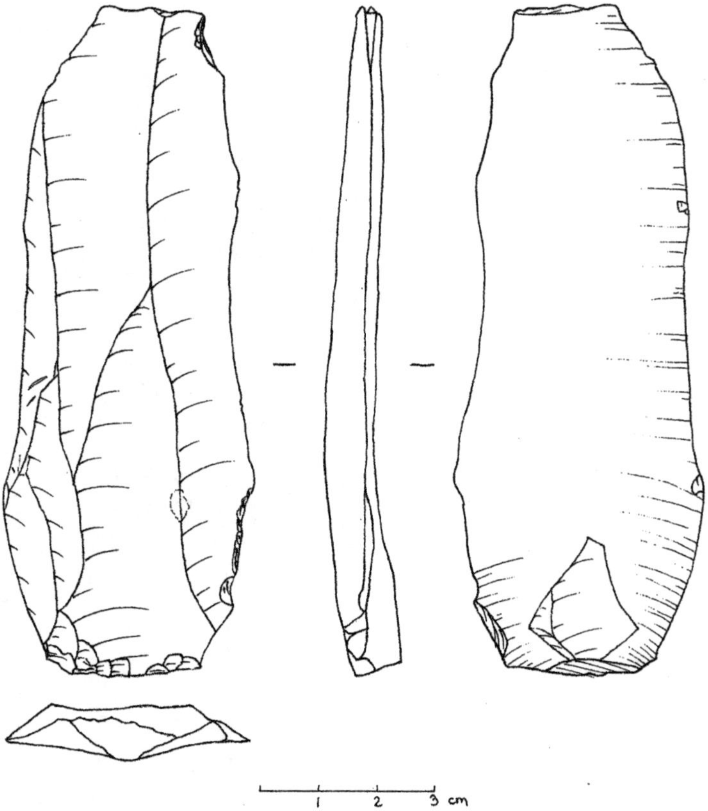

Fig. 6.9 A 'Canaanean element' from Tell 'Atij made from a wide and short shaping blade with a thick orthogonal plain butt: detached by indirect percussion

fragments that show a plain orthogonal thick butt without a crack, indicating the use of an antler punch (Fig. 6.9), and it is also suggested by some medial or distal elements that are irregular in form. Some of these could be 'secondary products' struck from the same core during the core shaping sequence (a few with residual cortex) or from its further reduction following the main production sequence by lever pressure. But one cannot rule out that some of these 'punch-detached' blades came from specific cores or even workshops that were primarily, if not exclusively, employing percussion techniques (this might particularly be the case if the raw material available is smaller in size).

From a similar context, one of us (J.P.) examined two regular proximal Canaanean elements from a group of about 20 pieces (4–7 cm long, most of them between 3 and 4 cm wide) from the site of Kutan in Northern Iraq (Nineveh V context, excavated by L. Bachelot), a collection that has been previously studied and published by Anderson and Inizan (1994, Fig. 3, pp. 91; here Fig. 6.10). Fortunately, the butts of these two pieces are well preserved (these Canaanean elements were often simply broken without preliminary retouch or truncation, and not further reused); here the platform preparation is somewhat different from that of Tell 'Atij and Tell Gudeda,

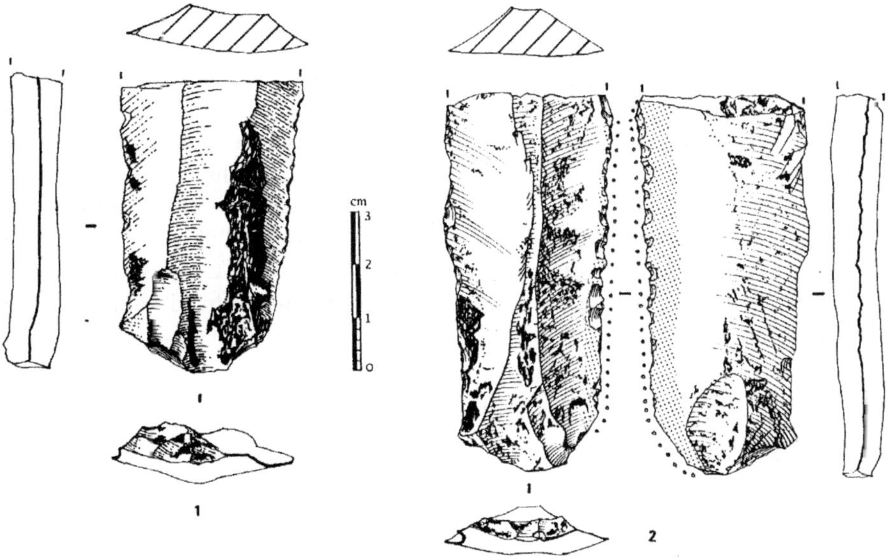

Fig. 6.10 Two proximal fragments from Kutan (K 258-23/10, K 249-9) (Courtesy of Anderson and Inizan 1994. Drawings: M. Reduron)

consisting mainly of faceting (except for a small removal towards the flaking surface for both of the pieces, in order to facilitate the platform faceting). Each butt thus looks convex facetted, forming an orthogonal edge angle, but the pressure occurred on an obtuse ridge in the middle of the butt surface (note that this ridge is much less prominent than that of the Tell 'Atij proximal blade fragments). The diagnostic character is that a minute crack is visible, forming a half circle or a '^' over the ridge only 2 mm in front of the back edge of the butt (Fig. 6.10). This indicates a limited contact over a few square millimetres, meaning the use of a hard and pointed material. This is subtle, but significant, evidence for the use of a copper point, because any organic material would spread on the platform and determine a much larger area of contact (and would not usually produce a crack, according to our experiments). In this case, with regard to wide flint blades, such a crack is indicative of a small contact area and is thus by itself an argument for a pressure technique because experiments proved that using a copper-tipped punch to detach large heavy blades by indirect percussion seems less efficient than using an antler punch.

We also note that, depending on the regions and possibly the periods, different varieties of flint are represented amongst these Canaanean blades and elements (as was originally reported in the Levant by Cauvin [1968]). This might signify the existence of different workshops or production areas, amongst which might have developed subtly (or overtly) different technical traditions. Presently, it seems that most of the Canaanean elements from Tell 'Atij and Tell Gudeda were made from a rather fine-grain flint, from light yellow to yellowish or pinkish grey in colour, which could possibly correspond to some of the varieties from the Upper Euphrates (200 km to the northeast) as worked in Hassek-Höjük (see colour photos in Behm-Blancke 1992), while those from Kutan and other Tigris valley sites were generally made from a coarse beige or whitish-grey flint, the origin of which is unknown to our knowledge.

Regarding Hassek-Hüyük, we now regard the claim that the Canaanean blades from the site had been produced by indirect percussion using a copper point as premature (Pelegrin and Otte 1992), the study having been undertaken at a time (1989) when J.P. had little practical experience concerning the detachment techniques for large blades and their technologically diagnostic features. Through a reconsideration of the well-illustrated blades and cores (Behm-Blancke 1992), we now believe that most of the blades were in fact detached by lever pressure. Discussing the material of the pressure point would, however, require a new first-hand examination of the archaeological material.

6.4 Conclusion

These observations, together with others (this volume), represent the building blocks for reconstructing a 'genealogy' of specific lithic technologies in this broad region during the Neolithic and Chalcolithic. We cannot, however, complete a general framework as long as so many pieces of the puzzle remain missing from the whole of the Near-Middle East and neighbouring regions.

At the very least, we can say that the origin of the South Caucasian obsidian pressure blade production techniques may well be found in Eastern Anatolia, amongst communities around the Bingöl and/or Nemrut Dağ volcanoes that lie only about 250 km southwest of the Araxe basin. However, it remains to be determined if the known period of Anatolian pressure with the lever technology, which at present ranges from the second half of the eighth millennium through the second half of the seventh (Altinbilek et al., this volume), continued until the beginning of the sixth millennium, a pre-requisite for this region being a viable candidate for the origin of this technique in the Southern Caucasus.

The phenomenon of the Canaanean blades was dealt with only briefly here. Long-lived and associated with several areas of production, it may take on an increasingly variable nature as more collections become the subject of specialized technological studies, documenting the modes of preparation and techniques of detachment. Currently, it is thought to have developed in Northern Mesopotamia, reaching the Southern Levant some centuries later, but it is premature to discuss with any confidence its origin and transfer/diffusion or indigenous development in other distant regions.

Acknowledgements The author would like to thank Ruben Badalyan (University of Erevan), Christine Chataigner (MOM/Université de Lyon 2), Michel Fortin (Laval University), Harry Lerner (Postdoc Laval University), James Woollett (Laval University) and Shirley Schreider.

Illustrations of Figs. 6.3, 6.4, 6.5, 6.6 and 6.9 were provided by Julie Leclerc under the supervision of J. Chabot and J. Pelegrin, with the participation of Gérard Monthel (UMR 7055).

The map of Fig. 6.1 was provided by Christine Chataigner and that of Fig. 6.7 by Andrée Héroux.

References

Anderson, Patricia C., and Marie-Louise Inizan 1994 Utilisation du tribulum au début du IIIe millénaire: des lames « cananéennes » lustrées à Kutan (Ninive V) dans la région de Mossoul, Iraq. *Paléorient* 20(2): 85–103.

Anderson, Patricia C., Jacques Chabot, and Annelou van Gijn 2004 The Functional Riddle of 'Glossy' Canaanean Blades: The Apogee of the Mesopotamian Threshing Sledge. *Journal of Mediterranean Archaeology* 17(1): 87–132.

Arimura, Makoto, Ruben S. Badalyan, Boris Gasparyan, and Christine Chataigner 2010 Current Neolithic Research in Armenia. *Neo-Lithics 1/10*: 77–85.

Badalyan, Ruben S., Pavel Avetisyan, and Adam T. Smith 2009 Periodization and chronology of southern Caucasia: from the Early Bronze Age through the Iron IV period. In *The Archaeology and Geography of Ancient Transcaucasian Societies I The foundations of Research and Regional Survey in the Tsaghkahovit Plain Armenia,* edited by Adam T. Smith, Ruben S. Badalyan and Pavel Avetisyan, pp. 33–93. Oriental Institute Publications 134. University of Chicago Press, Chicago.

Bar, Shay, and Haim Winter 2010 Canaanean Flint Blades in Chalcolithic Context and the Possible Onset of the Transition to the Early Bronze Age: A Case Study from Fazael 2. *TEL AVIV* 37: 33–47.

Behm-Blancke, Manfred R. (editor) 1992 *Hassek-Höyük: Naturwissenschaftliche Untersuchungen und lithische Industrie.* Istanbuler Forschungen 38. Ernst Wasmuth Verlag, Tübingen.

Caneva, I 1993 From Chalcolithic to Early Bronze Age III at Arslantepe: a lithic perspective. In *Between the rivers and over the mountains: Archaeologia Anatolica et Mesopotamica, Alba Palmieri Dedicata.* edited by Marcella Frangipane, Andreas Hauptman, Mario Liverani, Paolo Matthiae, Machted J. Mellink, pp 319–339. Universita di Roma "La Sapienza" Dipartimento di Scienze Storiche Archeologiche e Anthropologiche dell Antichita, Roma.

Cauvin, Jacques 1968 *Fouilles de Byblos IV: les outillages néolithiques de Byblos et du littoral libanais.* Maisonneuve, Paris.

Chabot, Jacques 2002 *Tell'Atij, Tell Gudeda : Industrie lithique; Analyse technologique et fonctionnelle. Avec une annexe de Jacques Pelegrin (CNRS, France) sur la reconnaissance du débitage par pression au levier.* Cahiers d'archéologie du CELAT no13, Série archéométrie n°3. Université Laval, Laval.

Chabot, Jacques, and Patrick Eid 2003 Le phénomène des lames cananéennes : état de la question en Mésopotamie du nord et au Levant Sud. In *Le traitement des récoltes : un regard sur la diversité du Néolithique au présent,* edited by Patricia C. Anderson, Linda S. Cummings, Thomas K. Schippers and Bernard Simonel, pp 401–415. Editions APDCA, Antibes.

Collin, Fernand 1992 Analyse tracéologique de l'industrie lithique d'Hassek Höyük. In *Hassek-Höyük: Naturwissenschaftliche Untersuchungen und lithische Industrie,* edited by Manfred R. Behm-Blancke, pp. 248–260. Istanbuler Forschungen 38. Ernst Wasmuth Verlag, Tübingen.

Crabtree, Donald E. 1968 Meso-American polyhedral cores and prismatic blades. *American Antiquity* 33(4): 446–478.

Edens, Christopher 2000 The chipped stone industry at Hacinebi: technological styles and social identity. *Paléorient* 25(1): 23–33.

Gijn, Annelou van 2003 The Ninevite 5 chipped stone assemblage from Tell Leilan: preliminary results. In *The origins of North Mesopotamian civilization: Ninevite 5 chronology, economy, society,* edited by Elena Rova and Harvey Weiss, pp. 401–416. Subartu IX. Brepols, Brussels.

Hartenberger, Britt, Steven A. Rosen, and Timothy Matney 1999 The Early Bronze Age Blade Workshop at Titris Höyük: Lithic Specialization in an Urban Context. *Near Eastern Archaeology* 63(1): 51–58.

Healan, Dan M., and Janet M. Kerley 1984 Edge Damage Induced by Core Immobilization in Prismatic Blade Manufacture. *Lithic Technology* 13(1): 1–10.

Hester, Thomas R. 1978 Preliminary notes on the technological analysis of obsidian artifacts from Villa Morelos, Michoacan, Mexico. In *Archaeological studies of Mesoamerican obsidian,* edited by Thomas R. Hester, pp. 131–158. Ballena Press Studies in Mesoamerican Art, Archaeology and Ethnohistory n°3. Ballena Press, Socorro.

Hirth, Kenneth G. (editor) 2003 *Experimentation and Interpretation in Mesoamerican Lithic Technology*. The University of Utah Press, Salt Lake City.

Inizan, Marie-Louise 1991 Le débitage par pression : des choix culturels. In *25 ans d'études technologiques en préhistoire. Bilan et perspectives. XIèmes Rencontres Internationales d'Archéologie et d'Histoire d'Antibes 18–20 octobre 1990*, pp. 367–377. A.P.D.C.A., Juan-les-Pins.

Inizan, Marie-Louise, Michèle Reduron-Ballinger, Hélène Roche, and Jacques Tixier 1999 *Technology and Terminology of Knapped Stone. Préhistoire de la Pierre taillée 5*. CREP, Nanterre.

Manolakakis, Laurence 1996 Production lithique et émergence de la hiérarchie sociale: l'industrie lithique de l'Enéolithique en Bulgarie (première moitié du IVème millénaire non cal.). *Bulletin de la Société Préhistorique Française* 93(1): 119–123.

Manolakakis, Laurence 2005 *Les industries lithiques énéolithiques de Bulgarie*. Coll. Internationale Archäologie, Bd 88. Rahden/Westf, Leidorf.

Matney, Timothy, Guillermo Algaze, and Steve A. Rosen 1999 Early Bronze Age Urban Structure at Titris Höyük, Southeastern Turkey: the 1998 season. *Anatolica* 25: 185–201.

Neuville, René 1930 Notes de Préhistoire palestinienne. *Journal of the Palestine Oriental Society* 10: 193–221.

Nishiaki, Yoshihiro 2003 Chronological developments of the Chalcolithic flaked stone industries at Tell Kosak Shamali. In *Tell Kosak Shamali,* edited by Yoshihiro Nishiaki and Toshio Matsutani, Vol. II, pp. 15–112. The University Museum and the University of Tokyo, Tokyo.

Pelegrin, J. 1988 Débitage expérimental par pression: "Du plus petit au plus grand". In *Technologie préhistorique*, edited by J. Tixier, pp. 37–53. Notes et Monographies Techniques du CRA No.25. CNRS, Paris.

Pelegrin, Jacques 2003 Blade making techniques from the Old World: insights and applications to Mesoamerican obsidian lithic technology. In *Experimentation and Interpretation in Mesoamerican Lithic Technology*, edited by Kenneth G. Hirth, pp. 55–71. The University of Utah Press, Salt Lake City.

Pelegrin, Jacques 2006 Long blade technology in the Old World: an experimental approach and some archaeological results. In *Skilled Production and Social Reproduction*, edited by Jan Apel and Kjel Knutsson, pp. 37–68. Stone Studies 2. Societas Archaeologica Upsaliensis, Upsalla.

Pelegrin, Jacques, and Marcel Otte 1992 Einige Bemerkungen zur Präparations- und Ausbeuttechnik der Kernsteine aus Raum 29. In *Hassek-Höyük: Naturwissenschaftliche Untersuchungen und lithische Industrie* edited by Manfred R. Behm-Blancke, pp. 219–224. Istanbuler Forschungen 38. Ernst Wasmuth Verlag, Tübingen.

Rosen, SA 1997 *Lithics after the stone age: a handbook of stone tools from the Levant*. AltaMira Press (A Division of Sage Publications Inc), Wallnut Creek, London & New Delhi.

Sheets, Payson D. 1978 A model of Mesoamerican obsidian technology based on Preclassic workshop debris in El Salvador. In *Archaeological studies of Mesoamerican obsidian,* edited by Thomas R. Hester pp 159–170. (Studies in Mesoamerican Art, Archaeology and Ethnohistory no3). Ballena Press, Socorro.

Shimelmitz, Ron 2009 Variability in Specialized Canaanean Blade Production of the Early Bronze Age Levant. In *Techniques and People. Anthropological perspectives on technology in the Archaeology of the Proto-historic and Early Historic periods in the southern Levant.* edited by Steve A. Rosen, Valentine Roux. pp. 135–156 (Mémoires et Travaux du Centre de Recherche Français à Jérusalem, Archéologie et Sciences de l'Antiquité et du Moyen Age, no 9), De Boccard, Paris.

Stein, Gil J. 2000 Material culture and social identity: the evidence for a fourth millennium BC mesopotamian Uruk colony at Hacinebi, Turkey. *Paléorient* 25(1): 11–22.

Tixier, Jacques 1984 Le débitage par pression. In *Préhistoire de la pierre taillée 2. Economie du débitage laminaire: technologie et expérimentation (IIIe table ronde de technologie lithique, Meudon-Bellevue, oct. 1982),* edited by Jacques Tixier, pp. 57–70. CREP, Paris.

Zohary, Daniel, and Maria Hopf 2004 *Domestication of Plants in the World - The origin and spread of cultivated plants in West Asia, Europe and the Nile Valley*. Third edition. Oxford University Press, Oxford.

Chapter 7
Pressure-Knapping Blade Production in the North-Western Mediterranean Region During the Seventh Millennium cal B.C.

Didier Binder, Carmine Collina, Raphaëlle Guilbert, Thomas Perrin, and Oreto Garcia-Puchol

7.1 Introduction

Following the research of Jacques Tixier and Marie-Louise Inizan (Tixier 1976, 1978; Inizan 1991), some scholars have suggested looking at pressure-knapping blade production as an original cultural marker from its first appearance, not only as a simple technique, but also as the component of a set of transferable methods.

In south-eastern France, in the beginning of the 1980s, the hypothesis was proposed that Castelnovian hunters had practised pressure-knapping technique (i.e. Châteauneuf-lès-Martigues – La Font des Pigeons), and that this indicated a rupture between Mesolithic technical tradition and the Cardial that was considered at that time as the first Neolithic culture of the western Mediterranean (Binder 1987, 2000). These observations were in contrast with the pattern proposed by M. Escalon de Fonton. He considered tool kits from both sets as similar and even identical, and developed a theory of the formation in the western Mediterranean of an original Neolithic techno-complex, independent from the near-eastern core of neolithization (Escalon 1956).

D. Binder (✉) • C. Collina • R. Guilbert
Université Nice Sophia-Antipolis, CNRS, UMR 6130, CEPAM, Nice, France
e-mail: didier.binder@cepam.cnrs.fr

T. Perrin
Université Le Mirail, Toulouse, France

O. Garcia-Puchol
Prehistory Department, Universitat de València, València, Spain

P.M. Desrosiers (ed.), *The Emergence of Pressure Blade Making: From Origin to Modern Experimentation*, DOI 10.1007/978-1-4614-2003-3_7,

The hypothesis of the use of pressure-knapping technique by Castelnovian groups was mainly based on the standardisation and regularity of the blades and trapezes. The situation in Provence would have therefore been similar to that described by M.L. Inizan for Upper Capsian industries in the Maghreb (Inizan 1976) where there was a relation between trapezes and pressure-knapping technique. Some aspects of the blade production fitted with this idea: regularity and parallel morphology of the arrises, development of the bulb and great variability of the butt morphologies, including particularly faceted butts canted towards the knapping surface of production, a detail that was supposed to exclude direct or indirect percussion. Here, one could recognise the characters defined by J. Tixier for the Capsian from the Aïn Dokkara (Tixier 1976).

During the last few years, the hypothesis of pressure-knapping technique within Castelnovian contexts has not been much discussed, and research concerning Mesolithic materials – for example the Tardenoisian industries from Northern France – suggested the use of indirect percussion for the production of laminar blanks whose general morphology was considered more or less similar to Castelnovian items (Pelegrin 2000; Valentin 1999).

Recent studies in Italy, France and Spain offer a new occasion for reassessing the laminar techniques and methods linked to Mesolithic trapeze production.

7.2 Dealing with Pressure-Knapping Blade Production, Techniques and Methods

After a first survey, the hypothesis that pressure was a common technique in use by Mesolithic groups during the seventh millennium cal B.C., at least in south-eastern France and in Italy, for producing bladelets as blanks for trapezoid microliths, becomes more and more solid.

Direct percussion, hard or soft, can for the most part be excluded, considering the absence of any evidence for localised impact and the development of the bulb. The main debate concerns pressure versus indirect percussion. Both generally provide more or less developed bulbs due to the inward-directed removal force; both provide a very small lip too, at the interface between the butt and the bulb, due to the outward-directed component of this force (Pelegrin 1988).

The indication for pressure is based upon the following characters:

- The bulbs are prominent, high and concentrated, and sometimes underlined by a thin ripple.
- The striking platform morphology is very diverse with many occurrences of faceted ones; there is great variability in the inclination towards the knapping surface of production as well as towards the sides of the core. Some form an obtuse *angle de chasse* between the platform and the dorsal surface of the bladelet; in several cases, the pressure is applied to a dihedral surface. Both of these characteristics would make it impossible to seat an intermediary antler tool for indirect percussion.

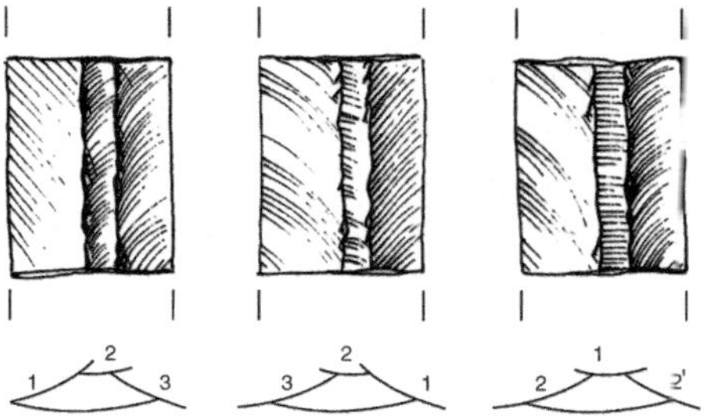

Fig. 7.1 Encoded designs ("schémas diacritiques") of the different types of trapezoidal central blades; from the *left* to the *right*: 123, 321, 212'

- The small or even very small dimensions of the cores and bladelets also argue against the general use of indirect percussion. Indirect percussion needs sufficient inertia: the cores have to be massive enough to be immobilised – between the thighs, under the foot or within a special clamp – and then to support the shock of the punch. Using a punch to remove bladelets from sub-spherical pebbles of ca. 3 cm diameter and even less is unrealistic. The removals, with a width usually around 7–8 mm and often less, are most of the time consistent with pressure applied in the hand, following Pelegrin's criteria.
- The last criteria concern the regularity, the parallel morphology of the arrises and the straightness of the bladelet profiles: these aspects are generally observed on the best products, often selected for making microliths.

Beyond the identification of the pressure-knapping technique, our first survey suggests to us that similar methods were in use throughout the area.

From Tixier's perspective, dealing with methods implicates, beyond the identification of techniques, a reconstruction of the chronology of technical gesture: sequencing the 'chaîne opératoire' and further identifying the knapping production 'rhythms' or sequences for the *plein débitage* (central blades whose scars only result from previous blade removals, unlikely lateral blades wearing cortex and/or traces of lateral sides of the preformed core [e.g. crossed removals from crests]) (Inizan et al. 1992). The observation of the chronology of previous blade removals on central trapezoidal blades let us differentiate two main designs (encoded 212' and 123/321) as depicted in Fig. 7.1. Encoding of removal sequence (i.e. *schemas diacritiques*; Dauvois 1976) is a quite productive method for differentiating pressure 'styles' (e.g. Early versus Late Chassey culture, Binder 1984; Early versus Middle and Late Pre-Pottery-Neolithic, Binder 2007).

Bladelets were often removed from small nodules, pebbles, prisms or core flakes. Complex shaping out of the cores, for example using crests, are exceptions. When pebbles were used – which is quite common – it is obvious that the bladelet detachment started in many cases just after the removal of a simple initial flake from the cortical knapping production surface:

- Most of the time, bladelets seem to be removed in small series (sequences of four to six items) sweeping the knapping production surface from both sides and converging towards its centre. This pattern provides a large number of trapezoidal sections with a large ratio of 212'designs (Fig. 7.1).

- Flaking on the wide face of the core provides a surface with low transversal curve, particularly at the end of the knapping production process. The cores showing such 'flat' surfaces and converging sequences are the most common ones, despite the shape and nature of the cores.
- The low transversal curve of the core is responsible for the removals spreading out, some irregularity of the blank sides and a high width versus thickness ratio.
- The wide faceted butt – sometimes quite close to the bladelet width – may result from a high force of pressure, but this has to be determined by further experimentation.
- Remodelling of the platforms between blade removals results in rapid reduction of the length of cores and bladelets and the production of many platform rejuvenation flakes.
- When the dimensions and shape of the raw materials and then of the cores are suitable, bladelet removals can be significantly reoriented, resulting in an increase of the transversal curve of the knapping production surfaces. This favours the production of much more regular bladelets with a lower width versus thickness ratio. Successive reorientations can also produce some twisted removals; the intersection of diversely oriented series creates specific designs of the arrises (converging features *en écharpe*) on the dorsal surfaces of the blanks.

Trapeze styles could also be influenced by the removal sequence: a marked transversal curve of the knapping production surface results in greater elongation of the blades as well as a higher relative thickness, the latter facilitating microburin fractures.

Knapping blades produced by small series of four to six items, produce include a high ratio (1/2–1/3) of lateral blades versus central ones. As the 'best' central products are selected for making microliths, the visibility of pressure is less if one only considers the unretouched material for dealing about techniques.

The raw materials, and particularly the original volume of the raw materials to be flaked, constitute a factor in blank design variability as soon as there is sufficient material for successive reorientations of the removals that contribute to increasing their regularity.

Another factor of the assemblage variability is obviously due to the territorial logistics and network complexity, as well as the raw material procurement strategies: the limited number of known settlements yet deprives us of any synthetic view.

In the Rhône basin, the eastern Po plain and the Alpine area, the location of Castelnovian sites, close to excellent raw materials sources is obvious. Nevertheless, no long distance diffusion of either cores or blanks have been demonstrated, save for a unique pressure bladelet from Vatte di Zambana (Verona), considered by R. Guilbert as very similar, and even identical, to some aspects, to the Cretaceous honey flint from Provence.

7.3 Some Examples Indicating Pressure-Knapping Blade Production Use at Key Mesolithic Settlements

7.3.1 Italy

L'Uzzo (Trapani, Western Sicily), a wide coastal cave found by E. Gobert and excavated by M. Piperno, has revealed a long sequence of deposits mainly corresponding to the Epipaleolithic-Mesolithic and Early Neolithic (Piperno et al. 1980) (Fig. 7.2). The Late Mesolithic or 'Transition' deposits from Trench F, layers 11–14, are dated from the early seventh millennium cal B.C. The lithic assemblages provided a very large set of blades and trapezes. All stages of the *chaînes opératoires* are present, from the rough material to the whole range of wastes and consumed tools (Collina 2009) (Fig. 7.3).

Bladelets were removed from small flint pebbles of high quality, collected on the beaches eastwards from the Uzzo promontory. Bladelets, with widths generally less than 8 mm, were used primarily for the production of trapezes, symmetrical or slightly asymmetrical, using the microburin technique (Fig. 7.3: 1, 2). The use of

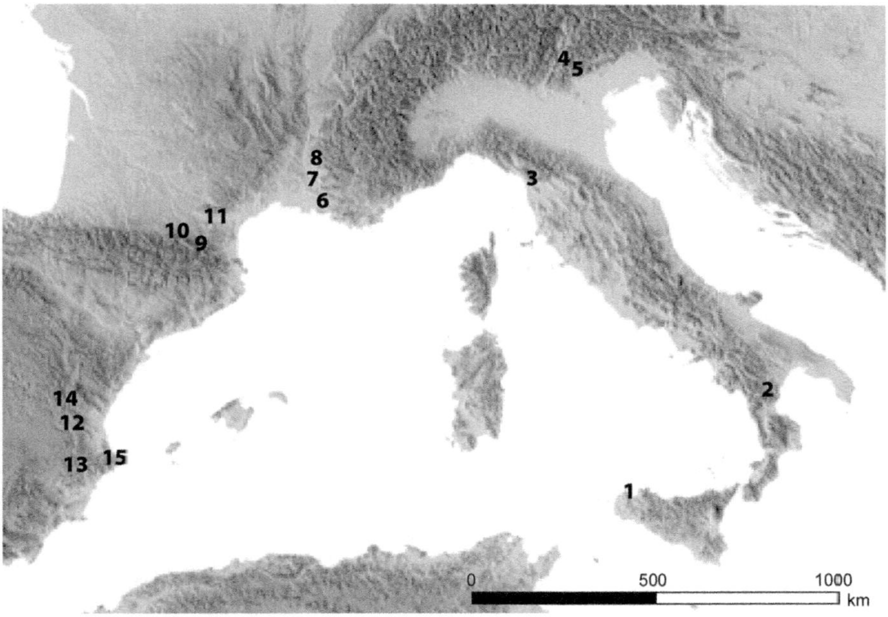

Fig. 7.2 Map of the key settlements from the Late Mesolithic cited in the text: *1* Grotta dell'Uzzo (Trapani); *2* Latronico 3 (Potenza); *3* Piazzana (Garfagnana); *4* Romagnano 3 (Trento); *5* Gaban (Piazzina di Martignano, Trento); *6* Grand abri de la Font-des-Pigeons (Châteauneuf-lès-Martigues, Bouches-du-Rhône); *7* Mourre du Sève (Sorgues, Vaucluse); *8* Lalo (Espeluche, Drôme); *9* Dourgne (Fontanès-de-Sault, Aude); *10* Gazel (Sallèles-Cabardès, Aude); *11* Buholoup (Montbéraud, Haute-Garonne); *12* Cueva de la Cocina (Dos Aguas, Valencia); *13* Abric de la Falguera (Alcoï, Alicante); *14* Covacha de Llatas (Andilla, Valencia); *15* Tossal de la Roca (Alcala, Castelon) (drawing Service cartographique-MOM-Lyon and S. Sorin-Mazouni)

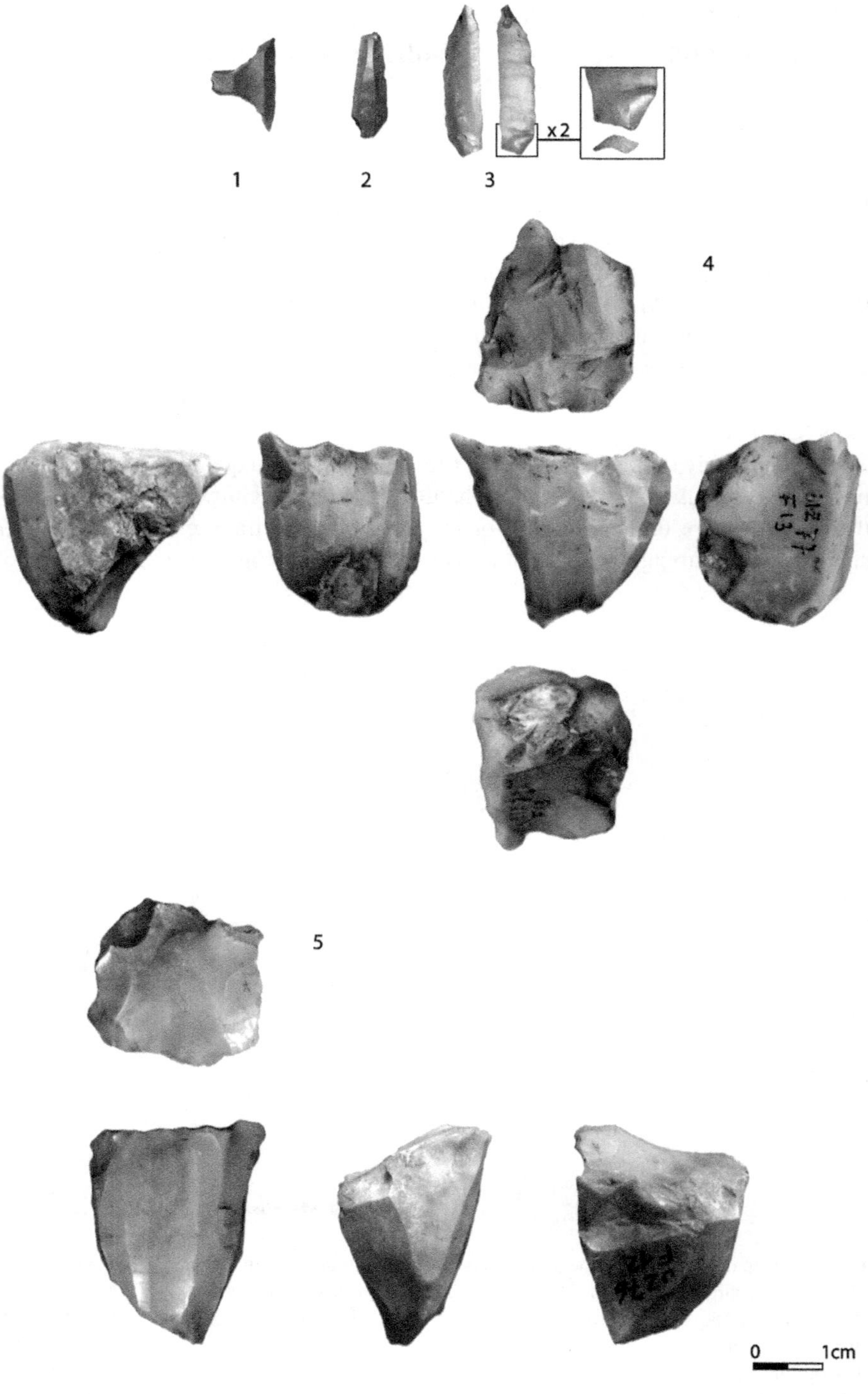

Fig. 7.3 Uzzo cave, industry from Trench F, layers 11–14: *1* trapeze; *2* microburin; *3* bladelet; *4–5* bladelet cores (Photo C. Collina)

pressure is indicated by several characters: the regular and parallel arrises (Fig. 7.3: 1–5), the marked inclination of the platform butt suggesting contact of the crutch on a dihedral (Fig. 7.3: 3–5), developed lips (Fig. 7.3: 3), high and well-delimited bulbs and wide faceted butts (Fig. 7.3: 3). Overhang of the butt is common (Fig. 7.3: 4).

The cores are very small, many with cortical residues. The knapping production surfaces have a low transversal curve. Bladelets are removed in short series, converging towards the centre of the knapping production surface with a systematic abasement of the pressure platform by faceting (Fig. 7.3: 4, 5).

The cave of Latronico 3 (Basilicate), excavated by G. Cremonesi, holds thick deposits from the Mesolithic and Neolithic periods (Cremonesi 1978) which have been recently studied from a typological perspective (Dini et al. 2008). Mesolithic layers 42–55 are dated from the whole seventh millennium cal B.C. The interpretation of the stratigraphic sequence is not very easy, because slumping may have caused possible replication. We examined lithic series from layers 41 to 42, situated in the core of sector 3 which we consider to be the more secure contexts most reliable.

Bladelet production (Fig. 7.4), similar to the L'Uzzo series, is preferentially realised on a very homogeneous grey flint. There is, at least for this phase, an excess of central products, as previously defined, compared to the lateral blades, waste and cores, indicating that the assemblage was selected from a larger production: we hypothesise that the selected blanks were flaked in another settlement or in another part of the cave for later use.

Trapezes are small, with direct retouch, symmetrical or slightly asymmetrical (Fig. 7.4: 1–17), with rare evidence for microburin facets, which fits with lack of microburins among the waste. The site of Piazzana (Garfagnana, Tosco-Emilian Apennine, Tuscany) revealed to C. Tozzi a Mesolithic layer dated from the late seventh millennium cal B.C. (Tozzi and Notini 1999).

The raw materials we examined are diverse, in the sense that they were probably collected at different stages of production and use, which could indicate a temporary settlement for specific activities such as hunting.

Pressure is visible on the blanks selected for the microliths (more or less elongated trapezes, symmetrical as well as asymmetrical) (Fig. 7.5: 1–5) and for the oblique truncations (Fig. 7.5: 6–9). When not in the rough (e.g. Fig. 7.5: 1, 4, 8), the *piquants trièdres* bear direct abrupt retouch.

A core on a pebble (Fig. 7.5: 10) shows a sequence of six pressure removals, converging from the sides towards the centre of the knapping production surface. The pressure platform is prepared for each blade removal, without any reduction of the overhang. The bladelet widths are small, in the same range as L'Uzzo or Latronico.

A second pressure core (Fig. 7.5: 11), obtained from a flake or from a prism, is also very illustrative of the removal sequence and methods in use at Piazzana. The knapping production surface shows the intersection of three laminar series (A, B, C); two of them (B, C) indicate a converging rhythm of the removals. The latter are flaked from an orthogonal faceted platform with a characteristic overhang (e.g. series B)

The Romagnano three rock shelters (Trento, Adige Valley) excavated by A. Broglio (Broglio 1971; Bagolini 1971; Alessio et al. 1984) provides one of the major sequences for the Epipaleolithic and Mesolithic of the eastern Po basin and

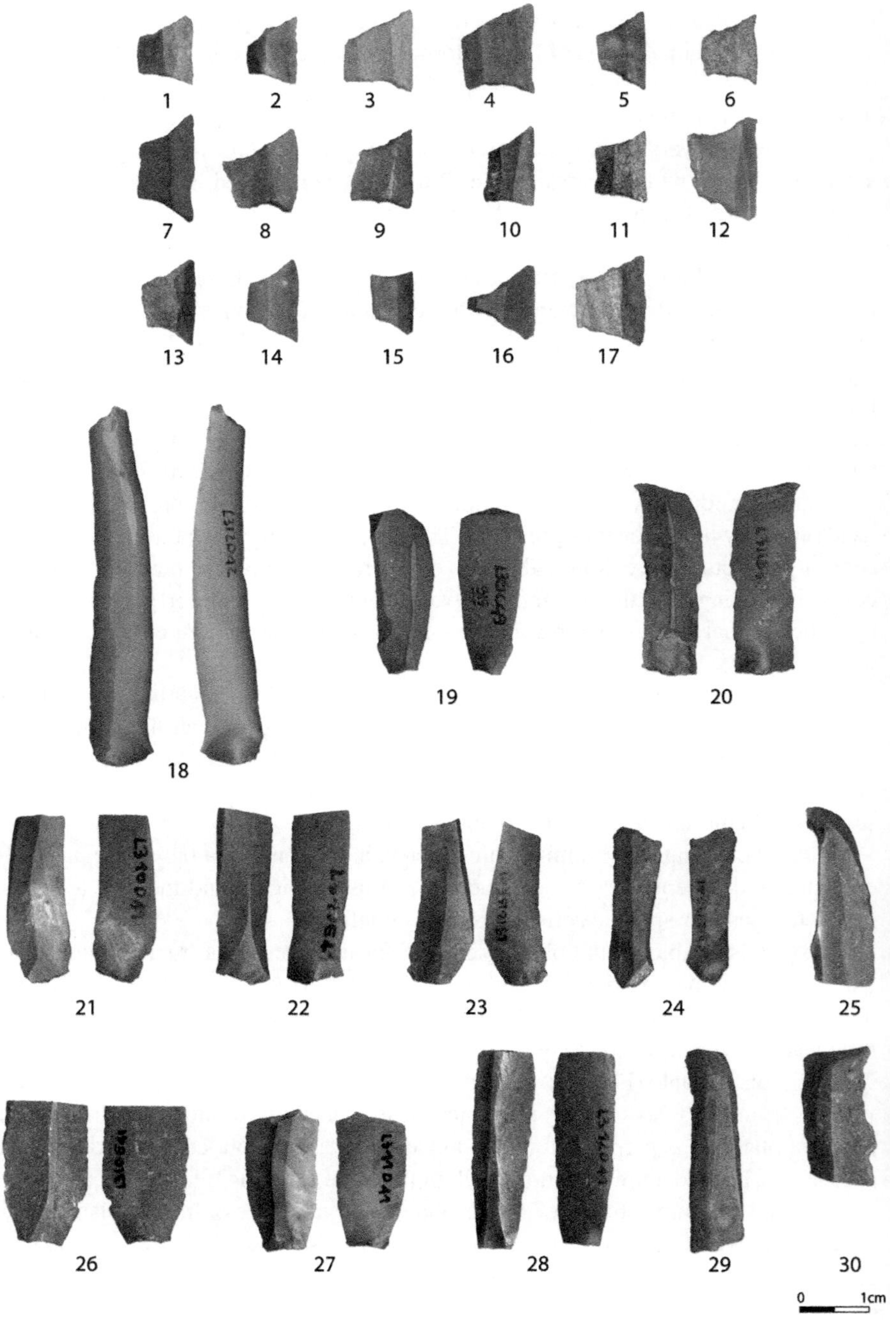

Fig. 7.4 Latronico 3 cave, industry from layers 41–42: *1–17* trapezes; *18–30* bladelets (Photo C. Collina)

Fig. 7.5 Piazzana, industry from layer 3A1: *1–5* trapezes; *6–9* truncations; *10–11* bladelet cores (Photo C. Collina)

Fig. 7.6 Romagnano 3 rock-shelter, industry from layers AB1-AB2: *1–2* bladelets (Photo R. Guilbert)

Italian Pre-Alps. Layers AB1 and AB2 are dated from the early seventh millennium cal B.C.

The assemblages are produced from local flint of the highest quality. The trapezes are diverse, symmetrical or asymmetrical, more or less elongated, with rough *piquants trièdres* or direct truncations. Pressure practices are illustrated here by two examples selected by R. Guilbert.

One of them (Fig. 7.6: 1) (RIII-743, Layer AB1) is a wide bladelet (ca. 15 mm) removed from a faceted platform on a Scaglia rossa flint core. The dorsal face presents three parallel blade removals graded from the left to the right (1,2,3) and cutting an anterior removal highly diverging (0); a fifth removal (4), parallel to 1, 2 and 3, is hinged, which indicates that the blank participates to the reparation of an accident. As far as pressure is concerned, the possibility of pressure 'in the hand' is clearly out of the question.

The second example (Fig. 7.6: 2) is provided by a blade removed from a Biancone flint core (RIII-667, Layer AB2). This twisted and plunging blank shows the intersection of two series of removals. On the right side, a set of five parallel removals bears a converging sequence towards the centre of the knapping production surface; on the left side, the latter are cut '*en écharpe*' by a sixth plunging removal. Here too, flaking in the hand is not realistic and the hypothesis of a crutch used in a sitting or even standing position is plausible.

The rock shelter of Gaban (Trento, Adige Valley) excavated by B. Bagolini (Bagolini and Biagi 1988; Kozlowski and Dalmieri 2000) is particularly known for its sequence of Mesolithic and Neolithic deposits; the latter is supposed to continue the Mesolithic lithic tradition, and these industries have been recently reassessed by Perrin (2006) as dating from the mid-seventh millennium cal B.C.

Perrin identified pressure removals for the more narrow bladelets, providing 212' designs and around 9 mm width, whereas wider ones, with faceted or concave butts, were considered to have been detached by punch. We wonder if this dichotomy is still consistent in light of more recent interpretations.

7.3.2 *South-Eastern France*

The large Font-des-Pigeons rock shelter (Châteauneuf-les-Martigues, Western Provence), excavated during the 1940s by M. Escalon de Fonton (Escalon 1956) and then by J. Courtin in 1979 (Courtin et al. 1985), is the eponymous settlement of the Castelnovian blade and trapeze late Mesolithic. Well-defined Castelnovian assemblages come from layers 18 F, 18 G1-G3 and 18 H of Courtin's excavations, dated to the seventh millennium cal B.C. The industry has been reassessed by Binder (1987) who identified the use of pressure for Castelnovian blade production at the site.

In the Provence hinterland, the Mourre-du-Sève rock shelter (Sorgues, Vaucluse), previously excavated by J. Marq (Paccard and Marq 1993), recently provided three Castelnovian units dated by AMS from the seventh millennium cal B.C. (Binder and Sénépart 2004). Cretaceous (Bedoulian) honey flint constitutes an excellent raw material for blade production. The use of pressure is evident on the geometric blanks: symmetrical or asymmetrical trapezes, sometimes very elongated (Fig. 7.7: 1–5); the *piquants trièdres* are most of the time modified by abrupt or crossed retouch.

Bladelets are usually regular with thin sections, but one can observe twisted lateral pieces. Central blades, as defined supra, seem to be over-represented compared to the wastes; that probably indicates a selection among the blade production. Their width is often more than 10 mm.

Proximal parts illustrate pressure removals: developed and well-delimited bulbs, well-expressed tearing-out lip (Fig. 7.7: 6–10, 10–12). The butts are small, generally flat or a little concave; the overhangs are reduced by abrasion and intensively smoothed. Some of the butts are canted towards the superior face, and one can note several dihedral butts as well.

The series also contains some more robust products:

- Considering the context, one of them has an exceptional width (18 mm) (Fig. 7.7: 18) that could be due to the spreading out of the removal on a surface with a weak transversal convexity. The metric and discrete characters (prominent bulb, tearing-out lip, dihedral butt) could suggest the use of pressure or of a punch.
- A second blade (Fig. 7.5: 19), passing over the right side of the core, indicates removals realised from an overhanging faceted platform and a surface with a low transverse convexity. This piece provides a marked S profile and a butt totally canted towards the knapping production surface that both indicate pressure. These characteristics suggest use of a short crutch, perhaps in a sitting position.

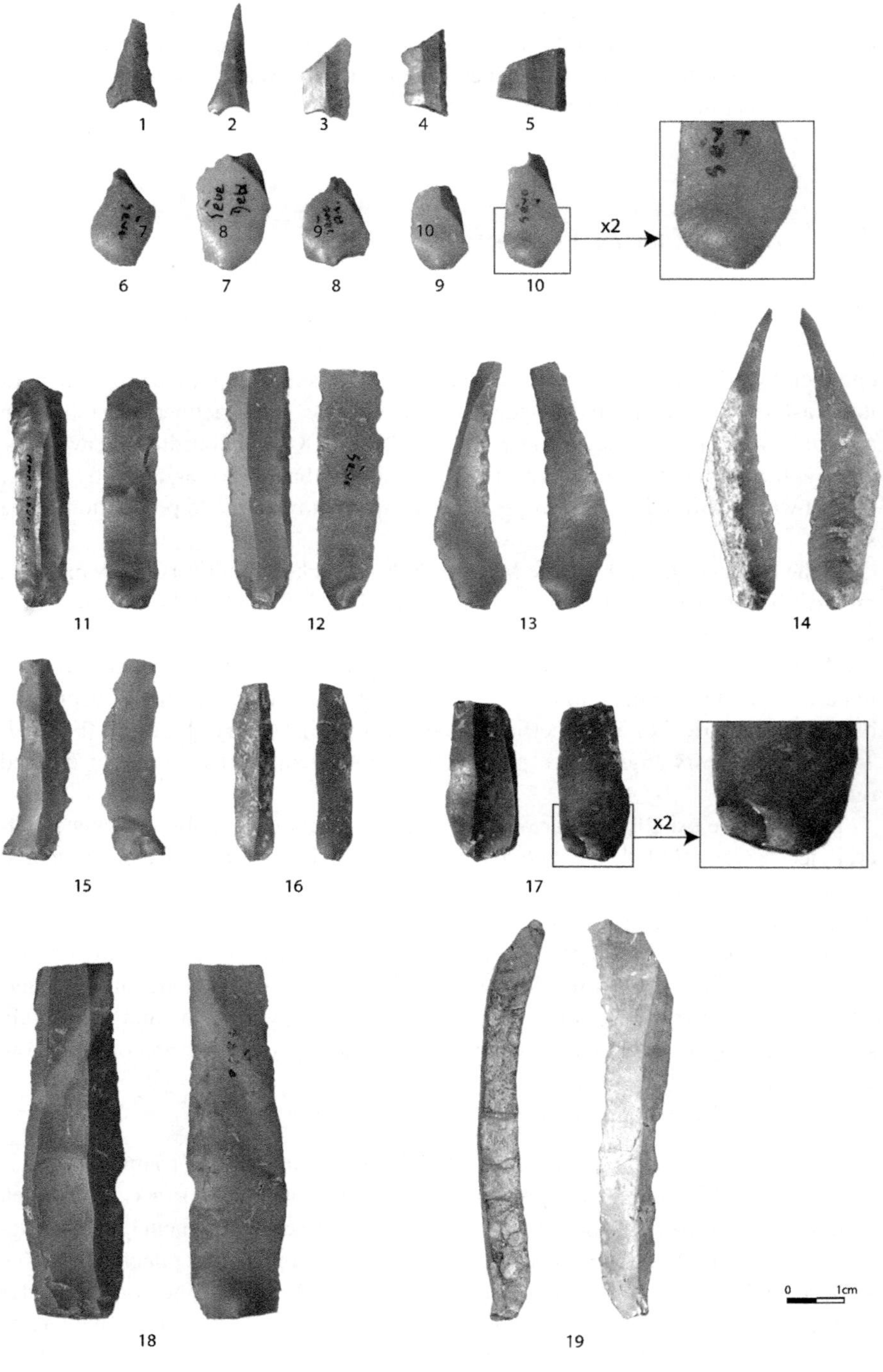

Fig. 7.7 Mourre du Sève rock-shelter, industry from Mesolithic layers: *1–5* trapezes; *6–10* microburins; *11–19* bladelets (Photo D. Binder)

In the Middle Rhône Valley, the stratified open air site of Lalo (Espeluche, Drôme), provided a Castelnovian settlement excavated by A. Beeching and dated to the late seventh millennium cal B.C. (Berger et al. 2002). The industry, currently being studied by R. Guilbert, was produced using the best available Cretaceous flints, acquired as little blocks or large flakes.

A single entire bladelet is one of the rare pieces that indicate a possible use of pressure: rectilinear converging arrises, with a 212' design, high and prominent bulb, marked lip, small flat butt, abraded overhang, etc.

On the other hand, the irregularity of most of the blades, with sinuous arrises and divergent edges, could mean that the best central blades were taken away. For instance, it is still difficult to conclude whether there was coexistence of pressure and punch techniques.

7.3.3 South-Western France and Spain

In eastern Languedoc and the Pyrenean piedmont, standard Castelnovian industries have still not been recognised. During the seventh millennium cal B.C., the caves and shelters of Dourgne, Gazel and Buholoup present specific assemblages, with triangular points. M. Barbaza (Guilaine et al. 1993), F. Briois (Briois 1997) or, more recently, T. Perrin did not identify any trace of blade production, except for rare single imported pieces. De facto, the raw material exploited, generally of a very poor quality, is not suitable for blade production.

In Spain, Levante and Aragon, this question is still open and needs investigation. Nevertheless, due to O. Garcia-Puchol's reassessment, the blade series of La Cocina, La Falguera, Llatas and Tossal de la Roca have shown, on diverse raw materials, similar characters to those that have been described above, with a dominance of wide faceted butts, inclined platforms, blade widths smaller than 10 mm, as well as a frontal bladelet removal and a low transverse convexity of the knapping production surface. A part of these sites are dated to the seventh millennium cal B.C. Nevertheless, the use of indirect percussion is also suggested (Garcia Puchol 2005).

7.4 Comments and Prospects

1. Beyond a common predilection for trapezoidal arrowheads, the Mesolithic industries discussed here present many similarities. These are due, not only to the common use of pressure techniques, but much more significantly to the methods involved. We suggest the sharing of a set of technical traditions within a wide area during the seventh millennium. In contrast to previous Mesolithic traditions, these practices are attested from Sicily to the Trentino and from the Provence hinterland to the Levante, beginning in the first half of the seventh millennium, and more securely since 6600 cal B.C. (Fig. 7.8).

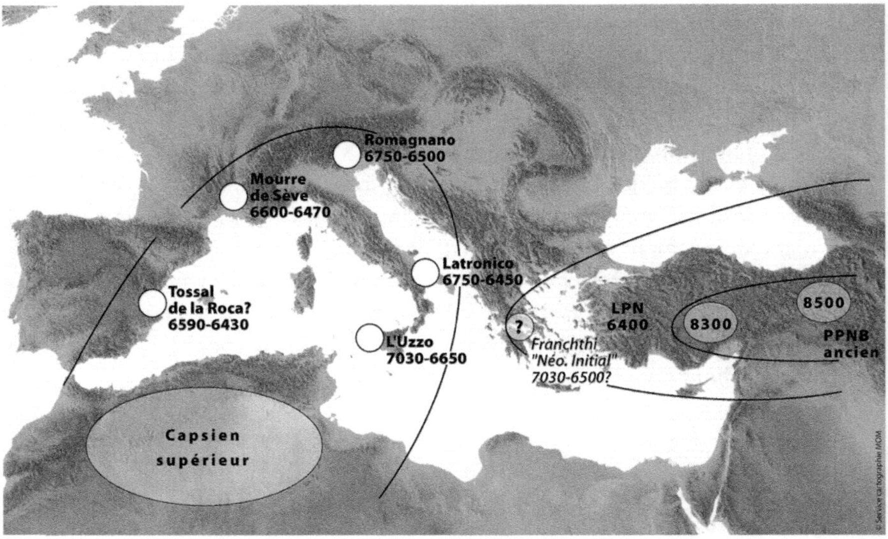

Fig. 7.8 Principal areas of diffusion of pressure within the Mediterranean with ranges of calibrate dates B.C. (Drawing Service cartographique-MOM-Lyon and S. Sorin-Mazouni)

The differences registered between the north (Veronese Pre-Alps and Provence) and the south of this zone concern particularly the removal sizes they are probably to be linked to the raw material quality, dimensions and availability. Actually, most of the raw materials in use to the south are small or very small pebbles, while the northern sites are generally close to huge outcrops of wide blocks of high quality flint. The Romagnano and the Mourre de Sève series, and maybe later Lalo, probably indicate use of a crutch in a sitting or even standing position, in addition to knapping production in the hand, which was much more broadly distributed in space.

2. The question of the origins of this technical tradition is still unresolved.

   In the eastern Mediterranean, pressure-knapping technique is known within the Anatolian and Aegean Neolithic. In Central Anatolia, the first evidence of pressure technique for blade removal comes from the Göllü Dağ obsidian workshops (i.e. Kömürcü-Kaletepe): these productions, linked to Early Pre-Pottery Neolithic B (EPPNB) contexts starting in the ninth millennium cal B.C., significantly decreased (if they did not totally disappear) with B.C. the emergence of the Aşıklı Hüyük complex around 8200 cal B.C. The first stages of Çatalhöyük do not show this technique, which does not seem to appear there before the mid-seventh millennium, due to the reactivation of eastern connections (Binder 2002; Carter et al. 2006). The reliable data today available suggest that the first spread of the Pottery Neolithic towards the Lake District and the Marmara has to be dated around 6400–6500 cal B.C. In these areas, most of the assemblages belonging to monochrome-red-slipped or painted ware contexts as well as impressed or scratched ware contexts provide evidence for pressure-knapping blade technique. Within these contexts, and particularly by the seventh millennium, pressure-knapping technique generally leads to conical or semi-conical core shapes (i.e. Çatalhöyük, Bialor 1962; Höyucek, Balkan-Atlı 2005).

The data used for suggesting 'Mesolithic' pressure-knapping technique in the Marmara region (i.e. Ağaşlı, Özdoğan and Gatsov 1998; Gatsov 2003) are not from reliable stratified contexts and belong to mixed surface series including Upper Paleolithic to Bronze Age lithic industries. The Pendik – Fikirtepe pressure blade production is more likely linked to the Central Anatolia and Lake District Neolithic traditions.

In Thessaly and the Argolid, pressure bladelets have been identified by Perlès (2001) in the Initial Neolithic phase at Franchthi Cave and Argissa Magoula. A few obsidian and flint pressure bladelets and symmetrical trapezes were identified at Franchthi. Nevertheless, three radiocarbon dates, which could situate this episode prior to 6400 cal B.C., need to be confirmed because two of them were obtained on ashy earth which is an unsuitable material (Perlès 2001: Table 5.3, 86); in addition, the stratigraphical context described as 'severely disturbed by more recent occupations' is still unclear (Perlès 2001: 46). Further details about the pressure method in use at Franchthi during the Initial Neolithic for blade production would be of a very high interest: could it be related to Anatolian traditions?

Currently, the probability for any connection between the western Mediterranean blade and trapeze complex and the eastern Neolithic package is weak and should require a new look at some more details from the Greek lithic assemblages.

On the other hand, the possibility of a link between 'Castelnovian-like' industries and the Upper Capsian (Inizan 1976; Tixier 1976; Rahmani 2003) has to be taken into account. Several authors previously discussed this point, for example Tixier commenting on La Cocina knapping production process as *débitage Capsien supérieur, en plus petit* (Jacques Tixier, personal communication) or J. Roche (Roche 1975). A lot of Upper Capsian dates were obtained on terrestrial or marine shells as well as ostrich eggs; most of them are to be considered suspiciously considering the complexity of carbon cycles within these materials (i.e. reservoir effect, post-depositional exchanges). Nevertheless, recent radiocarbon dates obtained on wood charcoal are reliable to Upper Capsian contexts where pressure techniques are attested at least during the second half of the eighth millennium cal B.C. (e.g. Kef Zoura (Algeria) – Unit IV; Jackes and Lubell 2008). These dates seem consistent with the presence of trapezes and bladelets at Medjez II (Camps-Fabrer 1975) and the generalisation of Upper Capsian dates mainly after 7300 cal B.C. (Rahmani 2003). Unfortunately, most of these measures have a large standard deviation and dates on short life material will be welcome. In addition, current research underlines (1) the long duration of Upper Capsian and the complexity of its internal chronology (Mulazzani et al. 2006) and (2) uncertainties due to the specific site formation processes (Lubell et al. 2009).

3. Considering the Impresso-Cardial complex from the Adriatic and the Tyrrhenian, pressure blade productions are definitely documented for the earliest Neolithic around 5900–5600 cal B.C.: for example l'Uzzo (Sicily), Favella (Calabria), Scamuzo and Rippa Tetta (Puglia), Torre Sabea (Basilicate), La Starza d'Ariano-Irpino (Campania), Arene Candide (Liguria) and Peyrosignado (Languedoc) (Barbaza and Briois 2003; Briois 2000; Collina 2009). In some cases punch technique is also attested (e.g. Torre Sabea). At Scamuzo, Favella and L'Uzzo, the bladelet modules seem to be very close to Mesolithic ones; on the contrary, for the Guadone contexts of Rippa Tetta or La Starza, a much more impressive blade production was probably spread from the Gargano promontory workshops.

Does the latter tradition proceed from Early Neolithic lever pressure, the use of which is attested at Early Neolithic Proto-Sesklo (Perlès 2004)? Further investigation is necessary in order to resolve the question of interaction between Mesolithic and Neolithic groups (continuity versus discontinuity; resistance versus adoption). Actually, the use of comparable techniques by Mesolithic and Neolithic knappers, even at different scales and in different ways, could have favoured adoption processes.

These big questions cannot be answered at the moment, and a large collective effort is still necessary in order to characterise and compare the different methods of pressure blade flaking on a large scale. To specify the diagnostics concerning knapping production techniques, methods and know-how, is still a pertinent issue for identifying aspects of craft specialisation that are supposed to have come from the eastern Neolithic package and for trying to compare them to practices which were more broadly shared.

Acknowledgements The authors thank Pr. R. Grifoni-Cremonesi and Pr. C. Tozzi (Pisa University), Pr. A. Broglio and Dr. A. Pessina (Trento Museum), Pr. M. Piperno (Roma-La Sapienza University) for their welcome and assistance in examining the lithic collections, J. Tixier who authorised access to his 1964 personal field and lab notebook and S. Sorin-Mazouni (CNRS-CEPAM) for illustrations.

References

Alessio, M., L. Allegri, F. Bella, A. Broglio, G. Calderoni, C. Cortesi, S. Improta, M. Preite-Martinez, V. Petrone, and B. Turi 1984 14 C datings of three mesolithic series of Trento Basin in the Adige Valley (Vatte di Zambana, Pradestel, Romagnano) and comparisons with mesolithic series of other regions. *Preistoria Alpina* 19: 245–254.

Bagolini, B. 1971 Considerazioni preliminari sull'industria litica dei livelli neolitici di Romagnano (Trento), (scavi 1969–1970). *Preistoria Alpina* 7: 107–133.

Bagolini, B., and P. Biagi 1988 The first Neolithic chipped stone assemblages from Northern Italy. In *Chipped stone industries of the early farming cultures in Europe*, edited by J. K. Kozlowski and S. K. Kozlowski, pp. 423–448. Archaologia interregionalis, Warsaw University Press, Warsaw.

Balkan-Atlı, N. 2005 Yontmataş Endüstrisi. In *Höyücek.1989–1992 yılları. Arasında Yapılan Kazıların Sonuçları*, edited by R. Duru and G. Umurtak, pp. 130–137. Türk Tarih Kurumu, Ankara.

Barbaza, M., and F. Briois 2003 L'industrie de pierre taillée de Torre Sabea. In *Torre Sabea. Un établissement du Néolithique ancien en Salento*, edited by J. Guilaine and G. Cremonesi, pp. 109–129. Collection de l'Ecole française de Rome, 315. Ecole française de Rome, Rome.

Berger, J.-F., C. Delhon, S. Bonté, S. Thiébault, D. Peyric, A. Beeching, and J. Vital 2002 Paléodynamique fluviale, climat, action humaine et évolution des paysages du bassin versant de la Citelle (moyenne vallée du Rhône, Drôme) au cours de l'Atlantique ancien (800–6,000 B.P.) à partir de l'étude de la séquence alluviale de l'Espeluche-Lalo. In *Les fleuves ont une histoire. Paléoenvironnement des rivières et des lacs français depuis 15,000 ans*, edited by J.-P. Bravard and M. Magny, pp. 223–238. Errance, Paris.

Bialor, P. A. 1962 The chipped stone industry of Çatal hüyük. *Anatolian Studies* 12: 67–110.

Binder, D. 1984 Systèmes de débitage laminaire par pression : exemples chasséens provençaux. In *Préhistoire de la pierre taillée, 2: Economie du débitage laminaire: Technologie et expérimentation*, edited by J. Tixier, M.-L. Inizan and H. Roche, pp. 71–84. CREP, Paris.

Binder, D. 1987 *Le Néolithique ancien provençal: technologie et typologie des outillages lithiques*. Supplément à Gallia-Préhistoire, 24. CNRS, Paris.

Binder, D. 2000 Mesolithic and Neolithic interaction in southern France and northern Italy: new data and current hypotheses. In *Europe's first farmers*, edited by T. D. Price, pp. 117–143. University press, Cambridge.

Binder, D. 2002 Stones making sense: what obsidian could tell about the origins of Central Anatolian Neolithic. In *The Neolithic of Central Anatolia*, edited by F. Gérard and L. Thissen, pp. 79–90. Ege Yayınları, Istanbul.

Binder, D. 2007 PPN Pressure Technology: views from Anatolia. In *Technical systems and PPN communities in the near East*, edited by L. Astruc, D. Binder and F. Briois, pp. 235–243. Association pour la promotion et la diffusion des connaissances archéologiques, Antibes.

Binder, D., and I. Sénépart 2004 Derniers chasseurs et premiers paysans de Vaucluse. Mésolithique et Néolithique ancien : 7000-4700 av. J.-C. In *Vaucluse Préhistorique. Le territoire, les hommes, les cultures et les sites*, edited by J. Buisson-Catil, A. Guilcher, C. Hussy, M. Pagni and M. Olive, pp. 131–162. Editions A. Barthélemy, Avignon.

Briois, F. 1997 *Les industries lithiques en Languedoc méditerranéen (6,000–2,000 av. J.-C.). Rythmes et évolution dans la fabrication des outillages de pierre taillée entre mer et continent.* Thèse de Doctorat, Ecole des Hautes Etudes en Sciences Sociales.

Briois, F. 2000 Variabilité techno-culturelle des industries lithiques du Néolithique ancien en Languedoc. In *Sociétés et Espaces. Actaulités de la Recherche. Rencontres Méridionales de Préhistoire Récente, Toulouse, 1998*, edited by M. Leduc, N. Valdeyron and J. Vaquer, pp. 43–50. Archives d'Ecologie Préhistorique, Toulouse.

Broglio, A. 1971 Risultati preliminari delle ricerche sui complessi epipaleolitici della Valle dell'Adige. *Preistoria alpina* 7: 135–241.

Camps-Fabrer, H. 1975 Le faciès sétifien du Capsien supérieur. In *L'Epipaléolithique méditerranéen : actes du colloque d'Aix-en-Provence, juin 1972,* edited by G. Camps, pp. 127–150. CNRS, Paris.

Carter, T., G. Poupeau, C. Bressy, and N. J. G. Pearce 2006 A new programme of obsidian characterization at Çatalhöyük, Turkey. *Journal of Archaeological Science* 33(7): 893–909.

Collina, C. 2009 *Evolution des industries lithiques du Néolithique ancien en Italie du sud.* Thèse de Doctorat, Université de Provence et Université Roma – La Sapienza.

Courtin, J., J. Evin, and Y. Thommeret 1985 Révision de la stratigraphie et de la chronologie absolue du site de Châteauneuf-lès-Martigues (Bouches-du-Rhône). *L'Anthropologie* 89(4): 543–556.

Cremonesi, G. 1978 Gli scavi nella Grotta N.3 di Latronico (nota preliminare). In *Atti della XX riunione scientifica dell'Istituto italiano di Preistoria e Protostoria, Basilicata ottobre 1976*, pp. 177–198. IPP, Firenze.

Dauvois, M. 1976 *Précis de dessin dynamique et structural des industries lithiques préhistoriques*. Pierre Fanlac, Périgueux.

Dini, M., R. Grifoni Cremonesi, S. K. Kozlowski, G. Molara, and C. Tozzi 2008 L'industria castelnoviana della Grotta di Latronico 3 (Potenza, Italia). *Preistoria Alpina (Trento)* 43: 49–74.

Escalon de Fonton, M. 1956 Préhistoire de la Basse-Provence. *Préhistoire* 12: 1–162.

Garcia Puchol, O. 2005 *El Proceso de Neolitización en la Fachada mediterránea de la Península Ibérica. Tecnología y Tipología de la piedra tallada.* British Archaeological Reports, International series 1430. Archaeopress, Oxford.

Gatsov, I. 2003 Chipped stone assemblages from Pendik. Technological and typological analysis. In *From village to cities. Early villages in the Near East. Studies presented to Ufuk Esin*, edited by M. Özdogan, H. Hauptmann and N. Basgelen, pp. 283–292. Vol. 1. Arkeoloji ve sanat yayinlari, Istanbul.

Guilaine, J., M. Barbaza and J. Gasco (editors) 1993 *Dourgne.* Centre d'Anthropologie des sociétés rurales, Archéologie en Terre d'Aude, Toulouse, Carcassonne.

Inizan, M.-L. 1976 *Nouvelle étude d'industries lithiques du Capsien.* Thèse de doctorat de 3ème cycle, Université Paris 10.

Inizan, M.-L. 1991 Le débitage par pression: des choix culturels. In *25 ans d'études technologiques en Préhistoire: bilan et perspectives*, pp. 367–378. XIe Rencontres internationales d'archéologie et d'histoire d'Antibes. APDCA, Juan-les-Pin

Inizan, M.-L., H. Roche, and J. Tixier 1992 *Technology of knapped stone.* Préhistoire de la pierre taillée, 3. CREP, Meudon

Jackes, M., and D. Lubell 2008 Early and Middle Holocene Environments and Capsian Cultural Change: Evidence from the Télidjène Basin, Eastern Algeria. *African archaeological revue* 25: 41–55.

Kozlowski, S. K., and G. Dalmieri 2000 Riparo Gaban: the Mesolithic layers. *Preistoria alpina* 36: 3–42.

Lubell, D., J. Feathers, and J. L. Schwenninger 2009 Post-Capsian occupation in the eastern Maghreb: implications of a revised chronological assessment for the adult burial at Aïn Misteheyia. *Journal of African Archaeology* 7(2): 175–189.

Mulazzani, S., R. Boussoffara, N. Aouadi, V. Azzarà, L. Belhouchet, A. Boujelben, A. Cavallari, E. R. Crema, A. Curci, C. Del Grande, E. Maini, M. A. Mannino, C. Mazzanti, and S. Scaruffi 2006 Entre Sahara et Méditerranée. Le peuplement de Sebkhet Halk el Menjel (Tunisie orientale) pendant l'Holocène (campagnes 2002–2005). *Africa* 21: 189–212.

Özdoğan, M., and I. Gatsov 1998 The aceramic Neolithic period in Western Turkey and in the Aegean. *Anatolica* 24: 209–232.

Paccard, M., and J. Marq 1993 L'abri Marq à Sorgues (Vaucluse). *Bulletin de la société d'étude des Sciences naturelles de Vaucluse* 1993: 17–32.

Pelegrin, J. 1988 Débitage expérimental par pression : du plus petit au plus grand. In *Technologie préhistorique*, edited by J. Tixier, pp. 37–53. Notes et monographies techniques du CRA, 25. CNRS, Paris.

Pelegrin, J. 2000 Les techniques de débitage laminaire au Tardiglaciaire: critères de diagnose et quelques réflexions. In *L'Europe centrale et septentrionale au Tardiglaciaire. Confrontation des modèles régionaux de peuplement*, edited by B. Valentin, P. Bodu and M. Christensen, pp. 73–86. Musée de Préhistoire d'Ile de France, Nemours.

Perlès, C. 2001 *The Early Neolithic in Greece.* Cambridge World Archaeology/University Press, Cambridge.

Perlès, C. 2004 *Les industries lithiques taillées de Franchthi (Argolide, Grèce). Tome 3: Du Néolithique ancien au Néolithique final.* Excavations at Franchthi cave, Greece, 13. Indiana university press, Bloomington and Indianapolis.

Perrin, T. 2006 Nouvelles réflexions sur la transition Mésolithique récent – Néolithique ancien à l'abri Gaban (Trento, Italie). *Preistoria Alpina* 41(2005): 89–146.

Piperno, M., S. Scali, and A. Tagliacozzo 1980 Mesolitico e Neolitico alla Grotta dell'Uzzo (Trapani). Primi dati per un'interpretazione paleoeconomica. *Quaternaria* 22: 275–300.

Rahmani, N. 2003 *Le Capsien typique et le Capsien supérieur: évolution ou contemporanéité. Les données technologiques.* British Archaeological Reports, International series 1187. Archaeopress, Oxford.

Roche, J. 1975 Les amas coquilliers de Muge (Portugal). In *L'Epipaléolithique méditerranéen*, pp. 79–81. CNRS, Paris.

Tixier, J. 1976 L'industrie lithique capsienne de l'Aïn Dokkara, région de Tebessa, Algérie. *Libyca* 24: 21–54.

Tixier, J. 1978 *Notice sur les travaux scientifiques*. Thèse de doctorat d'état ès lettres, Université de Paris X-Nanterre.

Tozzi, C., and P. Notini 1999 L'Epigravettien final et le Mésolithique de l'Apennin tosco-émilien et de la vallée du Serchio (Toscane septentrionale). In *L'Europe des derniers chasseurs. Epipaléolithique et mésolithique. Actes du 5e colloque international UISPP, commission XII, Grenoble septembre 1995*, edited by A. Thévenin and P. Bintz, pp. 483–488. Editions du Comité des travaux historiques et scientifiques, Paris.

Valentin, B. 1999 Techniques et cultures: les chasseurs cueilleurs du Tardiglaciaire dans le sud et l'ouest du Bassin parisien. In *L'Europe des derniers chasseurs. Epipaléolithique et mésolithique. Actes du 5e colloque international UISPP, commission XII, Grenoble septembre 1995*, edited by A. Thévenin and P. Bintz, pp. 201–212. Editions du Comité des travaux historiques et scientifiques, Paris.

Chapter 8
Origin and Development of Pressure Blade Production in the Southern Iberian Peninsula (6th–3rd Millennia B.C.)

Antonio Morgado and Jacques Pelegrin

8.1 Introduction

Crabtree (1968), based on Spanish chronicles, demonstrated pressure blade production in Mesoamerica. Tixier (1976) was the first to report on this technique in Northern Africa; following his contribution, the presence of the pressure knapping technique blade production has been increasingly acknowledged across a variety of cultural periods. Inizan (1991, 2003) compiled the available data from a number of Eurasian archaeological sites, providing a general framework on the origin, development, and dissemination of pressure blade manufacture. Although the distribution of pressure blade production in the Iberian Peninsula is likely to be consistent with that of neighboring Western Mediterranean regions, no systematic study has addressed the subject in depth.

The present study aims to provide a general outline of pressure blade production in the Southern Iberian Peninsula. For the purposes of technique identification, we will occasionally refer to experimental knapping studies (Tixier 1976; Pelegrin 1988, 2002). The analysis will be based on the assemblages of knapped lithic artifacts from the Late Prehistoric sites investigated by the Prehistory Department of the University of Granada (Martínez 1997; Martínez and Morgado 2005; Morgado 2008; Morgado et al. 2008, 2009; Ramos Millán 1997). Special attention will be given to the site of Los Castillejos de Montefrío (Arribas and Molina 1979; Afonso et al. 1996; Cámara et al. 2005) and Toro cave (Martín Socas 2004a, b), due to its broad stratigraphic sequence (6th–3rd millennia B.C.) (Fig. 8.1).

A. Morgado (✉)
Departamento Prehistoria y Arqueología, Universidad de Granada,
Campus universitario de Cartuja s/n; E- 18071 Granada, Spain
e-mail: morgado@ugr.es

J. Pelegrin
Laboratoire "Préhistoire et Technologie", CNRS et Université Paris Ouest Nanterre, MAE, Nanterre, France
e-mail: jacques.pelegrin@mae.u-paris10.fr

P.M. Desrosiers (ed.), *The Emergence of Pressure Blade Making: From Origin to Modern Experimentation*, DOI 10.1007/978-1-4614-2003-3_8,

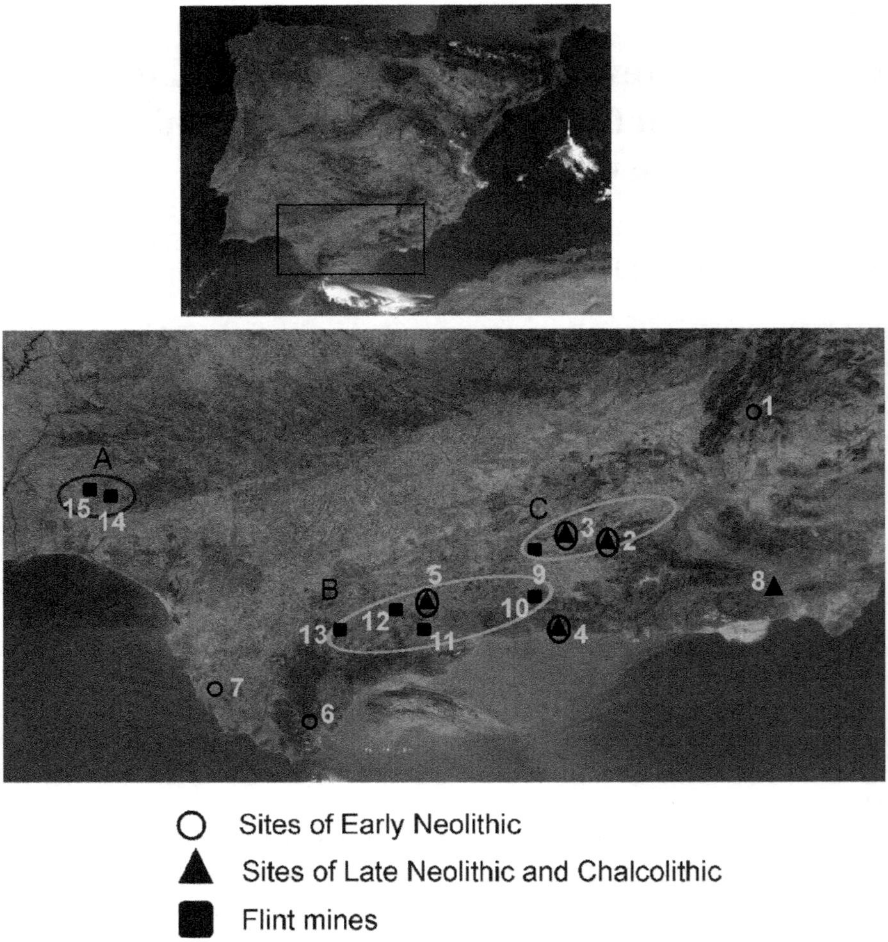

Fig. 8.1 Southern Iberian Peninsula, sites and flint mines. *1* Nacimiento cave; *2*: Carigüela cave; *3* Los Castillejos de Montefrío; *4* Nerja cave; *5* Toro cave; *6* Palmones; *7* Retamar; *8* Los Millares; *9* Los Gallumbares; *10* Cerro Alcolea; *11* Ardite/El Garrotal; *12* Valle del Turón; *13* Malaver; *14* Calañas; *15* Cerro de Andevalo

Pressure knapping blade production spread first among the Neolithic growers and shepherds of this region of the Mediterranean. The Neolithic in Southern Spain has been commonly regarded as an instance of a technical break due to the implementation of new productive systems absent from the last groups of hunter-gatherers (i.e., agriculture, cattle raising, pottery, and polished stone technology). However, among the lithic artifacts manufactured in this region, only knapped stone lithic assemblages have demonstrated typological differences (Fortea et al. 1987; Juan Cabanilles 1984). Moreover, the data supplied by a number of archaeological sites suggest that pressure blade production was widely disseminated in parallel to the systematic exploitation of abundant flint resources in the region.

Blade production experienced quite a unique evolution in the Southern Iberian Peninsula. This area of Western Europe developed a distinctive process of craft specialization for flint blade production beginning in the Late Neolithic. The technical features of blade production in Southern Spain became clearly distinguishable from elsewhere in the Western Mediterranean and seem technically closer to those of the Eastern Mediterranean and the Near East.

8.2 The Origin Issue and the Early Neolithic (ca. 5800–4000 B.C.)

Groups of farmers and cattle raisers of the Early Neolithic (ca. 6th millennium B.C.) were the first to make pressure knapping blade a common production, widespread technique in the Western Mediterranean (Binder 1984, 1987; Binder and Gassin 1988; Binder and Perlès 1990; Léa 2004; Pelegrin 2003; Terradas and Gibaja 2002). However, it should be noticed that some hunter-gatherers in Western Europe during the Upper Paleolithic were aware of the pressure flaking technique. Some groups used pressure to retouch tools, especially during the Solutrean period, and microblade production by pressure has been demonstrated in a few instances during the terminal Pleistocene (Alix et al. 1995). Some suggest that pressure knapping blade making may have emerged as a result of highly standardized bladelet production in Late Ice Age Western Europe. However, as posited by Pelegrin (2000), there were alternative technical procedures for blade production (e.g., direct percussion) used by late hunter-gatherers. The basic features of knapping by pressure technique use within flint material (i.e., detachment straightness, parallel arrises and edges, lightness; Tixier 1975) have been rarely reported in groups of the European Paleolithic (Alix et al. 1995). Pressure was unusual and limited to short periods and/or at a regional scale, thus holding a negligible role on the broad framework of blade and bladelet production.

Beginning in the Recent Mesolithic, regular pressure knapping bladelet production emerges in the last groups of hunter-gatherers in certain areas of the Western Mediterranean, the North African Upper Capsian (Inizan 1984; Tixier 1976), and some last groups of the European Mesolithic (Castelnoviense) (Binder 2000; Binder et al. this volume). These groups developed between the 7th and 6th millennia B.C. and coexisted short after with the Early Neolithic populations. Some of these groups developed bladelet standardization and geometric element manufacture (Inizan 1984). As a result of the arrival of typical Neolithic items and the neolithization of the Western Mediterranean, pressure blade production became a widespread technique.

Although a few sequences of hunter-gatherer groups of the Late Ice Age in Southern Spain have been reported, the evidence is still scarce. Although the available studies have described the basic typological characteristics of these assemblages, the analysis of the knapping production techniques has not been explored. The sequences from the Magdalenian and Epipaleolithic periods documented at the sites of Nerja Cave (Aura et al. 2009; Cava 1997; Jordá Pardo 1986), Nacimiento Cave (Asquerino and López 1981), Pirulejo site (Cortés 2008), and Palmones site

(Ramos Muñoz and Castañeda 2005) provide no evidence of the systematic use of pressure blade production. However, a few researchers have documented pressure knapping technique before the Neolithic expansion, for instance, at the site of Palmones River (Ramos Muñoz and Castañeda 2005), and the transitional period between the Epipaleolithic and Neolithic (Cava 1997; Ramos Muñoz and Lazarich 2002). However, both cases have been dated in times immediately preceding or concurrent with the earliest dates of the 6th millennium B.C. The situation changes dramatically when blade assemblages from Early Neolithic groups are analyzed. The first layers of the Early Neolithic are characterized by a high percentage of blades (60–80%), although in a few archaeological sites, a high percentage of geometric elements remain, suggesting continuity across Epipaleolithic groups (Asquerino and López 1981; Ramos Muñoz and Lazarich 2002). The percentage of geometric elements tends to decrease or disappear altogether over time. In regard to technology, the Neolithic brought about a more generalized pattern of heat treatment of flint together with pressure blade production. Blade production shows a low level of secondary modification: most of the time, bladelets have raw edges without retouch, a major typological feature of Neolithic blade assemblages. The systematic use of pressure after heat treatment lessens the need of secondary modifications. Pressure technique applied on heat-treated flint, together with knapping by pressure, thus became a distinctive feature of the Western Mediterranean Neolithic.

Although data are scarce, it can be assumed that the small production of microblades by means of handheld pressure does not demonstrate a widespread innovation by hunter-gatherer groups of the Late Pleistocene. As it has been suggested elsewhere, it is only at the beginning of the Neolithic that pressure knapping technique becomes part of the sociotechnological system of blade production, spreading across populations and with a variety of forms (blades and bladelets).

The technoeconomic and sociocultural context changed with the onset of the Neolithic in the 6th millennium B.C. The analysis of lithic production in the Early Neolithic Southern Spain has identified technical processes similar to those of other regions of the Western Mediterranean. Several studies describing the typology of these assemblages (Cava 1997; Ramos Muñoz 1988–89) have shown some geometric elements from the final hunter-gatherers (Ramos Muñoz and Lazarich 2002; Ramos Muñoz and Castañeda 2005; Aura et al. 2009). However, in purely Neolithic sites, geometric lithics are rare or absent (Cava 1997; Martín Socas et al. 2004a, b; Pellicer and Acosta 1997). On the other hand, from the very beginning of the Neolithic, there is a preeminence of bladelets and blades.

During the Middle Neolithic (5th millennium B.C.), geometric elements typical of preceding periods virtually disappeared along with backed blades and burins, while scrapers are rarely documented. On the other hand, a high percentage of bladelets with traces of use, both raw and retouched, have been found. Traces of use, notches, and denticulates are distinctive typological features of a variety of elements. There is also a strong development of blade technology along with an abundance of tools with high levels of secondary modification.

Early Neolithic strata from the sites of Cueva del Toro (ca. 5500–4600 B.C., Martín Socas et al. 2004a, b), Los Castillejos de Montefrío (ca. 5300–4800 B.C.,

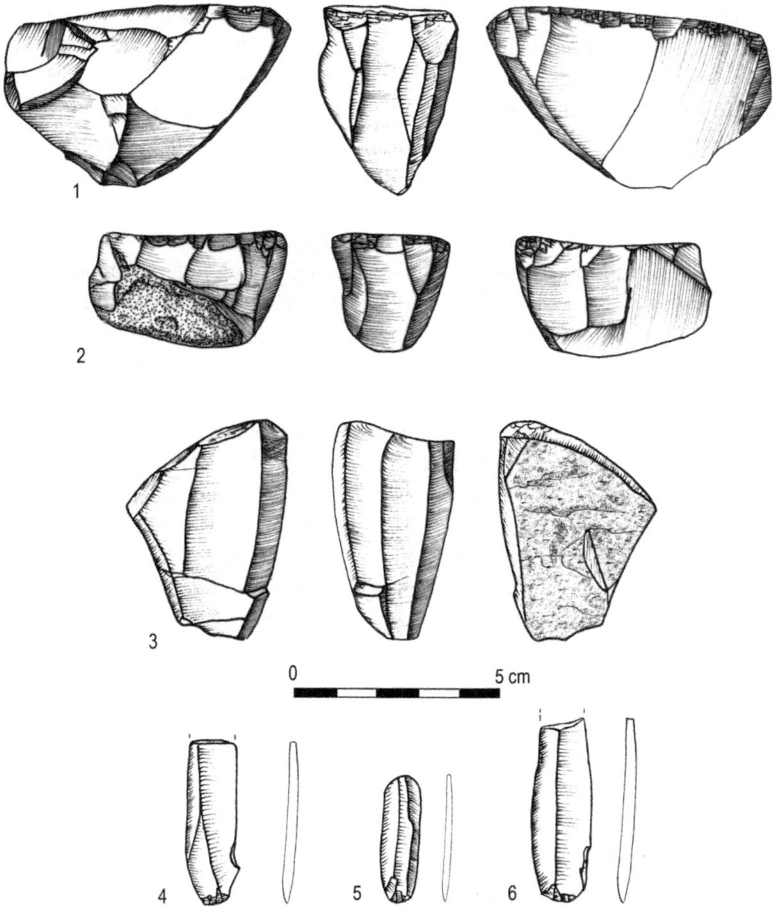

Fig. 8.2 Early Neolithic of Southern Iberian Peninsula. Pressure blades production. *1–2* Carinated performs, flint mine of Los Gallumbares; *3* Core, Los Castillejos site; *4–6* Blades and bladelets, Los Castillejos site

Cámara et al. 2005), and Carigüela Cave (Pellicer 1964) provide evidence of the use of pressure and preliminary heat treatment of the core for the production of bladelets (Inizan and Tixier 2001). Knapping by pressure technique applied on heat-treated material became a standard procedure from the Early Neolithic (ca. 5600/5500 cal. B.C.) until the 4th millennium B.C. The length of pressure-manufactured bladelets ranges from 20 to 40 mm, with almost no instances above 60 mm; less than 10% of bladelets are 13 mm or more in width. However, it should be noticed that for larger blades, indirect percussion without heat treatment was the customary practice. In summary, heat treatment was used in almost all pressure-manufactured blades, while very few had more than 13 mm in width. The domestic specialization of small blades may be a result of handheld pressure or pressure exerted with a short crutch in a sitting position (Fig. 8.2).

The preparation of the core prior to blade detachment requires carinated preforms with narrow striking platforms shaped by direct and indirect percussion (Fig. 8.2). Carinated preforms have been documented in a few flint deposits across the Betic Cordillera in Eastern Andalusia. These preforms are widely known in the Western Mediterranean and are also documented in domestic contexts indicating heat treatment (Binder 1987; Chauchat et al. 1996; Léa 2004).

Pressure blade production was persistent throughout the Early Neolithic. There are two major strategies for the preparation of the core prior to blade detachment: (a) abrasion and rubbing down of the overhang, following a movement directed toward the debitage surface, and (b) blades with a faceted butt and no edge abrasion. For the latter technique, the preparation of the pressure platform requires the detachment of minute flakes in order to place a point that will exert the pressure. We are still uncertain if these technical procedures for core preparation belong to distinguishable manufacture processes or geographical areas, or, on the other hand, are converging solutions for a single technical process. The site of Los Castillejos has one of the most abundant lithic assemblages of the advanced Early Neolithic, namely, over a thousand items and a collection of some 500 blades. Almost all blades at Los Castillejos follow a preparation of carinated cores with plain pressure platforms, while the proportion of blades with a faceted butt is low. This trend persists beginning in the 5th millennium B.C. until the Late Neolithic. During the early 4th millennium B.C., a technical change in blade production occurred. Average-size to large pressure-manufactured blades are characteristic of this period, while the domestic production of blades and bladelets of small size still remains with a variety of butt subtypes.

8.3 The Late Neolithic: The Technical Change (ca. 4000–3500/3400 B.C.)

Toward the first half of the 4th millennium B.C., a number of significant sociocultural changes took place. The construction of large villages in river valleys or inner lowlands and the appearance of collective burials are among several outstanding features of this period. The Late Neolithic has been divided into two phases: Early (ca. 4100–3800 B.C.) and Final Neolithic (ca. 3800–3400 B.C.) (Cámara et al. 2005; Molina and Cámara 2005; Pérez Bareas et al. 1999). The early phase displays continuity with the preceding period, but with a few significant changes, that will affect the next phase of village settlement consolidation.

These changes also parallel the lithic production of knapped stones. At the end of the period, geometric elements evolve to bifacial, knapped arrow points. As opposed to the Early Neolithic, these geometrics are larger due to their blade blanks. The increasing size of blades becomes obvious in the stratigraphic sequence of the region, as shown in natural caves occupied since the Early Neolithic

(Nerja, Carigüela, Toro, etc.) and in the new village settlements (Papa Uvas, Polideportivo de Martos, Llanete de Los Moros, etc.). The increasing size results from the increasing strength applied to the lithics. Additional features are the absence of heat treatment of flint and the trapezoidal sections of blades, resulting from the preparation of prismatic preforms and cores. Initially, these blades show a variety of butt subtypes (plain, plain faceted, convex) until butts with a sharp dihedron become the standard.

The stratigraphic sequence of the site of Los Castillejos de Montefrío may be a good example to illustrate the process. The site shows a technical change in pressure blade production during the third phase of Late Neolithic (Morgado et al. 2008). A number of enlarged blades with trapezoidal sections and sharp dihedral butts have been found from this period. These blades originated from prismatic cores and crests extending over the whole longitudinal length of the core. However, the new systems for blade extraction coexisted with the Neolithic systems of pressure bladelet manufacture from carinated cores and heat treatment.

The new system for blade production lasted until the Late Chalcolithic. This method required an extensive shaping of the core preforms by indirect percussion (Pelegrin and Morgado 2007; Morgado et al. 2009) (Fig. 8.3). There are two stages in the manufacture of preforms. In the first stage, the volume of the core and its crested edges are roughly shaped. In the second stage, the pressure platform and the crests, which will later guide the extraction of the first blades, are completed. There are abundant precores worked with three or four crests (two anterolateral and two posterolateral crests). A number of pieces (cores and crested blades) show that short and narrow transversal flakes were used to complete crests and eliminate imperfections. Several archaeological observations, compared to experimental stigmatas, suggest that a pointed punch that is made of some hard material was used for this purpose (Inizan et al. 1994; Méry et al. 2007; Pelegrin 1994, 2003; Pelegrin and Morgado 2007), most likely a punch armed with a point made of copper or a similar material (suggested by the small impact traces of about 2 mm in diameter). These marks resemble the blunt point of a pencil and are located in the hollow of a previous negative bulb. Thus, it might have been an enduring material harder than an antler point, which would quickly crack and split. Moreover, the flattened appearance of a few residual crests, crested blades, and neocrests also suggests that a pointy and enduring utensil was applied – a utensil that could be precisely located in the hollow of a negative bulb or over the crests.

Once the core preforms are completed, the crucial stage of blade detachment begins. The first extractions are obtained from the best delineated crests from the precore. Depending on the regularity of the first extractions, a steady pace of blade detachment will continue until the core's exhaustion. Given that a number of reshapings may occur during the process, the flint knapper may have to create new crests (i.e., neocrests) or use reserved ones (the posterior or lateroposterior crests of the core) to repair the core by a series of transversal flakes so that the blade extraction process can go on.

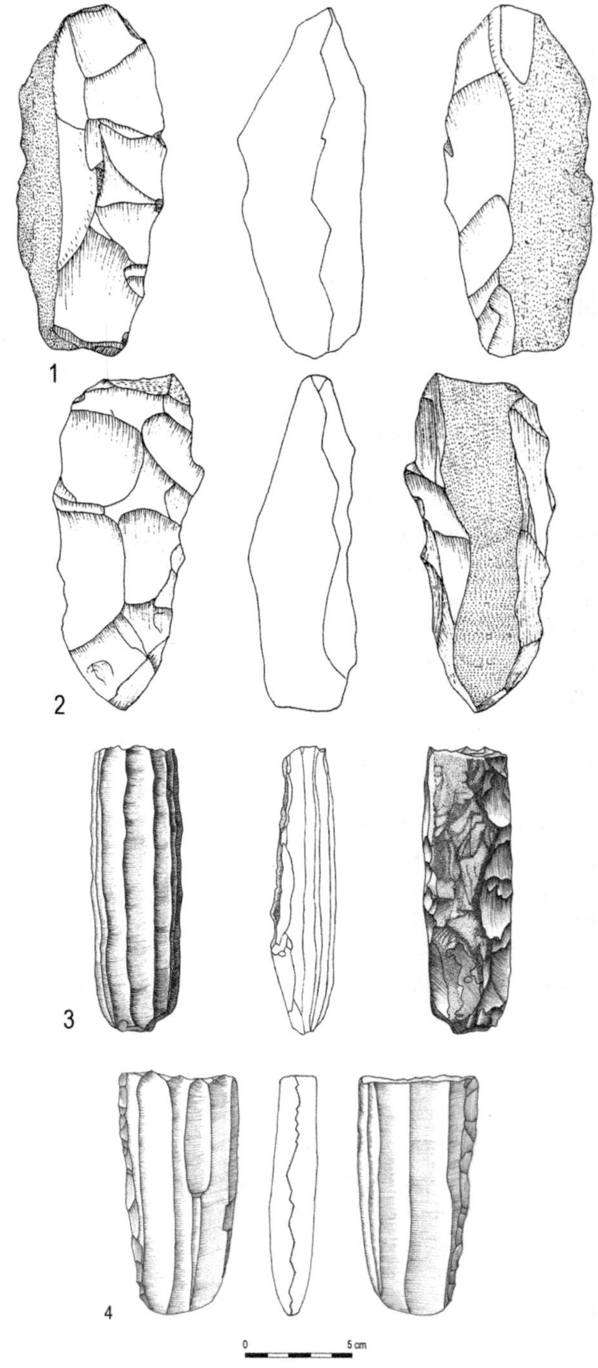

Fig. 8.3 Late Neolithic and Chalcolithic of Southern Iberian Peninsula. Pressure blade production. *1–2* Core performs, flint mine of Los Gallumbares; *3–4* Cores, flint mine of Los Gallumbares

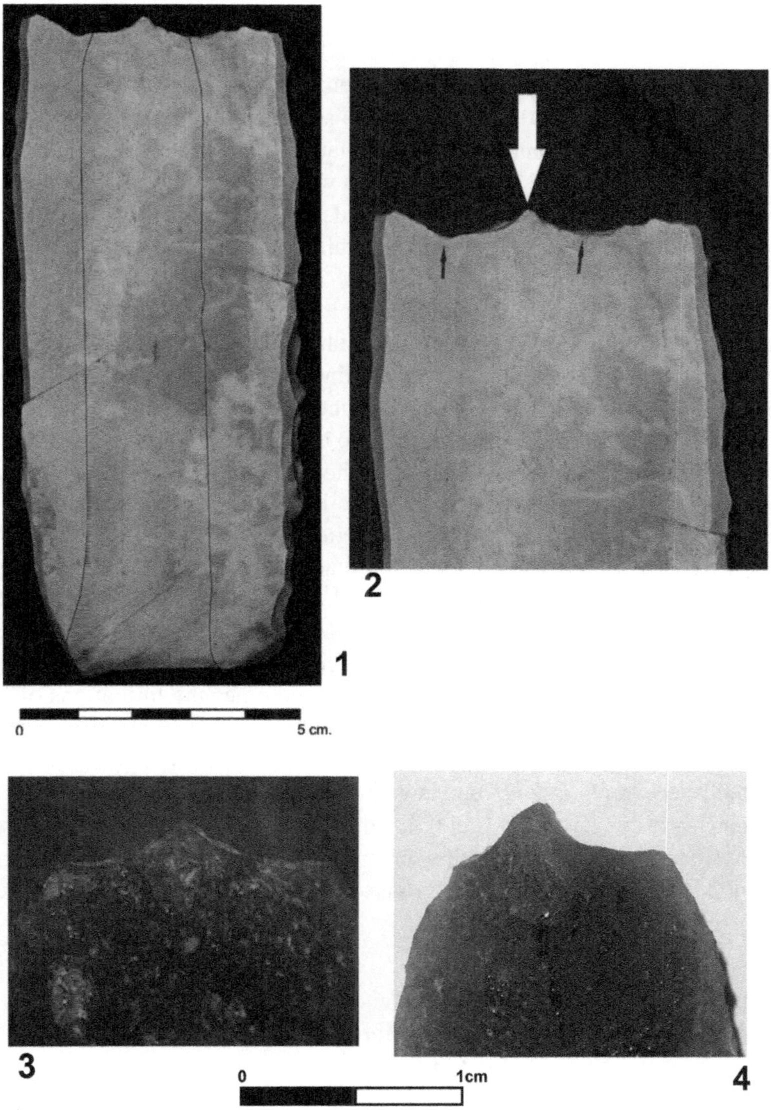

Fig. 8.4 Late Neolithic and Chalcolithic of Southern Iberian Peninsula. Pressure blade production. *1* Prismatic core, flint mine of Los Gallumbares; *2* The preparation of the pressure point; *3–4* Butts with a sharp dihedron

The blade's butt and the pressure platform of the core (observed during the last detachment) show that each blade has been extracted from a dihedron standing out of the core's edge (Fig. 8.4). This extraction procedure was highly standardized; therefore, most blade butts are consequently sharp dihedrons, while very few are asymmetric dihedrons, which may just be a version of the same design. A similar platform preparation has been already observed in Greece during the Late Neolithic (early 5th millennium B.C.), where it has been associated with copper-tipped pressure tools (Perlès 1984, 2004). A similar transition has been also demonstrated in Pakistan starting in the Chalcolithic period (late 5th millennium B.C.) (Inizan and Lechevallier 1990; Pelegrin 1994). Consistent with the cases of Greece and Pakistan, the butts extracted from sharp dihedrons found in Southern Spain also suggest a new knapping production element, namely, a metallic pressure point (copper) for blade detachment.

A number of hypotheses on the knapping production technique have been posed. Some of them propose the use of pressure, although they provided no insight on how the large blades were extracted. According to alternative hypotheses, indirect percussion may overcome the technical restrictions of pressure, allowing blades of nearly 40 cm in length in exceptional cases to be obtained.

The characteristics of these blades and their knapping traces strongly suggest the use of pressure technique. Moreover, the increased length and width of these products requires the exertion of higher strength and a higher control of the metallic point. As we have seen before, butts with a sharp dihedron entail the accurate location of the point where the pressure tool will be applied. On the other hand, as a number of experiments have shown, lever pressure systems could solve the issue of the increased need of pressure strength (Pelegrin 1988, 2002, 2003: 62–63; 2006; Pelegrin and Morgado 2007). Lever pressure overcomes the limitations of simple pressure (i.e., in a standing position), allowing the manufacture of larger blades (over 20 cm in length and 20 mm in width). In fact, various experimental studies have established a roof of 22–23 mm in width for flint blades detachment by means of pressure with a copper point (Pelegrin, this volume). Blades of smaller widths could have been extracted by standing pressure transmitted by a pectoral or abdominal crutch, while larger blades may have been manufactured applying pressure with a lever.

8.4 The Chalcolithic: The Skilled Production (ca. 3400–2400/2300 B.C.)

The Chalcolithic led to new forms of settlements and burials in the south of the Iberian Peninsula. Big fortified villages surrounded by stoned walls with strongholds attached emerged at this time. A similar phenomenon can be traced in Portugal (Zambujal, Lecia, Vila Nova de Sao Pedro, etc.) and the Western Mediterranean. A few instances of these settlements in Southern Spain are Los Millares in the Southeast, Marroquíes and Los Alcores de Porcuna in the Guadalquivir River Valley, and Valencina de la Concepción at the lower end of the Guadalquivir River. These sites emerge at the end of the 4th millennium B.C. For some of them the downfall comes at the end of the next millennium during the Late Copper and Early Bronze periods.

For instance, according to the absolute dating of Los Millares (Molina et al. 2004), its earliest foundation goes back to the Early Chalcolithic (ca. 3300 B.C.) expanding practically over all of the 3rd millennium B.C. until it was abandoned in the Late Copper period (ca. 2300/2200 B.C.).

The emergence of big villages during the Late Neolithic resulted in increasing territorial partitioning and social stratification during the Chalcolithic. The specialization of flint knapping artisans is coincident with the growing complexity of the society and the territorial structure. Flint knapping involved skilled production in different regions during the Chalcolithic. This process is manifest in Southern Spain, where the major necropolises of the period show knapped artifacts of high technical quality. These elements are of two main types: (a) large blades and (b) bifacial elements including arrow points and, more rarely, knives and "halberds" (very large and wide), all of them made of special materials (flint, jasper, rock crystal, etc.).

The production of large blades prevailed in this period, although a residual industry of bladelets remained during the Early Chalcolithic (end of the 4th millennium B.C.). These bladelets have sharp dihedral or convex-faceted butts, while cores have consistently a prismatic or conic-pyramidal morphology producing bladelets of less than 20 mm in length. The core is worked around a half or more of its perimeter. This production is typical of the Early Neolithic, but here they are manufactured with the new methods of crested preforms and pressure point preparation using sharp dihedrons. On the other hand, the preparation of the carinated cores with plain pressure platforms disappears during the 4th millennium B.C., suggesting that the old method was replaced.

Bladelet production is nevertheless residual, as there is a wide dominance of blades extracted with a standing crutch or lever from prismatic cores shaped with three or four crests. Craft specialization consolidates in three regions: the mountains of Eastern Andalusia in the areas of the Western and Central Subbético Mountain Range and the subvolcanic siliceous rocks of Huelva in Western Andalusia (Fig. 8.1). The specialization process for the blade production was consistent throughout Southern Spain.

As the evidence from several necropolis shows, blade sizes vary from 110 to 150 mm in length and 18 to 20 mm in width with outstanding cases reaching 400 mm in length and 40 mm in width (Fig. 8.5).

As mentioned before, the experimental tests conducted by Pelegrin strongly suggest that flint blades of 22–23 mm width or more required lever pressure to be manufactured. Lever pressure seems to be the main technique for large blade production in Southern Iberia during the Chalcolithic.

These large blades have been found far away from their manufacturing areas, in both domestic contexts and megalithic burials, suggesting that they were, to some extent, the subject of traffic and exchange. Their presence has been documented within the grave goods of the main megalithic necropolis (i.e., Los Millares, Valencina de la Concepción, Marroquíes). In these burials, the blades were left without retouch or with simple edge retouch, being fractured most of the time.

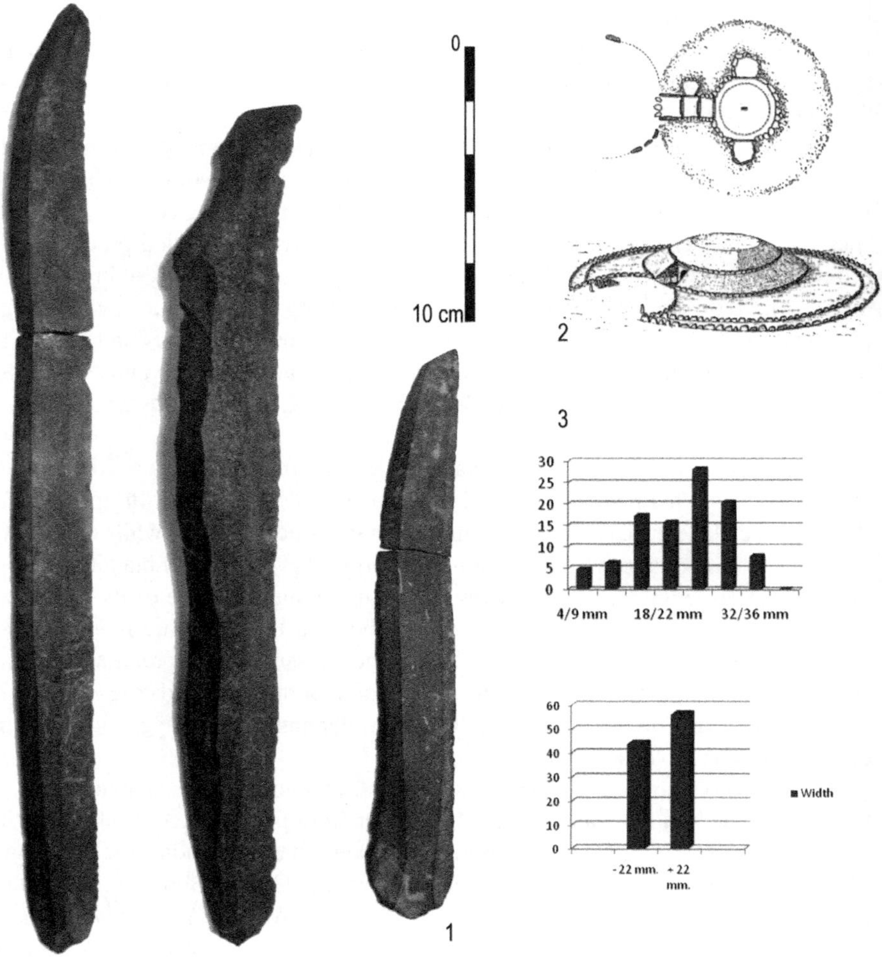

Fig. 8.5 The production of large blades, Chalcolithic of Southern Iberian Peninsula. *1* Long blades, Los Millares necropolis; *2* Tomb type tholos (reconstitution), Los Millares necropolis; *3* Statistical variability of the blades width, Los Millares necropolis

Although studies of the raw materials have just started, the technical features of the blades from Southern Spain (lever pressure, but with a sharp dihedron) are different from those of other regions such as Central Portugal (indirect percussion, thick-facetted butts) (Morgado et al. 2009; Pelegrin 2006) and the Ebro Valley (pressure technique knapping, butts without a sharp dihedron, also indirect percussion), clearly distinguishing these industries and establishing blade traffic routes. The presence of these blades in distant regions of the peninsula (from the Southeast to the West in Portugal and Galicia) is a clear sign of their circulation through long distances (Nocete et al. 2005; Morgado et al. 2011).

The second half of the 3rd millennium B.C. brought deep cultural changes. Incise Bell-Beaker pottery became a standard and was later followed by the rising of a new cultural period: the Bronze Age. These changes included new forms of settlements, changes in the burial rituals, and an expansion of metallurgy. Flint knapping specialization vanished, and consequently, pressure as a knapping technique for blade production also disappeared during the Late Chalcolithic. As the demand for large blades died down, so did the knapping production techniques related to blade specialization. From this moment, flint knapping was restricted to domestic settings for the purposes of local manufacture of some tools (primarily saw-toothed sickle elements extracted by simple direct stone percussion).

8.5 Conclusion

In the present state of our knowledge, the knapping pressure blade production appears for the first time, with the Neolithic in the south of the Iberian Peninsula, during the 6th millennium B.C. This technique of blade production was associated with heat treatment of flint. The preparation of the core prior to blade detachment required carinated preforms. These preforms have been documented in the flint outcrops of the Betic Cordillera in Eastern Andalusia and the Neolithic sites. They are also widely known in the Western Mediterranean (Southern France) and are also documented in domestic contexts indicating heat treatment, although the dates for the knapping pressure technique and the heat treatment are more recent. This production was maintained during the 5th millennium B.C. with little changes.

On the other hand, the Late Neolithic (early 4th millennium B.C.) is a period of deep social and technological change. There is evidence of a distinctive shift toward the use of pressure knapping in blade production in Southern Spain during this period. While Early Neolithic production systems are common across the Mediterranean (i.e., handheld pressure, mini crutch used in a sitting position), the Late Neolithic brings a new production system specific to Southern Spain. Here, blade production becomes based on preforms made from prismatic cores shaped with three or four antero- and posterolateral crests. In addition, the butt morphology is quite unique: butts with a sharp dihedron allowing pressure concentration over an acute arris. The experiments conducted by Pelegrin demonstrate pressure knapping blade production by means of a copper pressure point. Finally, lever pressure became a novel blade extraction method starting in the Late Neolithic and remained common throughout the Copper Age until the second half of the 3rd millennium B.C. The Late Copper Age is the end of the process of the development of pressure blade production in the Southern Iberian Peninsula (Fig. 8.6).

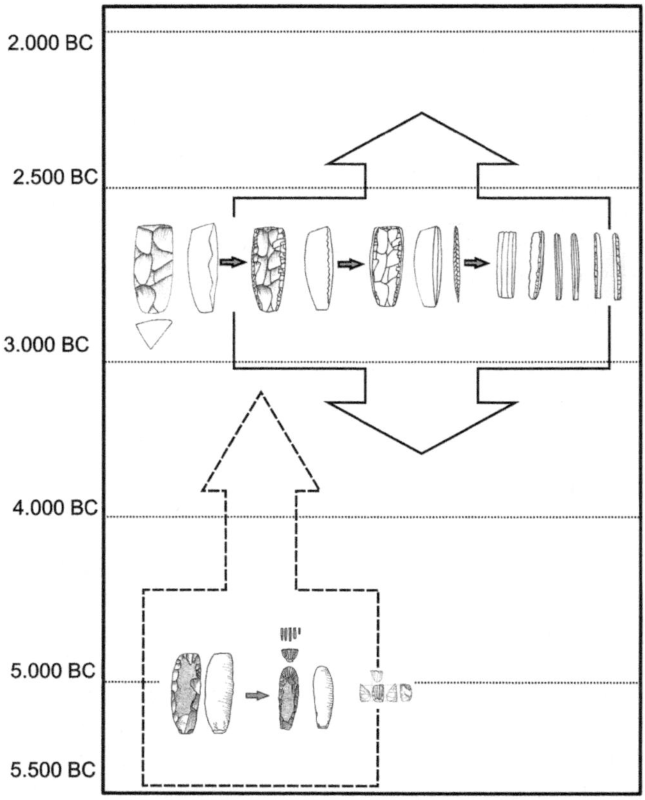

Fig. 8.6 Schematic evolution of pressure blade production, Southern Iberian Peninsula

References

Afonso, José A., Fernando Molina, Juan A. Cámara, Manuel Moreno, Ulises Ramos, and María O. Rodríguez 1996 Espacio y tiempo. La secuencia en Los Castillejos de las Peñas de Los Gitanos (Montefrío, Granada). *Rubricatum* 1 (I Congrés del Neolitic a la Peninsula Iberica. Formació e implantació de les comunitats agrícoles, Gavà-Bellaterra, 1995), Vol. I: 297–304.

Alix, Philippe., Jacques Pelegrin, and Huguette Deloge 1995 Un débitage original de lamelles par pression au Magdalénien du Rocher-de-la-Caille (Loire, France). *Paléo* 7: 187–199.

Arribas, Antonio, and Fernando Molina 1979 *El poblado de "Los Castillejos" en las Peñas de Los Gitanos (Montefrío, Granada). Campaña de excavaciones de 1971. El corte nº 1.* Cuadernos de Prehistoria de la Universidad de Granada, serie Monográfica no 3. Universidad de Granada, Granada.

Asquerino, María D., and Pilar López 1981 La Cueva del Nacimiento (Pontones): un yacimiento neolítico en la Sierra del Segura. *Trabajos de Prehistoria* 38: 94–148.

Aura Tortosa, José E., Juan F. Jordá Pardo, Manuel Pérez Ripoll, J.V. Morales Pérez, Oreto García Puchol, Javier González-Tablas, and B. Avezuela Aristu 2009 Epipaleolítico y Mesolítico en Andalucía oriental. Primeras notas a partir de los datos de la Cueva de Nerja (Málaga, España). In *El Mesolítico Geométrico en la Península Ibérica*, edited by Pilar Utrilla and Lourdes Montes, pp. 343–360. Universidad de Zaragoza, Zaragoza.

Binder, Didier 1984 Systèmes de débitage laminaire par pression: exemples chasséens provençaux. In *Préhistoire de la pierre taillée 2. Économie du débitage laminaire, technologie et expérimentation*, pp. 71–84. CREP, Paris.

Binder, Didier 1987 *Le Néolithique ancien provençal : technologie et typologie des outillages lithiques*. Supplément à Gallia-Préhistoire no. 24. Editions du CNRS, Paris.

Binder, Didier 2000 Mesolithic and Neolithic interaction in Southern France and Northern Italy: new data and current hypotheses. In *Europe's First Farmers*, edited by D. Price, pp. 117–143. Cambridge University Press, Cambridge.

Binder, Didier, and Bernard Gassin 1988 Le débitage laminaire chasséen après chauffe: technologie et traces d'utilisation. In *Industries lithiques, tracéologie et technologie*, edited by Sylvie Beyries, pp. 93–115. BAR International Series 411. British Archaeological Reports, Oxford.

Binder, Didier, and Catherine Perlès 1990 Stratégies de gestion des outillages lithiques au Néolithique. *Paléo* 2: 257–283.

Cámara, Juan A., Fernando Molina, and José A. Afonso 2005 La cronología absoluta de Los Castillejos en Las Peñas de Los Gitanos (Montefrío, Granada). In *III Congreso del Neolítico en la Península Ibérica* pp. 841–852. Universidad de Cantabria, Santander.

Cava, Ana 1997 La industria lítica tallada de la Cueva de Nerja (Cortes de las salas de la Mina 80-A y 80-B y de la Torca 82). In *El Neolítico y Calcolítico de la Cueva de Nerja en el contexto andaluz*, edited by Manuel Pellicer and Pilar Acosta, pp. 225–348. Trabajos sobre la Cueva de Nerja 6. Patronato de la Cueva de Nerja, Málaga.

Chauchat, Claude, Jacques Pelegrin, and Marie-Roger Seronie-Vivien 1996 Une pièce exceptionnelle dans le piémont pyrénéen: le nucléus à débiter par pression de Labastide-Villefranche (Pyr.-Atl.).- *Archéologie des Pyrénées Occidentales et des Landes* 15: 7–11.

Cortés, Miguel (editor) 2008 *El Pirulejo (Priego de Córdoba): Cazadores recolectores del Paleolítico superior en la Sierra Subbética*. Antiqvitas 20. Museo Municipal de Priego de Códoba, Córdoba.

Crabtree, Don E. 1968 Mesoamerican polyhedral cores and prismatic blades. *American Antiquity* 33(4): 446–478.

Fortea, Francisco J., Bernat Martí, and Joaquim Juan Cabanilles 1987 La industria lítica tallada del Neolítico antiguo en la vertiente mediterránea de la Península Ibérica. *Lucentum* 6: 7–22.

Inizan, Marie-Louise 1984 Débitage par pression et standardisation des supports: un exemple capsien au Relilaï (Algérie). In *Préhistoire de la pierre taillée 2. Économie du débitage laminaire, technologie et expérimentation*, pp. 85–92. CREP, Paris.

Inizan, Marie-Louise 1991 Le débitage par pression: des choix culturels. In *25 ans d'études technologiques en Préhistoire* (XI Rencontres Internationales d'Archéologie et d'Histoire d'Antibes), pp. 367–377. Éditions APDCA, Juan-les-Pins.

Inizan, Marie-Louise 2003 Tailler des roches par pression : émergence d'une technique, étapes de sa diffusion dans le monde. In *Matériaux, productions, circulations du Néolithique à l'Age du Bronze*, edited by Jean Guilaine, pp. 33–46. Éditions Errance, Paris

Inizan, Marie-Louise, and Monique Lechevallier 1990 A techno-economic approach to lithics: some examples of blade pressure debitage in the Indo-Pakistani subcontinent. In *South Asian Archaeology 1987*, edited by M. Taddei, pp. 43–59. ISMEO, Rome.

Inizan, Marie-Louise, and Jacques Tixier 2001 L'émergence des arts du feu : le traitement thermique des roches siliceuses. *Paléorient* 26(2): 23–36.

Inizan, Marie-Louise, Monique Lechevallier, and Jacques Pelegrin 1994 The use of metal in the lithics of Sheri Khan Tarakai, Pakistan. Evidence provided by the technological approach of pressure debitage. In *South Asian Archaeology 1993*, Vol. II, edited by A. Parpolla and P. Koskikallio, pp. 245–256. Annales Academiae Scientiarum Fennicae B 271, Helsinki.

Jodá Pardo, Jesús F. (editor) 1986 *Prehistoria de la Cueva de Nerja (Málaga)*. Patronato de la Cueva de Nerja, Málaga.

Juan Cabanilles, Joaquim 1984 El utillaje neolítico en sílex del litoral mediterráneo peninsular. Estudio tipológico-analítico a partir de los materiales de la Cova de l'Or y de la Cova de la Sarsa. *Saguntum* 18: 49–102.

Léa, Vanessa 2004 *Les industries lithiques du Chaséen en Languedoc oriental*. BAR International Series 1232. British Archaeological Reports, Oxford.

Martín Socas, Dimas, M. Dolores Cámalich, and Pedro González 2004a *La Cueva de El Toro (Sierra del Torcal-Antequera-Málaga). Un modelo de ocupación ganadera en el territorio andaluz entre el VI y II milenio a.n.e.* Junta de Andalucía, Consejería de Cultura, Sevilla.

Martín Socas, Dimas, M. Dolores Cámalich, Ramón Buxó, E. Chávez, Jean C. Echallier, Pedro González, Amelia Goñi, Manuel Mañosa, Teresa Orozco, María A. Paz, M.Oliva Rodríguez, Amelia Rodríguez, M. Tusell, and J.P.N. Watson 2004b Cueva de El Toro (Antequera, Malaga-Spain): a Neolithic stockbreeding community in the Andalusian region, between the 6th and 3th millennia B.C.. *Documenta Praehistorica* 31: 163–181.

Martínez, Gabriel 1997 Late Prehistory Blade Production in Andalusia (Spain). In *Siliceous Rocks and Culture*, edited by Antonio Ramos Millán and M. Angeles Bustillo, pp. 427–436. Universidad de Granada, Granada.

Martínez, Gabriel, and Antonio Morgado 2005 Los contextos de elaboración de hojas prismáticas de sílex en Andalucía Oriental durante el Neolítico Reciente. Aspectos técnicos, modelos de trabajo y estructuración social. In: *III Congreso del Neolítico en la Península Ibérica*, pp. 359–368. Santander.

Méry, Sophie, Patricia Anderson, Marie-Louise Inizan, Monique Lechevallier, and Jacques Pelegrin 2007 A pottery workshop with flint tools on blades knapped with copper at Nausharo (Indus civilisation, ca. 2,500 B.C.). *Journal of Archaeological Science* 34: 1098–1116.

Molina, Fernando, and Juan A. Cámara 2005 *Guía del yacimiento arqueológico de Los Millares*. Junta de Andalucía, Consejería de Cultura, Sevilla.

Molina, Fernando, Juan A. Cámara, Josefa Capel, Trinidad Nájera, and Leovigildo Sáez 2004 Los Millares y la periodización de la Prehistoria Reciente del Sureste. *II-III Simposio de Prehistoria Cueva de Nerja*: 142–158. Málaga.

Morgado, Antonio 2008 *Transformación social y producción de hojas de sílex durante la Prehistoria Reciente de Andalucía Oriental. La estrategia de la complejidad*, Ph.D., (digital edition, http://0-hera.ugr.es.adrastea.ugr.es/tesisugr/17486105.pdf), Universidad de Granada, Granada.

Morgado, Antonio, Jacques Pelegrin, Thierry Aubry, and José A. Lozano 2009 La producción especializada de grandes láminas del Sur y Oeste de la Península Ibérica. In *Les grans fulles de sílex. Europa al final de la Prehistòria*, edited by Juan F. Gibaja, Xavier Terradas, Antoni Palomo and Xavier Clop, pp. 89–97. Monografies 13. Museu d'Arqueologia de Catalunya, Barcelona.

Morgado, Antonio, Jacques Pelegrin, Gabriel Martínez, and José A. Afonso 2008 La production de grandes lames dans la Péninsule ibérique (IVe – IIIe mil. av.JC. cal.). In *Les industries lithiques taillées des 4e et 3e millénaires en Europe occidentale*, edited by Marie-Hélène Dias-Meirinho, Vanessa Léa, Karim Gernigon, Pierrich Fouéré, François Briois and Maxence Bailly, pp. 303–330. BAR International Series 1884. British Archaeological Reports, Oxford.

Morgado, Antonio, José A. Lozano, Carlos Rodríguez-Rellán and F. Rodríguez-Tovar 2011 De sur a norte de la Península Ibérica. Circulación a larga distancia del sílex tipo Turón (Málaga, España). Paper presented at the Congress Networks in the Neolithic. Exchange of raw materials, products and ideas in the Western Mediterranean (VII-III millennium B.C.). 2–4 February 2011, Gavà-Bellaterra Barcelona, Spain.

Nocete, Francisco, Reinaldo Sáez, José M. Nieto, Rosario Cruz-Auñón, Rosario Cabrero, Esther Alex, and Moisés R. Bayona 2005 Circulation of silicified oolitic limestone blades in South-Iberia (Spain and Portugal) during the third millennium B.C.: an expression of a core/periphery framework. *Journal of Anthropological Archaeology* 24: 62–81.

Pelegrin, Jacques 1988 Débitage expérimental par pression: "du plus petit au plus grand". In *Technologie préhistorique*, edited by Jacques Tixier, pp. 37–53. Notes et Monographies Techniques du CRA, n°25. Editions du CNRS, Paris.

Pelegrin, Jacques 1994 Lithic technology in Harappan times. In *South Asian Archaeology 1993*, Vol. II, edited by A. Parpolla and P. Koskikallio, pp. 585–598. Annales Academiae Scientiarum Fennicae B 271, Helsinki.

Pelegrin, Jacques 2000 Les techniques de débitage laminaire au Tardiglaciaire: critères de diagnose et quelques réflexions. In *L'Europe centrale et septentrionale au tardiglaciaire* (Table-ronde de Nemours, 13–16 mai 1997), pp. 73–86. Mémoires du Musée de Préhistoire d'Île de France, 7. Nemours.

Pelegrin, Jacques 2002 Principes de la reconnaissance des méthodes et techniques de taille. In J. *Tell Atij Tell Gudeda, industrie lithique. Analyse technologique et fonctionnelle*, edited by J. Chabot, pp. 215–224. Cahiers d'archéologie du CELAT 13, série archéométrie no. 3. Université Laval, Québec.

Pelegrin, Jacques 2003 Blade making techniques from the Old World: insights and applications to Mesoamerican obsidian lithic technology. In *Experimentation and Interpretation in Mesoamerican Lithic Technology*, edited by K. Hirth, pp. 55–71. The University of Utah Press, Salt Lake City.

Pelegrin, Jacques 2006 Long blade technology in the Old World: an experimental approach and some archaeological results. In *Skilled Production and Social Reproduction*, edited by Jan Apel and Kjel Knutsson, pp. 37–68. Societas Archaeologica Upsaliensis, Uppsala.

Pelegrin, Jacques, and Antonio Morgado 2007 Primeras experimentaciones sobre la producción laminar del Neolítico Reciente-Edad del Cobre del sur de la Península Ibérica. In *Arqueología Experimental en la Península Ibérica: investigación, didáctica y patrimonio*, edited by M. Luisa Ramos Sáinz, José E. González Urquijo and Javier Baena Preysler, pp. 131–139. Asociación Española de Arqueología Experimental, Sandander.

Pellicer, Manuel 1964 *El Neolítico y Bronce de la Cueva de la Carigüela de Píñar (Granada)*. Trabajos de Prehistoria 15. CSIC, Madrid.

Pellicer, Manuel and Pilar Acosta 1997 *El Neolítico y Calcolítico de la Cueva de Nerja en el contexto andaluz*. Trabajos sobre la cueva de Nerja, 6. Patronato de la Cueva de Nerja, Málaga.

Pérez Bareas, Cristobal, José A. Afonso, Juan A. Cámara, Francisco Contreras, and Rafael Lizcano 1999 Clasificación cultural, periodización y problemas de compartimentación en el Neolítico de la Alta Andalucía. *Saguntum extra-2 (II Congrès del Neolitico a la Peninsula Ibèrica)*: 485–492.

Perlès, Catherine 1984 Débitage laminaire de l'obsidienne dans le Néolithique de Franchthi (Grèce): technique et place dans l'économie de l'industrie lithique. In *Préhistoire de la pierre taillée 2. Économie du débitage laminaire, technologie et expérimentation*, pp. 129–137. CREP, Paris.

Perlès, Catherine 2004 *Les industries lithiques taillées de Franchthi (Argolide, Grèce: tome III); Du Néolithique ancien au Néolithique final*. Coll. Excavations at Franchthi cave, Greece. Indiana University Press, Bloomington & Indianapolis.

Ramos Millán, Antonio 1997 Flint Political Economy in a Tribal Society. A Material-Culture Study in the El Malagón Settlement (Iberian Southeast). In *Siliceous Rocks and Culture*, edited by Antonio Ramos Millán and M.Angeles Bustillo, pp. 671–711. Universidad de Granada. Granada.

Ramos Muñoz, José 1988-89 Las industrias líticas del Neolítico en Andalucía, sus implicaciones espaciales y económicas. *Zephyrus* 41–42: 113–148.

Ramos Muñoz, José, and María Lazarich 2002 *Memoria de la excavación arqueológica en el asentamiento del VI milenio a.n.e. de "El Retamar" (Puerto Real, Cádiz)*. Junta de Andalucía, Consejería de Cultura, Sevilla.

Ramos Muñoz, José, and Vicente Castañeda 2005 *Excavaciones en el asentamiento prehistórico del embarcadero del río Palmones (Algeciras, Cádiz): una nueva contribución al estudio de las últimas comunidades cazadoras y recolectoras*. Universidad de Cádiz, Cádiz.

Terradas, Xavier, and Juan F. Gibaja 2002 La gestión social del sílex melado durante el Neolítico Medio en el Nordeste de la Península Ibérica. *Trabajos de Prehistoria* 59: 29–48.

Tixier, Jacques 1976 L'industrie lithique capsienne de l'Aïn Dokkara, région de Tébessa. Fouilles L. Balout. *Lybica* 24: 21–54.

Chapter 9
The Arrival and Development of Pressure Blade Technology in Southern Scandinavia

Mikkel Sørensen

9.1 The Problem and the Research Field

In this paper, the following problem is discussed: When and why did pressure blade technology appear in Southern Scandinavia and how did the pressure blade concept evolve in this area?

This question is discussed from a technological point of view, thus the specific methods of pressure blade technology employed in prehistory within the area will be analysed, described and reconstructed. As technology today is defined not only as a science related to the technical transformation of raw materials but also to its social, cognitive and cultural aspects (e.g. Apel 2001; Audouze 1999; Leroi Gourhan 1964; Pelegrin 1990; Schiffer and Skibo 1987; Sørensen 2006c), the reasons for the arrival of the pressure blade technique and the meaning of this technology in the Maglemosian society will be discussed as well.

Pressure blade technology is, in this chapter, defined as the study of lithic production concepts in which pressure technique have been applied to produce blades, i.e. serially made detachments, from lithic cores, used as tools or preforms for tools (Sørensen 2006a: 289). The recognition of prehistoric pressure technique is based on macromorphological lithic criteria and stigmata found in experimental work and by analogy investigated and diagnosed in prehistoric lithic assemblages (Pelegrin 1984a, b, 1988, 2002, 2006; Sørensen 2006a). These criteria are extreme regularity, rare occurrence of ripples, straightness, a small bulb in combination with lip formation and, most importantly, the occurrence of small exhausted blade cores with negatives showing extreme regularity. These criteria have an overlap with blades made by indirect technique, and in many cases, single blades can therefore not be attributed with certainty; however, the careful investigation of large blade populations,

M. Sørensen, Ph.D. (✉)
Department of Archaeology, University of Copenhagen,
Njalsgade 76, Copenhagen, Denmark
e-mail: miksr@hum.ku.dk

P.M. Desrosiers (ed.), *The Emergence of Pressure Blade Making: From Origin to Modern Experimentation*, DOI 10.1007/978-1-4614-2003-3_9,

including blade cores, regarding the combination of the criteria can allow a reliable verification.

The Maglemose Culture, in this text termed the Maglemosian, can be defined as a complex of artefacts and technologies produced by hunter-gatherer groups in the Early Mesolithic and Early Holocene periods (9500–6500 B.C.). The Maglemosian complex is defined in Southern Scandinavia, Northern Germany and, in its early stages, in Eastern England and Western France. Towards the Baltic, assemblages with typical Maglemosian industries are known; however, these regions have traditionally been defined with other culture groups such as the Komornica Culture. The most important typological archaeological signifier of the complex is its microlithization of armature points as seen in the lithic assemblages. Bone point morphologies, lithic technology and bone technology are today also considered important for the definition of this complex. Early Holocene Southern Scandinavia was, due to the huge ice sheets of Northern Scandinavia, a period with a low seawater level characterized by vast low lands; for example a land bridge existed between Southern Scandinavia and the British Isles. The Maglemosian hunter-gatherer groups are generally interpreted to have employed this landscape depending largely on inland hunting and fishing.

The Maglemosian of Southern Scandinavia has been subject to a long and vast research history. It comprises a broad range of materials including an outstandingly well-preserved organic material due to the fact that most of the early excavated sites were found in anaerobic conditions (moors). During the early years of research, much effort was made to excavate, publish, define and date the archaeological material (Johansen 1919; Sarauw 1903). The Early Mesolithic, called the Mullerup Culture at that time, was the oldest known Mesolithic Stone Age culture in Northern Europe. Thus, the Maglemosian was filling out the chronological gap in Europe between the Ertebølle Culture in the North and the Upper Paleolithic found in the French cave sites until Late Paleolithic sites were documented in Northern Germany during the 1930s (see Fig. 9.1).

Later, research efforts merely addressed the typology and the chronology of the period, especially dealing with morphological analysis and seriation of microliths (Becker 1952; Petersen 1966, 1973; Skaarup 1979). The Maglemosian was divided into six phases from 0 to 5 mainly due to the seriation of microliths. This relative chronological ordering of the material has been criticized and scrutinized by several researchers since it was established. Critical points are that the microlith morphologies can reflect functional aspects as well as stylistic ones, and that solely defining a cultural chronology on the frequency of one artefact type is insufficient (e.g. Henriksen 1980; Sørensen 2006b). The latter argument is crucial when applying the chronology to smaller assemblages where the number of microliths is low, as only doubtful relative ages can be achieved in such situations.

During the last decade, new research methodologies have been employed in studying the Maglemosian artefact materials (David 2006; Sørensen 2006a, b; Toft 2006). As a result, a new chronological ordering has been put forward based on changes in the Maglemosian lithic blade industry (Sørensen 2006a, b). This chronology is now often preferred due to its broader application on the lithic material; however, the chronology based on microliths is generally still considered valid when the amounts of microliths are sufficient.

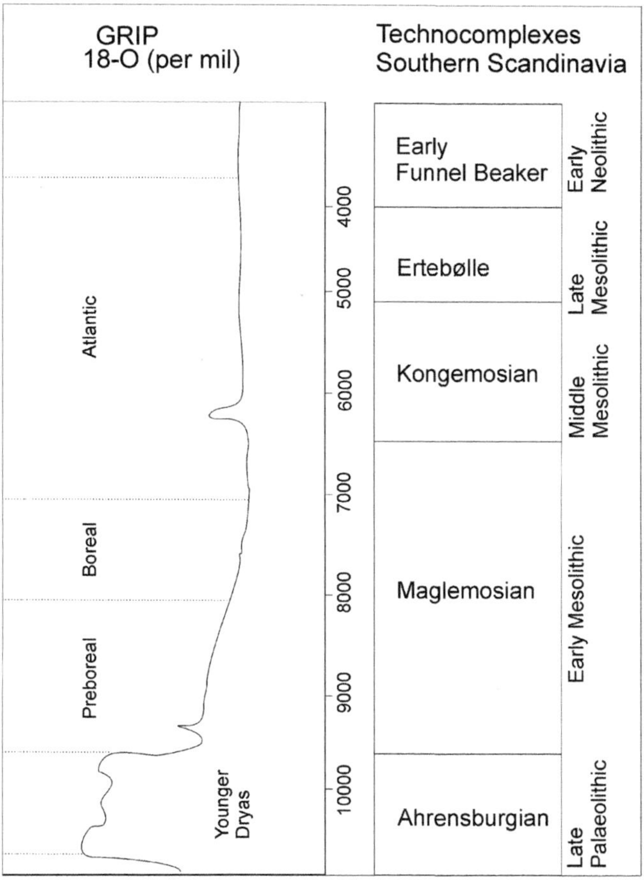

Fig. 9.1 Chronology, climate and chronozones of Mesolithic Southern Scandinavia (After Terberger 2006)

9.2 From Direct to Pressure and Punch Techniques in Southern Scandinavia: A Close Examination

Seven different concepts of blade production within the Maglemosian have been defined, representing four different diachronic techno-complexes (Sørensen 2006a, b) (Fig. 9.2).

During the Preboreal and the beginning of the Boreal period (9500–7000 B.C.), blades were made from unipolar cores. From techno-complex 2, dual platformed prismatic cores also appear. The blades from these periods have a generally irregular morphology. Typical blade attributes for the techno-complex 1 are triangular butts with impact cones, a variety of bulbs of percussion from mainly large to small and broken, with examples of split cone fractures. In the techno-complex 2, blades

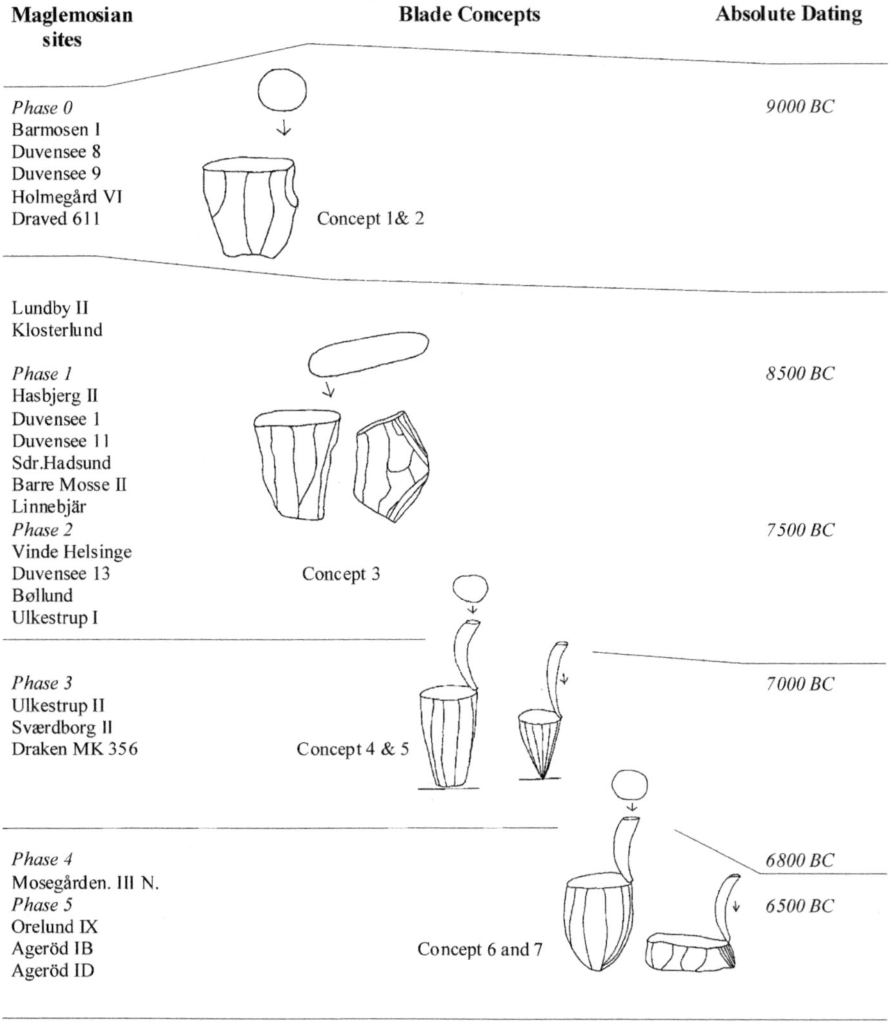

Fig. 9.2 The four technological defined complexes of the Maglemosian. The horizontal rows of the table below show the investigated sites classified according to the relative microlith chronology phases 0–5 with respect to the seven proposed concepts of blade production which are illustrated in the vertical columns. References: Barmosen I (Johansson 1990), Duvensee 8 (Bokelmann et al. 1981), Duvensee 9 (Bokelmann 1991), Holmegard VI (Becker 1945), Lundby II (Henriksen 1980), Klosterlund (Petersen 1966), Hasbjerg II (Johansson 1990), Duvensee 6 (Bokelmann 1971), Duvensee 13 (Bokelmann 1985), Sdr Hadsund (Petersen 1966), Barre Mosse (Skar 1987), Linnebjär (Salomonsson 1965), Vinde Helsinge (Mathiassen 1943), Bøllund (Petersen 1966), Ulkestrup I & II (Andersen et al. 1982), Sværdborg II (Petersen 1972), Mosegården III N (Andersen 1985), Orelund IX (Andersen 1985), Ageröd 1B & 1D (Larsson 1978)

are generally thinner with small bulbs and often small punctiformed butts or broken proximal ends (*fracture languette*). The blade attributes from the first two complexes thereby point to the use of respectively direct hard hammer percussion and direct soft stone hammer percussion when blades were produced.

Pressure blade technology arrived in Southern Scandinavia in the Maglemosian techno-complex 3 with the transition to the Atlantic period. Thereafter, this technology was in continuous use until the Ertebølle Culture. To understand the technological leap between direct percussion blade productions in techno-complex 2 to pressure and punch blade production in techno-complex 3, the contexts of blade production within the Maglemosian have to be investigated.

Substantial lithic material produced by pressure and punch technologies have been excavated as well as collected from several Maglemosian sites, especially in the large moors on Southern and Western Zealand and Eastern Denmark (Andersen 1983; Henriksen 1980; Johansson 1990; Schilling 1999). Most of these sites, however, were inhabited repeatedly through the Maglemosian (e.g. Sværdborg I and Lundby) and must be defined as mixed sites that are not suitable for contextual technological analysis. Among the few excavated sites, which are spatially defined and interpreted to have been used only within techno-complex 3, are the sites Ulkestrup II, hut 2 (Andersen et al. 1982) and Sværdborg II (Petersen 1972). Additionally a small, unmixed site named Draken MK 356 with an assemblage typical to techno-complex 3 was excavated recently in Scania (Sweden) (Gidlöf 2008; Sørensen 2007).

When blades and cores from the techno-complex 3 assemblages are analysed, extreme regularity is found compared to previous periods blade industries (Fig. 9.3). Techno-complex 3 blades range from 15 mm in width and 10 cm in length to 6 mm in width and 4 cm in length. Blades are prismatic, rather straight, and they generally show a combination of bulb and lip formation at their proximal ends. Butts are unfacetted and typically oval and unbroken. The dorsal faces of the blades display a careful trimming of the proximal edges by small feathering removals. A typical attribute is that most blades have broken distal ends and that blades generally seem to break during production. Fragmentation 'en languette' is seen on the larger blades, and this attribute appears in some cases also from the distal blade ends (Fig. 9.3k). Cores are single platform, unfacetted and can be both circular and single fronted. Generally, the cores display a series of regular blade removals (Fig. 9.3a, b)

Investigations of the blade production methods in techno-complex 3, employing a dynamical technological classification in which the blade assemblages are classified in accordance to the blade production process (*chaîne opératoire*) (Schild 1980; Sørensen 2006b) and series of lithic replicative experiments, by this author, in order to test, analyse and evaluate the specific technology (Sørensen 2006a, b) (Fig. 9.7), suggests that the blades were produced using two concepts: (1) a concept for production of macroblades by punch technique and (2) a concept for production of microblades by pressure technique. The appearance of regular bullet-shaped microblade cores up to 7 cm, together with microblades of extreme regularity and straightness being up to 9 mm in width, is evidence of the use of

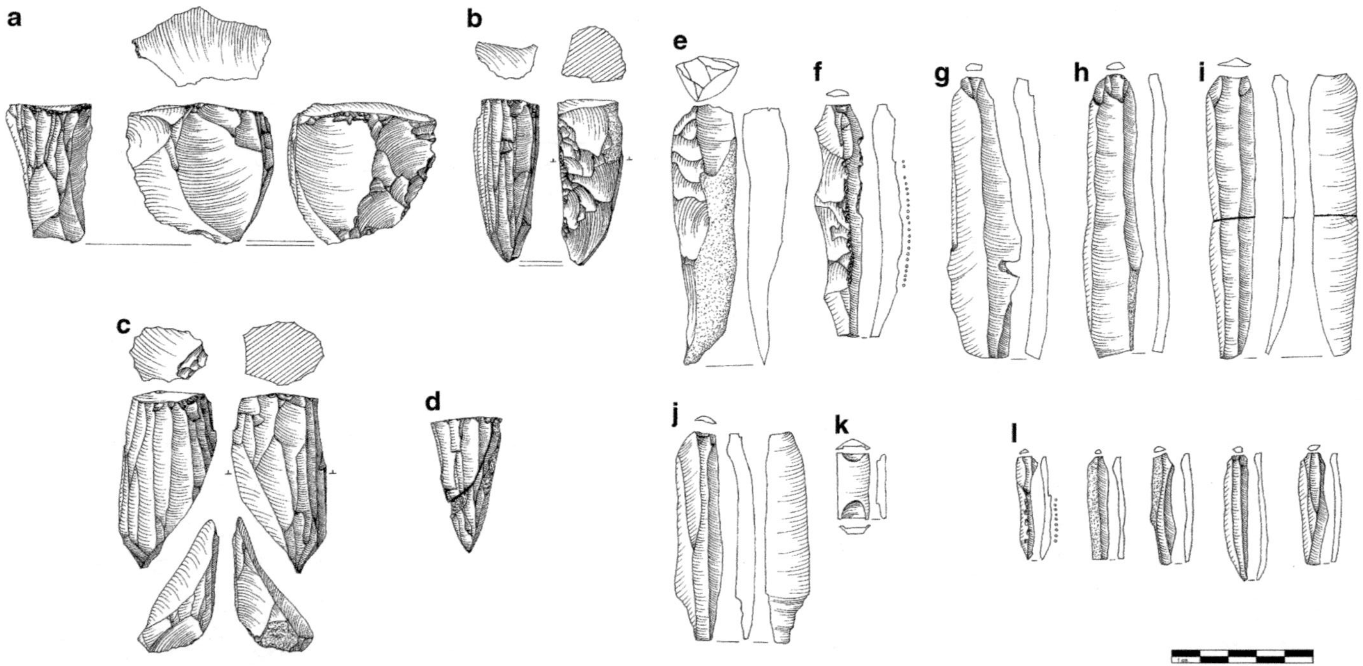

Fig. 9.3 Blades and cores from Maglemosian techno-complex 3. Site Ulkestrup II. (**a**) Single fronted microblade cores preform. (**b**) Exhausted single fronted microblade core. (**c**) Circular refitted core type. (**d**) Exhausted circular (bullet shaped) core type. (**e–h**) Macroblades typical of different production stages. (**j**) Blade with distal fracture *en languette* typical of punch technique. (**k**) Medial bladefragment with fracture *en languette* from both distal en proximal end typical of blades produced with distal support. (**l**) Microblades typical of an initial blade sequence (Drawing L. Johansen)

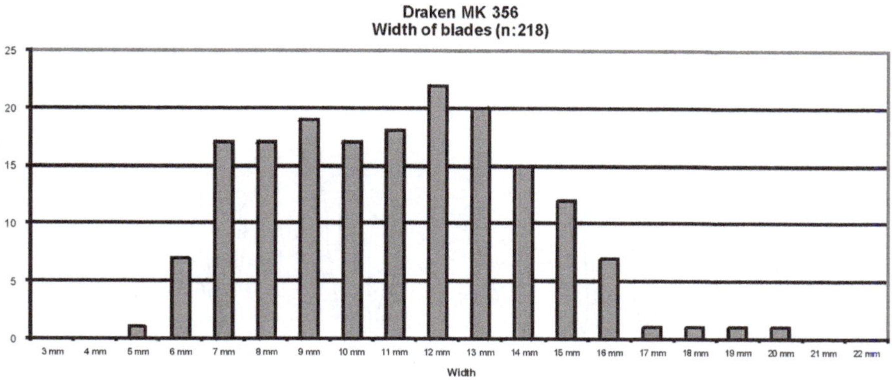

Fig. 9.4 Width measurements of the blade populations from the Scanian techno-complex 3 site Draken MK 356. It is seen that the blade populations of the concept 4 and 5 (punch and pressure techniques) have a metrical overlap. From the technological analysis it is suggested that the pressure blades have width up to 9 mm while punch made blades can have a greater width

pressure technique in techno-complex 3. Cortical, crested and regular thicker blades showing a width up to 15 mm are evidence of indirect percussion techniques. The blade industries from these two concepts, however, cannot be entirely isolated in the assemblages. The blades produced in the two concepts have many of the same morphological and technological attributes, and, as documented in the assemblage from Draken MK 356, there is a metric overlap between the two blade populations (Fig. 9.4).

Complete macroblade cores are seldom found at sites from techno-complex 3. The reason for this is probably that the macroblade cores were reused for the microblade production in a continuous production process. This continuous use can explain the metrical overlap between blades made by means of pressure and punch technique in techno-complex 3. One refitted blade core from Ulkestrup II can be interpreted as being in the stage when pressure technique was to be applied after the core initially had been exploited by punch technique (Fig. 9.3c).

An examination of the techno-complex 3 sites, for example Ulkestrup II and Sværdborg II, has revealed that two different methods for production of microblade core preforms are carried out at the sites (Fig. 9.5). Method A: Circular macroblade cores are reused as microblade cores resulting in circular conical (bullet shaped) microblade cores. Method B: (1) Oblong keeled microblade cores are produced and exploited from one front, resulting in exhausted single fronted cores. The two different methods for the production of pressure blades in techno-complex 3 constitute an important observation concerning the problem of arrival/innovation for the understanding of the development of the pressure blade technology in techno-complex 4, as shall be discussed below.

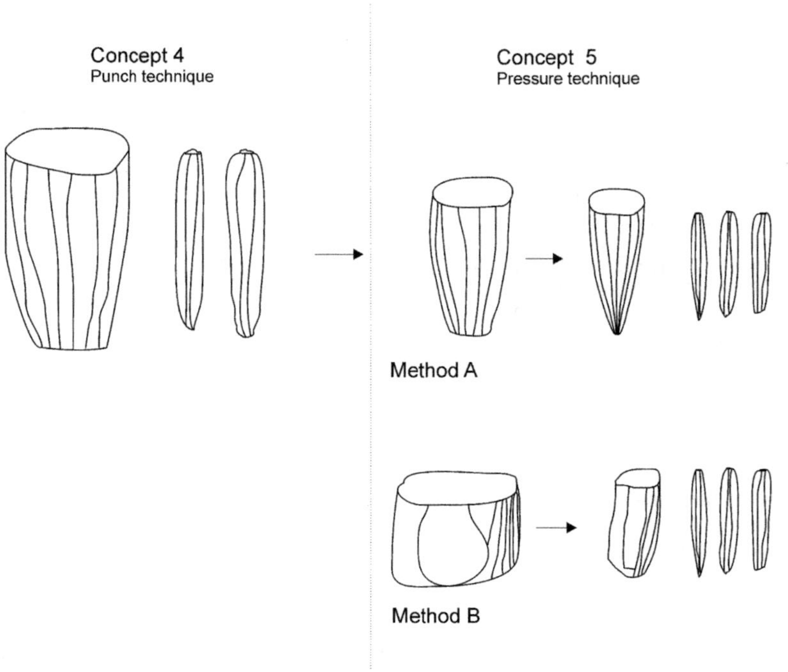

Fig. 9.5 Two methods for microblade production in Maglemosian techno-complex 3 (Drawing M. Sørensen)

Macroblade cores have, judging from the blade and core attributes, been distally supported during production in techno-complex 3. The straight blades of complex 3 point to this technological feature (Bordes and Crabtree 1969), and this hypothesis is strengthened by the fact that within the blade assemblage, there are fractures 'en languette' from the distal blade ends (Fig. 9.3k).

Several well-preserved elk antler tools were found at the Ulkestrup II site, which are interpreted as lithic pressure tools due to their morphology and use wear (Fig. 9.6) (Andersen et al. 1982: 79).

The bullet-shaped microblade cores must, due to their low inertia, have been fixed mechanically during blade production. Moreover, their circular morphology suggests that this fixation was flexible in a way which could allow the core to be easily turned after each blade detachment. Thus, a fixation system in which the core's lateral edges were held in a 'V'-shaped holder that also supported the distal end of the core seems most plausible (Pelegrin 1984c).

One of the observations made during the replicative experimental work was that the specialized single fronted microblade cores of techno-complex 3 (Fig. 9.3a, b) could be held as efficiently as the bullet-shaped cores in the 'V'-shaped holding system (Fig. 9.7).

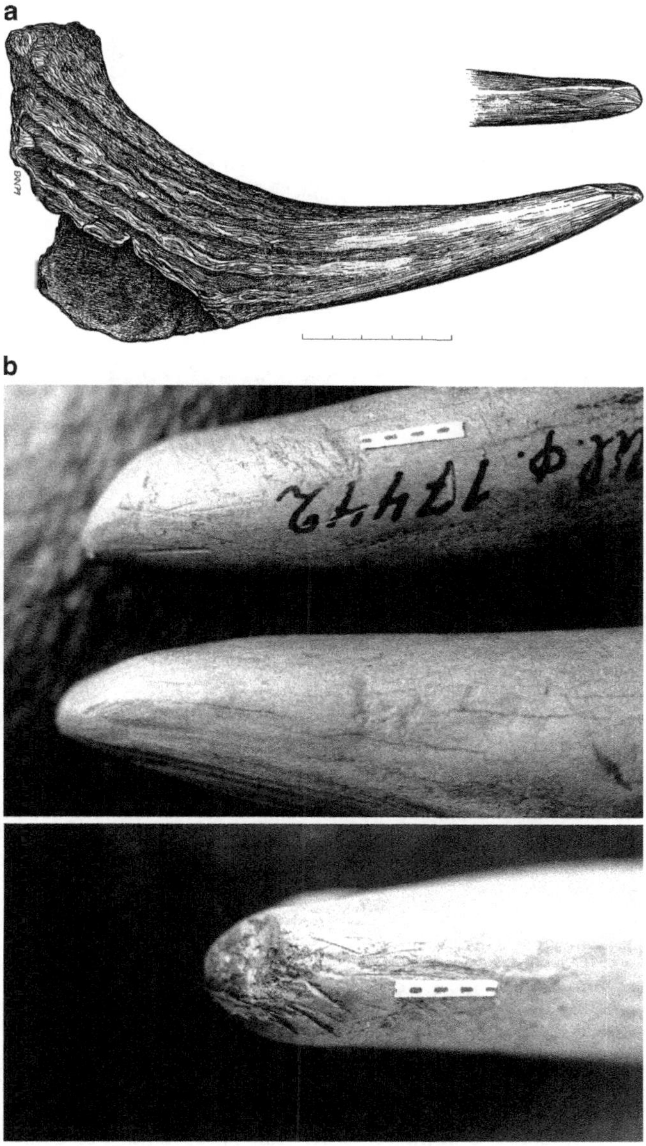

Fig. 9.6 Antler tools, possibly used for pressure blade production. (**a**) Pressure tool of elk antler from Ulkestrup II, length 21 cm (Drawing E. Koch). (**b**) Use wear on pressure tools from Ulkestrup II (Photo M. Sørensen)

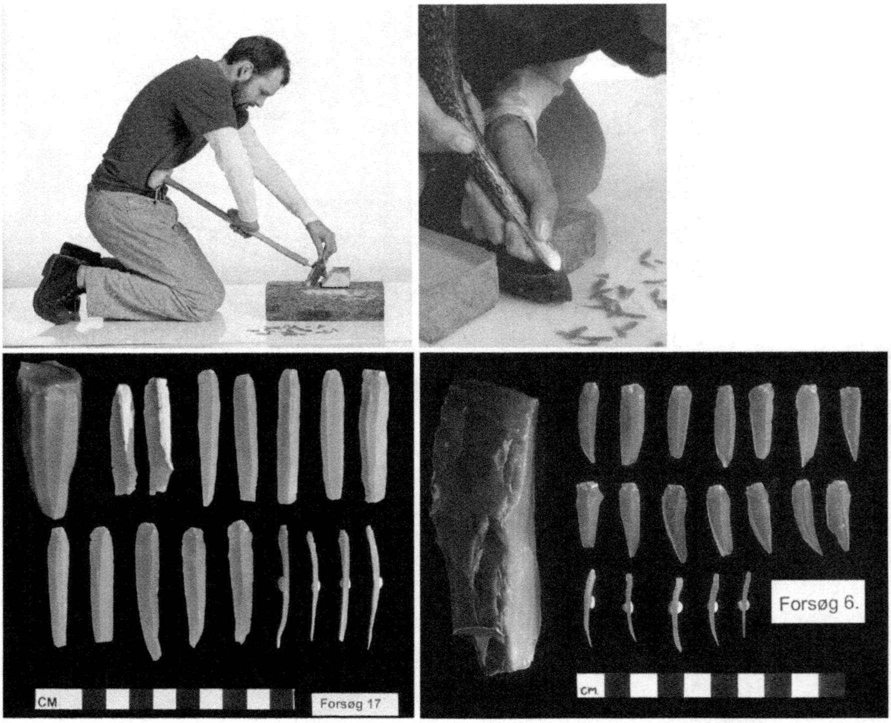

Fig. 9.7 Experiments the pressure micro blade concepts of techno-complex 3 (*to the left side*) and techno-complex 4 (*to the right side*) (Photos J. Sørensen)

9.3 The Absolute Dating of the First Pressure Blade Technology in Southern Scandinavia

Pollen analysis from the Sværdborg II and Ulkestrup II sites dates to the Boreal-Atlantic transition, 7500–7000 B.C. (Aaby 1993). Radiocarbon dating of the mentioned sites within techno-complex 3 is generally problematic as absolute dates are few and were made early in the history of radiocarbon analysis. From Ulkestrup II, hut 2, two dates have been obtained (K-1507 and K-2176) of which only the latter (8030 ± 140 B.P., calibrated to 7180–6690 B.C.) is considered viable due to a problematic site taphonomy. It thus seems possible that the pressure blade technology appears for the first time in Southern Scandinavia around 7000 B.C. Future AMS dating of artefacts from techno-complex 3, for example Ulkestrup II, can provide a more precise dating of this event.

9.4 The Development of Pressure Blade Technology in Southern Scandinavia

Concerning the subsequent Maglemosian techno-complex 4, many of the same problems existed with mixed assemblages as were seen for techno-complex 3. The sites Mosegården III and Orelunde IX from Åmosen on Zealand are among the few sites which can be considered unmixed (Andersen 1985). From Scania, the large sites Ageröd 1:B and 1:D belong to techno-complex 4 (Larsson 1978).

In techno-complex 4, a microblade and macroblade concept involving, respectively, pressure blade production and punch technique is maintained; however, the concepts are altered compared to techno-complex 3. Techno-complex 4 blades are generally very regular and prismatic, but metrically, the microblades generally decrease in size and thickness while the macroblades increase in size and thickness compared to techno-complex 3 (Fig. 9.8). Thus, in contrast to the techno-complex 3, two metrically different and distinguishable blade industries clearly exist in the techno-complex 4.

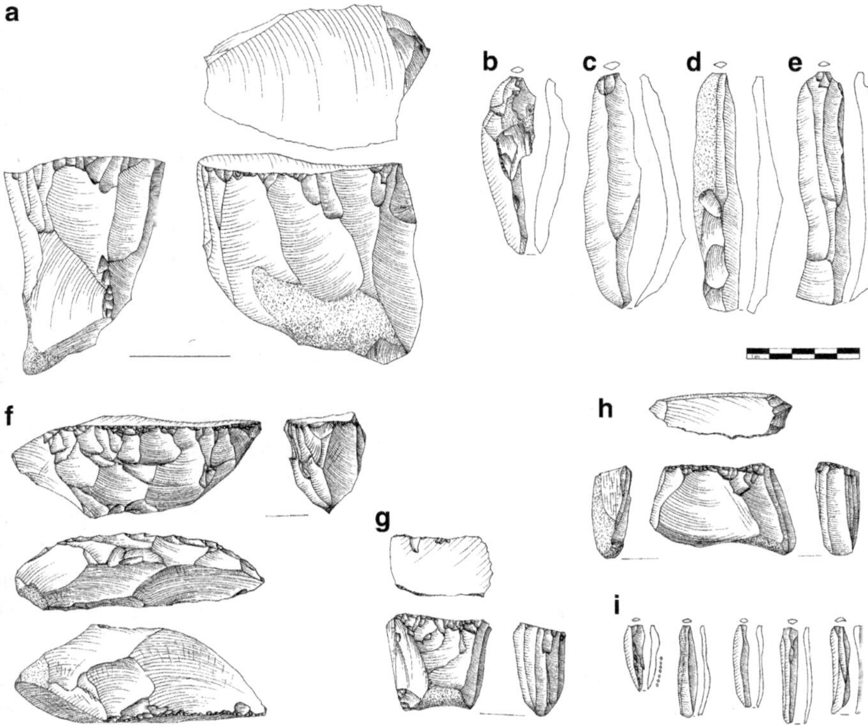

Fig. 9.8 Blades and cores from the Maglemosian techno-complex 4. Site Mosegården IIIN. (**a**) Preform for macroblade production. (**b–e**) Macroblades typical of different production stages. (**f**) Oblong keeled microblade core preform. (**g**, **h**) Exhausted oblong keeled microblade cores. (**i**) Microblades typical of different production stages (Drawing L. Johansen)

Concerning the regular macroblades, these are now generally more curved, larger and more robust than in the previous techno-complex 3. The difference can be interpreted as resulting from a change to distally unsupported macroblade cores, resulting in more curved subconical blade core morphologies and subsequently more curved and robust blades (Sørensen 2006a, b).

The microblade production undergoes a more substantial conceptual difference, in which the pressure blade method B from the previous techno-complex 3 seems to be developed. During techno-complex 4, a large keeled blade core morphology is produced (often termed the handle core). Such cores are produced from nodules or from large, specifically produced, thick flake blanks. The blade core type is mostly exploited from only one front, in few situations however, the core type is turned and reused from the opposite end.

A main difference between the microblade productions in techno-complexes 3 and 4 can be related to the morphology of the blade cores and how they were held during manufacture. While microblade production in techno-complex 3 probably had to take advantage of an open holding device, so that the knapper could be able to change the position of the circular (bullet shaped) cores after each blade detachment, the oblong keeled cores have a long rear end, that could be permanently fixed during blade production.

There are several attributes indicative of the change to a permanent fixation system of the rear ends of the oblong keeled core types in techno-complex 4. Firstly, this core type is generally discarded with a long rear end, as if the cores were not completely used up. This phenomenon can be explained if the cores were used in relation to a permanent fixation system, which 'occupied' the rear end of the blade cores. A second indication is based on microscopic analysis of the lateral blade core faces. A blade core from the Orelunde IX assemblage was examined under the microscope, and areas of use and wear, in the form of striations, were observed (Fig. 9.9). These striations are interpreted as resulting from a hard squeeze on the lateral faces of the core in combination with lithic dust and small movements of the core (Sørensen 2006b). This type of analysis was also conducted on keeled cores from the Scanian site Tågerup, belonging to the Kongemosian Culture, and showed wood polish on central arises on the lateral faces of the studied keeled cores. The interpretation is that a clamp made from wood had been in use to hold the core during blade production (Karsten and Knarrström 2003: 49).

A typical feature of the oblong keeled blade core type is that the lateral platform edges are heavily trimmed. These observations formerly lead to the belief that the core was a scraper, called the 'keeled scraper type' (Westerby 1927). However, in a series of replicative experiments, a permanent fixation system of the lateral sides of the keeled core types were tested, and the results can explain the heavy trimmed lateral core edges (Sørensen 2003; Sørensen 2006b). It was discovered that, if the clamp system rests on untrimmed lateral edges of the core, large damaging platform flakes will be detached when force is applied to the platform via the pressure tool. In order to avoid crucial damage to the blade core platform, the lateral edges should therefore be heavily trimmed before the core is mechanically fixed.

The two blade concepts and methods used in the Maglemosian techno-complex 4 were, with minor differences, continuously in use during the following

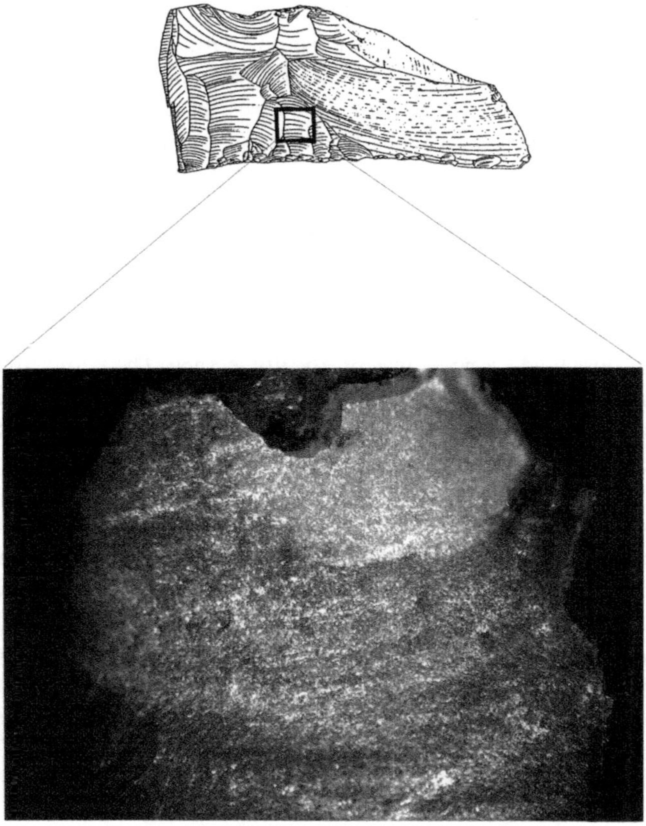

Fig. 9.9 Microwear seen as striations on the keeled blade core face from Orelund IX, interpreted as results of a holding device (Photo B. Knarrström)

Kongemosian period. Nonetheless, there is within concept 7 (Sørensen 2006a, b) (microblade production from keeled cores), during the Kongemosian, a tendency towards a decrease in the quality of the microblades and towards producing keeled core preforms increasingly on flake blanks rather than on small single nodules (Petersen 1993: 58). At the beginning of the Late Mesolithic Ertebølle Culture, the microblade concept 7 is discarded, and with the abandonment of this concept, the pressure blade technology in Southern Scandinavia 'dies out'.

9.5 Knowledge and Know-How in Relation to Pressure Blade Production

Pressure blade production can be described as a technology which, to be carried out, contains a great deal of knowledge in relation to know-how. The nature of knowledge is explicit, which means that it can be transmitted verbally, while know-how has an intuitive nature related to the body memory. Know-how can therefore mainly be achieved through practical training (Pelegrin 1990; Apel 2001).

Specific knowledge about pressure blade production in Mesolithic Scandinavia, appearing with the techno-complex 3, comprises information about several new techniques and knapping tool types in relation to earlier blade production techniques. These techniques are, for example, a mechanical core fixation system and a flexible, perhaps compound, pressure tool. The know-how needed to perform the pressure blade technology seems, on the other hand, to be limited and could thereby quickly have been achieved by Mesolithic flintknappers who already had substantial training from their day-to-day knapping experiences.

Personally, this paradox between knowledge and know-how was experienced when the pressure blade concept was practically learned. The knowledge needed to produce pressure blades, i.e. information about the construction of the pressure tool with the right flexibility and the construction of a holding device, was by far the most challenging part of the learning process. Once knowledge was gained about the device and the right tools were built, the blade production process itself was straightforward. In fact, with the system set up, it was possible to instruct (transmission of knowledge) a person without experience in flintknapping and have him or her produce pressure blades successfully. In contrast, this is generally not possible with blade concepts in which direct percussion is employed, as in techno-complex 1 and 2, because the knapper in this situation needs more practice (know-how). As discussed below, the relation between knowledge and know-how concerning pressure blade production is important when considering the crucial question of invention versus transmission of this technology.

9.6 Local Development or Diffusion of the Pressure Blade Technology in Southern Scandinavia

The Maglemosian is generally considered as a stable and conservative period with respect to typology, economy and settlement patterns. Despite the use of antler pressure tools and a possible holding device, no new tool types seem to have been introduced with the emergence of pressure technique in techno-complex 3. Thus, no new developments in subsistence and economy can be related to the use of pressure blade technology during the Maglemosian. The pressure and punch blade technology can be defined as a substitute for the former blade technologies, while the products (i.e. blades) are used for the same purposes as in the previous Early Mesolithic techno-complexes.

Economically, one could stress the raw material situation and argue that the increase in forests and coverage of the soils by vegetation during the Boreal periods resulted in a restricted access to lithic raw materials. This could have led to a more economical use of the lithic raw material and consequently the invention of the pressure blade technology. However, this argument does not seem to be valid as the assemblages from sites in techno-complex 3 often have large quantities of first quality flint materials and large surpluses of blade products, which are seemingly unused. Moreover, many areas in Scandinavia and Europe had a much more restricted access to lithic raw materials during the Early Mesolithic, which never resulted in the invention of pressure blade technology.

On these grounds, it is argued that the pressure blade technology did not fulfil a specific economic or functional need in the Maglemosian society which was not already fulfilled effectively or resolved by the previous blade production methods. To conclude, no functional, economical or environmental reasons have been found within the Southern Scandinavian area which can support an argument for a locally inspired innovation of the pressure blade technology.

9.7 The Problem of Studying Diffusion in the Maglemosian

The general idea within the Danish Mesolithic research has been that the Southern Scandinavian area yielded an independent development. In comparison, archaeologists from other nationalities in Northern Europe have, with some exceptions, described 'their' Early Mesolithic cultures independently and with a national terminology based on local site names. Regional and national research has often been focused on chronological studies, based on microlith morphology and frequency, and therefore not much attention has been paid to technology or the changes in blade production and the emergence of pressure blade technology within the period. This former research tradition of the Early Mesolithic complicates investigations of diffusion of ideas, technologies or the migration of people within the period.

9.8 A Search for Pressure Blade Technology in Areas Adjacent to Southern Scandinavia During the Early Mesolithic

In middle and Northern Scandinavia, the site Sujala in Northern Lapland, dated to around 8100 cal B.C., has clear traits of the pressure blade technology (Rankama and Kankaanpää 2008); however, this site is convincingly related to the post-Swiderian tradition and must be regarded as an example of a north-west penetration from the Eastern Baltic areas and Russia. From the West, i.e. the British Isles, pressure blade technology is not detected, as neither punch nor pressure seems to have existed in these areas during the Mesolithic (Costa et al. 2005). During the Preboreal and Boreal, when a land bridge existed between Southern Scandinavia and the British Isles, the blade technologies were similar in these areas; however, this similarity stopped when the sea level increased and the British Isles formed during the Boreal and Atlantic period, complicating contact between the two areas.

The Mesolithic of Northern Germany is part of the same cultural development as Southern Scandinavia from the Maglemosian onwards (Bokelmann 1999; Gerken 2001; Hartz et al. 2007; Terberger 2006), but in the adjacent Southern areas, there are seemingly no signs of pressure blade technology during the Early Mesolithic.

In contrast, the areas to the East, comprised of the Baltic states and Northern Poland, are yielding several Early Mesolithic assemblages from the post-Swiderian cultures, for example the Kunda Culture dated to the Preboreal and Boreal periods (9th–8th millennium B.C.), with artefacts indicating common use of pressure blade technology (Burov 1999; Sulgostowska 1999; Zhilin 1999). The Kunda Culture, as represented by the Pulli, Zvejnieki and Tlokowo sites situated in today's Estonia, Latvia and Eastern Poland, has an inventory and a technology which points towards an eastern Paleolithic origin (Sulgostowska 1996). Sites attributed to the Komornica Culture in Eastern Poland (e.g. the sites Lajty, Calowanie, Mszano and Chwalim) can, due to technology and typology, be regarded as part of the Early Mesolithic Western techno-complexes (Sulgostowska 1996, 1999), comparable to the Maglemosian techno-groups 1 and 2. Thus, in the eastern Baltic area, i.e. in today's Northeastern Polish lowland, there seems to have been an overlap between the Kunda Culture and Mesolithic groups of a Baltic/Western tradition.

9.9 Pressure Blade Technology in the Kunda Culture

A main lithic technology of the Kunda Culture is a blade concept in which extremely regular and straight blades are produced from single platform circular cores with facetted platforms. The blades are exploited so that the first sequence of relatively large blades are used for tanged points and large formal tool types, while the late blade sequence of smaller blades and microblades are used as inserts in slotted bone points (Sulgostowska 1996, 1999). If we focus on the Kunda blade technology in relation to the Maglemosian, it is evident that the concept of Kunda blade production is equal or strongly related to the blade production concepts introduced in the Maglemosian techno-complex 3. The main difference seems to be that within the Kunda Culture, platforms are currently facetted during the blade exploitation, while in the Maglemosian tradition, they are kept plain.

It is further evident that the technology of using snapped pressure blades as inserts into slotted bone points, a characteristic of the Late Maglemosian and Kongemosian in Scandinavia, is a technology which was used in the Baltic states and Western Russia, i.e. in the Kunda Culture prior to the Maglemosian. Thus, despite the fact that bone tools with lithic inserts appear before techno-complex 3 within the Maglemosian (Sarauw 1903), it is obvious that pressure blade production and the production of slotted bone tools are two connected technologies that are typical of an eastern Early Mesolithic tradition.

9.10 The Eastern Distribution of the Maglemosian Techno-complex 3

So far, only a little attention has focused on the relationship between the Maglemosian in Southern Scandinavia and the synchronous cultures in Poland (Bagniewski 1990; Domanska 1989). Galinski (2002) operates partly with a Maglemosian terminology

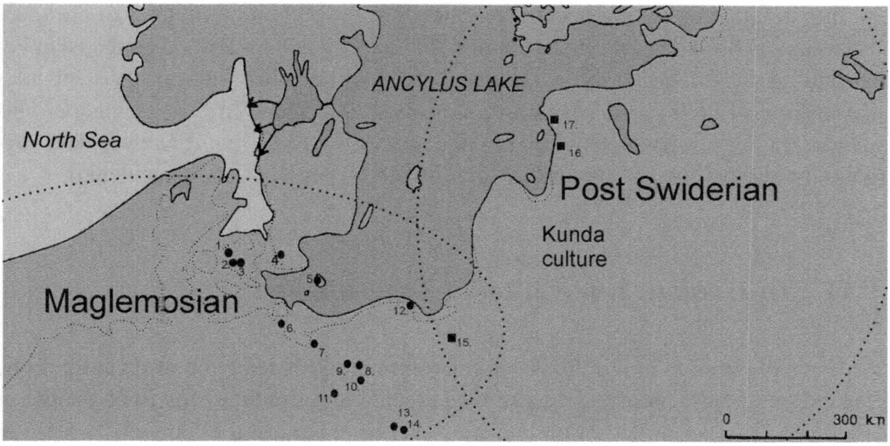

Fig. 9.10 The south Scandinavian and Baltic area with the Preboreal coastline (ca. 9000 uncal B.P.) (After Donner 1995). *Dotted line* represent present day coastline. Site mentioned in the text are numbered. *Round dots*: sites with pressure blade concepts typical of techno-complex 3. *1* Ulkestrup II (Andersen et al. 1982); *2* Sværdborg II (Petersen 1972); *3* Lundby 1 (Henriksen 1980); *4* Draken 356 (Gidlöf 2008); *5* Nr Sandegaard (Becker 1952); *6* Dobra (Galinski 2002); *7* Szczecin-Jezierzyce (Galinski 2002); *8* Wierzchow 6 (Bagniewski 1990); *9* Gudowo 3 (Bagniewski 1990); *10* Pomorski 3 (Domanska and Wąs 2009); *11* Trzebicz Mlyn (Domanska and Wąs 2009); *12* Jastrzebia Gora 4 (Domanska 1989); *13* Dąbrowa Biskupia 71 (Domanska and Wąs 2009); *14* Deby 29 (Domanska 1989); *square dots*: *15* Tlokowo (Sulgostowska 1999); *16* Zviejnieki (Sulgostowska 1999); *17* Pulli (Sulgostowska 1999)

for the different Early Mesolithic phases and complexes in Northern Europe and Poland, identifying, for example a 'Duvensee complex' and a 'Maglemose complex'. The complexes are defined solely on the basis of microlithic typology: The 'Duvensee complex' is typical of lanceolate microliths known primarily from techno-complexes 1 and 2, while the 'Maglemose complex' is defined by scalene triangular microliths typical of technology-complex 3; this typological horizon is sometimes also referred to as the 'Sværdborg phase'. Some of the 'Maglemose complex' sites belong, judging from published artefact drawings, to the Maglemosian techno-complex 3 (the sites Dobra Szcz and Szczecin-Jezierzyce) (Galinski 2002). Also, Bagniewski (1990), Domanska (1989) and Domanska and Wąs (2009) defined Maglemosian sites in Poland, of which several have to be ascribed to the techno-complex 3, often described as part of the 'Sværdborg Culture', including the Wierzchowo 6, Pomorski 3, Gudowo 3, Dobre 53, Trzebicz Mlyn and Dąbrowa Biskupia 71 assemblages from Northwest Poland. The Jastrzebia Góra site and the site Deby (Domanska 1989), situated respectively at the Baltic coast and in the Polish lowland, have blade industries and typologies that resemble the Maglemosian techno-complex 3 assemblages, for example, of the Baltic island Bornholm (Becker 1952). It can thus be concluded that the Maglemosian techno-complex 3, in which pressure blade technology appears for the first time in Southern Scandinavia, can be found from Southern Scandinavia through central parts of the Northern Polish lowland (Fig. 9.10).

It is comparatively interesting to notice that the techno-complex 3 in Southern Scandinavia have an eastern distribution. In fact, it was first defined on the island of Bornholm in the Baltic Sea (Becker 1952), while many of the classical sites (Sværdborg, Lundby and Ulkestrup) are found at Zealand in East Denmark. This picture might, of course, be biased by the research activity, but it nevertheless suggests a tendency towards the Baltic area, which is, perhaps, not coincidental.

9.11 Discussion: Innovation or Transmission

As argued above, pressure blade production mainly depends on knowledge to be carried out. Consequently, pressure blade technology 'only' needs to be shown or observed and transmitted orally before it can be reproduced effectively, while, on the other hand, it is a difficult technique to invent. Lithic technologies heavy in know-how, such as Upper Paleolithic blade concepts (Pigeot 1990) or Neolithic bifacial knapping (Apel 2001), involve training or even apprenticeship in order to be conducted, and they are therefore not quickly transmitted between people. In other words, pressure blade production (without reinforcement) is a technology which can rapidly be spread between people who already have practical know-how and knowledge about lithic fracture dynamics.

The second hypothesis concerns the lack of functional, economic or environmental explanations supporting the innovation of the pressure technique within the Maglemosian. The pressure blade technology replaces the former blade technologies and the blades function but does not fulfil new functional demands. An economical aspect related to pressure blade technology can thus be rejected as a cause for its use or invention in Southern Scandinavia. This does not, however, exclude the invention of pressure blade production during the Maglemosian, but the causes then have to be found within the social or ideological sphere of the society.

The third hypothesis concerns the areas adjacent to Southern Scandinavia. The Early Mesolithic tradition in the eastern Baltic area (Poland and Latvia, often termed the Kormonica Culture) overlaps geographically with the Kunda Culture, which employed the pressure blade technology during the Preboreal and Boreal periods (9th–8th millennium). This shared 'territory' suggests that knowledge concerning pressure blade production could have been transmitted between the two cultural traditions within the area, either with migrating people, or more possibly as transmitted knowledge during regional contacts between people. However, even though pressure was transmitted, the Kunda blade concept was not adopted completely. The Maglemosian tradition of maintaining the core platform's plane by avoiding facetting is unchanged within Southern Scandinavia, in contrast to the Kunda blade concept. The weakness of this hypothesis is the lack of sufficient data, since studies in the Polish area of pressure blade technology during the Early Mesolithic are only few (e.g. Płaza and Grużdź 2010). In order to understand the problem in depth, the original material needs to be studied from a technological perspective.

9.12 Conclusion

This paper has hopefully shed light on Maglemosian pressure blade technology and its development in Southern Scandinavia. On the basis of the technological analysis, it is suggested that the technology of producing pressure blades from single platform cores was transmitted from the Kunda Culture to Early Mesolithic hunter-gatherers of the Baltic and Scandinavian lowlands, known as the Kormonica Culture in Poland and the Maglemosian in Southern Scandinavia and Northern Germany. This transmission supposedly happened during the 9th millennium B.C. in the Southeast Baltic and in the Polish region and is observed in Southern Scandinavia in the 8th millennium. From only one absolute dating of techno-complex 3 assemblages within Southern Scandinavia, it can be suggested that pressure blade technology was carried out ca. 7000 B.C. AMS-radiocarbon dates need to be made on certain techno-complex 3 material before a more certain absolute age determination can be made on the arrival of the pressure blade technology in the region. The pressure blade concept of the Kunda and the Maglemosian differs concerning the preparation of the platforms, in that this preparation does not take place within the Maglemosian.

The pressure technique within flint-rich areas of the Maglemosian (Zealand, Denmark) was applied using two different methods of core exploitation (methods A and B) in techno-complex 3. The method A is equivalent to the Kunda Culture pressure blade concept, while method B employs single fronted oblong cores. It is suggested that during the following techno-complex 4, the single fronted core type is developed into a long oblong keeled core type (handle cores), while method A is abandoned.

So where did the Kunda Culture learn pressure blade technology? Was it a local invention from within the Kunda Culture? According to some researchers (Sulgostowska 1999), the Kunda Culture has an eastern origin in the Late Paleolithic of Siberia and Ural, with ties to sites such as Shikaevka, dated to 13000–12000 B.P. (Abramova 1984), or Mullino (Matiusin 1976). Skeletal material and anthropological data from sites related to the Kunda Culture, for example Zveinieki in Latvia and Popovo, do partly confirm this hypothesis (Potekhina 1999).

Seen from a technological perspective, pressure blades produced from single platform cores are found in the Butovo Culture in the upper Volga basin (Koltsov and Zhilin 1999), dated to the Preboreal period in the 9th millennium B.C. Thus, in a technological sense, there seems to be a link from the Kunda Culture towards an eastern area.

It is, as discussed by Inizan et al. (1992), possible that the pressure blade technology was transmitted as knowledge ('borrowed') across the central Russian plains and that this technology was invented during the Upper Paleolithic around 20000 B.C. in the Mongolian area. This hypothesis is supported by the fact that not only is pressure blade production as a technique 'arriving' in the hunter-gatherer societies of Northern and Eastern Europe (Butovo, Kunda and Maglemosian) but also almost the same concept of producing the blades, namely the use of single platform circular core types, is performed for the initial production of pressure blades. In this light, the Maglemosian can be understood as a cultural period, which receives knowledge about pressure blade production that has travelled across the continent from the Central Mongolian area.

There are many problems to be solved before more certain conclusions can be reached, especially concerning the relationship between the Maglemosian and the post-Swiderian cultures of the Baltic states and the Western Russian area. The national research traditions have so far prevented the Early Mesolithic of the North European lowland and Baltic area from being studied as a whole, i.e. a cultural phenomenon from Poland to Britain, and very few syntheses about the Early Mesolithic of Northern Europe are available. Secondly, the most chronological as well as regional studies of the Maglemosian are based on microlithic morphologies, a narrow perspective that does not facilitate, or in many cases permit, the study and discussion of cultural relations and cultural change within the Early Mesolithic Maglemosian. It is therefore time to leave the national focus and to study the Early Mesolithic internationally and interregionally and from new perspectives. One new perspective could involve detailed studies of specific technologies over large areas, as it has been clearly demonstrated that technology in prehistory, as in modern times, has strong social, traditional and cultural implications.

References

Aaby, B. 1993 Flora. In *Da klinger i muld,* edited by S. Hvass and B. Storgaard, pp. 24–30. Aarhus Universitetsforlag, København.

Abramova, Z.A. 1984 Pozdnij paleolit azjatskoj casti SSSR. In *Paleolit SSSR*, edited by F.I. Boriskovskij, pp. 302–346. Nauka, Moskva.

Andersen, K. 1983 *Stenalderbebyggelsen i den Vestsjællandske Åmose.* Fredningsstyrelsen, København.

Andersen, K. 1985 To Åmose bopladser med håndtagsblokke. *Aarbøger for Nordisk Oldkyndighed og Historie* 1984: 18–46.

Andersen, K., S. Jørgensen, and J. Richter 1982 *Maglemosehytterne ved Ulkestrup Lyng*. Det Kongelige Nordiske Oldtidsselskab, København.

Apel, J. 2001. *Daggers, knowledge & power: the social aspects of flint-dagger technology in Scandinavia, 2350–1500 cal. BC.* Uppsala University, Uppsala.

Audouze, F. 1999 New advances in French prehistory. *Antiquity* 73: 167–75.

Bagniewski, Z. 1990, Maglemose-Kultureinflüsse in Mitteleuropa. In *Contributions to the Mesolithic in Europe*, edited by P. Vermeersch and P.V. Peer, pp. 345–353, Leuven University Press. Leuven.

Becker, C.J. 1952 Maglemosekultur på Bornholm. *Aarbøger for Nordisk Oldkyndighed og Historie 1951*: 96–177.

Becker, C.J. 1945 En 8000-årig boplads i Holmegaards Mose. Foreløbige meddelelser. *Nationalmuseets Arbejdsmark 1945*: 61–72.

Bokelmann, K. 1971 Duvensee, ein Wohnplatz des Mesolithikums in Schleswig-Holstein, und die Duvenseegruppe. *Offa* 28: 5–26.

Bokelmann, K. 1985 Duvensee, Wohnplatz 13. *Offa* 42: 13–33.

Bokelmann, K. 1991 Duvensee, Wohnplatz 9. Ein präborealzeitlicher Lagerplatz in Schleswig-Holstein. *Offa* 48: 75–114.

Bokelmann, K. 1999 Zum Beginn des Spätmesolithikums in Südskandinavien. Geweihaxt, Dreieck und Trapez, 6100 cal BC. *Offa* 56: 183–197.

Bokelmann, K., F.R. Averdieck, and H.Willkomm 1981 Duvensee, Wohnplatz 8, Neue Aspecte zur Sammelwirtscharft im frühen Mesolithikum. *Offa* 38: 21–41.

Bordes, F., and D.E. Crabtree 1969 The Corciac Blade Technique and Other Experiments. *Tebiwa* 12(2): 1–20.

Burov, G.H. 1999 "Postswiderian" of the European North-East. In *Tanged Points Cultures in Europe,* edited by S. K. Kozlowski, J. Gurba and L.L. Zaliznyak, pp. 281–291. Wydawnictwo UMCS, Lublin.

Costa, L.J., S. Sternke, and P. Woodman 2005 Microlith to macrolith: the reasons behind the transformation of production in the Irish Mesolithic. *Antiquity* 79: 19–33.

David, E. 2006 Redskaber af ben og tak i tidlig Maglemosekultur. In *Stenalderstudier*, edited by B.V. Eriksen, pp. 77–100. Jysk Arkæologisk Selskab, Aarhus.

Domanska, L. 1989 Elements of a Food-Producing Economy in the Late Mesolithic of the Polish Lowland. In *The Mesolithic in Europe,* edited by C. Bonsall, pp. 447–455. Donald Press, Edinburgh.

Domanska, L., and M. Wąs 2009 Dąbrowa Biskupia 71: a specialised camp from the Maglemose culture. In *Mesolithic Horizons* Vol I, edited by S. McCartan, R. Schulting, G. Warren and P.C. Woodman, pp. 261–269. Oxbow Books, Oxford.

Donner, J. 1995 The Quaternary History of Scandinavia. In *World and Regional Geology* Vol. 7. Cambridge University Press, Cambridge.

Galinski, T. 2002 *Spolenczenstwa mezolityczne*. Museum Narodowe w Szczecine, Szczecin.

Gerken, K. 2001 *Studien zur jung- und spätpaläolithischen sowie mesolithischen Besiedlung im Gebiet zwischen Wümme und Oste.* Kommisionsverlag, Oldenburg.

Gidlöf, K. 2008 *Draken 1: Husie socken i Malmö stad, Skåne län: arkeologisk slutundersökning 2006.* Report, Enheten för arkeologi Malmö kulturmiljö, Malmö.

Hartz, S., H. Lübke, and T. Terberger 2007 From fish and seal to sheep and cattle: new research into the process of neolithisation in northern Germany. In *Proceedings of the British Academy. Going Over: The Mesolithic-Neolithic Transition in North-West Europe,* edited by A. Whittle and V. Cummings, pp. 567–597. Oxford University Press, Oxford.

Henriksen, B.B. 1980 *Lundby-holmen. Pladser af Maglemose-type i Sydsjælland.* Nordiske Fortidsminder Vol. 6. Det Kongelige Nordiske Oldtidsselskab, København.

Inizian, M.L., M. Lechevallier, and P. Plumet 1992 A technological marker of the penetration into North America: pressure microblade debitage. Its origin in the Paleolitihic of North Asia and its diffusion. In *Materials Issues in Art and Archaeology (III),* edited by P.B. Vandiver, J.R. Druzik, G.S. Wheeler and I.C. Freestone, pp. 661–681. Materials Research Society, Warrendale.

Johansen, F.K. 1919 En Boplads fra ældre Stenalder i Sværdborg Mose. *Aarbøger for Nordisk Oldkyndighed og Historie*: 106–235.

Johansson, A.D. 1990 *Barmose-gruppen. Præboreale bopladsfund med skiveøkser i Sydsjælland.* Århus Universitetsforlag, Århus.

Karsten, P., and B. Knarrström 2003 *The Tågerup Excavation.* National Heritage Board, Lund.

Koltsov, L.V., and M.G. Zhilin 1999 Tanged point cultures in the upper Volga basin. In *Tanged Point Cultures in Europe,* edited by S.K. Kozlowski, J. Gurba and L.L. Zaliznyak, pp. 346–360. Wydawnictwo UMCS, Lublin.

Larsson, L. 1978 Ageröd I:B, Ageröd I:D. A study of Early Atlantic Settlement in Scania. *Acta Archaeologica Lundesia* Vol. 12.

Leroi-Gourhan, A. 1964 *Le Geste et la Parole I – Technique et Langage.* Albin Michel, Paris.

Mathiassen, T. 1943 *Stenalderbopladser i Aamosen*. Nordiske Fortidsminder Vol. 3. Det Kongelige Nordiske Oldtidsselskab, København.

Matiusin, G.N. 1976. *Mezolit juznovo.* Nauka, Moskva.

Pelegrin, J. 1984a Debitage par pression sur silex: nouvelle experimentations. In *Préhistoire de la pierre taillée, 2: économie du débitage laminaire: technologie et expérimentation,* edited by J. Tixier, pp 117–127. CREP, Paris.

Pelegrin, J. 1984b Approche technologique expérimentale de la mise en forme des nucléus pour le débitage systématique par pression. In *Préhistoire de la pierre taillée, 2: économie du débitage laminaire: technologie et expérimentation,* edited by J.Tixier, pp. 93–103. CREP, Paris.

Pelegrin, J. 1984c Systèmes expérimentaux d'immobilisation du nucléus pour le débitage par pression. In *Préhistoire de la pierre taillée, 2: économie du débitage laminaire: technologie et expérimentation,* edited by J.Tixier, pp. 105–116. CREP, Paris.

Pelegrin, J. 1988 Débitage expérimental par pression: du plus petit au plus grand. In *Technologie préhistorique,* edited by J. Tixier. Notes et Monographies Techniques du CRA, (25), pp 37–53. CNRS, Paris.

Pelegrin, J. 1990 Prehistoric lithic technology: some aspects of research. *Archeological Review from Cambridge*, 9(1): 116–125.

Pelegrin, J. 2002 Principes de la reconnaisance des méthode et techniques de taille. In *Tell'Atij Tell Gueda. Analyse technologique et fonctionelle*, edited by J. Chabot, pp. 216–225. Cahiers d'Archéologie du CELAT 13, Quebec.

Pelegrin, J. 2006 Long blade technology in the Old World: an experimental approach and some archaeological results. In *Skilled Production and Social Reproduction. Aspects on Traditional Stone-Tool Technologies,* edited by J. Apel and K. Knutsson, pp. 37–69. SAU, Uppsala.

Petersen, E.B. 1966 Klosterlund – Sønder Hadsund – Bøllund. *Acta Archaeologica* (37).

Petersen, E.B. 1972 A Maglemose Hut From Sværdborg Bog, Zealand, Denmark. *Acta Archaeologica* 42: 43–78.

Petersen, E.B. 1973 A Survey of the Late Palaeolithic and the Mesolithic of Denmark. In *The Mesolithic in Europe,* edited by S.K. Kozlowski, pp.77–129. Warsaw University Press, Warszawa.

Petersen, P.V. 1993 *Flint fra Danmarks Oldtid.* Nationalmuseet, København.

Pigeot, N. 1990 Technical and social actors – flintknapping specialists and apprentices at Magdalenian Etoilles. *Archaeological Review from Cambridge* 9(1): 126–141.

Płaza, K., and W. Grużdź 2010 Experimental Study of Pomeranian Flint (Kugleflint) from Mesolithic Excavation Site in Jastrzębia Góra 4. *Materiały Zachodniopomorskie* IV/V (2007/2008): 5–21.

Potekhina, I.D. 1999 Postswiderian population of North-Eastern Europe as seen from the anthropological data. In *Tanged Points Cultures in Europe*, edited by S.K. Kozlowski, J. Gurba and L.L. Zaliznyak, pp. 333–337. Wydawnictwo UMCS, Lublin.

Rankama, T., and J. Kankaanpää 2008 Eastern arrivals in post-glacial Lapland: the Sujala site 10.000 cal BP. *Antiquity* 82: 884–900.

Salomonsson, B. 1965 Linnebjär. A Mesolithic Site in South-West Scania. *Meddelan Från Lunds Universitets Historiska Museum* 1964–65: 5–32.

Sarauw, G.F.L. 1903 En stenalders boplads i Maglemose ved Mullerup, sammenholdt med beslægtede fund. *Aarbøger for Nordisk Oldkyndighed og Historie* 1903: 148–315.

Schiffer, M.B., and M. Skibo 1987 Theory and Experiment in the Study of Technological Change. *Current Anthropology* 28: 595–611.

Schild, R. 1980 Introduction to Dynamic Technological Analysis of chipped stone assemblages. In *Unconventional Archeology*, edited by R. Schild, pp. 57–87. Ossolineum, Wroclaw.

Schilling, H. 1999 Maglemosekulturens bosættelse i Holmegårds Mose – miljøets diktat eller kulturelle valg. Unpublished thesis, Institute of Archaeology, University of Copenhagen, Copenhagen.

Skaarup, J. 1979 *Flaadet. En tidlig maglemoseboplads på Langeland.* Langelands Museum, Rudkøbing.

Skar, B. 1987 The Scanian Maglemose Site Barre Mosse II. A re-examination by refitting. *Acta Archaeologica* 58: 87–104.

Sulgostowska, Z. 1996 The Earliest Mesolithic Settlement of North-Eastern Polasnd. In *The Earliest Settlement of Scandinavia,* edited by L. Larsson. Acta Archaeologica Lundesia Series in 8, No. 24, pp. 297–305. Almquist & Wiksell International, Stockholm.

Sulgostowska, Z. 1999 Final Palaeolithic Masovian Cycle and Mesolithic Kunda Culture relations. In *Tanged Points Cultures in Europe,* edited by S.K. Kozlowski, J. Gurba and L.L. Zaliznyak, pp. 85–93. Wydawnictwo, Lublin.

Sørensen, M. 2003 Experiments with the "Tågerup pressure tool". In *The Tågerup Excavations* edited by P. Karsten and B. Knarström, pp 146–150. National Heritage Board, Lund.

Sørensen, M. 2006a Rethinking the Lithic Blade Definition – Towards a Dynamic Understanding. In *Skilled Production and Social Reproduction -Aspects on Traditional Stone Tool Technologies*, edited by J. Apel and K. Knutsson, pp. 277–299, Societas Archaeologica Upsaliensis, Stockholm.

Sørensen, M. 2006b Teknologiske traditioner i Maglemosekulturen. En diakron analyse af maglemosekulturens flækkeindustri. In *Stenalderstudier. Tidlig mesolitiske jægere og samlere i Sydskandinavien,* edited by B. V. Eriksen, pp. 19–77. Jutland Archaeological Society, Århus.

Sørensen, M. 2006c The Chaîne Opératoire Applied to Arctic Archaeology. In *Dynamics of Northern Societies: Proceedings of the SILA/NABO Conference on Arctic and North Atlantic Archaeology, Copenhagen, May 10th–14th, 2004,* edited by B. Grønnow and J. Arneborg, pp. 31–44. Aarhus University Press, Copenhagen.

Sørensen, M. 2007 Draken MK 356. En teknologisk undersøgelse at et Mesolitisk inventar med henblik på datering og arkæologisk potentiale. Report, Malmö Museer, Malmö.

Terberger, T. 2006 The Mesolithic Hunter-Fisher-Gatherers on the Northern German Plain. In *Across the western Baltic*, edited by K.M. Hansen and K.B. Pedersen, pp. 11–185. Sydsjællands Museum, Vordingborg.

Toft, P.A., 2006 Hvor går grænsen? Et studium af sociale territorier og ornamentik i Maglemosekulturen. In *Stenalderstudier. Tidlig mesolitiske jagere og samlere i Sydskandinavien*, edited by B. V. Eriksen, pp. 101–119. Jutland Archaeological Society, Arhus.

Westerby, E. 1927 *Stenalderbopladser ved Klampenborg. Nogle Bidrag til Studiet af den Mesolitiske Periode.* Reitzel, København.

Zhilin, M.G. 1999 New Mesolithic peat sites on the upper Volga. In *Tanged Point Cultures in Europe*, edited by S. K. Kozlowski, J. Gurba and L. L. Zaliznyak, pp. 295–311. Wydawnictwo, Lublin.

Chapter 10
Surface Pressure Flaking in Eurasia: Mapping the Innovation, Diffusion and Evolution of a Technological Element in the Production of Projectile Points

Kim Darmark

10.1 Introduction

During the archaeological excavations along the new stretch of the main arterial road, the E4, in Northern Uppland, Eastern Central Sweden, bifacially thinned arrowheads and associated waste by-products made out of flint, or flint-like materials, were found at several Late Neolithic and Bronze Age sites (Apel et al. 2005; Apel and Darmark 2007). A preliminary examination of the material suggested that Northern Uppland was a border area where two different traditions of making bifacial projectile points met (Fig. 10.1): a Northern tradition, in which projectile points were made from local raw materials through a combination of percussion flaking and pressure flaking, and a Southern tradition, in which projectile points were made from imported, South Scandinavian flint through edge-pressure flaking (Apel et al. 2005). These different traditions demarcate a classic cultural barrier between South and North Sweden with roots back to the Mesolithic. This cultural barrier is also a long-lasting division between hunter-gatherers/herders in the North and farming communities in the South. This realization triggered an interest in questions concerning the reasons behind the inclusion of surface pressure-flaking technologies in these economically and socially differentially situated populations.

In this chapter, we intend to present an attempt at mapping the chronological and spatial distribution of the use of surface pressure flaking in Eurasia. Such an endeavour will by necessity remain sketchy, but the arising pattern indicates that there is a strong selection towards the incorporation of this particular technological element across this vast geographical area, transgressing climatic and socioeconomic boundaries. We discuss our observations against specific attributes of surface pressure flaking as well as our conceptions of the principles of technological change.

K. Darmark (✉)
Åland Islands Museum, Mariehamn, Åland Finland
e-mail: kim.darmark@regeringen.ax

P.M. Desrosiers (ed.), *The Emergence of Pressure Blade Making: From Origin to Modern Experimentation*, DOI 10.1007/978-1-4614-2003-3_10,

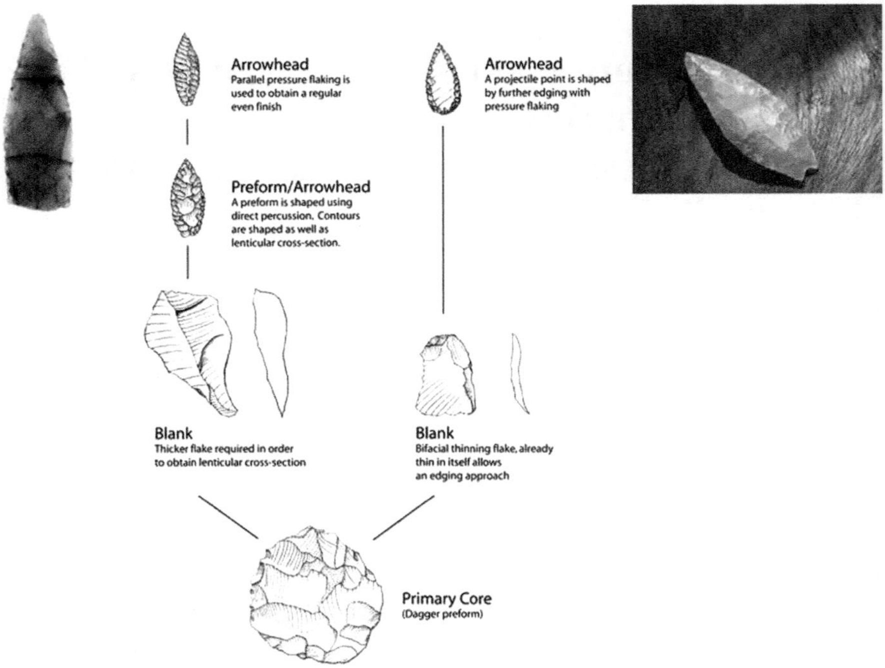

Fig. 10.1 Production chart illustrating two projectile point traditions in Central Middle Sweden (From Apel and Darmark 2007)

10.2 Pressure Flaking and the Technical Production Sequence

Below, we will discuss what type of archaeologically definable units are suitable to use in studies concerning the dynamics between on the one hand material cultural phenomena directly subjected to evolutionary processes, such as selection and drift, and on the other essential cultural phenomena which, due to the inherent cultural conservatism of humans, are reproduced almost intact through the centuries. A formulation of a relevant taxonomy of the cultural elements of a tool tradition should be based on a detailed mapping of the technological syntax, i.e. the ideas, materials and gestures included in every single production sequence constituting the technology (Darmark and Apel 2008). All of these features can be culture-specific. In this context, it is important to distinguish between individual technological elements and technological syntaxes. A technological element can be defined as an instant event consisting of a combination a gesture, a tool, a core and an intention. A technological syntax, on the other hand, consists of an ensemble of technical components that are chronologically structured into a sequence that ideally result in a finished artefact with the desired characteristics (Apel and Darmark 2007; Apel 2008). If a technology is complex enough, it is likely that such syntaxes will be transmitted vertically from parent to child – as the grammar of a language. The geographical diffusion of individual technical elements happens to a much greater extent horizontally between unrelated people – as the loanwords in a language. By articulating such a distinction regarding archaeological materials, tools are created which help us understand continuity and change over time as well as space.

A production sequence forms part of a greater complex of ideas that stretches from the *connaissance specifiques* (culture-specific knowledge) to the artefact and the waste by-products deriving from the making of the artefact (see for instance Pelegrin 1995; Sørensen 2006; Desrosiers and Sørensen 2008; Apel 2008). The cultural knowledge, which is accumulated and passed down a certain artefact tradition, constitutes the articulated prerequisite for new technical innovations.

An innovation, or the adoption of an already familiar technology, for that matter, can only take place if certain fundamental elements are available. When an intention has been defined – to make an object with the desired traits – a concept of the object in question is created. The operational scheme consists of the flintknapper's intentions, for instance, the intention to observe a certain reduction pattern (Eriksen 2000; Desrosiers and Sørensen 2008) required for the production of this particular object. It is important to take notice of the limiting factors existing partially outside of the cultural context in which the technique is performed, such as access to suitable raw materials, spatial limitations, climatic limitations, etc. All of these factors affect whether or not it is possible to put an operational scheme into practice. Such factors must always be considered in studies of specific stone technologies. If an artefact is produced in two separate areas where the access to raw material, as well as its quality, differs, the resulting artefact will display morphometric variability even if the same recipe of action is applied, as will the artefacts produced by craftsmen of different skill levels (Apel and Darmark 2007; Apel 2008).

Surface pressure flaking, in its simplest form, is not an exceedingly complicated technique. In relation to direct techniques, it requires the utilization of different tools (pressure flakers; Ishi sticks) as well as different body postures (Nunn 2006). Surface pressure flaking also seems to be connected to fine-grained raw materials. Advanced modifications to tools employing surface pressure, such as long edge-to-edge flaking, certainly involve a considerable amount of practice. There is, however, a logical transition between direct technique and pressure technique. Certain techniques of platform maintenance, e.g., can be characterized as a kind of pressure technique, when the platform is modified by gentle sliding motions, scraping off small flakes, or changing the angle of the platform. Even though the aim here is to prepare the platform for the removal of the principal flakes, the idea of using pressure as a means of releasing small thin flakes of a flaking surface comes easily to mind. If the manipulation of the surface becomes the desirable goal, special equipment and reduction strategies can be easily designed to meet this desire. The question thus is what the desired properties of surface pressure flaking are?

The most important feature of pressure flaking is that it facilitates the production of thinner and more evenly shaped flakes (Cotterell and Kamminga 1987: 681). Pressure flaking thus gives the knapper increased control over the reduction sequence. The amount of volume loss in relation to blank size is significantly lower when pressure flaking is employed than when direct technique is involved in the reduction sequence (Darmark and Apel 2008: 175f). This gives pressure flaking a position among risk-minimizing strategies since it reduces the risk of breakage and conserves the volume. This is a functionalistic property of the technique, and pressure flaking could thus be hypothesized to be closely correlated to environments of unreliable raw material access (cf. Elston and Brantingham 2002). Conceivably, pressure flaking could also have the effect of transforming a smooth ventral surface into a facetted and more lacerating projectile point (Ellis 1997: 51).

Another aspect of pressure flaking, also arising from the increased control on behalf of the knapper, is the possibility of enhancing the aesthetic qualities of the tool. The grinding and subsequent surface pressure flaking of the Late Neolithic daggers in Southern Scandinavia (Apel 2001) is a display of craftsmanship with universal appeal. The symmetrical finish obtained through a controlled reduction could well have great attraction value for a species among which symmetry is an important factor in signalling mate attractiveness (Grammer et al. 2003).

10.3 Principles of Material Culture Reproduction

Figure 10.2 illustrates different ways in which cultural, archaeological/technical elements remain stable or change over time. A craft element is reproduced according to the principle of generative materialism (Fig. 10.2a) if it is copied as soon as it becomes known because it brings a functional advantage compared to previous solutions to the same problem. Hence, the variation, which is represented by the grey areas surrounding the curves, is created by random or intentional discoveries and is greater at times when the natural and social environment is unstable, whereas optimal solutions are selected during periods of greater stability.

It has been suggested that artefacts that change because of a continuous adaptation should be of interest primarily to archaeologists who wanted to work within a Darwinian perspective (Dunnell 1978). The idea was that such functional types had a direct influence on people's ability to adapt and reproduce, and that they thereby affected the gene pool. Under ideal circumstances, no cultural conservatism operates on and delays this process of change. However, many archaeological phenomena have not had a clear function and consequently have not had any selective value. The combination of such elements, which can be compared to uncoded DNA, will not change through selection, but through cultural drift (Shennan 2002). The change that occurs within an ensemble of technological elements that do not have selective value will be stochastic, and thus, we have not felt the need to illustrate this process in a diagram. It is appropriate to name this principle "the principle of stochastic materialism" since chance determines which elements are forgotten and which survive.

It is important to make a distinction between cultural elements that are reproduced according to stochastic and generative materialism and those which are reproduced according to the principle of repetitive essentialism. It has been pointed

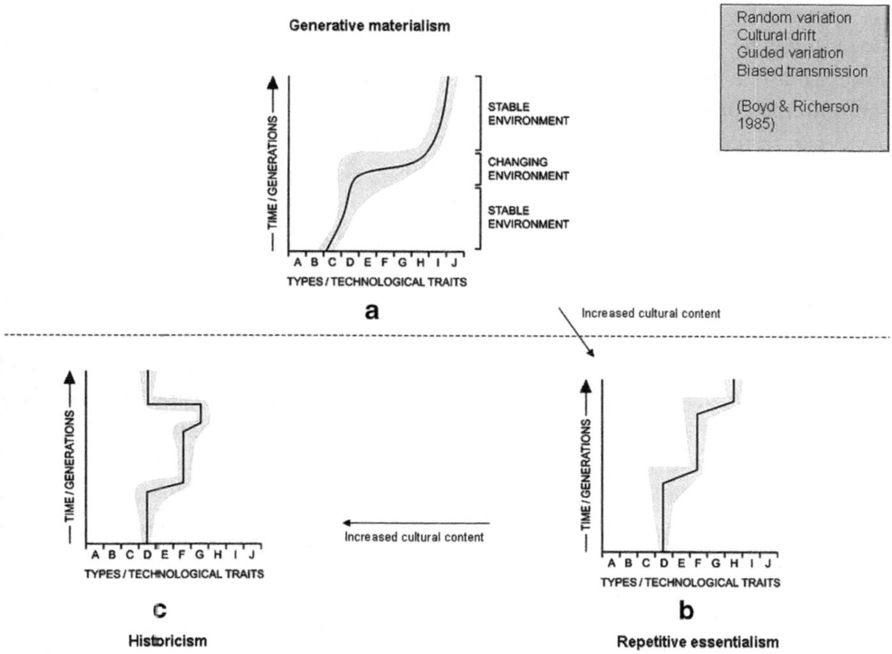

Fig. 10.2 Different principles by which material culture is reproduced (**a**) Generative materialism. (**b**) Repetitive essentialism. (**c**) Historicism. (From Apel and Darmark 2007)

out that human cumulative cultural evolution is dependent not only on the ability to emulate or imitate behaviour but also on inherent pedagogical resources that enable humans to make long-term educational investments in their children (Tehrani and Riede, in press). Neuropsychological research suggests that a theory of mind may have appeared as the result of primate tool use and that it projected humans from being passive niche constructors to active ones: "Once goal-directed intentional niche construction was introduced into the evolutionary process, biological and cultural processes became intertwined to an unprecedented degree" (Iriki and Sakura 2008: 11). A theory of mind made it possible to make a division between a subject (the mind) and an object (the body) and to acknowledge other minds, in other bodies, in different times. Thus, historical essence as a principle has to complement materialistic principles in the study of human cultural transmission processes.

An archaeological element reproduced according to the principle of repetitive essentialism (Fig. 10.2b) tends to remain almost unchanged through time as long as it constitutes an important part of the tradition, i.e. as long as it – in a social and cultural respect – carries a message which is important in order to create and maintain the identity of the group. Action patterns and explicit or implicit pedagogical strategies hidden in myths, structures, legends and traditions guarantee that the knowledge of the distinctive feature in question is copied in such a way as to make any deviations minimal. When it comes to the principal of repetitive essentialism, it is important to separate the form of a cultural phenomenon from its contextual meaning. The physical structure of the craft, ritual, law or institution will remain more or less constant over a longer period of time, whereas the historically situated cultural meaning connected to the phenomenon will vary.

As a result of repetitive essentialism, craft elements are reproduced almost unchanged over long periods of time, and this is also true for artefacts consisting to a great extent of such elements; the formal variation over time is small. When for some reason that idea is no longer of interest, the trait in question will be subjected to cultural drift to a greater extent. In other words, a type of artefact containing a great deal of essential elements will over time be subjected to a negligible amount of change, until the supporting cultural idea behind the craft (regardless of whether this idea is functional or identity-creating) becomes outdated. Perhaps one could picture that the variation occurring gradually in the margin during the lifetime of the trait will eventually result in the idea behind the trait losing its value. This will result in great variation for a short time, until the majority in the tradition have copied a new ideal, which then will constitute a new cultural norm. As we can see, this might happen as the result of several factors. In a stratified society, it is possible that an artefact which is initially exclusive to a certain social stratum, over time, is spread through copying to larger parts of the population, whereby it will eventually become devalued, and therefore replaced by a new type.

The diagrams in Fig. 10.2 only illustrate the change of the archaeological elements over time within one single tradition. The geographical distribution of a certain element is not taken into consideration. Consequently, when we speak of a historical reconnection in diagram c, outside influences are not intended. Instead, it is a matter of a human ability to reproduce a forgotten craft based on certain remains and traces combined with generative thinking. The point in this context is that humans are unique in the ability to, by way of dissociated examination and analysis, replicate old, historical artefacts and then use them in a partly new context. Such reconstructions occurred during prehistory as well (Knutsson 2006; Högberg 2006) and are consequently not solely the result of the impact of Enlightenment ideals over the past centuries (Knutsson 2006). Figure 10.2c illustrates such a process. Certainly, one could consider this principle as a variation of repetitive essentialism, but, according to us, there is a point in separating these two principles.

10.4 Tracing Surface Pressure Flaking: Methodological Aspects

In glass or obsidian, it is possible to calculate the crack velocity through the existence of Wallner lines on flat crack surfaces, a method which, however, is suggested as being of little use in distinguishing different rates of crack velocity in coarser materials (Rabinovitch et al. 2006; Cotterell and Kamminga 1987). Calculations of crack velocities based on Wallner lines and fracture wings on experimental material have shown the potential of these microscopic features to discriminate between different reduction techniques (Hutchings 1999). On a macroscopic level, the relationship between the thickness of the platform and the thickness of the percussion bulb on debris seems to distinguish pressure flaking from direct technique (Darmark and Apel 2008), which is connected to the operation of different fracture mechanics involved in the two techniques (Cotterell and Kamminga 1987). The fact that pressure technique produces thinner flakes can also be employed as a means of identifying pressure debitage (Darmark and Apel 2008).

Concerning the scope of this survey, however, we naturally did not have the possibility of rigidly analyzing the primary source material. Instead we have had to accept statements by other researchers and excavators as to the existence of pressure technique as well as in an impressionistic manner recognize pressure flaking in drawings and photographs of artefacts. The knowledge that pressure flaking can produce more regular flakes and a more symmetric parallel flaking is an important part of our preconceptions, even though we realize that there are problems with this notion, since there is considerable overlap in the diagnostic features attributed to different techniques (Callahan 1996). Since we see a conceptual difference between flaking in order to shape the contours of the objective in question (i.e. edging) and flaking in order to modify the surface of the object (i.e. flat retouch), even though both may utilize the technique of pressure, we have tried to present those assemblages that in our view explicitly are the result of an intention of modifying the surface characteristics of the tool.

10.4.1 Surface Pressure Flaking in Eurasia: Archaeological Data

10.4.1.1 ca. 35000–10000 B.C.

The earliest known examples of bifacial thinning using pressure flaking appear independently in two regions: within the Solutrean of Western Europe and the Dyuktai of easternmost Asia (Fig. 10.3). Within the Solutrean tradition, ranging from the 23rd to the 17th millennium B.C. (Smith 1966; Bordes 1968; Callahan, unpublished; Pelegrin 2006), pressure flaking is used mainly in the production of small projectile points (Aubry et al. 2003), but also in the final retouch of larger laurel-leaf points (Callahan, unpublished). In use-wear analyses of small, pressure-retouched shouldered points from various Solutrean sites, striations have been documented and subsequently replicated through experiments with spears (Geneste and Plisson 1990), and at the Le Combe Sauniére site, a spear-thrower has been found (Cattelain 1989; Sinclair 1995). In other words, the small, pressure-retouched points were probably fit into the shaft of a spear and used together with spear-throwers. It is likely that the origin of pressure flaking is to be found in the transitional phase between the Aurignacian period and Solutrean. It has been suggested that the surface pressure technique was first used on backed blades within the Gravettian tradition (Pelegrin 2006: 41) and that the idea then diffused to South-Western Europe

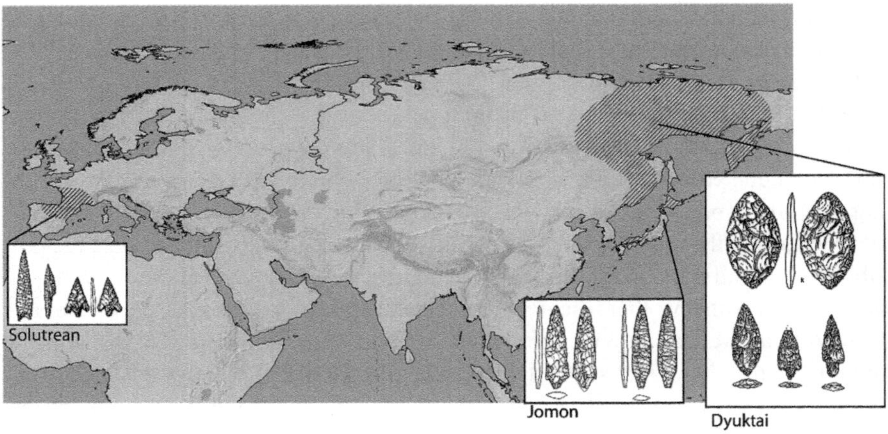

Fig. 10.3 Late Pleistocene surface pressure flaking in the Old World

(Tiffagom 2006). Regarding the morphometric appearance of the Solutrean-tanged projectile points on the Iberian Peninsula, this shape may depend on a North-African Late Aterian influence (Tiffagom 2006). Initially, small laurel-leaf points were made from blades with frequent and symmetrical retouch on the dorsal surface only. Gradually, the retouches are lengthened until eventually they are applied to both the dorsal as well as the ventral surface and cover the entire blade (Bordes 1968: 158; see also Chabas 1874: Plansch IV, 1; Callahan, unpublished). During the latter part of Solutrean, the use of heat treatment in pressure flaking, in order to improve the workability of the raw material, is fully evolved. Heat treatment is more common on the Iberian Peninsula, where quartzite is often used, than it is in France where the raw material is flint (Zilhão 1997; Tiffagom 1998; Aubry et al. 2003). As concerns the production of small bi- and unifacial artefacts, such as tanged points, shouldered points and points with a concave base, the production sequence includes percussion as well as pressure flaking.

One important observation regarding the bifacial thinning practised during Solutrean is that the technological syntax is practically identical over large areas, for instance, between Southern France and the Iberian Peninsula (Aubry et al. 2003), although there are some formal differences (Sinclair 1995: 52). At the same time, paleoclimatic models have indicated climatic variations between the areas (Zilhão 1997; Aubry et al. 2003), which have affected the fauna composition (Bayle 2000; Aubry et al. 2003). Consequently, the great technical similarities over large areas are not explained by an adjustment to similar environmental conditions but might instead reflect cultural conventions (Aubry et al. 2003; Tiffagom 2006). Bifacial thinning seems to disappear from Europe around 16000 B.C.

In North-Eastern Siberia, bifacial tools accompanied by wedge-shaped micro-blade cores are the hallmark of the Dyuktai culture. Sites with Dyuktai material are dated to as early as 35,000 years ago. Whether or not pressure technique was employed to fashion knives or projectile points at this early stage remains unclear. The earliest cultural strata at Ust-Mil 2, Verkhne-Troitskaya and Ikhine 2 contain sparse lithic material, with no irrefutable bifacials with traces of this technique (West 1996). What is clear, however, is that the later phases of the Dyuktai culture, dated between ca. 15000 and 10000 B.C., show clear evidence of the use of surface pressure technique. At the eponymous site of Dyuktai Cave, this forms an integral part of the technical repertoire in all cultural strata, where tools are produced not only using lithic material but mammoth tusk as well. The production sequence involves initial reduction by direct technique followed by fine pressure retouch along the edges often resulting in leaf-shaped or almond-shaped points. Evidence of bifacial pressure flaking has a wider geographical distribution during the Later Dyuktai phase. None of the sites outside the Aldan River Valley have dates stretching as far back as 30000 B.C. The dates indicate a widespread adoption of bifacial pressure technique over larger areas of Western Beringia beginning from about 15000 B.C., and in Japan, surface pressure flaking is employed to produce tools within the Incipient Jomon (Nagai 2007). Within the core area of the Dyuktai tradition, however, bifacial pressure flaking is not such a characteristic trait in the subsequent Sumnagin assemblages, which to a large extent lacks typical projectile points (Kol'tsov 1989: 191).

From other parts of Eurasia, evidence of surface pressure flaking is sparse. In the Caucasian Late Paleolithic, before 12000 B.C. at Mgvimevi and Gwardshilas-Klde, surface flaking is used to fashion projectile points on blades (Kol'tsov 1989: 96), and the finds from Gwardshilas-Klde include flakes released by pressure.

> ... Erstrangige Bedeutung hatte die Herstellung von Werkzeugen hauptsächlich direkt in den Höhlen. Grabungen zeigten, daß hier der Erst- zowie der Sekundärbearbeitungsprozeß vor sich ging. Davon zeugen zahlreiche Nukleus-Funde (manchmal tausende), aber auch Funde von Splittern und Spänen (auch schuppenartige Überbleibsel der Pressionsretusche) ....
>
> (Lordkipanidse 1991: 26)

10.4.1.2 ca. 10000–6000 B.C.

No surface pressure flaking technique is known to exist in Europe from the 17th to the 9th millennium B.C., even though it lives on in easternmost Asia. During the 9th millennium, however, its reintroduction over large parts of Eastern Europe and South-West Asia constitutes an interesting case (Fig. 10.4).

The Kunda culture, which is encountered in present-day Estonia, Latvia, Lithuania, Belorussia and parts of Poland and Finland, has many similarities with the contemporaneous Butovo culture in Western Russia, thus the term Kunda-Butovo complex. Sites within this complex are linked together by lithic provenience studies as far as 500–1,000 km apart (Zhilin 1999; Volokitin 2005). At the Pulli settlement in Estonia, belonging to the Kunda culture, arrowheads have been found which have been made from pressure-flaked blades whose distal fragment have been retouched with what would have been a pressure stick. The tang has also been worked (Jaanits et al. 1982: 31). Pulli has three radiocarbon dates to the period

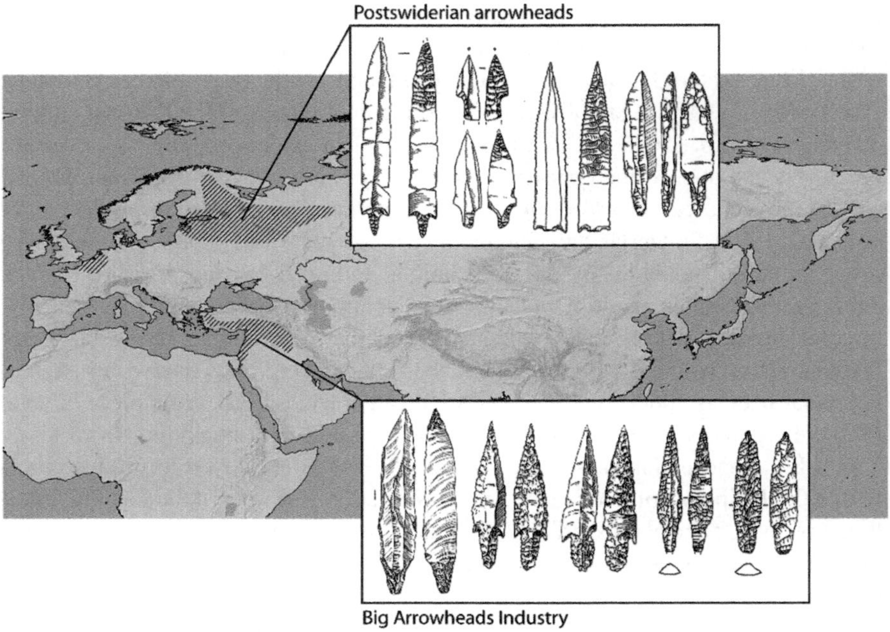

Fig. 10.4 Early Holocene surface pressure flaking in the Old World

from 9300 to 8600 B.C. (Kriiska 2001), but the earliest date of the Kunda/Butovo complex derives from Stanovoye 4, 10200 B.C. (Zhilin and Matiskainen 2000). Partial surface retouching of pressure-flaked blades is closely associated with the material culture of the post-Swiderian groups (Ostrauskas 2000). Arrowheads of this kind are known as post-Swiderian due to their morphological similarity with the tanged points of the East European Swiderian groups, as well as the idea of Swiderian groups migrating North during the Late Pleistocene/Early Holocene (Zaliznyak 1999). Fundamental differences between the industries of the Swiderian of the post-Swiderian (single platform cores, pressure blade debitage, inset technology, etc.) have been pointed out, possibly indicating Eastern influences on the post-Swiderian assemblages (Sulgostowska 1999). The origin of the post-Swiderian is thus a matter of discussion. We wish to point out, however, in concordance with Sulgostowska, that technologies are built up from technological elements, all of which can have very different genealogies. Surface pressure flaking constitutes one such technological element, which has greater visibility in the post-Swiderian than in the Swiderian.

Finds of post-Swiderian pressure-flaked tools were earlier known in Southern Finland (Edgren and Törnblom 1993: 29; Takala 2004), but recent finds show that a similar tradition is encountered in northernmost Finland as well (Kankaanpää and

Rankama 2005). The Parch settlements in the Vychegda river basin contain tanged arrowheads with evidence of ventral pressure flaking (Volokitin 2005).

In Anatolia and the Levant, the production of arrowheads following a similar action pattern – unifacial pressure retouching of blades – begins at the transition between Pre-Pottery Neolithic A (PPNA) and Pre-Pottery Neolithic B (PPNB) (ca. 9200–9000 B.C. [Aurenche et al. 2001]), i.e. roughly around the same time as the abovementioned post-Swiderian points. This industry is referred to as the BAI, or Big Arrowhead Industry (Koslowski 1999: 97). This period is characterized by the use of bipolar cores, the introduction of heat treatment, as well as the extensive production of pressure-retouched projectile points of Byblos or Amuq-type (Bar-Yosef 1981: 526). The sources of obsidian in central Turkey start to be used for the production of long, pointed blades which become widespread in the Near East (Balkan-Atli et al. 1999: 142; Özdogan 1999c: 229). Judging from the Anatolian materials, there seems to be a chronological line from the Byblos points where blades from bipolar cores have been partially pressure-retouched around the tang and base – e.g. at Çayönü and Cafer Höyük (Cauvin et al. 1999; Özdogan 1999a: 47) – to the Amuq points, where pressure retouching covers a larger surface, but where the blades are worked unifacially, but not yet by bifacial modification (Özbaşaran 1999). The Anatolian material thus seems to correspond well to the sequence proposed for the Levant, where there is an increase in the use of surface pressure flaking and surface covering; bifacial flaking is an important element of the PPNC (Koslowski 1999: 100, 124, 131). A thorough investigation of Levantine Neolithic arrowheads (Gopher 1994) reveals the chronological sequence of point types, which are fashioned using different technologies. Between ca. 9000 and 8000 B.C. the earlier el-Khiam and Helwan points disappear and are replaced by Byblos points, and in the Southern Levant, Jericho points. These are tanged points made on straight blades, on which the tang is fashioned using pressure flaking. Partial pressure flaking occurs at the body and the point as well, but the degree of retouch is sparse (Gopher 1994: 36). From ca. 8000 B.C., these types are complemented by the leaf-shaped or oval Amuq point, which is also made on a blade but lacks the distinct tang of the preceding types. The Amuq point is frequently pressure-flaked over large portions of the body, both dorsally and ventrally (Gopher 1994: 39). The Byblos and Amuq points are then produced simultaneously, with the Amuq point gradually becoming the more common variant. Regarding craft and idea, the Byblos-type points in particular are very similar to the abovementioned Baltic/Russian points since they are made from blades that have been partially pressure-retouched. The morphological difference between the points is mainly due to the differences in the blade technology behind them (bipolar versus unipolar cores, the removal of flakes by percussion flaking versus pressure flaking), as well as the placement of the retouches.

In Western Europe, surface pressure flaking is found on microliths from around 7500 B.C. in the Rhine-Meuse-Schilde Mesolithic of Belgium, France and the Netherlands. These have no obvious precursors and seem to be a geographically and chronologically isolated phenomenon (Heinen 2006; Otte and Noiret 2006).

10.4.1.3 ca. 6000–1000 B.C.

In Anatolia and the Levant, large Amuq-type points seem to be produced, until around 5500 B.C., well into the Pottery Neolithic (PN) (Gopher 1994; Gopher and Gophna 1993; Prausnitz 1970). By this time, small points (<4 cm), which have appeared already between ca. 6800 and 6600 B.C., become dominant. It could be a conceptual division in the flint craft, where the large blades formerly used as blanks for projectile points are henceforth used mainly within agriculture as tools for harvesting (Rosen 1997) or threshing (Anderson et al. 2004; Knutsson 2007), whereas small blades and flakes are used to produce arrowheads (Rosen 1997: 39; Copeland 1996: 332: 2, 337: 14–15). The most common shapes are the tanged Haparsa point, the oval, shouldered Nizzanim point, and the almond-shaped Herzeliya point (Bar-Yosef 1981: 560; Gopher 1994). Morphologically, they are similar to the large blade points but are made using careful bifacial pressure flaking and exhibit superb craftsmanship (Gopher 1994: 41). In the Levant, the production of these points ceases during the transition to Late Ceramic Neolithic Age, around 5000 B.C., and transverse arrowheads come to dominate subsequent assemblages (Gopher 1994) The production of bifacials survives only in the desert regions of Negev, Sinai and Southern Jordan (Rosen 1997: 43). At certain sites, such as Tel Eli in North-Eastern Israel, there is evidence of Pottery Neolithic leaf-shaped and rhomboid points living on well into the Chalcolithic (Prausnitz 1970). Prausnitz mentions that similar points are present at an earlier stage in Anatolia and that the same kinds of points reach Egypt during the Badarian stage, around 4500–4000 B.C. (Prausnitz 1970: 119). During the Late Ceramic phase, however, few arrowheads are found in the Anatolian area, from the Balkan Peninsula to the South-East.

The bifacial arrowhead tradition does not form part of the Neolithic package which moves up through the Balkan Peninsula – the Fikirtepe culture beginning around 6000 B.C. (Özdogan 1999b: 212) – nor does it occur within the Linear Pottery tradition (Gronenborn 1999: 169; Ošibkina 1996: 27). The Early Neolithic stone industry does not seem to be based on pressure retouching in Greece either (Perles 2001; Wijnen, unpublished), which is where the first European agricultural societies appear (Runnels 2004). Only with the Middle Neolithic (5900/5800–5400 B.C.) does retouching become more visible and assemblages start containing transverse arrowheads with invasive retouch (Demoule and Perlès 1993: 382). During the Late and Final Neolithic, symmetric, piercing bifacial arrowheads appear. With the Final Neolithic/Chalcolithic (ca. 4600–3300 B.C.), these are still found all over Greece, but in small numbers only, and it is suggested that they may be imports from the contemporary Gumelnitsa and Salcuta cultures in the Balkans (Demoule and Perlès 1993: 402).

According to the latest findings, agriculture is introduced in Egypt, in Fayium south of Cairo, between 6000 and 5000 B.C., possibly somewhat earlier, although not before 7000 B.C. (Wetterström 1996: 201; Hassan 2002: 63). Bifacial, pressure-flaked arrowheads, which are not known in the area before this period, are part of this agricultural package. There are factors indicating that the impulses for this first phase came from the Levant. In the E75-8 site in Napta Playa, sheep/goats that have not existed naturally in the area but have their origin in the Levant, have been dated to about 5500 B.C. (Smith 1989: 74). It has also been pointed out that the presence in Egypt of domesticated animals and plants and other ideas which have their origin in South-West Asia indicates the existence of contacts between these areas during this period (Trigger 1983; Smith 1989). These ideas include, for instance, sharpened stone axes and various kinds of bifacial arrowheads (Hassan 1988).

The conclusion of this line of reasoning is that agriculture is introduced in Egypt relatively late if one takes into consideration the early dates that exist of agriculture in the Levant and Turkey. Instead, the Neolithisation takes place almost simultaneously in many parts of the Mediterranean region (Wenke and Casini 1989: 141). However, surface pressure flaking does not reach the earliest Neolithic cultures immediately east of Mesopotamia; neither the Siyalk culture in present-day Iran nor the Jeitun culture in present-day Turkmenistan are associated with projectile points of any kind. Here, hunting is done using a bludgeon or a sling (Mellaart 1975: 187, 212).

In the western parts of North Africa, the introduction of bifacial thinning using pressure flaking is associated with the introduction of Neolithic, which is traditionally set at about 5000 B.C. (N. Rahmani, 2006, personal communication; Clarke 1970). According to Clarke (1970: 200), bifacial arrowheads are associated with the earliest Neolithic phase in the north-west of North Africa (*Neolithic of Capsian Tradition*). Therefore, it seems as if bifacial pressure flaking follows farming west during the first wave of distribution south of the Mediterranean. However, this is not true for the earliest spread of agriculture west along the north shores of the Mediterranean, which takes place earlier and can be placed in connection with the Cardial Ware tradition. In Greece, the Balkan Peninsula, Italy, France and the Iberian Peninsula, bifacial thinning using pressure flaking is not included in the introduction of agriculture. The production of bifacial arrowheads using pressure flaking does not emerge on the Iberian Peninsula until the middle fourth century B.C. (J. Zilhao, 2006, personal communication). A similar lack of bifacial technology applies for the spread of the agricultural package north along the dell of the Danube in connection with the Early Linear Pottery culture. Prior to 3500 B.C., bifacial thinning using pressure flaking is associated with agriculture only in the Levant, Anatolia, and Northern Africa. There are no indications of bifacial thinning using pressure flaking in the northern and central parts of Europe which are affected by the Linear Pottery complex. During the Early Neolithic phase, the stone craft in these regions is characterized by a microlith-based technique which relates to neighbouring Mesolithic traditions where transverse arrowheads, for instance, can be connected to the use of a bow.

In Central Europe, surface pressure flaking appears as early as 4000 B.C. Since it seems as if the surface pressure technique is reintroduced to Europe from the Levant and Anatolia along two different routes (Fig. 10.5), we must expect a fair amount of blending between different traditions. In Northern Europe, pressure flaking is introduced from two different areas which in turn share a common area of origin: on the one hand, from the East around 3000 B.C. in connection with an early expansion of the Corded Ware culture and, on the other, in connection with the expansion of the Bell-Beaker culture from the Iberian Peninsula which begins

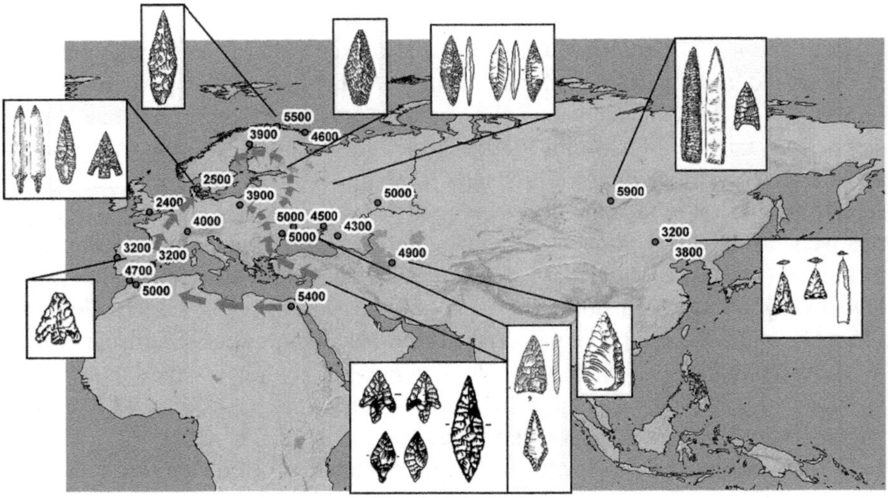

Fig. 10.5 Late Holocene surface pressure flaking in the Old World

around 3200 B.C. In the Netherlands, the Schipluiden site, which is situated by the coastline and has been dated to ca. 3500 B.C., 88 triangular, bifacial, flint arrowheads and large amounts of waste by-products have been gathered (Van Gijn et al. 2006: 142). Use-wear analyses of 41 points from the site show that 17 points have impact damage and a linear polish which is often associated with archery (Van Gijn et al. 2006: 158).

It is important to point out that the social context of the pressure-flaked, bifacially thinned arrowheads has changed in this later phase. If the bow was associated with hunting during the second phase, it is rather associated with individual graves starting with the appearance of the influences at the north of the coast of the Black Sea around 5000 B.C. and from there farther west into Central Europe. This could indicate that the bow should be regarded as a weapon and part of a warrior's equipment during this period, rather than anything else. Ethnographically, lithic projectile points have been used in connection with either warfare or large game hunting, while the hunting of small game is carried out using organic points (Ellis 1997).

The various Neolithic cultures of the former Soviet Union display ample evidence of the use of surface pressure flaking. From the Baltic states in the West to the Primorye in the East, surface pressure flaking is used in different ways: to shape projectile points, knives, microlithic insets and zoomorphic/anthropomorphic figurines. Several of these industries are based on blades, such as the cultures of the Upper Volga (Ošibkina 1996: 166), of the Volga-Kama river basins (Ošibkina 1996: 243) or the Novopetrovsk culture in the Priamur (Ošibkina 1996: 318), while flakes seem to constitute the primary blanks in other industries, such as is the case at the Starodubskoe II site on the Sakhalin Island (Ošibkina 1996: 328). The Baltic Narva culture is characterized as a "poor" lithic industry, and the blanks employed seem to have been flakes, modified by edging into small projectile points (Ošibkina 1996: 136).

Bifacial thinning using pressure flaking appears to reach Southern Russia and Ukraine during the Early Chalcolithic period. Triangular, bifacially retouched points constitute an important element within the Tripolye culture, which comprises present-day Romania (where the tradition is called Cucuteni), Moldavia and Western Ukraine (Klochko 2001: 21). The earliest phase, Tripolye A, is dated to the period between 5700 and 4200 B.C., based on 29 radiocarbon dates (Tjernych and Orlovskaja 2004). The development of Cucuteni-Tripolye is followed by the formation of similar groups farther East.

Between the rivers Dnieper and Don, pointed-bottom pottery and stone artefacts are found, belonging to the so called Skelya culture which has been dated to around 4550–4100 B.C. Blades and triangular, surface-retouched arrow- and spearheads can be noted among the stone artefacts (Rassamakin 1999: 76). Judging from the correlation between height and width, these are not blade arrowheads but are more likely to have been made from flakes.

Similar flake arrowheads with a triangular shape occur within the Khvalynsk culture further East, around the river Volga. The Khvalynsk culture seems to be influenced by Skelya and has been dated to ca. 5000–4500 B.C. (Rassamakin 1999: 61, 107, 111). A similar material culture is found east of the Black Sea as well, in the North Caucasian Zakubanskaya culture in which the same kind of triangular, bifacial points are found (Rassamakin 1999: 110).

In the middle of the Chalcolithic period (ca. 3800–3400 B.C.), the projectile points change their appearance somewhat. Several kinds of points are found in the Konstantinovka culture by the river Don, both in settlement contexts and as grave offerings. On the one hand, there are large, bifacial leaf-shaped points and, on the other, a kind of point with a slanting tang (Rassamakin 1999: 120).

Further East, in present-day Kazakhstan, the local Mesolithic tradition, influenced by the regions stretching from the Caspian Sea to the Aral Sea in the South-West, evolves into the first Neolithic culture of the region, the Atbasar culture (Kislenko and Tatarintseva 1999: 187). In connection with this, a production of triangular, bifacially thinned projectile points begins in the 5th millennium B.C. The dates correspond well with the Khvalynsk/Skelya cultures in the West. During the course of the 4th millennium, both the leaf- shaped points and those with a slanted tang appear in the area. Pressure-flaked bifacials constitute an important element within the subsequent Botai culture (Kislenko and Tatarintseva 1999: 203).

In the Chinese Hongshan culture (4000–3000 B.C.), there are pressure-retouched points, made from flakes, which are contemporary with long prismatic blades that are used in the Fuhe culture in Northern China. These have been partially pressure-retouched into points. Grinding also occurs (Da-shun 1995). There seems to be a gap in the use of pressure flaking between the Upper Paleolithic Era (Gao and Norton 2002), and this production of blade points, even though bifacially flaked projectile points, are found in Northern China during the Mesolithic phase (Chi 1999).

In Finland, bifacial points made from quartz or flint appear in connection with the transition to the typical Comb Ceramic period around 3900 B.C. However, their production seems to cease at the transition to the Late Comb Ceramic period 500 years later. During the Early Metal Age, the production of bifacials reemerges in a somewhat different shape, using flint, quartz and quartzite (Manninen et al. 2003). Simultaneously, around 4150 B.C. (Kriiska 2001), the same kinds of rhombic and almond-shaped points occur in Estonia as well as part of the typical Comb Ceramic period (Jaanits et al. 1982: 71). Comb Ceramic points made from Russian flint have been found also in Northern Sweden (Halén 1994). However, there are considerably earlier dates of pressure-retouched points on the North Calotte; on the Kola Peninsula, there are dates as far back as to 4600 B.C. (Gurina 1997). Even earlier dates have been obtained from the Early Northern Comb Ceramic culture at the Varangerfjord containing elements of bifacials made by pressure flaking which have recently been presented by Skandfer (2005). Here, the craft seems to exist as early as 5500 B.C. However, it should be pointed out that these points have been dated only according to the shoreline.

In Southern Scandinavia, pressure flaking is a part of the technological recipe within the younger Pitted Ware culture of Denmark, Western Sweden and Southern Norway (Vang Petersen 1999: 79), where projectiles are fashioned from blades employing surface pressure flaking. True, bifacially thinned points made with pressure flaking occur in Southern Scandinavia around 2500 B.C. It is a matter of at least two different kinds: an Eastern one which is related to the Corded Ware complex of Northern Europe and a Western one which has its origin in the Bell-Beaker culture of Western Europe. In Scania and the south-east of Denmark, a number of lancet-shaped Corded Ware points have been found which are common south of the Baltic Sea, in Mecklenburg and in Central Germany (Vang Petersen 1999: 92; Larsson 1999). In Western Scandinavia, bifacial points occur in graves around 2350 B.C., along with bell-beakers and slate wrist guards (Sarauw 2006). Bell-Beaker points are relatively common in Western Norway as well (Östmo 2006), but only a handful of these points have been found in Sweden, especially in the western parts of the country. The points which can be connected to the Bell-Beaker complex are mainly of two kinds: one kind, which is common in the archers graves on Jutland (Sarauw 2006), consists of triangular, bifacially thinned points with a deep indentation in the base (Vang Petersen 1999: 92), and the other, which in Scandinavia has been gathered primarily as stray finds, is made up of triangular, bifacially thinned points with a small tang. The latter type originated in Early Neolithic North Africa/Iberian Peninsula and is known in the Bell-Beaker contexts of Western Europe. After this, bifacial points made from Southern Scandinavian flint using pressure flaking are common in the south and central parts of Scandinavia, until the flint points are driven out of competition by metal points at the end of the Bronze Age. Northern Mälardalen constitutes the northern border of this tradition.

10.5 Summary

We realize that our attempt at mapping the chronological and spatial distribution of surface pressure flaking on this geographical and chronological scale is problematic, and the sweeping attitude towards the knowledge amassed by decades of thorough research has been a painful experience. However, we believe that this kind of approach has the potential to reveal information that contextual studies would not. Flawed as our study might be, we think that there are several interesting tendencies that merit further attention.

It is obvious that the use of surface pressure flaking becomes increasingly widespread with time. From the isolated Late Pleistocene origins through the Early Holocene to the Late Holocene, the technique seems to be applied in an increasing number of contexts, covering a larger geographical area. During the early part of the Holocene, the technique is to be found over large parts of Northern Eurasia, as well as in South-West Asia. In Western Europe, the distribution of surface pressure flaking is more patchy and isolated. The later part of the Holocene is a period during which surface pressure flaking is incorporated on a large scale into the technologies of Northern Africa, Western Europe and Scandinavia. In Eastern Europe, South-West and Northern Asia surface pressure flaking continues to be used even though there is evidence of a decline in the technology. To what extent this pattern is a result of analogous (local innovation) or homologous development (cultural transmission) is a matter of debate. However, the observed pattern certainly indicates that the technique must have had selective value.

Within several of the industries incorporating surface pressure flaking, there are similarities in the evolutionary sequence leading from unifacial, partial retouch to bifacial, surface-covering pressure flaking. This pattern has been described regarding the Solutrean, the BAI of South-West Asia, as well as in the Scandinavian Pitted Ware to Late Neolithic technologies. A similar pattern might apply to the Far Eastern technologies as well. Why there would be a selection towards more surface-covering, parallel pressure flaking is an intriguing question. This behaviour has no obvious functional advantage, in the sense that it would increase the efficiency of the projectile points. Since we also know that lithic projectile points are prone to break on impact (Ellis 1997), the functional aspects of surface retouch covering the tools seem all the more to be a costly behaviour. Attributing this simply to cultural preference also intuitively seems to be an unsatisfactory explanation. It is also notable that in several areas, a perceived change towards smaller projectile points often based on flakes instead of blades seems to take place.

Our survey of the use of surface pressure flaking has certainly shown that there are instances of independent innovation of the technique. There is no proximity in time and space between the Solutrean and the Dyuktai, and the Rhine-Meuse-Schilde Mesolithic seems to be an isolated occurrence of surface pressure flaking in the West European Mesolithic. Surely, however, the adoption of surface pressure flaking on a global scale is in no easily detectable way associated with mobility, arctic conditions, or scarcity of raw material, all of which would be situations under which an economic attitude towards tool production could be predicted. We believe that the impact of cultural transmission on the archaeological pattern has been of great importance.

As an example, it is unclear in what way the post-Swiderian and PPNB traditions are related. It has been claimed that the production of pressure-flaked blades constitutes a "natural stage of development" in the flint craft, and thereby the Kunda/

Butovo/Kama industries would have evolved from the preceding local Swiderian technologies (Zaliznyak 1999). The fact that chronologically contemporary and similar (albeit not identical) technical changes can be observed in the budding Neolithic society in the Middle East makes it unlikely that the two traditions would not be related in our view.

Judging from the data gathered, it seems as though the change which occurs in the Middle East during the Ceramic Neolithic period – when the production of macro blades is reserved for agriculture, whereas the points become smaller in size and are often made from flakes – has consequences in many parts of Eurasia and Northern Africa as well. Bifacial thinning using pressure flaking is part of the Neolithic package which reached Egypt around 5500 B.C. and which quickly covers the entire Maghreb region all the way to Spain. The Neolithisation of the Black Sea region, which occurs somewhat later (ca. 5000 B.C.), is also characterized by the existence of pressure-retouched points, not least in grave contexts. The dates collected indicate that there is certain slowness in the adoption of the technique in Central and Northern Europe, with relatively late dates. One obvious exception from the pattern are the early dates on the North Calotte, where the use of bifacial pressure flaking begins almost simultaneously with the change described in the Middle East.

According to us, there is cause to argue that even though surface pressure flaking is invented locally at certain points in time and space, cultural transmission is a major factor explaining the geographical distribution of the technique. Therefore, in this widely spread technology, which occurs in many different climatic zones and social systems, there is great potential to examine how cultural transmission takes place and how the evolutionary mechanisms accounted for in the theoretical chapter (random variation, cultural drift, guided variation and biased transmission) have influenced this process. From the material presented in this chapter, one interesting aspect of transmission to study is how it has occurred in areas populated by groups that can be presumed to have been characterized by high mobility versus areas where mobility has been low. In a mobile society, the individuals with whom one has close social relations are scattered over vast areas, which means that the rate of diffusion can be expected to be quicker in societies with high mobility than in societies with low mobility. We can also see that the Early Holocene dates related to surface pressure flaking appears in a triangular geographic area with sides stretching from 2,500 to 3,000 km, where the dates of the nodes are completely concordant (the Levant, Finland/Baltic, the Urals). From an archaeological perspective, the transmission has been extremely rapid. The transmission of pressure flaking during later Holocene exhibits certain inertia when it comes to Central and Northern Europe, where there appears to be a resistance to the adoption of pressure-retouched projectile points. This could be a consequence of the farming cultures established in the region in which the technology becomes ideologically (categorically) charged. During the same period, we can see how bifacial thinning through pressure flaking emerges on the North Calotte almost simultaneously as the changes in the Near East occur. If these processes are related, which we think they are, it means that ideas have travelled 4,000 km within a timeframe that cannot be measured archaeologically.

Acknowledgements We would like to thank the organizers of the session in Lisbon, Pierre M. Desrosiers and Noura Rahmani, who invited us to publish this chapter, the Lejre Research Centre in Denmark for their generous contribution to the experiments that form the basis of our definition of pressure flaking, as well as for a fantastic week in Lejre in the summer of 2005, and Mira Darmark for the English translation.

References

Anderson, P., A. Chabot, and A. van Gijn 2004 The functional riddle of 'glossy' Canaanean blades: the apogee of the Mesopotamian threshing sledge. *Journal of Meditaranean archaeology* 17(1): 87–132.

Apel, J., and K. Darmark 2007 Den flathuggna pilspetsens fylogeni. Mellansvenskt stenhantverk ur ett kulturhistoriskt perspektiv. I: Stenbäck, N. (ed.), Stenålder i Uppland. Uppdragsarkeologi och eftertanke. Arkeologi E4 Uppland – Studier volym 1. Uppsala.

Apel, J. 2001 *Daggers Knowledge & Power. The Social Aspects of Flint-Dagger Technology in Scandinavia 2350–1500 cal BC.* Uppsala University, Uppsala. 2008 Knowledge, Know-how and Raw Material – The Production of Late Neolithic Flint Daggers in Scandinavia. *Journal of Archaeological Method and Theory* 15(1): 91–111.

Apel, J. 2008 Knowledge, Know-how and Raw Material – The Production of Late Neolithic Flint Daggers in Scandinavia. *Journal of Archaeological Method and Theory* 15(1): 91–111.

Apel, J., K. Darmark, and L. Sundström 2005 Hantverkstraditioner som möts. Uppland 2005. 60–68.

Aubry, T., B. Walter, M. Almedia, and M.J. Neves 2003 Solutrean Laurel Leaves Production and Raw Material Procurement during L.G.M. in Southern Europe: Two examples from Central France and Portugal. In *Multiple Approaches to Bifaces Variability*, edited by M. Soressi and H.L. Dibble, pp. 165–182, University of Pennsylvania, Museum of Archeology and Anthropology, Pennsylvania.

Aurenche, O., P. Galet, E. Régagnon-Caroline, and J. Évin 2001 Proto-Neolithic and Neolithic cultures in the Middle East-the birth of agriculture, livestock raising, and ceramics: a calibrated 14 C chronology 12500–5500 cal BC. Radiocarbon 43(3): 1191–1202.

Balkan-Atli, N., D. Binder, M.C. Cauvin, E. Biçakçi, G. der Aprahamian, and C. Kuzucuoglu 1999 Obsidian: Sources, Workshops and Trade in Central Anatolia. In M. Özdogan & N. Basgelen (red.) *Neolithic in Turkey. The Cradle of Civilization*, edited by M. Özdogan and N. Basgelen, pp. 133–147. Ancient Anatolian Civilisations Series 3, New Discoveries, Istanbul.

Bar-Yosef, O. 1981 The "Pre-Pottery Neolithic" period in the Southern Levant. In *Préhistoire du Levant. Chronologie et organisation de l'espace depuis les origins jusqu'au Vie millénaire,* edited by J. Cauvin and P. Sanlaville, pp. 551–571. Colloques Internationaux du Centre National de la Recherche Scientifique No 598. Paris.

Bayle, G. 2000 Etude archéozoologique des niveaux solutréens de l'Abri Fritsch. La part de l'homme dans les accumulations de restes de grands mammifères. Mémoire de D.E.A. de l'Université de Paris I, Panthéon- Sorbonne, UFR 03 Préhistoire – Anthropologie – Ethnologie. Paris.

Bordes, F. 1968 *The Old Stone Age*. McGraw-Hill, London.

Callahan, E. 1996 *The Basics of Biface Knapping in the Eastern Fluted Point Tradition: A Manual for Flintknappers and Lithic Analysts.* 3rd ed. Eastern States Archaeological Federation, Lynchburg.

Cattelain, P. 1989 Un crochet de propulseur solutréen de la Grotte de Combe Sauniére. *Bulletin de la Société de la Préhistoire Française* 86: 213–216.

Cauvin, J., O. Aurenche, M.C. Cauvin, and N. Balkan-Atlı 1999 The Pre-Pottery Site of Cafer Höyük. In *Neolithic in Turkey. The Cradle of Civilization. New Discoveries*, edited by M. Özdogan and N. Basgelen, pp. 87–105. Ancient Anatolian Civilisations Series 3, Istanbul.

Chabas, F. 1874 Les silex de Volgu. *Rapport a la societe d'Histoire et d'Archaeologie de Charlon-sur-Saône*. Charlon-sur-Saône.

Chi, Z. 1999 The Mesolithic and Neolithic in China. In *Documenta Praehistorica* XXVI, edited by M. Budja, pp. 1–15. 6th Neolithic Studies.

Clarke, J. D. 1970 *The Prehistory of Africa*. Thames & Hudson, London.

Copeland, L. 1996 The Flint and Obsidian Industries. In *Tell Sabi Abyad. The Late Neolithic Settlement. Report on the Excavations of the University of Amsterdam (1988) and the National Museum of Antiquities Leiden (1991–1993) in Syria*, edited by P.M.M.G. Akkermans, pp. 285–339. l'Institut historique-archeologique neerlandais de Stamboul, Leiden.

Cotterell, B., and J. Kamminga 1987 The Formation of Flakes. *American Antiquity* 52(4): 675–708.

Darmark, K., and J. Apel 2008 The Dogma of Immaculate Perception. An experimental study of bifacial arrowheads and a contribution to the discussion on the relationship between personal experience and formalised analysis in experimental archaeology. In *Technology in Archaeology, Proceedings of the SILA workshop: The Study of Technology as a method for gaining insight into social and cultural aspects of Prehistory,* edited by M. Sørensen and P. Desrosiers. PNM, Publications from the National Museum Studies in Archaeology & History Vol. 14. Copenhagen.

Da-shun, G. 1995 Hongshan and related cultures. In *The Archaeology of Northeast China*, edited by S. Milledge Nelson, pp. 21–65. Routledge, London.

Demoule, J-P., and C. Perlès 1993 The Greek Neolithic: A New Review. *Journal of World Prehistory* 7(4): 355–416.

Desrosiers, P., and M. Sørensen 2008 Introduction. In *Technology in Archaeology,* edited by M. Sørensen and P. Desrosiers. Proceedings of the SILA workshop: The Study of Technology as a method for gaining insight into social and cultural aspects of Prehistory. PNM, Publications from the National Museum Studies in Archaeology & History Vol. 14. Copenhagen.

Dunnell, R.C. 1978 Style and Function. A Fundamental Dichotomy. *American Antiquity* 43: 192–202.

Edgren, T., And L. Törnblom 1993 Finlands Historia 1. Ekenäs.

Ellis, C.J. 1997 Factors Influencing the Use of Stone Projectile Tips: An Ethnographic Perspective. In *Projectile Technology*, edited by H. Knecht, pp. 37–79. Plenum Press. New York.

Elston, R.G., and P.J. Brantingham 2002 Microlithic Technology in Northern Asia: A Risk-minimizing Strategy of the Late Paleolithic and Early Holocene. *Archeological Papers of the American Anthropological Association* 12(1): 103–116.

Eriksen, B.V. 2000 Flintstudier. En håndbog i systematiske analyser af flintinventarier. Aarhus universitetsforlag. Aarhus.

Gao, X., and C.J. Norton 2002 A critique of the Chinese "Middle Palaeolithic". *Antiquity* 76: 397–412.

Geneste, J.-M., and H. Plisson 1990 Technologie fonctionnelle des points à cran Solutréennes: L'apport des nouvelles données de la Grotte de La Combe Sauniére. In *Feuilles de Pierre*, edited by J. Kozlowski, pp. 293–320, ERAUL 42. Liege.

Gopher, A. 1994 *Arrowheads of the Neolithic Levant*. Eisenbrauns, Winona Lake.

Gopher, A., and R. Gophna 1993 Cultures of the Eighth and Seventh Millennia BP in the Southern Levant: A Review for the 1990s. *Journal of World Prehistory* 7(3): 297–353.

Grammer, K., B. Fink, A.P. Møller, and R. Thornhill 2003 Darwinian aesthetics: sexual selection and the biology of beauty. *Biol* 78: 385–407.

Gronenborn, D. 1999 A Variation on a Basic Theme: The Transition to Farming in Southern Central Europe. *Journal of World Prehistory* 13(2): 123–210.

Gurina, N.N. 1997 Istorija kul'tury drevnego naselenija Kol'skogo poluostrova. Sankt Petersburg. (Гурина, Н.Н. 1997. История культуры древнего населения Кольского полуострова. Санкт Петербург.).

Halén, O. 1994 Sed*entariness during the Stone Age of Northern Sweden in the light of the Alträsket site 5000 BC and the Comb ware site Lillberget, c. 3900 BC*. Acta Archaeologica Lundensia Series 4, No. 20. Almqvist & Wiksell International, Stockholm.

Hassan, F.A. 1988 The Predynastic of Egypt. *Journal of World Prehistory* 2(2): 135–178.

Hassan, F.A. 2002 Holocene Environmental Change and the Transition to Agriculture in Southwest Asia and Northeast Africa. In *The Origins of Pottery and Agriculture*, edited by Y. Yasuda, pp 55–68, Lustre Press, New Dehli.

Heinen, M. 2006 The Rhine-Meuse-Schilde Culture in Western Europe. Distribution, Chronology and Development. In: *After the Ice Age. Settlements, subsitence and social development in the Mesolithic of Central Europe. Proceedings of the International Conference 9th to 12th of September 2003 Rottenburg/Neckar, Baden-Württemberg, Germany*, edited by Claus-Joachim Kind, pp. 75–87, Stuttgart.

Högberg, A. 2006 Continuity of Place: Actions and Narratives. In *Skilled Production and Social Reproduction*, edited by J. Apel and K. Knutsson, pp. 187–206, SAU Stone Studies 2, Uppsala.

Hutchings, W.K 1999 Quantification of Fracture Propagation Velocity Employing a Sample of Clovis Channel Flakes. *Journal of Archaeological Science* 26: 1437–1447.

Iriki, A., and O. Sakura 2008 The neuroscience of primate intellectual evolution: natural selection and passive and intentional niche construction. *Philosophical Transactions of the Royal Society B Biological Sciences* 363: 2229–2241

Jaanits, L., S. Laul, V. Lõugas, and E. Tõnisson 1982 *Eesti Esiajalugu*. Tallinn.

Kankaanpää, J., and T. Rankama 2005 Early Mesolithic Pioneers in Northern Finnish Lapland. In *Pioneer Settlements and Colonizations Process in the Barents Region*, edited by H. Knutsson, pp. 109–161. Papers on Hunter-gatherer Archaeology 1, Vuollerim.

Kislenko, A., and N. Tatarintseva 1999 The Eastern Ural Steppe at the End of the Stone Age. In *Late Prehistoric Exploitation of the Eurasian Steppe*, edited by M. Levine, Y. Rassamakin, A. Kislenko and N. Tatarintseva, pp. 183–216. McDonald Institute for Archaeological Research, Cambridge.

Klochko, V. 2001 *Weaponry of societies of the Northern Pontic Culture Circle: 5000–700 BC*. Poznan.

Knutsson, H. 2007 Spån och tidigt jordbruk i Mellansverige. In *Stenåldern i Uppland. Uppdragsarkeologi och eftertanke*, edited by N. Stenbäck, pp. 319–369 Uppsala.

Knutsson, K. 2006 A Geneaology of Reflexivity: The skilled Lithic Craftman as "Scientist". In *Skilled Production and Social reproduction*, edited by J. Apel and K. Knutsson, pp. 11–24, SAU Stone Studies 2. Uppsala.

Kol'tsov, L.V. 1989 (ed.). Arkheologiya SSSR. Mezolit SSSR. Moscow. (Кольцов, Л.В. 1989 (ред.). *Археология СССР. Мезолит СССР*. Москва.)

Koslowski, S.K. 1999 *The Eastern Wing of The Fertile Crescent. Late prehistory of Greater Mesopotamian lithic industries*. BAR International Series 760, Archaeopress, Oxford.

Kriiska, A. 2001 *Stone Age Settlement and Economic Processes in the Estonian Coastal Area and Islands*. University of Helsinki, Institute for Cultural Research, Archaeology, Helsinki.

Larsson, L. 1999 Fyrverkeri i eld och flinta. Ett massbrännoffer från stridsyxetid. Populär arkeologi 1999/1.

Lordkipanidse, O. 1991 *Archäologie in Georgien: von der Altsteinzeit zum Mittelalter*. Wiley VCH, Heidelberg.

Manninen, M., M. Tallavaara, and E. Hertell 2003 Subneolithic Bifaces and Flint Assemblages in Finland-Outlining the History of Research and Future Questions. In: *Uniting Sea. Stone Age Societies in the Baltic Sea Region. Proceedings from the First Uniting Sea Workshop at Uppsala University January 26–27, 2002*, edited by C. Samuelsson & N. Ytterberg, Uppsala University, Uppsala.

Mellaart, J. 1975 *The Neolithic of the Near East*. Scribner, London.

Nagai, K. 2007 Flake scar patterns of Japanese tanged points: toward an understanding of technological variability during the Incipient Jomon. *Anthropological Science* 115: 223–226.

Nunn, G. 2006 Using the Jutland Type IC Neolithic Danish Dagger as a model to replicate parallel, edge-to-edge pressure flaking. In *Skilled Production and Social Reproduction*, edited by J. Apel and K. Knutsson, pp. 81–113. SAU Stone Studies 2, Uppsala.

Ošibkina, S.V. 1996 Arkheologiya. Neolit Severnoy Evrazii. Moscow. (Ошибкина, С.В. 1996. Археология. Неолит Северной Евразии. Москва.).

Östmo, E. 2006 Oldsaksamlingens arsbok.

Ostrauskas, T. 2000 Mesolithic Kunda Culture. A Glimpse from Lithuania. In *De Temporibus Antiquissimis Ad Honorem Lembit Jaanits*, edited by V. Lang and A. Kriiska, pp. 167–181. Ajaloo Instituut, Tallinn.

Otte, M., and P. Noiret 2006 The Mesolithic of the Belgian Ardennes. In: *After the Ice Age. Settlements, subsistence and social development in the Mesolithic of Central Europe. Proceedings of the International Conference 9th to 12th of September 2003 Rottenburg/Neckar, Baden-Württemberg, Germany*, edited by Claus-Joachim Kind, pp. 95–101, Stuttgart.

Özbasaran, M. 1999 Musular: A general assessment on a new Neolithic site in Central Anatolia. In *Neolithic in Turkey. The Cradle of Civilization. New Discoveries*, edited by M. Özdogan and N. Basgelen, pp. 157–163. Ancient Anatolian Civilisation Series 3. Arkeoloji ve Sanat Yayinlari, Istanbul.

Özdogan, A. 1999a Çayönü. In *Neolithic in Turkey. The Cradle of Civilization. New Discoveries*, edited by M. Özdogan and N. Basgelen, pp. 35–65. Ancient Anatolian Civilisations Series 3. Arkeoloji ve Sanat Yayinlari, Istanbul.

Özdogan, M. 1999b Northwestern Turkey: Neolithic Cultures in Between the Balkans and Anatolia. In *Neolithic in Turkey. The Cradle of Civilization. New Discoveries,* edited by M. Özdogan and N. Basgelen, pp. 203–225. Ancient Anatolian Civilisations Series 3. Arkeoloji ve Sanat Yayinlari, Istanbul.

Özdogan, M. 1999c Concluding Remarks. In *Neolithic in Turkey. The Cradle of Civilization. New Discoveries*, edited by M. Özdogan and N. Basgelen, pp. 225–246. Ancient Anatolian Civilisations Series 3. Arkeoloji ve Sanat Yayinlari, Istanbul.

Pelegrin, J. 1995 Technologie Lithique le Chattelperronien de Roc-De-Combe (Lot) et de La Côte (Dordogne). Cahiers du Quarternaire 20, CNRS, Paris.

Pelegrin, J. 2006 Long blade technology in the Old World: an experimental approach and some archaeological results. In *Skilled Production and Social Reproduction. Aspects on Traditional Stone-Tool Technologies*, edited by J. Apel and K. Knutsson, pp. 37–69. SAU Stone Studies 2. Societas Archaeologica Upsaliensis, Uppsala.

Perles, C. 2001 *The Early Neolithic in Greece. The first farming communities in Europe.* Cambridge University Press, Cambridge.

Prausnitz, M.W. 1970 *From Hunter to Farmer and Trader. Studies in the lithic industries of Israel and adjacent countries (from the Mesolithic to the Chalcolithic age).* R. Habelt, Jerusalem.

Rabinovitch, A., V. Frid, and D. Bahat 2006 Wallner lines revisited. *Journal of Applied Physics* 99: 076102 – 076102-3.

Rassamakin, Y. 1999 The Eneolithic of the Black Sea Steppe: Dynamics of Cultural and Economic Development 4500–2300 BC. In *Late Prehistoric Exploitation of the Eurasian Steppe*, edited by M. Levine, Y. Rassamakin, A. Kislenko and N. Tatarintseva, pp. 59–183. McDonald Institute for Archaeological Research, Cambridge.

Rosen, S.A. 1997 *Lithics After the Stone Age. A Handbook of Stone Tools from the Levant.* Altamira Press, Walnut Creek/London/New Delhi.

Runnels, C. 2004 First Farmers of Europe. In *Ancient Europe 8000* B.C. *–A.D. 1000. Encyclopedia of the Barbarian World. Volume I. The Mesolithic to Copper Age (c. 8000–2000* B.C.*)*, edited by P. Bogucki and P.J. Cabtree, pp. 218–226. Thompson/Gale, New York.

Sarauw, T. 2006 Male symbols or warrior identities? The 'archery burials' of the Danish Bell Beaker Culture. *Journal of Anthropological Archaeology* 26: 65–87.

Shennan, S. 2002 *Genes, Memes and Human History. Darwinian Archaeology and Cultural Evolution.* Thames & Hudson, London.

Sinclair, A. 1995 The technique as a symbol in Late Glacial Europe. *World Archaeology* 27(1): 50–62.

Skandfer, M. 2005 Early, Northern Comb Ware in Finnmark: The Concept of Säräisniemi 1 reconsidered. *Fennoscandia Archaeologica* XXII: 3–27.

Smith, A.B. 1989 The Near Eastern Connection. Early to mid-Holocene Relations Between North Africa and the Levant. In *Late Prehistory of the Nile Basin and the Sahara*, edited by L. Krzyzaniak and M. Kobusiewicz, pp. 69–77, Polish Academy of Science, Poznan.

Smith, P.E.L. 1966 *Le Solutrean en France*. Publication de l'Institut de Prehistoire de l'Universite de Bordeaux. Memoire No. 5. Bordeaux.

Sørensen, M. 2006 Teknologi og Tradition I Østarktis 2500 BC–1200 AD. En dynamisk teknologisk undersøgelse af litiske inventarer i de palæoeskimoiske traditioner Vol. 1, Tekst. Det Humanistiske Fakultet, Københavns Universitet. København.

Sulgostowska, Z. 1999 Final Palaeolithic Masovian Cycle and Mesolithic Kunda Culture Relations. In *Tanged Points Cultures in Europe*, edited by S.K. Koslowski, J. Gurba and L.L. Zaliznyak, pp. 85–92. Maria Curie-Sklodowska University Press, Lublin.

Takala, H. 2004 *The Ristola Site in Lahti and the Earliest Postglacial Settlement of South Finland*. Lahti City Museum, Jyväskylä.

Tehrani, J.J., and F. Riede In press. The role of pedagogy in the transmission and maintenance of cultural traditions. *World Archaeology*.

Tiffagom, M. 1998 Témoignages d'un traitement thermique des feuilles de laurier dans le Solutréen supérieur de la grotte de Parpalló (Gandia, Espagne). *Paléo* 10: 147–161.

Tiffagom, M. 2006 *De La Pierre à l'homme. Essai sur une Paléoanthropologie solutréenne*. Eraul 113, Liège.

Tjernych, E.N., and L.B. Orlovskaja 2004 Radiouglerodnaja chronologija Eneolititjeskich kul'tur jogo-vostotjnoj Evropy: Rezultaty i Problemy Issledovanij. Rossijskaja Archeologija 4, 2004. 24–37. (Чёрных, Е.Н. & Орловская, Л.Б. 2004. Радиоуглеродная хронология Енеолитических культур юго-восточной Европы: Резултаты и проблемы исследований. Российская Археология 4, 2004: 24–37).

Trigger, B.G. 1983 The Rise of Egyptian Civilization. In *Ancient Egypt: A Social History*, edited by B. Trigger, D. Kemp, O'Connor and A.B. Lloyd, pp. 1–70, Cambridge University Press, Cambridge.

Van Gijn, A., V. Van Betuw, A. Verbaas, and K. Wentik 2006 *Schipluiden. A Neolithic Settlement on the Dutch North Sea Coast c. 3500 cal BC. Analecta Praehistorica Leidensia 37/38*. Leiden University, Leiden.

Vang Petersen, P. 1999 *Flint fra Danmarks Oldtid*. Nationalmuseet, Copenhagen.

Volokitin, A.V. 2005 Some peculiarities of colonization of the European north-east in Mesolithic. In *Pioneer settlements and colonization process in the Barents region*, edited by H. Knutsson, pp. 11–18, Vuollerim.

Wenke, R.J., and M. Casini 1989 The Epipaleolithic-Neolithic transition in Egypt's Fayum depression. In *Late Prehistory of the Nile Basin and the Sahara*, edited by L. Krzyzaniak and M. Kobusiewicz, pp. 139–155. Poznán Archaeological Museum, Poznán.

West, F.H. (editor) 1996 *American Beginnings: The Prehistory and Palaeoecology of Beringia*. University of Chicago Press, Chicago.

Wetterström, W 1996 Foraging and farming in Egypt: the transition from huntering and gathering to horticulture in the Nile valley. In *The Archaeology of Africa. Food, metals and towns*, edited by T. Shaw, P. Sinclair, B. Andah and A. Okpoko, pp. 165–226, Routledge, London.

Wijnen, N. Unpublished The Early Neolithic I settlement at Sesklo. 061211. http://www.unige.ch/lettres/archeo/introduction_seminaire/neolithique/wijnen.html.

Zaliznyak, L.L. 1999 Tanged point cultures in the Western part of Eastern Europe. In *Tanged Points Cultures in Europe*, edited by S.K. Koslowski, J. Gurba and L.L. Zaliznyak, pp. 202–218. Maria Curie-Sklodowska University Press, Lublin.

Zhilin, M.G., and H. Matiskainen 2000 Deep in Russia, deep in the bog. Excavations at the Mesolithic sites Stanovoje 4 and Sakhtysh 14, Upper Volga region. In *Mesolithic on the Move. Papers presented at the Sixth International Conference on the Mesolithic in Europe, Stockholm 2000, edited by* L. Larsson, H. Kindgren, K. Knutsson, D. Loeffler and A. Åkerlund, pp. 694–702. Oxbow Books, Oxford.

Zhilin, M.G. 1999 New Mesolithic Peat Sites on the Upper Volga. In *Tanged Points Cultures in Europe*, edited by S.K. Koslowski, J. Gurba and L.L. Zaliznyak, pp. 295–310, Maria Curie-Sklodowska University Press, Lublin.

Zilhão, J. 1997 O Paleolítico superior da Estremadura portuguesa. Lisbonne: Colibri. Unpublished Ph.D. dissertation, Lisbon.

Chapter 11
Emergence and Development of the Pressure Microblade Production: A View from the Upper Paleolithic of Northern Japan

Jun Takakura

11.1 Introduction

As one of the most sophisticated innovations in prehistoric lithic technologies, the pressure technique has attracted considerable scholarly attention. Several experimental studies demonstrate that the pressure technique requires complicated knowledge concerning the repertoire of gestures as well as well-developed know-how which can only be acquired through repeated practice (e.g. Apel 2008; Pelegrin 2003, 2006). Studies of the pressure blade and microblade production open undeniably important insights into the dynamics of interaction between prehistoric society and technology by directing explicit attention towards the skill and craft learning underlying the technological practice (e.g. Migal 2006). Also, understanding the timing of its appearance has been of notable archaeological interest in evaluating the temporal changes to the chaîne opératoire of lithic assemblages and the socio-economic conditions in relation to the adoption of such technique (e.g. Rahmani 2004, 2007).

Some archaeologists have proposed that the pressure technique used for producing blades or microblades emerged in the Upper Paleolithic of Eurasia (Flenniken 1987), found in large areas of Northern Asia, including Northern Japan. Inizan et al. (1992) claimed that the pressure technique in these areas was linked to the production of microblades detached from wedge-shaped microblade cores, and its appearance probably occurred around 20,000 years B.P. Recent results of archaeological research in these areas certainly support their suggestions in terms of new AMS radiocarbon dates and the technological re-evaluation of the microblade reduction sequences. In this regard, understanding the microblade technology of Northern Asia is critical to research into the origin of pressure microblade production during the Late Pleistocene.

J. Takakura (✉)
Archaeological Research Center, Hokkaido University, Sapporo, Hokkaido, Japan
e-mail: jun-ta@let.hokudai.ac.jp

P.M. Desrosiers (ed.), *The Emergence of Pressure Blade Making: From Origin to Modern Experimentation*, DOI 10.1007/978-1-4614-2003-3_11,

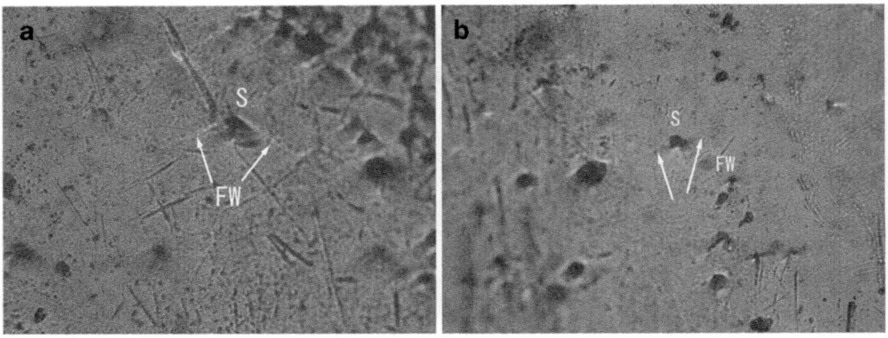

Fig. 11.1 Fracture wings observed on the archaeological samples from the Kamihoronai-moi site (field view width, 1 mm) (After Takakura 2008). (**a**) Observed on the ventral surface of spall detached from the platform of the Sakkotsu type microblade core. These indicate the employment of direct percussion using stone for platform formation. (**b**) Observed on the ventral surface of microblade detached from the Sakkotsu type microblade core. These indicate the employment of pressure for microblade knapping

Despite its importance to structuring research questions concerning the Upper Paleolithic in Northern Asia, there has so far been only a limited examination of the processes of widespread adoption and development of the pressure microblade production. Indeed, it is unclear what technological and socio-economic conditions of the Upper Paleolithic in this region have affected the employment of the pressure technique. Obstacles to addressing this issue may stem not only from the scarcity of explicit research aimed at determining techniques of detachment based on a systematic and reliable method of identification but also from the lack of attention to explanation of the associations between the various microblade reduction sequences and specific applications of the pressure microblade technique. Considering these associations can assist in ordering the employment of the pressure technique into the chaîne opératoire of the various microblade reduction sequences, thereby moving beyond just the identification of pressure microblade production to understanding the technological and socio-economic conditions in relation to the development of pressure microblade production.

In order to deal with such challenges, I identified techniques of detachment for blades and microblades in the Upper Paleolithic assemblages of Northern Japan by focusing on the analysis of fracture wings found on the fracture surfaces of a wide range of brittle solids including obsidian (Takakura 2007a, 2008, n.d.). Fracture wings (see Fig. 11.1) are microscopic fracture surface ripple markings which show distinctive 'V'-shape. They are very reliable registers of crack velocity according to the model of fracture mechanics (Cotterell and Kamminga 1979; Hutchings 1999; Tomenchuk 1988). Crack velocity can be determined by measuring the effective angle of divergence of fracture wings. Several of our experiments demonstrate that a strong association between crack velocity and detachment techniques does in fact exist, and establish three groups of detachment techniques ((1) pressure; (2) indirect percussion, direct percussion using antler or wood; (3) direct percussion using stone or metal) which are strongly dependent on the difference in crack velocity (Takakura and Izuho 2004, 2005). In turn, this makes it possible to identify detachment techniques of archaeological samples made of obsidian. Analysing fracture wings offers a systematic and explicit method of identification of the detachment techniques, although most archaeologists are seemingly unaware of the method's analytical potential.

In this chapter, the goal of my assessment is to confirm the temporal change in various microblade reduction sequences in the Upper Paleolithic assemblages of Japan and then to discuss the roles of the pressure technique among them. Additionally, I attempt to address the issues on the emergence and development of the pressure microblade production in Japan. In particular, I focus on the variable microblade technologies of the middle Upper Paleolithic and late Upper Paleolithic assemblages from Hokkaido, located at the northern tip of the Japanese islands (Fig. 11.2). The microblade technology of Hokkaido spans 9,000 radiocarbon years, from approximately 20,000 to 11,000 radiocarbon years B.P. The radiocarbon dates and stratigraphical contexts of lithic assemblages reveal that the microblade technology of Hokkaido emerged several thousand years earlier than on Honshu, the central part of the Japanese islands (Sano 2007; Sato and Tsutsumi 2007). A few microblade assemblages of Hokkaido assigned to the middle Upper Paleolithic have been found that appear to document the emergence of the pressure microblade production in the Japanese islands. Furthermore, there has been some vital research into reconstructing the lithic reduction sequences (particularly microblade reduction sequences) and the behavioural patterns in relation to the technological organization of hunter-gatherers (Kimura 1992; Shiraishi 1993; Tsurumaru 1979; Yamada 2006). Thus, it is indisputable that these can contribute to our understanding of when the pressure technique appeared and how it was adopted and developed.

This chapter is divided into three main sections. First, I begin with a review of the archaeological record concerning the microblade assemblages of Hokkaido, Northern Japan, especially the available radiocarbon dates and the technological variability recognized in the microblade reduction processes. Second, I attempt to assess the archaeological record before the appearance of the microblade assemblage and discuss the emergence of the pressure microblade production in Hokkaido. Third, I deal with the development of the pressure technique in Hokkaido, with a consideration of the temporal change in relationships between microblade reduction processes and the pressure technique.

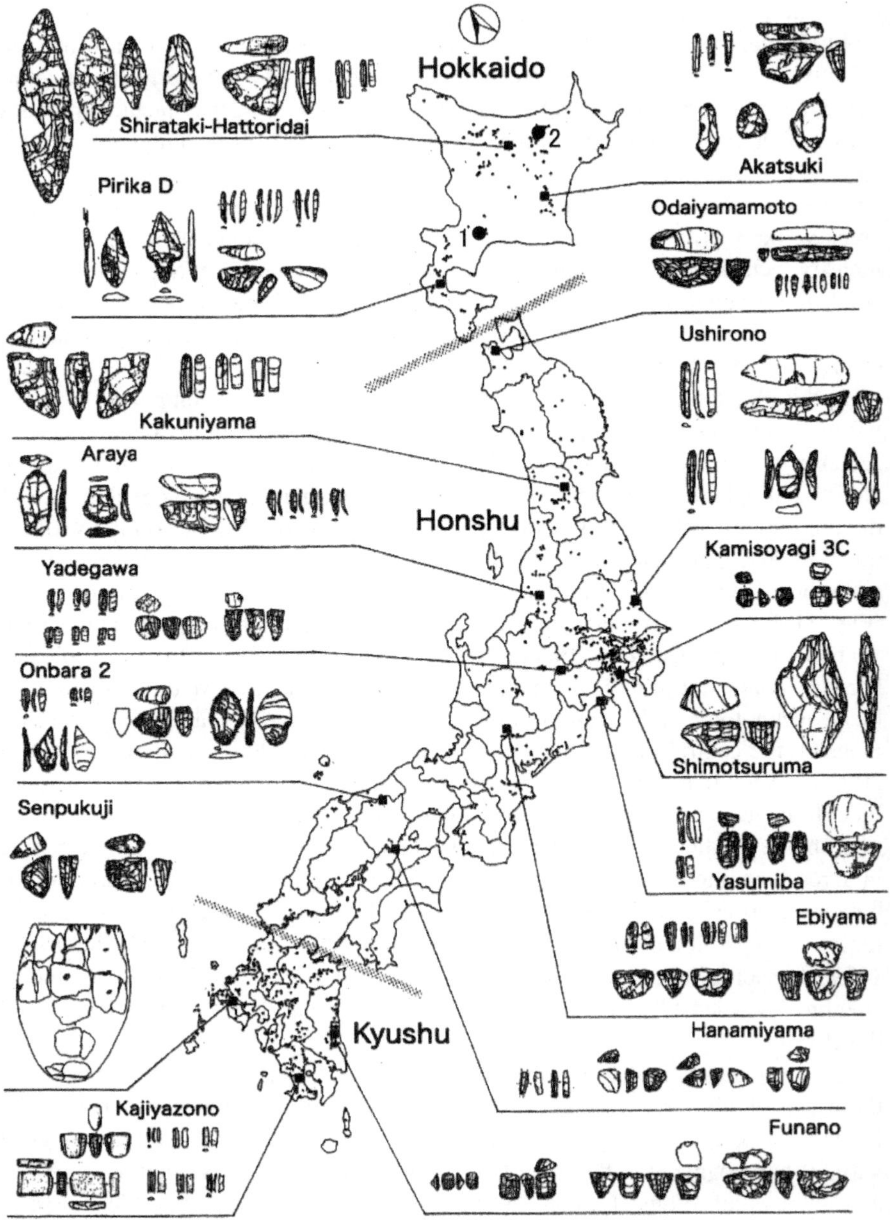

Fig. 11.2 Distribution of the microblade assemblages on the Japanese islands in 2003 (Modified from Sato and Tsutsumi 2007) *1* Kashiwadai-1 site; *2* Nakamoto site

11.2 Microblade Assemblages in Hokkaido: Technological Variability and Dating

The microblade assemblages of Hokkaido are characterized by the presence of variable microblade reduction sequences. Many of the technological features observed in the microblade reduction sequences tend to resemble those of the microblade assemblages distributed across Northern Asia during the Late Pleistocene (e.g. Tsurumaru 1979; Vasilevski 2006). The microblade assemblages of Hokkaido are always associated with curated flake tools, such as end scrapers, side scrapers and gravers, which were often altered by repeated edge resharpening (Takakura 2007b). Blanks of these tools were generally obtained from either blade cores or bifacial cores. Broadly speaking, techno-typologically similar flake tools are known from sites in Northern Asia. We should therefore fully consider the connections with the Asian continent in order to understand the appearance and development of microblade assemblages in Hokkaido.

Due to a large number of general surveys and excavations, many archaeological sites containing microblade assemblages have been uncovered in Hokkaido. Apparently, these investigations have specifically concentrated upon the areas surrounding sources of lithic raw materials, such as obsidian and hard shale, which are largely used in the microblade assemblages of Hokkaido. It is necessary to note that distribution of these sources is definitely limited in Hokkaido. Therefore several sites in these areas have yielded a large number of lithic artefacts and their refitted materials. Such archaeological remains are suitable for reconstructing the lithic reduction sequences and, thus, have allowed us to undertake detailed technological analyses of the various microblade reduction sequences (e.g. Kimura 1992).

Microblades were produced with interestingly complex processes, from preparing blanks of cores to detaching microblades (Bleed 1996, 2002). As presented in our previous discussion (Nakazawa et al. 2005), an outline of the microblade reduction methods and microblade core types observed in the archaeological record of Hokkaido is given below (see Figs. 11.3, 11.4).

Yubetsu Method: This method involves preparing mainly bifacial or boat-shaped core blanks with symmetrical cross sections and forming platforms by removing spalls from the lateral edge of a blank. The Yubetsu method generates the Sakkotsu type and Shirataki type microblade cores. While the microblade cores of the Sakkotsu type are relatively large and wide, the microblade cores of the Shirataki type tend to be smaller and narrower and in the case of obsidian have obvious traces of scratching on the platform. The so-called Pirika type is a variety of the Sakkotsu type.

Togeshita Method: This method involves preparing unifacial blanks on flakes or blades with asymmetrical cross sections and forming platforms generally by removing spalls. The Togeshita method generates the Togeshita type microblade cores.

Oshorokko Method: This method involves preparing relatively small bifaces as blanks and forming platforms generally by removing short spalls. The Oshorokko method generates the Oshorokko type microblade cores.

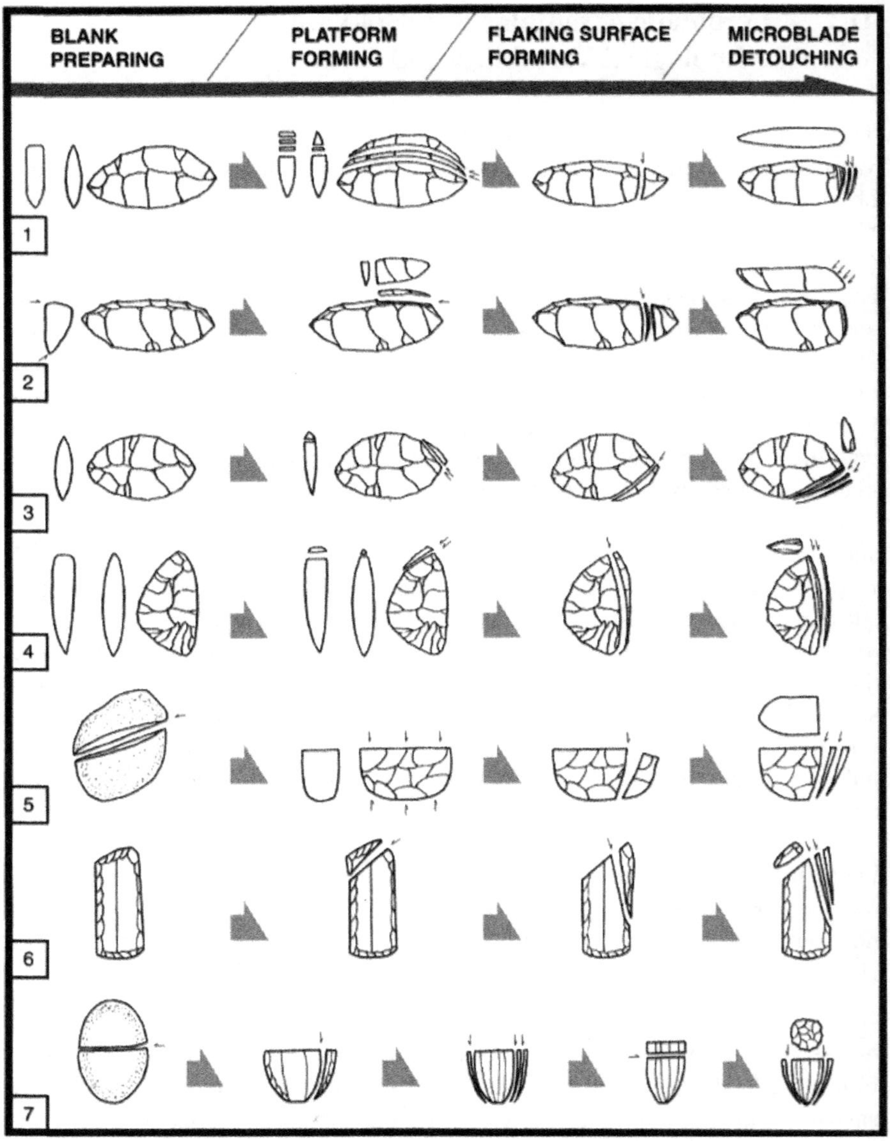

Fig. 11.3 Various microblade reduction methods in Hokkaido (*1–4*: Modified from Tsurumaru 1979; *5–7*: after Nakazawa et al. 2005). *1* Yubetsu method; *2* Togeshita method; *3* Oshorokko method; *4* Rankoshi method; *5* Horoka method; *6* Hirosato method; *7* Oketo method

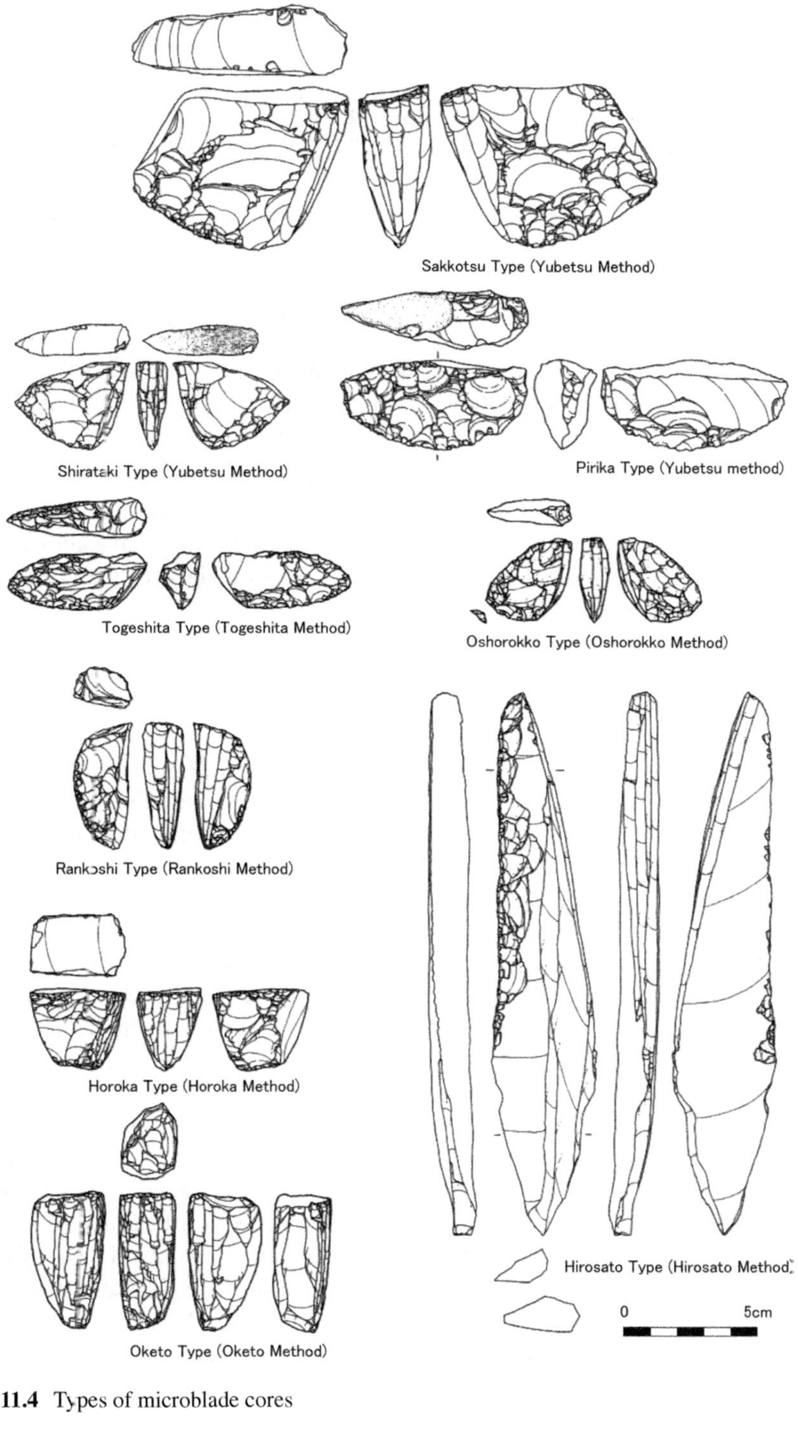

Fig. 11.4 Types of microblade cores

Rankoshi Method: This method involves forming platforms at the ends of elongated wedge-shaped blanks and detaching blades or microblades parallel to the long axis. The refitted materials from the Rankoshi method show that blades and microblades were detached from the same cores in the progress of reduction sequence, with the repeated core rejuvenations. The Rankoshi method generates the Rankoshi type microblade cores.

Horoka Method: This method begins with a split cobble or an angular mass of lithic material. Boat-shaped blanks are prepared by detaching in one direction away from a single flat surface and then microblades are detached from the sharp ends of platforms. The Horoka method generates the Horoka type microblade cores.

Hirosato Method: This method involves forming platforms by preparing the ends of large blades and detaching microblades roughly parallel to the long axis. The Hirosato method generates the Hirosato type microblade cores.

Oketo Method: This method is sometimes called 'the Momijiyama method'. Platforms of cores were made early in the reduction process and then blades and microblades were progressively detached with the repeated core rejuvenations. This method generates the conical-shaped microblade cores defined as the Oketo type and other microblade cores.

While each reduction method or microblade core type was previously used as a diagnostic chronological trait, some researchers are currently highlighting the significance of lithic raw material, especially its morphology and quality, which affected the variability of microblade reduction sequences (Kimura 1992). It demonstrates that constructing the chronology of microblade assemblages in Hokkaido is much more complicated than originally claimed. We therefore need to consider the many causal factors resulting in inter-assemblage variability, in relation to the context of overall socio-economical settings and the behavioural patterns of hunter-gatherers. Such issues have been discussed in different ways more recently (e.g. Nakazawa et al. 2005).

Nevertheless, it is at least possible to say that the microblade assemblages of Hokkaido are divided into two chrono-cultural sub-divisions: Early period and Late period. The distinction between the two sub-divisions mainly lie in the diagnostic stone tool classes and the combination of microblade core types, even though there is no abrupt change in the overall technological features of the lithic assemblages between the Early and Late periods, in terms of the presence of blade and bifacial technologies. The microblade assemblages of the Late period are often accompanied by new tool types such as bifacial leaf-shaped points, bifacial stemmed points, flake adzes and bifacial axes, which are generally not seen in the Early period. On the other hand, microblade cores of the Early period consist mainly of the Rankoshi, Pirika, Togeshita, Sakkotsu and Horoka types, while those of the Late period consist mainly of the Shirataki, Oshorokko and Hirosato types. Acquisition of chronometric dates has been limited at most sites, but recent progress in obtaining AMS radiocarbon dates enables us to re-examine the chronology of the microblade assemblages in Hokkaido (Ono et al. 2002). Such dates show that the Early period probably lasted from 20,000 to 13,500 radiocarbon years B.P., when it was succeeded by the Late period, which lasted until approximately 11,000 radiocarbon years B.P.

Only a few data have been reported on discrete and well-preserved lithic assemblages including the Oketo microblade core type. Additionally, recent radiocarbon dates from the lithic assemblages associated with the Hirosato microblade core type range between 16000 and 12000 B.P., even though the techno-typological assessment suggests that these assemblages are assigned to the Late period (Nakazawa et al. 2005; Terasaki 2006; Yamada 2006). Therefore, it is likely that archaeologists do not have a complete consensus on the development of the microblade assemblages in Hokkaido.

For an understanding of the appearance of microblade assemblages in Hokkaido, the case of the Kashiwadai-1 site, located in Central Hokkaido (Fig. 11.2), is important. At the Kashiwadai-1 site, microblade assemblages including the Rankoshi type and the Pirika type microblade cores were recovered from below the primary Eniwa-a (En-a) pumice fall deposit which dates to approximately 17,000 radiocarbon years B.P. (Figs. 11.5, 11.6). AMS radiocarbon dates obtained from discrete hearths associated with lithic concentrations including the Rankoshi type range from 20790 ± 160 (Beta-126175) to 18830 ± 150 B.P. (Beta-126177), but cluster more tightly around 20500 B.P. (Fukui 1999). These dates suggest that the lithic assemblage from the Kashiwadai-1 site is the earliest appearance of microblade assemblages in Hokkaido, and the emergence of microblade technology in Hokkaido dates back to the Last Glacial Maximum.

In contrast, the case of the Nakamoto site, located in Eastern Hokkaido (Fig. 11.2), provides an essential insight into the end of the microblade assemblages in Hokkaido (Fig. 11.6). At the Nakamoto site, AMS radiocarbon dates on four charcoal samples obtained from discrete hearths or hearth-related features (dense charcoal) associated with lithic concentrations including the Hirosato type microblade cores range from 12580 ± 90 (Beta-111878) to 12280 ± 170 B.P. (Beta-111880) (Nakazawa et al. 2005). It is clear that these recent AMS radiocarbon dates from the Nakamoto site support the conclusion that the microblade assemblages of the Late period in Hokkaido persisted during the terminal Pleistocene.

Research into the terrestrial and marine sediments in Japan and its adjacent regions reveals that cold and dry conditions lasted until the abrupt warming known as the Bölling-Alleröd event (e.g. Lea et al. 2000; Nakagawa et al. 2002, 2008; Prokopenko et al. 2001). Based on the dates of these environmental events and the archaeological record, the Early period roughly corresponds to the phase of cold and dry conditions, while the Late period corresponds to the phase of warm conditions during the terminal Pleistocene (Yamada 2006). In this regard, the difference of microblade assemblages between two periods is not only restricted to representing the chrono-cultural units but also perhaps related to the dynamics of human behaviour to adapt such environmental change.

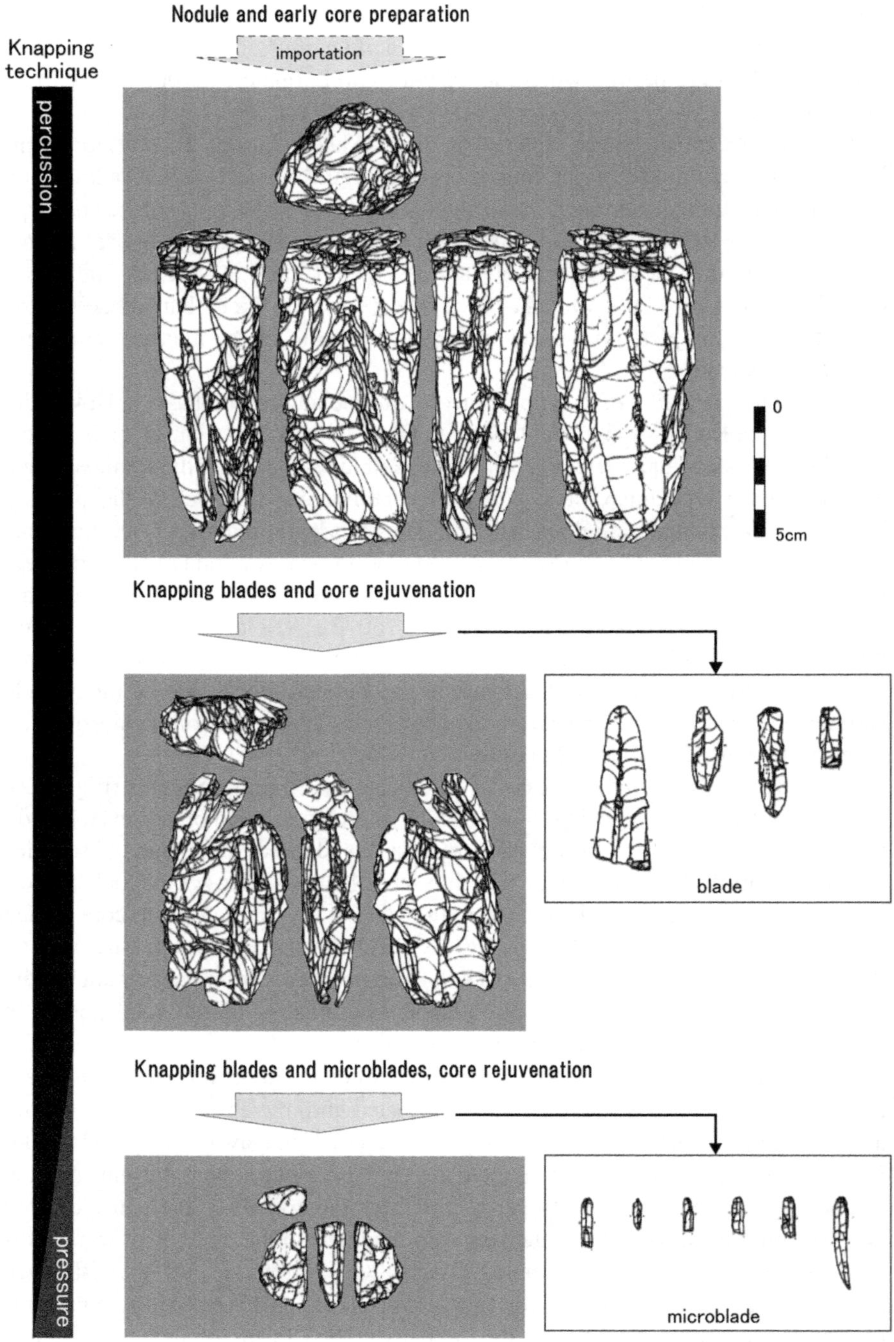

Fig. 11.5 Lithic reduction sequence related to the Rankoshi method. This shows one of the refitted materials from the Kashiwadai-1 site

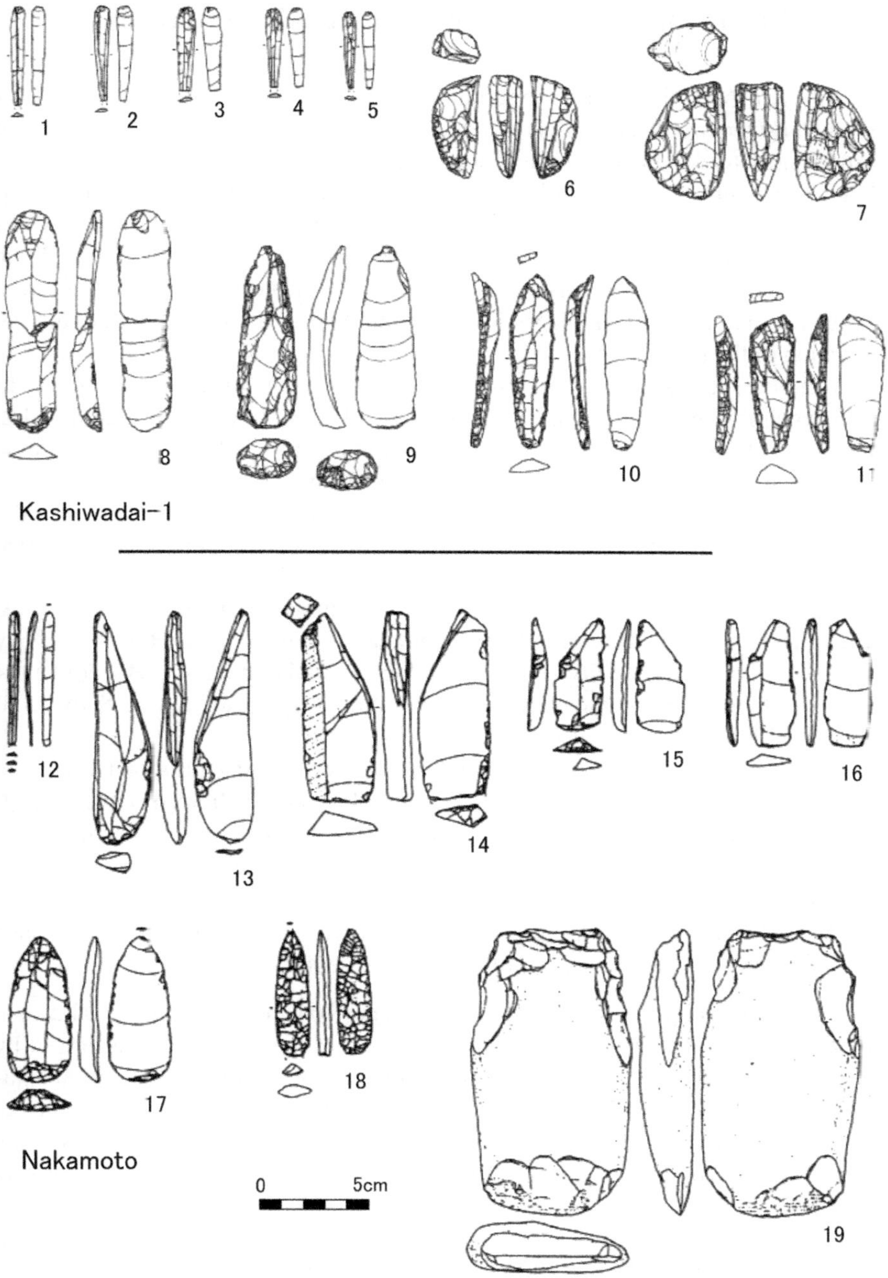

Fig. 11.6 Microblade assemblages in Hokkaido. *1–11* The microblade assemblages of the early period from the Kashiwadai-1 site (Fukui 1999); *12–19* The microblade assemblages of the late period from the Nakamoto site (Nakamoto site research group, n.d.)

11.3 Emergence of the Pressure Microblade Production

In Northern Asia, the earliest appearance of the pressure technique, used not for retouching tools but for producing blanks of tools, has been associated with the origin and spread of microblade technology (Inizan et al. 1992). Inasmuch as microblade production is intimately bound with the pressure technique, their interpretation is supported by my recent identifications based on the analysis of fracture wings observed in some microblade assemblages of Hokkaido (Takakura 2007a, 2008, n.d.). At present, archaeologists are forced to re-evaluate the processes of emergence of microblade assemblages in their respective areas in order to explain the adoption of the pressure technique.

To address this issue, it is essential to explore the technological characteristics of lithic assemblages prior to the microblade assemblages. In the past three decades, much effort has been expended on demonstrating the evidence for these lithic assemblages in Hokkaido. Most of them have been uncovered in the Tokachi plain, Eastern Hokkaido, and in the southern part of the Ishikari lowland, Central Hokkaido. The primary En-a pumice fall deposit which was erupted approximately 17,000 radiocarbon years B.P. is distributed across these areas. It is possible to estimate the chronology of such lithic assemblages from a tephrochronological point of view. Thus, the recent increase in available records allows us to discuss the chronology and technological variability among the lithic assemblages recovered from below the En-a tephra and to interpret their behavioural significance (e.g. Sato 2003; Terasaki 2006; Yamahara 1996).

Various lithic assemblages recovered from below the En-a tephra, and thought to be older than the microblade assemblages, can be roughly divided into four sub-divisions (Groups A–D) based on the technological features of the lithic reduction sequences and the typological features of the stone tools. The characteristics of each sub-division are described below (Fig. 11.7). Note that my classification does exclude a few assemblages probably belonging to the stage prior to the microblade assemblages because of scarcity of the related materials.

Group A: This group includes the lithic assemblages from the Shukubai-Sankakuyama site, the Wakabanomori site and many of other sites. Group A is characterized by small, irregular flakes and amorphous cores defined as multi-directional, unstandardized, expedient and rotated cores. Blades and microblade-like flakes are completely absent. The stone tools in these assemblages are dominated by small, irregular flake tools, which are roughly edge-trimmed. Sometimes, a few side scrapers, end scrapers and drills are seen in these assemblages. At the Wakabanomori site, AMS radiocarbon dates on five charcoal samples obtained from hearths range from 27640 ± 230 (Beta-174960) to 23930 ± 220 B.P. (Beta-162683) (Kitazawa 2004).

Group B: This group includes the lithic assemblages from the Shimaki site and other sites. Group B is characterized by an abundance of end scrapers and round scrapers. Sometimes a few side scrapers, drills and gravers are associated with these assemblages. Blanks of these tools are derived from the multi-directional, rotated cores or

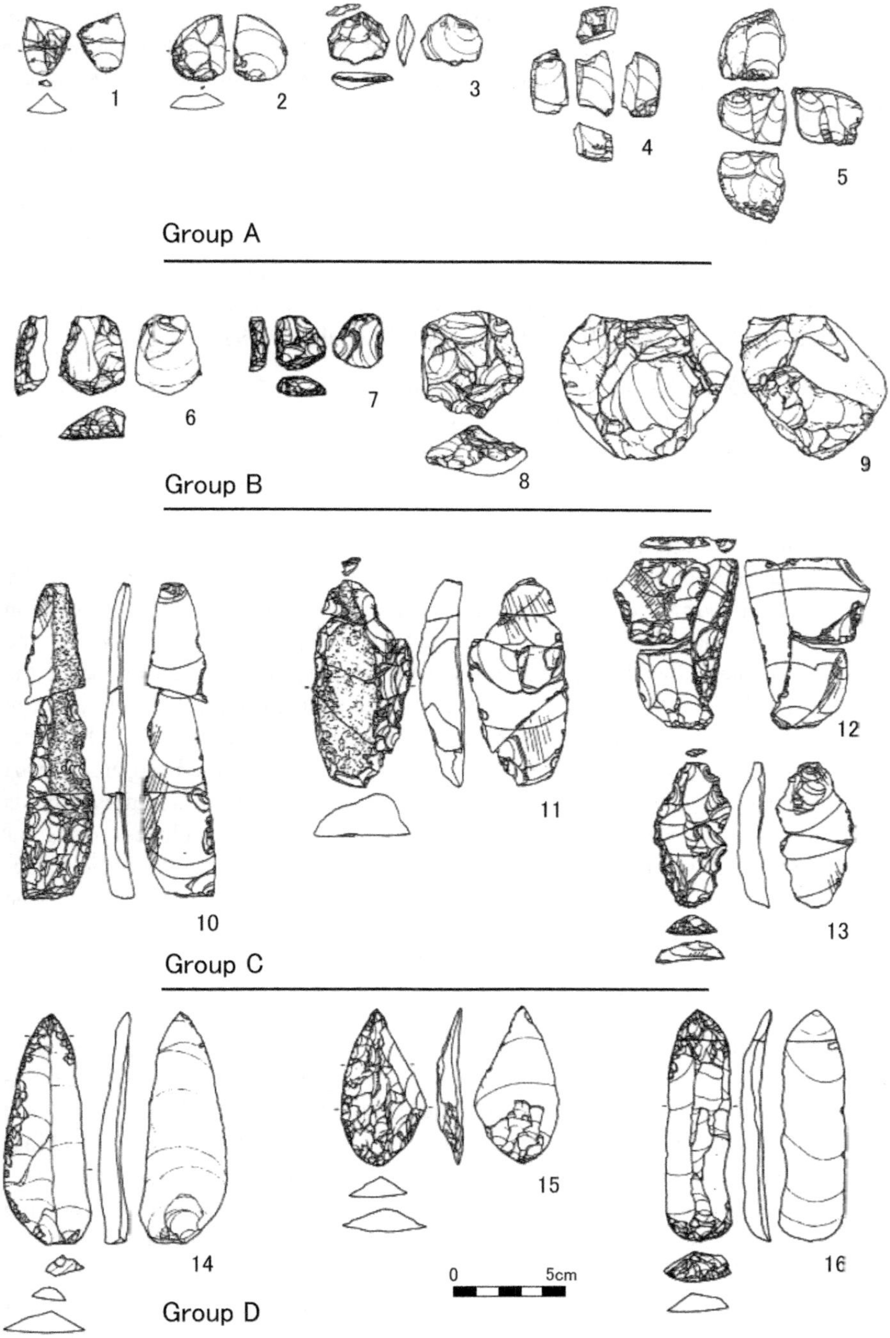

Fig. 11.7 The lithic assemblages older than the microblade assemblages. *1–5* Materials from the Wakabanomori site (Kitazawa 2004); *6–9* Materials from the Kashiwadai-1 site (Fukui 1999); *10–13* Materials from the Kawanishi-C site (Kitazawa 1998); *14–16* Materials from the Kamishirataki-7 site (Naganuma 2000)

radial cores. Lithic raw materials used for these cores include a large variety of sizes of round gravel. Blades and microblade-like flakes are scarce excluding a few materials yet to be discussed, for example the materials from the Shimaki site (Kato and Yamada 1988). At the Kashiwadai-1 site, the lithic assemblage assigned to group B is found in the stratigraphic unit below the En-a tephra, which also yields the microblade assemblages including the Rankoshi type and the Pirika type microblade cores. However, the spatial distribution of it is incongruent with that of the microblade assemblage within the site, which suggests that they are not contemporary. AMS radiocarbon dates on eight charcoal samples obtained from hearths associated with the group B lithic assemblage at the Kashiwadai-1 site range from 22550 ± 180 (Beta-126171) to 20390 ± 70 B.P. (Beta-120880) (Fukui 1999).

Group C: Various side scrapers made on blades are abundantly represented in the lithic assemblages belonging to group C. End scrapers made on blades are also often found. Microblade-like flakes are completely absent. Depending on the morphological features of the lithic raw materials and the sizes of the stone tools, group C is divided into two sub-groups: C1 and C2. The lithic assemblage from the Kawanishi-C site represents sub-group C1. Lithic raw materials used for the sub-group C1 are dominated by large round gravel. Cortex was not removed specifically before blade making. Core shaping began with the removal of platform preparation flakes on large mass. The result was the creation of large blades with cortex on one or more of their dorsal surfaces and several roughly faceted platforms. At the Kawanishi-C site, AMS radiocarbon dates on two charcoal samples obtained from hearths are 21780 ± 90 (Beta-106506) and 21400 ± 190 B.P. (Beta-107731) respectively (Kitazawa 1998). In contrast, the lithic assemblage from the Kukouminami-A site and other sites represent the sub-group C2. The round gravels mainly used as the lithic raw materials and the stone tools that dominate the assemblages are smaller than those of sub-group C1, whereas the technological features of the blade reduction sequences substantially resemble those of sub-group C1. The quantity of stone tools made on blades exhibits great diversity between sites. No radiocarbon dates obtained from samples associated with reliable archaeological features, such as hearths, have yet been reported for sub-group C2.

Group D: This group includes the lithic assemblages from the Hirosato-8 site, Kami-shirataki-7 site and other sites. Group D is characterized by a variety of unifacial points made on blades. In general, the striking platforms and the ridges of the dorsal surfaces of blade cores were not prepared intensively before blade making. Microblade-like flakes are not recognized at all. Sometimes end scrapers and gravers are associated with these assemblages. Adequate radiocarbon dates related to the chronological position of these assemblages have not yet been obtained.

Although the number of archaeological remains assigned to the stage prior to the microblade assemblages has increased recently, their typological and technological relationships to each other remain obscure, and the number of chronometric dates is definitely limited. More detailed studies are needed to determine the chronological sequence and the technological inter-relationships of these assemblages. Thus, we cannot say whether each group signifies a chronological unit during the early and middle Upper Paleolithic in Hokkaido. In particular, there has been little consensus on the chronological positions of the groups B, C and D. The question as to whether each existed simultaneously or not is still debated. However, based on the radiocarbon dates and the typological assessment, it is clear that group A is older than groups B, C and D, as has been previously suggested (Sato 2003; Terasaki 2006; Yamahara 1996).

In Hokkaido, most lithic assemblages older than the microblade assemblages virtually lack microblades or microblade-like flakes. In addition, blades are scarce in these lithic assemblages, except in groups C and D. The lithic assemblages belonging to groups A and B are characterized by the production of flakes derived from multi-directional, rotated cores or radial cores. Blades are found in the lithic assemblages belonging to groups C and D, but the characteristics of the blade reduction sequences in these assemblages are clearly different from those of the microblade assemblages. In general, the blade cores in the microblade assemblages of Hokkaido were prepared and rejuvenated intensively, and parallel-sided blades were largely produced.

In contrast with these lithic assemblages in Northern Japan, the lithic assemblages dated from 40,000 to 20,000 radiocarbon years B.P. in Northern Eurasia are characterized by the presence of typical blade technology, with intensive core preparations and rejuvenations, and formal stone tools made from blade such as graver, side scraper and end scraper (e.g. Brantingham et al. 2004; Derevianko et al. 1998; Goebel 2004). Also, microblades or microblade-like flakes are often recognized in the early and middle Upper Paleolithic assemblages of Siberia (Derevianko et al. 2000; Derevianko and Shunkov 2002). Consequently, it bears emphasizing that the lithic assemblages prior to the microblades assemblages of Hokkaido, Northern Japan, are strikingly different from many Northern Eurasian Upper Paleolithic occurrences both preceding and contemporaneous with them.

The lithic assemblage from the Kashiwadai-1 site is the earliest representative of microblade assemblages in Hokkaido, as mentioned above. In this lithic assemblage, highly regular, parallel-sided and very thin microblades were detached from the wedge-shaped microblade cores defined as the Rankoshi type and the Pirika type. These can perhaps be interpreted as the earliest examples in Northern Japan of using pressure technique to produce blanks of tools, such as microblades or blades. In addition, the refitted materials from the Rankoshi method show that both blades and microblades were detached from the same cores in the reduction sequence, with intensive core preparations and rejuvenations (Fig. 11.5). These reduction sequence characteristics raise the possibility that the detachment technique was converted from direct percussion to pressure during the progress of blades and microblades production.

Obviously, there is a remarkable difference in the overall technological and typological characteristics of the lithic assemblages between before and after the appearance of the microblade assemblages in Hokkaido. The observations presented here suggest, on the whole, that both microblade technology and the pressure technique appear suddenly around 20,000 radiocarbon years B.P. and without local precedent in Hokkaido. It is probable that these represent an intrusive phenomenon from some adjacent area.

The hypothesis of Inizan et al. (1992) that the pressure microblades production appeared around 20,000 years B.P. in Northern Asia finds empirical support in the recent evidence presented here. However, the conclusion that the emergence of the microblade technology and the pressure technique was an intrusive phenomenon from some adjacent area of Hokkaido may prompt us to rethink why and how the pressure microblade production dispersed across the large areas of Northern Asia, including Northern Japan. To exploit fully what the record can tell us, it is necessary to understand the socio-economic settings in relation to the adoption of the pressure microblade production.

It is interesting to note that the appearance of microblade technology and the pressure technique in Hokkaido roughly corresponds to the beginning of the severe cold and dry conditions in the Last Glacial Maximum. Since the available mammal resources were sparsely scattered over a large area under such climatic conditions, people may have needed to increase the frequency and magnitude of residential moves. It seems that the changes of residential moves favoured lightening of portable tool kits and effective use of raw materials designed to minimize stone transport costs (Elston and Brantingham 2002; Nishiaki 2001), in the case of when the lithic raw material of the needed quality and morphology was spatially limited. Availability of the lithic raw materials such as obsidian and hard shale, as well as the distribution of food resources in Hokkaido, may also have strengthened the behavioural scheduling of highly mobile hunter-gatherers. The abrupt adoption and development of microblade technology associated with the pressure technique in Northern Japan may be closely related to these requirements.

This inference has important implications as to precisely why the pressure microblade technique was able to disperse across the overall area of Northern Asia. Severe environmental conditions across large areas during the Last Glacial Maximum probably caused abrupt depopulation by hunter-gatherers, who were obliged to use the dispersed resources on the landscape effectively. As a result, the highly mobile hunter-gatherers who met these socio-economic conditions were able to disperse across large areas. This may have caused the spread of the pressure microblade production across Northern Asia.

To test this hypothesis, we would need to confirm that the emergence of the pressure microblade production occurred almost simultaneously in the different areas of Northern Asia. In future, more detailed studies comparing the earliest appearances of the pressure microblade production in Northern Japan and some adjacent areas, for example Siberia, Mongolia, Northern China, Korea and the Russian Far East, are needed to evaluate and improve current understanding.

11.4 Microblade Reduction Sequences and the Pressure Technique

One of the most intriguing aspects of the refitted materials from the Kashiwadai-1 site is that the blades and microblades were produced from the same cores through a series of reduction processes (see Fig. 11.5). A similar characteristic is often noted in other reduction sequences of microblade core types, such as the Sakkotsu and the Oketo types (Sato 2003; Takakura 2007a). However, this characteristic is not seen in the various microblade core types of the Late period, such as the Shirataki, Oshorokko and Hirosato types. Among the microblade assemblages of the Late period, the blanks of microblade cores were always made on bifaces or blades.

In addition, as some researchers have already suggested (e.g. Naganuma 2008; Shiraishi 1993), the bifacial blanks of microblade cores of the Sakkotsu type also produced the blanks of various flake tools, such as gravers, end scrapers and side scrapers. This is the so-called bifacial reduction strategy. On the other hand, the blanks of flake tools in the microblade assemblages of the Late period were mainly produced from prismatic blade cores. The bifacial reduction strategy is recognized exclusively in the Early period of the microblade assemblages.

It is clear that the relationships between the production of flake tools (made on blades or bifacial flakes) and microblades changed considerably from the Early period to the Late period. This is important for at least two reasons. First, this enables us to infer the temporal change in behavioural patterns through the association between the lithic reduction strategy and the technological organization of the hunter-gatherers. Indeed, the temporal change in the lithic reduction strategies from the Early period to the Late period may demonstrate that minimizing transport costs was not the primary concern for the hunter-gatherers in the Late period. Further, the appearance of new tool types such as bifacial leaf-shaped points, bifacial stemmed points, flake adzes and bifacial axes in the Late period was probably related to this alteration of lithic reduction strategies. Second, this also raises a new question as to why the microblade technology persisted under such alteration and how the temporal change in lithic reduction sequences was related to the development of the pressure microblade production.

Presumably not every microblade production on the Japanese islands, including Hokkaido, necessarily depended on the employment of the pressure technique. In this regard, the possibility that some microblades were detached by direct or indirect percussion cannot be ruled out. Identification of the detaching technique in the microblade assemblages must be an important subject. Yet most of the microblade reduction processes in Hokkaido probably resulted from the employment of the pressure technique. This is shown by the result of experimental study (Ohnuma 1992) aimed at identifying microblade techniques among some microblade assemblages in Hokkaido. My analysis focused on the fracture wings also offers empirical support for this inference (Takakura 2007a, 2008, n.d.).

If this inference is valid, it means that the pressure microblade production lasted approximately 9,000 radiocarbon years in Northern Japan. We then should discuss the development of pressure technique among the microblade assemblages of Hokkaido through a comparison of the various microblade methods. For the sake of dealing with this question, it is necessary to explore the specific applications of pressure technique used for various microblade reduction methods and to examine their relationships to the overall lithic reduction sequences.

The results of some experimental studies (e.g. Flenniken and Hirth 2003; Pelegrin 2003; Tabarev 1997) suggest that choices of the type of pressure tools, body position, method of holding and use of devices for stabilizing cores are fundamentally important to ensuring successful microblade removal. However, it is very difficult to determine these specific applications of pressure technique in the archaeological record. More systematic experimental studies focusing on an examination of particular archaeological remains are needed to address this issue. It is possible, at least, to infer that the difference in the form of microblade cores significantly affects the choice of the specific applications of pressure technique.

Since the microblade cores of the Rankoshi, Sakkotsu, Shirataki, Togeshita, Horoka and Oshorokko types are generally triangular in cross section, elongated in shape with a flat striking platform, these are obviously included in the wedge-shaped microblade cores. Experimentation shows that microblades can be detached from these wedge-shaped microblade cores by the handheld technique (e.g. Pelegrin 2003), occasionally with a grooved piece of wood as a stabilization device for the pressure microblade production. On the other hand, the form of the Hirosato and Oketo types differs from that of the wedge-shaped microblade cores by not having the elongated flat striking platform and the so-called keel that give the microblade cores its distinctive wedge-shaped appearance. It probably indicates that there is some variation in the method of holding and use of a device for the pressure microblade production among the assemblages. The chronological position of the Oketo type is still debated, while it is indisputable that the Hirosato type does not date back to the Last Glacial Maximum according to the available AMS radiocarbon dates. Therefore, the Hirosato type potentially suggests the existence of a diversity of applications of pressure technique in terms of one or several parameter such as the pressure tools, the body position and the method of holding.

Unfortunately, the conditions responsible for this diversification of applications remain to be identified. To address this issue, it is necessary to determine the specific applications among the pressure microblade techniques, based on the analysis of archaeological remains, and to examine the significance of them in the context of the behavioural patterns of hunter-gatherers.

11.5 Conclusion

This review of archaeological records before and after the appearance of microblade assemblages in Hokkaido reveals that the pressure microblade production appeared suddenly in this area around 20,000 years B.P. This is confirmed by the well-preserved stratigraphical context and several radiocarbon dates from the Kashiwadai-1 site. In addition, it is possible that the emergence of the pressure microblade production was associated with diffusion from some area adjacent to Hokkaido. Such an intrusive phenomenon may be closely related to the large-scale migrations of hunter-gatherers who attempted to adapt to drastic environmental fluctuations, suggested by comparisons of the chronometric dates of environmental events and the archaeological record.

Radiocarbon dates show that the microblade assemblages of Hokkaido persisted over a period of approximately 9,000 radiocarbon years. During this time, various microblade reduction methods, affected by the abundance of lithic raw material resources, were distributed throughout most areas of Hokkaido. The development of pressure microblade techniques seems to have been related to the diverse use of specific applications among the microblade assemblages of the Late period, along with the difference in the form of the microblade cores.

Acknowledgements Special thanks to Dr. Ichiro Yamanaka for his encouragement and valuable suggestions on my research. I am also grateful for the insights of one anonymous reviewer whose comments improved this paper. Additionally, my thanks extend to Pierre M. Desrosiers and Noura Rahmani who organized the session on pressure blade and microblade production. Of course, all errors are my responsibility.

References

Apel, Jan 2008 Knowledge, Know-how and Raw Material: The Production of Late Neolithic Flint Daggers in Scandinavia. *Journal of Archaeological Method and Theory* 15: 91–111.

Bleed, Peter 1996 Risk and Cost in Japanese Microblade Production. *Lithic Technology* 21: 95–107.

Bleed, Peter 2002 Cheap, Regular and Reliable: Implications of Design Variation in Late Pleistocene Japanese Microblade Technology. In *Thinking Small: Global Perspectives on Microlithization*, edited by Robert G. Elston, and Steven L. Kuhn, pp. 95–102. Archaeological Papers of the American Anthropological Association 12, American Anthropological Association, Washington, DC.

Brantingham, Jeffrey P., Xin Gao, David B. Madsen, Robert L. Bettinger, and Robert G. Elston 2004 The Initial Upper Paleolithic at Shuidonggou, Northwestern China. In *The Early Upper Paleolithic beyond Western Europe*, edited by Jeffrey P. Brantingham, Steven L. Kuhn, and Kristopher W. Kerry, pp. 223–241. University of California Press, Berkeley and Los Angeles.

Cotterell, Brian, and Johan Kamminga 1979 The Mechanics of Flaking. In *Lithic Use-wear Analysis*, edited by Brian Hayden, pp. 97–112. Academic Press, New York.

Derevianko, Anatoly P., and Michael V. Shunkov 2002 Middle Paleolithic Industries with Foliate Bifaces in Gorny Altai. *Archaeology, Ethnology and Anthropology of Eurasia* 1(9): 16–42.

Derevianko, Anatoly P., Demitri B. Shimkin, and Roger W. Powers (editors) 1998 *The Paleolithic of Siberia: New Discoveries and Interpretation.* University of Illinois Press, Chicago.

Derevianko, Anatoly P., Valeriy T.Petrin, and Evgeny P. Rybin 2000 The Kara Bom Site and the Characteristics of the Middle-Upper Paleolithic Transition in the Altai. *Archaeology, Ethnology and Anthropology of Eurasia* 1(2): 33–52.

Elston, Robert G., and Jeffrey P. Brantingham 2002 Microlithic Technology in Northern Asia: A Risk-minimizing Strategy of the Late Paleolithic and Early Holocene. In *Thinking Small: Global Perspectives on Microlithzation*, edited by Robert G. Elston and Steven L. Kuhn, pp. 103–116. Archaeological Papers of the American Anthropological Association 12, American Anthropological Association, Washington, DC.

Flenniken, Jeffrey J. 1987 The Paleolithic Dyuktai Pressure Blade Technique of Siberia. *Arctic Anthropology* 24: 117–132.

Flenniken, Jeffrey J., and Kenneth G. Hirth 2003 Handheld Prismatic Blade Manufacture in Mesoamerica. In *Mesoamerican Lithic Technology: Experimentation and Interpretation*, edited by Kenneth G. Hirth, pp. 98–107. The University of Utah Press, Salt Lake City.

Fukui, Jun-ichi (editor) 1999 *Kashiwadai 1 Iseki (Kashiwadai 1 Site).* Hokkaido Center for Archaeological Operations, Ebetsu.

Goebel, Ted 2004 The Early Upper Paleolithic of Siberia. In *The Early Upper Paleolithic beyond Western Europe*, edited by Jeffrey P. Brantingham, Steven L. Kuhn, and Kristopher W. Kerry, pp. 162–195. University of California Press, Berkeley and Los Angeles.

Hutchings, Karl W. 1999 Quantification of Fracture Propagation Velocity Employing a Sample of Clovis Channel Flakes. *Journal of Archaeological Science* 26: 1437–1447.

Inizan, Marie-Louise, Monique Lechevallier, and Patrick Plumet 1992 A Technological Maker of the Penetration into North America: Pressure Microblade Debitage, Its Origin in the Paleolithic of North Asia and Its Diffusion. In *Materials Issues in Art and Archaeology III*, edited by Pamela B. Vandiver, James R. Druzik, Gary S. Wheeler and Ian C. Freestone, pp. 661–681. Materials Research Society Symposium Proceedings 267. Materials Research Society, Pittsburgh.

Kato, Shimpei, and Masahisa Yamada 1988 Hokkaido Katou-gun Kamishihoro-cho Shimaki Iseki no Sekki Bunka (A Palaeolithic Culture at Shimaki: The Oldest Palaeolithic Components in Hokkaido). *Rekishi to Jinrui (History and Anthropology)* 16: 1–64. (In Japanese).

Kimura, Hideaki 1992 *Reexamination of the Yubetsu Technique and Study of the Horokazawa Toma Lithic Culture*. Archaeological Museum of Sapporo University, Sapporo. (In Japanese and English).

Kitazawa, Minoru (editor) 1998 *Kawanishi-C Iseki (Kawanishi-C site)*. Board of Education, Obihiro City.

Kitazawa, Minoru (editor) 2004 *Wakabanomori Iseki (Wakabanomori site)*. Board of Education, Obihiro City, Obihiro. (In Japanese).

Lea, David W., Dorothy K. Pak, and Howard J. Spero 2000 Climate Impact of Late Quaternary Equatorial Pacific Sea Surface Temperature Variations. *Science* 289: 1719–1724.

Migal, Witold 2006 The Macrolithic Flint Blades of the Neolithic Times in Poland. In *Skilled Production and Social Reproduction*, edited by Jan Apel and Kjel Knutsson, pp. 387–398. Societas Archaeologica Upsaliensis, SAU Stone Studies 2, Uppsala.

Naganuma, Masaki 2008 Ryoumen Sekki Ridakushon no Jirei (1): Sakkotsu Saisekijin-kaku ni Kanrensuru Sekkigun (A Case Study of Biface Reduction (1): The Microblade Assemblages with the Sakkotsu Type Microbladecores). *Ronshu Oshorokko (The Memoirs of the Oshorokko Reseach Group)* II: 125–150. (In Japanese).

Naganuma, Takashi (editor) 2000 *Shirataki Isekigun I (Shirataki Sites I)*. Hokkaido Center for Archaeological Operations, Ebetsu (In Japanese).

Nakagawa, Takeshi, Pavel E. Tarasov, Kotoba Nishida, Katsuya Gotanda, and Yoshinori Yasuda 2002 Quantitative Pollen-based Climate Reconstruction in Central Japan: Application to Surface and Late Quaternary Spectra. *Quaternary Science Reviews* 21: 2099–2113.

Nakagawa, Takeshi, Masaki Okuda, Hitoshi Yonenobu, Naoya Miyoshi, Touru Fujiki, Katsuya Gotanda, Pavel E. Tarasov, Yoshimune Morita, Keiji Takemura, and Shoji Horie 2008 Regulation of the Monsoon Climate by Two Different Orbital Rhythms and Forcing Mechanisms. *Geology*, 36: 491–494.

Nakamoto Site Research Group. n.d. *Nakamoto Iseki (Nakamoto Site)*. Nakamoto Site Research Group, Kitami.

Nakazawa, Yuichi, Masami Izuho, Jun Takakura, and Satoru Yamada 2005 Toward an Understanding of Technological Variability in Microblade Assemblages in Hokkaido, Japan. *Asian Perspectives* 44: 276–292.

Nishiaki, Yoshihiro 2001 Zenki Kyusekki to Ouatsu Hakuri (The Early Paleolithic and Pressure Knapping). *Rekishi Hyouron (Journal of Historical Review)* 617: 100–107. (In Japanese).

Ohnuma, Katsuhiko 1992 Study by Replication on Manners of Micro-blade Detachment. In *Origin and Dispersal of Microblade Industry in Northern Eurasia*, edited by Hideaki Kimura, pp. 207–216. Organizing Committee of International Symposium, Sapporo.

Ono, Akira, Hiroyuki Sato, Takashi Tsutsumi, and Yuichiro Kudo 2002 Radiocarbon Dates and Archaeology of the Late Pleistocene in the Japanese Islands. *Radiocarbon* 44: 477–494.

Pelegrin, Jacques 2003 Blade Making Techniques from the Old World: Insights and Applications to Mesoamerican Obsidian Lithic technology. In *Mesoamerican Lithic Technology: Experimentation and Interpretation*, edited by Kenneth G. Hirth, pp. 55–71. The University of Utah Press, Salt Lake City.

Pelegrin, Jacques 2006 Long Blade Technology in the Old World: An Experimental Approach and Some Archaeological Results. In *Skilled Production and Social Reproduction*, edited by Jan

Apel and Kjel Knutsson, pp. 37–68. Societas Archaeologica Upsaliensis, SAU Stone Studies 2, Uppsala.

Prokopenko, Alexander A., Eugene B. Karabanov, Douglas F. Williams, Mikhail I. Kuzmin, Galina K. Khursevich, and Alexander A. Gvozdkov 2001 The Detailed Record of Climate Events during the Past 75,000 yrs BP from the Lake Baikal Drill Core BDP-93-2. *Quaternary International* 80–81: 59–68.

Rahmani, Noura 2004 Technological and Cultural Change among the Last Hunter-Gatherers of the Maghreb: The Capsian (10,000–6000 B.P.). *Journal of World Prehistory* 18: 57–105.

Rahmani, Noura 2007 From Mitred Cores to Broken Microliths: In Search of Specialization during the Capsian. *Lithic Technology* 32: 79–97.

Sano, Katsuhiro 2007 Emergence and Mobility of Microblade Industries in the Japanese Islands. In *Origin and Spread of Microblade Technology in Northern Asia and North America*, edited by Yaroslav V. Kuzmin, Susan G. Keates and Chen Shen, pp. 79–90. Archaeology Press, Simon Fraser University, Montreal.

Sato, Hiroyuki 2003 Hokkaido no Kouki Kyusekki Jidai Zenhanki no Yousou (The Early Upper Paleolithic in Hokkaido). *Kodai Bunka (Cultura Antiqua)* 55: 181–194. (In Japanese with English Abstract).

Sato, Hiroyuki, and Takashi Tsutsumi 2007 The Japanese Microblade Industries: Technology, Raw Material Procurement, and Adaptations. In *Origin and Spread of Microblade Technology in Northern Asia and North America*, edited by Yaroslav V. Kuzmin, Susan G. Keates and Chen Shen, pp. 53–78. Archaeology Press, Simon Fraser University, Montreal.

Shiraishi, Noriyuki 1993 Hokkaido niokeru Saisekijin Sekkigun no Tenkai (Development of Microblade Assemblages in Hokkaido). *Bushitsu Bunka (Material Culture)* 56: 1–22. (In Japanese).

Tabarev, Andrei V. 1997 Paleolithic Wedge-Shaped Microcores and Experiments with Pocket Devices. *Lithic Technology* 22: 139–149.

Takakura, Jun 2007a Hokkaido Monbetsu-gun Engaru-cho Okushirataki-1 Iseki Shutsudo Sekkigun niokeru Hakuri Houhou no Doutei (Identification of Blade and Microblade Flaking Techniques in the Lithic Assemblage of the Okushirataki-1 Site, Hokkaido, Japan). *Kodai Bunka (Cultura Antiqua)* 58(IV): 98–109. (In Japanese with English Abstract).

Takakura, Jun 2007b New Evidence of Endscraper Reduction in the Upper Paleolithic Japan. *Current Research in the Pleistocene*, 24: 40–43.

Takakura, Jun 2008 Hokkaido Yufutsu-gun Atsuma-cho Kamihoronai-Moi Iseki no Kyusekki Jidai Sekkigun niokeru Hakuri Houhou no Doutei (Identification of the Flaking Techniques in the Paleolithic Assemblages of the Kamihoronai-Moi Site, Hokkaido (Japan)). *Ronshu Oshorokko (The Memoirs of the Oshorokko Reseach Group)* II: 41–48. (In Japanese with English Abstract).

Takakura, Jun, and Masami Izuho 2004 Fracture Wing niyoru Hakuri Houhou no Doutei Kenkyu (Identification of Flaking Techniques: From the Analysis of Fracture Wings). *Daiyonki Kenkyu (The Quaternary Research)* 43: 37–48. (In Japanese with English Abstract).

Takakura, Jun, and Masami Izuho 2005 Identification of Flaking Techniques through the Analysis of Lithic Crack Velocity. Paper presented at the 70th Annual Meeting of the Society for American Archaeology, Salt Lake City.

Takakura, Jun n.d. Microblade and Blade Knapping Techniques in the Upper Paleolithic of Hokkaido, Northern Japan: Identification through the Analysis of Fracture Wings and Its Implications. In preparation.

Terasaki, Yasufumi 2006 Hokkaido no Chiiki Hennen (Regional Chronology of Hokkaido). In *Kyusekki Jidai no Chiiki Hennen Teki Kenkyu (A Study of Regional Chronology in the Paleolithic)*, edited by Masato Anzai and Hiroyuki Sato, pp. 277–314. Dousei-sha, Tokyo. (In Japanese).

Tomenchuk, John 1988 Effects of Loading Rate on the Reliability of the Engineering Use-wear Models. In *Industries Lithiques: Tracéology et Technologie*, edited by Sylvie Beyries, pp. 99–113. BAR International Series 411(II). British Archaeological Reports, Oxford.

Tsurumaru, Toshiyuki 1979 Hokkaido Chihou no Saisekijin Bunka (The Microlithic Culture in Hokkaido District). *Sundai Shigaku (Sundai Historical Review)* 47: 23–50. (In Japanese).

Vasilevski, Alexander 2006 The Upper Palaeolithic of Sakhalin Island. In *Archaeology of the Russian Far East: Essays in Stone Age Prehistory*, edited by Sarah M. Nelson, Anatoly P. Derevianko, Yaroslav V. Kuzmin and Richard L. Bland, pp. 75–100. BAR International Series 1540. British Archaeological Reports, Oxford.

Yamada, Satoru 2006 *Hokkaido niokeru Saisekijin Sekkigun no Kenkyu (A Study of Microblade Assemblages in Hokkaido, Japan)*. Rokuichi-Shobou, Tokyo. (In Japanese).

Yamahara, Toshirou 1996 Hokkaido niokeru Saisekijin Bunka Izen no Sekkigun (The Stone Industries before Micro-blade Culture in Hokkaido). *Obihiroshi Hyakunen-kinenkan Kiyou (Report of the Obihiro Centennial City Museum)* 14: 1–28. (In Japanese).

Chapter 12
The Technique of Pressure Knapping in Central Asia: Innovation or Diffusion?

Frédérique Brunet

12.1 Introduction

Details concerning the adoption and use of the pressure knapping technique in Central Asia are still relatively poorly understood by researchers in comparison with Southwestern Asia (especially the Near East) and Europe; moreover, it remains a marginal aspect in the technological studies focused upon this region. This chapter aims to review and clarify this issue through a consideration of the period when the pressure technique took place in Central Asia (Fig. 12.1): from the Final Paleolithic/Mesolithic to the Chalcolithic (or Eneolithic). The specific context of the processes of neolithization, a time characterized by multifaceted (cultural, social, economic, technical, symbolic) transformations, is particularly significant for understanding the development of pressure blade technology in Central Asia, as well as the reasons linked to its adoption and application in different cultural entities. The Neolithic transition and the development of a new way of life did not occur simultaneously across the entire region. The additional information provided here will enrich the discussion underway in this field for the neighboring regions of Russia, Caucasus, Iran, and Afghanistan.

The chronological framework for these periods varies by region (Fig. 12.2) and suffers from a paucity of secure radiocarbon dates. The lack of precise dating for many archaeological sites results mainly from poor organic preservation due to the extensive erosion in these desert and steppe areas. This poses particular challenges to archaeological studies through a scarcity of stratigraphic contexts and the localization of most of the records on the surface.

F. Brunet (✉)
CNRS-UMR 7041 ArScAn, équipe Asie centrale, Maison René-Ginouvès de l'Archéologie et de l'Ethnologie, Nanterre, France
e-mail: frederique.brunet@mae.u-paris10.fr

P.M. Desrosiers (ed.), *The Emergence of Pressure Blade Making: From Origin to Modern Experimentation*, DOI 10.1007/978-1-4614-2003-3_12,

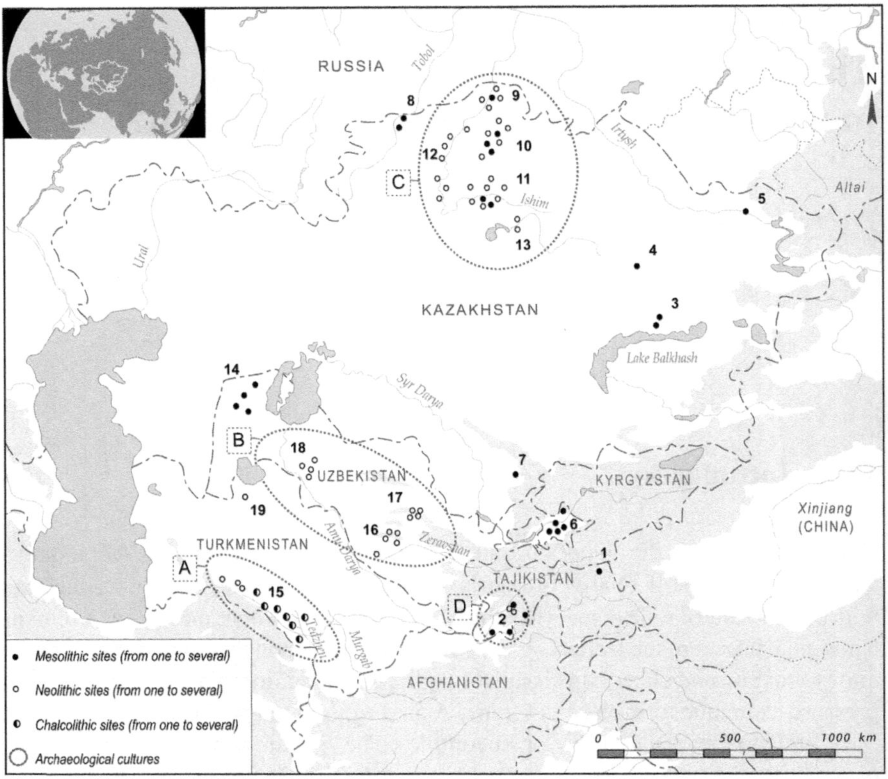

Fig. 12.1 The main archaeological sites and cultures of Central Asia mentioned (*map drawn by the author*). *1* Oshkhona; *2* Beshkent, Javan, Beshkent, Mullo-Nijaz, Makoni-Mor (Mesolithic), and Tutkaul, Saj-Sajёd (Neolithic); *3* Semizbugu, and group of Northern Pribalkhash; *4* Akimbek; *5* Shul'binka; *6* Ferghana (Obishir, Jangikadam, Zambar, Shorkul', Dorazkul'); *7* Bugun'; *8* group of Ubagan; *9* group of Javlenka; *10* group of Vinogradov; *11* group of Tel'man; *12* group of Akkan; *13* group of Kyrgal'dzhino; *14* Ustjurt (group of Ajdabol); *15* Jeitun, Bami, Chopan-depe, Togolok-depe, Chagylly-depe (Neolithic), and Dashlyzhyj-depe, Ak-depe, Anau, Kara-depe, Namazga-depe, Ulug-depe, Geoksijur, Altyn-depe, Ilgynly-depe (Eneolithic/Chalcolithic); *16* Zeravshan (Uchashchi, Ajakagytma, Khodzhagumbas); *17* Inner Kyzyl-Kum (group of Ljavljakan, Beshbulak); *18* Chorasmia (Tolstov, Dzhanbas-Kala, Dzhingel'dy, Kavat); *19* Uzboj. (**A**) Neolithic and Chalcolithic Southern Turkmenistan sedentary farmers. (**B**) Neolithic "culture of Kel'teminar". (**C**) Neolithic "culture of Atbasar". (**D**): Neolithic "culture of Hissar"

In Central Asia, the main data for the period considered here derives from archaeological research undertaken during the Soviet period. Since then, the excavation of several Stone Age sites in Central Asia has significantly added to this field. I attempted to document, with as much detail as possible, the archaeological evidence for the technical processes of stone artifact production. This new approach consists of a technological study of the major lithic assemblages recovered from Upper Paleolithic to Chalcolithic contexts across dispersed parts of Central Asia. The complete analysis of these assemblages, in particular the detailed reconstruction

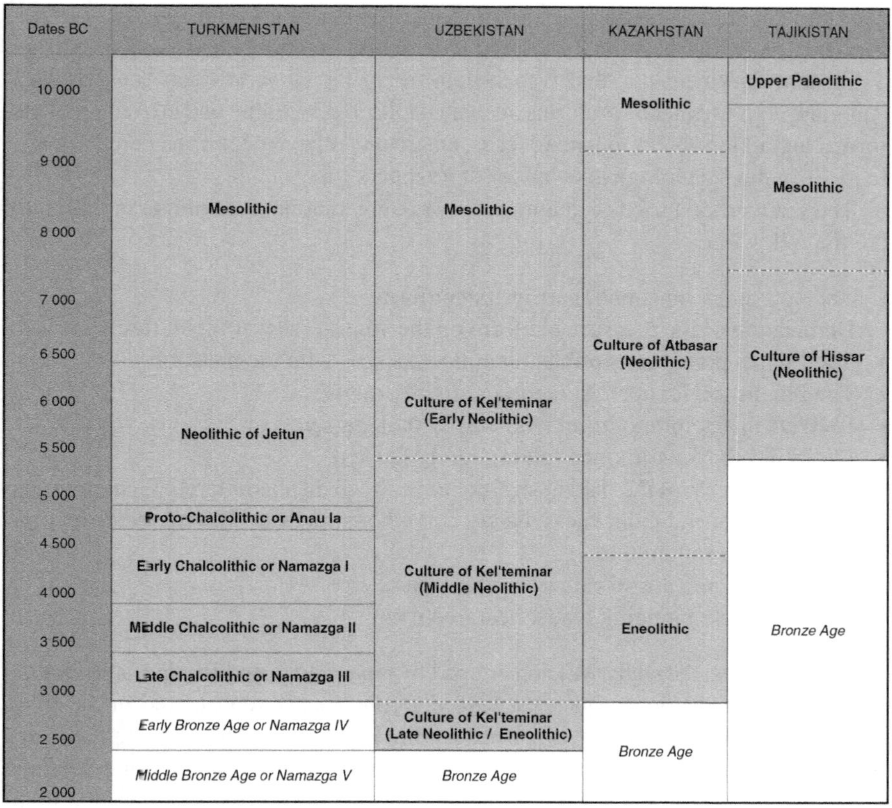

Fig. 12.2 A chronological outline of the Central Asia cultures and archaeological periods

of reduction sequences, is beyond the scope of this chapter. What I propose to examine here are those results concerned with the pressure knapping technique. In addition, I highlight avenues for future archaeological research, particularly the implications that stone knapping activities and tool use may have for the wider social, cultural, and economic dimensions. Finally, regional intersite comparisons and overviews are used to illustrate both synchronic and diachronic technical variability and thus provide an insight into the appearance and the use of the pressure blade technology in this part of Eurasia.

12.2 Identification of the Pressure Knapping Technique in Central Asia

The first problem that arose was the recognition of the pressure knapping technique in lithic assemblages. Some researchers have suggested that, given the extreme regularity of some blades and bladelets, this technique was probably used during the

Mesolithic and the Neolithic in Central Asia. In order to test this hypothesis and to further explore this question, I analyzed several lithic assemblages from Kazakhstan, Tajikistan, Turkmenistan, and Uzbekistan. All of the stone artifacts (i.e., nodules, cores and core fragments, preforms, technical falls, flakes, entire and broken products, formal and nonformal retouched pieces, and chunks) have been considered in order to identify the technological behavior of knappers.

The main criteria used to diagnose the pressure knapping technique in this study are the following:

- The butt: thick but small, narrow, punctiform
- Delineation of the pressure platform on the ventral face: straight, rarely curved
- The impact point: perceptible but almost merges with the platform
- The flaking angle: open, sometimes slightly obtuse
- Bulb attributes: pronounced with a lip, small, presence of fissuring, squat
- Platform preparation: none (plain), rarely faceted
- Preparation toward the flaking surface: none, frequent slight (rarely high) abrasion
- Blade profile: equal thickness, barely curved or straight with a curved distal part
- Blade cross section: thin
- Blade edges and dorsal ridges: straight, parallel
- Specific mark: plunging blade, heat treatment

In some cases, the study was constrained by taphonomic and stratigraphic factors, such that full reduction sequences were not consistently represented in the lithic assemblages. When dealing with biased assemblages or open air, nonstratified sites with only surface finds, refitting became nonproductive, and the pressure debitage processes could not be specified in detail. In this instance, the goal became more elusive when the cores were exhausted and their final stage obscured the previous reduction sequences. Furthermore, it was sometimes difficult to discriminate the pressure technique from very well-controlled indirect percussion during the early stages of blade removal, especially if the knapper used indirect percussion for shaping the core. Lastly, some aspects of the knapping methods (nature and morphology of tools, gesture, mode of holding or gripping the core, and posture of the knapper) will be clarified through future replication and experimentation analysis.

Three categories of intended products resulting from different uses of the pressure knapping technique have emerged from this study. As experiments have shown (Pelegrin 1988; Texier 1982, 1984), the most significant metric element is the length rather than the width of products, as the width categories generally overlap.

- Microblades (2–7 mm wide and <5 cm long): the core is held in one hand, usually using a grooved device, and the knapper, from a seated position, applies pressure with a short hand crutch
- Bladelets (6–10 mm wide and <8 cm long): the core is either held in one hand after being enclosed in a device or fixed, then set on a hard slab so as to use a crutch braced against the shoulder or abdomen
- Blades (12–20 mm wide and up to 18 cm long): the knapper, in a standing position, uses a pectoral crutch

The pressure knapping technique requires access to high-quality, internally homogenous raw materials which were (and are) available and widely distributed throughout the territory of Central Asia, i.e., various flints, jasper, and chalcedony. There was an obvious relationship between raw material types and production systems during Mesolithic and Neolithic since these materials have always been chosen for these types of core reduction strategies. Other raw materials were used on a more occasional basis. Archaeological sites are located rather close to the sources of high-quality raw material. When they are not, local and bad-quality raw material (diverse types of limestone, sandstone, volcanic pebbles) was usually used for the production of flakes. In the mountainous parts of Central Asia, volcanic pebbles were sometimes transformed into flake tools but were more commonly exploited for pebble tools. In this case, the exotic high-quality raw material was procured either by direct exploitation of a remote source or through either import or exchange. The nearest obsidian outcrops are located in the Caucasus, and, as far as we know, they have hardly been used in Central Asia: only one or two pieces that appear to have been imported as "finished products" have been found at the site of low Uzboi (Turkmenistan) during Neolithic (Tolstov 1958). Another source of obsidian in the region of Ghazni, near Dasht-i-Nawur (Afghanistan), may have been used at the nearby Epipaleolithic sites of G.P.2 and G.P.4 (Davis and Dupree 1977) where the pressure knapping technique was also identified.

12.3 Introduction of Bladelet and Microblade Assemblages: The Dawn of the Mesolithic

The Upper Paleolithic stone assemblages consist mainly of two components: a blade technology, which had just appeared, and an older but still prominent flake production strategy. Most of these blades are irregular and rather short because the direct percussion blade technique was the only method used at this time. The first clear evidence for pressure microblade and bladelet technologies takes place at the end of the Pleistocene and during the Early Holocene. These technologies appeared at different times in various regions in Central Asia and differed in several crucial ways from Upper Paleolithic period. We can summarize these differences as follows:

- Use of more homogeneous, elastic, fine-grained raw materials.
- Introduction and development of microblade and bladelet production which then became prevalent.
- Use of new techniques such as pressure and indirect percussion.
- Appearance of regular and even standardized blanks for manufacturing various tools, particularly those displaying projectile and hafting elements. During the Upper Paleolithic, any regularity in form apparent in these tool types, when it existed, was the result of substantial retouch. Also, there is no longer systematic reduction in blank size or shape through retouch; energy was invested in the blank production rather than retouch manufacturing.

- Presence of new or more refined tools fashioned on these new types of blanks; among them, I note particularly the presence of large geometric microliths (triangles, trapezes, and, more rarely, crescents). The last tool type has been suggested for use in hunting spears or as a weapon; however, this remains to be proven: they may just have easily been served as cutting implements.

So there are important changes apparent in the methods, techniques, skills, and conceptual knowledge, as well as the finality of the lithic reduction processes. It is therefore necessary to define these industries as belonging to a new entity called the Mesolithic. This is contrasted with the Epipaleolithic, which is defined as the continuation of the earlier Upper Paleolithic tradition into the Early Holocene; this period is illustrated by the discoveries from Oshkhona, Tajikistan (Ranov 1975). At this site, the stone artifacts show some convincing similarities with the lithic technical system of the Upper Paleolithic layer two from Shugnou (Ranov 1973, 1988; Ranov et al. 1976) in the same region. This point has been discussed in detail elsewhere (Brunet 2002; Ranov 1988, 1996).

12.3.1 Tajikistan, Eastern and Central Kazakhstan, and Xinjiang

In Southern Tajikistan, the end of the 9th and the beginning of the 8th millennia B.C. (9530 ± 130 B.P.; Ranov 1984) are characterized by two distinct situations. The first one is represented by the settlement of Oshkhona (Figs. 12.1, 12.3). Here, the microblade component of the assemblage, which was obtained by the pressure technique, has integrated with a lithic industry combining two types of flaking techniques using hard-hammer percussion for the removal of ordinary flakes and a soft-hammer organic percussor for bladelet making. The second approach is observed through the existence of several workshops (Beshkent, Makoni-mor, Javan, Mullo-Nijaz; Amosova et al. 1991; Jusupov and Solov'ev 1973; Ranov 1992) where only regular microblades, obtained by the pressure technique, were produced by the reduction of the local high-quality raw material at the source (Figs. 12.1, 12.3). Detailed debitage analysis has testified to a preference for a specific stoneworking process to reduce cores and generate tool blanks. The microblades were removed only from the frontal and narrow surface of single platform cores after detachment of a frontal crested blade that was generally associated with a dorsal crest. Sometimes, the process of microblade removals has progressed laterally onto one of the sides of the core. The same operation was usually carried out at the other end of the pressure platform, resulting in a core with a cylindroconical shape at the end of its use life. It is striking that this kind of debitage (methods, technique) is clearly similar in the two categories of sites described here, but the relation between them remains unclear. The most obvious (but admittedly simplistic) possibility is that there was a partial "supply" of cores and blanks by the workshops toward the settlements. Alternatively, another possible explanation is that knappers themselves were mobile.

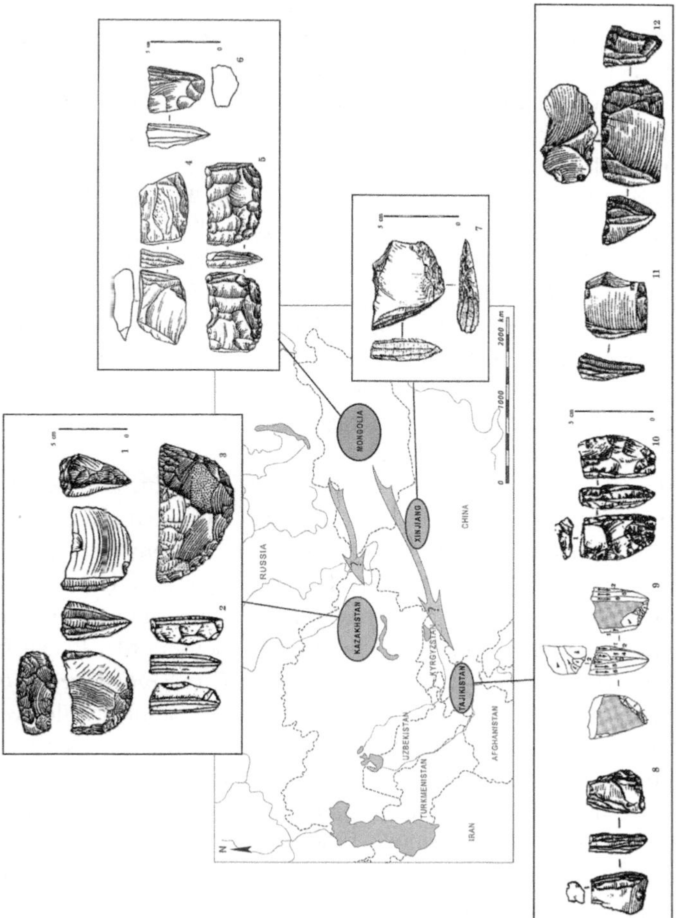

Fig. 12.3 The earliest occurrences of the use of pressure knapping technique in Central Asia employing the "Yubetsu" method (map drawn by the author). *1* Semizbugu (After Medoev 1970); *2* Akimbek (After Chindin 1989); *3* Northern Pribalkhash (After Medoev 1970); *4* Ulan-Bator (After Okladnikov and Abramova 1994); *5* Dno-Gobi (After ibid.); *6* Mojl'tyn-Am (After Okladnikov 1981); *7* Qijiaojing (After Debaine-Francfort 1988); *8–9* Beshkent (After Amosova et al. 1991; Brunet 2003); *10* Makoni-Mor (After Jusupov and Solov'ev 1973); *11–12* Mullo-Nijaz (After Ranov 1985)

These lithic assemblages exhibit the evidence of rules for the core reduction strategy, which, while not extremely complex, are well defined. This set pattern is characterized by the (almost) bifacial preparation of the core, its exploitation along the frontal surface following the removal of one or two (opposing) crests, the (generally) unidirectional and continuous (or not) detachment of microblades, and the use of the pressure technique with a short crutch. However, in these Tajik sites, this seems to be an additional method, although one that was clearly used and assimilated to the earlier Upper Paleolithic tradition. It appears that the knappers developed their own variants through the adoption of single aspects of this technique.

A more typical application of this method is seen in another part of Central Asia, suggesting that there was a strong knapping tradition linked to this particular use of the pressure technique. Indeed, the study has led me to consider the possibility that this microblade technology had also penetrated Kazakhstan, particularly in the eastern and central territories (Figs. 12.1, 12.3): at Semizbugu, Shul'binka, and the group of Northern Pribalkhash in the east (Artjukhova et al. 2001; Derevjanko et al. 1993; Medoev 1970, 1982; Petrin and Tajmagambetov 2000) and at Akimbek in the central region (Chindin 1989). Moreover, this method can be identified not far from this part of Kazakhstan, in Xinjiang (e.g., the *naviform cores* at the sites of Qijiaojing and Chaiwopu; Debaine-Francfort 1988) (Fig. 12.3). New investigations should be focused upon the few river valleys in these two territories and along the Ili valley in particular, which remains the main and perhaps the only way to connect these regions in Kazakhstan and in Western China.

This "technological style," as well as the absence of microlithic tools, especially those of geometric form, is significant (Inizan 1991) and reminds me of a very well-known microblade production technique identified not far from there in Siberia (Eastern and Russian Far East) and Mongolia. It could be generically referred to as *Yubetsu*. There are many variants of this type of reduction system, but it remains necessary to better define these variants according to region in order to clarify the core classification (*wedge-shaped core*, *klinovidnyj nukleus*, *Gobi core*, *tortsovyj nukleus*, *end cores*, etc.). This last point deserves a specific treatment which will be addressed in a future article. Research appears to have confirmed that the pressure knapping technique associated with this specific method originated at the crossroads of Northern China, Mongolia, and Eastern Siberia, where it appeared about 28000 B.P. or even 35000 B.P. (Abramova 1986; Inizan 1991; Inizan et al. 1992; Goebel 2002)[1] and then spread both eastward (Japan, Korea, Northern America) and westward from approximately 20000 B.P. This hypothesis can probably also be applied to the sites of Eastern and Central Kazakhstan (through the Xinjiang territory?), and even reached Tajikistan on the path from the Far East (Fig. 12.3). The last scenario echoes Vadim Ranov's suggestion that Tajikistan belonged to a "Sibiro-Mongolian"-influenced group (Litvinskij and Ranov 1998).

---

[1] For more details about the radiocarbon dating of Siberian Paleolithic and Mesolithic sites, see Lisitsyn and Svezhentsev (1997); Kuzmin and Orlova (1998).

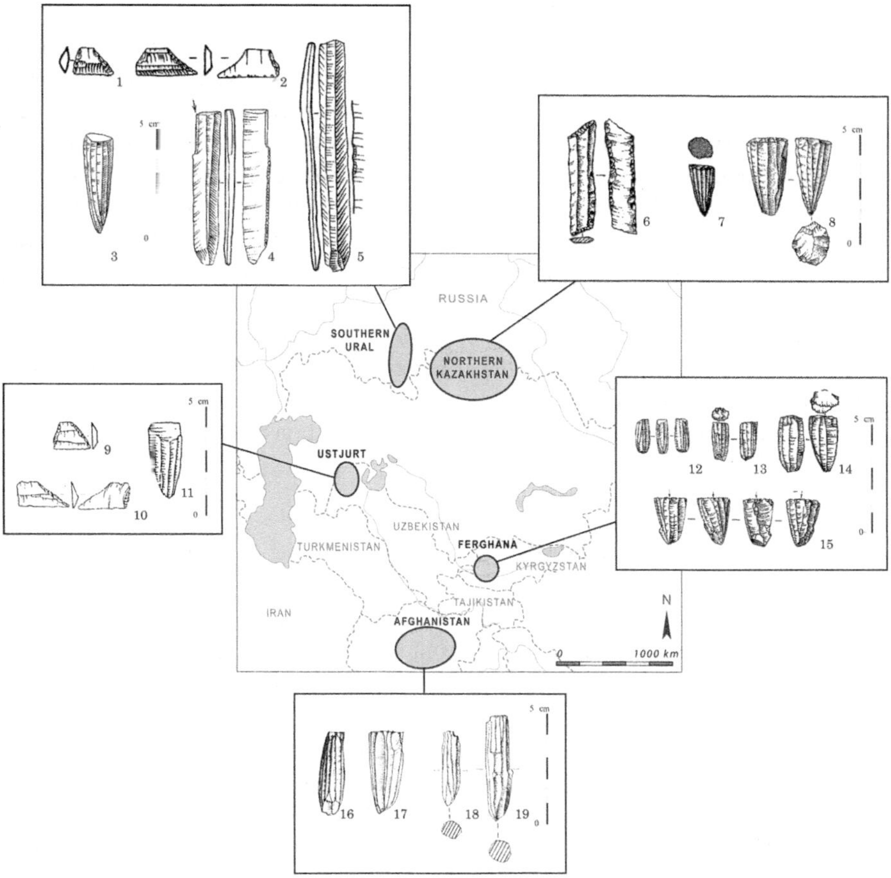

Fig. 12.4 The earliest occurrences of the use of pressure knapping technique in Central Asia employing the "bullet-shaped core" method (*map drawn by the author*). *1–2, 4–5* Jangel'ka (After Matjushin 1989); *3* Dolgyj El'nik II (After ibid.); *6* Tel'mana IXa (After Zajbert and Potemkina 1981); *7* Verkhnaja Alabuga (After ibid.); *8* Tel'mana VII (After ibid.); *9–10* Ajdabol 25 (After Avizova 1990); *11* Ajdabol 6 (After Bizhanov 1982); *12* Jangikadam 26 (After Islamov and Timofeev 1986); *13* Tajpak 15 (After ibid.); *14* Achchik-Kul' 3 (After ibid.); *15* Obishir I (After Islamov 1980); *16–17* Aq-Kupruk (After Dupree and Davis 1972); *18–19* Darra-i-Kalon (After Mussi 1979)

12.3.2 Eastern Uzbekistan (Ferghana) and Southern Kazakhstan

Evidence for use of the pressure technique in microblade production was also found in Uzbekistan, particularly in the region of Ferghana (Islamov and Timofeev 1986) (Figs. 12.1, 12.4). Here, the core reduction method consisted principally in the removal of products from its wide main surface following a unidirectional method. I noted the presence of gradually invasive flaking on the lateral sides of the single platform core; also this flaking usually has covered the back of the core. In the context of the pressure knapping technique, this method, tightly linked to the *bullet-shaped* cores, can be characterized here as "classical" as compared to the *Yubetsu* method. At the end of the reduction process, the core remnant was discarded when it was either too small or offered only step-fractured working surfaces. Some of the open air sites in Ferghana were probably ambush or kill sites and as such yielded mainly finished projectile points and the last stages of the flintworking process; the by-products of manufacturing are absent. The selected microblades were modified into tools, mainly microliths sometimes showing a geometric form (triangles and crescents). Close to this area, new surveys in Southern Kazakhstan have revealed the existence of a similar pattern of microblade production in the site of Bugun' (Aleksandr N. Podushkin, 2002, personal communication), suggesting a probably extensive development of this specific use of the pressure knapping technique in this territory. Sometimes, evidence of microblade production was found in other sites such as at the possible residential site of Obishir located in Ferghana (Islamov 1980), where bladelets (approximately 7 mm wide and 10–12 mm long) were produced by indirect percussion and two additional items (irregular blades and flakes) were made by a direct, hard-hammer percussion technique.

12.3.3 Northern Kazakhstan

Another version of the "classical" method was identified at several groups of sites in Northern Kazakhstan: at Ubagan, Vinogradov, and Tel'man (Zajbert and Potemkina 1981) (Figs. 12.1, 12.4). Indeed, a particularly intense focus upon the production of narrow, elongated, and straight products can be noted. In most cases, these blanks have been used without having been retouched; only some of them were modified by fine retouch to create geometric forms such as *trapeze-rectangles* and parallelograms. The two types of blanks observed, microblades and bladelets, were produced by separate methods: the use of a short hand crutch with or without a holding device or by means of a shoulder crutch. These *bullet-shaped* cores (*conical* and *pencil* cores in the scientific literature) were wholly exploited following a continuous detachment of bladelets and microblades from around the entire perimeter of the core. The relationship between this tradition and the bladelet assemblages of Southern Siberia and the Ural regions (e.g., at Jangel'ka and Dolgyj El'nik; Matjushin 1989), where the early use of the pressure knapping technique continued during the Neolithic period (e.g., at Chebarkul; Krizhevskaja 1968) and even up to the third millennium, appears significant. However, the Mesolithic Southern Ural lithic industries (Fig. 12.4) reveal a high level of flintknapping skills and production processes so complex that they imply to the presence of near knapping specialists.

The situation is quite different in Northern Kazakhstan. If a technical transfer between this last territory and the Ural region ever happened, it must have been confined to the methods linked to the production of microblades and bladelets. Indeed, at the Ural sites, many regular blades (12–14 mm wide and 17 cm long) emphasize a peculiar technological feature specific to this group: a pressure knapping technique that appears to have involved the use of a pectoral crutch. This method has not yet been observed in these Northern Kazakhstan sites. In fact, this last technique implies a complex transition from the use of a shorthand device to a pectoral crutch, which would require a long-term learning process.

12.3.4 Western Uzbekistan (Ustjurt)

The Ustjurt plateau in Uzbekistan (Avizova 1990; Bizhanov 1982, 1996) is the last place where this early introduction to pressure knapping occurred in Central Asia. Some of the many sites here (e.g., the Ajdabol group of sites) (Figs. 12.1, 12.4) have been identified as hunting stations, others as settlements. The lithic assemblages suggest some connections with the Mesolithic Southern Ural cultures (Matjushin 1973, 1976, 1989; Serikov 1998). Indeed, similar elements are particularly well represented at the Ustjurt sites by the technological tradition of knapping (bladelets and microblades obtained by a "classical" method using the pressure technique) and also by the manufacture of geometrical microliths, especially the trapezes easily recognizable by their specific shape (elongated, right-angled, oblique truncation) and retouch (direct and inverse truncation).

12.3.5 A Brief Synthesis

Before moving on to the Neolithic, it is worthwhile to summarize and briefly discuss some of these results about knapping strategies and techniques employed in Central Asia during the Early Holocene. Firstly, the emergence of the use of the pressure knapping technique in this part of Asia was associated with the appearance of microblade technology and, to some extent, bladelet productions. The pressure technique appeared in hunter-gatherer groups that contrast sharply with the previous Paleolithic stone reduction traditions; henceforth, these are referred to as Mesolithic. Two concepts have been identified: the first one, called here the *Yubetsu* method, is closely related to the technical tradition from the Far East (Sibero-Sino-Mongolia area), and the second one, linked to a *bullet-shaped* core and the more "classical" method, is most often associated with geometrical microliths. The absence of the pressure knapping technique in other new cultures during this period in Central Asia invites to question the reasons and the modalities (diffusion or invention?) of its appearance in the sites described here.

The *Yubetsu* method (Fig. 12.3) appears to have been adopted in Tajikistan by "Paleolithic groups," as seen at Oshkhona. The introduction of the pressure technique into local traditions may have stemmed from contact with experienced flintknappers from the nearby workshop sites, where the microblade reduction technology has been observed. In Eastern and Central Kazakhstan, this study confirms the penetration of the pressure knapping technique into this area from the Far East. But a question remains: did this westward transmission occur through displacements of human groups, or by adoption, or as a consequence of diffusion? The pressure flaking equipment linked to this specific microblade technology, which is characterized by the handheld technical mode with a short crutch, supports the suggestion of mobile knappers. Alternatively, this *Yubetsu* method, which shows a single and relatively simple conceptual scheme, could have been reproduced after some contact, even indirect, with experienced knappers (Inizan 1991).

With the respect of the *bullet-shaped* core method (Fig. 12.4) from the Northern Kazakhstan and the Ustjurt plateau (Uzbekistan), the hypothesis of a local invention is conceivable. However, the close relationship of these sites with those of Ural cultures, where the pressure knapping technique was very refined and included several variations, suggests that this technique developed first in the Ural region. If we accept the idea of diffusion toward Northern Kazakhstan and fully acknowledge the complexity of this learning process, especially for the shaping out of the core, then three modes of transmission can be envisaged: (a) the diffusion of preformed cores, implying a contact between the expert and apprentice knappers; (b) the transmission of this specific knowledge by itinerant experienced knappers; or (c) the technique was carried out in situ by foreign experienced knappers. Thus, this northern region of Central Asia might be another core area of the pressure knapping technique, distinct from the one known in Southern Asia (Iran–Afghanistan). More precise radiocarbon dates are necessary to consolidate this assumption. Likewise, concerning several Epipaleolithic/Mesolithic sites in Afghanistan (e.g., Aq Kupruk, Darra-i-Kalon III-Ia, Kara-Kamar I, and G.P.2 and G.P.4) (Davis and Dupree 1977; Dupree and Davis 1972; Mussi 1979), the precise origin (either local or in connection with Iran?) and the mode of transmission for the pressure knapping technique using the *bullet-shaped* core method remain unclear.

12.4 Introduction of the Blade Technology: The Emergence of Neolithic Societies

With the appearance of agropastoral Neolithic societies in Central Asia, particularly in Southern Turkmenistan, new industries arose that exhibited a main intended blank: the blade. The principal features of these societies consisted of settled village life, an agropastoral economy (farming and animal husbandry), and a rich material culture (lithic and bone industries, painted handmade pottery with geometric and zoomorphic patterns, terracotta figurines, beads). These well-known sites, especially those belonging to the *Jeitun culture* (7th–6th millennia B.C.) (Harris et al. 1993, 1996; Masson 1960, 1971), reveal the use of the pressure knapping technique for the production of regular blades employing the *bullet-shaped* core method. Within the lithic toolkit, the hafted elements include, among others, tools with backed edges, many truncations, and various trapezes.

A more interesting and specific case in Central Asia is found among societies involved in the process of neolithization.

12.4.1 The Kel'teminar Culture (Uzbekistan)

This culture, located in Uzbekistan and especially in the Kyzyl Kum desert (Fig. 12.5), illustrates the beginning of the settlement process in a landscape then characterized by

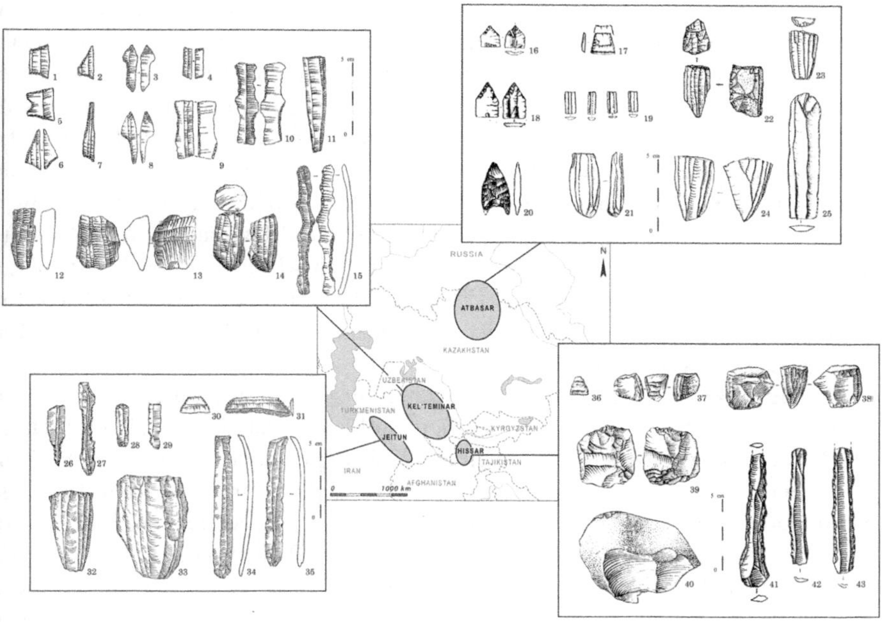

Fig. 12.5 Some significant features of the lithic assemblages from the main Neolithic cultures of Central Asia (*map drawn by the author*). *1–15* the "culture of Kel'teminar"; *1, 2* trapezes (Uchashchi 131, after Vinogradov 1981b); *3* Kel'teminar arrowhead (Tolstov, after Vinogradov 1981a); *4* parallelogram (Tolstov, after ibid.); *5* horned trapeze (Khodzhagumbas 5, after ibid.); *6* triangle (Dzhanbas-Kala 4, after Vinogradov 1981b); *7* triangle (Uchashchi 131, after ibid.); *8* Kel'teminar arrowhead (Ljavljakan 26, after Vinogradov and Mamedov 1975); *9* denticulate on blade (Khodzhagumbas 5, after Vinogradov 1981a); *10* denticulate on blade (Uchashchi 131, after ibid.); *11* blade (Ljavljakan 120, after Vinogradov and Mamedov 1975); *12, 13* cores (Uchashchi 131, after Vinogradov 1981a); *14* core (Ljavljakan 120, after Vinogradov and Mamedov 1975); *15* denticulate on blade (Tolstov, after Vinogradov 1981a); *16–25* the "culture of Atbasar"; *16, 18* point (Tel'mana I, after Zajbert 1992); *17* trapeze (Tel'mana I, after ibid.); *19* microblades (Vinogradovka II, after ibid.); *20* bifacial point (Tel'mana XII, after ibid.); *21* core (Vinogradovka X, after ibid.); *22* core (Vinogradovka XIV, after Kislenko and Pleshakov 1998); *23* core (Tel'mana, after Zajbert 1992); *24* core (Vinogradovka X, after ibid.); *25* blade (Tel'mana I, after ibid.); *26–35* the "culture of Jeitun" (site of Jeitun, after Masson 1971); *26–27, 29* denticulate on bladelet; *28* end scraper on bladelet; *30–31* trapezes; *32–33* cores; *34–35* blades. *36–43* the "culture of Hissar"; *36* trapeze (Tutkaul 2–1, after Ranov and Korobkova 1971); *37–38* cores (Tutkaul 2–1, after ibid.); *39* core (Tutkaul 2–1, after ibid.); *40* chopper (Tutkaul 2–1,after ibid.); *41–43* backed edges on blades (Tutkaul/Saj-Sajëd, after Ranov 1982)

Tugais forests[2] and typical steppe close to river deltas and lakes; its technical tradition came mainly from the local Mesolithic background (Brunet 2006; Dzhurakulov and Kholmatov 1991; Guljamov et al. 1966; Szymczak and Khudzhanazarov 2006a, b; Tolstov 1958, 1959; Vinogradov 1960, 1963, 1968, 1981a; Vinogradov and Mamedov 1975).

2 *Tugai* are the riparian forests along the rivers in the desert regions of Central Asia. They mainly consist of typical floodplain vegetation: trees (poplar, tamarisk, maple, ash, and elm), shrubs, reed, and grass communities.

The subsistence strategies were marked by a focus not only on hunting and gathering but also with the appearance of domestic cattle. The period, which spanned the 7th millennium B.C. to the early 4th millennium B.C., can be divided in three broad chronological phases that led to the blossoming of the *Kel'teminar* tradition's main features. The sites of Uchashchi 131 and Ajakagytma (Vinogradov 1981a) provide a good illustration of the early stage of this culture (end of the 7th–6th millennia B.C.) in the region of Zeravshan (Fig. 12.1). New excavations were carried out at the site of Ajakagytma initially by a Polish-Uzbek archaeological expedition (Szymczak and Khudzhanazarov 2006a) and now by a joint French-Uzbek team (MAFANAC)[3] in which Prof. Dr. K. Szymczak (Institute of Archaeology of Warsaw University, Poland) is an important collaborator. During the second phase, distinctive regional variants, principally in Zeravshan and in Choresmia, have been observed.

During the early stage of the *Kel'teminar culture*, the lithic industry has evidence of several production systems (Fig. 12.5): microblades, bladelets, and blades. These blanks were often broken by the microburin blow technique in order to transform them into specific geometrical morphotypes such as triangles and trapezes. Microblades and bladelets were essential elements for composite tools; blades were the main blanks for side scrapers and denticulates. Reduction sequence analysis indicates that there were at least two techniques employed for obtaining these blanks. The most common is a very well-controlled indirect percussion. The majority of the crested blades, second-generation bladelets, and striking platform rejuvenation flakes were produced using indirect percussion. The second technique is represented by the *bullet-shaped* core method using a pressure technique that implies a complex reduction system. It took place mainly in the production of narrow blanks (bladelets and microblades). However, there is also little evidence for the production of regular blades through the use of the pressure knapping technique. Therefore, it seems that in comparison with the Mesolithic period, this last technique was no longer restricted in the *Kel'teminar culture* to the production of very small supports, suggesting that new skills had been acquired. This change might have been introduced by Mesolithic Ural groups that were in contact with this culture, as some other lithic elements (e.g., geometrical microliths) indicate.

During the second stage (5th–4th millennia B.C.), the lithic technological pattern remained quite the same except for one aspect. This temporal stage saw the significant development of blade production (Fig. 12.5), which requires the involvement of elaborated skills and, to some extent, bladelet production as well. Both of these blank types were selected for various tools (geometrical microliths, notches, denticulates, burins on truncations, scrapers, and end scrapers). Among these, the *Kel'teminar arrowhead* and the *horned trapeze*, which have wide distribution in many parts of Central Asia (Kazakhstan, Russia, Uzbekistan, Turkmenistan,

[3] This archaeological expedition, codirected by Dr. M. Khudzhanazarov (Institute of Archaeology of the Academy of Sciences of the Republic of Uzbekistan, Samarkand) and the author, is financially supported by the French Foreign Office; we owe our deep thanks to these two institutions. We express also our greatest gratitude to Prof. Dr. K. Szymczak for his financial and scientific support.

Northern Afghanistan), could have been seen as having not just functional but also symbolic value, as a way to define social identity. It is interesting to note that from this period onward, the place of the pressure knapping technique becomes more prominent in the blade production. On that subject, one hypothesis I am exploring is the suggestion of the new relationship that developed between the *Kel'teminar culture* (especially the regional variant located in Chorasmia) and the agropastoral community of Southern Turkmenistan (Fig. 12.1), which is indicated in other spheres of the material culture, particularly the decoration of the handmade pottery (Vinogradov 1957; Itina 1959; Brunet 2007). The *Kel'teminar culture* appears as a key culture located at the crossroads of the north and the south of Central Asia, in connection with several groups from steppe and oasis areas.

12.4.2 The Atbasar Culture (Kazakhstan)

In the steppe-forest zone of Northern Kazakhstan, the Neolithic *Atbasar culture* (5th–4th millennia B.C.) is represented through an abundance of functionally variable sites (Figs. 12.1, 12.5), including both settlements and workshops (Kislenko and Pleshakov 1998; Pleshakov 1993; Zajbert 1992). It developed from the local Mesolithic, retaining microblade production using the pressure knapping technique (*bullet-shaped* cores) and the production of very regular narrow blanks that remained almost unmodified by retouch. However, the introduction of few regular blades (detached by indirect percussion or pressure knapping technique?) and new formal tools can be observed – points, trapezes, triangular arrowheads with a basal notch, bifacial pieces, and leaf-shaped bifacial points (Fig. 12.5). It seems possible that these tools were operated in quite different functional, even socioeconomic, contexts. Innovations appear simultaneously in other spheres, in particular the appearance of handmade pottery with incised or combed decoration, and the domestic horse at the very end of this period (Zajbert 1993; Meshcherjakov and Morgunova 1996; Kalieva 1998; Levine 1999; Anthony and Brown 2000). These new implements, which enrich the Mesolithic cultural background, are quite similar to those found in the neighboring regions of Eastern Kazakhstan, Altai (Kirjushin and Kljukin 1985; Molodin 1977) and Eastern Siberia (Khlobystyn 1996), where most related sites have been dated to the 5th–4th millennia B.C. Due to the lack of precise radiocarbon dates, it is currently not possible to identify the exact origin of these new features and, consequently, to explain the reasons and the source of their appearance in Northern Kazakhstan (invention? acculturation? diffusion? borrowing?).

12.4.3 The Hissar Culture (Tajikistan)

In the foothills of Tajikistan, the *Hissar culture* (7th–4th millennia B.C.) (Litvinskij and Ranov 1998; Ranov 1982, 1984, 1985; Ranov and Korobkova 1971) reveals a similar situation to *Atbasar culture* (Figs. 12.1, 12.5). Has the food producing economy been developed from the local Mesolithic just as the material culture did? The faunal remains from the sites show the exploitation of both domestic (sheep/goat) and wild animals, with a higher proportion of the latter, suggesting a short-distance form of mobile pastoralism. The lithic assemblage, especially at the site of Tutkaul, shows the continuation of the earlier Mesolithic tradition (pressure microblade technology according to the *Yubetsu* method) together with the introduction of new Neolithic components such as a blade production using the indirect percussion, flake production by direct percussion (hard hammer), and the presence of trapezes and of polished axes (Fig. 12.5). These new elements seem to be linked with intrasite evidence of domestic activities inside as revealed by the use-wear studies. These include tasks related mainly to leather, skin, and woodworking (Ranov and Korobkova 1971). However, the presence of some very regular blades raises the question of the use of the pressure knapping technique and, consequently, of the origin of this new technological skill.

12.4.4 Toward the Bronze Age

During the Chalcolithic/Eneolithic period, pressure knapping tends to disappear gradually from Central Asia. Following the emergence of the first Bronze Age communities, it is seen only in the shaping process of bifacial tools and projectile points. From then on, stone knapping leads mainly to the production of flakes using hard-hammer percussion. It would be too simplistic and certainly premature to attribute this considerable transition, characterized by the abandonment of the pressure knapping technique and blade technology, to the arrival of new population. If we cannot deny the existence of new cultural features tightly linked to Bronze Age communities and thus perhaps of new groups, in particular in the steppe zone of Central Asia, it seems to us that a more complex situation, changing according to regions, took place during the Eneolithic period. Additional characterizations are necessary in order to better appreciate this transitional period, which suggests the beginning of not only economic (marked especially by the introduction of new tools and the development of domestication processes) but also symbolic and cultural transformations.

12.5 Conclusion

It must be recognized that further research needs to be conducted in order to confirm some of the interpretations presented here and to better understand the factors surrounding the development of the pressure knapping technique in Central Asia. Nevertheless, this investigation has yielded three significant results.

Firstly, the pressure knapping technique appeared in several areas of Central Asia at the very beginning of the Holocene. It was used by mobile foragers to produce microblades and bladelets that generally served as for projectile elements (weapon components?) or hafted tools. Except in the case of sites of Tajikistan and Kazakhstan (eastern and southern), some of these tools show a geometric form. Indeed, in these regions of Tajikistan and Kazakhstan, the stone knapping system echoes the Final Paleolithic tradition of the Far East (Southern Siberia, Russian Far East, Mongolia, and Northern China), which is characterized by the use of the hand crutch. The documented penetration of the pressure technique linked to this method into eastern and southern Central Asia suggests that this innovation was adopted either through cultural contact with the Far East or by means of migration of the bearers of this technique across Siberia, Mongolia, or Xinjiang.

The second method, which is linked with the pressure knapping technique in Central Asia, involves a more "classical" style of knapping technology leading to the development of *bullet-shaped* cores with their corresponding narrow bladelets and microblades. Depending upon the culture, these products were detached using either a short crutch or a shoulder crutch. This observation raises one important issue: was the appearance of this method a result of a transmission of knowledge and technologies, or a local invention? Actually, the existence of several core areas, notably in the northern (the Kazakh–Ural area) and southern (the Irano–Afghan area) parts of Central Asia, has to be considered as a distinct possibility. Moreover, the data for the northern region suggests the particular influence or the Ural Mesolithic lithic tradition on other groups in Kazakhstan and Uzbekistan.

Secondly, and with reference to the early sedentary farmers of Southern Turkmenistan (for instance, the *Jeitun culture*), the pressure knapping technique changes radically with the appearance of these first Neolithic societies: it is now exclusively used for the production of regular blades.

Lastly, microblade production using the pressure technique is still found in cultures in the process of neolithization in Uzbekistan (*Kel'teminar*), Tajikistan (*Hissar*), and Kazakhstan (*Atbasar*), a fact that highlights the persistence of the local Mesolithic tradition in these cultures. However, it seems that blade production became widespread in parallel with the appearance of Neolithic features although these new blanks remain rare in comparison with the other types (e.g., the *Hissar culture*). The picture given by the *Kel'teminar culture* is quite different, with the existence of blade production, probably based upon detachment using the pressure technique, dating back to the early phase (6th millennium B.C.). This production mode becomes significant in the following phase, a factor that may reflect the development of a relationship between the two major cultural areas: *Kel'teminar* in Uzbekistan and the Eneolithic agropastoral societies in the Southern Turkmenistan. Likewise, I can infer that the early connection with the Mesolithic Southern Ural cultures, where a blade production using the pressure technique is observed, acquainted the *Kel'teminar* groups with this method which they later adopted. Indeed, it could be argued that these different cultures were in mutual contact through exchange systems.

Therefore, the situation of Central Asia's territory is partly similar to that known in other Eurasian areas since the shift toward the blade technology occurs at the time of the emergence of the Neolithic with major changes in diet as well as in living conditions. However, in Central Asia, the situation remains complex, and we must be careful about taking this scenario too far, given the preliminary nature of the technological investigations on this territory. Moreover, the chronological (i.e., reliable radiocarbon dates), economical (i.e., information on subsistence behavior), and paleoenvironmental database for this period is not sufficiently representative and abundant to offer a secure comparative framework for the study of the use of pressure knapping technique inside various cultural contexts and its evolution through time. This paucity of data complicates interpretations of the archaeological materials. Lastly, many issues remain to be addressed: not least of all are those concerning the reasons for the use or invention of the pressure knapping technique in Central Asia. Indeed, several possible explanations can be proposed – reducing the rates of errors in the manufacturing process (Elston and Brantingham 2002), the development of production efficiency, conservation of high-quality raw material, standardization of production or of the toolkits, or the result of considerable interaction that affected cultural areas during the process of neolithization.

Acknowledgments The results discussed in this article would not have been possible without the support of many colleagues from various research institutes: the Institute of Archaeology of the Academy of Sciences of Uzbekistan (Samarkand), the Al-Farabi Kazakh National University (Almaty), the Margulan Institute of Archaeology of the Academy of Sciences of Kazakhstan (Almaty), the Central State Museum of Kazakhstan (Almaty), the Turco-Kazakh Ahmed Yesevi University (Turkestan), the History Museum (Shymkent), the Institute of Archaeology, History and Ethnology of the Academy of Sciences of Tajikistan (Dushanbe), the Institute for the History of Material Culture of the Russian Academy of Sciences (St. Petersburg), and the Institute of Archaeology of the Russian Academy of Sciences (Moscow). I would like to express here my deep gratitude to all of them for providing me with full access to their scientific material. My grateful thanks are due to the reviewers for helping to improve noticeably the English language quality of the manuscript.

References

Abramova, Z. A. 1986 Klinovidnye nukleusy v paleolite Severnoj Azii. In *Paleolit i Neolit*, edited by V. P. Ljubin, pp. 11–16. Nauka, Leningrad

Amosova, A. G., S. P. Lomov, and S. A. Nesmejanov 1991 *Istorija formirovanija i drevnejshego zaselenija Beshkentskoj doliny, Juzhnyj Tadzhikistan*. Donish, Dushanbe

Anthony, David W., and Dorcas R. Brown 2000 Eneolithic Horse Exploitation in the Eurasian Steppes: Diet, Ritual and Riding. *Antiquity* 74(283): 75–86.

Artjukhova, O. A., A. P. Derevjanko, V. T. Petrin and Zh. K. Tajmagambetov 2001 *Paleoliticheskie kompleksy Semizbugu, punkt 4*. Institute of Archaeology and Ethnography of the Siberian Division of the Russian Academy of Sciences, Novosibirsk.

Avizova, A. K. 1990 K voprosu o kul'turnom svoeobrazii neolita Ustjurta, po dannym tipologo-statisticheskogo analiza kamennoj industrii. In *Arkheologija Priaral'ja*, edited by V.N. Jagodin, pp. 3–28. FAN, Tachkent.

Bizhanov, E. B. 1982 Mezoliticheskie i neoliticheskie pamjatniki severo-zapadnogo Ustjurta. In *Arkheologija Priaral'ja*, Vol. 3, edited by S. K. Kamalov, pp. 14–39. FAN, Tashkent.

Bizhanov, E. B. 1996 Kamennyj vek Ustjurta. Unpublished abstract of dissertation for the degree of Doctor of Historical Sciences, Tashkent.

Brunet, F. 2002 Asie centrale: vers une redéfinition des complexes culturels de la fin du Pléistocène et des débuts de l'Holocène. *Paléorient* 28(2): 9–24.

Brunet, F. 2003 De l'Epipaléolithique au Néolithique en Asie centrale : quelle néolithisation? Synthèse critique des données et étude technologique d'industries lithiques. Unpublished Ph.D. dissertation, University of Paris I Panthéon-Sorbonne, Paris.

Brunet, F. 2006 Pour une nouvelle étude de la culture néolithique de *Kel'teminar*, Ouzbékistan. *Paléorient* 31(2): 87–106.

Brunet, F. 2007 De l'imitation à l'emprunt dans les sociétés néolithiques et chalcolithiques d'Asie centrale: Ouzbékistan–Turkménistan–Iran. In *Mobilités, immobilismes. L'emprunt et son refus*, edited by Pierre Rouillard, Catherine Perlès and Emmanuel Grimaud, pp. 253–266. De Boccard, Paris.

Chindin, A. Ju 1989 Otnositel'naja khronologija kremnevogo inventarja poselenija Akimbek. In *Voprosy arkheologii Tsentral'nogo i Severnogo Kazakhstana*, pp. 12–19. National University of Karaganda, Karaganda.

Davis, Richard S., and L. Dupree 1977 Prehistory Survey in Central Afghanistan. *Journal of Field Archaeology* 4(2): 139–148.

Debaine-Francfort, C. 1988 Archéologie du Xinjiang: des origines aux Han. *Paléorient* 14(1): 5–30.

Derevjanko, A. P., B. Zh. Aubekerov, V. T. Petrin, Zh. K. Tajmagambetov, O. A. Artjukhova, V. N. Zenin, and V. G. Petrov 1993 *Paleolit Severnogo Pribalkhash'ja (Semizbugu punkt 2, rannij-pozdnij Paleolit)*. Institute of Archaeology and Ethnography of the Siberian Division of the Russian Academy of Sciences, Institute of Archeology of the Academy of Sciences of the Republic of Kazakhstan, Novosibirsk.

Dupree, L., and R. S. Davis 1972 The Lithic and Bone Specimens from Aq-Kupruk and Darra-i-Kur. *Transactions of the American Philosophical Society, new series* 62(4): 14–32.

Dzhurakulov, M. D., and N. U. Kholmatov 1991 *Mezolit i Neolit srednego Zarafshana, Sazaganskaja kultura*. FAN, Tashkent.

Elston, R. G., and P. J. Brantingham 2002 Microlithic Technology in Northern Asia: A Risk-Minimizing Strategy of the Late Paleolithic and Early Holocene. In *Thinking Small: Global Perspectives on Microlithization*, edited by Robert G. Elston and Steven L. Kuhn, pp. 103–116. Archeological Papers of the American Anthropological Association, Arlington.

Goebel, T. 2002 The 'Microblade Adaptation' and Recolonization of Siberia during the Late Upper Pleistocene. In *Thinking Small: Global Perspectives on Microlithization*, edited by Robert G. Elston and Steven L. Kuhn, pp. 117–131. Archeological Papers of the American Anthropological Association, Arlington.

Guljamov, JA. G., U. I. Islamov, and A. Askarov 1966 *Pervobytnaja kul'tura i voznik'novenie oroshaemogo zemledelija v nizov'jakh Zarafshana*. FAN, Tashkent.

Harris, D. R., V. M. Masson, Y. E. Berezkin, M. P. Charles, C. Gosden, G. C. Hillman, A. K. Kasparov, G. F. Korobkova, K. Kurbansakhatov, A. J. Legge, and S. Limbrey 1993, Investigating early agriculture in Central Asia: new research at Jeitun, Turkmenistan. *Antiquity* 67(255): 324–338.

Harris, D. R., C. Gosden, and M. P. Charles 1996, Jeitun: Recent Excavations at an Early Neolithic Site in Southern Turkmenistan. *Proceedings of the Prehistoric Society* 2: 423–442.

Inizan, M.-L. 1991 Le débitage par pression: des choix culturels. In *25 ans d'études technologiques en Préhistoire: bilan et perspectives (Actes des XIe Rencontres internationales d'archéologie et d'histoire d'Antibes, 18-20 octobre 1990)*, edited by Catherine Perlès, pp. 367–377. Ed. APDCA, Juan-les-Pins.

Inizan, M.-L. Monique Lechevallier, and Patrick Plumet 1992 A technological marker of the penetration into North America: Pressure Microblade debitage. Its Origin in the Paleolithic of North Asia and its Diffusion. In *Materials Research Society Symposium Proceedings*, edited by Pamela M. Vandiver, James R. Druzik, George Segan Wheeler and Ian C. Freestone, pp. 661–681. Materials Issues in Art and Archaeology III, Vol. 267. Materials Research Society, Pittsburgh.

Islamov, U. I. 1980 *Obishirskaja kul'tura*. FAN, Tashkent.

Islamov, U. I., and V. I. Timofeev 1986 *Kul'tura kamennogo veka Tsentral'noj Fergany*. FAN, Tashkent.

Itina, M. A. 1959 Pervobytnaja keramika Khorezma. In *Keramika Khorezma. Trudy khorezmskoj arkheologo-etnograficheskoj ekspeditsii*, Vol. IV, edited by S. P. Tolstov and M. G. Vorob'eva, pp. 5–62. Academy of Sciences of the USSR, Moscow.

Jusupov, A. Kh., and V. S. Solov'ev 1973 Novye arkheologicheskie otkrytija v Javanskoj doline. *Arkheologicheskie raboty v Tadzhikistane* 10: 62–78.

Kalieva, S. S. 1998 *Poselenie Khozaj I*. Academy of Sciences of the Republic of Kazakhstan, Almaty

Khlobystin, D. P. 1996 Vostochnaja Sibir' i Dal'nij Vostok. In *Neolit Severnoj Evrazii*, edited by S. V. Oshibkina, pp. 270–305. Nauka, Moscow.

Kirjushin, Ju. F., and G. A. Kl'jukin 1985 Pamjatnika neolita i bronzy jugu-zapadnogo Altaja. In *Altaj v epokhe kamnja i rannego metala*, pp. 73–117. National University of Altaï, Barnaul.

Kislenko, A. M., and A. A. Pleshakov 1998 Neoliticheskaja stojanka Vinogradovka 14. In *Voprosy arkheologii Kazakhstana*, edited by Z. Samachev, pp. 25–37. Gylym: Almaty-Moscow.

Krizhevskaja, L. Ja. 1968 *Neolit Juzhnogo Urala*. Nauka, Leningrad.

Kuzmin, Y. V., and L. A. Orlova 1998 Radiocarbon Chronology of the Siberian Paleolithic. *Journal of World Prehistory* 12(1): 1–53.

Levine, M. A. 1999 The origins of horse husbandry on the Eurasian Steppe. In *Late Prehistoric Exploitation of the Eurasian Steppe*, edited by Marsha A. Levine, Yuri Rassamakin, Aleksandr M. Kislenko and Nataliya S. Tatarintseva, pp. 5–58. McDonald Institute for Archaeological Research, Cambridge.

Lisitsyn, N. F., and Ju. S. Svezhentsev 1997 Radiouglerodnaja khronologija verkhnego paleolita Severnoj Azii. In *Radiouglerodnaja khronologija paleolita Vostochnoj Evropy i Severnoj Azii. Problemy i perspektivy*, edited by A. A. Sinitsyn and N. D. Praslov, pp. 67–108. Russian Academy of Sciences, Sankt-Petersburg.

Litvinskij, B. A., and V. A. Ranov (editors) 1998 *Istorija tadzhikskogo naroda*. Surush, Dushanbe.

Masson, V. M. 1960 Dzhejtunskaja kul'tura. In *Trudy Juzhno-Turkmenistanskoj Arkheologicheskoj Kompleksnoj Ekspeditsii*, Vol. X, edited by M. E. Masson, pp. 37–109. Academy of Sciences of Turkmenistan, Ashgabat.

Masson, V. M. 1971 *Poselenie Dzhejtun (problema stanovlenija proizvodjashchej ekonomiki). Materialy i issledovanija po arkheologii SSSR* n°180. Nauka, Leningrad.

Matjushin, G. N. 1973 Mnogoslojnoe poselenie Mysovoe na oz. Karabalykty na Juzhnom Urale. *Kratkie soobshchenija* 136: 67–76.

Matjushin, G. N. 1976 *Mezolit Juzhnogo Urala*. Nauka, Moscow.

Matjushin, G. N. 1989 Mezolit Juzhnogo Zaural'ja. In *Mezolit SSSR*, edited by L. V. Kol'tsov, pp. 144–148. Nauka, Moscow.

Medoev, A. G. 1970 Arealy paleoliticheskikh kul'tur Sary-Arka. In *Po sledam drevnikh kul'tur Kazakhstana*, edited by M. K. Kadyrbaev, pp. 200–216. Nauka, Alma-Ata.

Medoev, A. G.1982 *Geokhronologija paleolita Kazakhstana*. The XI INQUA Congress. Nauka, Alma-Ata.

Meshcherjakov, D. V., and N. L. Morgunova 1996 K probleme proiskhozhdenija konevodstva na Juzhnom Urale. In *Voprosy arkheologii zapadnogo Kazakhstana. Sbornik nauchnykh trudov*, Vol. 1, edited by V. V. Rodionov and E. Zh. Kurmanbekov, pp. 46–55. Dialog, Samara.

Molodin, V. I. 1977 *Epokha neolita i bronzy lesostepnogo Ob'-Irtysh'ja*. Nauka, Novosibirsk.

Mussi, M. 1979 Darra Kalon Rock Shelter (Northern Afghanistan): Levels I-V. *East and West, new series* 29(1–4): 189–220.

Okladnikov, A. P. 1981 *Paleolit Tsentral'noj Azii: Moil'tyn am, Mongolija*. Nauka, Novosibirsk.

Okladnikov, A. P., and Z. A. Abramova 1994 *Paleolit Tsentral'noj Azii-Mongolii*. Nauka, Sankt-Pertersburg.

Pelegrin, J. 1988 Débitage expérimental par pression: du plus petit au plus grand. In *Technologie Préhistorique*, edited by Jacques Tixier, pp. 37–53. Notes et Monographies Techniques du CRA 25. Editions du CNRS, Valbonne.

Petrin, V. T., and Zh. K. Tajmagambetov 2000 *Kompleksy paleoliticheskoj stojanki Shul'binka iz verkhnego Priirtysh'ja*. Al'-Farabi Kazakh National University, Almaty.

Pleshakov, A. A. 1993 *Kharakter i razvitie orudij truda naselenija kamennogo veka epokhi golotsena Priishim'ja*. Unpublished abstract of dissertation for the degree of Candidate of Historical Sciences, Almaty.

Ranov, V. A. 1973 Shugnou, mnogoslojnaja paleoliticheskaja stojanka v verkhov'jarkh r. Jakhsu (raskopki 1969-1970 gg.). *Arkheologicheskie raboty v Tadzhikistane* 10: 42–61.

Ranov, V. A. 1975 Pamir i problema zaselenija vysokogornoj Azii chelovekom kamennogo veka. *Strany i narody Vostoka* 16: 136–157.

Ranov, V. A. 1982 Gissarskaja kul'tura: rasprostranenie, khronologija, ekonomika. In *Kul'tura pervobytnoj epokhi Tadzhikistana, ot Mezolita do Bronzy*, pp. 22–41. Donish, ushanbe.

Ranov, V. A. 1984 Douze années de recherches sur la préhistoire au Tadjikistan méridional et au Pamir (1971-1982). *Paléorient* 10(2): 5–22.

Ranov, V. A. 1985 Gissarskaja kul'tura, neolit gornykh oblastij Srednej Azii (proiskhozhdenie, rasprostranenie, osobennosti). In *Kamennyj vek Severnoj, Srednej i Vostochnoj Azii*, edited by V. E. Larichev, pp. 10–35. Nauka, Novosibirsk.

Ranov, V. A. 1988 *Kamennyj vek Juzhnogo Tadzhikistana i Pamira*. Unpublished abstract of dissertation for the degree of Doctor of Historical Sciences, Novosibirsk.

Ranov, V. A. 1992 Genezis i periodizatsija pamjatnikov kamennogo veka v Tadzhikistane. In *Problemy istorii kul'tury Tadzhikskogo naroda*, pp. 28–48. Hisor, Dushanbe.

Ranov, V. A. 1996 Mogut li geometricheskie mikrolity byt' pokazatelem migratsionnykh protsessov v Srednej Azii? *Drevnosti* 19: 25–26.

Ranov, V. A., and G. F. Korobkova 1971 Tutkaul, mnogoslojnoe poselenie Gissarskoj kul'tury v Juzhnom Tadzhikistane. *Sovetskaja Arkheologija* 2: 133–147.

Ranov, V. A., A. A. Nikonov, and M. M. Pakhomov 1976 Osadnistvo epokhi kamnja u podnozhi Pamiru. *Acta Archaeologica Carpathica* 16: 5–20.

Serikov, Ju. B. 1998 Geometricheskie mikrolity Srednego Zaural'ja. *Rossijskaja Arkheologija* 1: 117–125.

Szymczak, K., and M. Khudzhanazarov (and with the contribution of Michel Fontugne) 2006a *Exploring the Neolithic of the Kyzyl-Kums: Ayakagytma "the Site" and other collections*. Swiatowit, Supplement Series P: Prehistory and Middle Ages Vol. XI, Central Asia, Prehistoric Studies 2. Institute of Archaeology, Warsaw University, Warsaw.

Szymczak, K., and M. Khudzhanazarov 2006b Bullet-shaped core reduction in Kelteminar culture (Neolithic of Central Asia). In *The Stone: Technique and Technology*, edited by A. Wiśniewski, T. Płonka and J. M. Burdukiewicz, pp. 191–198. Institute of Archaeology, University of Wrocław, Wrocław.

Texier, P.-J. 1982 Le Débitage par pression pectorale à la béquille: une approche expérimentale au plus près des paramètres à l'origine de la rupture fragile des roches dures. In *Tailler ! Pour quoi faire: Préhistoire et technologie lithique II. Recent progress in microwear studies,* edited by Daniel Cahen, pp. 57–64. Studia Praehistorica Belgica 2.Koninklijk Museum voor Midden-Afrika, Tervuren.

Texier, P.-J. 1984 Un débitage expérimental de silex par pression pectorale à la béquille. *Bulletin de la Société préhistorique française* 81(1): 25–27.

Tolstov, S. P. 1958 Raboty Khorezmskoj arkheologo-etnograficheskoj ekspeditsii AN SSSR v 1949–1953 gg. In *Arkheologicheskie i etnograficheskie raboty khorezmskoj ekspeditsii 1949–1953,* edited by S. P. Tolstov, pp. 7–258. Trudy khorezmskoj arkheologo-etnograficheskoj ekspeditsii Vol. II. Academy of Sciences of USSR, Moscow.

Tolstov, S. P. (editor) 1959 *Polevye issledovanija Khorezmskoj ekspeditsii v 1954–1956 gg.* Materialy Khorezmskoj ekspeditsii Vol. 1. Academy of Sciences of USSR, Moscow.

Vinogradov, A. V. 1957 K voprosu o juzhnykh svjazjakh kel'teminarskoj kul'tury. *Sovetskaja Arkheologija* 1: 25–45.

Vinogradov, A. V. 1960 Novye neoliticheskie nakhodki Khorezmskoj ekspeditsii AN SSSR 1957 g. In *Polevye issledovanija Khorezmskoj ekspeditsii v 1957 godu*, edited by S. P. Tolstov, pp. 63–81. Materialy Khorezmskoj ekspeditsii Vol. 4. Academy of Sciences of USSR, Moscow

Vinogradov, A. V. 1963 Novye materialy dlja izuchenija Kel'teminarskoj kul'tury. In *Polevye issledovanija khorezmskoj ekspeditsii v 1958-1961 gg,* edited by S. P. Tolstov, pp. 91–102. Materialy Khorezmskoj ekspeditsii vol. 6. Academy of Sciences of USSR, Moscow

Vinogradov, A. V. 1968 *Neoliticheskie pamjatniki Khorezma*. Nauka, Moscow.

Vinogradov, A. V. 1981a *Drevnie okhotniki i rybolovy Sredneaziatskogo Mezhdurechija*. Nauka, Moscow.

Vinogradov, A. V. 1981b Novye materialy po periodizatsii i khronologii neolita Kyzylkumov. In *Kul'tura i iskusstvo drevnego Khorezm*, edited by M. A. Itina, Ju. A. Rapoport, N .S. Sycheva and B. I. Vajnberg, pp. 88–98. Nauka, Moscow.

Vinogradov, A. V., and E. D. Mamedov 1975 *Pervobytnyj Ljavljakan, etapy drevnejshego zaselenija i osvoenija Vnutrennikh Kyzylkumov*. Nauka, Moscow.

Zajbert, V. F. 1992 *Atbasarskaja kul'tura*. Ural Division of the Russian Academy of Sciences, Ekaterinburg.

Zajbert, V. F. 1993 *Eneolit Uralo-Irtyshskogo mezhdurech'ja*. Academy of Sciences of the Republic of Kazakhstan, Petropavlovsk.

Zajbert, V. F., and T. M. Potemkina 1981 K voprosu o Mezolite lesostepnoj chasti Tobolo-Irtyshkogo mezhdurech'ja. *Sovetskaja Arkheologija* 3: 107–130.

Chapter 13
Blades and Microblades, Percussion and Pressure: Towards the Evolution of Lithic Technologies of the Stone Age Period, Russian Far East

Andrei V. Tabarev

13.1 Introduction

The Final Paleolithic–Early Neolithic period (15000–6000 B.P.) in the Russian Far East is represented by a series of cultures with developed production of blades and microblades for various types of tools. Typological and experimental analysis suggests a *multi-linear model* for the evolution of these industries including local and external factors (raw material availability, adaptation to new climatic conditions, influence of economy, trade and exchange networks, etc.). Until recently, not much information was published in western languages about these archaeological materials. During the last 10 years, a new series of excavations throughout the region, including several joint projects (Russian-Japanese, Russian-Korean), new carbon dating and a series of publications and presentations in the international conferences have made it possible to attract a wide range of specialists to the collections and to discuss the structure of lithic industries (Table 13.1).

Thanks to recent archaeological research in the Russian Far East, we have a much more detailed picture and sequence of archaeological cultures than was the case during the initial stage of investigations in the 1960–1970s (Fig. 13.1).

A.V. Tabarev (✉)
Division of Foreign Archaeology, Institute of Archaeology and Ethnography,
17, Lavrentieva Ave., Novosibirsk 630090, Russia
e-mail: olmec@yandex.ru

P.M. Desrosiers (ed.), *The Emergence of Pressure Blade Making: From Origin to Modern Experimentation*, DOI 10.1007/978-1-4614-2003-3_13,

Table 13.1 Carbon dates on the Fareastern cultures

Region	Cultures or sites	Data (ybp)	Period, technology
Middle Amur	Selemjinskaya	22530 ± 320 (SNU03-365) 19350 ± 65 (SOAN-2619) 16460 ± 170 (SNU03-366)	Late Paleolithic Wedge-shaped microblade industry, blade percussion industry
	Gromatukhinskaya	13310 ± 110 (AA-20940) 13240 ± 85 (AA-20939) 12340 ± 60 (AA-36079) 11320 ± 150 (SNU02-002) 9895 ± 50 (AA-36447)	Final Paleolithic – Early Neolithic Wedge-shaped microblade industry, blade percussion industry, microprismatic industry
	Novopertovskaya	12720 ± 130 (AA-38103) 10400 ± 70 (AA-20938) 9765 ± 70 (AA-20937) 9740 ± 60 (AA-38109)	Early Neolithic Pressure-blade industry
Lower Amur	Osipovskaya	13260 ± 100 (AA-13392) 12960 ± 120 (LE-1781) 12500 ± 60 (LLNL-102169) 10875 ± 90 (AA-13393) 9890 ± 230 (GaK-18981)	Wedge-shaped microblade industry, blade percussion industry
	Mariinskaya	8565 ± 65 (SOAN-4869) 6180 ± 60 (SOAN-4109)	Early Neolithic Microprismatic industry
Maritime Region	Ustinovka	15900 ± 120 (AA-36626) 15340 ± 90 (AA-36625) 15300 ± 140 (Ki-3502) 15105 ± 100 (AA-9463) 11550 ± 240 (GEO-1412) 11750 ± 620 (SOAN-3538)	Final Paleolithic Wedge-shaped microblade industry, blade percussion industry
	Vetka	6010 ± 90 (SOAN-6146) 5860 ± 55 (SOAN-6306) 5830 ± 95 (SOAN-6145)	Neolithic Pressure-blade industry
Sakhalin Island	Ogon'ki	19320 ± 145 (AA-20864) 18920 ± 150 (AA-25434) 17860 ± 120 (AA-23137)	Late Paleolithic Wedge-shaped microblade industry, blade percussion industry
Kamchatka Peninsula	Ushki	14300 ± 200 (MAG-550) 13600 ± 250 (GIN-167) 10860 ± 400 (MAG-400) 10360 ± 350 (MO-345)	Final Paleolithic Wedge-shaped microblade industry
	Avacha	6180 ± 50 (GIN-8144a)	Neolithic Pressure-blade industry

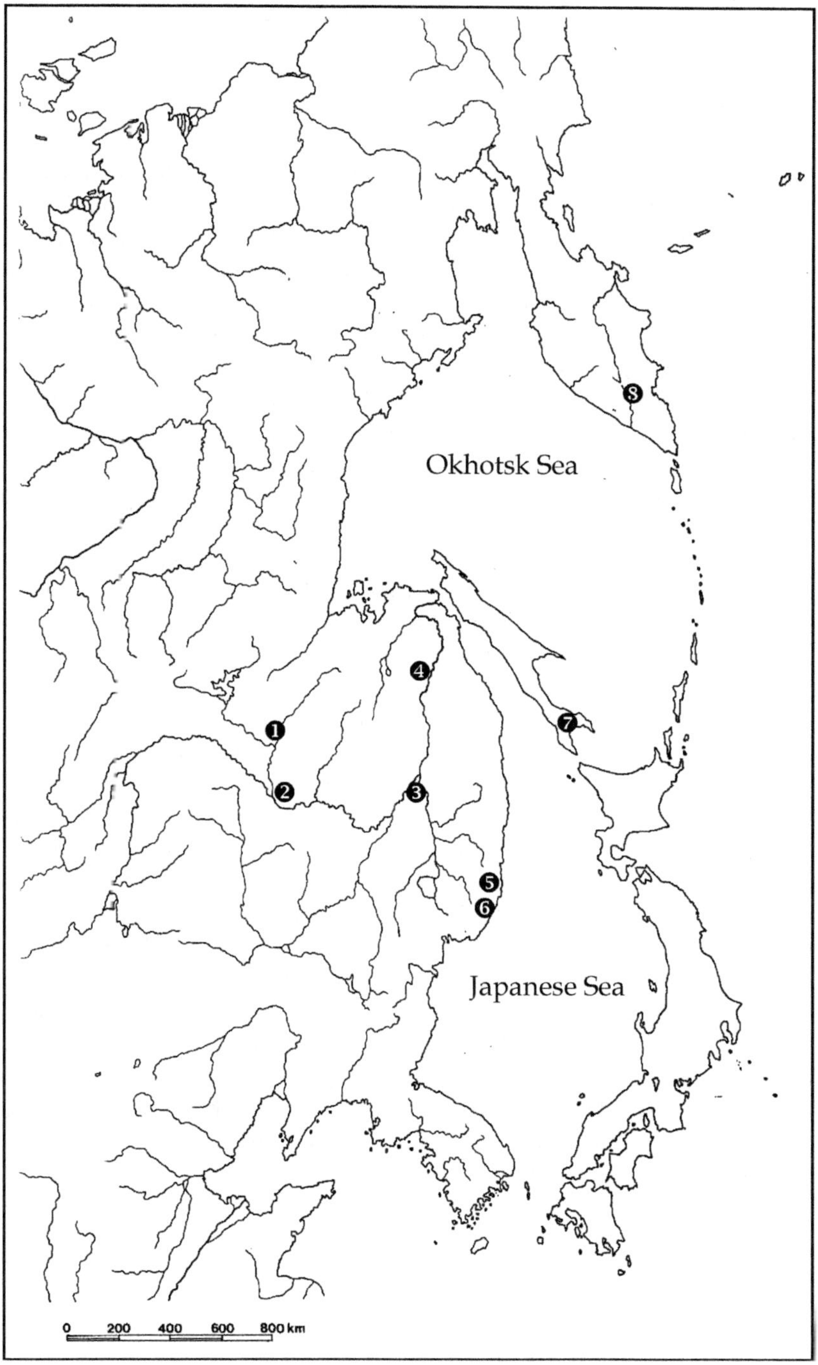

Fig. 13.1 Russian Far East. Cultures and sites locations. *1 Selemja Culture*; *2 Gromatukhinskaya and Novopetrovskaya Cultures*; *3 Osipovskaya Culture*; *4 Mariinskaya Culture*; *5 Ustinovka Culture*; *6 Vetka Culture*; *7* Ogonki Sites; *8* Ushki Lake sites

13.2 Geography and Chronology of Cultures

13.2.1 Middle Amur Region

Selemjinskaya Culture (24000–12000 B.P.) is the first culture with microblade technology known in the region. This culture was studied on a series of multilevel sites which can be interpreted as seasonal camps of forest hunters and fishers located on river banks. Microblade cores of wedge-shaped configuration, microblades and ski-spalls were found in great quantities and in a variety of raw materials starting from the lower horizons with an age of 24000–22000 B.P. up to the upper horizons (13000–12000 B.P.) (Derevianko et al. 2006). Microblade technology (pressure methods), along with the flake technology (direct percussion of pebble cores), provided the economy with all types of blanks for tools. Interestingly, these Selemja craftsmen never developed a blade technique with prismatic or sub-prismatic blade cores. The Terminal Pleistocene–Early Holocene continuation of this tradition is represented in the *Gromatukhinskaya Culture* (13000–4000 B.P.) (Okladnikov and Derevianko 1977). The sites of this culture are of the same economic orientation but include the earliest evidence of pottery which is dated to about 14000–13000 B.P. Looking at the lithic materials, we see that the wedge-shaped microcores were replaced by conical cores around 9000 B.P. It looks like the local inhabitants moved from one type of portable device to another one using the same principles of pressure techniques and the same raw material base (Fig. 13.2).

A different technology existed in the Middle Amur during the same period in the form of the so-called *Novopetrovskaya Culture*. Several sites with subterranean dwellings, pottery shreds, and rich lithic materials were excavated on the Amur River tributaries in the transitional type of landscape between forest and open plains (Derevianko 1970; Derevianko et al. 2005). Using fine-grained flinty tuff and chert, the people of this culture produced big blade cores and regular prismatic blades. Some of the blanks are about 12–15 cm long and, after additional edge retouching, were modified into points, burins, scrapers and knives. Unfortunately, we do not have any remains of devices or clamps used by Novopetrovka flintmakers but technologically, such blades are of the highest quality and skills. It should be mentioned that while working with pressure blade techniques, these people never tried to explore wedge-shaped or other versions of microblade technique. The origin of *Novopetrovskaya Culture* is also problematic and may be tentatively linked with the territory of Northern China (Figs. 13.3, 13.4).

13.2.2 Lower Amur Region

The Middle Amur region is one of the possible centres of origin for the impulse of migrations of ancient tribes to the Lower Amur territory. This can be shown with the

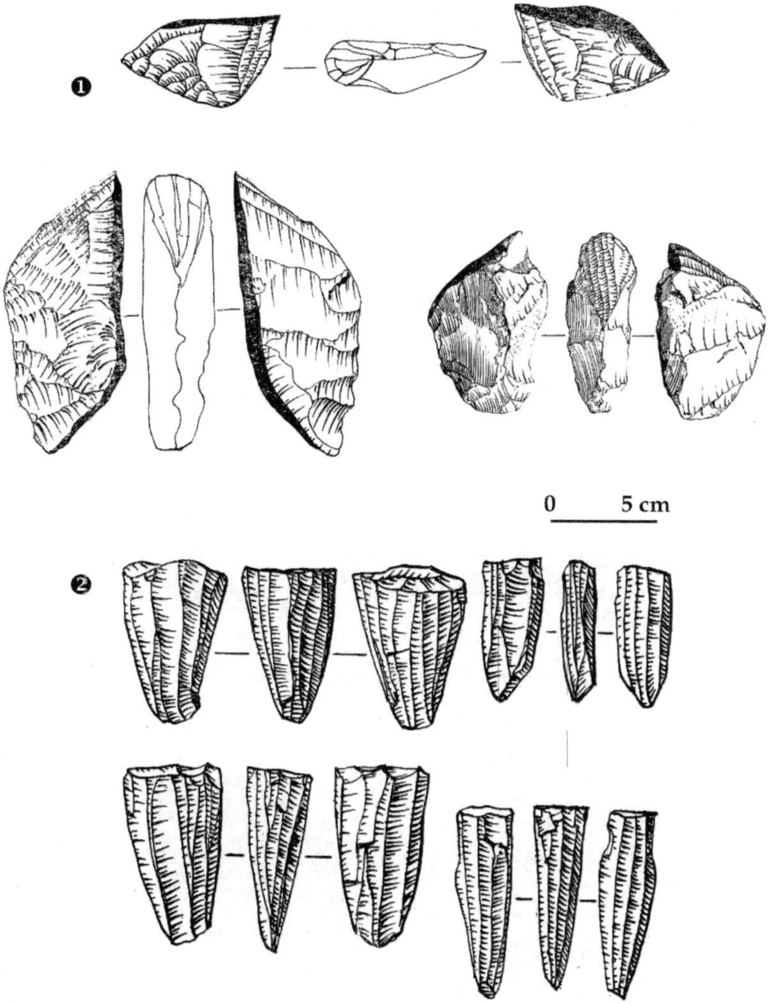

Fig. 13.2 *Gromatukhinsakaya Culture*. *1* Final Paleolithic industry; *2* Early Holocene industry (By Okladnikov and Derevianko 1977)

materials of the *Osipovskaya Culture* (about 30 sites around the city of Khabarovsk), which are technologically and chronologically very close to *Gromatukhinskaya Culture*. The lithic industry includes two types of cores: big ones for flakes and elongated flakes, and wedge-shaped microcores. The culture as a whole is also very similar, having seasonal camps with fishing activities and also hunting and gathering in a forest zone. After 10000–9500 B.P., the lithic industry was transformed following another scenario: wedge-shaped microcore techniques disappeared, and

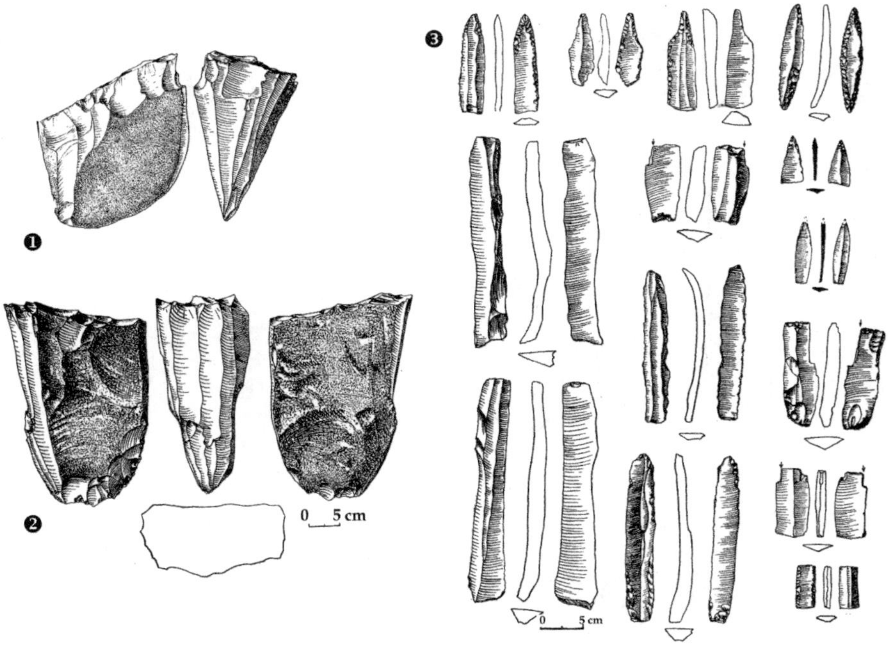

Fig. 13.3 *Novopetrovskaya Culture*. *1* Cores; *2* Blades (By Derevianko 1970)

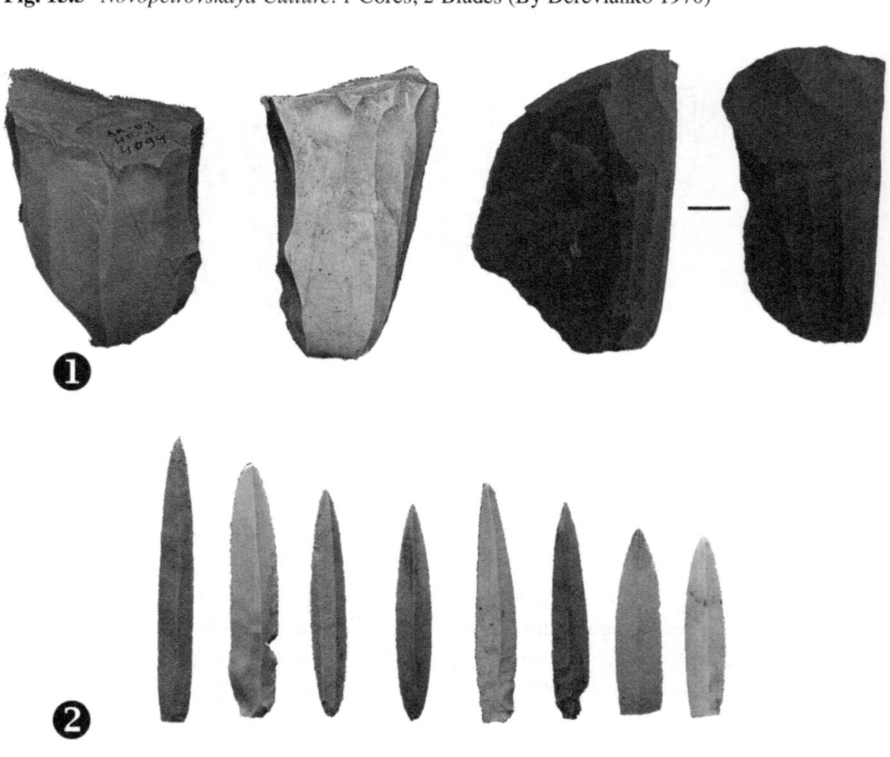

Fig. 13.4 *Novopetrovskaya Culture*. *1* Cores; *2* Tools on blades (By Derevianko et al. 2005)

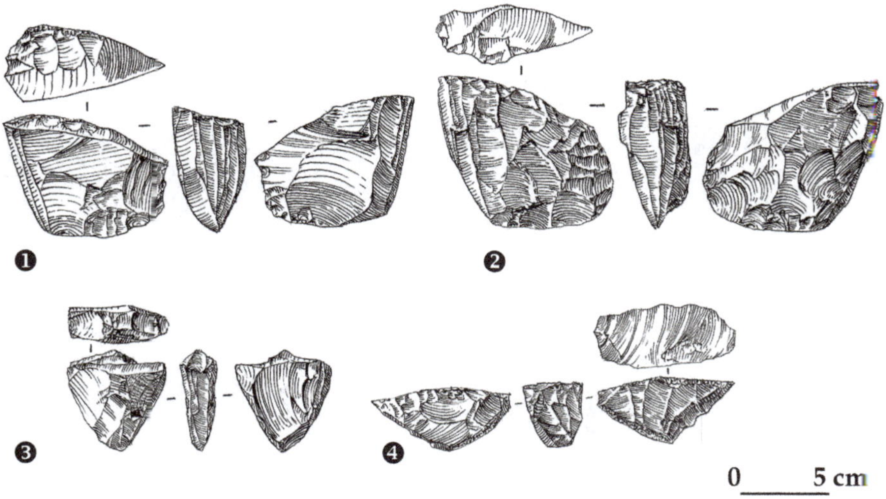

Fig. 13.5 *Osipovskaya Culture*. 1–4 Microblade cores (By Derevianko et al. 2006)

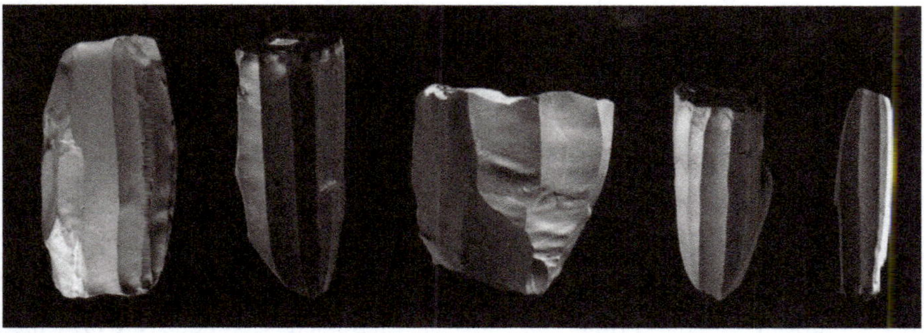

Fig. 13.6 *Mariinskaya Culture*. Microconical and microprismatic cores (By National Research Institute of Cultural Heritage 2006)

there are just slight traces of microconical, microprismatic and bullet-shaped cores such as those in the Middle Amur region (Fig. 13.5).

Only during a brief period from 8500 to 8000 B.P. do we see rare evidence of excellent pressure blade techniques in some parts of the Lower Amur Region; for example we see evidence of excellent pressure blade technique with the recently located *Mariinskaya Culture*. The first materials were found on Suchu Island along with distinctive pottery and the possible remains of dwelling constructions (National Research Institute of Cultural Heritage 2006) (Fig. 13.6).

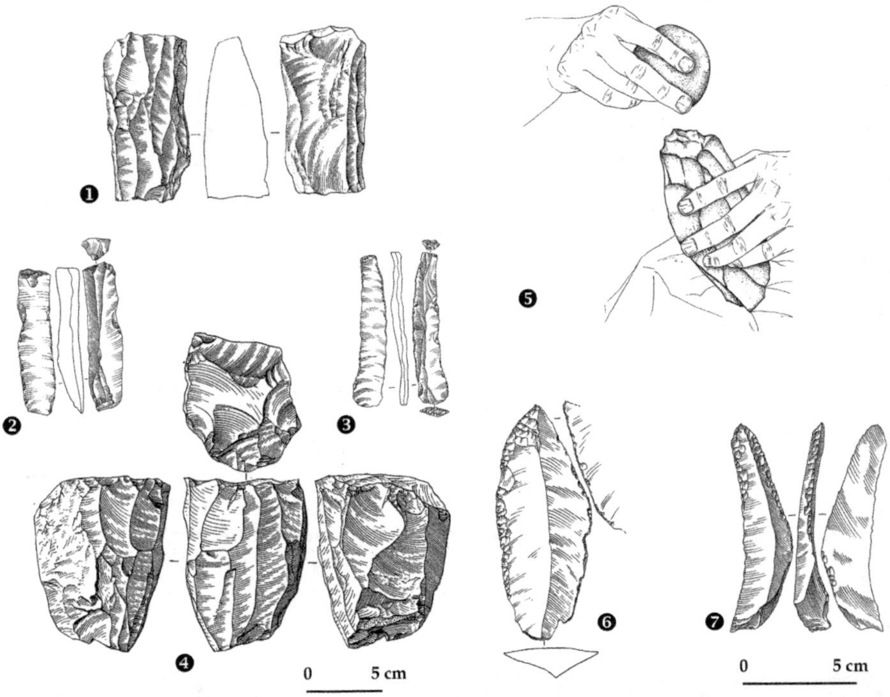

Fig. 13.7 *Ustinovka Culture*. 1, 4 Blade cores; 2, 3, 6, 7 Blades; 5 Technique of direct percussion (By Krypianko and Tabarev 2001)

13.2.3 Maritime Region

In the Maritime Region (Primorye), blade and microblade techniques are presented in the archaeological complexes starting at 16000–15000 B.P. and are connected with the *Ustinovka Culture* (Derevianko and Tabarev 2006; Kononenko et al. 1995; Tabarev 1994, 2003; Tabarev et al. 1999 Vasilievsky and Gladyshev 1989). These sites are located both in the coastal zone (mostly in the Zerkal'naya River basin) and in the continental parts of the region. In the Zerkal'naya River basin, the lithic industry demonstrates a high level of direct percussion blade technology. Based on the rich local raw material sources (flinty tuff), direct percussion blade technology was the dominant technology for at least 5,000 years, and only during the change from Pleistocene to Holocene did it eventually dwindle and disappear. Microblade technology (wedge-shaped version) was a minor part of the local industry, and it existed in several modifications (on bifacial blanks or on unifacial blanks). Cores, microblades and tools from exotic materials, for example obsidian, are extremely rare and appeared in the coastal zone no earlier than 11000–11500 B.P. (Fig. 13.7).

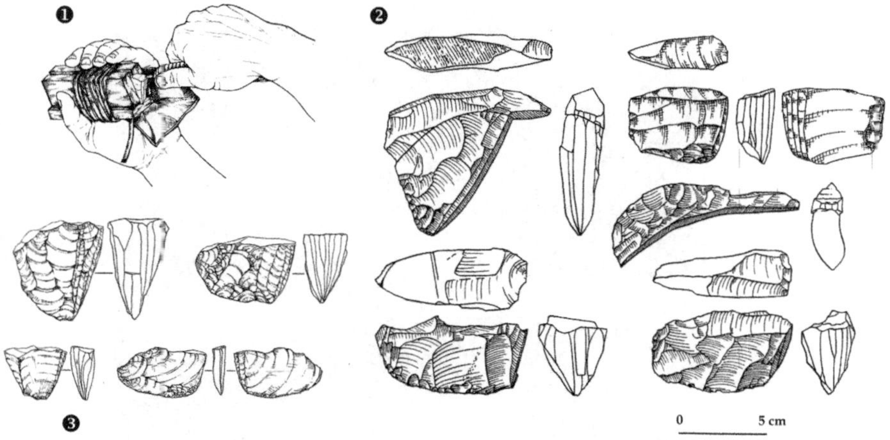

Fig. 13.8 Ustinovka culture. *2–3* Microblade cores (*wedge-shaped*) and *1* Portable device for microblades production by pressure (By Krypianko and Tabarev 2001)

In contrast, the lithic industry of the inland part of the Maritime Region demonstrates a higher percentage of microblade cores and microblades. Several local sources of volcanic glass and exchange with the territories of Korea and China (Gillam and Tabarev 2004) provided people with high-quality material for various types of microblade cores and tools. Thanks to the presence of obsidian during the final stage of the *Ustinovka Culture* (10500–9000 B.P.), wedge-shaped technology in the continental part switched to microconical, but the period of its existence was very short, and we do not see any evidence of microconical cores after this time. The Neolithic period is associated only with a simple flake technology (Figs. 13.8, 13.9).

Up until recently, it seemed that microblade technology and pressure blade technology as a whole did not exist in the Maritime Region after the Final Pleistocene. Recent excavations, however, on the Vetka site and Ustinovka-8 sites between 2004 and 2006, along with some separate finds in the coastal zone (Krypianko 2006; Popov and Tabarev 2008), have demonstrated that pressure blade technology was successfully used by people during the 8000–5000 B.P. interval (*Vetka Culture*). This technology has no roots in the previous Paleolithic cultures in the coastal and inland zones and may have originated in the adjacent territories, possibly the Lower Amur Region (Fig. 13.10).

13.2.4 Sakhalin Island

Evidence from the Final Paleolithic and Early Neolithic cultures of Sakhalin Island has strongly confirmed the influence and implication of raw materials in the development

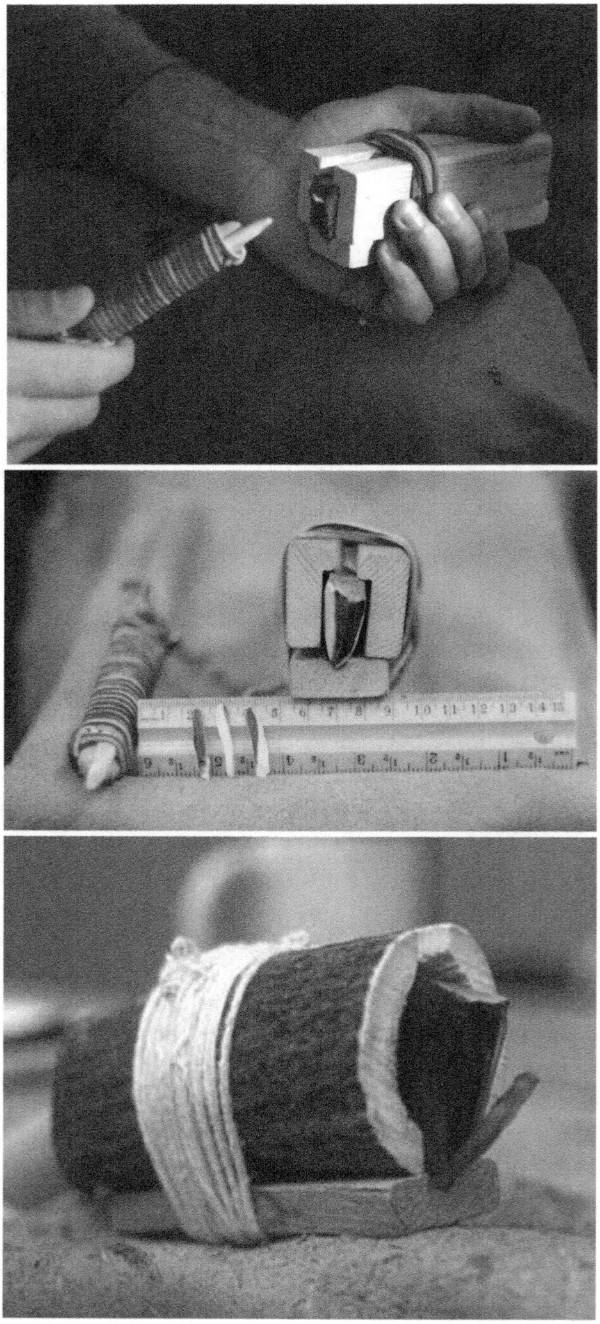

1
2
3

Fig. 13.9 Experimental production of microblades in portable devices *1–2* From hard wood and *3* Antler (Photos by Andrei V. Tabarev)

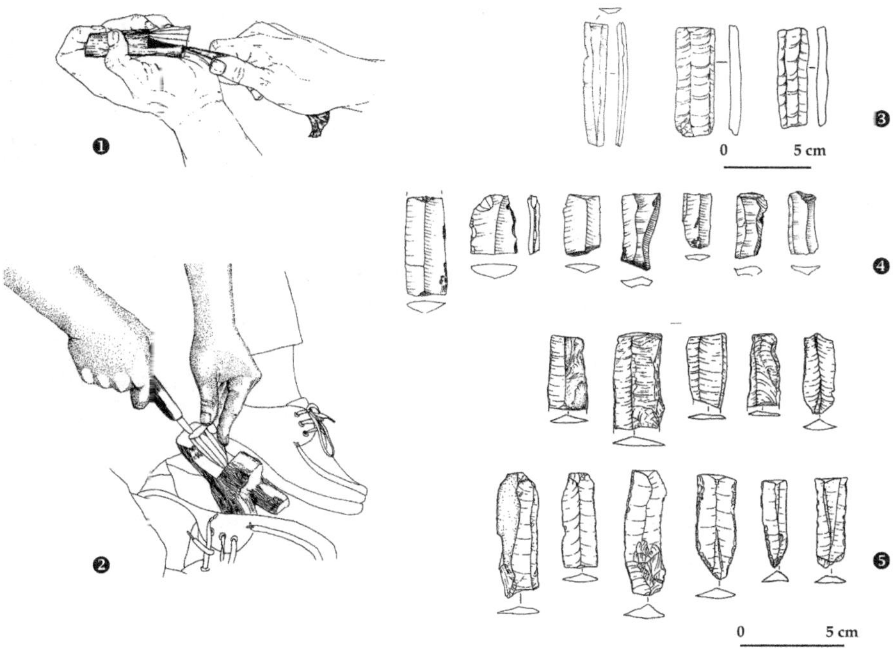

Fig. 13.10 *3–5* Early Holocene blades and *1–2* Possible methods of pressure techniques (By Krypianko and Tabarev 2001)

of lithic technologies. Extensive contact with the northern part of Hokkaido Island since the Upper Paleolithic times was demonstrated on the basis of the obsidian artefacts found there. Obsidian was regularly transported from the sources and used in blade and microblade production, with very similar designs to the distinctive types of cores found in Japan (Vasilevsky 2006). The transition from wedge-shaped microcores to conical cores associated Pleistocene–Holocene border and the Early Holocene from 10500 to 8000 B.P. (Sokol and Ogonki Sites) (Vasilevsky and Shubina 2006) (Figs. 13.11, 13.12).

13.2.5 *Kamchatka Peninsula*

The first traces of microblade production (series of microblades) on the Kamchatka Peninsula are known from the earliest level (Level VII, 13000–12000 B.P.) of the famous Ushki Lake sites (Dikov 1977). Microblade cores with a wedge-shaped modification appear in the next level and may be dated from as early as 12000 to 11000 B.P. The following stages (Neolithic period) of this tradition are connected with micro-conical technology, which was developed from previous traditions of flintknapping.

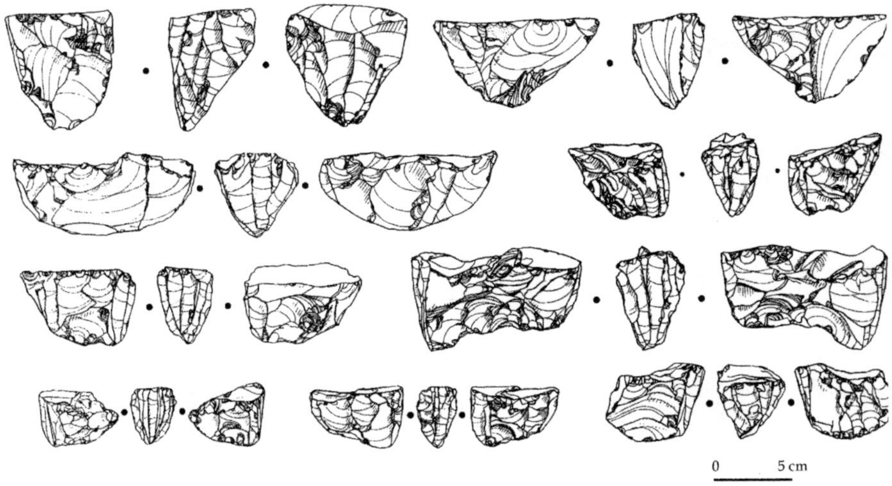

Fig. 13.11 Sakhalin Island. *1–9* Obsidian microblade cores (*wedge-shaped*) (By Vasilivsky 2006)

According to N.N. Dikov, microconical and microprismatic cores were found in Level IV, which was roughly dated to between 6000 and 4000 B.P. (Fig. 13.13).

Other types of blade technologies were not well documented before the very end of twentieth century. A number of big prismatic blades made from local obsidian and flint were found on the surface, in disturbed contexts and in local museum collections. Thanks to new field research conducted by archaeologists in 2000–2001, several sites with very interesting materials were found on the coastal zone in the southern part of the Kamchatka Peninsula (Lebedintsev 2006).

The Avacha localities are of significant interest because archaeological materials are represented by a great number of obsidian blade cores and prismatic blades which were used as knives and scrapers, with or without additional retouch. The preliminary chronology of the sites (7000–6000 B.P.) is based on typology and was confirmed by the carbon dating (6180 ± 50 B.P. – GIN-8144a). These new materials open a very interesting perspective to research into the origin of early blade industries in the Northern Pacific including the Anangula blade site on the Aleutian Islands (Fig. 13.14).

13.3 Technological and Experimental Interpretations

Even this preliminary picture of the microblade and blade industries from five regions of the Russian Far East (Middle Amur Region, Lower Amur Region, Maritime Region, Sakhalin Island and Kamchatka Peninsula) demonstrates that we have several models of technological evolution during the Terminal Pleistocene to Early Holocene period. Wedge-shaped microblade technology seems to be the basic

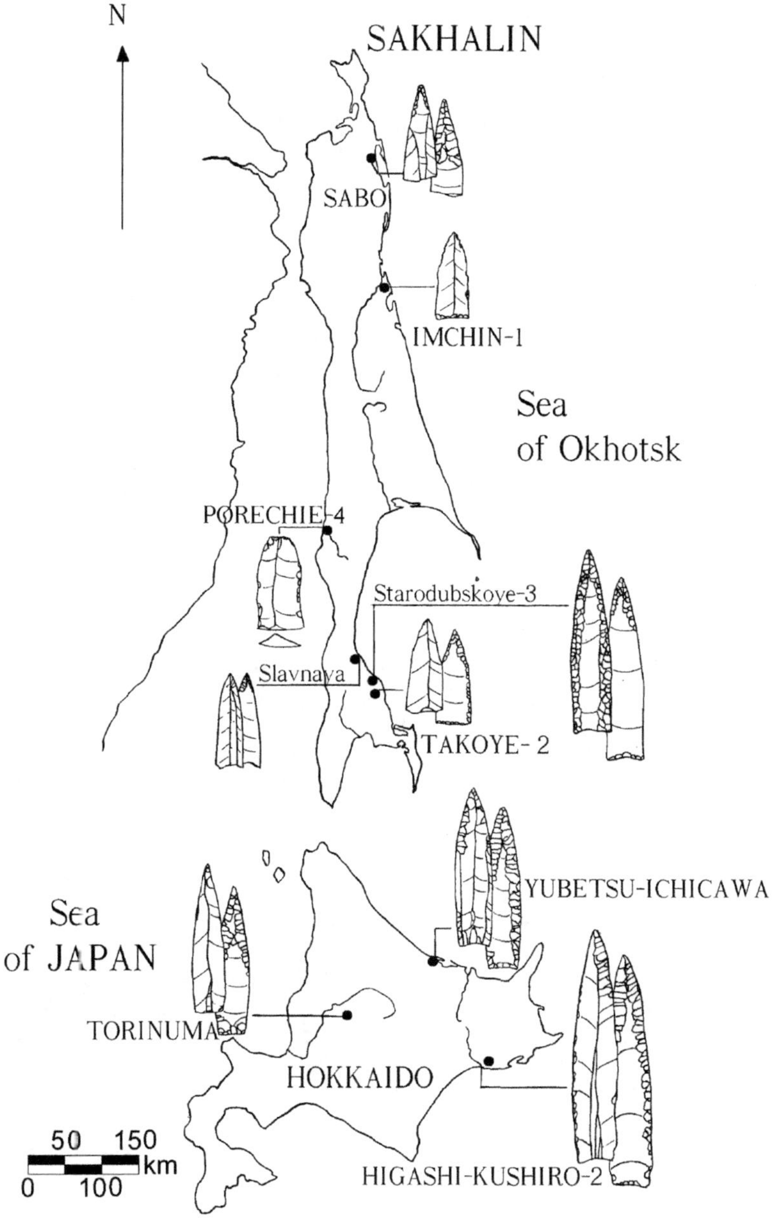

Fig. 13.12 Sakhalin and Hokkaido Islands in Early Holocene. Sites with tools on blades (By Vasilevsky and Shubina 2006)

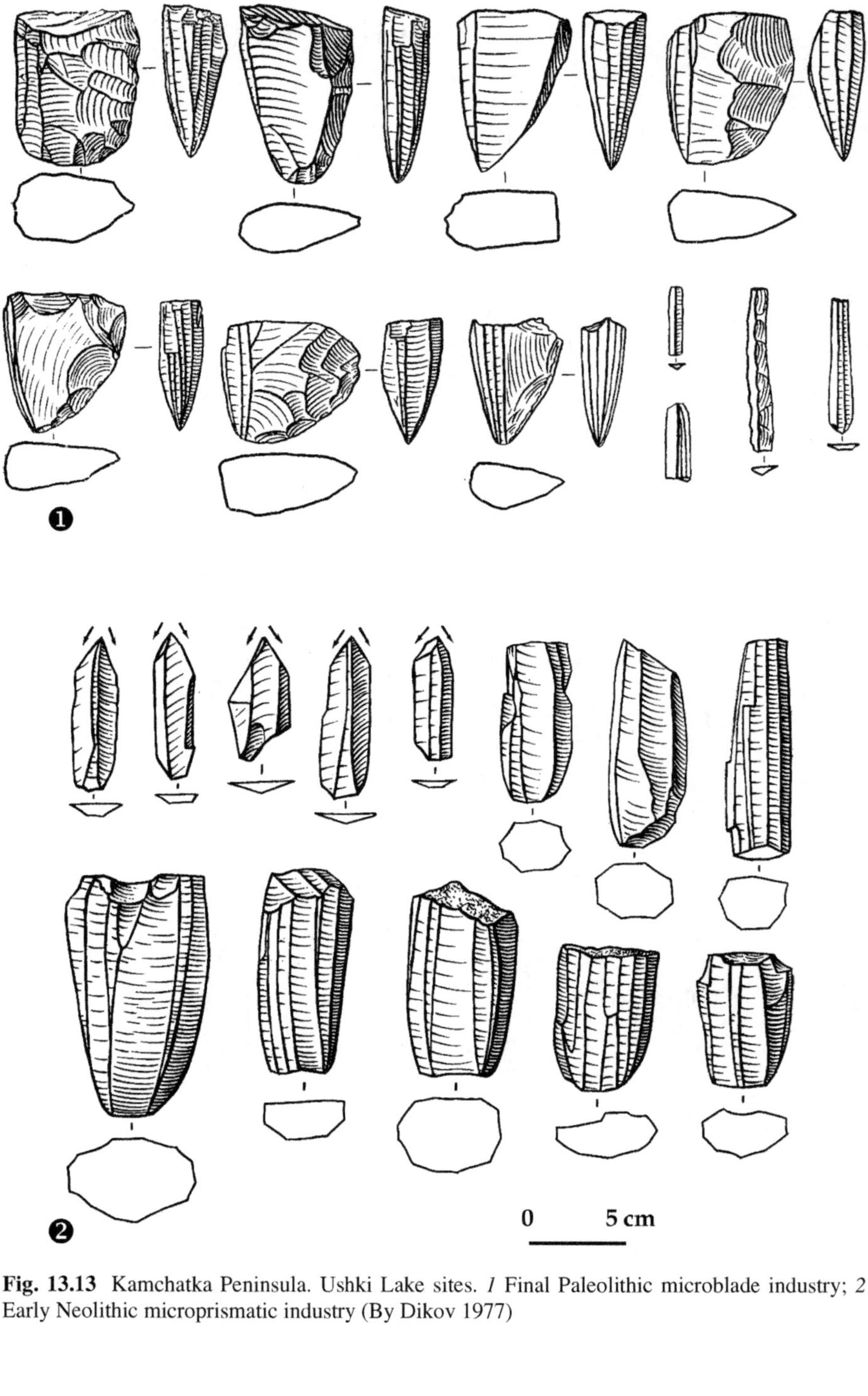

Fig. 13.13 Kamchatka Peninsula. Ushki Lake sites. *1* Final Paleolithic microblade industry; *2* Early Neolithic microprismatic industry (By Dikov 1977)

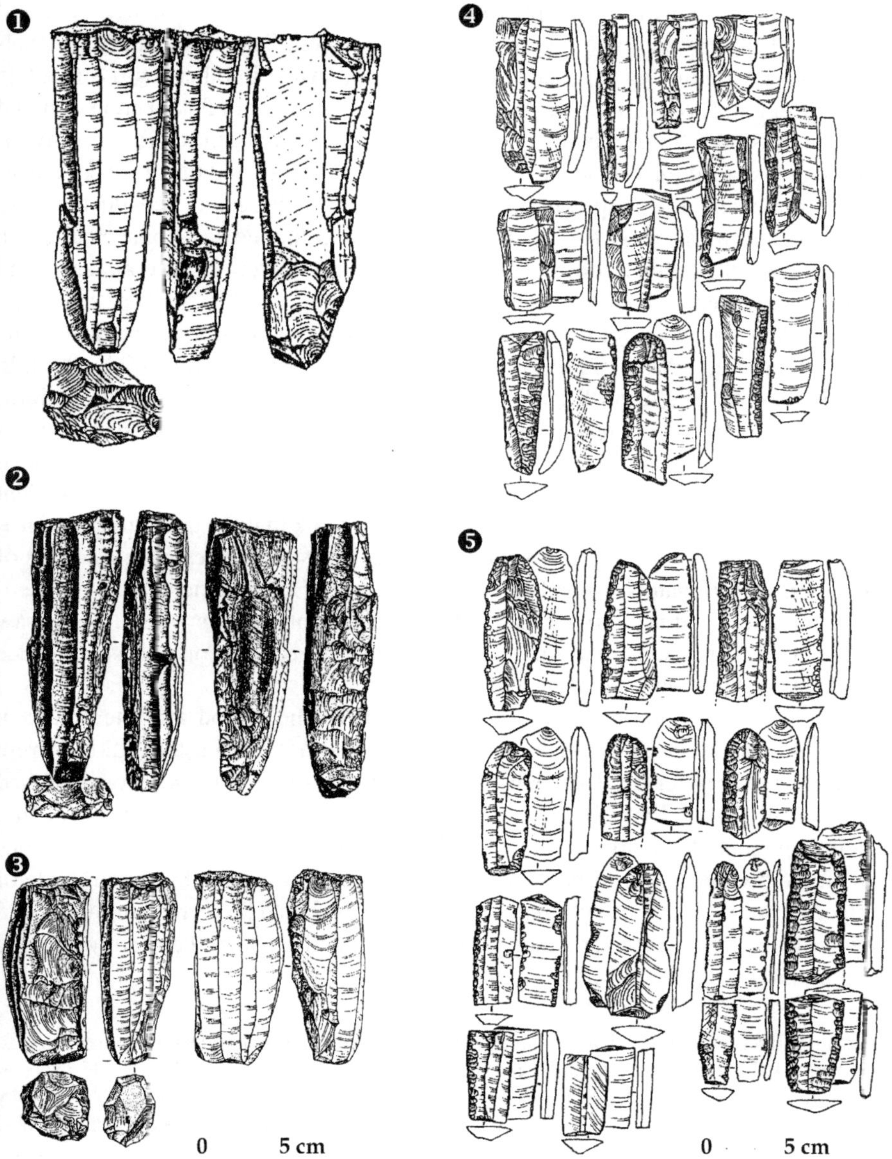

Fig. 13.14 Kamchatka Peninsula. *1–3* Neolithic blade cores from obsidian and 4–5 Tools on blades (By Lebedintsev 2006)

cultural component for all the regions in spite of raw material resource base and ecological factors. In some regions (Maritime Region, Sakhalin Island), it developed along with the big prismatic blade technique, while in other regions (Middle Amur Region, Lower Amur Region), it developed along with flake percussion and amorphous cores. So far, detailed experimental works have been carried out only for the wedge-shaped microblade technique (pressure method) and the blade core technique (direct and indirect percussion) (Tabarev 1997). After having conducted these experiments, we strongly suggest that microblade technique was connected with a wide range of portable compact devices where cores were tightly attached and flaked or reduced with short or long pressure flakers.

The Pleistocene to Holocene transition in the Far East was accompanied by the changes in lithic industries (Krypianko and Tabarev 2001; Tabarev 2008, 2001). In some cases (Middle Amur Region, Sakhalin Island, Kamchatka Peninsula), microprismatic and microcore techniques appeared out of a previous wedge-shaped tradition, whereas in other territories, this transition was more complicated and depended on external influences (Middle Amur Region, Maritime Region) or independent local innovations (Lower Amur Region). We also think that the transition to microprismatic cores represents the transition to other types of devices and principles of pressure. Unfortunately, we still have no evidence of such devices in an archaeological context and need to conduct further experimental works with the local raw materials. This is also very useful for an understanding of the economic significance of blades and microblades in ancient cultures.

Traditionally, blade technologies for the Paleolithic period are interpreted in terms of hunting activities. Since Final Pleistocene times, Far Eastern cultures were oriented towards seasonal salmon fishing, and the role of this activity increased dramatically in the Early Holocene. This had a strong effect on all aspects of life, including technology, art and rituals (Tabarev 2006). Our preliminary conclusions about the disappearance of blade and microblade industries and the leading role of biface technologies in the cultures of salmon fishers should be corrected. Blades and microblades were in use for a long period of time and satisfied the needs of fishing communities.

Acknowledgements I would like to express my gratitude to my Siberian and Far Eastern colleagues: Dr. Sergei P. Nesterov (Institute of Archaeology and Ethnography, Novosibirsk), Dr. Alexander A. Krypianko and Dr. Alexander N. Popov (Far Eastern State University), Dr. Andrei V. Ptashinsky (Kamchatka State Pedagogical University) and Dr. Alexander A. Vasilevsky (Sakhalin State University) for the opportunity to work with archaeological collections, access to recent publications and for sharing their ideas and concepts on the origins of blade and microblade industries in the Far East. Supported by grant #09-06-00006-a from the Russian Foundation for Basic Researches.

References

Derevianko, Anatoly P. 1970 *Novopetrovskaya Culture, Middle Amur.* Nauka, Novosibirsk.

Derevianko, Anatoly P., Sergei P. Nesterov, Sergei V. Alkin, Vladimir G. Petrov, Pavel V. Volkov, Olga S. Kudrich, Kang Chan Hgwa, Lee Hon Jeon, and Kim Ken Chzu 2005 *The Excavation Report of the Novopetrovka Site on the Amur River.* Novosibirsk-Cheju.

Derevianko, Anatoly P., and Andrei V. Tabarev 2006 Palaeolithic of the Primorye (Maritime) Province. In *Archaeology of the Russian Far East: Essays in the Stone Age Prehistory*, edited by Saragh M. Nelson, Anatoly P. Derevianko, Yaroslav V. Kuzmin and Richard L. Bland, pp. 41–54. BAR International Series 1540. British Archaeological Reports, Oxford.

Derevianko, Anatoly P., Vasily N. Zenin, and Igor Y. Shevkomud 2006 Palaeolithic of the Priamurye (Amur River Basin). In *Archaeology of the Russian Far East: Essays in the Stone Age Prehistory*, edited by Saragh M. Nelson, Anatoly P. Derevianko, Yaroslav V. Kuzmin and Richard L. Bland, pp. 55–73. BAR International Series 1540. British Archaeological Reports, Oxford.

Dikov, Nikolay N. 1977 *Archaeological Sites of Kamchatka, Chukotka and Upper Kolyma*. Nauka, Moscow (In Russian).

Gillam, John C., and Andrei V. Tabarev 2004 On the Path of Upper-Paleolithic Obsidians in the Russian Far East. *Current Research in the Pleistocene* 21(1): 3–6.

Kononenko, Nina A., Alexander A. Krypianko, and Andrei V. Tabarev 1995 Ustinovka-VI Site: Recent Investigations of the Microblade Industries in the Maritime Region, Russian Far East. *The Wyoming Archaeologist* 39(1–2): 1–5.

Krypianko, Alexander A. 2006 New Materials on the Archaeology of Eastern Sikhote-Alin (Russian Far East). In *Archaeological Education of the Japanese Fundamental Culture in East Asia*, Vol. 7, edited by Tatsuo Kobayashi, pp. 89–109. 21 COE Program Archaeology Series. Kokugakuin University, Tokyo. (In Russian and Japanese).

Krypianko, Alexander A., and Andrei V. Tabarev 2001 *Archaeological Complexes of the Stone Age Period in the Eastern Maritime Region*. Siberian University Press, Novosibirsk (in Russian).

Lebedintsev, Alexander I. (editor) 2006 *Neolith and Paleometal of the North of the Far East*. NEISRI FEB RAS, Magadan (in Russian).

National Research Institute of Cultural Heritage 2006 *The Treasures of Primorie and Priamurie*. New Results of Korea-Russia Joint Excavations. National Research Institute of Cultural Heritage, Daejeon.

Okladnikov, Alexei P., and Anatoly P. Derevianko 1977 *Gromatukhinskaya Culture*. Science Press, Novosibirsk (in Russian).

Popov, Alexander N., and Andrei V. Tabarev 2008 Neolithic Cultures of the Russian Far East: Technological Evolution and Cultural Sequence. *Turkish Academy of Sciences Journal of Archaeology* 2008(11): 41–62.

Tabarev, Andrei V. 1994 The Ustinovka Culture in the Stone Age of the Russian Far East: 40 Years of Discoveries. *Lithic Technology* 19(1): 21–34.

Tabarev, Andrei V. 1997 Paleolithic Wedge-Shaped Microcores and Experiments with Pocket Device. *Lithic Technology* 22(2): 139–149.

Tabarev, Andrei V. 2001 Russian Far East in the Final Paleolithic: Peopling, Migrations, Maritime, and Riverine Adaptation. In On Being First: Cultural Innovation and Environmental Consequences of First Peopling. *Proceedings of the 31st Annual Chacmool Conference*, edited by Beau Cripps and Ruth Dickau, The Archaeological Association of the University of Calgary, pp. 511–525. University of Calgary Press, Calgary.

Tabarev, Andrei V. 2003 Problems of the Stone Age in the Russian Far East. *Cultura Antiqua* 55(10): 17–24.

Tabarev, Andrei V. 2006 People of Salmon: Technology, Art and Ritual of the Stone Age Cultures, Russia Far East. In *Archaeological Education of the Japanese Fundamental Culture in East Asia*, Vol. 7, edited by Tatsuo Kobayashi, pp. 111–124. 21 COE Program Archaeology Series. Kokugakuin University, Tokyo.

Tabarev, Andrei V. 2008 Late and Final Paleolithic in the Continental Part of the Russian Far East: Current Situation and Some Perspective Directions of Research. In *Human Ecosystem Changes in the Northern Circum Japan Sea Area (NCJSA) in Late Pleistocene*, edited by Hiroyki Sato, pp. 106–114. Tokyo University, Tokyo.

Tabarev, A. V., Alexander A. Krypianko, Laura B. Niven, and Craig M. Lee 1999 The Final-Paleolithic Sites of Suvorovo III and IV in the Maritime Region, Russian Far East. *Current Research in the Pleistocene* 16: 73–75.

Vasilevsky, Alexander A. 2006 The Upper Palaeolithic of Sakhalin Island. In *Archaeology of the Russian Far East: Essays in the Stone Age Prehistory,* edited by Saragh M. Nelson, Anatoly P. Derevianko, Yaroslav V. Kuzmin and Richard L. Bland, pp. 75–100. BAR International Series 1540. British Archaeological Reports, Oxford.

Vasilevsky, A. A., and Olga A. Shubina 2006 Neolithic of the Sakhalin and Southern Kurile Islands. In *Archaeology of the Russian Far East: Essays in the Stone Age Prehistory*, edited by Saragh M. Nelson, Anatoly P. Derevianko, Yaroslav V. Kuzmin and Richard L. Bland, pp. 151–166. BAR International Series 1540. British Archaeological Reports, Oxford.

Vasilievsky, Ruslan S., and Sergei A. Gladyshev 1989 *Upper Paleolithic of the Southern Maritime Region.* Nauka, Novosibirsk (in Russian).

Chapter 14
Pressure Microblade Industries in Pleistocene-Holocene Interior Alaska: Current Data and Discussions

Yan Axel Gómez Coutouly

14.1 Introduction

Late Pleistocene and Early Holocene sites in Alaska have obvious ties with the Siberian Late Paleolithic based on the presence of pressure microblade production. The study of these sites and their corresponding lithic assemblages is essential to our understanding of the peopling of the New World, especially when considering the significance of Swan Point and its lower microblade-bearing layer (currently the earliest reliably dated human occupation documented in Alaska).

The aim of this chapter is to present the context for the emergence of pressure microblade technique in Late Pleistocene – Early Holocene interior Alaska. I will do so by illustrating some of the technological variability from different assemblages in Alaska using the examples of Dry Creek Component II (in the Nenana River Valley) and Swan Point Cultural Zone 4 (in the Tanana River Valley). Both sites are located in the Alaskan interior, where some of the oldest sites with clear evidence of pressure microblade production have been unearthed. The two main methods of microblade production that have been identified in early sites in interior Alaska, the Yubetsu method and the Campus method, will be detailed.

Because the early prehistory of Alaska is so closely tied to the archaeology of Beringia, especially with regard to microblade industries, references to other regions and sites in Northeast Asia and northwestern North America will be made. In the discussion section, I will describe some general technological aspects concerning the affiliation between Northeast Asian and Alaskan microblade assemblages and present a theoretical model to explain the emergence and disappearance of the Yubetsu method.

Y.A. Gómez Coutouly (✉)
Prehistory and Technology Laboratory, University of Paris 10, Paris, France
e-mail: yanaxel@northpacificprehistory.com

P.M. Desrosiers (ed.), *The Emergence of Pressure Blade Making: From Origin to Modern Experimentation*, DOI 10.1007/978-1-4614-2003-3_14,

14.2 Historical Context

Alaska and the Yukon Territory represent the easternmost regions of the Beringian continent, which is traditionally defined as the area extending from the Lena River in Central Siberia to the Mackenzie River in the Western Northwest Territories. The peopling of the New World has long been tied with the study of microblade assemblages given that it provides the clearest technological and cultural link between Northeast Asian and North American Paleolithic assemblages. The exact birthplace and timing of the first appearance of microblade industries is hotly debated. While the exact birthplace of microblade industries remains unknown, all the evidence seems to indicate that they originated somewhere around southern Siberia, northern China, Mongolia or the Far East (e.g. Goebel et al. 2000; Graf 2009; Ikawa-Smith 2007; Inizan et al. 1992; Yesner and Pearson 2002). Most researchers have suggested that microblade industries first appeared around 35000 B.P. in southern Siberia (e.g. Kuzmin et al. 2007), while others contend the emergence occurred later, around 20000 B.P. in the Far East (e.g. Gómez Coutouly 2011b; Graf 2009). These differences in interpretation are due not only to a different assessment of the published archaeological data but also to a different conception of what microblade cores and microblades are. For the partisans of an early appearance, the assemblages taken into account are generally bladelet cores not necessarily manufactured with pressure technique or the Yubetsu method. For the partisans of a later appearance, the assemblages taken into account are of the Yubetsu type and exhibit pressure technique. Based on the current evidence, it appears that the Pacific Northwest Coast was the last region in America colonized by the Paleolithic microblade-bearing populations migrating from Northeast Asia.

Paleolithic microblade assemblages can be associated with various tool types such as burins, bifaces, end-scrapers, organic projectile points, blade technology and lithic projectile points. While some of these tool types are common to all regions and virtually present in every component, others are restricted to some geographical areas. Burins (mainly dihedral), bifacial projectile points and end-scrapers are the main tool types found across Beringia, from the Far East to Alaska. Slotted organic projectile points (mainly made of antler) are found in various microblade-bearing assemblages and are very characteristic since it is widely accepted that microblades were mainly produced to be mounted on such tools. The scarcity of these organic tools in some regions (e.g. Primorye, Japan and Alaska) is more likely due to preservation reasons than to cultural ones. Finally, while virtually absent of Paleolithic microblade components in Siberia and Alaska, blade technology is common in various regions of the Far East such as Korea, Primorye and Japan (e.g. Bae 2010; Derevianko and Kononenko 2003; Gómez Coutouly 2007). Of course, all of this toolkit is only representative for Paleolithic sites since in latter times and in other contexts, microblades are known to be associated with other tool types.

Over the last few decades, the study of pressure microblades has become increasingly important in our efforts to understand the first peopling of the New World. Indeed, 'if it's still difficult to accurately link the other elements of the Siberian assemblages, especially points and other foliate bifacial points, to American equivalents, there is no doubt left that microblade industries from both sides of the Bering strait have a common origin in Asia' (Plumet 2004: 266). Pressure microblade production is thus one of the main diagnostic elements that can be easily identified across greater Beringia.

In 1935, based on his observations of Alaskan and Mongolian microcores, N.C. Nelson suggested a possible cultural connection between Asia and America when he noticed that 'about twenty small semi-conical flint cores and several end-scrapers [from the Campus site] … are identical in several respects with thousands of specimens found in the Gobi desert … [and] furnish the first clear archaeological evidence we have of early migration to the American continent' (Nelson 1935: 356, cited in Mobley 1991: 1). This statement is reiterated a couple of years later when he says that 'in one of these collections are certain specimens of more than ordinary significance because they appear to suggest definite cultural relations between Alaska and Mongolia' (Nelson 1937: 267), 'suggesting that we have here a possible specific proof of culture connection between Asia and America' (ibid.: 268). These comments were largely based on his analysis of two site assemblages: Shabarakh Usu in Mongolia and the Campus site in Alaska (Fig. 14.1).

Since N.C. Nelson's publications, microblade assemblages and pressure technique have received a great deal of attention from archaeologists. To date, evidence for microblade industries has been found in all regions of greater Beringia, including Siberia (e.g. Flenniken 1987), the Russian Far East (e.g. Derevianko and Kononenko 2003; Tabarev 1997), Japan (e.g. Kobayashi 1970; Morlan 1976), China (e.g. Chen and Wang 1989), Korea (e.g. Seong 1998), Alaska (e.g. West 1996b), the Yukon Territory (e.g. Clark 1992), British Columbia (e.g. Carlson 1996; Magne and Fedje 2007) and Canada (e.g. Desrosiers 2009, this volume; Owen 1988).

The second half of the twentieth century witnessed a parade of terms used to cluster together Northeast Asian and northwest North American sites with microblade assemblages. It should be noted, however, that these different complexes and traditions have their own peculiarities and do not always deal with the same exact regions and/or time periods. Some of the terms used in North America in recent decades include (in chronological order of publication): the *Northwest Microblade Tradition* (defined by MacNeish in 1954), the *Arctic Small Tool Tradition* (defined by Irving in 1957), the *Denali Complex* (defined by West in 1967), the *American paleoarctic Tradition* (defined by Anderson in 1968) and the *Northeast Asian Northwest American Microblade Tradition*, or *NANAMT* (defined by Smith in 1974). In 1981, F.H. West proposed a larger construct for the early prehistory of Beringia that encompassed a wider variety of technologies but still included the early microblade assemblages: the *Beringian Tradition* (West 1981, 1996a). More recently, C.E. Holmes (2001) modified F.H. West's concept to *East Beringian Tradition*.

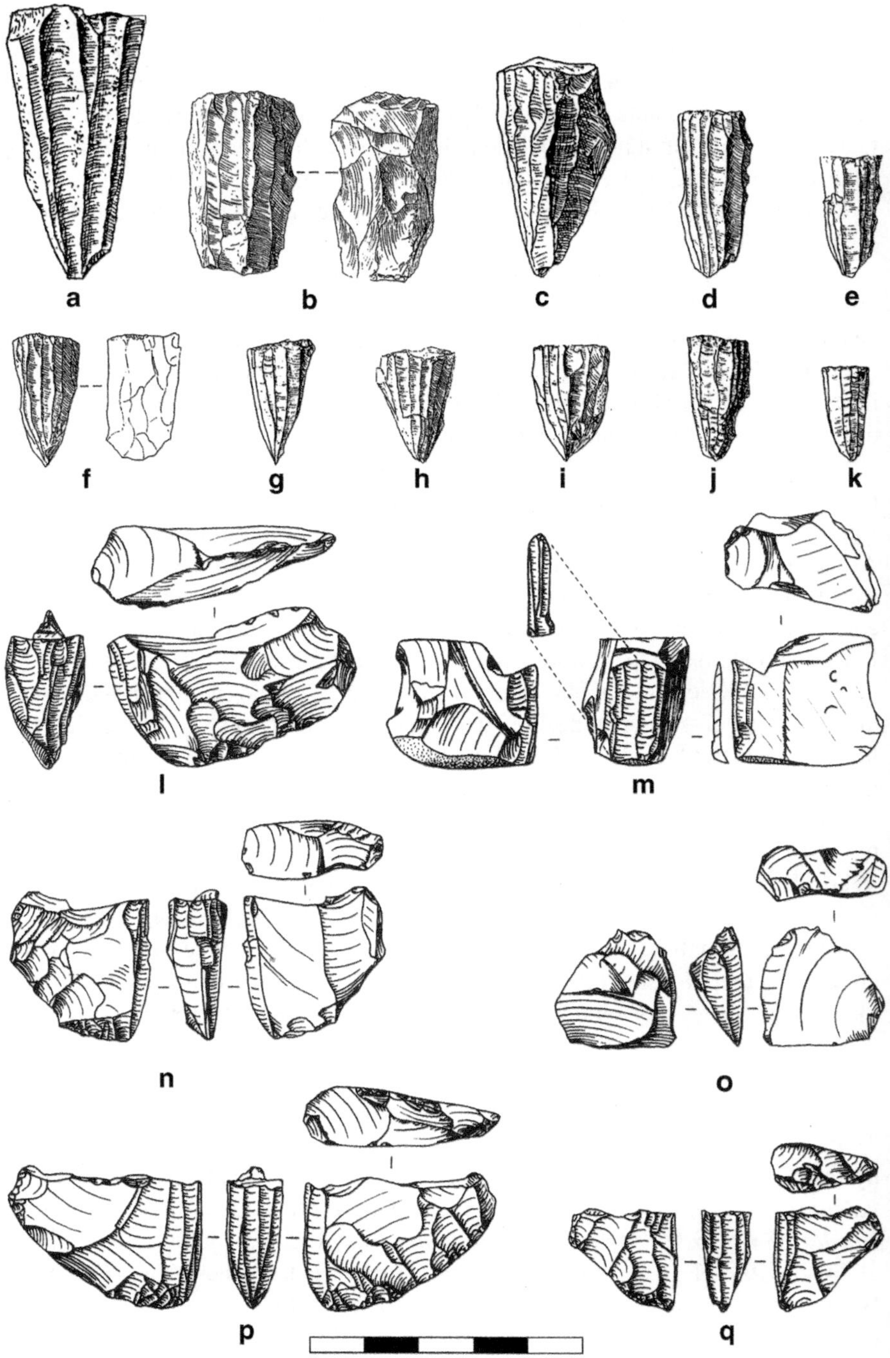

Fig. 14.1 (**a–k**) Microcores from Shabarakh Usu, Mongolia (Adapted from Fairservis 1993). (**l–q**) Microcores from Campus site, Alaska (Adapted from Mobley 1991)

14.3 Over 10,000 Years of Pressure Microblade Industries

Based on the current data from Swan Point, pressure microblades manufacture first appeared in interior Alaska around 12000–13000 uncal B.P. (Holmes 1996, 2001, 2011). According to C.E. Holmes, 'once this [microblade] technology became established in Alaska it appears to have remained throughout almost the entire Holocene, at least in the inter-montane regions of central Alaska and as well in Yukon and Northwest Territory' (Holmes 2001: 167). As research progresses and new sites are discovered, 'data continue to accumulate suggesting that microblade technologies, including wedge-shaped cores, persisted throughout most of the Holocene' (Bowers 1999: 12) (Fig. 14.2).

There is a succession of cultural traditions in Alaska that have employed pressure microblade production as part of their lithic toolkits. From oldest to youngest, they are the Denali Complex, the Northern Archaic Tradition and the Arctic Small Tool Tradition, as well as other late Holocene occurrences. Within these distinct cultural and chronological entities, different knapping methods and techniques as well as microcore morphologies can be identified. Yet, the sole discovery of pressure microblades in an assemblage can hardly inform us on the chronological and/or cultural period of a given site without the presence of other diagnostic artefacts and confirmation by absolute dating. Too many sites in Alaska have initially been interpreted as belonging to the Late Pleistocene based on typological comparisons only to be

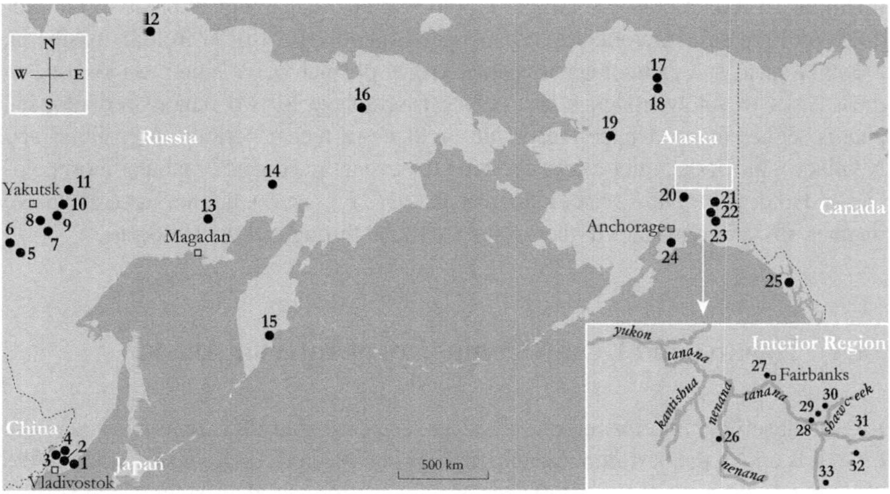

Fig. 14.2 Map of greater Beringia and Interior Alaska showing the location of mentioned sites in the article. *1* Ustinovka-6, *2* Gorbatka-3, *3* Molodiojna-1, *4* Risovoe-1, *5* Leten Novyy-1, *6* Kurung-2, *7* Ust-Timpton, *8* Tumulur, *9* Dyuktai Cave, *10* Verkhne-Troitskaya, *11* Ezhantsy, *12* Berelekh, *13* Kheta, *14* Druchak-V, *15* Ushki Lake sites, *16* Tytylvaam-4, *17* Lisburne, *18* Mesa, *19* Nogahabara-1 *20* Ravine Lake locality, *21* Phipps site, *22* Reger site, *23* Sparks Point, *24* Round Mountain microblade locality, *25* Hidden Falls, *26* Dry Creek, *27* Campus site, *28* Broken Mammoth, *29* Swan Point, *30* Mead site, *31* Healy Lake, *32* Gerstle River Quarry, *33* Donnelly Ridge

re-assigned to the Holocene following radiocarbon dating. The Campus site is a classic example of this problem. Based on typological comparisons, the microblade assemblage was first believed to be part of the final Pleistocene Denali Complex (West 1967). During a re-analysis of the microblade assemblage, C.M. Mobley (1991) dated this component to the Late Holocene (ca. 3500 B.P.). Finally, the latest re-investigation carried out at the site by G.A. Pearson and W.R. Powers placed the microblade assemblage around 7000 B.P., which they felt represented 'either early to mid-Holocene occupations of the Denali Complex and Northern Archaic tradition, or one or more early Northern Archaic occupation(s)' (Pearson and Powers 2001: 100). Another recent example is the Lisburne site in northern Alaska, which offers new 'evidence for a mid-late-Holocene persistence of a widespread Beringian lithic technology characterized primarily by wedge-shaped cores, microblades, and distinctive burins' and illustrates the hazardous task of making a 'de facto assumption of a Late Pleistocene/Early Holocene age for such artifacts' (Bowers 1999: 13).

On the contrary, in other regions of Beringia, such as in Yakutia (and more generally in Siberia), there are sometimes very distinct technological and typological differences between the microblade techniques and methods of the Upper Paleolithic (Dyuktai Complex) and those of the following periods (Sumnagin Late Paleolithic/Mesolithic and the Neolithic). On the one hand, the Dyuktai Complex microcores maintain morpho-technological links with the Yubetsu method of the Japanese Paleolithic (Fig. 14.3a–d). On the other hand, from the Mesolithic onwards, microcores are not of the wedge-shaped type, but will almost exclusively exhibit the morphology of conical, tabular or pyramidal cores (Fig. 14.3e–h). The appearance of the Sumnagin Complex in Northeast Asia represents a major typological and technological break in the production of pressure microblades (e.g. Gómez Coutouly 2011b). Therefore, in Siberia, morphological and technological characteristics of microcores alone can be used in most cases to safely assign a microblade assemblage to two major chronological phases: either to the Upper Paleolithic or to more recent periods (Mesolithic and Neolithic). In Alaska, microcores also tend to have more conical or tabular morphologies in Holocene toolkits (e.g. Holmes et al. 1996: Figs. 6–9), but not in an exclusive manner, since wedge-shaped microcores are found throughout the Holocene.

14.4 Nenana and Denali Complexes of Interior Alaska

I have chosen the interior region of Alaska as a case study for three main reasons. First, it is one of the best-documented regions in eastern Beringia where some of the oldest sites and the larger assemblages have been discovered, especially concerning early microblade-bearing occupations. Second, distinctive methods for producing microblades have been identified at several of these sites. Finally, there are specific discussions in this region concerning the chrono-cultural sequence of archaeological complexes and traditions (non-microblade and microblade) that are essential for understanding some of the current debates on Beringian archaeology. Hence, this region has produced the necessary data to present a good case study for the understanding of Alaska's early prehistory specifically and Beringia more generally.

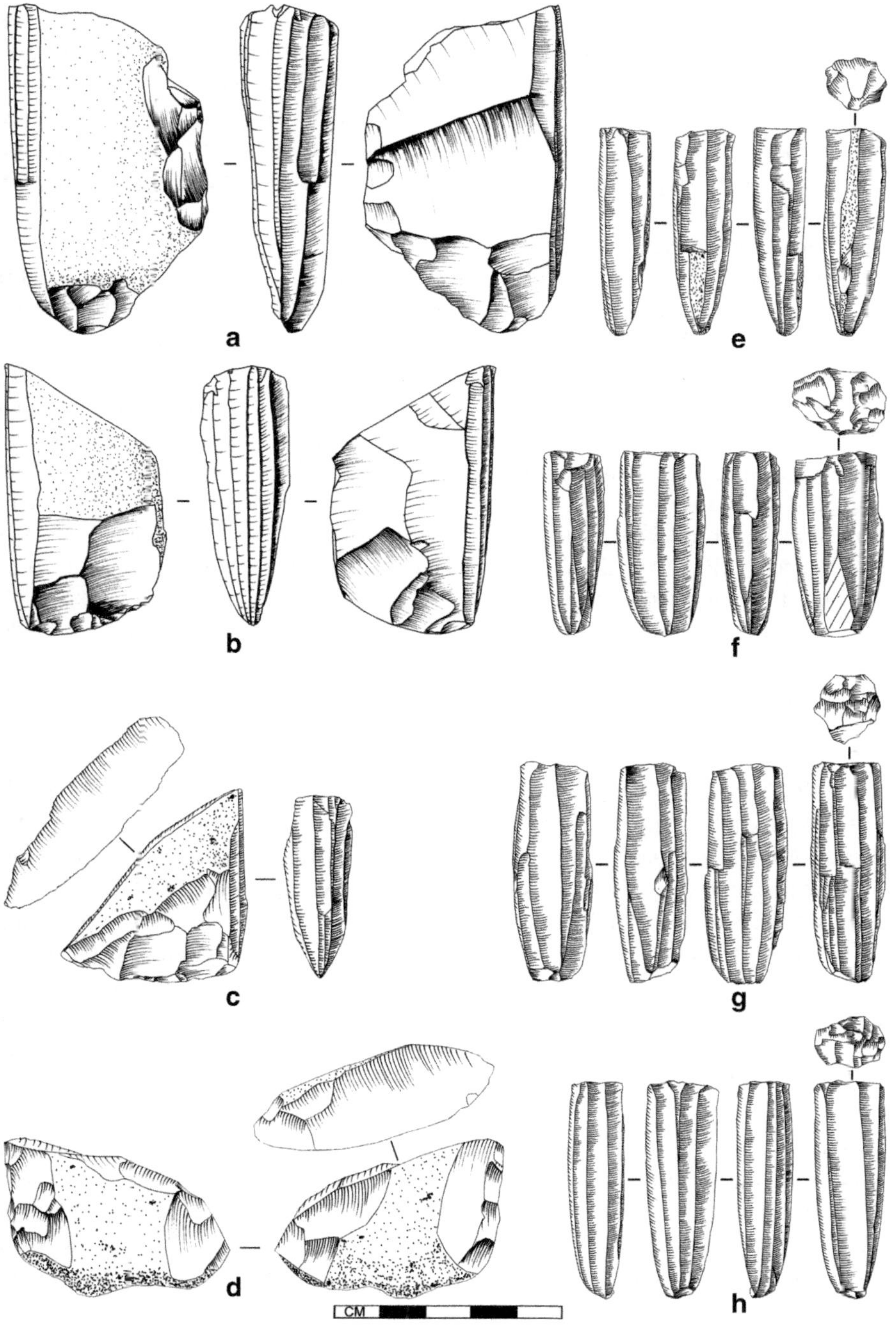

Fig. 14.3 (**a–d**) Microcores from Dyuktai Cave (Dyuktai Complex). (**e–h**) Microcores from Ust-Timpton site (Sumnagin Complex)

When discussing this region, I am mainly referring to the Nenana and Tanana River Valleys. The archaeological data unearthed in these two areas, though relatively close to one another, is contradictory. As J.F. Hoffecker (2001: 139) notes, 'the archaeological record in the Nenana River Valley suggests that two separate and temporally successive industries were present in central Alaska during the Late Pleistocene and Early Holocene, but the Tanana Basin sites indicate a more complex picture'. The two 'separate and temporally successive industries' refer to the Nenana and Denali Complexes.

The Nenana Complex was defined by W.R. Powers and J.F. Hoffecker (1989) and refers to an archaeological horizon that recurs in Pleistocene sites of the Nenana Valley. This complex is primarily characterized by the absence of microcores and microblades and is always stratigraphically below the Denali Complex (microblade-bearing levels). Apart from the absence of microcores, the Nenana Complex is also defined by a macro-blade technology, tools on blades, end-scrapers and bifaces. The teardrop-shaped Chindadn points are also considered characteristic tools of this complex, although some authors have reported the co-occurrence of Chindadn points and microblade industries (Holmes 2008). Dating between ca. 11500 and 10500 uncal B.P. (Dixon 2001), the Nenana Complex was originally considered as the oldest human occupation in Alaska.

The Denali Complex was defined by F.H. West in 1967 based on microblade sites from central Alaska. It is characterized by the presence of microcores and pressure microblades, but also by other tools such as bifaces and burins. This complex dates from ca. 10500 to 8000 uncal B.P. (Dixon 2001) and has clear ties with Siberian Paleolithic microblade industries (Dyuktai Complex and related sites).

In the Nenana Valley, Nenana Complex occupations are chronologically followed by Denali Complex occupations. However, this is not always the case in the Tanana Valley, especially at the Shaw Creek sites (Broken Mammoth, Swan Point and Mead sites). C.E. Holmes (2001) reminds us that up to the 1970s, only a few sites where known in this region such as Donnelly Ridge (West 1967), Campus (Mobley 1991; Rainey 1939, cited in Holmes 2001) and Healy Lake (Cook 1975, 1996). It was not until the 1990s that new major sites were discovered and investigated such as Swan Point (Holmes et al. 1996) and Broken Mammoth (Holmes 1996) or reinvestigated such as at the Gerstle River Quarry site (Potter 2002). These sites, especially Broken Mammoth and Swan Point, changed our understanding of what was a relatively clear and simple vision of the first peopling of Alaska.

The lowest level of Broken Mammoth could represent a pre-Nenana human occupation or, at least, a much earlier Nenana-related occupation. Some archaeologists are indeed reluctant to 'assign these components [lowest level of Broken Mammoth] to the Nenana Complex on negative evidence alone [lack of microcores and microblades]' (Holmes 2001: 165). However, it is the Swan Point site that really questions the previously established picture, due to its microblade component dated to over 12000 B.P., making it the earliest evidence of human occupation in Alaska (Bever 2006; Holmes 2001). It is also the first clear proof of a pre-Nenana occupation with a microblade component. The Mead site has not produced evidence of microblade production thus far (Hamilton and Goebel 1999).

An ongoing debate concerns the exact relationship between the microblade and non-microblade Paleolithic sites. It revolves around trying to understand whether they represent separate archaeological entities or whether it is simply the consequence of different activities within the same complex, thus producing what might be perceived as different toolkits. In Alaska, this discussion is due to 'the weak and uncertain anteriority of the Nenana Complex to the Denali Complex that archaeologists thought they had perceived' and that is now 'tending towards a chronological overlap of both complexes as slightly older dates for the paleoarctic appear, as in Swan Point'* (Plumet 2004: 270). C.E. Holmes, discussing the microblade and non-microblade assemblages of the Tanana Valley, states that 'differences in assemblage compositions may reflect differences in site habitat, function, or seasonality of occupation' (Holmes 2001: 156) 'or most importantly, the sampling methods used at the site' (ibid.: 162).

A recently discovered site in northwestern interior Alaska, Nogahabara-1, has created new discussions into this debate given 'the co-occurrence ... of a number of tools and technologies considered by some to be characteristic of distinct archaeological cultures' (Odess and Rasic 2007: 708). This assemblage is considered by the authors as a single coherent Late Pleistocene assemblage corresponding to a brief occupation of the site and contains microcores and microblades, burins, lanceolate projectile points and side-notched projectile points. Nonetheless, a response to this article (Holmes et al. 2008) has countered D. Odess and J. Rasic's arguments and hypothesis based on a critical interpretation of the geological and cultural data available for the site and its surroundings. The results of their analysis led them to propose an alternative hypothesis according to which the site would most probably be a palimpsest or a mixed assemblage. Also contradicting the original authors, they believe that 'a review of relevant data on regional archaeology suggests a mid to Late Holocene age for these assemblage(s)' (ibid.: 782), and they consider erroneous the statement according to which the different Nogahabara-1 tool types 'are not currently encompassed by any single analytical construct for interior Alaska' (Odess and Rasic 2007: 708). This debate on the degree of interrelationship between microblade and non-microblade assemblages has been extrapolated to other regions of Alaska and of Beringia. Such is the case in a recent study discussing the differences in raw material procurement and selection between the non-microblade and microblade components of the Ushki Lake site in Kamchatka and of the Dry Creek site in Alaska (Graf and Goebel 2009).

14.4.1 Campus Method Versus Yubetsu Method

The two main methods that I have identified in early sites in interior Alaska are the Yubetsu method and the Campus method. Although not always named 'Campus

*Translated by the author

method', the presence of a distinct method other than Yubetsu in Alaska has been noticed by a variety of archaeologists (e.g. Chen 2007; Clark 1992; Morlan 1976; Holmesf 2008, 2011; Plumet 2004; West 1967). For example, F.H. West (1967: 367–368) described the difference in these terms:

> Platform preparation is distinctive [in the Denali Complex]. The technique employed is similar, and probably related to, the Yubetsu technique characteristic of Shirataki core burins of northern Japan ... There is an important difference, however: where the Shirataki artisan aimed at removing the entire top (as used here), thus creating a continuous smooth surface, the platform in the Donnelly core was prepared by a distinctive partial removal of the top.

The Yubetsu method (Fig. 14.4) was first described in the context of the Japanese Paleolithic by M. Yoshizaki in 1961 (Kobayashi 1970), although T. Sato had previously mentioned the existence of wedge-shaped cores made on split bifacial blanks in Mongolia (Sato 1960, cited in Kobayashi 1970). The Yubetsu method is found throughout greater Beringia and as far away as Central Asia (Brunet 2002) and Turkey (Balkan-Atli et al. 1999; Binder and Balkan-Atli 2001). There are obvious differences in core morphology from one region to the other due to the size, quality and type of raw materials available. There are also differences in the type of preparation and maintenance of the core, in the way tablets are removed and in the reduction sequence. However, the main characteristics of the Yubetsu method are the following (based on the definitions of Inizan et al. 1999; Kobayashi 1970; Tixier 1984 and on my own observations):

Step 1: Shaping of a leaf-shaped, often asymmetrical, bifacial preform.
Steps 2 and 3: Removing the less convex ridge of the biface (ridge-spall) followed by one or several tablets (ski-spalls) to create a flat pressure platform.
Step 4: Removal of a crested blade.
Step 5: Pressure microblade production is then carried out at one end of the bifacial core along the width of the biface (in some rare cases, microblades will be removed at both ends of the biface).

The Campus method (Fig. 14.5) is well documented in Alaska. Historically, Alaskan microcores have been labelled Campus, Denali or Gobi, but I propose to use the term Campus since C.M. Mobley (1991) already presented a detailed technological analysis of the microblade *chaîne opératoire* for the Campus site. D.W. Clark (1992) also described a very similar method for the microcores from the KbTx-2 site in the Yukon. It is important to note that the terms *Campus*, *Denali* and *Gobi* have not always been employed with the same meaning, and depending on the authors, they can refer to different methods. The main characteristics of the Campus method are the following (based on the definition of Mobley 1991 and on my own observations of the Alaskan material):

Step 1: The Campus method is mainly characterized by the use of flake blanks. Morphologies of the cores are quite varied due to the wide diversity of selected blanks and due to the variations in preform shaping.

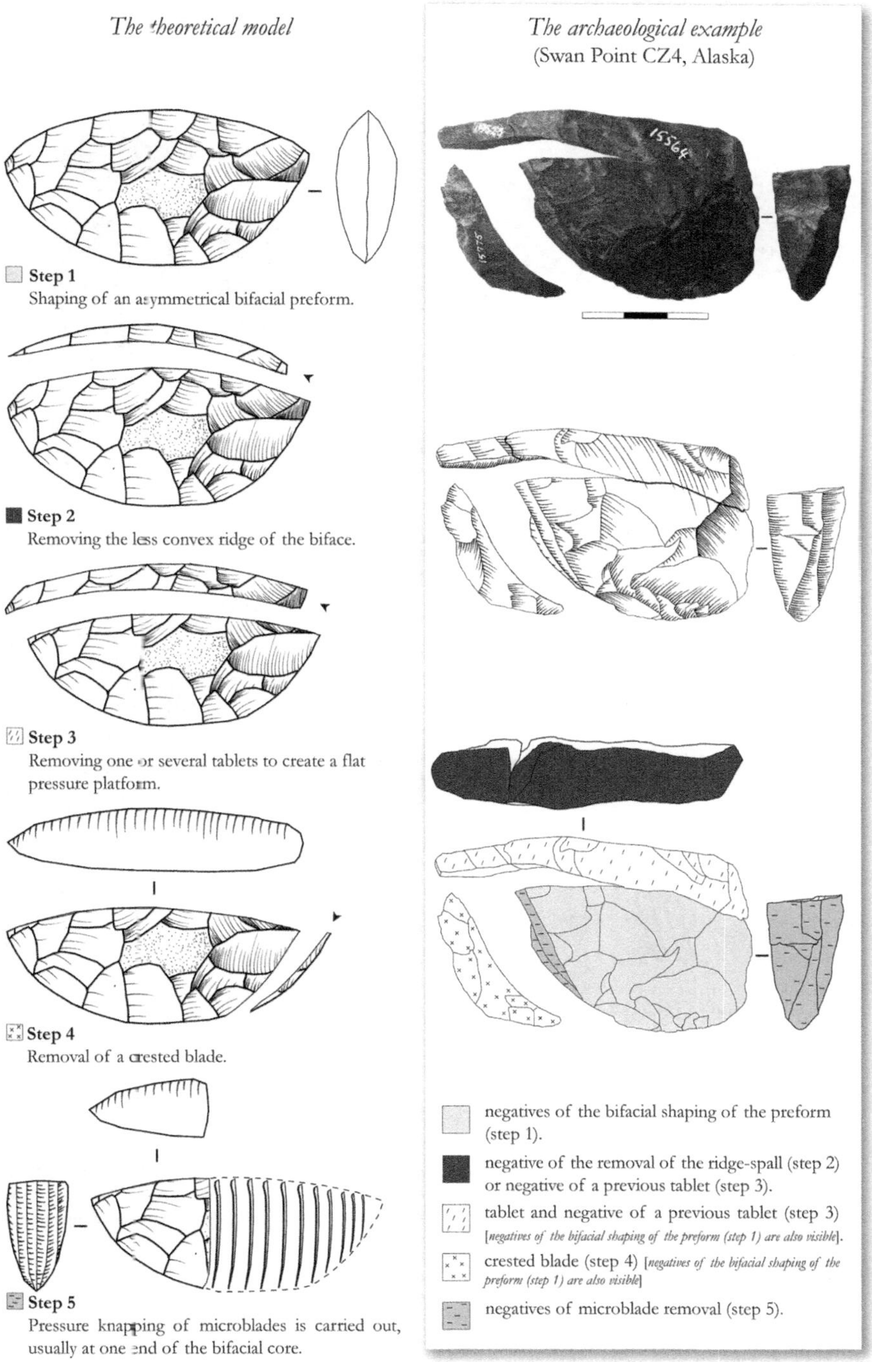

Fig. 14.4 The Yubetsu method

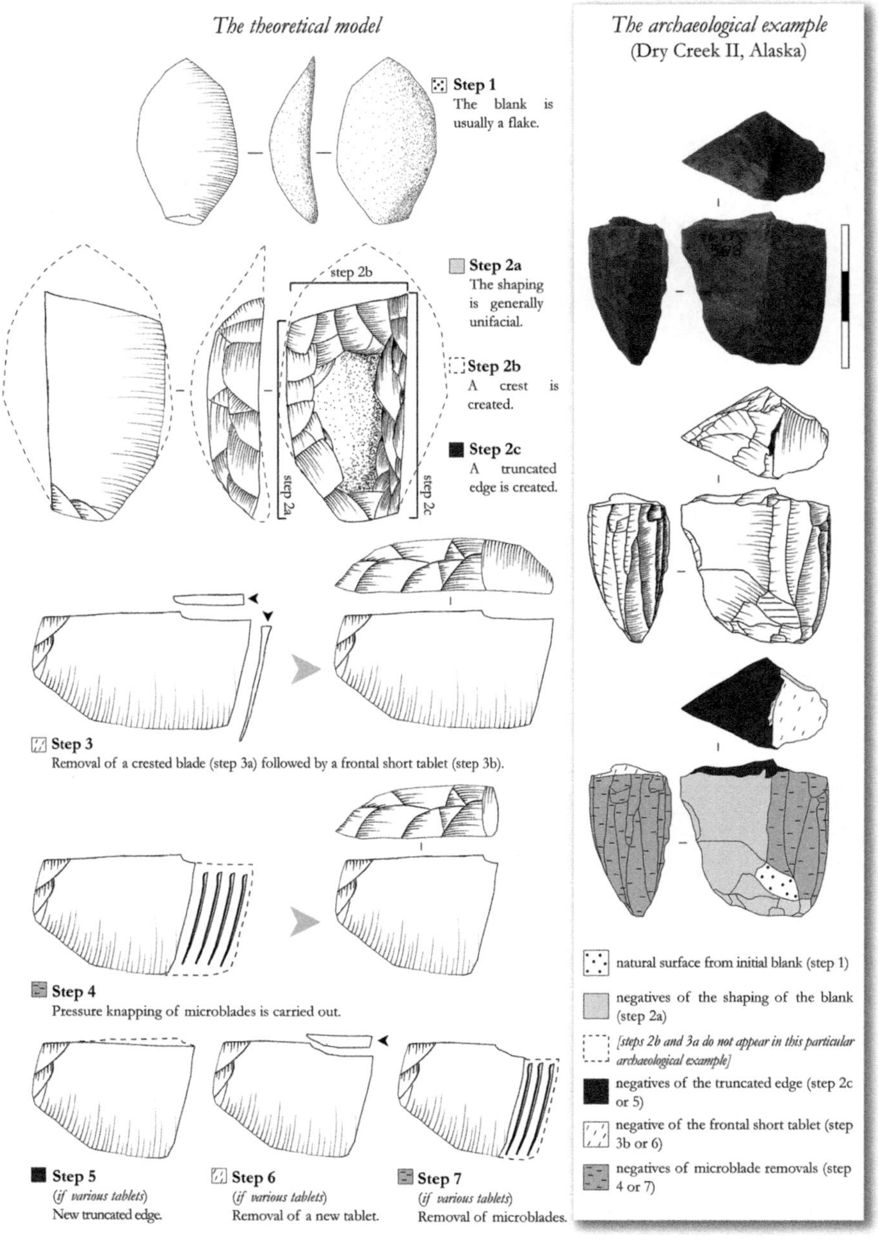

Fig. 14.5 The Campus method

Step 2a: The shaping of the blank is done by either unifacial or partially bifacial flaking, although in some cases, fully bifacial preforms have been observed (see below for a discussion on Campus method on bifacial preforms). The Campus method is also characterized by the relatively common presence of unretouched areas (natural surfaces, ventral surfaces of flakes, etc.). The Campus method therefore differs from the Yubetsu method, in that the latter usually has a complete bifacial preform and therefore lacks large portions of natural or unretouched surfaces.

Step 2b: During the shaping of the blank, a crest is also created.

Step 2c: Preparation of the platform is also very distinctive in the Campus method. Instead of preparing a ridge, as in the Yubetsu method, platforms were prepared with lateral blows to produce what have been called 'side struck platform flakes' by J.E. Mauger (1971: 9, cited in Mobley 1991: 32). These side-struck platform flakes create a straight truncated edge used as a platform.

Step 3 and 4: Once the platform is prepared, a short tablet and a crested blade are removed, followed by the removal of microblades.

Step 5, 6 and 7: If several tablets are removed during the reduction sequence, a new series of side-struck flakes to flatten the platform will be necessary. Furthermore, given that 'the point of force [is] deliberately aligned below the negative bulb of the previous removal' (Mobley 1996: 299), this technical procedure will sometimes produce characteristic 'gull-wing'-shaped flakes. In most cases, side-struck platform flakes will be seen on the core during most of the microblade production phase Then, a new short tablet will be removed, followed by microblades.

There are other less common variants, such as preparing the platform with side-struck flakes to be subsequently removed in its whole length by a long tablet (similarly to ski-spalls removals in the Yubetsu method). In this case, if the removal of the tablet was successful, no negatives of side-struck flakes will be visible on the platform (only through re-fitting will we be able to see the platform previously prepared with side-struck flakes).

However, the presence of side-struck platform flakes removal on a core should not be automatically considered as definitive evidence of the Campus method. This technical procedure can also be used to repair a failed platform removal (such as a Yubetsu ridge-spall) and should not be confused with the platform preparation of the Campus method. A miss-struck ridge-spall or ski-spall removal is often characterised by deviating to one side of the Yubetsu microcore, thus creating an unbalanced and sloped platform. Side-struck flake removals can then be used to even the platform into a flat surface. Only at that point can a new spall be detached. Therefore, although they will have very similar morphological attributes, they are two different technical procedures, with different intentions, and should not be confused.

Campus method on a bifacial preform. Out of context, the distinction between the Yubetsu and Campus methods is minimal when Campus microcores are shaped into bifacial preforms. Even though I am aware of the delicate issue of creating two different terms for two very similar cores, I propose to distinguish one from the other on the basis of platform preparation and tablet removals. Yubetsu platforms are characterized by the removal of a long ridge-spall, while Campus platforms are characterized by the removal of short spalls that have been previously prepared on a truncated edge with the detachment of side-struck platform flakes. However, the distinction should also be based on the overall assemblage (e.g. predominance of Yubetsu or Campus microcores).

14.4.2 Dry Creek Site

Dry Creek is located about 5 km from the town of Healy, in the Nenana River Valley. First discovered and initially investigated by C.E. Holmes in 1973, large-scale excavations were carried out afterwards by W.R. Powers until the late 1970s (Hoffecker 2001). In 1992, geoarchaeological investigations were carried out at the site by N. Bigelow and W.R. Powers (Bigelow and Powers 1994).

Dry Creek has two main components dating to the Late Pleistocene and Early Holocene. The earliest occupation level, Component I, is characterized by the presence of flake and blade cores, retouched blades, transverse scrapers, notches, gravers, cobble tools, side-scrapers, end-scrapers, bifaces and bifacial points, including the typical Chindadn points (Graf and Goebel 2009; Hoffecker 2001). Dry Creek Component I 'represents the "type-assemblage" for the central Alaskan Nenana Complex - a blade-and-biface industry lacking any signs of microblade and burin technologies' (Graf and Goebel 2009: 57). This component has been assigned to the non-microblade Nenana Complex and is presently dated by a single radiocarbon date of 11120±85 uncal B.P. (Powers and Hamilton 1978). Overlying this occupation is Component II, which is assigned to the Denali Complex, with three radiocarbon dates ranging from 10060±75 to 10690±250 uncal B.P. (Bigelow and Powers 1994). It is this prehistoric occupation, the oldest evidence of a microblade component in the Nenana Valley so far and one of the oldest from Alaska, which I will now discuss in more detail.

Dry Creek Component II (DC II) has yielded a wide variety of lithic artefacts, such as microcores, microcore preforms and tablets, microblades, burins, bifaces, bifacial points, end-scrapers, side-scrapers and cobble tools. It is one of the most complete and characteristic Denali assemblages recovered to date. The DC II assemblage contains almost 29,000 artefacts (Hoffecker et al. 1996), among which there are about 1,800 pieces directly associated with microblade technology (microcores, tablets, microblades, etc.) including 1,772 microblades (Hoffecker 2001).

Raw materials found at DC II include degraded quartzite, eight varieties of cryptocrystalline silicates (CCS), four varieties of fine-grained volcanics (rhyolite, diabase, basalt and dacite), obsidian, quartzite and argillite (Graf and Goebel 2009). Degraded quartzite and CCS are by far the main type of raw material used at DC II, making up nearly 78% of the total, while obsidian artefacts only amount to 2.7% of the total assemblage (ibid.). Although no obsidian microcores have been found, we know that obsidian was used for pressure microblade production based on the presence of obsidian microblades. One of the obsidian microblades (a fragment) is

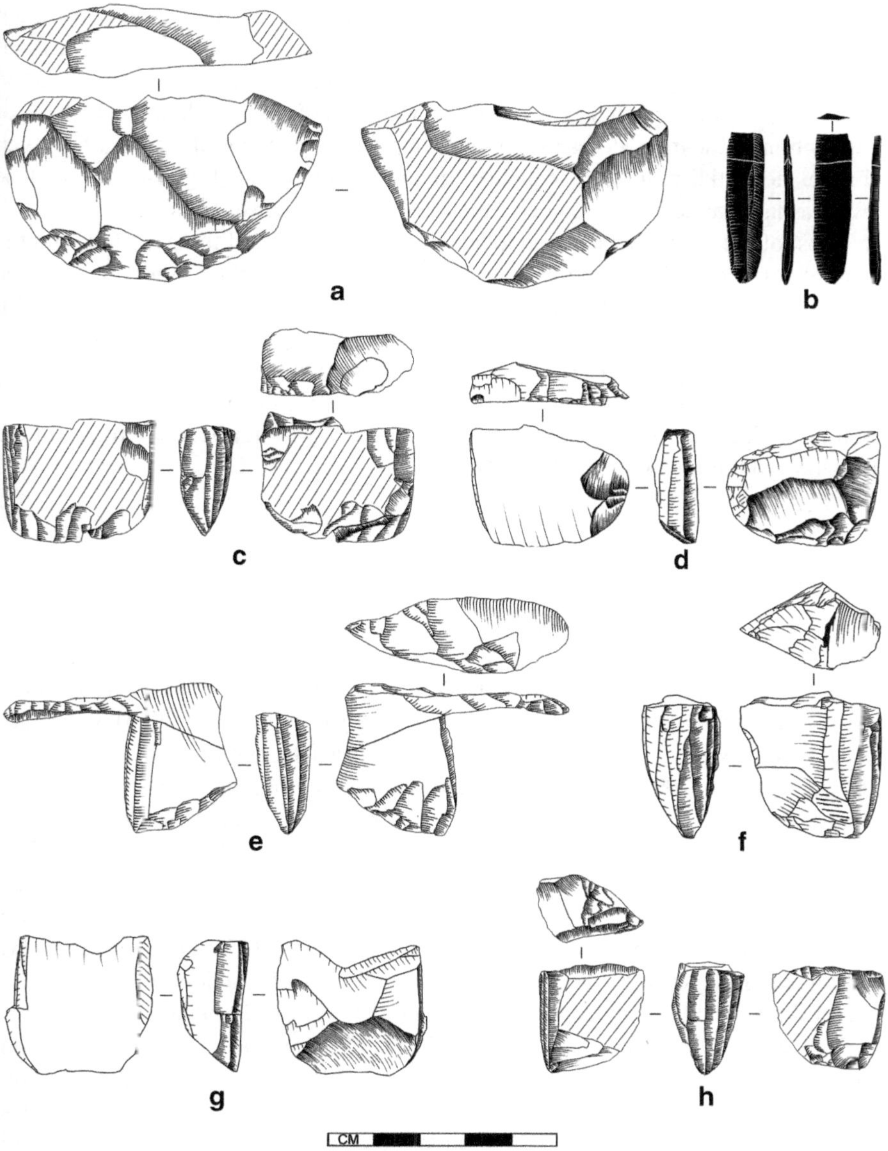

Fig. 14.6 Selected artifacts from Dry Creek component II, Nenana valley, Alaska. (**a**) Microcore preform. (**b**) Obsidian microblade. (**c**–**h**) Microcores

interesting because its length and width are actually larger than most of the other microblades and microcores (Fig. 14.6b). The nearest obsidian sources known are from *Batza Téna* (300 km from Dry Creek) and from the Wrangell Mountains (350 km from Dry Creek), and at least some of the obsidian artefacts have been brought from these sources (Cook 1995).

The characteristics seen on the microblades and on the microcores leave no doubt about the use of pressure technique in this assemblage. Microcores at DC II are usually made according to the Campus method. Microcores are essentially produced on small cobbles or thick flakes removed from such cobbles. In their final phase of production, these microcores usually have a maximum length of 3–4 cm; however, during the initial reduction sequence, microcore preforms (Fig. 14.6a) could be twice as large as seen from some of the remaining preforms and tablets.

Most of the microcores are made on flake blanks. Some of them are left with the ventral surface of the flake unretouched (Fig. 14.6d, g), while others retain large portions of their natural or cortical surfaces (Fig. 14.6c, h), although a few exhibit bifacial shaping (Fig. 14.6f). One refitted microcore (Fig. 14.6e) shows the removal of at least one short tablet on a truncated platform (with side-struck flakes), subsequently removed in its whole length by a long tablet. This variant can only be identified through re-fitting since the rejected microcores will not bear any traces of the previously mentioned chain of events.

14.4.3 Swan Point Site

Swan Point is located in the Shaw Creek Flats in the central Tanana Valley. First discovered by R. VanderHoek and T.E. Dilley (under the direction of C.E. Holmes) in 1991, the site is still being excavated periodically (Holmes et al. 1996). Its archaeological levels span 14,000 years from the Late Pleistocene up to historical times (Speakman et al. 2007). It is an important site not only because a whole succession of chronological periods are well represented but mainly because its oldest component has provided numerous radiocarbon dates that makes it, based on current data, the oldest site in Alaska (and thus the oldest evidence of pressure microblade production in the Americas). The discussion and analysis to follow is on Swan Point's earliest microblade component, Cultural Zone 4 (CZ4). The earlier microblade-bearing layer at Swan Point has produced about 14 radiocarbon dates older than 10000 uncal B.P., and ten of these are around or over 12000 uncal B.P. (Holmes 2011). Among those dates, one of them (12060 ± 70 uncal B.P.) was obtained from mammoth ivory, another one (11770 ± 140 uncal B.P.) on carbon residue recovered from a chert microcore platform rejuvenation flake (Holmes 2001), and the oldest one so far is 12360 ± 60 uncal B.P. (Holmes 2011). As recently as a few years ago, some authors were still sceptical about the archaeological reality of the earliest microblade component at Swan Point, given that they were only associated with 'a couple of amorphous' microcores and considering that the microblades recovered from that early layer 'might best be labeled "blade-like flakes"' (Yesner and Pearson 2002: 136). Nevertheless, new radiocarbon dates and new artefacts have been obtained that firmly establish a Late Pleistocene age for CZ4 that is undeniably related with the production of microblades.

The assemblage of the early microblade-bearing level is composed of microcores, crested blades, ski-spalls, microblades, dihedral burins and blades (Holmes 1996). In Alaska, microblade and burin tools are considered typical of the Denali

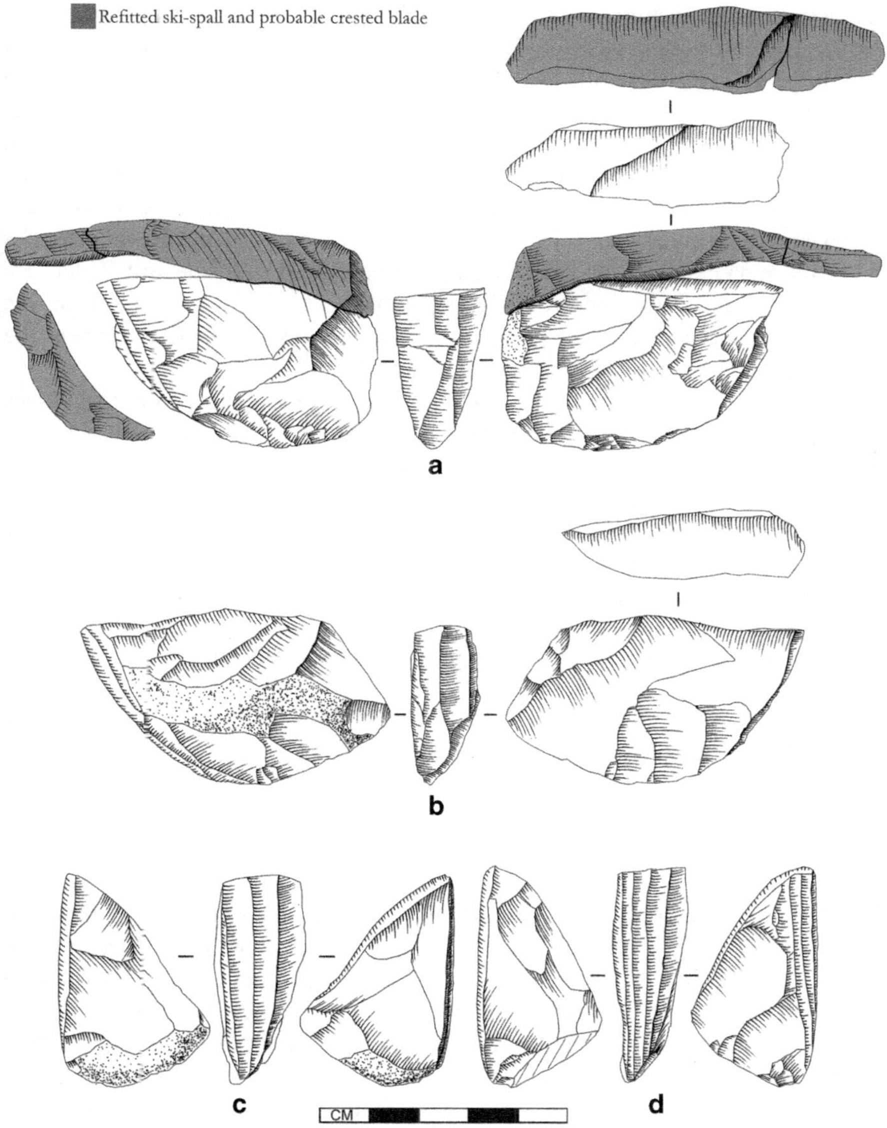

Fig. 14.7 Selected artifacts from Swan Point Cultural Zone 4, Tanana valley, Alaska. (**a**) Microcore with refitted ski-spall and crested blade. (**b–d**) Microcores

Complex when dated to the Pleistocene-Holocene boundary. Bifaces are another major tool of the Denali Complex, but are uncommon in the early Swan Point microblade assemblage (except, of course, for microblade cores themselves which are predominantly bifacial).

The microblade assemblage from CZ4 (Fig. 14.7a–d) is represented by microcores, various tablets, preforms as well as microblades. New microcores, microcore preforms and various tablets have been recovered at Swan Point since I did the analysis of the collection in 2006 (Holmes, 2007, personal communication; Tedor 2010). The microblade assemblage is highly standardized and coherent from a techno-typological point of view. The manufacturing process of almost all the microcores is in accordance with the typical Yubetsu method. Ridge-spalls and ski-spalls (i.e. tablets) found at the site clearly indicate that the technique for platform preparation was almost exclusively through the removal of long longitudinal blows to create flat pressure platforms (at least in one instance, multiple ski-spalls were removed from a microcore). There is however one reported microcore (Holmes 2008) made on a flake blank that has a platform prepared with side-struck platform flakes that would correspond to the Campus method, according to the definition above. However, this microcore was discovered after my first-hand analysis of the collection in 2006; therefore, I cannot comment on this latest find.

CZ4 microcores are not only coherent based on the method and techniques employed, but they are also standardized in almost every aspect and clearly show that one main *chaîne opératoire* was being used at the site for the production of pressure microblades. A greenish CCS is the main raw material used for the microcores and tablets recovered, although chert, rhyolite and obsidian were also used for the production of microblades (Holmes 2001). Obsidian is however quite rare in this early component (Speakman et al. 2007). Preform size (usually in the shape of bifaces) is also standardized, having an average size of about 8 cm long and 4–5 cm tall, based on the recovered preforms, ridge-spalls and ski-spalls.

14.5 Discussion and Hypothesis

14.5.1 Variability Among Alaskan Sites

Based on the study of the two key assemblages discussed in this article, there are two general methods for the production of microcores in Late Pleistocene – Early Holocene Alaska. On the one hand, we have microcores produced with the Campus method (Dry Creek Component II being one of the most characteristic assemblages), while on the other hand, at Swan Point Cultural Zone 4, the main method for producing microblades is Yubetsu (as we will see below, C.E. Holmes (2008, 2011) has already distinguished these two methods). The recognition of these two methods should not lead archaeologists to systematically fit every Alaskan wedge-shaped microcore within one of these two categories. Other methods yet to be identified and numerous variants of these two main methods most probably exist and co-exist.

Based on both my observations and published figures, it is clear that the Campus method is present in many Alaskan microblade sites, including Dry Creek, Campus (Mobley 1991, 1996), Donnelly Ridge (West 1967, 1996c), Phipps (West et al. 1996a), Sparks Point (West et al. 1996b), Reger (West 1996c), Hidden Falls (Davis 1989, 1996), Round Mountain microblade locality (Reger and Pipkin 1996), Ravine Lake locality (Robinson et al. 1996) and Mesa (Bever 2008), among others (Fig. 14.2). The Campus method is also known in sites from the Yukon Territory (e.g. Clark 1992). However, it is currently scarce at Swan Point CZ4, with only one reported microcore (Holmes 2008, 2011) that could be classified as the Campus method. Also, Swan Point is the only site to my knowledge that has a clear and consistent use of the Yubetsu method in Alaska, information which has already been observed before by C.E. Holmes. It is undeniable that the Denali Complex has clear ties with the Siberian Upper Paleolithic. Yet, from a technological and typological point of view, Swan Point CZ4 might be an even closer candidate for a direct link with the Siberian Dyuktai Complex. The fact that Swan Point is the earliest known human occupation in Alaska is important, but it is not a definitive factor. C.E. Holmes (ibid.) has argued that Swan Point CZ4, which has clear ties with the 'Dyuktai/Yubetsu technique' (i.e. Yubetsu method), should be viewed as the eastern branch of the Dyuktai Complex, while later microblade components of Swan Point and other Denali-related sites are more similar to the 'Campus technique' (i.e. Campus method). My analysis of Russian and American microblade collections supports his conclusion, as evidenced by the review of the archaeological data from Northeast Asia presented below.

14.5.2 Yubetsu and Campus Methods in Northeast Asia

The Campus method is widespread in Alaska but almost nonexistent in Siberian assemblages. My first-hand analysis of microblade sites from Siberia (including Dyuktai Cave, Verkhne-Troitskaya, Ezhantsy, Ushki Lake sites, Druchak-Vetrenikh and Kheta), from a number of sites in the Russian Far East (including Ustinovka-6, Molodiojna-1, Risovoe-1 and Gorbatka-3) and from bibliographic references of various sites, has shown that the Campus method and its specific platform preparation is quite limited in Northeast Asian microblade assemblages. In published data, drawings do not often show the platform, thus making its preparation difficult to understand. In all of these Siberian assemblages, the main method employed (usually the only one) is the Yubetsu method, and thus the platform is prepared through the removal of long longitudinal tablets covering the whole length of the core.

I have personally observed the presence of platform preparation with side-struck flakes on one single core from the Dyuktai Cave (Gómez Coutouly 2011a, b; Mochanov and Fedoseeva 1996b: Fig. 3.5a) and on the unique microcore found at Berelekh (Siberia) (Gómez Coutouly 2011b; Mochanov and Fedoseeva 1996a: Fig. 4.2, n) (Fig. 14.2). Based on bibliographic references, similar microcores have been found at Tumulur (Mochanov and Fedoseeva 1996e: Fig. 3.23), Kurung 2 (Mochanov and Fedoseeva 1996c: Fig. 3.31) and Leten Novyy-1 (Mochanov and Fedoseeva 1996d: Fig. 3.33) (Fig. 14.2). Interestingly enough, Mochanov and Fedoseeva consider that some of these sites (Leten Novyy-1, Tumulur and Berelekh) relate 'to the final stage of the Dyuktai culture and have an age of 13,000–10,500 years' (ibid.: 209). In the Tytylvaam site, one of the very rare dated Paleolithic sites of Chukotka (Kiryak 2004; Kiryak et al. 2003), platform preparation is usually made with side-struck flakes, typical of the Campus method. Nonetheless, this assemblage is mainly represented by microcores on bifacial preforms, not typical of the Campus method (Gómez Coutouly 2011b).

Therefore, it can safely be said that even if there are rare examples of Campus-like microcores in Siberia and in the Russian Far East, the Campus method (and its characteristic platform preparation with side-struck flakes) is a peculiarity from Alaska and the Yukon, until demonstrated otherwise. This leads us to the next question: how does the Campus method relate with the Yubetsu method?

14.5.3 Campus Method: Adaptability of the Yubetsu Mental Template?

The shaping of a biface in order to prepare a Yubetsu microcore is a demanding procedure. As described above, the platform preparation of the Yubetsu method is carried out by removing a ridge-spall followed by ski-spalls. In order to remove the ridge-spall, the bifacial preform has to be very carefully shaped: its ridges must be rectilinear and lined up with the axis of the biface to avoid knapping accidents. If struck incorrectly, the ridge-spall can deviate from the horizontal plane (i.e. towards one side of the biface), creating an unbalanced and sloped platform (Pelegrin, 2008, personal communication). Thus, the removal will be unsuccessful and repairing of the platform will be necessary (as described earlier).

The fact that side-struck platform flakes are a technical procedure seen on microcores on flakes or on irregular wedge-shaped cores should not be viewed as a coincidence. It is in fact quite the opposite. On one hand, if a crude preform is used, there is a high risk of failure when removing the long ridge-spall that creates the platform. On the other hand, if the intention is to prepare the platform using side-struck flakes, there is no need to prepare a bifacial preform and a ridge, and, thus, a wider variety of crude preforms (i.e. flake blanks with minimal shaping) can be used. Therefore, I consider that the Campus method widely used in Alaska corresponds to a more flexible adaptation (i.e. adaptability) of the Yubetsu mental template in that it provides more liberty and inflicts fewer restrictions when it comes to the selection of microcore blanks. In summary, the advantages of the Campus method are the following:

1. It permits the use of more varied blanks (such as flakes) that require much less shaping.
2. It is time effective since there is no thorough bifacial preform needed.
3. It is raw material efficient, since preparing the platform with side-struck flakes consumes less raw material than when preparing a symmetrical ridge (for the Yubetsu method).
4. It is risk effective through the use of short tablet removals.

This leads us to our final and most problematic question: which factor(s) caused this technological evolution (from Yubetsu to Campus)?

14.5.4 The Importance of Obsidian and Good-Quality Raw Material

The availability of good-quality raw material is important for pressure flaking and pressure blade production (Inizan et al. 1999; Pelegrin and Yamanaka 2007; Whittaker 1994; and many others). Concerning pressure microblade assemblages in greater Beringia, many authors have already mentioned, explicitly or implicitly, the importance of raw material in the development of this new technology. In Japan 'it is presumed that the development of the various microblade techniques of Hokkaido was caused by the technological adaptation to exploit obsidian or oil shale. For example, at the Shirataki sites, located at the Shirataki obsidian sources, bifacial blanks were consistently produced with the Yubetsu method' (Sato and Tsutsumi 2007: 58). For Korea, 'one possible explanation for the origin of blade and microblade technology in Korea is based on the quality of raw material ... With the increased need to produce more standardized stone implements through time it has been suggested that higher quality raw materials were utilized (e.g., obsidian, shale and tuff)' (Norton et al. 2007: 99). In Mongolia, the importance of high-quality raw material is also encountered such as in the Moil'tyn Am site, where the Yubetsu method is present: 'the almost totality of worked material ... are coming from strictly local sources ... An exception could be made by a very thin grained flint ... whose origin could locate farther. Most of the pressure bladelets were by the way made from that high quality material' (Bertran et al. 1998: 221–222).

In Alaska (and neighbouring regions), there are various identified obsidian sources including the *Batza Téna* in northwestern Alaska (Clark and Clark 1993), the Wrangell Mountains in eastern Alaska (Cook 1995), Suemez Island in Southeastern Alaska (Moss and Erlandson 2001) and Mount Edziza in British Columbia, Canada (Fladmark 1984). But there are no obsidian sources identified in interior Alaska (Graf and Goebel 2009), which leaves *Batza Téna* and the Wrangell Mountains as the closest known sources. And even though some authors consider that some of the unidentified obsidian sources could be located somewhere nearer to the Nenana and Tanana Valleys (Graf and Goebel 2009; Speakman et al. 2007), I consider the possibility of a local or near-local source in these regions highly improbable because it would be clearly reflected in the raw material consumption of microblade-bearing sites located in the region. There is no doubt that obsidian is the most suitable raw material for pressure microblade production, and if it had been easily acquired, it would have quickly become a preferential raw material, as seen in many microblade-bearing sites located near obsidian sources in the Russian Far East, Japan, northern Alaska, southern Alaska and British Columbia.

14.5.5 Hypothetical Model: The Rise and Fall of the Yubetsu Method

The debate concerning the birthplace and chronology of the first appearance of microblade industries is outside the scope of this article (for more information on this issue, see Gómez Coutouly 2011b; Kuzmin et al. 2007). The following hypothetical model focuses on the technological processes and factors that might be involved in the rise and fall of the Yubetsu method. The objective is to argue that the Campus method should be viewed as a technological and cultural variant of the widespread Asian Yubetsu method and was probably influenced by raw material availability.

Based on my critical analysis (Gómez Coutouly 2011b) of the published data, I believe that the appearance of the Yubetsu method is to be found in the Far East. Some of the earliest reliably dated sites with clear data proving the presence of the Yubetsu method or one of its variants are to be found in Korea (such as at the Hopyeong-dong site, ca. 24000–20000 B.P.) or in Hokkaido (e.g. at the Kashiwadai-1 site, ca. 20000–19000 B.P.). The available raw material probably played an important role in the emergence of the Yubetsu method when taking into consideration that pressure technique and the demanding procedure of carefully shaping an asymmetrical biface (for the Yubetsu method) require high-quality raw material. Therefore, an ideal environment in terms of raw material would most certainly be a good candidate for the birthplace of the Yubetsu method. And obsidian and other high-quality raw materials are indeed abundant in Hokkaido and Korea.

Upon arrival in Alaska, the Yubetsu method evolved into the Campus method, based on our current knowledge. In some regions of Alaska, high-quality raw material and large nodules of chert are found. Nonetheless, the oldest known sites are in interior Alaska where raw material is found mainly in the form of cobbles and locally available obsidian sources appear to be lacking. C.M. Mobley (1991: 25) describes that 'few of the chert cobbles available locally in alluvial deposits of the Tanana River are larger than one's fist' and the same is true for the Nenana River (Graf and Goebel 2009). Thus, interior Alaska, where high-quality obsidian and large blocks of chert are not as easily accessible as in other areas of Alaska could be considered as the broad region where the simplified and raw material–efficient version of Yubetsu (i.e. the Campus method) materialized. In this scenario, Swan Point CZ4 could be viewed as the first assemblage yet discovered to represent the 'transitional' period between the fall of the Yubetsu method and the rise of the Campus method.

14.6 Conclusion

The model above does not take into account methods other than Yubetsu and Campus that have been identified in other areas of Northeast Asia (such as the Horoka method) given that my purpose was to propose a model for the development of the Campus method in Alaska, as well as its technological and cultural relationship with the Asian Yubetsu method. This is obviously a broad hypothetical model that still needs to be demonstrated and argued in detail as concrete evidence and new dated microblade sites are discovered in Alaska and Northeast Asia.

It is possible that new sites, older than Swan Point, located near obsidian or other high-quality raw material sources and already producing microcores with the Campus method will be discovered in Alaska or in neighbouring Chukotka. If so, it will probably invalidate the assumption that the Campus method arose from the lack of high-quality raw material. However, the fact that the Campus method is risk effective, time effective and raw material efficient (when compared to Yubetsu) cannot be considered a minor matter and will have to be taken into account when explaining the external factors that could have caused this technological change.

Acknowledgements I am grateful to Dr. Jacques Pelegrin (CNRS, French National Center for Scientific Research) for our discussions concerning methodological and technological issues relevant to this article. I would like to thank the following researchers for their help during my stay in Alaska and for giving me access to their collections: Prof. David Yesner (University of Alaska Anchorage), Dr. Chuck Holmes (Office of History and Archaeology, Anchorage), Dr. Jim Whitney and Dr. Dan Odess (Museum of the North, Fairbanks) and Robert Gal (National Park Service, Anchorage). My gratitude also goes to the following Canadian colleagues, who have also permitted me to study their collections: Dr. Roy Carlson and Dr. Knut Fladmark (Simon Fraser University, British Columbia), Dr. Quentin Mackie (University of Victoria, British Columbia) and Dr. Daryl Fedje (Parks Canada, Victoria, British Columbia). For giving me the opportunity to study the Siberian collections, I am most grateful to Dr. Yuri A. Mochanov and Dr. Svetlana Fedoseeva (Center for the Archaeology and Paleoecology of Arctic People, Russian Academy of Sciences, Yakutsk); Dr. Sergey Slobodin, Dr. Alexander I. Lebedintsev and Dr. Margarita Kyriak (North-East Interdisciplinary Scientific Research Institute, Russian Academy of Sciences, Magadan); Prof. Irina Y. Ponkratova (Northern International University, Magadan) and Dr. Igor Vorobei (Magadan Oblast Museum of Local Studies). I would also like to thank the colleagues from Vladivostok who have welcomed me and have given me access to their microblade collections from the Russian Far East: Dr. Nikolay Kluyev, Dr. Nina A. Kononenko, Dr. Alla V. Garkovik (Institute of History, Archaeology and Ethnography, Russian Academy of Sciences, Vladivostok) and Dr. Alexander Krupyanko and Dr. Anatoly Kuznetsov (Far Eastern University, Vladivostok). Finally, I would like to thank Randy Tedor (University of Alaska Anchorage, USA), Dr. Kelly Graf (Center for the Study of the First Americans, College Station, USA) and Dr. Caroline Renard (University of Paris 1, France) for their revisions and comments of earlier drafts of this article

References

Anderson, Douglas D. 1968 A Stone Age Campsite at the Gateway to America. *Scientific American* 218(6): 24–33.

Bae, Kidong 2010 Origin and patterns of the Upper Paleolithic industries in the Korean Peninsula and movement of modern humans in East Asia. *Quaternary International* 211: 103–112.

Balkan-Atli, Nur, Didier Binder, and Catherine Kuzucuoğlu 1999 L'Atelier Néolithique de Kömürcü-Kaletepe: fouilles de 1998. *Anatolia Antiqua* 7: 231–243.

Bertran, Pascal, Jacques Jaubert, Monique Olive, Valéry Sitlivy, and B. Tsogtbaatar 1998 The Palaeolithic Site of Moil'tyn Am (Harhorin, Mongolia), Thirty Years after A.P. Okladnikov. In *Paleoekologia Pleistotzena i Kulturi Kamienovo Beka Srednei Asii i Sopredelnikh Teritorii (Materiali Mejdunar Simposiuma)*. Vol. 2, pp. 210–226. Institute of Archaeology and Ethnography, Russian Academy of Sciences, Novosibirsk.

Bever, Michael R. 2006 Too Little, Too Late? The Radiocarbon Chronology of Alaska and the Peopling of the New World. *American Antiquity* 71(4): 595–620.

Bever, Michael R. 2008 Distinguishing Holocene Microblades from a Paleoindian Component at the Mesa Site, Alaska. *Journal of Field Archaeology* 33(2): 133–150.

Bigelow, Nancy H., and W. Roger Powers 1994 New AMS dates from the Dry Creek Paleoindian Site, Central Alaska. *Current Research in the Pleistocene* 11: 114–116.

Binder, Didier, and Nur Balkan-Atli 2001 Obsidian Exploitation and Blade Technology at Kömürcü-Kaletepe (Cappadocia, Turkey). In *Beyond Tools: Reconsidering Definitions, Counting and Interpretation of Lithic Assemblages, Workshop PPN Chipped Lithic Industries*, edited by C. Caneva, C. Lemorini, D. Zampetti and P. Biagi, pp.1–16. Ex oriente, Berlin.

Bowers, Peter M. 1999 AMS Dating of the Area 22 American PaleoArctic Tradition Microblade Component at the Lisburne Site, Arctic Alaska. *Current Research in the Pleistocene* 16: 12–15.

Brunet, Frédérique 2002 Asie Centrale: vers une Redéfinition des Complexes Culturels de la Fin du Pléistocène et des Débuts de l'Holocène. *Paléorient* 28(2): 9–24.

Carlson, Roy L. 1996 Early Namu. In *Early Human Occupation in British Columbia*, edited by Roy L. Carlson and Luke Dalla-Bona, pp. 83–102. University of British Columbia, Vancouver.

Chen, Chun 2007 Techno-typological Comparison of Microblade Cores from East Asia. In *Origin and Spread of Microblade Technology in Northern Asia and North America*, edited by Y.V. Kuzmin, S.G. Keates and C. Shen, pp. 7–38. Archaeology Press, Simon Fraser University, Burnaby.

Chen, Chun, and Xiang-Qian Wang 1989 Upper Paleolithic Microblade Industries in North China and their Relationships with Northeast Asia and North America. *Arctic Anthropology* 26(2): 127–156.

Clark, Donald W. 1992 A Microblade Production Station (KbTx-2) in the South Central Yukon. *Canadian Journal of Archaeology* 16: 3–23

Clark, Donald W., and A. McFadyen Clark 1993 *Batza Téna Trail to Obsidian. Archaeology at an Alaskan Obsidian Source*. Archaeological Survey of Canada, Paper No. 147, Mercury Series. Canadian Museum of Civilization, Hull.

Cook, John P. 1975 Archaeology of Interior Alaska. *The Western Canadian Journal of Anthropology* 5(3–4): 125–133.

Cook, John P. 1995 Characterization and Distribution of Obsidian in Alaska. *Arctic Anthropology* 32(1): 92–100.

Cook, John P. 1996 Healy Lake. In *American Beginnings: the Prehistory and Palaeoecology of Beringia*, edited by Frederick H. West, pp. 323–327. University of Chicago Press, Chicago.

Davis, Stanley D. 1989 *The Hidden Falls site, Baranof Island, Alaska*. Aurora monograph series. Alaska Anthropological Association, Anchorage.

Davis, Stanley D. 1996 Hidden Falls. In *American Beginnings: the Prehistory and Palaeoecology of Beringia*, edited by Frederick H. West, pp. 413–424. University of Chicago Press, Chicago.

Derevianko, Anatoly P., and Nina A. Kononenko (editors) 2003 *Foraging Population of the Sea of Japan during the Late Pleistocene - Early Holocene*. Institute of History, Archaeology and Ethnography of the Peoples of the Far East, Russian Academy of Science, Far Eastern Branch, Novosibirsk.

Desrosiers, M. Pierre 2009 *A l'origine du dorsétien. Apports de la technologie lithique des sites GhGk-63 et Tayara (KbFk-7) au Nunavik*. Unpublished Ph.D. dissertation, UFR d'Histoire de l'Art et d'Archéologie, Université de Paris 1, Paris.

Dixon, E. James 2001 Human Colonization of the Americas: Timing, Technology and Process. *Quaternary Science Reviews* 20(1–3): 277–299.

Fairservis, Walter Ashlin 1993 *Archaeology of the Southern Gobi of Mongolia*. Carolina Academic Press, Centers of civilization, Durham.

Fladmark, Knut R. 1984 Mountain of Glass: Archaeology of the Mount Edziza Obsidian Source, British Columbia, Canada. *World Archaeology* 16(2): 139–156.

Flenniken, J. Jeffrey 1987 The Paleolithic Dyuktai Pressure Blade Technique of Siberia. *Arctic Anthropology* 24(2): 117–132.

Goebel, Ted, Michael R. Waters, Ian Buvit, Mikhail V. Konstantinov, and Aleksander V. Konstantinov 2000 Studenoe-2 and the Origins of Microblade Technologies in the Transbaikal, Siberia. *Antiquity* 74: 567–575.

Gómez, Coutouly, Yan Axel 2007 Rethinking the Ustinovka Complex: Lithic Technology and Raw Material in Palaeolithic Microblade Industries of Primorye (Russian Far East). *North Pacific Prehistory* 1: 65–110.

Gómez, Coutouly, Yan Axel 2011a Identifying Pressure Flaking Modes at Dyuktai Cave: A Case Study of the Siberian Upper Paleolithic Microblade Tradition. In *From the Yenisei to the Yukon: Interpreting Lithic Assemblage Variability in Late Pleistocene/Early Holocene Beringia*, edited by Ted Goebel and Ian Buvit. Texas A&M University Press, College Station, in press.

Gómez, Coutouly, Yan Axel 2011b Industries lithiques à composante lamellaire par pression du Nord Pacifique de la fin du Pléistocène au début de l'Holocène : de la diffusion d'une technique en Extrême-Orient au peuplement initial du Nouveau Monde. Unpublished Ph.D. dissertation, Institute for Archaeology and Anthropology, Paris West University Nanterre La Défense, Nanterre, in press.

Graf, Kelly E. 2009 Modern Human Colonization of the Siberian Mammoth Steppe: A View from South-Central Siberia. In *Sourcebook of Paleolithic Transitions: Methods, Theories and Interpretations*, edited by Marta Camps and Parth R. Chauhan, pp. 479–501. Springer, New York.

Graf, Kelly E., and Ted Goebel 2009 Upper Paleolithic Toolstone Procurement and Selection across Beringia. In *Lithic Materials and Paleolithic Societies*, edited by Brian Adams and Brooke S. Blades, pp. 55–77. Blackwell Publishing, Oxford.

Hamilton, Thomas D., and Ted Goebel 1999 Late Pleistocene Peopling of Alaska. In *Ice Age People of North America: environments, origins, and adaptations*, edited by Robson Bonnichsen and Karen L. Turnmire. Center for the Study of the First Americans, Oregon.

Hoffecker, John F. 2001 Late Pleistocene and Early Holocene Sites in the Nenana River Valley, Central Alaska. *Arctic Anthropology* 38(2): 139–153.

Hoffecker, John F., W. Roger Powers, and Nancy H. Bigelow 1996 Dry Creek. In *American Beginnings: the Prehistory and Palaeoecology of Beringia*, edited by Frederick H. West, pp. 343–352. University of Chicago Press, Chicago.

Holmes, Charles E. 1996 Broken Mammoth. In *American Beginnings: the Prehistory and Palaeoecology of Beringia*, edited by Frederick H. West, pp. 312–318. University of Chicago Press, Chicago.

Holmes, Charles E. 2001 Tanana River Valley Archaeology Circa 14,000 to 9,000 B.P. *Arctic Anthropology* 38(2): 154–170.

Holmes, Charles E. 2008 Lithic Technology of the East Beringian Tradition: Sustaining the Cutting Edge. Paper presented at the 73 rd Annual Meeting of the Society for American Archaeology, Vancouver.

Holmes, Charles E. 2011 The Beringian and Transitional Periods in Alaska: Technology of the East Beringian Tradition As Viewed From Swan Point. In *Explaining Lithic Assemblage Variability across Beringia*, edited by Ted Goebel and Ian Buvit. Texas A&M University Press, College Station, in press.

Holmes, Charles E., Ben A. Potter, Joshua D. Reuther, Owen K. Mason, Robert M. Thorson, and Peter M. Bowers 2008 Geological and Cultural Context of the Nogahabara I site. *American Antiquity* 73(4): 781–790.

Holmes, Charles E., Richard VanderHoek, and Thomas E. Dilley 1996 Swan Point. In *American Beginnings: the Prehistory and Palaeoecology of Beringia*, edited by Frederick H. West, pp. 319–323. University of Chicago Press, Chicago.

Inizan, Marie-Louise, Monique Lechevallier, and Patrick Plumet 1992 A Technological Marker of the Penetration into North America: Pressure Microblade Debitage, its Origin in the Paleolithic of North Asia and its Diffusion. In *Material Issues in Art and Archaeology III*, edited by P.B. Vandiver, J.R. Druzik, W.G.S. Freestone and I.C. Freestone, pp. 661–681. Symposium Proceedings, Vol. 267. Material Research Society, Pittsburgh.

Inizan, Marie-Louise; Michèle Redurin-Ballinger; Hélène Roche, and Jacques Tixier 1999 *Technology and Terminology of Knapped Stone*. CREP, Nanterre.

Ikawa-Smith, Fumiko 2007 Conclusion: in Search of the Origins of Microblades and Microblade Technology. In *Origin and Spread of Microblade Technology in Northern Asia and North*

America, edited by Y.V. Kuzmin, S.G. Keates and C. Shen, pp. 189–198. Archaeology Press, Simon Fraser University, Burnaby.

Irving, William N. 1957 An Archaeological Survey of the Susitna Valley. *Anthropological Papers of the University of Alaska* 6(1): 37–52.

Kiryak, Margarita 2004 Upper Paleolithic Sites in Western Chukotka (the Valley of the River Titilvaam). In *"Beringian Days" International Scientific and Practical Conference Papers, September 18–21, 2003, Anadyr*. Sovetsky Sport, Moscow.

Kiryak, Margarita A., Olga Y. Glushkova, and Thomas A. Brown 2003 Upper Paleolithic Sites in the Tytylvaam River Valley (Polar Chukotka). *Archaeology, Ethnology and Anthropology of Eurasia* 3(15): 2–15.

Kobayashi, Tatsuo 1970 Microblade Industries in the Japanese Archipelago. *Arctic Anthropology* 7(2): 38–56.

Kuzmin, Yaroslav V., Susan G. Keates, and Chen Shen (editors) 2007 *Origin and Spread of Microblade Technology in Northern Asia and North America*. Archaeology Press, Simon Fraser University, Burnaby.

MacNeish, Richard S. 1954 The Pointed Mountain site near Fort Liard, Northwest Territories, Canada. *American Antiquity* 19(3): 234–253.

Magne, Martin, and Daryl Fedje 2007 The Spread of Microblade Technology in Northwestern North America. In *Origin and Spread of Microblade Technology in Northern Asia and North America*, edited by Yaroslav V. Kuzmin, Susan G. Keates and Chen Shen, pp. 171–188. Archaeology Press, Simon Fraser University, Burnaby.

Mauger, J.E. 1971 *The Manufacture of Campus site Microcores*. Unpublished manuscript, Washington State University, Pullman.

Mobley, Charles M. 1991 *The Campus Site: A Prehistoric Camp at Fairbanks, Alaska*. University of Alaska Press, Fairbanks.

Mobley, Charles M. 1996 Campus site. In *American Beginnings: the Prehistory and Palaeoecology of Beringia*, edited by Frederick H. West, pp. 296–302. University of Chicago Press, Chicago.

Mochanov, Yura A., and Svetlana A. Fedoseeva 1996a Berelekh, Allakhovsk Region. In *American Beginnings: the Prehistory and Palaeoecology of Beringia*, edited by Frederick H. West, pp. 218–222. University of Chicago Press, Chicago.

Mochanov, Yura A., and Svetlana A. Fedoseeva 1996b. Dyuktai Cave. In *American Beginnings: the Prehistory and Palaeoecology of Beringia*, edited by Frederick H. West, pp. 164–174. University of Chicago Press, Chicago.

Mochanov, Yura A., and Svetlana A. Fedoseeva 1996c. Kurung 2 (Stratum, VI). In *American Beginnings: the Prehistory and Palaeoecology of Beringia*, edited by Frederick H. West, pp. 206–208. University of Chicago Press, Chicago.

Mochanov, Yura A., and Svetlana A. Fedoseeva 1996d. Leten Novyy 1 (Stratum IV). In *American Beginnings: the Prehistory and Palaeoecology of Beringia*, edited by Frederick H. West, pp. 208–211. University of Chicago Press, Chicago.

Mochanov, Yura A., and Svetlana A. Fedoseeva 1996e. Tumulur. In *American Beginnings: the Prehistory and Palaeoecology of Beringia*, edited by Frederick H. West, pp. 195–198. University of Chicago Press, Chicago.

Morlan, Richard E. 1976 Technological Characteristics of Some Wedge-shaped Cores in Northwestern North America and Northeast Asia. *Asian Perspectives* 19: 96–106.

Moss, Madonna L., and Jon M. Erlandson 2001 The Archaeology of Obsidian Cove, Suemez Island, Southeast Alaska. *Arctic Anthropology* 38(1): 27–47.

Nelson, Neis C. 1935 Early Migration of Man to America. *Natural History* 35(4): 356.

Nelson, Neis C. 1937 Notes on Cultural Relations between Asia and America. *American Antiquity* 2(4): 267–272.

Norton, C.J., K. Bae, H. Lee, and J.W.K. Harris 2007 A Review of Korean Microlithic Industries. In *Origin and Spread of Microblade Technology in Northern Asia and North America*, edited by Y.V. Kuzmin, S.G. Keates and C. Shen, pp. 91–102. Archaeology Press, Simon Fraser University, Burnaby.

Odess, Daniel, and Jeff Rasic 2007 Toolkit Composition and Assemblage Variability: The Implications of Nogahabara I, Northern Alaska. *American Antiquity* 72(4): 691–718.

Owen, Linda R. 1988 *Blade and microblade technology: selected assemblages from the North American Arctic and the Upper Paleolithic of southwest Germany.* BAR international series 441. Archeopress, Oxford.

Pearson, Georges A., and W. Roger Powers 2001 The Campus Site re-Excavation: New Efforts to Unravel its Ancient and Recent Past. *Arctic Anthropology* 38(1): 100–119.

Pelegrin, Jacques, and Ichiro Yamanaka 2007 From the Smallest to the Largest: Experimental Flaking by Pressure Techniques. *Cultura Antiqua* 58(4): 1–16. (In Japanese)

Plumet, Patrick 2004 *Peuples du Grand Nord: Des Mythes à la Préhistoire (tome I).* Editions Errance, Paris.

Potter, Ben A. 2002 A Provisional Correlation of Stratigraphy, Radiometric Dates, and Archaeological Components at the Gerstle River Site, Central Alaska. *Anthropological Papers of the University of Alaska, New Series* 2(1): 73–93.

Powers, W. Roger, and John F. Hoffecker 1989 Late Pleistocene Settlement in the Nenana Valley, Central Alaska. *American Antiquity* 54(2): 263–287.

Powers, W. Roger, and Thomas D. Hamilton 1978 Dry Creek: A Late Pleistocene Human Occupation Site in Central Alaska. In *Early Man in America from a Circum-Pacific Perspective*, edited by Alan Lyle Bryan. University of Alberta Occasional Papers 1. University of Alberta, Edmonton.

Rainey, F. 1939 Archaeology in Central Alaska. *Anthropology Papers of the American Museum of Natural History* 36(4): 355–405.

Reger, Douglas R., and Mark E. Pipkin 1996 Round Mountain Microblade Locality. In *American Beginnings: the Prehistory and Palaeoecology of Beringia*, edited by Frederick H. West, pp. 430–433. University of Chicago Press, Chicago.

Robinson, Brian S., Frederick H. West, and Douglas R. Reger 1996 Ravine Lake Locality. In *American Beginnings: the Prehistory and Palaeoecology of Beringia*, edited by Frederick H. West, pp. 438–443. University of Chicago Press, Chicago.

Sato, Tatsuo 1960 Horonbairu no Saisekki-bunka (The Microlithic Culture of Hulun-Buir). *Kokogaku Zasshi* 46(3): 252–260.

Sato, Hiroyuki, and Takashi Tsutsumi 2007 The Japanese Microblade Industries: Technology, Raw Material Procurement, and Adaptations. In *Origin and Spread of Microblade Technology in Northern Asia and North America*, edited by Y.V. Kuzmin, S.G. Keates, and C. Shen, pp. 53–78. Archaeology Press, Simon Fraser University, Burnaby.

Seong, Chuntaek 1998 Microblade Technology in Korea and Adjacent Northeast Asia. *Asian Perspectives* 37(2): 245–278.

Smith, Jason W. 1974 The Northeast Asian-Northwest American Microblade Tradition (NANAMT). *Journal of Field Archaeology* 1(3/4): 347–364.

Speakman, Robert J., Charles E. Holmes, and Michael D. Glascock 2007 Source Determination of Obsidian Artifacts from Swan Point (XBD-156), Alaska. *Current Research in the Pleistocene* 24: 143–145.

Tabarev, Andrei V. 1997 Paleolithic Wedge-Shaped Microcores and Experiments with Pocket Devices. *Lithic Technology* 22(2): 139–149.

Tedor, Randy 2010 Ice Patches, Mammoth Tusks, and the East Beringian Paleolithic. Paper presented at the roundtable Préhistoire de l'Alaska, 9 and 16 Septembre 2010, Institute for Archaeology and Anthropology, Nanterre.

Tixier, Jacques 1984 Le débitage par pression. In *Préhistoire de la pierre taillée. 2, Economie du débitage laminaire*, edited by Jacques Tixier, Marie-Louise Inizan and Hélène Roche, pp. 57–70. CREP, Paris.

West, Frederick H. 1967 The Donnelly Ridge Site and the Definition of an Early Core and Blade Complex in Central Alaska. *American Antiquity* 32: 360–382.

West, Frederick H. 1981 *The Archaeology of Beringia.* Columbia University Press, New York.

West, Frederick H. 1996a Beringia and New World Origins: II. The Archaeological Evidence. In *American Beginnings: the Prehistory and Palaeoecology of Beringia*, edited by Frederick H. West, pp. 537–559. University of Chicago Press, Chicago.

West, Frederick H. 1996b Donnelly Ridge. In *American Beginnings: the Prehistory and Palaeoecology of Beringia*, edited by Frederick H. West, pp. 302–307. University of Chicago Press, Chicago.

West, Frederick H. 1996c Reger site. In *American Beginnings: the Prehistory and Palaeoecology of Beringia*, edited by Frederick H. West, pp. 399–403. University of Chicago Press, Chicago.

West, Frederick H., Brian S. Robinson, and Mary Lou Curran 1996a Phipps site. In *American Beginnings: the Prehistory and Palaeoecology of Beringia*, edited by Frederick H. West, pp. 381–386. University of Chicago Press, Chicago.

West, Frederick H., Brian S. Robinson, and R. Greg Dixon 1996b Sparks Point. In *American Beginnings: the Prehistory and Palaeoecology of Beringia*, edited by Frederick H. West, pp. 394–399. University of Chicago Press, Chicago.

Whittaker, John C. 1994 *Flintknapping: Making and Understanding Stone Tools*. University of Texas Press, Austin.

Yesner, David R., and Georges A. Pearson 2002 Microblades and Migrations: Ethnic and Economic Models in the Peopling of the Americas. In *Thinking Small: Global Perspectives on Microlithization*, edited by Robert G. Elston and Steven L. Kuhn, pp. 133–161. Archaeological Papers of the American Anthropological Association, Arlington.

Chapter 15
Eastern Arctic Under Pressure: From Paleoeskimo to Inuit Culture (Canada and Greenland)

Pierre M. Desrosiers and Mikkel Sørensen

15.1 Introduction

Pressure microblade production appeared with the arrival of the Paleoeskimo people (4500–800 B.P.) in the Eastern Arctic, long after the technique was established in other areas of the world (Fig. 15.1). Previous assumptions have all too quickly proposed that the Paleoeskimo produced microblades by 'pressing them off' from the core. As a result, there was no real attempt made to analyse the techniques employed to detach microblades in later studies. In addition, early studies did not focus on lithic technology in any great detail, which likely explains why our present knowledge is limited with regard to detachment techniques in the Arctic. This study seeks to improve upon our current knowledge of the detachment technique for microblade production employed by the Paleoeskimo.

In this chapter, we define the current state of knowledge of pressure microblade production as well as the context in which lithic technology developed in the study area. Analysis and observation of microblade collections from Canada and Greenland allow us to identify the place of pressure technique in lithic tool production. Initially, we will look at the origin and diffusion of pressure microblade production in the Eastern Arctic until its disappearance at the arrival of Thule/Inuit people (Neoeskimo).

P.M. Desrosiers (✉)
Avataq Cultural Institute, Westmount, Québec, Canada
e-mail: servicearch@avataq.qc.ca

M. Sørensen
Department of Archaeology, University of Copenhagen, Copenhagen, Denmark
e-mail: miksr@hum.ku.dk

P.M. Desrosiers (ed.), *The Emergence of Pressure Blade Making: From Origin to Modern Experimentation*, DOI 10.1007/978-1-4614-2003-3_15,

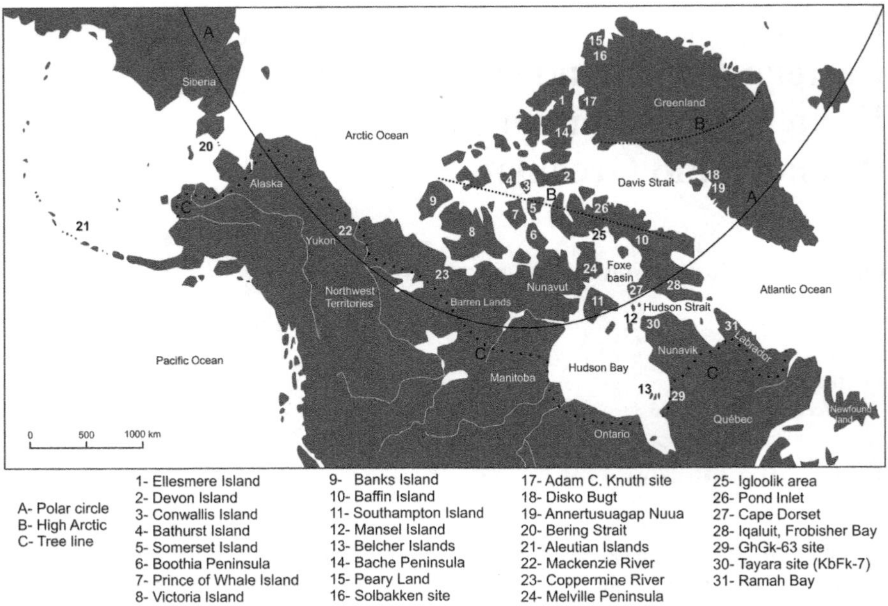

Fig. 15.1 Map of the Eastern and Western Arctic showing locations of the sites discussed in the chapter (Map prepared by Mikkel Sørensen)

In our discussion, we present the pressure tool as well as the use of pressure-flaking techniques employed by the Thule/Inuit people. Furthermore, examining the relationship between pressure tools, pressure flaking for bifacial production, and pressure-produced microblades expands upon our understanding of Paleoeskimo technological organization.

15.2 Early Assumptions Regarding Detachment Techniques

Archaeological investigations in the Arctic increased significantly in the 1950s with the onset of the Cold War and the resulting development of Arctic infrastructure such as airports and the DEW line. Such developments, and the establishment of policies related to Arctic sovereignty, gave archaeologists better access to remote regions and the sites they contained. Initial research focused on defining the cultural history, chronology, material culture, and origin of the Arctic's past inhabitants. Figure 15.2 depicts the broad chronological and cultural sequences of the North American Arctic and Greenland. These are imperfectly defined since differences between subregions reflect not only a broad diversity of environments (i.e. from the tree line to the High Arctic) but also the influence of different archaeologists in each area.

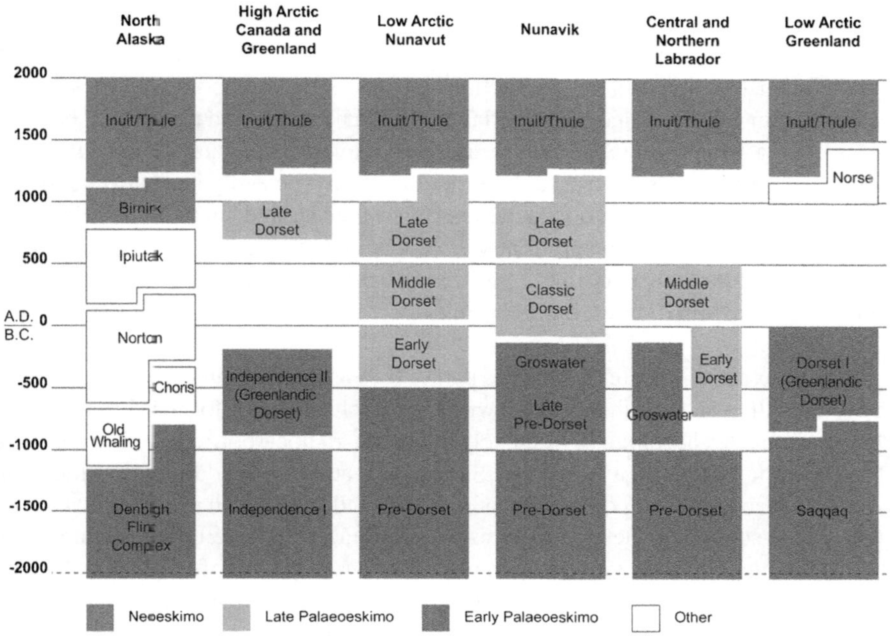

Fig. 15.2 The broad chronological sequence of Arctic culture in North America and Greenland (http://www.avataq.qc.ca/en/Institute/Departments/Archaeology/Discovering-Archaeology/Arctic-Chronology). According to Grønnow and Sørensen (2006) the Greenlandic Dorset (formerly defined as the Independence II and the Dorset I) should be associated with the Late Paleoeskimo. According to Schledermann (1990) and Grønnow and Sørensen (2006) the High Arctic 'North Water' region was, from 2500 to 0 B.C. used by many of the Paleoeskimo cultural groups from both Canada and Greenland: Independence I, Predorset, Saqqaq and Late Predorset/Transitional Canadian Dorset, and Greenlandic Dorset groups. The dating of the Independence I culture is from 2500 to 1900 B.C. In low Arctic Greenland the dating of the Saqqaq is from 2500 to 700 B.C., while the Greenlandic Dorset is dated to 800–0 B.C., thus an overlap in absolute dating appear in Central West Greenland. According to Desrosiers (2009) the situation in Low Arctic Nunavut and Central and Northern Labrador is similar to Nunavik with regards to the so called 'Early and Middle Dorset' periods. Instead, these should be placed within the 'Classic Dorset' period and its chronology

The Dorset was the first Paleoeskimo culture to be distinguished from the Thule culture (the direct ancestors of the present-day Inuit). Jenness (1925) identified Dorset culture, which became better defined in the following decades (Rowley 1940; Wintemberg 1939, 1940). Shortly thereafter, in the 1950s, the simultaneous discovery of spalled burins across the Arctic led to the definition of earlier Paleoeskimo cultures (Giddings 1949, 1951; Irving 1951; Knuth 1952; Meldgaard 1952; Solecki 1950; Solecki and Hackman 1951). Authors such as Meldgaard (1952: 223) proposed that a pressure technique was used to detach the spall from the resultant burin, while Collins (1956: 70) suggested spalls were either 'struck or pressed off'.

By the 1960s and 1970s, this idea extended to 'tip fluting' of harpoon end blades and the flaking of chert points (Giddings 1967: 230; Meldgaard 1960b: 592). It also extended to microblade production: 'At this site no true microblades in the sense of thin parallel-sided lamellar flakes, driven by controlled pressure or percussion flaking from a prepared polyhedral core, have been found' (Maxwell 1962: 28); '… Denbigh artisan pressed his antler flaker tip repeatedly against small flinty stones, turning out the burins, scrapers, microblades, and marvellous miniature arrowpoints and edging blades that we recognize at a glance as the hallmarks of Denbigh culture …' (Giddings 1967: 247); and 'Microblades are undoubtedly removed from microcores by well-controlled pressure flaking process' (Wyatt 1970: 100).

This interest in lithic technology as part of the earlier cultural historical archaeology gradually vanished in the 1970s with the establishment of formal chronological frameworks. Archaeologists often rely more on radiocarbon dates than artefact typologies to assign an assemblage to a specific culture. In addition, the focus of Arctic archaeology shifted towards more diversified topics such as ecological adaptations, the study of settlement patterns and site function, as well as the documentation of architecture and zooarchaeology. The non-recognition of the archaeological context problems associated with occupation admixture led many to adopt a limited view of material culture for distinguishing Paleoeskimo groups (Desrosiers 2009: 120). This inhibited the development of research into lithic technology, with some distinct exceptions. One of them is the microblade study conducted by Owen (1988). If Owen's study did not aim at documenting detachment techniques, it did include interesting general observations based on her meticulous analysis based on a large quantity of microblades: 'The microblade assemblages from the Independence I of Port Refuge and the Early Pre-Dorset are remarkably similar and clearly belong to the same microblade technology. On the basis of microblade form and attributes, it seems likely that they were produced with a well controlled pressure technique' (1988: 122).

It was noted that the size of microblades varies according to the type of raw material used for their manufacture (McGhee 1970: 95–96). Owen also noted: 'It seems likely that the quartz crystal microblades were produced with the same general technique as the other Dorset microblades, probably with pressure. In contrast, the pieces of Ramah chert are larger, more irregular in form, have less carefully prepared platforms and fewer ridge blades. They were probably manufactured with a different technique, i.e. with indirect or direct percussion' (1988: 127). By extension, we can propose that the detachment techniques would vary according to raw material type used by the Paleoeskimos.

As a result of the previous assumptions with regard to microblade production, it would be of interest to better document the removal techniques employed by the Paleoeskimo. The following questions have motivated our research into the subject: Did microblade production techniques vary in space and over time? And were these variations the result of different factors such as the availability of raw material?

15.3 The Origins of Pressure Microblade Production in the Eastern Arctic

Before its appearance in the Arctic, pressure microblade production developed much earlier outside North America (Inizan, et al. 1992). Three theories exist for the origin of Arctic cultures. The first (e.g. Bogoras 1925; Cranz 1770; Dawkins 1874; Markham 1865; Mathiassen 1927; Sapir 1916; Thalbitzer 1914) proposes that the Bering Strait was the point of origin of the Thule/Inuit people. With the discovery of earlier cultures, it was also treated as the region from which the Paleoeskimo cultures had emerged (Collins 1940; De Laguna 1946; Harp 1964: 159–161).

The second theory (e.g. Boas 1888; Murdoch 1892; Rink 1873) proposes that certain Arctic cultures at their origin represent a progressive adaptation of interior Amerindian peoples to coastal Arctic regions. This model was later proposed to explain the origins of Paleoeskimos (Collins 1934: 311; Hoffman 1952; Mathiassen 1935: 421–422; 1936: 130; Meldgaard 1960a, b, 1962). Since the 1950–1960s, archaeological research has demonstrated that there is no clear relationship between the Amerindian peoples in the Subarctic and the development of the Eastern Arctic cultures (e.g. Harp 1964).

Finally, the third theory (i.e. McGhee 1983) asserts the challenging position that the prehistoric cultures of the Eastern Arctic originated in North-Western Russia or possibly Northern Europe. The archaeological record does not present any evidence to lend support to this theory however.

Thus, only the first theory explains the origins of pressure techniques employed for microblade manufacture in the Eastern Arctic. The exact circumstances under which Paleoeskimo pressure techniques had evolved in the production of microblades in the Western Arctic and Bering Strait is still poorly understood. The Westernmost Paleoeskimo culture from Alaska is the Denbigh Flint culture (Fig. 15.2). The Denbigh Flint culture is roughly contemporaneous with other Early Paleoeskimo cultures and is believed to be their direct ancestor due to its Western position (e.g. Taylor 1968).

Both authors have observed Denbigh chert and obsidian microblades from the Iyatayet site of Alaska. During this quick overview, we noted an overall lack of edge regularity. This would hardly attest to the systematic employment of pressure techniques. This overall lack of regularity is also distinguishable from illustration in Giddings' book (1964: 207). In fact, only midsized and small microblades show a significant degree of regularity and other characteristics associated with the use of pressure technique. At present, we can only suggest that pressure was not the only technique employed, particularly for production of the largest microblades, which occasionally get larger towards their distal ends (Fig. 15.3).

The exact detachment technique employed by Denbigh for the manufacture of microblades remains to be fully studied. As well, the relationship between Denbigh and other Paleoeskimo cultures needs to be better understood before a convincing argument on the origin of the pressure microblade technique in the Eastern Arctic can be put forth.

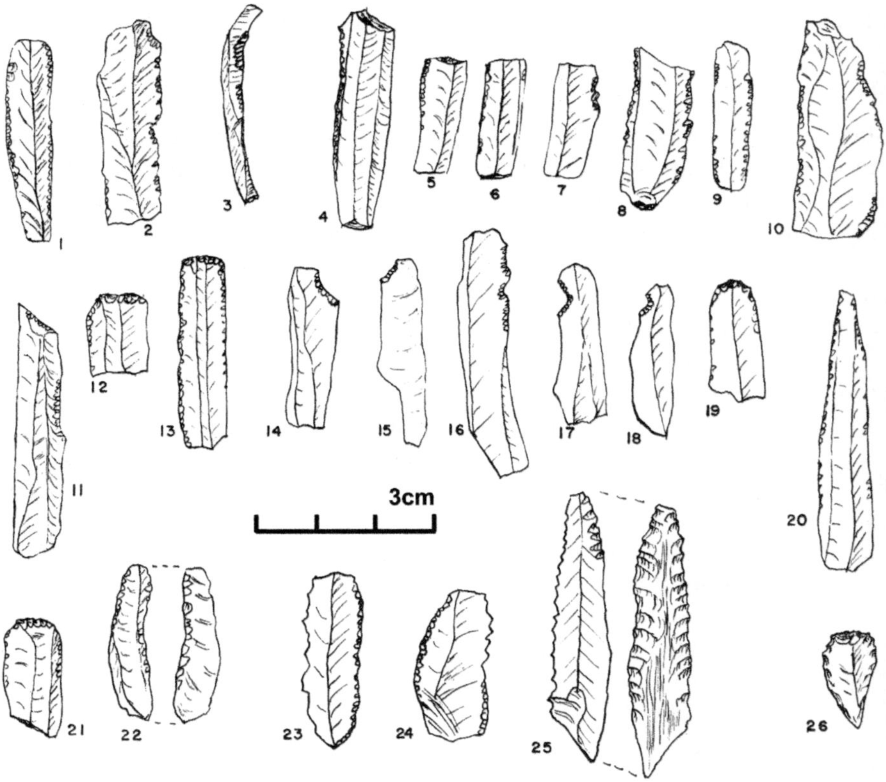

Fig. 15.3 Sample of microblades from the Denbigh Flint Complex (Giddings 1964: 207)

15.4 A Case Study from the Canadian Arctic

Desrosiers (2009) documents the techniques of detachment for several classic Dorset collections from Nunavik (Fig. 15.2). These sites represent lithic technology from the beginning of the Dorset period in the Eastern Arctic and include GhGk-63 site and level II of the Tayara site (KbFk-7) (Fig. 15.1). Both are dated between 2100 and 1800 B.P. These collections demonstrate that relatively similar raw materials were used for microblade production – primarily small-sized pieces of chert and quartz crystal.

The *chaîne opératoire* involved in microblade production for both sites has been described previously (Desrosiers 2007, 2009). Only the results for the identification of the exact detachment techniques are mentioned below. Desrosiers derived the diagnostic criteria through the observation of modern flintknappers (i.e. Jacques Pelegrin, Éric Boëda, Sylvain Sorriano, Mikkel Sørensen, and others), personal experience, and from lithic technology seminars. Descriptions of certain criteria are also provided in the literature (Crabtree 1968; Marchand 1999; Pelegrin 2000, 2002; Texier 1984; Tixier et al. 1980).

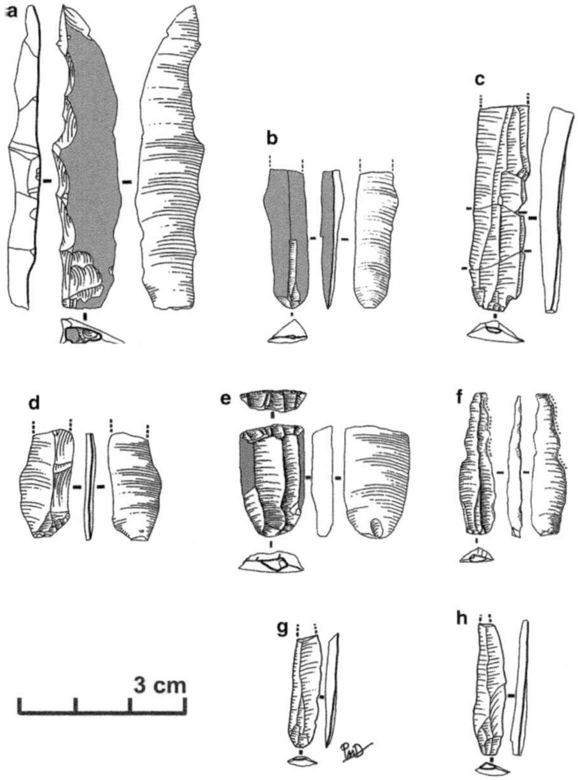

Fig. 15.4 Chert microblades from GhGk-63 site: (**a**) crested, (**b**) with natural surfaces, (**c–d**) refitted fragments, (**e**) end-scraper, (**f**) lateral edge retouched (or use-wear) and (**g–h**) unretouched. (Drawing: Pierre M. Desrosiers)

A definite combination of criteria can rarely prove that a particular technique of detachment was employed over another since identifying a specific technique is a delicate task (Pelegrin 1995: 20–23; Tixier 1982). At times, a particular diagnostic feature could be used to eliminate a detachment technique, for instance a concave butt is unlikely to relate to the use of direct percussion. Our diagnosis of a given detachment technique by and large represents tendencies as opposed to an absolute fail-safe identification. These tendencies demonstrate that microblades would show a combination of characteristics that suggest a given technique of detachment was employed as opposed to others.

The GhGk-63 site dates from the first half of the Dorset period and is situated 100 km within the tree line, on the Eastern coast of Hudson Bay near Kuujjuarapik (Avataq Cultural Institute 1991, 1992; Bernier 1997; Desrosiers 1999, 2009; Desrosiers and Gendron 2004, 2006; Desrosiers and Rahmani 2003). Microblades from this site indicate the use of a variety of detachment techniques. A total of 175 chert microblades were studied. The techniques identified can be summarized as follows: 9 by direct soft hammer percussion, 41 by indirect percussion, 11 by pressure, and 114 undetermined (Fig. 15.4). As for quartz crystal microblades the results are 3 by soft hammer direct

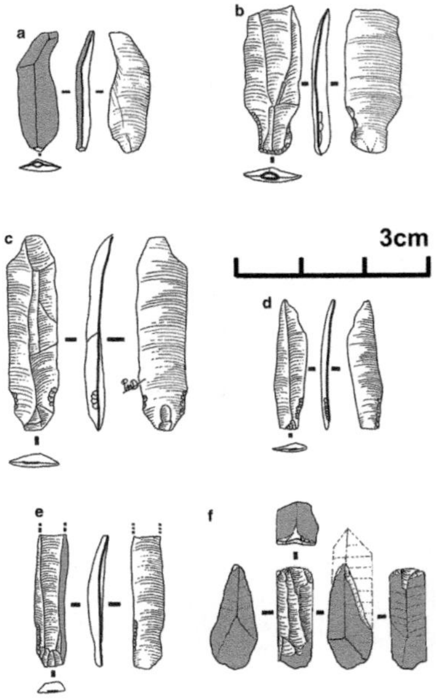

Fig. 15.5 Quartz crystal microblade production at Dorset GhGk-63 site: (**a**) natural surfaces microblade, (**b–e**) tanged microblades and (**f**) microblade core. (Drawing: Pierre M. Desrosiers)

percussion, 3 by indirect percussion, 11 by pressure, and 21 undetermined (Fig. 15.5). The undetermined samples relate to degrees of fracture, the size of the microblades, and the degree of retouch, particularly on the proximal end. Furthermore, on the smallest microblades, it is difficult to distinguish diagnostic attributes (Desrosiers 2009).

The Tayara site (KbFk-7) presents exceptionally deep stratigraphy, which is unusual among sites in the Arctic. The central area of the site includes three layers close to the permafrost. The archaeological context and other aspects of the site are well documented (Desrosiers 2009; Desrosiers et al. 2006, 2007; Desrosiers et al. 2008; Houmard 2006; Avataq Cultural Institute 2002, 2003, 2004, 2006, 2007; Todisco 2008; Todisco and Bhiry 2007, 2008a, b; Todisco et al. 2009; Todisco and Monchot 2008). Chert microblades from the Dorset level II are numerous, with a total of 413 having been identified. Detachment techniques are as follows: 12 by direct soft hammer percussion, 21 by indirect percussion, 115 by pressure, and 265 undetermined (Fig. 15.6). On the other hand, 440 quartz crystal microblades were also produced by various techniques: 12 direct percussion soft hammer, 12 indirect percussion, 187 pressure, and 229 undetermined (Desrosiers 2009).

Considering the fact that the smallest microblades incorporate mainly unidentifiable features in both assemblages, the tendency appears to be that pressure was the main technique employed for the detachment of small- and medium-sized microblades by the Dorset people. This is not only visible on the microblades themselves but on the microblade cores as well, which exhibit regular parallel scars on their knapping surface upon abandonment. This is especially true among quartz crystal cores (Figs. 15.5f, 15.7j–k).

The largest microblades result from indirect or soft hammer percussion. More specifically, they often include crested microblades (Figs. 15.4a, 15.6a–c, 15.7d).

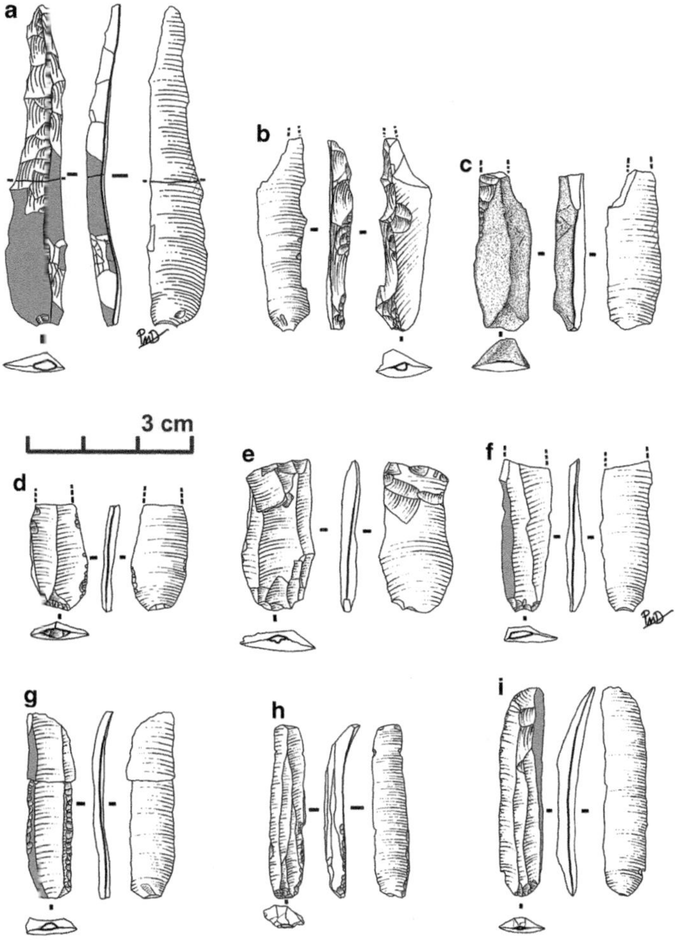

Fig. 15.6 Chert microblades from Tayara site (KbFk-7): (**a–b**) crested, (**c**) with natural surfaces, (**d**) probably detached by soft hammer percussion, (**e**) retouched, (**f**) with concave butt and getting larger toward distal end, (**g**) lateral retouched edges, (**h**) tanged and (**i**). (Drawing: Pierre M. Desrosiers)

The *chaîne opératoire* of microblade production suggests that the crest was shaped by direct and indirect percussion flaking (Figs. 15.4a, 15.6a, b, 15.7a, b) or consists of selecting the intersection of two natural flat surfaces (Figs. 15.4b, 15.5a, 15.6c, 15.7c, d). The crested microblades tend to be much thicker and their edge regularity is poor. On some of the crested microblades, the butt is deeply concave, a characteristic almost incompatible with the use of direct percussion. Those characteristics suggest the use of indirect percussion with a punch (Fig. 15.6a, b).

There was a logical application of various detachment techniques that relate closely to the steps of the *chaîne opératoire* and the microblade size. It seems that pressure

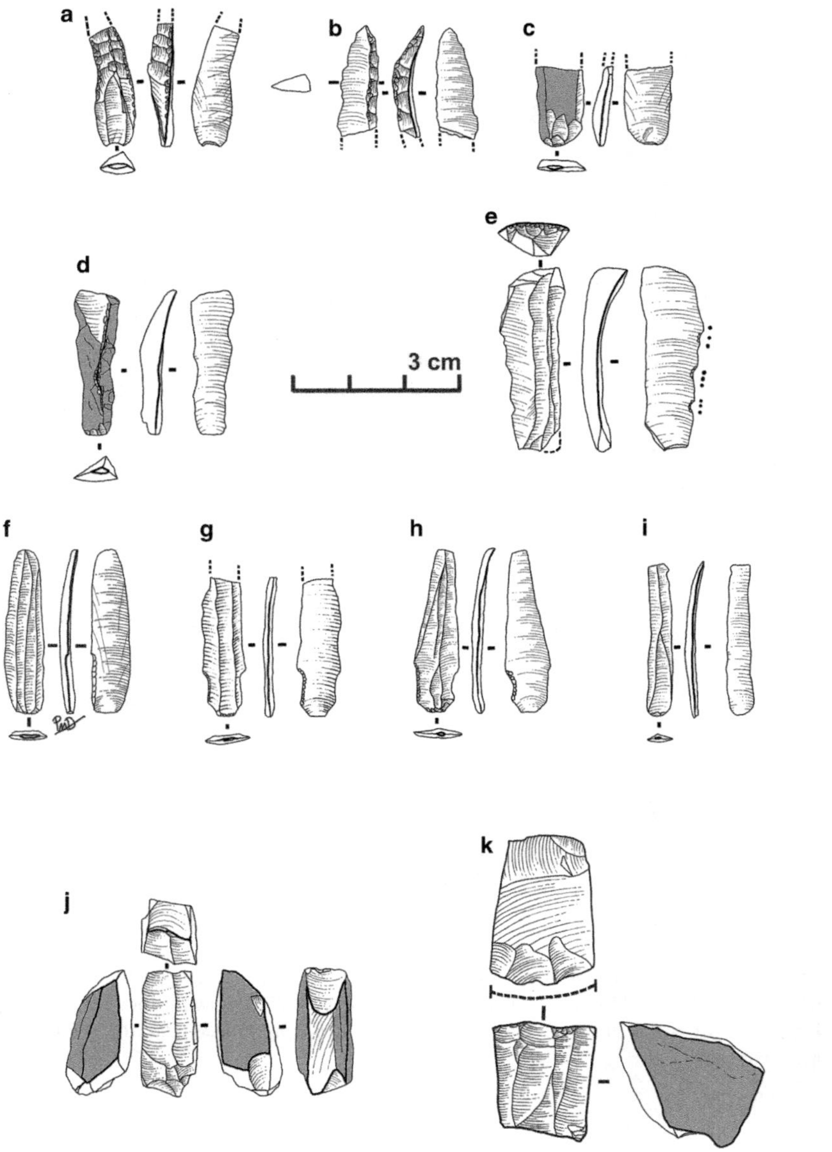

Fig. 15.7 Quartz crystal microblade production at Tayara site (KbFk-7): (**a–b**) crested microblades, (**c–d**) natural surfaces microblades, (**e**) end-scraper, (**f–h**) tanged microblades, (**i**) microblade detached by pressure and (**j–k**) microblade cores. (Drawing: Pierre M. Desrosiers)

could only be applied within a limited range of force (i.e. producing small- and middle-sized microblades), a feature linked to the type of pressure tool and the way it was manipulated. When larger microblades were required, namely, at the beginning of the production sequence, soft hammer and indirect percussion were the preferred techniques. The biggest microblades (most likely not produced by pressure techniques)

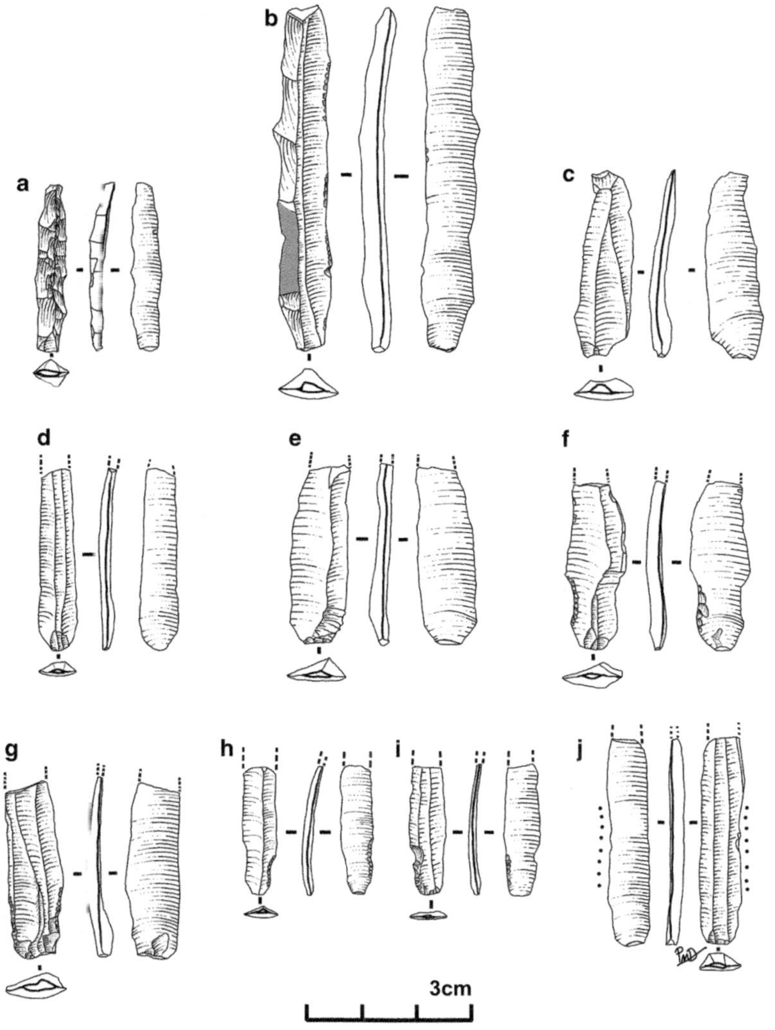

Fig. 15.8 Sample of microblades from T1 site (KkHh-3 [**a**–**g**]) on Southampton Island and Alarnerk site NhHd-1 (**h**–**j**) in Igloolik area. All microblades are in chert with the exception of one (H) in quartz crystal (Drawing: Pierre M. Desrosiers)

were often selected and transformed into tools (Desrosiers 2009). This suggests that regularity was not the main criterion for microblade blank selection.

We compared the results of the microblade study of GhGk-63 and level II of Tayara site (KbFk-7) with collections from Hudson Strait (KkHh-3, NhHd-1, and NjHa-1) and Dorset sites from Labrador (IdCr-6 and JaDb-10). The microblades from Hudson Strait are particularly similar to those of the GhGk-63 and Tayara (level II) sites. A rapid overview indicated that the small- and medium-sized microblades were most likely produced by pressure, while the largest ones were detached by either direct or indirect percussion (Fig. 15.8). By contrast, Ramah

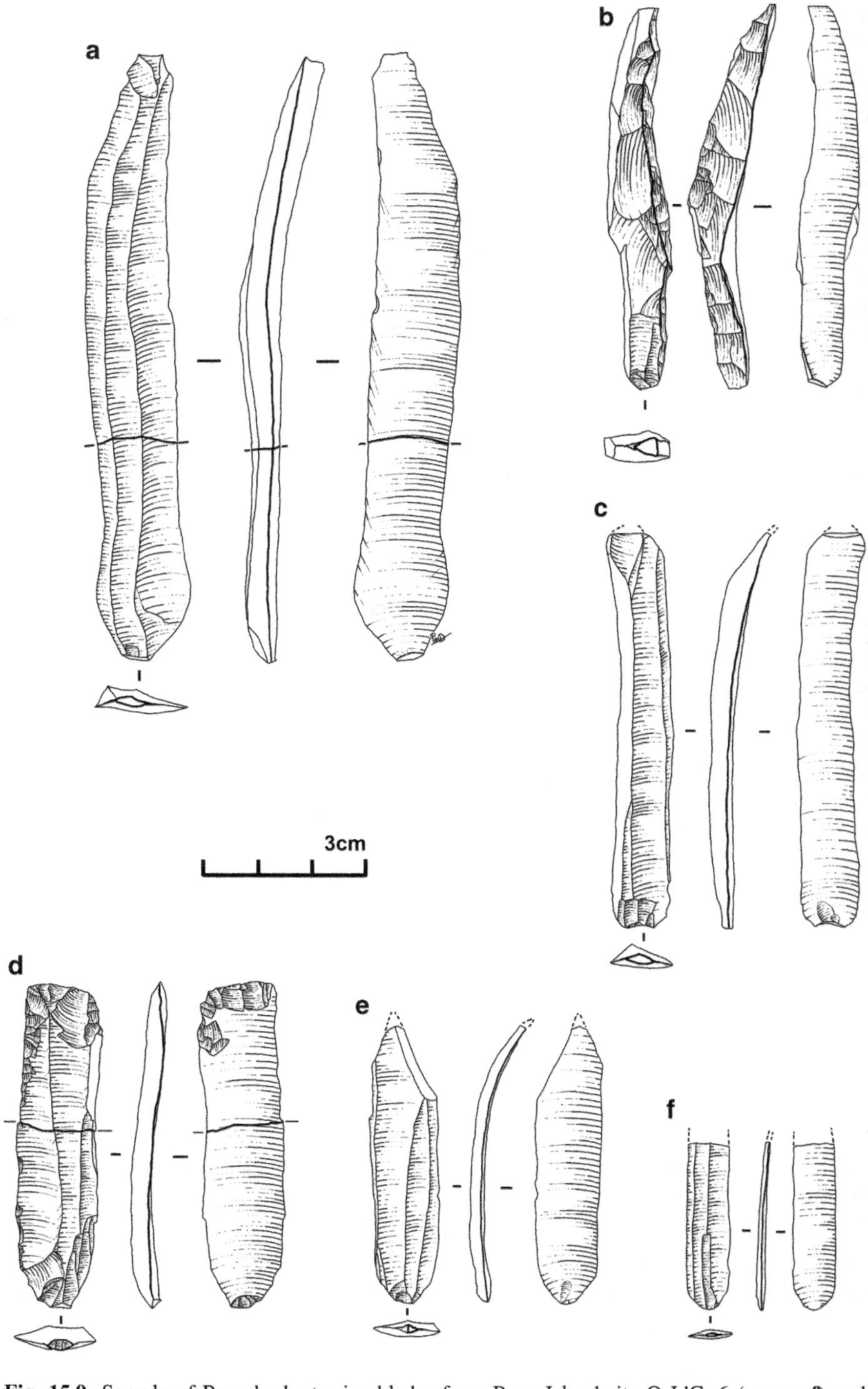

Fig. 15.9 Sample of Ramah chert microblades from Rose Island site Q IdCr-6 (**a**–**c**, **e**–**f**) and Avayalik site JaDb-10 (**d**)

chert microblades from Labrador are often larger and would easily be classified as blades as opposed to microblades. The brief overview of the Labrador collections does not permit a complete understanding of the *chaîne opératoire* involved in their production and whether Ramah chert blades represent a different aim of production than the other regular-sized microblades (Fig. 15.9). Like Owen (1988), we note that it is unlikely that pressure would have been used to detach the large and roughly regular blades. Conversely, the production of regular-sized microblades from the same region seems to follow the same succession of different detachment techniques than other Dorset sites.

15.5 A Case Study from Greenland

Sørensen has analysed the lithic technology of the Paleoeskimo cultures of Greenland (Sørensen 2006a, 2012). Among the Paleoeskimo cultures of Greenland, Independence I (2500–1900 B.C.), Saqqaq (2500–800 B.C.), Greenlandic Dorset (formerly Independence II/Dorset I [Grønnow and Sørensen 2006]) (800–0 B.C.), and Late Dorset (A.D. 800–1400) artefacts reveal the use of pressure microblade production and pressure-flaking techniques (Fig. 15.2). These techniques were employed with some variation over time.

The production from the Independence I Adam C. Knuth site (Jensen and Pedersen 2002; Knuth 1983) and Solbakken site (Grønnow and Jensen 2003) in Northern Greenland is considered here. With the addition of the Greenlandic Dorset site Annertusuagap Nuua, situated in Disko Bugt, they illustrate the differences between the two groups (Fig. 15.1 for location), both of which employed pressure in the production of microblades (Sørensen 2012). In the following, the main focus will be on the blade production method and concept.

At Adam C. Knuth site (Fig. 15.10), microblades are produced from keeled, single-fronted cores with small-faceted platforms prepared from tabular nodules of high-quality microcrystalline quartz (MCQ). One of the narrow faces of the nodule is selected as the platform, and the core is shaped with respect to this platform. The front of the core and its cross section is often shaped by the production of a single central crest on the front. The bottom of the core sometimes has a crest from which the cross section of the core can be controlled. In other situations, the bottom is flat and left unworked. The width of the core is constantly between 20 and 25 mm, and the height of the core can be up to 70 mm during the initial step of production. The angle between the front and platform is generally right-angled during all steps. Prismatic regular and relatively straight microblades are produced from these cores, and possibly up to 50 microblades can be produced from a single core. The platform of the core is repeatedly faceted during microblade production. Microblades from the Solbakken site ($n=56$) have an average width from 7 to 9 mm and a thickness of approximately 2 mm. The length of the microblades is normally between 40 and 60 mm ($n=14$). The butts are usually oval; 28% have a smooth butt ($n=9$), and 68% are facetted ($n=17$). The method employed in their production and modification steps can be described as the Independence I microblade concept (Sørensen 2012: 178).

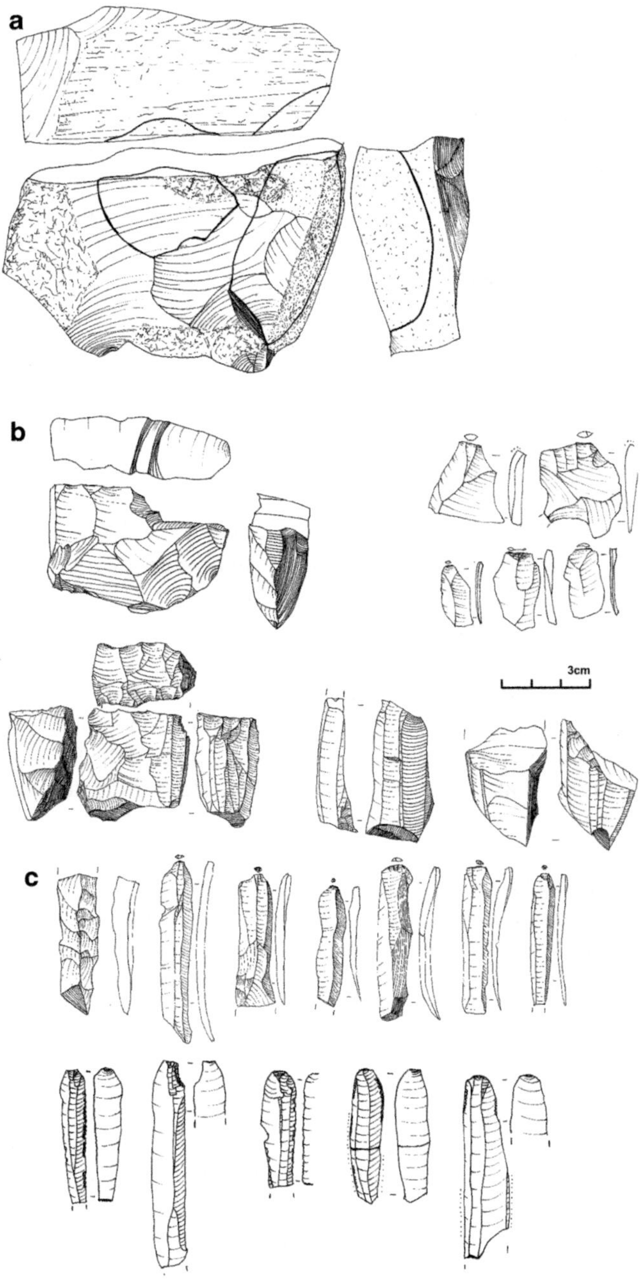

Fig. 15.10 Chert microblade production at the Independence I Adam C. Knuth Site. (**a**) Refitted core preform, (**b**) cores and by-products, (**c**) upper row: a crested microblade and six microblades; lower row: microblade with hafting retouch on proximal ends and use-wear retouch on lateral edges (Drawing: Mikkel Sørensen)

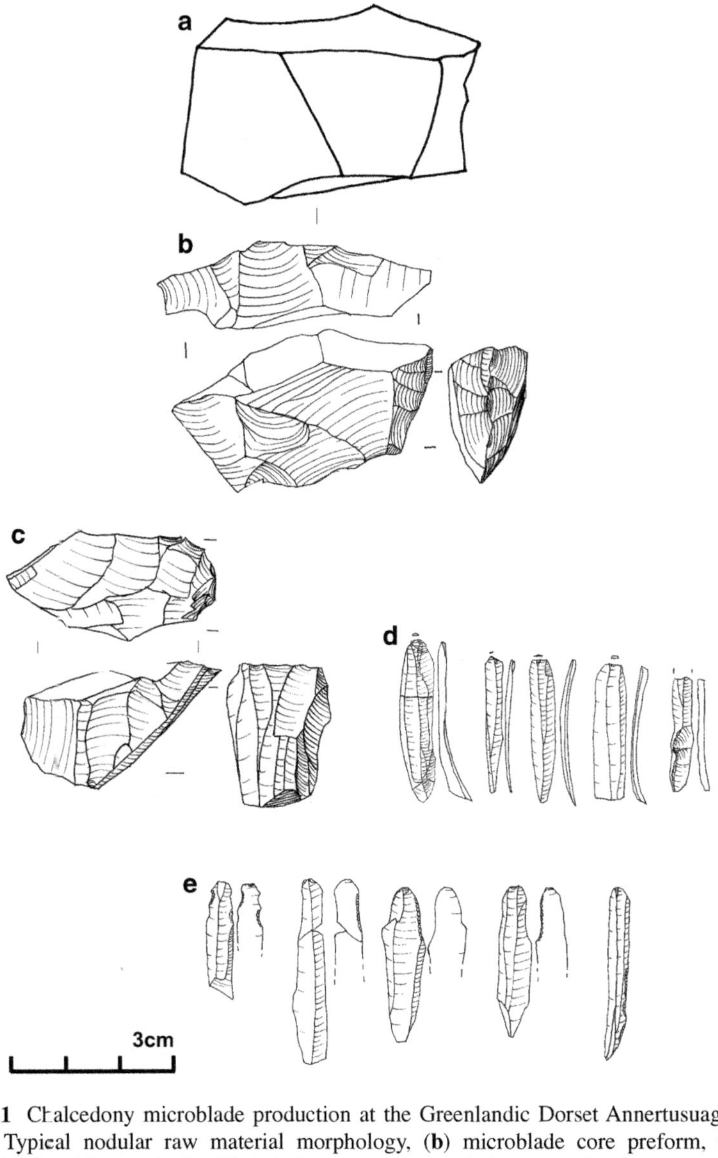

Fig. 15.11 Chalcedony microblade production at the Greenlandic Dorset Annertusuagap Nuua site. (**a**) Typical nodular raw material morphology, (**b**) microblade core preform, (**c**) core, (**d**) microblades, (**e**) tanged microblade knives, retouched at their proximal ends, presumably for a hafting system (Drawing: Mikkel Sørensen)

At Annertusuagap Nuua (Fig. 15.11), the microblades are produced from unipolar and unifacial wedge-shaped cores using MCQ as a raw material as well as quartz crystals. Tabular blanks of MCQ are selected as preforms for microblade cores. The core preform is transformed into a wedge shape. The width of the cores ranges from 15 to 20 mm. The core front is only occasionally created by a crest, as it is more commonly left unprepared before the production. Quartz crystal cores are produced either by creating a platform at the top by a platform flake or by exploiting the crystal from the bottom. The production method and technique employed in the detachment of quartz crystal microblades are the same as for other MCQ types. The core platform is prepared from the side, and the angle between the front and platform is only 50–60°, which is typical for the Greenlandic Dorset. The complete microblades investigated ($n=68$) have a mean width of 5 mm and a mean length of approximately 35 mm. The butts of the microblades ($n=99$) are generally oval; 80% have a smooth butt, and 20% are facetted. During production, the core platform is adjusted by small flake removals from the front. The core's width is not reduced during its stage of blade exploitation. In several cases, a second front is established at the core's rear end using the same platform; in these cases, the core will become triangular. Microblade production generally stops when the core is too small for further detachments and platform preparation. Generalized methods can be described as the Greenlandic Dorset microblade concept (Sørensen 2012: 220).

At both sites, considering the regularity, straightness, and butt as well as bulb attributes, the microblades are generally perceived as produced by means of a pressure technique. Due to the low inertia of microblade cores, it seems most likely that they were mechanically fixed. The size and width of the microblades in conjunction with the results of modern experiments (Pelegrin 1988; Sørensen 2006b) suggests that approximately 20–30 kg of pressure was required for microblade detachment, and possibly more in Independence I production, due to the larger size of the microblades produced and their facetted butts. The analysis of raw material types for microblade and biface production reveals that in many instances, these may have been heat-treated before the pressure technique was applied, especially in the case of the Greenlandic Dorset (Sørensen 2012: 310). The heat-treated raw materials are typically agate and chalcedony-like types of MCQ.

According to the archaeological record, it appears that the Independence I culture arrived in Northern Greenland with a well-developed, pressure-produced microblade technology around 4500 B.P. However, the Saqqaq group that arrived at the same time in Central Western Greenland does not prioritize microblade production as much as Independence I, as they favour killiaq, a metamorphosed slate which is not appropriate for this process. The Greenlandic Dorset culture appeared at around 2800 B.P. and brought with it a pressure microblade concept that is somewhat different. Their cores are narrower, which results in the production of narrower microblades, and the front platform angle is more acute when compared with the cores of earlier cultures. The pressure tool now had a square cross section (Fig. 15.13b) and different hafting, and for microblade detachment, it was often placed on a smooth part of the platform instead of a facet.

The Late Dorset people had a similar microblade concept as the one performed by the Greenlandic Dorset, but the results were lower in quality. For instance, these microblades are generally shorter and irregular; the cores exhibit less preparation. At the same time, larger microblade types, most likely detached by indirect percussion, also appear (Sørensen 2012: 296).

15.6 Pressure Techniques from Paleoeskimo to Thule/Inuit Culture

Did the detachment techniques employed in microblade production evolve from the earliest Paleoeskimo cultures through to the end of the Late Dorset Period? Owen's (1988) study of a large quantity of microblades from the earliest to the latest Paleoeskimo periods provides a starting point for research into this topic. She observed no major changes from Early Pre-Dorset to the beginning of the Dorset period (Owen 1988: 124–126). However, she stated that 'Microblades decrease in frequency in the Middle and Late Dorset and production becomes less carefully controlled … There is also a corresponding rise in the number of irregularly shaped pieces' (Owen 1988: 126). It appears that the most significant changes happened towards the end of the Dorset period. On the other hand, Paleoeskimo sites in Nunavik indicate that towards the end of the Pre-Dorset, a period sometimes referred to as Groswater-like (Gendron and Pinard 2000), there are an unusually large proportion of microblades in these assemblages; however, this phenomenon remains to be better understood.

In Greenland, it has been noted that pressure blade production was more common in the Independence I and Greenlandic Dorset groups, as opposed to the Saqqaq and Late Dorset groups (Sørensen 2012). Moreover, considerable differences in the pressure microblade concept have been observed between the Early Paleoeskimo groups (Saqqaq and Independence I) and the Dorset groups (Greenlandic Dorset and Late Dorset) (Sørensen 2012).

Throughout the Eastern Arctic, microblade production vanished with the disappearance of the Dorset culture, which was marked by the arrival of the Thule people. The question one may ask is why microblade production was abandoned following the arrival of a new culture into this region? This is mainly explained by the fact that Thule culture is not contiguous with the previous Paleoeskimo cultures.

We know from the ethnographic work that Inuit had employed pressure-flaking techniques for bifacial reduction: 'Selecting a log of wood, in which a spoon-shaped cavity was cut, they placed the splinter to be worked over it, and by pressing gently along the margin vertically, first on one side, then the other, as one would set a saw, they splintered off alternate fragments until the object, thus properly outlined, presented the spear or arrow-head form, with two cutting serrated sides' (Belcher 1861: 138–139). In Greenland, pressure-flaking techniques employed by the Thule only occurred upon their first arrival in the High Arctic, a period otherwise termed as the Ruin Island phase, where few examples of chipped points demonstrate this technology (Holtved 1944; McCullough 1989). However, this was abandoned probably when meteoritic iron began to be exploited for tool manufacture (Sørensen 2010).

Despite the fact that the Thule brought with them the knowledge of pressure flaking, they did not, however, use it to produce microblades. The origin of the Thule people remains to be better understood in the Bering Sea region in order to know why microblade production was abandoned in the Eastern Arctic.

15.7 Paleoeskimo Pressure Flaking and Pressure Tools

The Paleoeskimo people used pressure flaking for the production of bifaces. Their finely chipped points are among the most characteristic elements that demonstrate the well-controlled use of pressure flaking in Early Paleoeskimo (Fig. 15.12a–d). The later Dorset people also used pressure flaking (Fig. 15.12e–i). When the tip-fluting spall method was first identified on Dorset harpoon head end blades, it was believed to have been produced by pressure: 'The end-blade has first been chipped equally on both sides, like the ordinary type, whereafter two long flakes were pressed off from the pointed end on the same side, each removing approximately one half of the chipped surface' (Meldgaard 1960b: 592). This interpretation of the method used to detach spalls survives up until today (e.g. Maxwell 1985: 152). Plumet and Lebel's (1991, 1997) elaborate attribute analysis reached the same conclusion; however, this understanding of the situation is not entirely convincing.

The main problem with this interpretation is that the spalls get progressively wider towards their distal end (Fig. 15.12h–i), a characteristic usually incompatible with the use of pressure techniques (Pelegrin, personal communication). Preliminary experimental tests on detaching tip-fluting spalls were briefly conducted by Mikkel Sørensen and Jacques Pelegrin. These attempts did not succeed in producing by pressure the characteristic spalls that become larger towards the distal end. They reveal that elaborate experiments must be conducted before we can reach any sound conclusion.

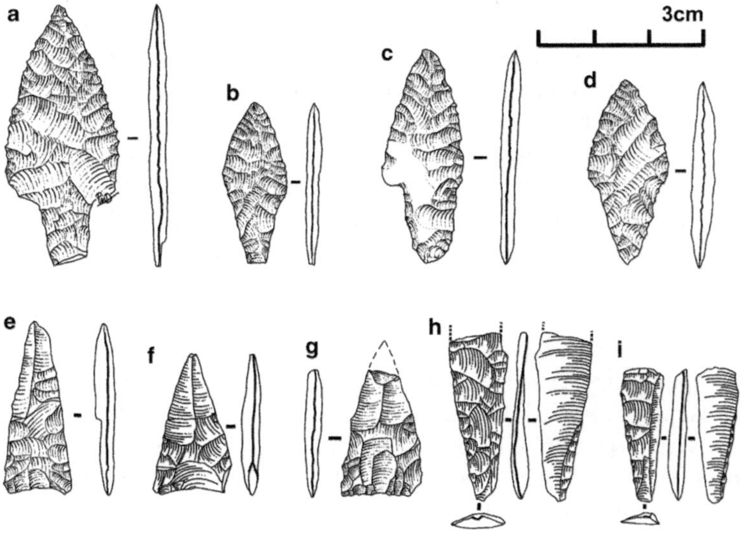

Fig. 15.12 Sample of finely chipped chert points (**a**–**g**) from Pre-Dorset occupation of KcFr-5 (**a**, **b**), GhGk-4 (**c**) and IgDj-2 (**d**) sites. Also, examples of Dorset tip fluted chert points (**e**–**g**) and tip fluting spalls (**h**, **i**) from Dorset occupation of T1 (**e**), GhGk-63 (**f**–**i**) and Tayara (**g**) sites (Drawing: Pierre M. Desrosiers)

Fig. 15.13 Presumed pressure tool tips (**a**–**f**) and one complete pressure tool (**g**). (**a**) Tayara site (KbFk-7), (**b**, **c**) Malmquist site, (**d**) Solbakken site, (**e**, **f**) Den blå flints boplads (*blue flint site*), (**g**) Qerqertaaraq site. (Photo: Pierre M. Desrosiers (**a**) and Mikkel Sørensen (**b**–**g**))

Another issue worth examining would be to determine if different tools were used for pressure microblade production and for pressure flaking. The tool for pressure flaking was most likely handheld and can only produce a limited amount of force. The tendencies noted for microblades indicate that pressure tools involved in their production could probably only detach small- and midsized microblades. Therefore, the use of a long pressure tool, such as the one employed in the production of Mesoamerican blades upon which a knapper's full body weight can be applied, seems unlikely.

It has been proposed by Maxwell (1985: 151–152) that short wooden handles and punches carved from walrus baculum would be lashed together and used as a pressure tool. Unfortunately, both elements have not been found in direct association by him. Moreover, observations of probable punch tips, such as the one found at the Tayara site and in other collections, often do not exhibit any evidence of use wear (Fig. 15.13a).

In Greenland, tips of presumed pressure tools are known from several Independence I sites, e.g. the Solbakken site and 'Den blå flints boplads' (blue flint site) (Grønnow and Jensen 2003), and at Greenlandic Dorset sites such as the Malmquist site. Preserved Independence I pressure tips are typically made from walrus tusk and have oval cross sections, are up to 5 cm in length and 1 cm wide with a rounded distal end characterized by extensive use wear (Fig. 15.13d–f). In order to make such a small tip function as a pressure tool, it would need to be fixed into a handle in order to properly direct the force. Grooved wooden handles that fit pressure tips with oval cross sections have been recovered from the Saqqaq site Qeqertasussuk. These suggest that the pressure tips were fixed to a rather short handle (Grønnow 1996: 24). Modern experiments with this hafting type demonstrate that it functions rather well when employed in pressure flaking. One reason for this is that lashing provides a minimal degree of flexibility in the application of pressure, which increases the contact time during detachment.

At the Greenlandic Dorset sites, and contemporary sites found in the Central Canadian Arctic regions (Meldgaard 1962), a difference is observed in the design of presumed pressure tool tips when compared with the Early Paleoeskimo groups. These tips have a square or rectangular cross section and may approach 6 cm in length and 1 cm in width (Fig. 15.13b). Preserved specimens are made from bone, possibly walrus baculum. Due to their specific design, they must have functioned within a different hafting system when compared to Early Paleoeskimo cultures. From Late Dorset contexts at the site of Qeqertaaraq (Appelt and Gulløv 1999), a pressure tool in which the handle and tip were produced from a single piece of walrus baculum has been identified (Fig. 15.13g). Similar pressure tools have been documented in Late Dorset contexts on Ellesmere Island such as the Shelter site (Schledermann 1990: 275) and in the Captain Comer collection acquired from Sadlermiut people of Southampton Island (Boas 1901: 63).

At present, we do not have definitive proof that the artefacts currently identified as pressure tool or punch tips necessarily functioned as such. It is tempting to make an analogy to the well-documented hafted pressure tools used by Thule/Inuit people; however, the characteristic spoon-shaped handle of such implements (Holmes 1919: 319; Murdoch 1892: 287–289; Nelson 1899: 91) has not been identified in Paleoeskimo assemblages.

From the experiments conducted by Sørensen, we know that pressure tips combined with short hafts identified in Paleoeskimo assemblages would constitute an efficient tool for pressure flaking; however, it would hardly explain the whole range of microblade size produced by pressure (small and midsize). Using the same tips, two different hafting methods relating respectively to microblade manufacture and pressure flaking may have existed.

Finally, another problem to solve is the holding method of the small-sized and low-inertia microblade cores. They were almost certainly held in a fixation device during pressure detachment. The volumetric concept of Paleoeskimo cores implies that microblades were produced from a single narrow surface at a time. Among other possibilities, this would have allowed the side of the core to be held in some sort of pliers-like device. The holding device remains to be identified in an archaeological context.

15.8 Concluding Remarks

According to current knowledge, pressure microblade production was introduced to the Eastern Arctic with the first eastward migration of Paleoeskimo people from the Bering Strait region around 4,000–4,500 years ago. These Early Paleoeskimo peoples spread southwards as far as the treeline and northwards to the High Arctic, inhabiting a vast territory that spans from Alaska to Greenland (more than 5,000 km from west to east and more than 3,000 km from north to south). In other words, pressure technique was carried from the West and not independently invented in the Eastern Arctic.

Paleoeskimo technology developed until the end of the Dorset period. The role of the pressure technique and the exact nature of its evolution remain to be fully documented. Based on the cases studied in this chapter, we can state that different detachment techniques were employed and used in combination to produce microblades. Moreover, pressure microblade manufacture, so long assumed to have been the technique used in the Eastern Arctic, has now been demonstrated as a fact that needs to be studied in relation with other detachment techniques for both microblades and bifacial tools.

The roughly regular microblade blanks that result from this type of production appear to have fulfilled the needs of the Paleoeskimo. The largest microblades that were most likely not detached by pressure were often selected to be transformed into tools. Consequently, from the Western Arctic Denbigh culture to the Eastern Arctic Late Dorset culture, it is difficult to conceive that the pressure technique was employed to respond to a need for the production of very regular microblades with parallel sides. In fact, the pressure technique was employed in a standard sequence. This sequence involved direct and/or indirect percussion in the first steps of the *chaîne opératoire*, when the core permitted the production of long microblades. This was followed by the use of pressure as the core was gradually reduced to a smaller size. In instances where a small quartz crystal was used, the whole production sequence involved the use of pressure.

The flintknapper most likely wanted to produce the longest microblades possible; however, this was limited by the properties of the raw material and the knapping tools, as well as the skill and strength of the knapper. For instance, the homogeneous Ramah chert of Labrador permits the production of much longer microblade blanks. The choice of detachment technique was made according to size and the possibilities offered by raw material.

Improvement of our present understanding of pressure detachment techniques is related to our understanding of poorly preserved pressure tools in the archaeological record of the Paleoeskimo. Our current knowledge excludes the likely use of pressure tools capable of generating great amounts of force for the detachment of microblades. It seems likely that the pressure tools of the Paleoeskimo were shorter and more portable than those of the Mesoamericans, which stand as one of the best known examples (Clark 1982; Crabtree 1968).

A crucial effort remains to be invested in the study of different techniques used to detach microblades. During our preliminary study, one of the major problems encountered was the particularly small size of microblades in the Eastern Arctic. The smaller the microblades, the smaller the associated attributes and, therefore, the greater the difficulty in recognizing and identifying associated techniques of detachment. Considering the fact that lithic technology research is still in its infancy in this part of the world, we expect this situation to improve greatly in the future.

Acknowledgements We are grateful to Max Friesen for carefully reviewing this paper and would like to thank David Howard and Benjamin Pateneaude who greatly helped us to improve the English.

References

Appelt, M. and H.C. Gulløv 1999 *Late Dorset in high Arctic Greenland, Final Report on the Gateway to Greenland Project.* Danish Polar Center Publication 7. Dansk Polar Center, Copenhague.

Avataq Cultural Institute 1991 *Archaeological Salvage Excavation of the GhGk-63 site, 1990, Kuujjuarapik, Nunavik.* Report presented to the Municipality of Kuujjuarapik. Copies available from Avataq Cultural Institute AR 010.

Avataq Cultural Institute 1992 *Archaeological Salvage Excavation of the GhGk-63 site, 1991, Kuujjuarapik, Nunavik.* Report presented to the Municipality of Kuujjuarapik. Copies available from Avataq Cultural Institute AR 011.

Avataq Cultural Institute 2002 *Rapport de la fouille archéologique 2002 sur le site de Tayara (KbFk-7), Nunavut, dans le cadre du Programme de Recherche ARUC "Des Tuniit aux Inuits".* Présenté au Government of Nunavut, Inuit Heritage Trust, Prince of Whales Northern Heritage Centre, Canadian Museum of Civilization et à la Municipalité de Salluit. Copies available from Avataq Cultural Institute No. AR 062.

Avataq Cultural Institute 2003 *Rapport de la fouille archéologique 2002 sur le site de Tayara (KbFk-7), Nunavut, dans le cadre du Programme de Recherche ARUC "Des Tuniit aux Inuits". Volume 2: Catalogue.* Présenté au Government of Nunavut, Inuit Heritage Trust, Prince of Whales Northern Heritage Centre, Canadian Museum of Civilization et à la municipalité de Salluit. Copies available from Avataq Cultural Institute No. AR 063.

Avataq Cultural Institute 2004 *CURA 2001–2003 from Tuniit to Inuit: Three years of research at the Tayara site (KbFk-7), Qikirtaq. Summer fieldwork 2003 and preliminary synthesis.* Présenté au Government of Nunavut, Department of Cultural Heritage, Inuit Heritage Trust, Prince of Whales Northern Heritage Centre, Canadian Museum of Civilization et à la municipalité de Salluit. Copies available from Avataq Cultural Institute No. AR 064.

Avataq Cultural Institute 2006 *Tayara Site (KbFk-7), Qikirtaq, Summer Fieldwork 2005.* Rapport présenté au Government of Nunavut, Inuit Heritage Trust, Prince of Wales Northern Heritage Centre, Canadian Museum of Civilization, Municipality of Salluit, et au Qaqqalik Landholding. Copies available from Avataq Cultural Institute No. AR 241.

Avataq Cultural Institute 2007 *Tayara site (KbFk-7), Qikirtaq, summer Fieldwork 2006.* Rapport présenté au Government of Nunavut, Inuit Heritage Trust, Prince of Wales Northern Heritage Centre, Canadian Museum of Civilization, Municipality of Salluit et au Qaqqalik Landholding. Copies available from Avataq Cultural Institute No. AR 248.

Belcher, E. 1861 On the Manufacture of Works of Art by the Esquimaux. *Transactions of the Ethnological Society of London* 1: 129–146.

Bernier, L. 1997 Analyse archéologique de quatre habitations paléoesquimaudes du site GhGk-63, Kuujjuarapik, Arctique québécois, Université de Montréal, Montréal.

Boas, F. 1888 The Central Eskimo, pp. 399–669. Annual Report of the Bureau of Ethnology. Vol. 6, J. W. Powell, general editor. Government Printing Office, Washington D.C.

Boas, F. 1901 *The Eskimo of Baffin Land and Hudson Bay. From Notes Collected by Capt. George Comer, Capt. James S, Mutch, and Rev. E. J. Peck.* Bulletin of the American Museum of Natural History XV, Part I. Knickerbocker Press, New York.

Bogoras, W. 1925 Early Migrations of the Eskimo between Asia and America. *Proceedings of the XXIième Congrès International des Américanistes*. Göteborg.

Clark, J.D. 1982 Manufacture of Mesoamerican Prismatic Blades: An Alternative Technique. *American Antiquity* 47(2): 355–376.

Collins, H.B. 1934 Eskimo Archaeology and Somatology. *American Anthropologist* 36(2): 309–313.

Collins, H.B. 1940 Outline of Eskimo Prehistory. *Smithsonian Miscellaneous Collections* 100: 533–592.

Collins, H.B. 1956 The T1 Site at Native Point, Southampton Island, N.W.T. *Anthropological Papers of the University of Alaska* 4(2): 63–90.

Crabtree, D.E. 1968 Mesoamerican Polyhedral Cores And Prismatic Blades. *American Antiquity* 33(4): 446–478.

Cranz, D. 1770 *Historie von Grönland.* Barby.

Dawkins, W.B. 1874 *Cave Hunting Researches On The Evidence Of Caves Respecting The Early Inhabitants Of Europe.* Macmillan & Co, Londre.

De Laguna, F. 1946 The Importance of the Eskimo in Northeastern Archaeology. In *Man in Northeastern North America*, edited by F. Johnson, pp. 106–142. Papers of the Peabody Foundation for Archaeology 3. Phillips Academy, Andover.

Desrosiers, P. 1999 Les traditions lithiques au Paléoesquimau. Approche technologique préliminaire de l'industrie du Dorsétien moyen de Kuujjuarapik (GhGk-63) Nunavik, Canada. Caractérisation du débitage lamellaire, Université Paris 1, Panthéon-Sorbonne, Paris.

Desrosiers, P. 2007 Palaeoeskimo Lithic Technology: Constraints and Adaptation. *Lithic Technology* 32(1): 17–38.

Desrosiers, P. 2009 A l'origine du Dorsétien, apport de la technologie lithique des sites GhGk-63 et Tayara (KbFk-7) au Nunavik, Université Paris 1, Sorbonne, Paris.

Desrosiers, P., and D. Gendron 2004 The GhGk-63 Site: A Dorset Occupation in Southeastern Hudson Bay, Nunavik. *Canadian Journal of Archaeology* 28(1): 75–99.

Desrosiers, P., and D. Gendron 2006 Étude de la production d'outils en schiste dorsétiens et thuléens au Nunavik, Hypothèses préliminaires sur les schémas techniques. In *Préhistoire de l'Amérique. Actes of the XIVth UISPP Congress (section 17)*, edited by S.D. Congrès. British Archaeological Reports. Vol. 1524. Archeopress, Oxford.

Desrosiers, P., D. Gendron, and N. Rahmani 2006 Harpoon Head Seriation and the Dorset Phases: About the Tayara Sliced and the Other Types. In *Dynamics of Northern Societies. Proceedings of the SILA/NABO Conference on Arctic and North Atlantic Archaeology, Copenhagen, May 10th–14th, 2004*, edited by J. Arneborg and B. Grønnow, pp. 131–144. Studies in Archaeology & History. Vol. 10. National Museum of Denmark, Copenhague.

Desrosiers, P., D. Gendron, and N. Rahmani 2007 Trois ans de recherche sur le site de Tayara (KbFk-7), 2001–2003. In *Des Tuniit aux Inuits. Patrimoine archéologique et historique au Nunavik*, edited by D. Arsenault and D. Gendron, pp. 27–44. Nunavik Archaeology Monograph Series. Vol. 2. CELAT, Université Laval, Avataq Cultural Institute, Québec.

Desrosiers, P., D. Gendron, D. Todisco, H. Monchot, N. Rahmani, N. Bhiry, and C. Houmard 2008 Tayara (KbFk-7) et le Dorsétien: recherche pluridisciplinaire sur un site-clé du Paléoesquimau du détroit d'Hudson (Nunavik, Canada) *L'Anthropologie* 112(4–5): 757–779.

Desrosiers, P., and N. Rahmani 2003 Analyse spatiale du site GhGk-63 (Kuujjuarapik, Nunavik): apports à la compréhension de l'architecture dorsétienne. *Études Inuit Studies* 27: 131–153.

Gendron, D., and C. Pinard 2000 Early Palaeo-Eskimo Occupations in Nunavik: A Re-Appraisal. In *Identities and Cultural Contacts in the Arctic, Proceedings from a Conference at the Danish National Museum, Copenhagen, November 30 to December 2 1999*, edited by M. Appelt, J. Berglund and H. C. Gulløv, pp. 129–142. The Danish National Museum, Danish Polar Center, Copenhague.

Giddings, J.L.J. 1949 Early Flint Horizons on the North Bering Sea Coast. *Journal of the Washington Academy of Science* 39(3): 85–90.

Giddings, J.L.J. 1951 The Denbigh Flint Complex. *American Antiquity* 16(3): 193–203.
Giddings, J.L.J. 1964 *The Archaeology of Cape Denbigh*. Brown University Press, Rhode Island.
Giddings, J.L.J. 1967 *Ancient Men of the Arctic*. Alfred A. Knopf, New York.
Grønnow, B. 1996 The Saqqaq Tool Kit - Technological and Chronological Evidence from Qeqertasussuk, Disko Bugt. In *The Paleo-Eskimo Cultures of Greenland, New Perspectives in Greenlandic Archaeology*, edited by B. Grønnow, pp. 17–34. Vol. 1. Danish Polar Center, Copenhague.
Grønnow, B., and J.F. Jensen 2003 *The Northernmost Ruins of the Globe. Eigil Knuth's Archaeological Investigations in Peary Land and Adjacent Areas of High Arctic Greenland.* Meddelelser om Grønland, Man & Society 29. National Museum of Denmark, Copenhague.
Grønnow, B., and M. Sørensen 2006 Paleo-Eskimo Migrations into Greenland: the Canadian Connection. In *Dynamics of Northern Societies. Proceedings of the SILA/NABO Conference on Arctic and North Atlantic Archaeology, Copenhagen, May 10th–14th, 2004*, edited by J. Arneborg and B. Grønnow, pp. 59–74. Studies in Archaeology & History. Vol. 10. National Museum of Denmark, Copenhague.
Harp, E.J. 1964 *The Cultural Affinities of the Newfoundland Dorset Eskimo*. National Museum of Canada Bulletin 200.
Hoffman, B.G. 1952 Implications of Radiocarbon Datings for the Origin of the Dorset Culture. *American Antiquity* 18(1): 15–17.
Holmes, W.H. 1919 *Handbook of Aboriginal American Antiquities. Part I Introductory. The Lithic Industries*. Bureau of American Ethnology Bulletin 60. Smithsonian Institution, Washington.
Holtved, E. 1944 Archaeological Investigations in the Thule District. *Meddelelser om Grønland* 141(1–2).
Houmard, C. 2006 The Tayara Site (KbFk-7, Dorset Period, Nunavik): Preliminary Results of a Typo-Technological Study *Proceedings of The View from Here: History and Ecology of the North Atlantic Region, NABO (The North Atlantic Biocultural Organisation) Conference.* Québec.
Inizan, M.-L., M. Lechevallier, and P. Plumet 1992 A Technological Markers of the Penetration into North America: Pressure Debitage in the Indo-Pakistani Subcontinent. In *Materials Issues in Art and Archaeology III*, edited by G.S. Wheeler and I.C. Freestone, pp. 661–681. Materials Research Society symposium proceedings. Vol. 267. Materials Research Society, Pittsburgh.
Irving, W.N. 1951 Archaeology in the Brooks Range of Alaska. *American Antiquity* 17: 52–53.
Jenness, D. 1925 A New Eskimo Culture in Hudson Bay. *Geographical Review* 15(3): 428–437.
Jensen, J.F., and B. Pedersen 2002 Jordens nordligste stenalder. *Nationalmuseets Arbejdsmark* 71–85.
Knuth, E. 1952 An Outline of the Archaeology of the Peary Land. *Arctic* 5(17–33).
Knuth, E. 1983. The Northmost Ruins of the Globe. *Folk* 25: 5–21.
Marchand, G. 1999 *La néolithisation de l'ouest de la France: caractérisation des industries lithiques*. International Series 748. British Archaeological Reports.
Markham, C.R. 1865 On the origin and migrations of the Greenlander Eskimaux. *Journal Royal Geographical Society of London* 35.
Mathiassen, T. 1927 *Archaeology of the Central Eskimos II, The Thule Culture and its Position Within the Eskimo Culture*. Report of the Fifth Thule Expedition 1921–24 IV (2). Gyldendalske Boghandel, Nordidsk Forlag, Copenhague.
Mathiassen, T. 1935 Eskimo Migrations in Greenland. *Geographical Review* 25(3): 408–422.
Mathiassen, T. 1936 The Eskimo Archaeology of the Julianhaab District. *Meddelelser om Grønland* 118(1).
Maxwell, M.S. 1962 Pre-Dorset and Dorset Sites in the Vicinity of Lake Harbour, Baffin Island, N.W.T., A Preliminary Report. *National Museum of Canada Bulletin* 180: 20–44.
Maxwell, M.S. 1985 *Prehistory of the Eastern Arctic*. Academic Press, New York.
McCullough, K.M. 1989 *Ruin Islanders. Early Thule Culture Pioneers in High Eastern Arctic*. Mercury Series 141. Canadian Museum of Civilization, Hull.
McGhee, R. 1970 A Quantitative Comparison of Dorset Culture Microblade Samples. *Arctic Anthropology* 7(2): 89–96.

McGhee, R. 1983 Eastern Arctic Prehistory: the Reality of a Myth? *Musk-Ox* 33: 21–25.
Meldgaard, J. 1952 A Paleo-Eskimo Culture in West Greenland. *American Antiquity* 17(3): 222–230.
Meldgaard, J. 1960a Origin and Evolution of Eskimo Cultures in the Eastern Arctic. *Canadian Geographical Journal* 60: 64–75.
Meldgaard, J. 1960b Prehistoric Culture Sequences in the Eastern Arctic As Elucidated by Stratified Sites at Igloolik. In *Men and Cultures: Selected Papers of the Fifth International Congress of Anthropological and Ethnological Sciences, 1956*, edited by A.F.C. Wallace, pp. 588–595. University of Pennsylvania Press, Philadelphia.
Meldgaard, J. 1962 On the Formative Period of the Dorset Culture. In *Prehistoric Cultural Relations between the Arctic and Temperate Zones of North America*, edited by J.M. Campbell, pp. 92–95. Technical Paper. Vol. 11. Arctic Institute of North America, Montréal.
Murdoch, J. 1892 Ethnological Results of the Point Barrow Expedition. *Annual Report of the Bureau of American Ethnology* 9: 1–441.
Nelson, E.W. 1899 *The Eskimo About Bering Strait*. Annural Report of the Bureau of American Ethnology 18. Smithsonian Institution, Washington.
Owen, L. 1988 *Blade and Microblade Technology. Selected Assemblages from the North American Arctic and the Upper Paleolithic of Southwest Germany*. BAR International Series 441. B.A.R., Oxford.
Pelegrin, J. 1988 Débitage expérimental: «Du plus petit au plus grand». In *Journée d'études technologiques en Préhistoire*, edited by J. Tixier, pp. 37–53. CNRS, Paris.
Pelegrin, J. 1995 *Technologie lithique: le Châtelperronien de Roc-de-Combe (Lot) et de la Côte (Dordogne)*. Cahier du Quaternaire 20. CNRS, Paris.
Pelegrin, J. 2000 Les techniques de débitage laminaire au Tardiglaciaire: critères de diagnose et quelques réflexions. In *L'Europe centrale et septentrionale au Tardiglaciaire. Table-ronde de Nemours, 13–16 mai 1997*. Mémoires du musée de Préhistoire de l'Île-de-France. Vol. 7, Nemours.
Pelegrin, J. 2002 Principes de la reconnaissance des méthodes et techniques de taille. In *Tell'Atij Tell Gudeda industrie lithique*, edited by J. Chabot. Cahier d'archéologie du CELAT. Vol. 13. CELAT & Université Laval, Québec.
Plumet, P., and S. Lebel 1991 Les pointes à cannelures distales dorsétiennes, seconde invention américaine. In *25 ans de technologie en préhistoire. XIe rencontre internationales d'archéologie et d'histoire, octobre 1990: Bilan et perspectives.*, pp. 337–355. Éditions APDCA, Juan-les-Pins.
Plumet, P., and S. Lebel 1997 Dorset Tip Fluting: A Second "American" Invention. *Arctic Anthropology* 34(2): 132–162.
Rink, H. 1873 On the Descent of the Eskimo. *Journal of the Anthropological Institute of Great Britain and Ireland* 2: 104–108.
Rowley, G. 1940 The Dorset Culture of the Eastern Arctic. *American Anthropologist* 42(3): 490–499.
Sapir, E. 1916 *Time Perspective in Aboriginal American Culture, A Study in Method*. Anthropological Series 90. Canada Geological Survey.
Schledermann P. 1990 *Crossroads to Greenland, 3000 Years of Prehistory in the Eastern High Arctic*. Komatik Series 2. The Arctic Institute of North America of the University of Calgary, Calgary.
Solecki, R.S. 1950 A Preliminary report of an archaeological reconnaissance of the Kukpowruk and Kokolik rivers in northwestern Alaska. *American Antiquity* 16(1): 66–69.
Solecki, R.S. and R.J. Hackman 1951 Notes on two archaeological discoveries in northern Alaska, 1950. *American Antiquity* 17(2): 55–57.
Sørensen, M. 2006a The Chaîne Opératoire Applied to Arctic Archaeology. In *Dynamics of Northern Societies. Proceedings of the SILA/NABO Conference on Arctic and North Atlantic Archaeology, Copenhagen, May 10th–14th, 2004*, edited by J. Arneborg and B. Grønnow, pp. 31–44. Studies in Archaeology & History. Vol. 10. National Museum of Denmark, Copenhagen.

Sørensen, M. 2006b Rethinking the Lithic Blade Definition - Towards a Dynamic Understanding. Skilled Production and Social Reproduction - Aspects on Traditional Stone Tool Technologies. In *Skilled Production and Social Reproduction - Aspects on Traditional Stone Tool technologies*, edited by J. Apel and K. Knutsson, pp. 277–299. Societas Archaeologica Upsaliensis, Stockholm.

Sørensen, M. 2010 Thulekulturens anvendelse af bjergarter til fremstilling af redskaber i Nordøstgrønland. *Tidskriftet Grønland* 58(4): 288–297.

Sørensen, M. 2012 *Technology and Tradition in the Eastern Arctic, 2500 BC–AD 1200: A Dynamical Technological Investigation of Lithic Inventories in the Paleo-Eskimo Traditions*. Meddelser om Grønland 350. Museum Tusculanum Press, Copenhagen.

Taylor, W. E. 1968 *The Arnapik and Tyara Sites, an archaeological study of Dorset culture origins*. Memoirs of the Society for American Archaeology 22, Salt Lake City.

Texier, P.-J. 1984 Le débitage par pression et la mécanique de la rupture fragile. Initiation et propagation des fractures. In *Préhistoire de la pierre taillée 2. Éconimie du débitage laminaire.*, pp. 139–147. Cercle de recherche et d'études préhistoriques, Paris.

Thalbitzer, W. 1914 The Ammassalik Eskimo: Contributions to the Ethnology of the East Greenland Natives. *Meddelelser om Grønland* 39: 1–755.

Tixier, J. 1982 Techniques de débitage: osons ne plus affirmer. In *Tailler ! pour quoi faire: Préhistoire et technologie lithique II. Recent progress in microwear studies*, edited by D. Cahen and URA28duCRA, pp. 13–22. Sudia Praehistorica Belgica. Vol. 2. Koninklijik Museum voor Midden-Afrika, Tervuren.

Tixier, J., M.-L. Inizan, and H. Roche 1980 *Préhistoire de la pierre Taillée 1. Terminologie et technologie*. Cercle de recherche et d'études préhistoriques, Antibes.

Todisco, D. 2008 Géoarchéologie et processus de formation d'un site paléoesquimau en contexte périglaciaire: l'exemple de Tayara (KbFk-7), île Qikirtaq, Nunavik (Canada), Département de géographie, Université Laval, Québec.

Todisco, D., and N. Bhiry 2007 Genèse et évolution du site paléoesquimau de Tayara (KbFk-7), île Sugluk, détroit d'Hudson, Canada: données préliminaires. In *Des Tuniit aux Inuits. Patrimoine archéologique et historique au Nunavik*, edited by D. Arsenault and D. Gendron, pp. 167–184. Nunavik Archaeology Monograph Series. Vol. 2. CELAT, Université Laval, Avataq Cultural Institute, Québec.

Todisco, D., and N. Bhiry 2008a Micromorphology of Periglacial Sediments From the Tayara Site, Qikirtaq Island, Nunavik (Canada) *Catena* 76(1): 1–21.

Todisco, D., and N. Bhiry 2008b Palaeoeskimo Site Burial by Solifluction: Periglacial Geoarchaeology of the Tayara Site (KbFk-7), Qikirtaq Island, Nunavik (Canada) *Geoarchaeology* 23(2): 177–211.

Todisco, D., N. Bhiry, and P. Desrosiers 2009 Palaeoeskimo Site Taphonomy: An Assessment of the Integrity of the Tayara Site, Qikirtaq Island (Nunavik, Canada). *Geoarchaeology*.

Todisco, D., and H. Monchot 2008 Bone weathering in a Periglacial Environment: the Tayara Site (KbFk-7), Qikirtaq Island, Nunavik (Canada). *Arctic* 61: 87–101.

Wintemberg, W.J. 1939 Eskimo Sites of the Dorset Culture in Newfoundland. *American Antiquity* 5(2): 83–102.

Wintemberg, W.J. 1940 Eskimo Sites of the Dorset Culture in Newfoundland, Part II. *American Antiquity* 5(4): 309–333.

Wyatt, D. 1970 Microblade Attribute Patterning: A Statistical Examination. *Arctic Anthropology* 7(2): 97–105.

Chapter 16
The Organizational Structures of Mesoamerican Obsidian Prismatic Blade Technology

Kenneth G. Hirth

16.1 Introduction

Complex state-level societies flourished throughout Mesoamerica for more than 1,500 years before the Spanish Conquest. What is important and interesting from the perspective of technological development is that complex societies in Mesoamerica developed and spread using stone-based technologies. Although copper-bronze metallurgy was an emerging technology in Western and Central Mexico at the time of the Spanish Conquest, it was used primarily for ornamentation and ritual rather than utilitarian purposes. Stone tools, particularly those fashioned from obsidian volcanic glass, provided the cutting edges used by Mesoamerican states. Of particular importance was the development and spread of obsidian pressure blade technology that supplied cutting tools for both domestic and state-level consumption needs. Although obsidian use varied in intensity from region to region based on its availability and the quality of alternative types of flakeable stone, it was a major component of long-distance, interregional trade for nearly 3,000 years before European contact.

This chapter presents a broad outline of the organization of obsidian blade economy in pre-Hispanic Mesoamerica. Its primary goal is to identify the organizational structures of Mesoamerican obsidian blade technology and how it was incorporated into the societies that used it. As a result, this chapter focuses on the structural aspects of production and distribution systems that moved obsidian tools over space. While the development of obsidian blade technology is an important topic in and of itself (Sheets 1975; Parry 1994), I do not attempt to deal with it in depth here. Instead, I address dimensions of the scale, complexity, and integration of obsidian blade production at different points in its developmental sequence. Its utility within

K.G. Hirth (✉)
Department of Anthropology, Penn State University, University Park, PA, USA
e-mail-AMIL: kgh2@psu.edu

P.M. Desrosiers (ed.), *The Emergence of Pressure Blade Making: From Origin to Modern Experimentation*, DOI 10.1007/978-1-4614-2003-3_16,

the broader discussions of this volume is that it provides a point of comparison for percussion and pressure blade industries found in the Old World.

Any attempt to paint with a broad brush blurs the line between fact and inference. Eleven characteristics are identified below that represent the key organizational structures of Mesoamerican pressure blade technology. While our knowledge about the technology of production is relatively secure, a great deal of research is still needed to clarify the form, scale, and complexity of the production and distribution mechanisms that moved finished goods from craftsman to consumer.

16.2 Prismatic Blades: 4,000 Years of Technological Development

Obsidian prismatic blades begin to appear in secure archaeological contexts in Mesoamerica around 2500 B.C. Examples of possible earlier blades have been recovered in the Basin of Mexico at Zohapilco during the Zohapilco phase (Neiderberger 1976, 1987) and in the Valsequillo, Puebla during the Texcal II phase (5000–2500 B.C.) (García Moll 1977: 87). Obsidian blades with unprepared platforms may appear in the Tehuacan Valley as early as the Abejas phase (3400–2300 B.C.) and with prepared platforms during the Purron and Ajalpa phases (2300–800 B.C.) (MacNeish et al. 1967: 22–23). Despite these occurrences, obsidian pressure blades are rare occurrences in sites before 1200 B.C. Their appearance is important, however, because they document the antiquity of pressure blade technology and the importance of obsidian as a long-distance trade item during the Late Archaic (ca. 2500 B.C.).

Obsidian pressure blades begin to appear with regularity at archaeological sites in Mesoamerica after 1000 B.C. They are definitely present in the Coapexco and Ayotla phases in the Basin of Mexico (Boksenbaum et al. 1987; Neiderberger 1987), during the San Lorenzo A phase on the Gulf Coast (Coe and Diehl 1980:248), in San Jose phase deposits in Oaxaca (Parry 1987, 1994), and in Conchas phase households at La Blanca, Guatemala (Jackson and Love 1991). This technology is remarkably uniform over space and produced standardized prismatic blades. Although the origin of Mesoamerican pressure blade technology remains obscure (Cassiano 2005), the distribution of early blades manufactured of obsidian from the Zinapecuaro, Otumba, and Paredon sources (Boksenbaum et al. 1987; Cobean et al. 1971) suggests that the technology was already widespread by 1200–1000 B.C. (Fig. 16.1).

Obsidian pressure blades have a prismatic cross section with two parallel arrises on their dorsal surface. Obsidian blades were produced unidirectionally from a specially prepared polyhedral core. The specific technology used in blade production evolved over time with different foot-held and hand-held techniques used at different places in Mesoamerica (Clark 1982; Flenniken and Hirth 2003; Titmus and Clark 2003). The structure of obsidian pressure blade technology is remarkably stable over time, with most of the variation across Mesoamerica a result of the type

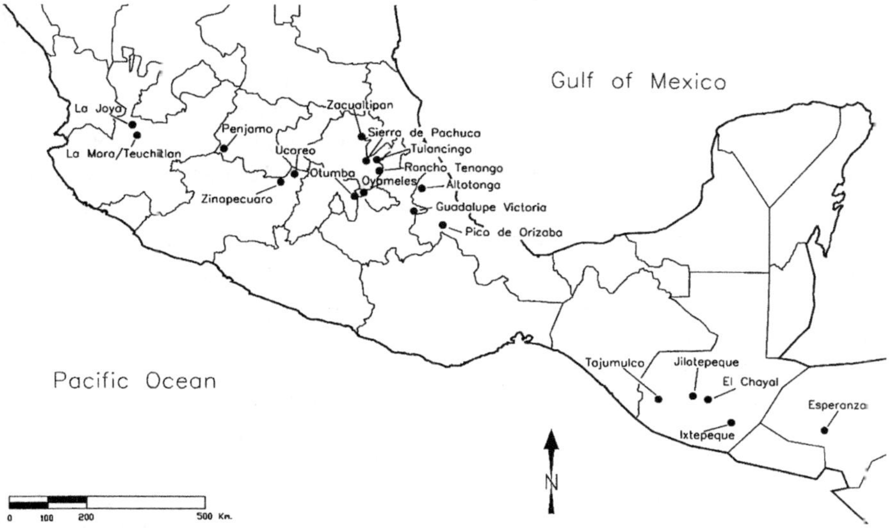

Fig. 16.1 The location of obsidian sources in Mesoamerica

and form of raw material available to local producers. One of the few major changes in pressure blade technology occurred between A.D. 600 and 700 where faceted platforms are replaced by pecked and ground ones. This change made blade removal faster, easier, and enabled the use of hand-held techniques that prolonged the use of blade cores and kept them in production longer (Crabtree 1968; Flenniken and Hirth 2003; Hirth et al. 2003).

16.3 Technology, Sedentism, and Cultural Complexity

While obsidian pressure blade technology was not included in the list of traits defining the cultural limits of Mesoamerica (Kirchoff 1952), it is often identified as a key feature of its core states. It is an example of a basic technology that developed with hunting and gathering groups in Central Mexico during the Late Archaic period (4000–1500 B.C.), and continued in use with minor modification through the formation of complex states. Obsidian prismatic blades increased in frequency with the appearance of sedentary agricultural populations during the Early Formative period (1200–1000 B.C.). Obsidian, however, was an important trade item in Mesoamerica even before prismatic blades made their appearance, with raw material moving as small nodules used in an expedient flake technology employed in a wide variety of domestic tasks (Boksenbaum et al. 1987; Clark 1987; Cobean et al. 1971; Parry 1987).

Obsidian blade technology spread widely and increased in frequency during the Middle Formative period between 900 and 500 B.C. This corresponds with the development of ranked societies, the growth of regional populations, and an increase in interregional interaction networks through which obsidian certainly was traded. What the role of elites was in the spread of obsidian technology remains unclear. Clark (1987) has suggested that emerging elites in Mesoamerica fostered the spread of blade technology through the sponsorship of blade production for local distributions. However, whether obsidian blade technology spread throughout Mesoamerica as a result of elite initiatives or the normal pull of demand from the household economy remains unclear. The earliest evidence for specialized obsidian blade production comes from the Middle Formative site of Chalcatzingo (Burton 1987), where production does not appear to have been under the direct control of elite households (Hirth 2008).

Pressure blades and the technology to produce them provided the important cutting tools needed by complex state-level societies in Central Mexico, Oaxaca, West Mexico, and the Maya region. Copper-bronze metallurgy made its appearance in West Mexico around A.D. 600–800. (Hosler 1994: 12) but never was used widely to produce cutting tools. Obsidian remained the primary material for cutting tools and continued in use during the Colonial period well after Spanish steel was introduced in the New World in A.D. 1519 (Pastrana and Fournier 1998).

16.4 Ubiquitous Consumption of Obsidian Blades

The spread of this technology made obsidian prismatic blades the cutting tool of choice in areas where obsidian was readily available. Despite the occurrence of natural obsidian outcrops in only a few regions (Fig. 16.1), it is not an exaggeration to say that the *majority* of households in greater Mesoamerica had access to obsidian or was regularly consuming obsidian prismatic blades by the end of the Formative period (ca. 150 B.C.–A.D. 200). In some areas, pressure blades and related blade tools dominate normal household assemblages comprising 75–80% of all recovered lithics. Nevertheless some areas of West Mexico remained committed to expedient flake technology in obsidian and resisted the adoption of pressure blade technology longer (Darras 1999). Other areas like the Maya region had access to high-quality chert and chalcedony and readily used it for tools in lieu of obsidian (McAnany 1989; Shafer and Hester 1983, 1991). Even here, however, obsidian pressure blades still moved into the Maya region on a regular basis (Dreiss and Brown 1989).

What is important is that obsidian pressure blades were in high demand throughout Mesoamerica from the Formative period onward. This helped to structure and maintain interregional trade routes through prehistory. While obsidian moved freely throughout Mesoamerica, specific sources dominated some regional networks. Obsidian sources in Central Mexico supplied the Northern and Central highlands and obsidian from the Guatemala Highlands provisioned the greater Maya region. While obsidian was one of *many* commodities moving between regions, it was also one of the most important because of the large number of consumers who used it on a daily basis.

16.5 Pressure Blade Production: A Specialized Activity

The high demand for obsidian prismatic blades required high output, and this provided the basis for obsidian pressure blade production to develop into a specialized craft. Most experts agree that obsidian pressure blade production involved special training and some degree of labor specialization. From their earliest appearance, there is no evidence for ad hoc production of obsidian blades at the household level. In many parts of Mesoamerica, ad hoc obsidian tool production at the household level occurs early as bipolar or expedient flake industries (Boksenbaum et al. 1987; Clark 1987; Coe and Diehl 1980; Jackson and Love 1991; Parry 1987). In contrast, when obsidian pressure blades began to appear in domestic industries they did so as well formed finished pieces without associated production debris or elevated error rates indicative of infrequent production (Boksenbaum et al. 1987; Coe and Diehl 1980).

Research indicates that obsidian blade production was a specialized craft activity from at least the Middle Formative period onward (900–500 B.C.) (Clark 1987; Hirth 2008). Experimental replication reinforces this conclusion suggesting that training and considerable practice was necessary to maintain high levels of craftsman skill (Clark 2003). The Middle Formative workshop at Chalcatzingo is an example of early craft specialization (Burton 1987), although it is unclear whether this was a full-time or part-time activity. Obsidian pressure blades continued to be produced by craft specialists throughout the remainder of Mesoamerican prehistory with production intensifying with the emergence of state societies.

16.6 Specialization and Craftsmen Interdependencies

Technological analyses have enabled investigators to reconstruct production sequences for pressure blade manufacture under different sets of conditions (Clark and Bryant 1997; Hirth and Andrews 2002; Sheets 1978). What is particularly interesting is that the complete production sequence is usually never identified at one location. Instead, reduction activities are differentially distributed over space with mining and core shaping taking place at or near source locales and blade production often located hundreds of kilometers away. A structural interdependency developed over time between craftsmen who specialized in different types of activities. Craftsmen tend to separate into two groups: (1) those located near the source who mined, prepared cores, and produced some blades and (2) craftsmen residing in more distant areas who received shaped cores and produced finished blades for their constituent consumers. The relationship between supplier craftsmen and receiver craftsmen varied over time, but it is a dynamic that channeled access to obsidian through local craftsmen.

How these linkages were established and maintained remains unclear and requires further research. Recent research at Xochicalco suggests that linkages between craftsmen were idiosyncratic and personal rather than centralized or controlled (Hirth 2006). Disruptions in supplier-receiver linkages between craftsmen resulted in major changes in the sources of obsidian exploited at different times. This is apparently what occurred at the Sierra de las Navajas obsidian source near Pachuca, Hidalgo around A.D. 600. The Sierra de las Navajas source had been the primary source used for pressure blade production in Central Mexico between A.D. 150 and 600. A disruption in exploitation and core processing at the quarries around A.D. 600 led to a decline in the use of Pachuca obsidian in Central Mexico for the next 300 years (Garcia et al. 1990; Healan 2003; Hirth 2006; Santley et al. 1995). It was not until focused exploitation resumed around A.D. 900 (Pastrana 2002: 16–18) that Sierra de las Navajas once more became a primary source for obsidian blade production in Central Mexico.

16.7 Obsidian Craft Production as a Political Process

While researchers concur that obsidian craft production was a specialized activity, there is less agreement about how production was organized. Some investigators believe that centralized control of the production and distribution of prismatic blades was important in the political economies of many Mesoamerican states (Santley 1983, 1984). It has been argued, for example, that the key to Teotihuacan's expansion during the second century A.D. was the control of the important obsidian source of Sierra de las Navajas at Pachuca, Hidalgo (Sanders and Santley 1983; Spence et al. 1984). Likewise, the subsequent control of this source by the Toltecs is proposed as one of the reasons for the growth and expansion of the large city at Tula, Hidalgo (Sanders and Santley 1983). Other instances of monopolized political control over obsidian sources include Kaminaljuyu's domination of the obsidian source of El Chayal, Guatemala (Sanders and Santley 1983), Tzintzuntzan's control over obsidian mines at Zinapecuaro and Ucareo, Michoacan (1993), and the role of Cantona in the exploitation of obsidian deposits at Oyameles, Puebla (Garcia; García Cook 2003).

There are two problems, however, with political models of obsidian production. First, we do not know if or how natural resource zones were controlled by political authorities. The limited data suggest that natural resources within the community commons were open to all potential users. The distribution of obsidian production around the El Chayal obsidian source in the Valley of Guatemala suggests that no attempt was made to restrict access to it in any way (Hurtado de Mendoza 1977; Hurtado de Mendoza and Jester 1978). Similarly, Darras (2006) has documented direct extraction and exploitation of obsidian by craftsmen in rural households 20–80 km away from source locales in West Mexico. A similar pattern is found for jade exploitation in the Rio Motagua. Although we would anticipate that the extraction and manufacture of jade objects would be carefully controlled because of its value in Mesoamerican society, this was not the case. Instead, jade was available to and worked by households at all levels of society (Rochette no date). If access to jade was not controlled, then it is likely that the same was true for less valued items like obsidian.

Second, there is little evidence for either elite control or state supervision of blade production that models of political control imply. None of the workshops thought to be state-controlled at Teotihuacan (Spence 1981, 1984) or Cantona (García Cook 2003) have been systematically excavated (cf. Clark 1986). Recent excavations at Teotihuacan uncovered refuse from biface production in an area thought to be a state-controlled workshop (Carballo 2005). Unfortunately, this material was recovered in refuse pits, and the context of production remains unclear. Without question, all Mesoamerican states consumed obsidian blades and bifaces in large quantities to equip their armies. This was accomplished through corvee labor and by imposing a form of tax-in-kind on craftsmen who worked goods needed by the state (Carrasco 1978). This provided an effective way of meeting the state's obsidian needs without necessitating *direct* control of resource areas or the operation of state-sponsored workshops.

16.8 Obsidian Craft Production as a Commercial Process

The evidence indicates that the vast majority of all craft production in ancient Mesoamerica took place in domestic contexts by independent craft producers (Feinman 1999). The same was true for obsidian blade production. Except for the possibility of a small workshop in the Tlatelolco palace (García Velázquez and Cassiano 1990), all of the obsidian craft areas excavated thus far in Mesoamerica are domestic workshops (Hirth 2006: Table 13.2). Excavations at Xochicalco reveal that obsidian craftsmen were independent specialists with a high level of entrepreneurial skill; they obtained obsidian through individual trade partnerships and sold finished goods in its central marketplace (Hirth 2006). Ethnohistoric sources indicate that obsidian craftsmen did not hold a special place in society despite the importance of the goods they produced. Obsidian craftsmen like other craftsmen were part of the broad commoner (*macehualli*) class without special rank or privilege.

The production of obsidian blades by independent specialists raises the question of how production was organized in domestic contexts. Was it a full or part-time activity? Although the data are far from clear, I suspect that part-time production was much more common than full-time production (Hirth 2006). There are two reasons for this. First, the production capacity for obsidian craftsmen was high. A single full-time craftsman reducing one core of 150 blades per day would produce between 48,000 and 54,000 blades per year; this is enough to supply 2,400–5,000 families if they consumed between 10 and 20 blades per year (cf. Clark 1986: 36–38). Second, from the point of view of the craftsman, most full-time craft production is risky endeavor. The ability to defer the purchase of craft goods during food shortages place specialist households at risk if they do not include food production in their normal activities. Craft diversification is one way to defer risk, and we are beginning to recognize multi-crafting as a common practice throughout Mesoamerica (Hirth 2006: Table 12.2).

Several scholars have argued that craft producers were organized in craft guilds as a means to foster training, maintain product quality, and increase production efficiency (Berdan 1982; Katz 1966). Although guild-like organizational structures are found in merchant groups, a similar structure for craftsmen did not exist. Craftsmen certainly collaborated with one another and were organized into tribute cadres for purposes of mobilizing resources used by the state, but there is no indication that they were organized by ward or community into commercial craft guilds like we find in medieval Europe (Epstein 1991).

16.9 The Distribution of Obsidian over Space

Obsidian moved primarily as cores or finished goods (blades, bifaces) rather than as raw material. The reason for this is a simple energetic one: finished goods weigh less than unworked rock. Nevertheless, the distance over which prismatic cores and blades moved is impressive. Obsidian deposits are found in only two areas of Mesoamerica: in the highlands of Guatemala and throughout the transverse volcanic axis of Central Mexico (Fig. 16.1). As a result, obsidian goods often were transported 200–400 km to arrive at their final destination. Analysis of obsidian from Chichen Itza revealed that its major source of obsidian was Ucareo, Michoacan, more than 1,200 km from the site (Braswell and Glascock 2002). Mesoamerica largely lacks navigable rivers, beasts of burden, and wheeled transportation, making the costs of transportation very high. Instead, Mesoamerica had a tumpline economy, and all cargos were moved overland on the backs of human porters. The fact that obsidian moved over long distances despite high costs of transportation underscores the importance of obsidian blades in the everyday life of pre-Hispanic societies.

The question of how obsidian moved and who moved it over space remains unanswered. We know that among the Aztecs, long-distance merchants (*pochteca*) took finished obsidian tools with them to trade in local markets (Sahagún 1981: 3:30–31). The *pochteca*, however, dealt primarily in high-value goods and were not the primary agents by which obsidian was distributed throughout Mesoamerica. This task primarily fell either to merchant peddlers (*oxtomeca*, *tlacocoalnamacac*, *tlanecuiloque*) who specialized in one or more types of utilitarian goods, or to itinerant craftsmen who produced blades as they traveled over space (Carrasco 1978; Katz 1966: 67; Sahagún 1961: 91).

16.10 Craftsman in the Marketplace

The marketplace was the central institution for converting and distributing goods in Mesoamerica. Ethnohistoric sources indicate that independent craftsmen were required by law to sell their wares in the marketplace which enabled the state to collect a small market tax on the goods sold. Marketplaces appear early in Mesoamerica and are established in the highlands by 300–200 B.C. (Feinman et al. 1984). They were the principal mechanism through which obsidian blades were distributed to consuming households. They also provided a network through which itinerant craftsmen could move, produce, and distribute obsidian blades.

I personally favor the view that itinerant craftsmen were the primary means by which obsidian moved and prismatic blades were distributed throughout the greater part of Mesoamerican prehistory. There are excellent ethnohistoric descriptions of craftsmen working in the marketplace where blades were produced for immediate resale (Clark 1989; Diaz del Castillo 1956; Sahagún 1981: 148). At Coyoacan, the obsidian craftsmen listed in market records are specialists from outside the region (Anderson et al. 1976: 149). The best explanation for these specialists is that they are itinerant craftsmen who moved from market to market to produce stone tools. Archaeological evidence from Xochicalco, Mexico, indicates that local craftsmen either obtained obsidian cores from itinerant craftsmen and not from merchants or procured it directly at the source themselves (Hirth 2006). Likewise, excavations in Xochicalco's public marketplace recovered microdebitage from prismatic blade manufacture recovered directly from the earthen plaza floor where craftsmen worked (Hirth 2006). Markets and obsidian craft production were closely linked and it is difficult to imagine how prismatic blades could become such a ubiquitous commodity in Mesoamerica without them.

16.11 Specialization and the Development of Mesoamerican Obsidian Systems

Obsidian craft production needs to be examined in terms of the larger *industry system* of which it was a part. In Mesoamerica, lithic *industries* are defined in technological terms as the manufacture of tools from a common raw material using a shared body of techniques (Sheets 1978: 3). Examples include expedient flake, biface, and pressure blade industries, each of which employed different combinations of percussion and pressure techniques to manufacture stone tools. A lithic *system* refers to the production relationships and social mechanisms that move raw material and products over space. These systems are networks of interacting craftsmen and merchants through which obsidian from a specific source was worked and distributed. Multiple lithic systems developed and operated simultaneously throughout Mesoamerica over time. These systems could (1) employ the same or different production techniques, (2) overlap or be completely separate, and (3) compete with or complement one another depending on the framework of supply and demand

In Central Mexico, craftsmen utilized obsidian sources in ways that matched their physical properties. Obsidian sources with more impurities (e.g., Otumba, Pizarrín-Tulancingo, Pico de Orizaba, and Paredon) were often used for bifaces and simple percussion flakes. Obsidian sources with few impurities (e.g., Pachuca, Ucareo, and Zacualtipan) could be used for any production task, but more often were selected by craftsmen for producing prismatic blades. Selective use of obsidian in this way probably developed over time as a result of increased circulation of obsidian from different sources with different working properties.

Since obsidian systems represent interconnected networks of suppliers and consumers, it is not surprising to find them overlapping in space and growing or shrinking over time. Hammond (1972) first recognized this phenomenon for the Maya lowlands during its Classic period (A.D. 300–900), where he observed obsidian from the Ixtepeque source circulating along the coast of Belize while obsidian from the El Chayal source was more common in inland sites. In Central Mexico, Pachuca obsidian was the dominant material used to produce pressure blades during its Classic period (A.D. 150–650), while Otumba obsidian was used for bifaces. This changed during the Epiclassic period (A.D. 650–900) when obsidians from the Ucareo, Pachuca, Oyameles, and Zacualtipan sources were all used to make pressure blades and Otumba and Pizarrin-Tulancingo sources were used for bifaces. What is particularly interesting is that these obsidian systems developed during the Epiclassic which was a period of intense political competition. Nevertheless, obsidian moved easily across political boundaries underscoring the organization of obsidian systems as commercial networks rather than politically organized procurement systems.

16.12 Obsidian Use in the Colonial Period

Copper-bronze metallurgy appeared in northwestern Mesoamerica around A.D. 600–800 (Hosler 1994). While it was used to produce utilitarian tools such as needles, axes, tweezers, and awls, it never competed with obsidian as the primary cutting tool at any time before the Conquest. In the Old World, some of the most extensive evidence for stone tool craft production dates to the Middle to Late Bronze Age (Hartenberger et al. 2000; Rosen 1997; Torrence 1986). The same is true of Mesoamerica, with obsidian and metallurgical craftsmen specializing in the manufacture of different types of products.

In Mesoamerica, obsidian continued to be used well into the Colonial period, after knives and axes became available in Spanish steel after A.D. 1519. The Pachuca obsidian mines continue to be exploited, and the presence of a sixteenth century church at the quarries indicates that they remained the focal point of regional socioeconomic life (Pastrana and Fournier 1998). Two factors probably helped to extend the use of obsidian into the Colonial period. The first was the high cost of imported steel tools vis-a-vis the less expensive obsidian. The second, and perhaps more important reason, was the long-standing use of obsidian at the domestic level and the established operation of distribution systems that met household needs. Household economies tend to be conservative and resistant to change as long as they can reliably provision themselves with the resources they need at a lower cost (Hirth 2009). It may have been only after inexpensive steel tools were available in local market settings that obsidian began to lose its appeal to traditional users.

16.13 Discussion

Mesoamerica is an important area for researchers interested in the study of prehistoric stone tool technology because it represents the development of state-level societies using obsidian for all its cutting needs. Metallurgy was present 700 years before the Spanish Conquest, but unlike the Old World, it did not compete with or replace the obsidian craft specialist as the source of primary cutting tools. Obsidian pressure blade technology was part of the infrastructure on which complex society in Mesoamerica would develop. Although it first appeared in hunting and gathering societies during the Late Archaic period (ca. A.D. 2500), obsidian pressure blade technology expanded in use with the appearance of sedentary agricultural communities, early chiefdoms, and the increased demands of early states.

From 1000 B.C. onward, obsidian prismatic blade production appears to have been in the hands of craft specialists. There is little evidence for attempts of ad hoc prismatic blade production at the household level. Instead, blades in early deposits are well formed, efficiently made, and have the same low error rates as blades produced in later specialized craft workshops. Obsidian prismatic blade production remained in the hands of specialized craftsmen throughout the remainder of Mesoamerican prehistory. Obsidian blades were the cutting tool of choice in many households and led to the development of an extensive system of interdependent craft specialists across Mesoamerica. Multiple competing obsidian systems formed over time linking craftsmen and merchants producing and distributing obsidian from different source locales.

While the early development and spread of this technology remains incompletely understood, most obsidian craft production was in the hands of independent craft specialists before the appearance of early states. Obsidian craft production was primarily a commercial rather than a political activity. While it may have played an important role in the development of large urban centers such as Teotihuacan, Cantona, or Kaminaljuyu, production and distribution were not controlled through political processes. Obsidian blade production was too important for the millions of Mesoamerican households that consumed tens of millions of blades each year to allow that to happen. Instead, merchants and entrepreneurial craftsmen met the demand of their consumers through broad-based and diversified domestic production systems spread widely across the rural and urban landscape. Instrumental to this commercial system was the development of an integrated market system that fostered the emergence of craft specialists and linked them with the consumers of their finished products. The importance of the marketplace for obsidian blade production is evident in its role as a place of blade production by itinerant craftsmen. At the time of the Conquest, obsidian craftsmen regularly produced blades and other tools in major marketplaces where households provisioned themselves with all kinds of utilitarian goods.

Mesoamerica was economically challenged. The absence of all but a few navigable rivers and beasts of burden made it difficult to move heavy utilitarian goods like obsidian very far over space. Despite these difficulties, obsidian cores, blades, and finished goods moved and were traded over hundreds of kilometers. That this was possible is a function to its desirability as a product and the efficiency of the socioeconomic systems that produced and distributed it. Mesoamerica contained state-level societies organized using Stone Age technologies, and it serves as an excellent case study for contemplating what stone tool technology is capable of when metal tools are not available.

References

Anderson, Arthur, Frances Berdan, and Jane Lockhart 1976 *Beyond the codices. The Nahua view of colonial Mexico*. University of California Press, Berkeley.

Berdan, Frances 1982 *The Aztecs of Central Mexico. An imperial society*. Holt, Rinehart & Winston, New York.

Boksenbaum, Martin, Paul Tolstoy, Garman Harbottle, Jerome Kimberlin, and Mary Nivens 1987 Obsidian industries and cultural evolution in the Basin of Mexico before 500 B.C. *Journal of Field Archaeology* 14:65–75.

Braswell, Geoffrey, and Michael Glascock 2002 The emergence of market economies in the ancient Maya world: Obsidian exchange in Terminal Classic Yucatan, Mexico, In *Geochemical evidence for long-distance exchange*, edited by M. Glascock, pp. 33–52. Bergin and Garvey, Westport.

Burton, Susan 1987 Obsidian blade manufacturing debris on terrace 37. In *Ancient Chalcatzingo*, edited by D. Grove, pp. 321–328. University of Texas Press, Austin.

Carballo, David 2005 State political authority and craft production at the moon pyramid, Teotihuacan, Mexico. Unpublished Ph.D. dissertation, Department of Anthropology, University of California, Los Angeles.

Carrasco, Pedro 1978 La economía del México prehispánico. In *Economía política e ideología en el México prehispánico*, edited by P. Carrasco and Johanna Broda, pp. 13–74. Editorial Nueva Imagen, México.

Cassiano, Giofranco 2005 La technología de navajillas prismáticfas en la perspectiva histórica. Paper presented at the INAH Homenaje to Guadalupe Alba Mastache, Mexico City.

Clark, John 1982 Manufacture of Mesoamerican prismatic blades: An alternative technique. *American Antiquity* 47: 355–376.

Clark, John 1986 From mountains to molehills: A critical review of Teotihuacan's obsidian industry. In *Research in economic anthropology, supplement No. 2. Economic aspects of prehispanic highland Mexico*, edited by B. Isaac, pp.23–74, JAI Press, Greenwich.

Clark, John 1987 Politics, prismatic blades, and Mesoamerican civilization. In *The organization of core technology*, edited by J. Johnson and C. Morrow, pp. 259–285. Westview Press, Boulder.

Clark, John 1989 Obsidian: The primary Mesoamerican sources. In *La Obsidiana en Mesoamérica*, edited by M. Gaxiola G. and J. E. Clark, pp. 299–319. INAH, Colección Científica 176. Mexico.

Clark, John 2003 Craftsmanship and craft specialization. In *Mesoamerican lithic technology: Experimentation and interpretation,* edited by K. Hirth, pp. 220–238. University of Utah Press, Salt Lake City.

Clark, John, and Douglas Bryant 1997 A technological typology of prismatic blades and debitage from Ojo de Agua, Chiapas, Mexico. *Ancient Mesoamerica* 8: 111–136.

Cobean, Robert, Michael Coe, Edward Perry, Karl Turekian, and Dinkar Kharkar 1971 Obsidian Trade at San Lorenzo Tenochtitlan, Mexico. *Science* 174: 666–671.

Coe, Michael, and Richard Diehl 1980 *In the land of the Olmec. The archaeology of San Lorenzo Tenochtitlan*. University of Texas Press, Austin.

Crabtree, Donald 1968 Mesoamerican polyhedral cores and prismatic blades. *American Antiquity* 33: 446–478.

Diaz del Castillo, Bernal 1956 *The discovery and conquest of Mexico 1517–1521*. Farrar, Straus and Cudahy, New York.

Darras, Véronique 1999 *Tecnologías prehispánicas de la obsidiana: Los centros de producción de la región de Zinápero-Prieto, Michoacán*. Cuadernos de Estudios Michoacanos, No. 9. Centre Français d' Études Mexicaines et Centraméricaines, Mexico City.

Darras, Véronique 2006 Obsidian strategy and power in the Tarascan society from the Malpais de Zacapu, Michoacan. Paper presented at the 71st annual meeting of the Society for American Archaeology, San Juan, Puerto Rico.

Dreiss, Meredith, and David Brown 1989 Obsidian Exchange Patterns in Belize. In *Research in Economic Anthropology*, Supplement 4, edited by P. McAnany and B. Issac, pp. 57–90. JAI Press, Greenwich.

Epstein, Steven 1991 *Wage labor and guilds in medieval Europe*. University of North Carolina Press, Chapel Hill.

Feinman, Gary 1999 Rethinking our assumptions: Economic specialization at the household scale in ancient Ejutla, Oaxaca, Mexico. In *Pottey and people*, edited by J. Skibo and G. Feinman, pp. 81–98. University of Utah Press, Salt Lake City.

Feinman, Gary, Richard Blanton, and Stephen Kowalewski 1984 Market system development in the prehispanic Valley of Oaxaca, Mexico. In *Trade and exchange in early Mesoamerica*, edited by K. Hirth, pp. 157–178. University of New Mexico Press, Albuquerque.

Flenniken, J. Jeffrey, and Kenneth Hirth 2003 Handheld prismatic blade manufacture in Mesoamerica. In *Mesoamerican lithic technology: Experimentation and interpretation,* edited by K. Hirth, pp. 98–107. University of Utah Press, Salt Lake City.

Garcia, Chávez, Michael Glascock, J. Michael Elam, and Harry Iceland 1990 The INAH salvage archaeology excavations at Azcapotzalco, Mexico: An analysis of the lithic assemblage. *Ancient Mesoamerica* 1: 225–232.

García Cook, Angel 2003 Cantona: The city. In *El urbanismo en Mesoamérica. Urbanism in Mesoamerica*, edited by W. Sanders, A. Mastache and R. Cobean, pp. 311–343. INAH/Penn State University, Mexico City/University Park.

García Moll, Roberto 1977 *Análisis de los materialies arqueológicos de la cueva del Texcal, Puebla*. Colección Cientifica 56. INAH, Mexico.

García Velázquez, Jorge, and Gianfranco Cassiano 1990 La producción de navajillas prismáticas en el Postclásico tardío: El caso de la plaza de la Banca Nacionalizada. In *Etnoarqueología: Primer coloquio Bosch-Gimpera*, edited by Y. Sugiura and M. Carmen Serra, pp. 513–526. UNAM, Mexico City.

Hammond, Norman 1972 Obsidian Trade Routes in the Maya Area. *Science* 178: 1092–1093.

Hartenberger, Brit, Steven Rosen, and Timothy Matney 2000 The Early Bronze age blade workshop at Titris Hoyuk: Lithic specialization in an urban context. *Near Eastern Archaeology* 63: 51–58.

Healan, Dan 2003 From quarry pit to the trash pit: Comparative core-blade technology at Tula, Hidalgo and the Ucareo obsidian source region. In *Mesoamerican lithic technology: Experimentation and interpretation,* edited by K. Hirth, pp. 153–169. University of Utah Press, Salt Lake City.

Hirth, Kenneth 2006 *Obsidian Craft Production in Ancient Central Mexico.* The University of Utah Press, Salt Lake City

Hirth, Kenneth 2008 Unidad doméstica, comunidad y artesanía en un cacicazgo del Formativo Medio: Revalorando la importancia del Proyecto Chalcatzingo, In *El Formativo en Mesoamerica*, edited by A. Cyphers and K. Hirth, pp. 93–125. UNAM, Mexico City.

Hirth, Kenneth 2009 Craft production, household diversification, and domestic economy in prehispanic Mesoamerica, In *Housework: Craft Production and Domestic Economy in Ancient Mesoamerica*, edited by K. Hirth, pp. 13–32 Archaeological Publications of the American Anthropological Society No. 19.

Hirth, Kenneth, and Bradford Andrews 2002 *Pathways to prismatic blades*, edited by K. Hirth and B. Andrews, pp. 1–14. Cotsen Institute of Archaeology Monograph 45, University of California, Los Angeles.

Hirth, Kenneth, Bradford Andrews, and J. Jeffrey Flenniken 2003 The Xochicalco production sequence for obsidian prismatic blades: technological analysis and experimental inferences. In *Mesoamerican lithic technology: Experimentation and interpretation,* edited by K. Hirth, pp. 182–196. University of Utah Press, Salt Lake City.

Hosler, Dorothy 1994 *The sounds and colors of power.* MIT Press, Cambridge.

Hurtado de Mendoza, Luis 1977 Obsidian studies and the archaeology of the Valley of Guatemala. Unpublished Ph.D. dissertation, Penn State University, University Park.

Hurtado de Mendoza, Luis, and William Jester 1978 Obsidian sources in Guatemala: A regional approach. *American Antiquity* 43: 424–435.

Jackson, Thomas, and Michael Love 1991 Blade running: Middle Preclassic obsidian exchange and the introduction of prismatic blades at La Blanca, Guatemala. *Ancient Mesoamerica* 2: 47–59.

Katz, Friedrich 1966 *Situación social y económica de los Aztecas durante los siglos XV y XVI.* Instituto de Investigaciones Históricas, UNAM, Mexico City.

Kirchoff, Paul 1952 Mesoamerica: Its geographic limits, ethnic composition and cultural characteristics, In *Heritage of Conquest*, edited by S. Tax, pp. 17–30. Cooper Square Publishers, New York.

MacNeish, Richard S., Antoinette Nelken-Terner, and Irmgard Weitlaner de Johnson 1967 *The prehistory of the Tehuacan Valley, Vol. 2: Non-Ceramic artifacts.* University of Texas Press, Austin.

McAnany, Patricia A. 1989 Stone Tool Production and Exchange in the Eastern Maya Lowlands: The Consumer Perspective from Pulltrouser Swamp, Belize. *American Antiquity* 54: 332–346.

Neiderberger, Christine 1976 *Zohapilco: Cinco Milenios de Ocupación Humana en un Sitio Lacustre de la Cuenca de México.* Colección Científica 30, INAH, Mexico.

Neiderberger, Christine 1987 *Paleopaysages et archaeologie pre-urbaine du Bassin de México.* Col. etudes mesoamericaines 1–11. CEMCA.

Parry, William 1987 *Chipped Stone Tools in Formative Oaxaca, Mexico: Their Procurement, Production and Use.* Memoirs of the Museum of Anthropology, No. 20. University of Michigan, Ann Arbor.

Parry, William 1994 Prismatic blade technologies in North America. In *The organization of North American prehistoric chipped stone tool technologies*, edited by P. Carr, pp. 87–98. International Monographs in Prehistory, Ann Arbor.

Pastrana Cruz, Alejandro 2002 Variation at the source: Obsidian exploitation at Sierra de las Navajas, Mexico. In *Pathways to prismatic blades*, edited by K. Hirth and B. Andrews, pp. 15–28. Cotsen Institute of Archaeology Monograph 45, University of California, Los Angeles.

Pastrana, Alejandro, and Patricia Fournier 1998 Explotación Colonial de Obsidiana en el Yacimiento de Sierra de Las Navajas. In *Primer Congreso Nacional de Arqueología Histórica: Memoria Oaxaca, 1996*, edited by E. Fernández, D. Gómez and S. Gómez, pp. 486–496. CONACULTA-INAH, Mexico.

Rochette, Erick No date Archaeological research in the Rio Motagua Valley, Guatemala. (Manuscript)

Rosen, Steven 1997 *Lithics after the stone age.* Altamira Press, Walnut Creek.

Sahagún, Fray Bernardino de 1961 *Florentine Codex. General history of the things of New Spain, book 10, the people.* Translated by D. Dibble and A. Anderson. Monographs of the School of American Research and the University of Utah, No. 14, Part 11, Santa Fe.

Sahagún, Fray Bernardino de 1981 *Historia general de las cosas de Nueva EspaZa.* 4 Volumes, Editorial Porrua, Mexico City.

Sanders, William, and Robert Santley 1983 A tale of three cities: Energetics and urbanization in pre-hispanic central Mexico. In *Prehistoric settlement patterns*, edited by E. Vogt and R. Leventhal, pp. 243–291. University of New Mexico Press/the Peabody Museum of Archaeology and Ethnology, Albuquerque/Cambridge.

Santley, Robert 1983 Obsidian trade and Teotihuacan influence in Mesoamerica. In *Highland-lowland interaction in Mesoamerica*, edited by A. Miller, pp. 69–124. DORLC, Washington.

Santley, Robert 1984 Obsidian exchange, economic stratification, and the evolution of complex society in the Basin of Mexico. In *Trade and exchange in early Mesoamerica*, edited by K. Hirth, pp. 43–86. University of New Mexico, Albuquerque.

Santley, Robert, Janet Kerley, and Thomas Barrett 1995 Teotihuacan period lithic assemblages from the Teotihuacan Valley, Mexico. In *The Teotihuacan Valley project final report - Volume 3, The Teotihuacan period occupation of the valley, part 2. Artifact analyses*, edited by W. Sanders, pp 466–497. Occasional Papers in Anthropology No. 20. Matson Museum of Anthropology, The Pennsylvania State University, University Park.

Shafer, Harry, and Thomas Hester 1983 Ancient Maya chert workshops in Northern Belize, Central America. *American Antiquity* 48: 519–543.

Shafer, Harry, and Thomas Hester 1991 Lithic craft specialization and product distribution at the Maya site of Colha, Belize. *World Archaeology* 23: 79–97.

Sheets, Payson 1975 Behavioral analysis and the structure of a prehistoric industry. *Current Anthropology* 16: 369–391.

Sheets, Payson 1978 Artifacts. In *The prehistory of Chalchuapa, El Salvador*, edited by R. Sharer, 2: 2–107. University of Pennsylvania Press, Philadelphia.

Spence, Michael 1981 Obsidian production and the state in Teotihuacan. *American Antiquity* 46: 769–788.

Spence, Michael 1984 Craft production and polity in early Teotihuacan. In *Trade and exchange in early Mesoamerica*, edited by K. Hirth, pp. 87–114. University of New Mexico Press, Albuquerque.

Spence, Michael, Jerome Kimberlin, and Garmon Harbottle 1984 State-controlled procurement and the obsidian workshops of Teotihuacan, Mexico. In *Prehistoric quarries and lithic production*, edited by E. Ericson and B. Purdy, pp. 97–105. Cambridge University Press, Cambridge.

Titmus, Gene, and John Clark 2003 *Mexica* blade making with wooden tools: Recent experimental insights. In *Mesoamerican lithic technology: Experimentation and interpretation,* edited by K. Hirth, pp. 72–97. University of Utah Press, Salt Lake City.

Torrence, Robin 1986 *Production and exchange of stone tools*. Cambridge University Press, Cambridge.

Chapter 17
Development of Pressure Blade Technology in North-Central and Western Mexico

Véronique Darras

17.1 Introduction

Among the numerous Mesoamerican studies on obsidian blade production, several stress as a particularly significant factor the sociopolitical complexity of the societies in which it developed (for examples, see Santley 1984; Santley et al. 1986; Spence 1981, 1987; Clark 1987, 1989). The latest publications led by Kenneth Hirth (Hirth and Andrews 2002; Hirth 2003, 2006), which describe the mechanisms regulating obsidian blade production, distribution, and consumption in several regions of Mesoamerica at diverse periods of pre-Hispanic history, make clear the large diversity of social and political contexts in which the technology developed.

Despite uncertainty as to its place and moment of origin, the prismatic blade is present in most regions of Mesoamerica from the Early Pre-Classic, more precisely from 1200 B.C. However, some regions, especially in North-Central and Western Mexico, are noticeably different (Fig. 17.1). In these areas, the prismatic blade was an imported product that was not introduced until the end of the Pre-Classic period, and its technology then developed along various paths without – a priori – any spatial logic. On the other hand, during the Proto-Classic (A.D. 1–250; Table 17.1), percussion blade manufactures acquired increasing importance in the lithic systems alongside flake and bifacial industries. Pressure blade technology was only introduced at the end of the Epi-Classic (A.D. 750–900), for the Mexican Far West, and the end of the Early Post-Classic (A.D. 900–1100), for the Northern Michoacan, replacing the older tradition of obtaining percussion blades.

V. Darras (✉)
Archéologie des Amériques, CNRS-Université Paris 1, Nanterre, France
veronique.darras@mae.u-paris10.fr

P.M. Desrosiers (ed.), *The Emergence of Pressure Blade Making: From Origin to Modern Experimentation*, DOI 10.1007/978-1-4614-2003-3_17,

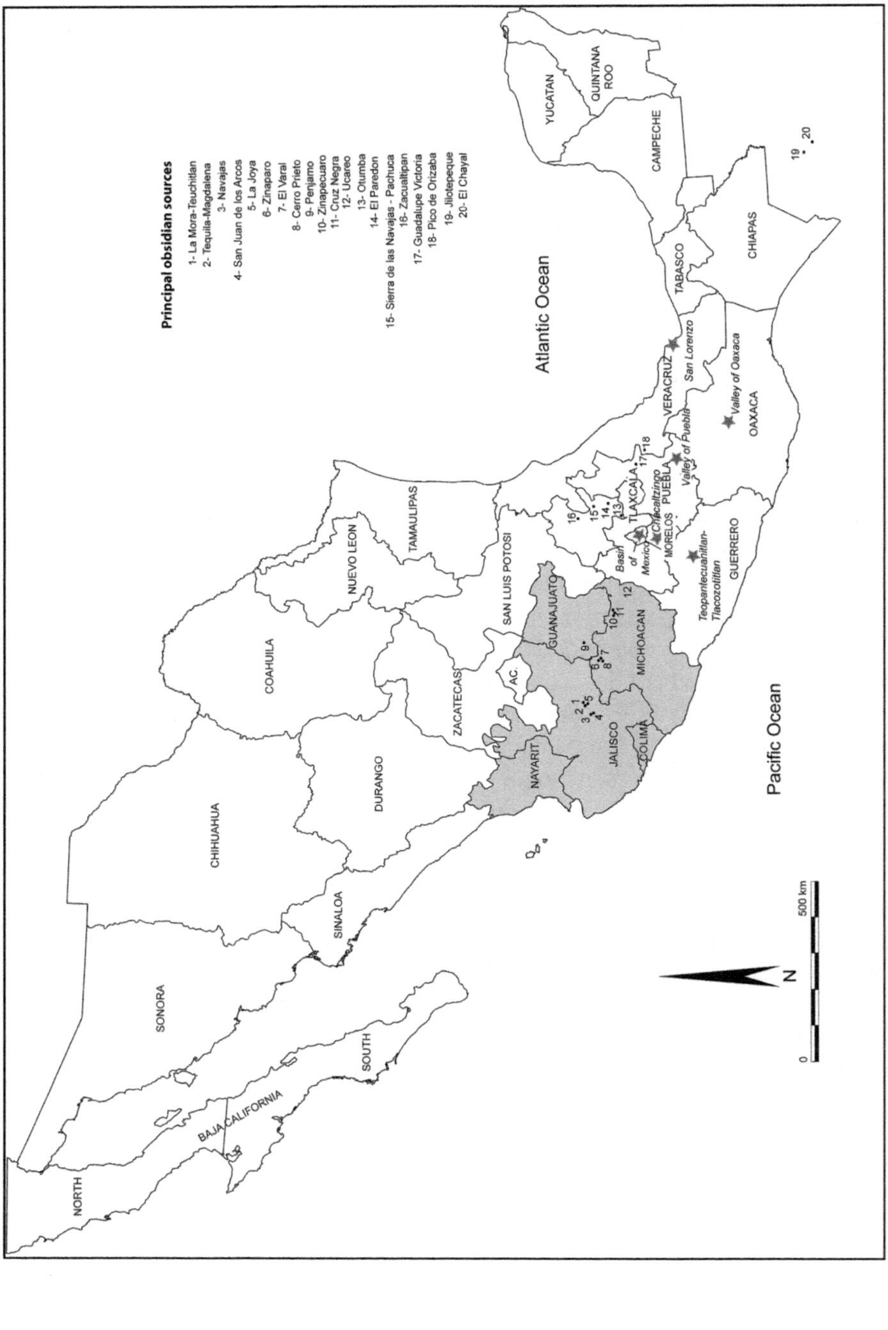

Fig. 17.1 Western area and principal obsidian Mesoamerican sources mentioned in the chapter (Drawing by V. Darras)

Table 17.1 Chronological chart of Mesoamerica and regions mentioned in the chapter

	Period	Jalisco central highlands (Beekman and Galvan 2006) Nayarit/ Colima	Jalisco highlands (Beekman and Weigand in press)	Zacapu region (Michelet 1992) Northeast Michoacan (Oliveros 2004)	Lerma/ Acambaro (Darras and Faugère 2005; Hernandez 2006)	Basin of Mexico
1500						
1450						Azteca IV
1400						Azteca III
1350	Late post-classic				Late Acámbaro	Azteca I
1300				Milpillas *Development of the prismatic blade technology*	Early Acámbaro	
1250						
1200						
1150	Middle post-classic		Aztatlan			
1100						
1050						
1000						Mazapan
950			*Development of the prismatic blade technology*			
900	Early post-classic			Palacio	Perales Terminal	
850				La Joya		
800			Teuchitlan II			
750		El Grillo				
700				Late Lupe		
650	EPI-classic			*Ucareo*		Coyotlatelco
600				*Prismatic blades*	Perales	Metepec
550			Teuchitlan I	Jarácuaro		
500	Middle classic					
450		Late Tabachines		*Pachuca and Ucareo*		
400				*Prismatic blades*		Xolalpan
350		?		Loma Alta 3	Choromuco	
300			Ahualulco			
250	Early classic			*Pachuca and Ucareo*	Mixtlán 2	Tlamimilopa
200				*Prismatic blades*		

(continued)

Table 17.1 (continued)

	Period	Jalisco central highlands (Beekman and Galvan 2006) Nayarit/ Colima	Jalisco highlands (Beekman and Weigand in press)	Zacapu region (Michelet 1992) Northeast Michoacan (Oliveros 2004)	Lerma/ Acambaro (Darras and Faugère 2005; Hernandez 2006)	Basin of Mexico
150		Middle Tabachines				
100						Miccaotli
0				Loma Alta 2	Mixtlán 1	Tzacualli
50	Proto-classic or	Early Tabachines	Late El Arenal			Cuicuilco
100	Terminal pre-classic			Loma Alta 1	Transition	
200			Early El Arenal			
300						
400	Late pre-classic				Late Chupicuaro	Ticoman
500						
600			San Felipe		Early Chupicuaro	
700	Middle pre-classic					
800						
900		Capacha				
1000						
1100						*Presence of the prismatic blade technology*
1200						
1300	Early pre-classic					
1400						
1500				El Opeño		

We put forward here a synthesis of the available data for two regions of North-Central and Western Mexico: one set from the Jalisco Highlands and the other from Northern Michoacan and the Middle Lerma Valley, also referred to as Bajio. Studying the lithic systems developed by the populations living in these regions allows us to focus on the conditions under which pressure blade technology appeared and explore various hypotheses. In what way did social and political factors interact with its development? Why such a delay for its adoption, despite the abundance of high-quality obsidian sources and the early existence of particularly dynamic cultural centers?

17.2 The Beginnings of the Prismatic Blade in Mesoamerica

Though the origin of the technique, the moment in which it appeared, and how it expanded are not perfectly known (see the works of Parry 1994; Hirth and Flenniken 2002; Darras 2005a), the studies on the Pre-Classic period in the Mesoamerican highlands and lowlands show the obsidian prismatic blade was present at several settlement sites in the Early Pre-Classic from 1200 B.C. The most precise data concerns the Olmec site of San Lorenzo (Cobean et al. 1971, 1991), on the Gulf coast, the Mexico Valley sites (Niederberger 1976; Boksenbaum 1978; Boksenbaum et al. 1987), and, finally, several establishments in the Tehuacan and Oaxaca valleys, in the highlands of South-Central Mexico (MacNeish et al. 1967; Pires-Ferreira 1975, 1976; Elam et al. 1994) and the Maya lowlands. Prismatic blades are also observed during this period in the State of Guerrero (Niederberger 1976, 1986, 1987). In Morelos, prismatic blades appear between 1050 and 900 B.C. (Grove 1974, 1987; Burton 1987). They are always found as finished products, and without any other artifacts, such as cores or preparation flakes and blades, to indicate on-site production. Cores, although still rare, begin to appear in a few settlement sites in the Basin of Mexico toward 1000 B.C., such as those of Zohapilco (Niederberger 1976), Tlapacoya (Narez 1990), Tlatilco (Niederberger 1987), or Coapexco (Boksenbaum 1978; Boksenbaum et al. 1987). At the site of Chalcatzingo (Morelos), workshops producing prismatic blade appear about 700 B.C. (Grove 1987; Burton 1987).

From as early as 800 B.C., the whole of Central and Southern Mesoamerica was using prismatic obsidian blades and most of the regions were embedded in long-distance exchange networks. Geographically, prismatic blades covered a geocultural area affected, directly or indirectly, by the Olmec phenomenon. Characterization analyses have revealed the diversity of supply sources, but the two sources supplying the bulk of the raw material for making prismatic blades have been found to be Otumba and, more significantly, Ucareo-Zinapecuaro (Michoacan) in the Eastern part of our research area (Figs. 17.1, 17.2).

In general, the data available for the Early and Middle Pre-Classic period is insufficient to determine how prismatic blades were produced; the earliest sites to manufacture blades are still unknown. Boksenbaum (1978); Boksenbaum et al. (1987) proposed a scenario in which traveling craftsmen prepared cores close to obsidian deposits and then produced and distributed the blades throughout the settlements, thereby creating a consumption market and an extended network of dependence. Overall, the information available today designates Highland Central Mexico as a crucial region, which, owing to the abundance and quality of its obsidian deposits, seems to have been the prime mover involved in the invention of pressure blade technology and the organization of production systems with long-distance circulation networks.

But whereas most of Mesoamerica is involved in the prismatic blade phenomenon, the regions in North-Central and Western Mexico remained outside of the process, despite the fact that Ucareo – one of the main deposits to have played a leading role in this economy – is far to the West.

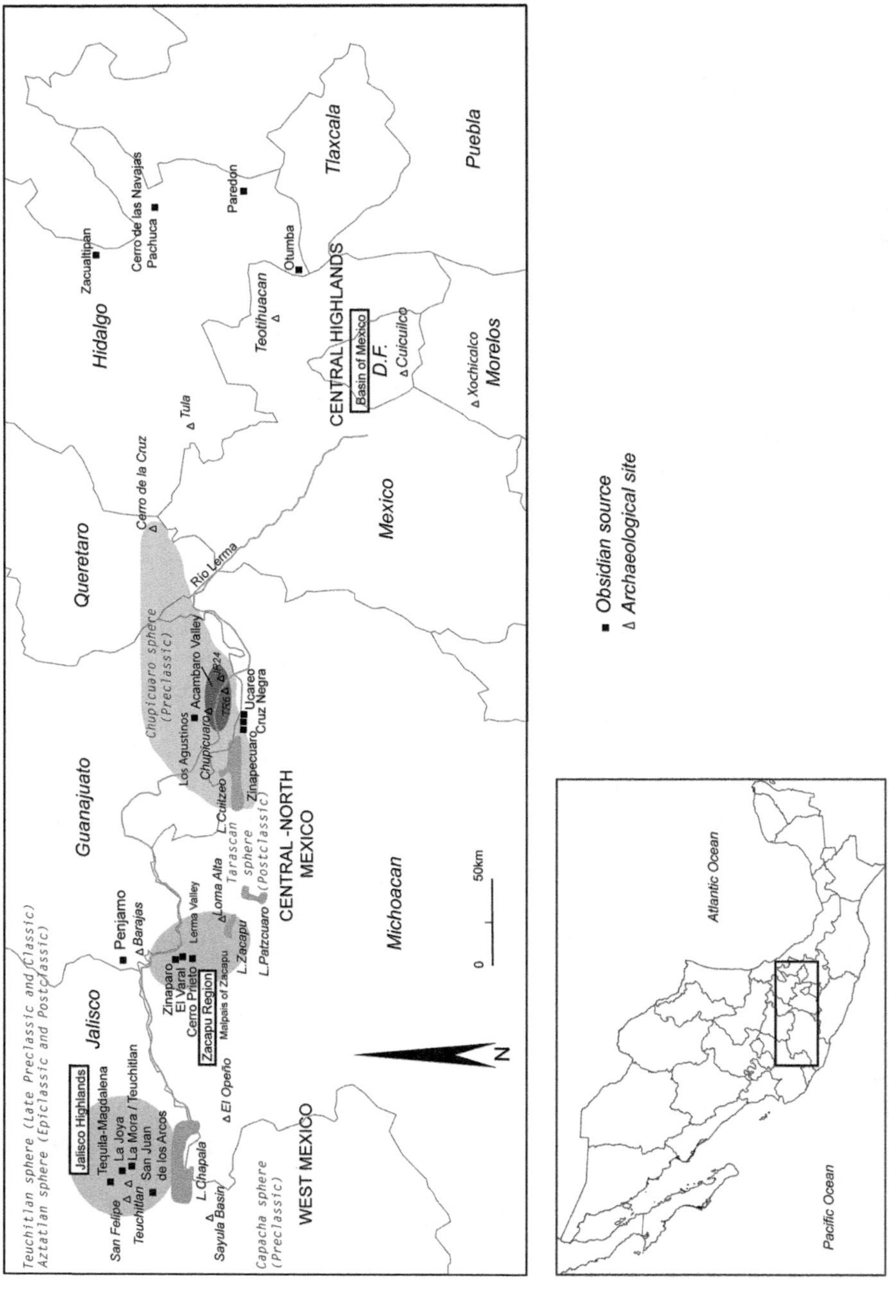

Fig. 17.2 West and North Central Mexico (Drawing by V. Darras)

17.3 Obsidian in North-Central and Western Mesoamerica

The regions in North-Central and Western Mexico are integrated as one of the ten geocultural areas defined by Paul Kirchhoff in his 1943 definition of Mesoamerica (Western Mexico). This area brings together the modern States of Michoacán, Jalisco, Colima, and Nayarit and a part of the States of Guanajuato, Guerrero, Zacatecas, Durango, and Sinaloa (Figs. 17.1, 17.2). Traversed by the trans-Mexican neovolcanic axis, the whole region is rich in lithic raw material, especially obsidian, since it contains various sectors with excellent quality deposits that were all systematically exploited during the pre-Hispanic era (Fig. 17.2). For the purposes of this chapter, we will focus upon Teuchitlan, in the State of Jalisco, which comprises numerous deposits and mine workshops, including those of La Mora-Teuchitlan, Tequila-Magdalena, San Juan de los Arcos, and La Joya and Navajas; the Zinaparo region in North-Western Michoacan, with three different sources – El Cerro Varal, El Cerro Zinaparo, and El Cerro Prieto; and finally, in the North-East corner of the same State, the Ucareo-Zinapecuaro complex consists of three sources – Ucareo, Zinapecuaro, and Cruz Negra.

Most of the regions in this cultural area have long been considered marginal, as they do not meet all of the criteria required for being really "Mesoamerican," and so have not been the subject of extensive long-term archaeological studies. As a result of this lack of interest, the archaeological information available for the area is uneven in both quantity and quality and is clearly behind the advances achieved in the other regions of Mesoamerica. Numerous efforts undertaken over the last 20 years have now mitigated this notion of marginality, and today, research as a whole tends to stress the idea of multiple trajectories, certainly characterized by their originality, but fully involved in the more global Mesoamerican dynamics. Nevertheless, despite a definite renewed interest in the issues, research on the lithic industries plays a minor, even nonexistent, role. In reality, technological studies have not been developed, and only those carried out in Northern Michoacan and Guanajuato (Darras 1993, 1994, 1999, 2005a, 2005b, 2008; Lodeho 2007; Healan 1997, 2002, 2003, 2004, 2005) are relevant to our study. Data on the Jalisco Highlands is still incomplete, despite several publications on the subject (Soto de Arachevaleta 1982, 2005; Weigand and Spence 1982; Spence et al. 2002; Esparza and Ponce Ordaz 2005).

17.4 Early and Middle Pre-Classic Obsidian Industries in Western Mexico

Archaeological knowledge of Early and Middle Pre-Classic occupations in all the regions of Western Mexico comes exclusively from the evidence of burials. This period is known through two cultural traditions called "El Opeño" and "Capacha" (Fig. 17.2). The first tradition developed between 1500 and 1000 B.C. and the second between 1000 and 800 B.C.: they were thus contemporaries of the main cultural centers of Central Mesoamerica, notably the Gulf coast (Olmecs) and the Basin of Mexico. The characteristics of burial furnishings indicate that they were agrarian societies with complex sociopolitical and religious organization (Kelly 1980; Oliveros 2004). The observations on the lithic industries come from artifacts recovered from fill deposits or burial offerings. At El Opeño in North-Western Michoacan, a region where the nearest obsidian sources are approximately 40 km away, the obsidian material appears to consist of flake industries and bifacial tools; the latter artifacts are generally found as burial offerings[1] (Oliveros 2004). As for the Capacha tradition archaeological contexts, found in the modern States of Colima and Jalisco – zones where the nearest obsidian sources are roughly 30 km away – they generally yield reduced quantities of this material, mainly in the form of flakes and projectile points (Kelly 1980; Mountjoy 2004). Obsidian blades, prismatic or nonprismatic, are completely absent (Kelly 1980: 83). Finally, in the Jalisco Highlands, the sparse information available for the San Felipe phase (about 1000–300 B.C., although still poorly dated) (Weigand 2000: 65) is also remarkable for the absence of blade industries and the exclusive existence of flake industries and bifacial production.

17.5 Blade Industries in the Jalisco Highlands

Blade production in the Mexican West raises a certain number of issues – chronology in particular – owing to the extreme complexity of its archaeological contexts.[2] Today, we benefit from only one really thorough study on the obsidian technologies used in the specialized workshops at the Teuchitlan site[3] (Soto de Arachevaleta 1982, 2005), occupied mainly from the beginning of our era to A.D. 700, but afterward also occupied by Aztatlan tradition populations[4] (about A.D. 900; Fig. 17.2).

[1] The obsidians' provenance has not been determined.

[2] We mean this situation is linked to the complexity of the stratigraphic contexts.

[3] The Teuchitlan site has given its name to a cultural phenomenon that developed in the modern States of Jalisco, Colima, and Nayarit between 300 B.C. and A.D. 900. It is mainly defined by circular public architectural complexes: a circular patio was bordered by a circular platform backed by several rectangular buildings and surrounding a central circular pyramid (these circular complexes are named Guachimonton). The other distinctive features of the Teuchitlan tradition are shaft tombs and the production of large clay and hollow anthropomorphic figures. The peak of the Teuchitlan tradition may be placed between A.D. 400 and 700 (Beekman and Weigand, 2000, 2010).

[4] The Aztatlan tradition refers to a cultural phenomenon that developed in the Mexican West and North-West, in the modern States of Jalisco, Nayarit, Durango, and Sinaloa, between A.D. 900 and 1,350/1,400. Based on the characteristics of certain features of its material culture (its pottery above all), several authors have associated it with the Mixteca-Puebla complex (which refers to a particular ceramic style and iconography), while others have found links with the Toltecs. In general, these authors agree on attributing the Aztatlan tradition to foreigners coming from the Central Highlands (Mountjoy 2000). Their success seems to have been due to prolific and diversified craftsmanship – copper working, work in shells, pottery, obsidian debitage – and a very structured widespread trading system (see Mountjoy 1990; Kelley 2000).

Based on studies of more than 200,000 artifacts recovered from surface collections and excavations in the obsidian workshops, this work describes two types of blade debitage carried out simultaneously: percussion and pressure, although without reconstituting the details of their respective reduction sequence (Soto de Arachevaleta 2005: 274). This major work has acted as the starting point for other publications suggesting the presence of prismatic blade technology in the Teuchitlan region from the Proto-Classic, i.e., the beginning of our era (Weigand 2000; Spence et al. 2002). All the same, these suggestions do not accord with recent archaeological data from stratigraphic excavations. These underline the systematic absence of prismatic blades from layers previous to the Epi-Classic, i.e., before the ninth century A.D. (Beekman, 2007, personal communication; Calgaro 2007; Lopez personal communication in 2007; Reveles 2005; Liot et al. 2006). On the other hand, the same research attests to the presence of tools made from obsidian macroblades. This finding may even apply to the Teuchitlan site: although it is true that prismatic blades are found in abundance there, they always appear on the surface, and Esparza and Ponce Ordaz stress that their presence occurs above all in Epi-Classic and Early Post-Classic contexts together with a diminution of macroblade artifacts (2005: 150). These "anomalies" have made us take a second look at the issue of pressure blade debitage chronology and revise the available literature, in order to date its beginnings more precisely. This revision, enhanced by exchanges with several specialists in Jalisco highland archaeology (in particular Phil Weigand, Chris Beekman, Lorenza Lopez, Catherine Liot, and Javier Reveles), has enabled us to reach the conclusion that the prismatic blade most probably remained a very rare artifact until the Epi-Classic period and supports the hypothesis of a late development for its technology.

17.5.1 The Prismatic Blade During the Final Pre-Classic and Classic Periods

So far, no archaeological records dated with certainty to these periods have yielded prismatic blades. In reality, when they are found, they come from long-distance exchanges. Phil Weigand notes the presence of a pressure blade of green obsidian from Pachuca (a deposit in the State of Hidalgo more than 500 km away) in the Capilla sector at Teuchitlan (Spence et al. 2002: 71). Other examples in green obsidian, also from Pachuca, have been found in the Basin of Sayula in contexts thought to date from the beginning of the Sayula phase (A.D. 550–1000, Reveles 2005: 358; 2007, personal communication). These blades, which are mainly surface finds, have a single-facet punctiform platform (Reveles, 2005: 359, personal communication). These few foreign artifacts, associated with rare "Thin Orange" sherds, are evidence of the existence of circulation routes, albeit little-used, between the Mexican West and the Central Highlands (Weigand 1990, 1993). In reality, in these very Western regions, the prismatic blade's manifestation in lithic assemblages coincides with the local development of its technology.

17.5.2 Percussion Blade Industries: A Regional Tradition Widely Predominant During the Classic Period

In spite of the unanswered questions connected with pressure blade debitage, all of the available information for the Western regions of Mexico confirms that, as early as the Proto-Classic period, obsidian working had acquired an essential place, as seen in jewelers' ornamental work in the form of pendants, pectorals, beads, mirrors, anthropomorphic figurines, etc. Similarly, tools made from macroblades, such as scrapers and large bifacial knives, find an important place in the archaeological assemblages. In addition, the fact that the greatest Teuchitlan civic-ceremonial centers were placed next to high-quality obsidian sources is evidence of the material's strategic importance.

The first signs of percussion blade production can be perceived in the Arenal phase (300 B.C.–A.D. 200). In their 2002 publication, Spence et al. mention two blades designated as "flake blades,[5]" as part of a surface collection gathered on the El Arenal site in the Basin of Eztatlan. This collection of 75 artifacts is made up of scrapers, bifacial pieces, and the two irregular blades whose description is not given. According to the same authors, surface collections gathered from at least five sites dated without any possible distinction from the Arenal to the succeeding phase – Ahualulco (A.D. 200–400), located in the Tequila Valley – attest to the existence of an industry producing both large irregular blades (flake blades), generally having a single-facet platform, as well as more regular blades (fine blades). They also mention the occurrence of ground platforms and ask whether the abrasion technique used for ornamental objects was transferred to blade technology and the preparation of core platforms (Spence et al. 2002: 66). According to the same authors, unlike the Arenal phase shaft tombs, which contained particularly rich obsidian furnishings of ornaments (pectorals, beads, pendants, etc.) and so-called ritual objects (bifacial knives) (ibid.: 65), certain shaft tombs from the Ahualulco phase seem to have yielded blade macrocores. One of them – Las Cuevas – yielded seven macrocores, the largest of which measured 45 cm long and 25 cm wide (ibid.: 66). Other shaft tombs excavated by Galvan Villegas (1991) in the Atemajac Valley (Jalisco) also contained blade products – some retouched bifacially: burial 17 yielded an incomplete blade in black obsidian with bifacial retouches 12×2.4 cm, which must have measured about 18 cm (ibid.: 181); a large blade of $19 \times 4.2 \times 1.3$ cm made from red obsidian was found in grave 18 (ibid.: 184). These tombs, according to a revision by Beekman (Beekman and Galvan 2006; Beekman 2006), seem to date from the Tabachines phase (300 B.C.–A.D. 600), and most probably from the Middle Tabachines subphase (100 B.C.–A.D. 200).

[5] To describe their material, the authors divided the blades into two large categories: "flake blades and fine blades." The former are "more roughly formed and generally broader, with somewhat irregular edges and dorsal arrises" (in English) (Spence et al. 2002: 63). The latter are "narrower and highly regular in form, with linear dorsal arrises" (in English). No indication is given about the proximal parts.

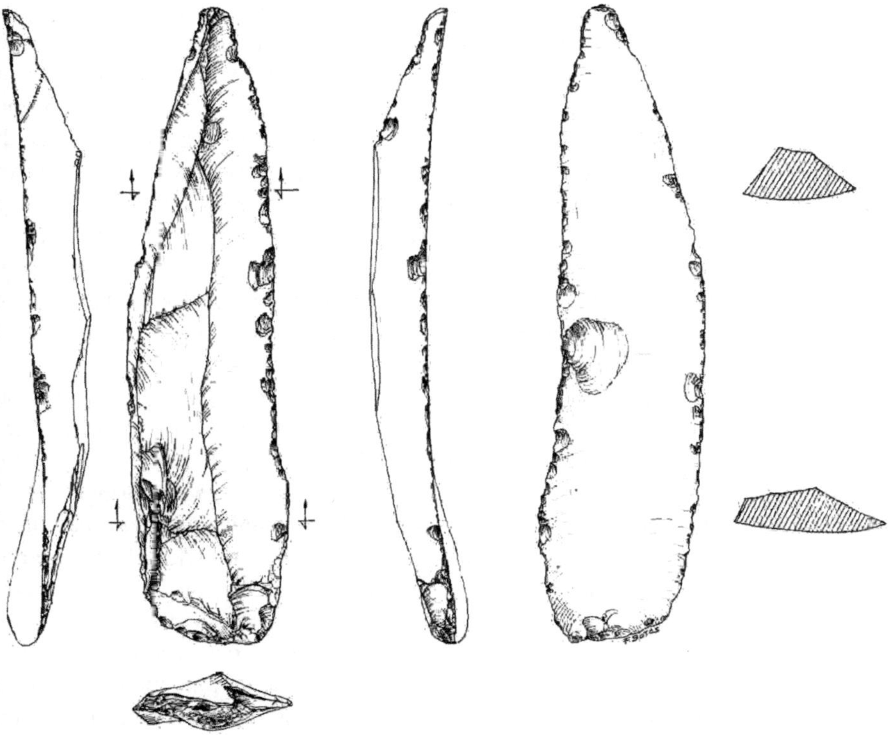

Fig. 17.3 Macroblade found in the Teuchitlan workshop (in Soto de Arachevaleta 2005: 147, Fig. 2: picked and ground platform – no scale)

As for research at the site of Teuchitlan – considered to be the regional capital – it showed the artifacts made from obsidian macroblades are widely prevalent in the lithic assemblages, alongside nonspecialized flake industries. Surface collections generally supply blade products from at least two *chaînes opératoires*, one using percussion and the other pressure (see, e.g., Esparza and Ponce Ordaz 2005). As we have mentioned, Soto's research examined the technological aspects of this blade production, studying a production workshop 200 m from the ceremonial center (1982, 2005). Besides the evidence for pressure blade production, his work casts light on the manufacture of macroblades by direct percussion from conical cores,[6] of which the average dimensions are 6.3 cm wide (±2.2 cm), 19.7 cm long (±5.8 cm), and 3.5 cm thick (±0.4 cm) (Soto de Arachevaleta 2005: 145; Fig. 17.3). The author mentions the predominance of faceted or ground platforms, both for percussion and

[6] Of the 229,421 artifacts collected from this workshop, only 52 correspond to cores or core fragments. Twenty-three of them have a prepared platform, 3 smooth, 7 faceted, and 11 ground. But these cores are not differentiated nor connected to a particular *chaîne opératoire*.

for pressure blades.[7] These macroblades were used as blanks for making a variety of tools, mostly distal scrapers and bifaces. Soto's studies do not enable the precise reconstruction of the *chaînes opératoires* associated with the two techniques, and integrating certain categories of artifacts, for example, macroblades, in one or the other production process. For this reason, several questions remain unanswered: What were the stages followed for one or the other process? What was the final form of the cores used for percussion? How can the very small proportion of this category of objects in the workshop be explained? How is it possible to explain the coexistence of three types of platforms – implying three types of preparation – for a single category of artifacts? Was the pecking and grinding technique really used to prepare the striking platforms for supplying percussion macroblades? Or was it rather used on the cores for pressure blade production?

Lastly, in the Sayula Basin, some 60 km South of the Tequila Valley in a region without obsidian deposits, evidence for the existence of percussion blade production is found from the Late Usmajac phase (200 B.C.–A.D. 300; Valdez et al. 2005). More precisely, macroblades detached by direct percussion appear during the first century A.D. (Reveles 2005: 359; 2005, personal communication) and may have been used in connection with exploiting salt. According to Reveles, they were not manufactured in the Sayula Basin but imported as finished products. Furthermore, characterization analyses indicate the deposits of Las Navajas and San Juan de los Arcos, about 40 km to the North, were the main supply sources. Little information is available on these obsidian deposits, but surface observations made at several quarry workshops associated with that of San Juan de los Arcos indicate production mainly consisted of percussion macroblades. A lot of blade refuse has been found there, including numerous exhausted macroblade cores with average dimensions varying between 15 and 22 cm long, 10 and 16 cm wide, and 3 and 5 cm thick. These percussion cores are tabular in form with a totally flat face and rectangular cross section. They have a single-facet and oblique platform, more rarely faceted, and the blades were removed only from one of the two principal faces, the unused face with cortex or bearing percussion scars.

The data from the westernmost regions thus seems to suggest the existence of percussion blade production from the very beginning of our era. Moreover, this technology seems to have been predominant throughout the Classic period. The final purpose of these reduction sequences seems to have been to produce macroblades or blades as blanks for instruments such as scrapers or bifacial knives. The *chaînes opératoires* used for this technique are still not very well understood, and numerous questions remain unanswered, such as when the pecking and grinding technique was first developed and how it was used. In any case, we think the exploitation of certain deposits, such as San Juan de Los Arcos, could have been connected

[7] Of the 7,327 macroblade platforms it has been possible to examine, 3,287 were grinding and 2,595 had a multifaceted surface – the others being with a single facet or cortical. Among the complete blades or proximal fragments obtained with the pressure technique, which number 13,394, 7,562 have ground platform, 3,791 multifaceted platforms, and 1,856 single-facet platform.

to these industries. Although the *chaînes opératoires* cannot be known in detail, most of the production from the workshops of this deposit is highly similar to that from the Zinaparo – El Varal workshops 200 km to the East – where we have worked – and which we shall discuss below.

17.5.3 Pressure Blade Technology: A Late Development Connected to the Aztatlan Tradition (A.D. 800/900–1350)

Just as for percussion, the conditions under which pressure blade production was developed in are still not understood in the most Western regions of Mexico. Notwithstanding a few publications proposing that the technology was present in the Teuchitlan zone from the beginning of the Classic (Soto de Arachevaleta 1982; Spence et al. 2002), it now appears established that it did not arise until later during the Epi-Classic, probably from the ninth century, and really only spread from the Early Post-Classic, i.e., from A.D. 900. Recent archaeological studies associate its development with the rise of the Aztatlan tradition (Beekman personal communication in 2007); Lopez personal communication in 2007; Reveles 2005; Mountjoy 2004). When it appears, whether in ceremonial or residential areas, the prismatic blade is made from regional obsidians – the La Joya deposits above all, but also Teuchitlan – La Mora and Magdalena –Tequila. The La Joya source, producing mainly green-colored obsidian, is especially interesting since the start of its systematic exploitation coincided with the development of pressure blade workshops on the island of Las Cuevas-Atitlan in the Teuchitlan Lake Basin. According to Weigand (1993: 220; Weigand and Spence 1982; Spence et al. 2002: 72–73), this island went in for large-scale production of prismatic blades – workshops spread over 15 ha having been identified. The obsidian arrived from the La Joya deposits across the lake in polyhedral cores. Nothing is known about production organization or the *chaînes opératoires*, but according to available information, the whole production process was carried out there in order to produce finished blades. The blades extracted were quite large – on average 2 cm wide, about 15–20 cm long, and 5 mm thick. Weigand mentioned the largest prismatic blade found measured 30 cm long and 2 cm wide (personal communication in 2007), and in the Sayula Basin, a blade made from La Joya obsidian was 22.5 cm long with a width less than 2 cm (Reveles, personal communication in 2007). All these blades have a ground linear but well-defined platform and are also slightly curved (Fig. 17.4).

Most of the production from the Las Cuevas workshops is dated from the Early and Late Post-Classic period and should be related to the development of the Aztatlan cultural complex. This was the time when most sites in the West began to make regular use of prismatic blades – all coming from the main Jalisco deposits, La Joya in particular. According to various authors, the circulation of certain varieties of obsidian and artifacts like prismatic blades was controlled by the elites (Liot et al. 2006, 2007; Lopez-Mestas 2007). In the Sayula Basin, for example, the site of La Peña – occupied from the Early Post-Classic by Aztatlan tradition populations

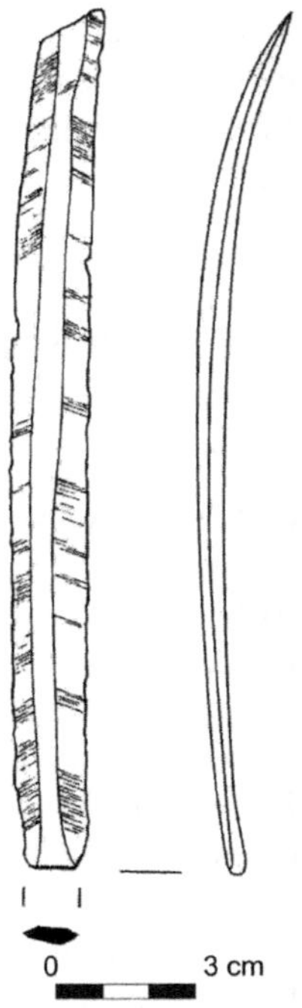

Fig. 17.4 Prismatic blade found in the Sayula basin (in Reveles 2005: 365, Fig. 5d)

foreign to the local substratum – seems to have monopolized a certain number of specialized activities, including pressure blade production, and have supplied the rest of the basin (Liot et al. 2006, 2007).

So, according to information available today, blade pressure technique first would appear in the Highlands of Jalisco right at the end of the Classic or at the start of the Early Post-Classic, i.e., between A.D. 800 and 900. Prismatic blade production progressively supplanted percussion blade, and its development seems to have been closely connected to the rise of the Aztatlan cultural tradition.

17.6 Blade Industries in Northern Michoacan and Middle Valley of the Lerma

Let us now take a close look at the situation in North-Central Mexico, more precisely North Michoacan and the Middle Valley of the Lerma, between the Jalisco Highlands and the Central Highlands (Fig. 17.2). Owing to its geographic situation, between two particularly remarkable influential regions (Teuchitlan and Basin of Mexico), but also to the intersection of other cultural spheres (hunter gatherers, Mazahua, Otomis, etc.), this region has always been considered a strategic communication route, a corridor in which certain political entities destined to play an essential role in the history of Central Mexico crystallized (see, e.g., Brambila and

Crespo 2005). Looking beyond remains such as monumental architecture, ceramics, or mortuary features, less prestigious evidence such as lithic industries also contributes to the debate about the regional cultural identity between the Pre-Classic and the Late Post-Classic. The many archaeological studies that have been conducted in this region since the mid-1980 enable us to have a global view and reconstruct behavior relating to the lithic economy quite clearly.

17.6.1 The Prismatic Blade During the Pre-Classic and Early Classic: A Rare Product Obtained Through Medium and Long-Distance Exchange

As in most Mesoamerican regions, when the prismatic blade appears in North-Central Mexican lithic assemblages, it does so as a foreign product, acquired through medium and long-distance exchange networks. Its presence is only attested from the Late Pre-Classic (400–100 B.C.).

The first archaeological data on the region's trajectories concerns the Chupicuaro culture, which developed during the Middle and Late Pre-Classic, between 600 B.C. and A.D. 250 (Porter Noé 1956). An interdisciplinary project has been ongoing since 1999 in the Acambaro Valley, considered the culture's heartland (Darras and Faugère 1999, 2007). Its objective is to reconstruct the population dynamics with the aim of obtaining a clearer idea of the role of Chupicuaro in the regional and supraregional processes. The three main concerns are to identify the origins of the Chupicuaro peoples, who formed agrarian societies with a complex sociopolitical organization; to better understand the cultural content at the peak period, between 400 and 100 B.C.; and to comprehend the nature of its relations with neighboring populations, especially in the Basin of Mexico (Darras and Faugère 2007; Darras 2006). In the Chupicuaro region (Fig. 17.2), the technical systems developed using obsidian were structured around flake industries, producing expedient instruments and a few categories of standardized tools. The populations of the Acambaro Valley mainly supplied themselves from the three regional sources[8] (Cruz Jimenez, personal communication in 2005; Lodeho 2007); geographic proximity seems to have been a significant, but not decisive, criterion for the strategies adopted.[9] The high proportions of eroded cortex flakes, as well as the identification of

[8] The three deposits of the nearby region are Los Agustinos, Ucareo, and Zinapécuaro. Some distant deposits are also found (Cerro Zinaparo and Cerro Varal, 150 km to the West and Pachuca 200 km to the East).

[9] Variability from one site to another is found: site JR 24, on the right bank, was massively supplied from the Los Agustinos deposit (77.5%), 20 km as the crow flies, and next from Ucareo (12.5%), on the left bank and at the same distance; the deposit of Zinapecuaro, also on the left bank but 40 km away, only represents 5%. On the other hand, for site TR6, on the left bank, 63.8% of the obsidians came from Los Agustinos, only 16 km as the crow flies, but on the other bank of the river; 13.3% came from Ucareo – the closest deposit: 15 km – and 9% from Zinapécuaro (32.5 km away). The rest comes from distant sources. The Ucareo results will be discussed below.

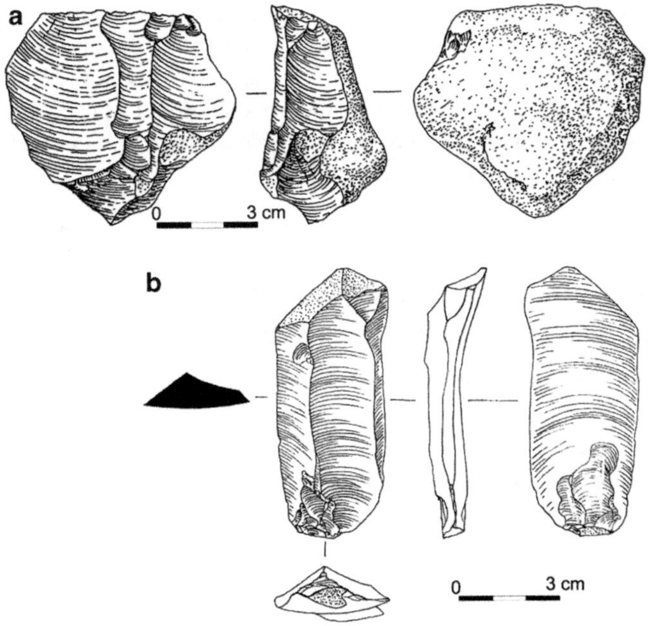

Fig. 17.5 (**a**) Unipolar core (JR 24–319) and (**b**) percussion short blade with cortical platform in Chupicuaro (JR 24–56) (Drawing by F. Bagot)

unworked nodules of moderate weight, between 0.5 and 2 kg, show that the raw material was obtained from surface collections and that it reached the sites without having been prepared at the quarry. Typo-technological study of assemblages collected at two settlements in the valley[10] (Lodeho 2007) reveals three types of production using distinct methods: multidirectional debitage, "salami-slicing" with nodules long in shape, and direct unipolar percussion debitage, to produce short unstandardized blades (Fig. 17.5a, b). These different types of production needed only a limited technical investment, but produced a varied range of flakes capable of being used either directly, without intentional modifications as an instrument, or to make various categories of specialized tools. Unipolar flakes were sought to meet the aim of obtaining blanks more suitable for making pedunculate scrapers. The blanks made with these debitage systems seem to have met the needs of these populations and allowed tool sets to be made for the whole range of tasks required.

Alongside these local industries, some imported artifacts have been identified, including very rare prismatic blade segments at site TR 6 (Fig. 17.6). These specimens were collected from layers dated between 300 and 100 B.C. and constitute 0.2% of the Late Chupicuaro phase obsidian collection (2,494 artifacts). They consist of four mesial fragments and one distal, made with translucent gray-black obsidian

[10] Sites JR 24 and TR 6 excavated as part of the Chupicuaro project (directed by V. Darras and B. Faugère).

Fig. 17.6 Prismatic blade fragments found at Chupicuaro (TR 6–79) (Photo by V. Darras)

from the Ucareo source.[11] It is interesting to observe that only the excavations at TR 6 yielded artifacts of this kind as neither the excavations at JR 24, about 10 km away as the crow flies, nor those at Chupicuaro, between 1946 and 1947, recovered prismatic blades. These five examples are the only evidence of pressure blades coming from stratigraphic contexts that can be confidently assigned to the Late Pre-Classic in this region of Mexico.

Not until the transition period between the Pre-Classic and the Classic – commonly called Terminal Pre-Classic or Proto-Classic (100 B.C.–A.D. 250) – does the presence of the prismatic blade become more evident in the lithic assemblages in the Western and North-Central regions. This artifact is then found next to flake industries made from regional obsidians, or basalts and andesites also of regional origin.

The sites occupied during this period developed in the lake basins of the North-Central region (Acambaro Valley, Cuitzeo Basin, Patzcuaro Basin, Zacapu Basin, Fig. 17.2). Archaeological investigations have yielded green obsidian prismatic blades from the source of Pachuca at a distance ranging from 200 km (Acambaro Valley) to 320 km (Zacapu Basin, Fig. 17.2) depending on the site (Darras 1993; Darras and Faugère-Kalfon 2007; Carot 2001; Filini and Cardenas 2007; Macias Goytia 1990; Pereira 1999). Additionally, several of these sites, such as Loma Alta, also acquired blades in translucent gray-black obsidian from Ucareo (Manzanilla Lopez 1984; Darras 1999; Pereira 1999, personal communication in 2007). The first instances of green prismatic blades in these various regions were associated with the rise of Teotihuacan in the Valley of Mexico. But, although these products were circulated, they are found in only moderate proportions and in particular contexts – such as burials (Fig. 17.7; Darras 1993; Carot 2001). The blade represented a special and

[11] Analyses by B. Gratuze and S. Boucetta, IRAMAT, Orleans (*Institut de Recherche sur les Archéomatériaux*).

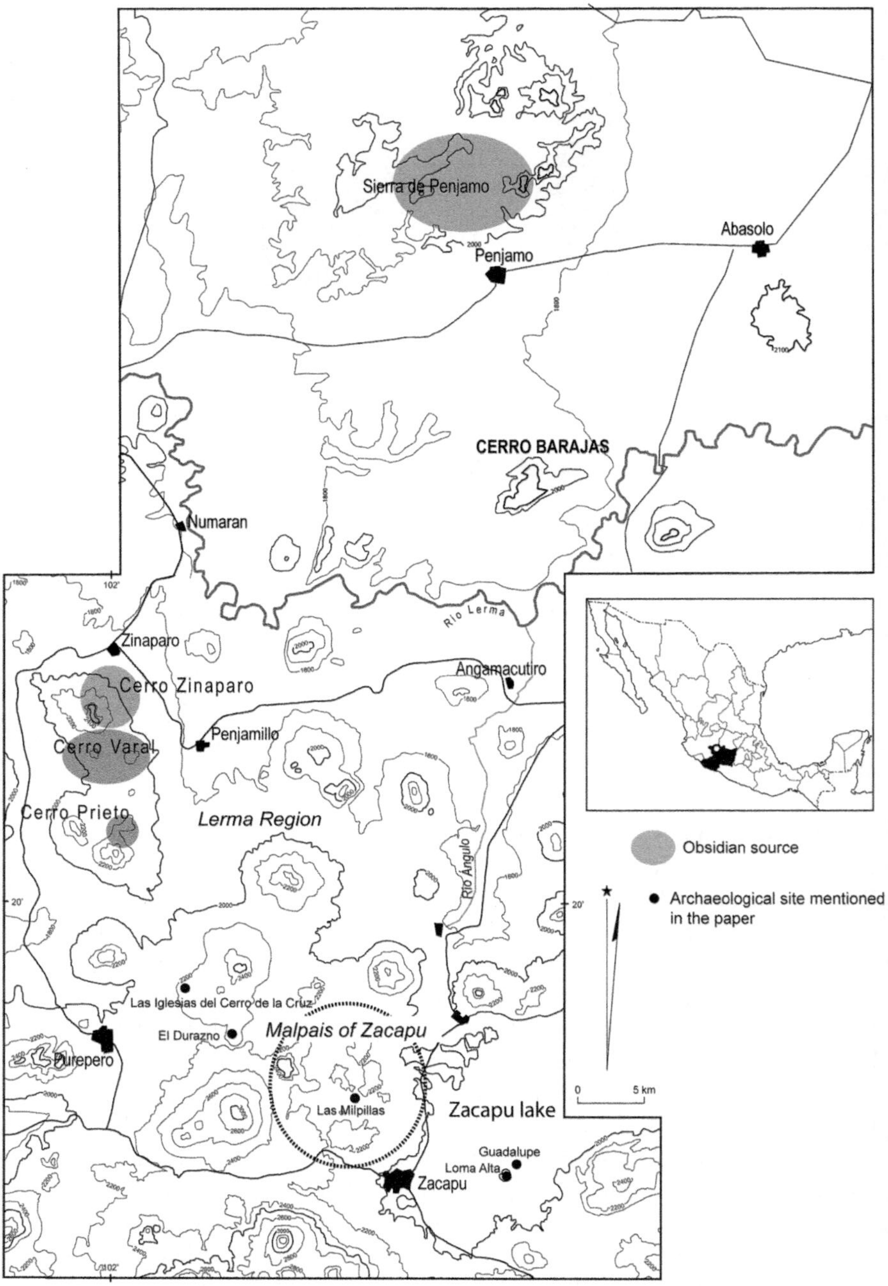

Fig. 17.7 Pachuca green prismatic blades found at Loma Alta (S3 C.7) (Drawing by F. Bagot)

unusual object, perhaps acquired and used by a certain fringe of the population involved in the interaction networks emanating from Teotihuacan. The contexts that they are found in, and the fact that they show no intentional retouching, nor obvious usewear, seem to indicate a specific use – perhaps linked to bloodletting rituals (Darras 1993, 1998; Pereira 1999; Carot 2001).

During the Late Pre-Classic and Early and Middle Classic – and in the whole region of Central-North and Western Mexico – the obsidian prismatic blade evidently remained a rare object acquired through exchange. But its relatively constant proportions in the lithic collections indicate a real and stable presence in the long- and middle-range circulation networks. There were two networks, which were probably independent of each other. The first was directly linked to the Basin of Mexico and Teotihuacan, covering a distance between 200 and 350 km depending upon the context being studied; this network provided the green blades. The second was linked to the Ucareo deposit North-East of Michoacan, covering an area between 20 (Acambaro Valley) and 120 km (Zacapu Basin); this network provided the black-gray blades.

17.6.2 The Particularity of Ucareo, Michoacan

This obsidian source has a special place in the history of the prismatic blade technology. Located in the North-East of Michoacan, it seems to have played a leading role since the Early and Middle Pre-Classic within the obsidian circulation networks in Central and Eastern Mesoamerica. The provenance analyses of the obsidian artifacts recovered at the Olmec sites of San Lorenzo or La Venta, and contemporary sites in the Oaxaca Valley and the Basin of Mexico, show that this deposit was an essential source, comparable to Barranca de Los Estetes (Otumba) or El Chayal (Guatemala) (Boksenbaum et al. 1987; Cobean et al. 1971, 1991; Elam et al. 1994). Furthermore, Ucareo ranks, alongside Otumba, as one of the first sources to supply the raw material used to make the prismatic blades recovered from these consumer sites. This importance increased over the following centuries as it became the main supply source for sites in Central Mexico, such as Xochicalco, between A.D. 650 and 900 (Hirth 1995, 2002, 2006), and Tula, between A.D. 900 and 1200 (Healan 1997, 2002, 2003, 2004, 2005). Healan's research conducted in the Ucareo-Zinapecuaro region since 1994, investigating the nature of obsidian exploitation between the Epi-Classic and Post-Classic, and the region's relations with Tula, has not been able to reveal evidence for obsidian extraction and transformation during this early period. The earliest archaeological evidence in this region for the systematic exploitation of obsidian sources and workshops connected to pressure blade technology dates from the Epi-Classic or Early Post-Classic period. If Healan's research results cast more light today on how Ucareo obsidian was produced and how it reached Tula, it must be admitted we are still in the dark as to how and by whom the deposit was exploited during the Pre-Classic to Middle Classic period. What is certain, however, is that those who exploited the resource from the Pre-Classic to the Early Post-Classic must

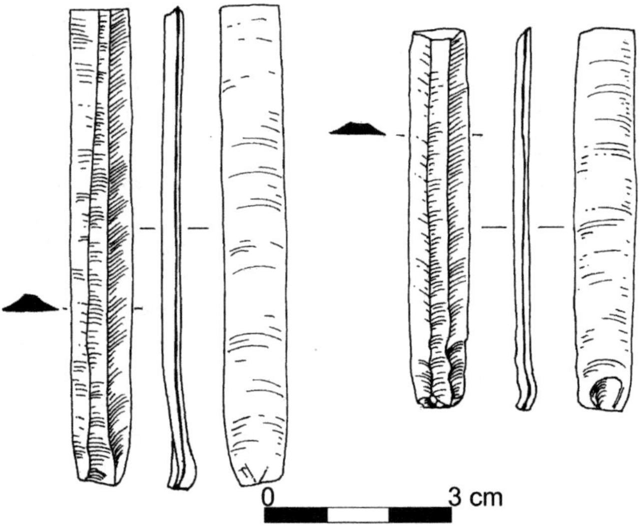

Fig. 17.8 The Zacapu region, Michoacan (Drawing by V. Darras)

have taken part in more than merely local – or even regional – dynamics and that they were involved in the long-distance circulation and interaction networks that remained foreign to the people of Western Mexico.

17.6.3 Blade Industries in the Classic Period (Middle Classic and Epi-Classic)

In the North-Central region, particularly in Zacapu, the end of the Classic was marked by two concordant phenomena: on the one hand, the prismatic blade became noticeably scarcer in the lithic assemblages and, on the other, the percussion blade industries became more essential.

17.6.3.1 The Prismatic Blade: An Increasingly Rare Artifact

The various projects carried out by the CEMCA[12] in the Zacapu region (Fig. 17.8) between 1983 and 1997 (Michelet 1992; Arnauld et al. 1993; Darras 1993; Pereira

[12] Archaeological research has been carried out here by the Centre of Mexican and Central American Studies (CEMCA, Mexico) and the CNRS (*Centre National de la Recherche Scientifique* – National Scientific Research Centre) between 1984 and 1997. This research has resulted in several doctoral theses and various publications.

1999) showed that from the Jarácuaro phase (A.D. 550–600) – and especially from the Early Lupe (A.D. 600–700) – the quantities of green blades decreased significantly whereas the black blades from Ucareo continued to be acquired, albeit moderately. This trend grew stronger over the three following centuries: during this time period, the green Pachuca blades disappeared and only a few gray-black specimens have been recovered. For example, the research realized south of the Middle Río Lerma River shows prismatic blades as the exception there (Faugère-Kalfon 1989). Surface collections and the 19 stratigraphic pits undertaken in 12 of the most important sites characterized by a civic-ceremonial architecture and dating to the Late Lupe to Palacio phases (A.D. 700–1200) yielded 13 prismatic blade segments, including only two from the excavations (0.1% of the obsidian collection, which comprises a total of 1,739 artifacts), all on material from Ucareo.

The information available for the sites dated to the same period in the North-Eastern region of Michoacan, near the Ucareo deposit, describes a similar situation (Healan 2002: 33; Healan 2005: 175–177). According to Healan, the production from the Ucareo workshops, consisting of polyhedral cores for pressure blade production, was reserved exclusively for Central Mexico, more specifically for the site of Tula. The populations living close to Ucareo consumed very few or no prismatic blades, probably because they had no interest in this type of artifact. All the same, Healan suggests that the archaeological site which may have been involved in the exploitation activities of the Ucareo deposit, named Las Lomas and located in the valley of the same name, seems to have had workshops carrying through the entire prismatic blade production process (Healan 2004, 2005). These products may have been destined for a closer consumption market; the few prismatic blade segments found on the Epi-Classic and Early Post-Classic sites in the Middle Valley of the Lerma could have come from workshops of Las Lomas.

The near disappearance during the sixth century of the Pachuca prismatic blade, the presence of which in the regions studied is known to have been tied to the influence of Teotihuacan, seems to be due to the evolutions taking place in the Basin of Mexico in the middle of the sixth century. It is known that the fire which destroyed the ceremonial center of Teotihuacan in A.D. 550 appears to have accelerated the metropolis' decline (Manzanilla 2003), causing the administrative and economic power structures to fall apart and, consequently, a probable disruption of certain long-distance circulation networks.

17.6.3.2 The Rise of Percussion Blade Industries During the First Millennium A.D.

While the pressure blade industries remained marginal in the second part of the Classic period, the percussion blade technology rose to a dominant position. However, unlike the Jalisco Highlands, where this type of industry was well established at the beginning of our era, it developed later in the regions under discussion. The lack of technological data on the San Juan de los Arcos production processes does not allow the two regions to be compared rigorously. However, the surface observations we made at the latter site allow a possible morpho-technological correlation with Northern Michoacan production processes, which will be worth confirming through further research. If this technological correspondence was to be confirmed, it could be surmised that the technology we shall describe below could be the result of technological borrowing from the populations of the Jalisco Highlands.

In fact, the archaeological work carried out in the Zacapu region, in the North of Michoacan, indicates the existence of percussion blade industries at least from the sixth century A.D. (Jarácuaro phase A.D. 500–550). Found alongside flake debitage – which made possible the development of expedient and unstandardized tooling – and prismatic blades acquired through the exchange discussed above, the first evidence was found at the sites of Loma Alta and Guadalupe (Darras 1993; Pereira 1999; Carot 2001). It is interesting to note the coexistence of the two types of blades for this period. The excavations in 1986 at Loma Alta yielded a total of 377 lithic artifacts, including 173 in obsidian, nine macroblade fragments from Cerro Varal, one point from an irregular blade, and three knives with basal fixation notches made from a large blade (Darras 1993: 170–181). The burials at the Guadalupe site, excavated by Pereira, resulted in the collection of 293 lithic artifacts – 234 of which were obsidian. Among the latter are six pedunculate points from large blades and five bifacial knives made from large-sized blanks (Pereira 1999: 127–128). Physicochemical analyses have confirmed Cerro Varal as the obsidian source. The blades, all fragmented, are wide (between 2.5 and 3 cm) and thick (0.5–0.75 cm), and the platforms, noticeable on proximal samples, are oblique, wide, and single faceted. The presence of these blades in contexts dated to the Late Classic indicates that the obsidians from Cerro Varal and Cerro Zinaparo were used from this time. But as we shall see, the first evidence for systematic exploitation of the Cerro Varal and Cerro Zinaparo deposits, as well as the development on site of specialized workshops, dates from the beginning of the eighth century A.D.

The same archaeological work has brought to light important changes during the eighth century, at which time an expansion took place from the Zacapu Basin toward a more Northern region on the Southern side of the Lerma Valley. This expansion could be explained by significant demographic pressure in the Zacapu Basin and a resulting need to colonize new agricultural land (Faugère-Kalfon 1989, 1996; Faugère in press). This research has shown a dispersed settlement pattern, characterized by agricultural sites organized around small civic-ceremonial centers with autonomous political and economic organization. These developments in the regional settlement pattern coincide with the earliest exploitation of the two obsidian deposits in the Zinaparo massif (Cerro Varal and Cerro Zinaparo) and the one in Cerro Prieto. Prior to this period, the deposits were used to meet the needs of local populations. The obsidian exploitation was accompanied by the establishment of a population of farmers and craftsmen in the Zinaparo massif, who carried out the production logistics from mining activities to large-scale manufacture of obsidian blades. The earliest evidence for human occupation of the Zinaparo massif dates only from the late seventh to the early eighth century A.D. The population reached its peak toward A.D. 800, with 21 farming and/or artisanal settlements established at the obsidian deposits or in their immediate vicinity (Darras 2008).

The 14 identified production centers consist of major extraction areas – both subterranean and opencast – and quarry workshops, usually adjacent and extending up to several hectares (Darras 1994, 1999, 2008). The thousands of tons of obsidian waste that resulted from this activity pile up on depths of up to two meters.[13] The purpose of 13 of these production centers was to obtain by percussion, using andesite hammerstones, blades of varied shapes and sizes – often transformed on the spot into tool preforms. This production process was carried out by means of two debitage methods, neither involving heavy technological investment in core preparation.[14] The first was performed on angular blocks of variable dimensions, sometimes on macroflakes, on which a single-facet or multifaceted strike platform – generally very oblique – was prepared straight away. Use was made most often of the blocks' natural edges to detach the first blades; the following were then removed alternately from a single production face. The cores were generally thrown away when they were too flat or thin to allow further blades to be extracted. These cores show – in their residual state – a tabular morphology, rectangular or subtriangular in form and rectangular in cross section, with a totally flat debitage surface most often opposite an unworked cortical surface (Fig. 17.9). The removed blades showed a certain morphological variability and were not really standardized. Their dimensions could vary considerably depending on the initial dimensions of the obsidian blocks (Fig. 17.10a–c). The residual cores have a length varying from 10 to 23 cm, a width between 11 and 17 cm, and an average thickness of 4 cm. As for the blades, they have a width of between 2 and 5 cm, are 10–25 cm long, and are between 5 and 10 mm thick. The other production method started with smaller nodules or by using very thick macroflakes. Decortication was followed by preparing a horizontal striking platform, from which unipolar flakes were detached, so as to create longitudinal arises and finally obtain more regular blades. Production took place on the circumference of a conical core. The aim of this method was to produce small blades, no more than 10 cm long and between 1.5 and 2.5 cm wide.

The macroblades produced by the first method were generally retouched as tool preforms in the quarry workshops and were used as blanks for scrapers and unifacial or bifacial tools (Fig. 17.11a–c). Some were used to make bifacial knives, carefully formed by means of pressure retouching. The smallest blades, obtained with one or the other method, were more rarely preformed in the workshops. In general, these blades, when found in settlement sites, turn out to have small basal fixation notches and were used as knives.

The studies on the distribution of these products demonstrate the regional importance of the quarries and workshops of the Zinaparo massif over at least four centuries, since the percussion blades that were produced were distributed to settlement

[13] The study concerned a collection of 74,994 obsidian artifacts collected from a total of five investigatory excavations in the workshops of four production centers (Darras 1999).

[14] The last production center is on the Cerro Prieto, 3 km south of the Zinaparo massif. The obsidian found here is of inferior quality. The zones of activity extend for 8 ha and include opencast extraction sectors and workshops specialized in the manufacture of unipolar cores and bifacial preforms.

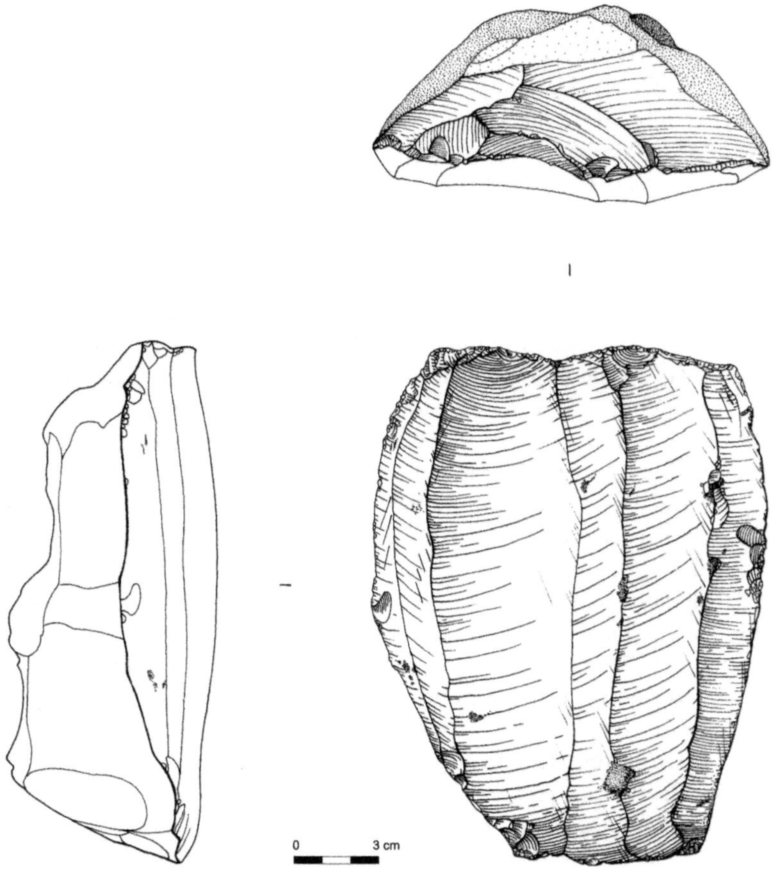

Fig. 17.9 Obsidian blade percussion core (Mich. 156 surface) (Drawing by F. Bagot)

sites in the region of the Lerma Valley and Zacapu (Darras 2008). Healan (2005) observed for this same period that the populations of the Ucareo-Zinapecuaro region, culturally close to the populations in the Middle Lerma Valley, also relied on percussion blade debitage. Nonetheless, we have no further information about the production of this kind in these obsidian deposits, although Healan (personal communication in 2006) mentions the presence of cores with a similar morphology to those of Zinaparo.

17.6.3.3 Pressure Blade Technology: A Late Development Related to the Rise of Tarascan Culture

The lithic traditions of this region are thus dominated, during the Classic (A.D. 250–900) and Early Post-Classic periods (A.D. 900–1200), by flake production in

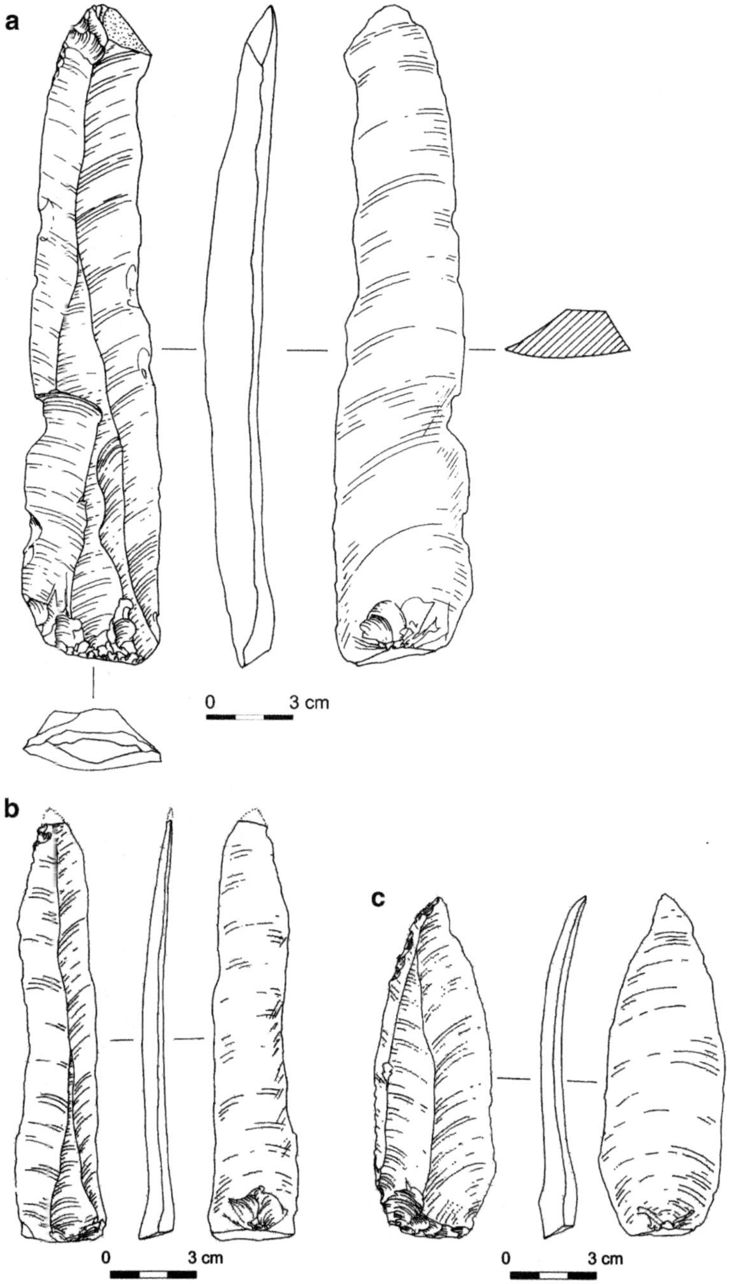

Fig. 17.10 (**a**) Macroblade (Mich.156, CNC). (**b**) Blade (Mich.117, S1.C.1.). (**c**) wide and short blade (Mich.117,S1.C.1.) (Drawing by F. Bagot)

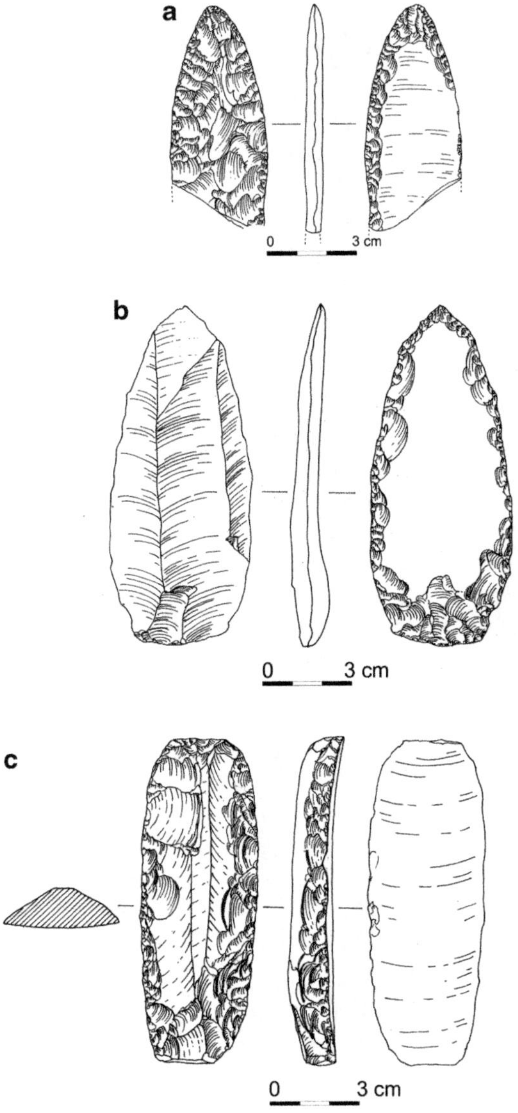

Fig. 17.11 Tools blanks on percussion blades: (**a**) Mich.117 C.1. (**b**) Mich.158, surface. (**c**) Mich.151, surface (Drawing by F. Bagot)

the domestic sphere, on obsidian and on andesite or basalt, and, as we have just explained, by percussion blade production.

Between A.D. 1100 and 1200, Northern Michoacan was the theater for important changes observed in the spatial and social reorganization. The Lerma Valley region was abandoned by most of its population, and the obsidian mine workshops of the Zinaparo region ceased their activity (Faugère-Kalfon 1996; Darras 1998). At the same time, spectacular population growth became apparent in the Zacapu region – most especially in the Malpaís sector, about 20 km South of the Lerma Valley region: new settlements appeared while the existing sites grew significantly (Migeon

1990; Michelet 1998). The Zacapu Malpaís at this time had a concentration of 18 settlements – several of urban character – harboring a numerous population. The population estimates for only four of them (representing a surface of about 4 km²) vary between 10,000 and 12,000 persons (Michelet 1998). These regional transformations in settlement patterns represent important social and political evolutions within Tarascan society, culminating in the fourteenth century in the creation of a centralized state with its capital in the Patzcuaro Basin[15] (Pollard 1993, 2003; Migeon 1998; Michelet et al. 2005; Michelet 1998; Darras 2005b).

The study of obsidian artifacts from the region's Post-Classic sites has shown that the lithic systems had also undergone deep transformations. The most radical change lay in the disappearance of percussion blade tooling and the local development of pressure blade production, with significant consequences on the populations' consumption habits: the prismatic blade became a very commonplace artifact, consumed in great quantities by every level of Tarascan hierarchy. What happened, and how did this change in blade technologies take place?

While the Zinaparo massif hamlets and mine workshops were abandoned, five settlements, with high concentrations of obsidian waste on the surface, appeared halfway between the Zinaparo obsidian deposits and the major Tarascan sites in the Malpais of Zacapu (Fig. 17.8). The work carried out there (Darras 2009) has shown that two of them were thirteenth century creations, while the three others give evidence of an occupation going back to the Early Post-Classic (A.D. 900–1200). In both cases, however, the activity of the obsidian workshops developed during the Milpillas phase (A.D. 1200–1450). Nonetheless, the few differences observed in the morphological characteristics of the various localities – and certain strategies adopted for the production, notably in choice of raw materials, workshop localizations, and general level of organization – seem to indicate a slight chronological variation: prismatic blade production seems to have developed in two of the oldest settlements initially and then to have been continued by the others in slightly different ways (Darras 2009).

The *Las Iglesias del Cerro de la Cruz* site is installed on the slopes of the volcano of the same name at an altitude of 1,800 m. It is composed of a small civic-ceremonial center located at the higher level, of a residential zone with a system of agricultural terraces on the slope, and three sectors for obsidian transformation. Two sectors are below and immediately adjacent to the residential zone, and the third is isolated about 1,000 m North-East of the site. The three concentrations have a surface of between 150 and 200 m² and cover irregular terrain, characterized by small basalt outcrops.

[15] The Tarascans formed an ethnic group mainly occupying the modern State of Michoacan. At the time of the Spanish conquest, the Central Highlands of Mesoamerica were dominated by two rival powers: the Aztecs in Central Mexico and the Tarascans in the West. Just as for the Aztec empire, the Tarascan kingdom was a late creation, dating from the Late Post-Classic. It was centralized politically, administratively, and economically. The Michoacan project, carried out by the CEMCA between 1982 and 1996, was concerned, in part, with understanding the beginnings of the social and political processes that resulted in the kingdom's consolidation. The choice of the Zacapu region was made on the basis of ethnohistorical evidence from the only sixteenth century account relating the official history of the Tarascan people and which designated Zacapu as their place of origin.

The *El Durazno* site, which was only occupied during the Milpillas phase, is located on a basalt plateau close to Malpaís of Zacapu. It is a hamlet composed of two main platforms with the remains of several houses surrounded by agricultural zones, today including high quantities of andesite tools. Of the four obsidian concentrations, three seem to be directly associated with the residential zones with dimensions of 20–150 m^2, while the fourth, of more than 250 m^2, is located in a sector characterized by small basalt outcrops 75 m from the second residential platform.

The contents of the two well-analyzed obsidian concentrations refer to the production of prismatic blades, and all the stages of reduction are represented, from the decortication flakes and microdebitage up to the residual cores; the finished products, i.e., the third series prismatic blades, are extremely rare. The average depth of the deposit varies between 15 and 22 cm depending on the workshop, and the density of the waste can also vary from one workshop to another: for instance, 19 kg of obsidian waste were collected within a perimeter of 3×4×0.15 m at the Las Iglesias de La Cruz workshop no. 2, and 31 kg from a pit 2×2×0.15 m from the Durazno workshop no. 2.

The optical and physicochemical analyses show only one variety was used at the Las Iglesias del Cerro de la Cruz site, coming from Cerro Varal, and 20 km to the North. On the other hand, analysis of the El Durazno obsidians indicates two varieties were mainly used – one coming from Cerro Varal and the other, grayish-green in color, from Penjamo, a source 80 km to the North. In spite of the variations in distance from one deposit to the other, the acquisition strategies were the same: the high proportion of decortication flakes and blades indicates the craftsmen were supplied with blocks either unprepared or freed of a part of their cortex. The supply was probably direct, and the nature of the cortex could be an indication that the blocks had been extracted, although we have not been able to prove the existence of extraction mines dating to the Post-Classic.

The *chaîne opératoire* followed to obtain the prismatic blades was the same for the two localities. No raw material or polyedrical cores have been found in these workshops, which suggests that all of the blocks selected at the deposits and brought back to the workshops had been reduced. The dimensions of the cortical flakes and other sizeable products (particularly residual cores) indicate that these blocks were small, not exceeding 15 cm long and weighing 1.5–2 kg; they were generally angular blocks easy to start working. Without going into the details of the *chaîne opératoire,* it may be said that the final objective was to make prismatic blades of dimensions varying between 0.7 and 1.8 cm, with an average width of 1.2 cm and between 0.2 and 0.4 mm thick. The residual cores come in varying sizes: at the site of Durazno, the smallest core is 6×3.4×1.7 cm and the largest 9.2×2.6×1.7 cm, whereas at the Las Iglesias de la Cruz site, slightly bigger dimensions are found, the smallest measuring 10.2×2.6×2.1 cm and the largest 11.5×3.2×2 cm (Fig. 17.12a). Core fragments found at the site indicate that some may have reached up to 13 cm in length and thus would have allowed the production of blades of a similar length (Fig. 17.12b). Globally, it can be seen that the preparation blades were equally moderate in size – between 6 and 13 cm – which seems to indicate that the initial blocks were no larger (in their debitage axis) than the residual cores.

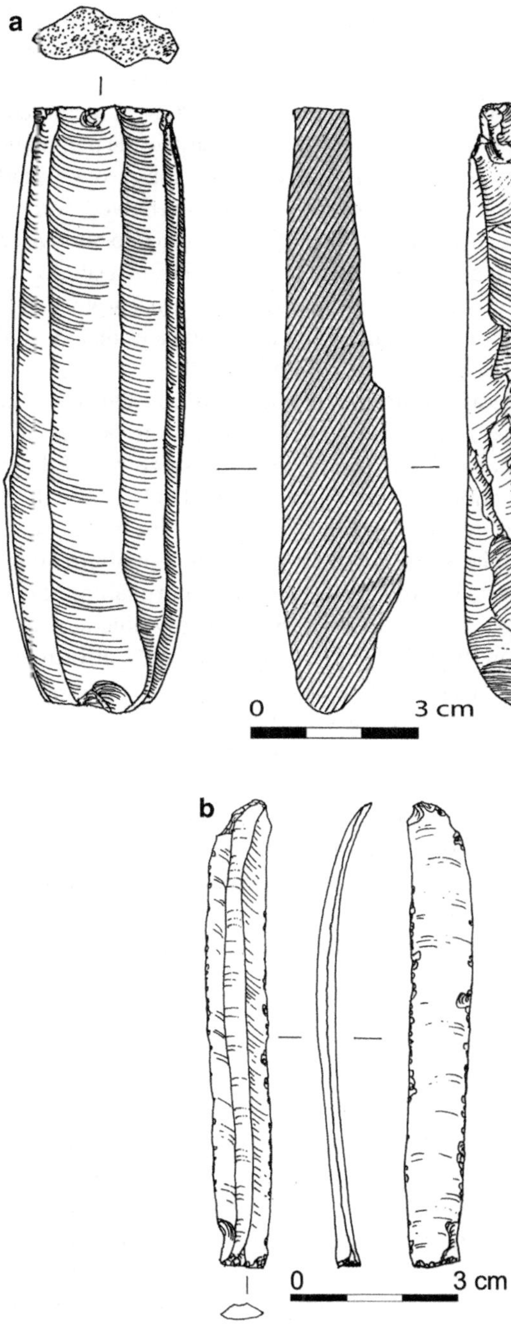

Fig. 17.12 (**a**) Tarascan pressure cores, Mich. 101. (**b**) Prismatic blade, Mich. 95 (955-I20 niv1) (Drawing by F. Bagot)

Broadly speaking, the reduction stages were as follows: blocks were shaped with decortication flakes and brought to final polyhedral form after blade makers created a multifaceted platform that was subsequently pecked and ground. It is important to stress that a number of flakes and percussion blades also were identified with ground platforms (7.3%), indicating that grinding occurred early in the preparation sequence, and that the percussion reduction progressed using this ground platform. Blade makers usually took advantage of natural ridges to removed cortical flakes and blades; when these were absent, a crested ridge was created to remove crested blades (4.2%). Percussion and pressure blades were removed from a single face of the core leaving its opposite side in cortex or with multidirectional negative scars (Fig. 17.12a). The size of the exhausted cores and finished blades indicates that most blades ranged from 9 to 12.5 cm in length and 0.8 to 1.4 cm in width.

The research at the Tarascan sites of the Malpais of Zacapu and, in particular, the analysis of the collection of obsidian recovered at the site of Milpillas, one of the largest Tarascan settlements in this sector, have shown 48.2% of the obsidian artifacts found in the excavations consisted of prismatic blade segments (Fig. 17.12b). These blade segments were used as blanks for tools, most often expedient, used in cutting soft materials. The study of their spatial distribution among the various excavated structures does not reveal any quantitative and qualitative differences that would imply difference of access. On the contrary, everything indicates undifferentiated mass consumption, both for domestic and probably also more ritual tasks, such as self-bloodletting (Darras 1998, 2005).

In the North of Michoacan, and most especially in the region of Zacapu, the Middle Post-Classic period is thus marked by an important technological change: the introduction of pressure blade technology. This technique was developed in small settlements, distant from the supplying deposits (between 20 and 80 km), but near the urban settlements of the Malpais of Zacapu. Our research has revealed specialized part-time activity by families of farmers/craftsmen, who were involved in the whole production cycle – from obtaining the raw material from the deposits up to distributing the products among the consumer sites.

17.7 Discussion

The examination of several archaeological contexts in two key regions in Western Mexico clearly indicates that the obsidian prismatic blade was not a commonplace item. Today, all of the archaeological data points to the same conclusion: in spite of a precocious production center in the Ucareo area, exclusively connected with the populations of Central and Southern Mexico, the populations living farther to the West adopted prismatic blade technology at a particularly late date. This adoption occurred at a variable pace and did not follow a clear spatial logic progressing from the traditional production places.

The conditions for the development of pressure blade technology in the Jalisco Highlands are visibly different from the pattern evident in Northern Michoacan.

In the first region, the available data seems to show that this technology was acquired two centuries before the Northern Michoacan, though the latter is closer to Central Mexico. In the same way, when the technology developed, the sociopolitical contexts showed significant differences. In one case, prismatic blade production and distribution may have been controlled by elite groups (Liot et al. 2006, 2007); in the other, they were assured by independent craftsmen pursuing their specialized activity within a very flexible framework. Whichever the case, none of these variations contradict the following observations: a priori, all of these populations satisfied the conditions required for acquiring the technology[16] – “unlimited” high-quality obsidian reserves, sociopolitical stratification, complex economic organization, demographic density assuring a potential market, and contacts with the populations possessing the skills and expertise. But then, why did it not develop before? We will examine below the various scenarios that could explain the behavior of the pre-Hispanic populations of these regions between the Pre-Classic and Late Post-Classic.

17.7.1 The Pre-Classic: Technocultural Choices or Ethnic Control?

During the Early Pre-Classic period and the beginning of the Middle Pre-Classic (1500–800 B C.) – for which archaeological remains have been found in North-West Michoacan and the States of Jalisco and Colima – ignorance of the prismatic blade, as a manufactured product, was probably due to the absence of well-established commercial networks between Western Mexico and the Central Highlands. When these artifacts started to circulate, i.e., during the third century B.C. in some Chupicuaro sites, it was in very small quantities and made with obsidian from Ucareo, a deposit very close to the geographic core of the Chupicuaro culture. In fact, the regional situation cannot be discussed without reflecting on a comparison with the role played by the sources of Ucareo-Zinapecuaro throughout the Pre-Classic.

First, the hypothesis that during the Middle and Late Pre-Classic period these sources were controlled can be put forward. During the Middle and Late Pre-Classic, the Ucareo deposit was on the Southern margins of Chupicuaro territory,[17] on the frontier of a region culturally related to the Valley of Mexico.[18] The research of Healan and (Hernandez 2000) in the zone of Ucareo-Zinapecuaro has revealed that sites occupied by the Chupicuaro were concentrated on the shores of the Lake of Cuitzeo, near

[16] In fact, Clark (1987) suggests the abundance of raw material and access facilities to the deposits, as well as the degree of complexity of the societies' social organization (strong hierarchization), are the conditions for development of prismatic blade technology.

[17] The limits of this territory still have to be defined – the Northern limits in particular. They are determined by ceramic, technological, and stylistic criteria.

[18] Ceramic characteristics of the Ticoman or Cuicuilco I to IV phases of the Basin of Mexico.

the Zinapecuaro obsidian deposit[19] located farther to the West, and were about 10 km from the Ucareo Valley (Hernandez 2000) – although this does not seem to have yielded any traces of Pre-Classic occupation. This information suggests that Ucareo was not integrated into Chupicuaro territory (Hernandez 2006) and was exploited by other populations, of distinct ethnic origin. Physicochemical analyses, however, have shown that the Chupicuaro groups of the Acambaro Valley acquired a part of their raw material from Zinapecuaro and Ucareo, even if the Los Agustinos deposit was their main source. Still, the nodules acquired were small and came from surface collecting: Were these strategies guided by technocultural choices or were these "constrained" strategies, adapted to a context of restricted access? There is no doubt the populations living near Ucareo could pick up their raw material from the thousands of nodules exposed across a wide area through erosion, without having access to the concentrations of the better-quality large blocky material located at the deposit's core. But is the hypothesis that non-Chupicuaro populations controlled access to the Ucareo deposit really reasonable? Actually, the absence of Pre-Classic settlements (and non-Chupicuaro) on the spot or nearby would seem to rule this idea out.

On the other hand, the scenario in which Ucareo was at the boundary between two cultural regions – not apparently controlled by either group – could be much more relevant. In this case, the deposit would have been used by two groups of different origins with different interests, and, consequently, each would have exploited the deposit in a different way. The raw material acquisition strategies of the Chupicuaro groups may have been guided by precise requirements. The adoption of prismatic blade technology may have served no purpose, owing to the existence of local well-established lithic systems that perfectly fulfilled the Chupicuaro groups' needs.

On another level, a consideration of the forms of organization of prismatic blade production at the Ucareo deposit may help us to understand why the local populations did not acquire the technology. Data on the Pre-Classic exploitation of obsidian at Ucareo is still limited; however, the fact that mines and quarry workshops for this period have not been identified is not proof that they did not exist (Healan 1997). Nonetheless, some indications suggest that it is improbable that the whole manufacturing process took place on the spot: firstly, as stated earlier, there is no Pre-Classic settlement in the immediate vicinity of the Ucareo deposit; secondly, the prismatic blades were mainly destined for a supraregional market, in the regions of the East and South-East.

Additionally, for the Epi-Classic period, Healan (2002: 33) notes that only the first part of the *chaîne opératoire* took place at Ucareo, and that it consisted in configuring polyhedral cores, which were then transported over long distances toward their places of manufacture, such as Tula. This form of organization was by far the most common in Mesoamerica, since prismatic blades were generally manufactured in secondary workshops, sometimes far from the obsidian source (see Clark 1987,

[19] While still being of good quality, this deposit does not present the same potential as Ucareo, especially as far as the size of the blocks is concerned. Healan has shown that its systematic exploitation dated above all to the Late Post-Classic and was by the Tarascans (2005).

1988, 1989; Hirth 2006; Hirth and Andrews 2002; Parry 2002; Spence 1981, etc.) Thus, this information suggests a similar pattern for the Pre-Classic and the Early Classic. The groups involved in the production of prismatic blades from Ucareo obsidian could not have lived nearby. Their strategies for obsidian procurement probably involved cyclical trips and short-term stays that would have enabled blocks to be acquired and/or transformed into polyhedral cores, which were then transported back to the residential areas where they produced pressure blades.

No part of this scenario, however, explains how the technology was acquired. Why did the groups in possession of the savoir faire not transmit it to the local populations? It is difficult to imagine a lack of contact, but what was the nature of their relations? Could the cyclical travels and short stays on the spot and the segmentation of the *chaîne opératoire* be sufficient explanatory factors? Could it be that those with the job of procuring the material did not have all the skills to manufacture prismatic blades and were merely responsible for acquiring the raw or shaped blocks? It is possible that, if the reduction sequences were not entirely realized on the spot, and if the workers only came periodically for restricted periods, transmission of the technical knowledge could have been limited – the locals having only a partial and approximative understanding of the technological process, its purpose, and its usefulness.

The rarity of prismatic blades in the lithic assemblages of the Chupicuaro populations appears to support this hypothesis.[20] In this scenario, if the blades were manufactured far to the East, far from the Ucareo sources, if this production was not mainly for the Chupicuaro populations, and if the latter did not have relationship with the producers, then it is predictable that few blades would actually reach them. This would have been especially true if the artifact held no practical or symbolic meaning to the Chupicuaro people.

Consequently, we suggest that the absence of the prismatic blade technology in the Chupicuaro region could have resulted from a combination of three phenomena: on the one hand, the ways in which the producer groups were organized did not ease transmission of the technical knowledge to the region supplying the raw material, on the other, the absence of demand on the spot, probably owing to firmly rooted lithic traditions which were well adapted to the locals' needs, and, lastly, the absence of well-established interaction with the places in which the blades were manufactured.[21] In the heart of Chupicuaro territory, obsidian was a plentiful and easily accessible resource, just like basalt and andesite, and the various *chaînes opératoires* used – particularly unipolar debitage to obtain small short blades – made accessible the whole range of tools required by the people who lived there. Without rejecting the

[20] The variations observed from one Chupicuaro site to another also support this view: the Pre-Classic occupation levels of site TR 6 are for the moment the only ones to yield prismatic blades.

[21] This agrees with what has been found during our research: throughout the early and during the first half of the Late Chupicuaro phase, little archaeological evidence has been found showing well-established contacts with Central Mexico, commercial or otherwise (Darras 2006; Darras and Faugère 2007)

possibility of restricted access to the prismatic blade market, or deliberate retention of technical expertise by the groups possessing it, the hypothesis of a technocultural choice led by autosufficiency seems most reasonable to us today.

17.7.2 The Situation from the Proto-Classic to the Middle Classic: The Prismatic Blade – An Article for the Elite?

In contrast to the situation described for the Pre-Classic, the archaeological data for the extreme end of this period followed by the Early and Middle Classic shows the regional consumption of black and green obsidian prismatic blades was part of the general cultural evolution in North-Central Mexico, within which the interactions with Central Mexico – in particular with Teotihuacan – became more dynamic. In the Jalisco Highlands, Teotihuacan's influence is attested but seems definitely more discreet. In the Center-North, the metropolis's influence is perceptible to varying degrees in architectural patterns, pottery decoration techniques, iconography, and certain prestige goods (Carot 2001; Filini and Cardenas 2007; Filini 2004; Gomez and Gazzola 2007; Saint Charles 1996). Green prismatic blades were thus among the items circulating in most of the Center-North sites touched by the aura of Teotihuacan. However, based upon the present state of knowledge, we suggest the regional use of the prismatic blade was limited to special circumstances – ritual uses in particular – since the contexts in which the artifact is found give it a strong symbolic value. Its presence in particular seems to represent adhesion to an ideological model incarnated by Teotihuacan. As the prismatic blade became a useful and meaningful artifact, the question is raised as to why pressure blade technology was not adopted then, given that high-quality obsidian deposits were plentiful in the region. However, since the consumption market was restricted and probably reserved for special activities, the technology's local development may not have been justified. Also, if the use of the artifact was the privilege of a restricted number of people, for economic and/or symbolic reasons, the latter had no interest in promoting its local development. Finally, and clearly apart from the artifact itself, the variety of obsidian – in this instance, translucent green obsidian – could have had a particular symbolic value and have made the elites dependent on this circulation network.

17.7.3 Raw Material Abundance: A Brake on the Development of Technology During the Classic Period?

Several authors working in Mesoamerica see the development of pressure blade technology as a technical solution to the problem of managing the raw material efficiently, thus maximize the use of the obsidian blocks' potential (Clark 1982, 1987; Healan 2002, 2005; Hirth 2006; Hirth and Andrews 2002). It appears clearly that "the distance between obsidian sources and consumers sites … and the transportation costs affected how obsidian was worked in different regions and that a considerable amount of technological variation may be a response to economizing scarce resources in areas of high demand" (Hirth and Andrews 2002: 9).

In Central and Eastern Mesoamerica, during the Pre-Classic and Classic, prismatic blade technology development and logistics seem to have been strongly driven by leading regions distant from the obsidian sources (Olmec area, Puebla, Morelos, and Oaxaca). The consolidation of the web formed during the Early and Middle Pre-Classic in any case created a permanent substratum: most of the regions that were significantly touched by prismatic blades during this period remained as such. Later, during the Classic, very important populations centers emerged in the Central Highlands (see, e.g., Teotihuacan, Xochicalco, and Tula), which were also far from the obsidian sources that they used.[22] On the other hand, if we take the case of the populations in the West, we see that those of the Teuchitlan heartland are established at the hub of a vast system of deposits of excellent quality with an incalculable and inexhaustible supply of obsidians – black, gray, green, and red. In the Zinaparo region, deposits are located within a territory that was relatively well populated but characterized by a dispersed settlement pattern. So in these two regions, the obsidian sources and their surroundings were permanently occupied by a relatively numerous population of farmers and craftsmen. Settling close to sources may have been the result of a techno-economic strategy.

The direct percussion blade traditions produced unstandardized macroblades and blades of variable dimensions and thicknesses. Owing to the debitage methods used, this technique did not allow a maximized use of the obsidian cores – only a limited number of blades could be extracted (Darras 1999). Meanwhile, the craftsmen's decision to install themselves right next to the extraction areas meant that they had all the raw material they needed close at hand. Could the abundance of raw material and ease of access be enough to explain their technological traditions' perpetuity? Research done in the workshops of the Zinaparo region has shown the artisans did not need to be economical with their raw material and that they could choose to be highly profitable without exploiting cores efficiently (Darras 1999). This way of organizing production may thus have made it unnecessary to adopt a new technology, which certainly gave higher yields but was far more exacting in time and technical investment.

Lastly, we may recall percussion blade technologies enabled the production of a whole range of blanks for manufacturing certain types of tools, such as macroblade scrapers, bifacial knives, finely retouched with pressure, or again knives with basal fixation notches. Since these specialized products completed nonspecialized flake industries able to produce one-off nonstandard instruments, one may wonder if the adoption of the new technology was justified – given the well-established lithic traditions perfectly adapted to the local populations' needs.

[22] Teotihuacan is located 25 and 50 km, respectively, from its two favorite sources – Otumba and Pachuca. Xochicalco and Tula functioned synergetically with Ucareo at a distance of 250 km and more than 150 km, respectively.

17.7.4 Organization of Bajio Classic Cultures

Quite apart from these pragmatic reasons, the characteristics of the Classic cultures of the Lerma Valley or Bajio may open fruitful ways of thinking. At the end of the Classic period and in the Early Post-Classic (A.D. 800–1100), all the Center-North of Mexico was affected by two phenomena: several population movements occurred and autonomous regional developments were reinforced. The social and political context was then marked by a strong segmentation of the power structures, whether political, economic, or religious. The patterns of implantation in the macroregional space reflect this segmentation: on one hand, neighboring centers are found crystallizing an important population, embodying a politico-religious power (e.g., the site of Plazuelas or the massif of Barajas; Migeon and Pereira 2007; Pereira et al. 2005), and, on the other hand, vast rural zones with dispersed settlements run by small civic-ceremonial sites (Faugère, in press). In this macrocontext, and particularly in the region on the South margin of the Lerma River, the organization of the sites – both spatial and architectural – seems to reflect intercommunal competition and tensions (Faugère, in press). It is interesting to note that the systematic exploitation of the obsidian sources of Cerro Varal and Zinaparo, as well as the development of percussion blade production, occurred within this particular regional framework. The blade makers were culturally affiliated with the populations occupying the rest of this side of the Lerma Valley and the region of Zacapu, who were their main clients, while restricted circulation of obsidian from Zinaparo into more Northern regions – such as the Barajas massif – has been found (C. Andrieu, personal communication, 2007). Actually, it seems that blade production from the Zinaparo region mine workshops had a strictly regional impact going no farther than about 50 km East, South, and West.

Accordingly, can ignorance of pressure blade technology be explained by the nature of the social and political structures of populations in the Bajio? Did their social organization and territorial implantation prevent the acquisition of this technology? As their region, interconnected to the obsidian deposits in the Zinaparo massif, was not very densely populated, might the effective lack of a market of consumption explain this absence?

17.7.5 The Political and Social Conditions for Its Appearance in the Post-Classic

Studies carried out in the Jalisco Highlands and Center-North of Michoacan have shown that prismatic blade expertise and production appeared at the same time as a series of social and political phenomena. We are not able to discuss how the technology appeared in the first region, except to repeat that it was associated with a new tradition, called Aztatlan; so we shall focus on what we know best: the North of Michoacan.

For this region, one has every right to wonder whether the technological changes could have something to do in one way or another with the social and political evolutions. Certainly, only in the twelfth century did a certain number of elements favorable to this technology's development come together: concentration of a very numerous population in a restricted territory – which meant the prospect of a dense, stable, and regular market of consumption – and progressive political changes tending toward more complex power structures – concluding in the fourteenth century with the consolidation of a centralized state governed by a Tarascan sovereign (Darras 2005b). By the end of the twelfth century, the unification of the territory was still incomplete, and the regions of Zacapu and Patzcuaro seem to have experimented with an organization involving the cohabitation of several rival lineages (Darras 2008, 2009; Michelet 1998). Curiously, it was in this rather unstable political situation that the technology developed: at the time it was practiced within a flexible framework, in which control upstream – access to raw material deposits, production – and downstream – distribution and consumption – seem to have been nonexistent or insignificant (Darras 2008, 2009). Control by the Tarascan authorities is imperceptible in the archaeological record, although it may have existed indirectly through tribute payments. Our research suggests that the artisans produced the prismatic blades alternately with other unspecialized subsistence activities, such as agricultural activities (Darras 2008, 2009). Responsible for the whole production process from procuring the raw material to selling the goods, and with no intermediaries, the Tarascan blade makers of the Zacapu region were independent. Distribution was probably through market networks, mentioned in ethnohistorical documents (Relación de Michoacan 1977; Pollard 1993; Pollard and Vogel 1994), but the identification of residual prismatic cores and some preparation blades in several consumer sites could also suggest the craftsmen were itinerant as well and reduced their prismatic cores where they sold the blades.[23] The absence of a control infrastructure and the ways in which the profession was practiced could also explain the vulgarization of the prismatic blade, which became the most widespread artifact in the Tarascan peasant's toolkit. In this way, a very clear correlation can be established between the evolution of Tarascan social and political structures and the development of pressure blade technology: probably the authorities favored its development and the very wide diffusion of its products. In contrast, however, while the process of centralization and unification could be expected to be accompanied by augmented supervision of certain craft productions, a great freedom can be seen in the ways the technology was put into practice. This situation differs from the model upheld by Clark (1987), who insists the technology's logistics could only have been managed by elite groups, but also from the situation described by Pollard for the Patzcuaro Basin, the core of the Tarascan kingdom, during the fifteenth

[23] The presence of residual cores and a few preparation blades is not enough, however, to infer the passage of itinerant craftsmen. These residual cores may have been acquired deliberately to be recycled and used for other purposes.

century. This suggests the Tarascans exploited Ucareo and the prismatic blades were distributed under the control of the reigning dynasty (Pollard 2003: 232).[24] This is explicitly described in the *Relación de Michoacan*, in which the prismatic blade makers appear as members of the Assembly of the Uris, directly subordinated to the authority of the sovereign (*Relación de Michoacan* 1977). The situation we find in the Zacapu region may then reflect a complex situation, with great upheavals, in which the process of unification and centralization was only beginning and still had no impact on the organization of craft activities.

17.7.6 The Economic Variable: A Technical Choice Guided by the Quest for Profit?

During the Late Post-Classic, the abandonment of the area along the South bank of the Lerma and the subsequent redevelopment in Zacapu moved the people farther away from the obsidian deposits and forced them to reorganize their activities. This distance (which remained moderate) could have favored the adoption of a new production logistic and thus prismatic blade technology. The distance that they had to travel implied regular movements, and the transport of the raw material to the workshops could have resulted in a concern for using raw material efficiently, and justified a heavier technical investment. These observations would be in line with the results of other authors' research (Healan 2005: 177; Hirth and Andrews 2002).

Lastly, the very significant widespread of the prismatic blade – which characterized all of the Tarascan settlements in the region of Zacapu, but also in Ucareo – could be explained by improvements in technical knowledge. According to Healan (2002: 35; 2009: 110), the systematic grinding of pressure platforms, which became generalized during the Late Post-Classic, would have facilitated the detachment of prismatic blades and made the skills easier to acquire. In addition to this technical bonus – which, according to this author, would have reduced many production errors – one may wonder if the general simplification of the *chaîne opératoire*[25] did not favor and accelerate skill learning and ensure better efficiency. In any case, the development of the technology in the region studied was accompanied by an immediate widespread of the prismatic blade (Darras 2008, 2009).

[24] Healan's research confirms the Ucareo deposit came under Tarascan control during the Late Post-Classic.

[25] Two factors could have favorite the simplification of the "*chaîne opératoire*": all the stages of the reduction sequence were carried out in one place, and the blade makers had the possibility to select small angular blocks producing small prismatic blades.

17.7.7 How Were the Skills Transmitted?

The study of the prismatic blade workshops in the Zacapu region has permitted a technological correspondence to be made with other contemporary collections. It is remarkable that the *chaîne opératoire* followed was, broadly speaking, identical to that followed at the Aztec site of Otumba and at Tenochtitlan (Cassiano 1991; Garcia Velázquez 1990; Parry 2001, 2002): blade removal on a single-core face, abrasion of the future pressure platform during percussion stages, rarity of pressure platform rejuvenation, and production of short blades no more than 13 cm long – on average between 10 and 11 cm. These similarities suggest that pressure blade making at Zacapu followed the same principles as the Aztecs of the same period. However, the work of Healan (2003, 2005, 2009) in the Ucareo deposits has proven the long tradition of pressure blade production and quite clearly shown the generalization of prismatic blade production during the Middle Post-Classic, i.e., with the rise of the Tarascan culture. It is, therefore, probable that the development of the technology in the Tarascan region started at Ucareo in contact with populations already in possession of it. The skills were then transmitted to the populations in the Zacapu region from this initial core between A.D. 1100 and 1200. What is certain is that the prismatic blade makers of Zacapu shared the same material culture and were indeed Tarascans.

17.8 Conclusion

The obsidian prismatic blade can be considered a technocultural marker for the Mesoamerican identity of the peoples who made or used it. Thus, its rarity in the Western archaeological records could be an argument in favor of their marginal and unique character.

In this way, changes in the prismatic blade's status may illustrate the role of the ideological factor. During the Pre-Classic period, it clearly was not economically useful and on the ideological level was meaningless for these populations. It was not until the Early and Middle Classic – with the exportation throughout nearly all Mesoamerica of an ideological and religious model crystallized by Teotihuacan – that the prismatic blade acquired a meaning for the elites of many groups. However, at the end of the Early Classic phase, the decline of the metropolis's supremacy and the prevalence of regional developments – characterized by their withdrawn and rural character – entailed the disappearance of this artifact, which had no techno-functional advantage over the local industries. For the elites, the prismatic blade was now devoid of symbolic connotations. Finally, the blade's widespread use during the Middle Post-Classic could be related to improvements in technical knowledge, thereby facilitating its generalization and associated loss of value – both economic and symbolic.

However, if the material and social criteria necessary for the implementation of the technology are examined, there can be no doubt that all of these populations met – at least by the Late Classic – the conditions that permitted its development. But neither the abundance and quality of the obsidians, nor the level of social and political organization, nor the demographic mass, nor the excellence of local technical skills, acted as triggers. On the contrary, the abundance of raw material may well have prevented the adoption of a foreign technology perceived to be less than essential. The percussion blades technologies, directly depending for their implementation and viability on this abundance, enabled a whole range of standardized instruments to be created that were well adapted to the populations' needs.

Based upon the present state of knowledge, the absence of a market – due to technocultural preferences sustained by the advantages of easy resource access – may thus be a reasonable hypothesis for explaining ignorance of the technology. Its later adoption can be linked to the conjunction of two phenomena: radical transformation of the political and social structures, and technical simplification of the skills.

However, numerous questions remain unanswered, making it difficult to grasp the phenomenon of the prismatic blade in all its complexity. Clearly, to answer them, it is vital to understand how Ucareo-Zinapecuaro was exploited throughout the Pre-Classic and by whom. For the Jalisco Highlands, further research must be undertaken: first of all, the exact reconstitution of the *chaînes opératoires* associated with percussion blade productions, especially those realized at San Juan de los Arcos; then, studying the origin and forms of development of prismatic blade technology at the end of the Epi-Classic. The ways by which the skills were transmitted may enable an understanding of how the populations interacted; a rigorous reconstitution of the technology could reveal the links to other cultural groups – thereby helping to discern the identity of the Aztatlan peoples better.

Acknowledgments This chapter was made possible thanks to the constructive discussions with many colleagues, and I especially gratefully acknowledge Dominique Michelet, Chris Beekman, Phil Weigand, Catherine Liot, Javier Reveles, Dan Healan, and Lorenza Lopez.

References

Arnauld, Marie C., Patricia Carot, and Marie F. Fauvet-Berthelot 1993 *Arqueología de Las Lomas en la Cuenca Lacustre de Zacapu, Michoacán, México*. Cuadernos de estudios Michoacanos 5, CEMCA, México.

Beekman, Christopher 2006 The Chronological context of the central Jalisco Shaft tombs. *Ancient Mesoamerica* 17(2): 239–249.

Beekman, Christopher, and Luis J. Galvan Villegas 2006 The shaft tomb of the Atemajac Valley and their Relation to Settlement. *Ancient Mesoamerica* 17(2):259–270

Beekman, Christopher, and Phil W. Weigand 2010 La secuencia cronológica de la tradición Teuchitlan. In *El Eje Lerma-Santiago durante el Formativo Terminal y Clásico temprano: precisiones cronológicas y dinámicas culturales,* edited by Laura Solar, INAH, México.

Beekman, Christopher, and Phil W. Weigand 2000 *La cerámica arqueológica de la tradición Teuchitlan, Jalisco*. El Colegio de Michoacan/Secretaria de Cultura de Jalisco.

Boksenbaum, Martín W. 1978 *Lithic Technology in the basin of Mexico during the Early and Middle Pre-classic*. Unpublished Ph.D. dissertation, University of New York.

Boksenbaum, Martin W., Paul Tolstoy, Garman Harbottle, Jerome Kimberlin, and Mary Neivens 1987 Obsidian Industries and Cultural Evolution in the Basin of Mexico before 500 B.C. *Journal of Field Archaeology* 14(1): 65–76.

Brambila, Rosa, and Ana Maria Crespo 2005 Desplazamientos demográficos y creación de territorios en el Bajío. In *Reacomodos demográficos del Clásico al Posclásico en el centro de México*, edited by Linda Manzanilla, pp. 155–174. UNAM/INAH, México.

Burton, Susan S. 1987 Obsidian Blade Manufacturing Debris on Terrace 37. In *Ancient Chalcatzingo*. edited by David Grove, pp. 321–328. University of Texas Press, Austin.

Calgaro, B. 2007 Sección 9: Análisis Preliminar de la Lítica. In *Informe Final de la Temporada de 2007, Proyecto Arqueológico Regional Valles de Tequila*. Informe científico del INAH.

Carot, Patricia 2001 *Le site de Loma Alta, lac de Zacapu, Michoacan, Mexique*. Paris Monographs in American Archaeology 9, Archaeopress BAR International Series 920, Oxford.

Cassiano, Gianfranco 1991 La tecnología de navajillas prismáticas. *Arqueología* 5: 107–118.

Clark, John 1982 Manufacture of Mesoamerican prismatic blades: An alternative technique. *American Antiquity* 47: 355–376.

Clark, John 1987 Politics, prismatic blades, and Mesoamerican civilization. In *The organization of core technology*, edited by Jay K. Johnson and Carol A. Morrow, pp. 259–285. Westview Press.

Clark, John 1988 *The lithic artifacts of La Libertad, Chiapas, Mexico. An economic perspective*. Papers of the New World Archaeological Foundation 52. Provo, UTA, Brigham Young University, Utah

Clark, John 1989 Hacia una definición de talleres. In *La obsidiana en Mesoamérica*, editado por Margarita Gaxiola y John Clark, pp. 213–217, Colección Científica 176, INAH, México.

Cobean Robert H., Michael Coe, Edward A. Perry, and Dinkar P. Kharkar 1971 Obsidian Trade at San Lorenzo Tenochtitlan, Mexico. *Science* 174: 666–671.

Cobean Robert H., James R. Vogt, Michael D. Glascock, and Terry L. Stocker 1991 High precision Trace Element Characterization of Major Mesoamerican Obsidian Sources and further Analyses of Artifacts from San Lorenzo Tenochtitlan, Mexico. *Latin American Antiquity* 2(1): 69–91.

Darras, Véronique 1993 La lítica tallada de las Lomas. In Arqueología de las Lomas en la Cuenca lacustre de Zacapu, Michoacán, México, edited by Marie Charlotte Arnauld, Patricia Carot, and Marie France Fauvet-Berthelot, pp. 168–190, Cuadernos de Estudios Michoacanos 5, CEMCA, México.

Darras, Véronique 1994 Les mines-ateliers d'obsidienne de la région de Zinaparo-Prieto, Michoacan, Mexique. *Bulletin de la société préhistorique française* 91(4–5): 290–301, Paris

Darras, Véronique 1998 La obsidiana en la Relación de Michoacan y en la realidad arqueológica: del símbolo al uso o del uso de un símbolo. In *Génesis, culturas y espacios en Michoacan*, edited by Véronique Darras, pp. 61–88, CEMCA, México.

Darras, Véronique 1999 *Tecnologías prehispánicas de la obsidiana: los centros de producción de la región de Zináparo-Prieto, Michoacán*. Cuadernos de Estudios Michoacanos 9, CEMCA, México.

Darras, Véronique 2005a La tecnología de la navaja prismática: una singular invención mesoamericana. In *Reflexiones sobre la industria lítica*, edited by Leticia González Arratia and Lorena Mirambell, pp. 111–134, Colección Científica 475, INAH, México.

Darras, Véronique 2005b Economía y poder: la obsidiana entre los tarascos de Zacapu, Michoacán (fase Milpillas 1200 a 1450 d.C.). In *Reflexiones sobre la industria lítica*, edited by Leticia González Arratia and Lorena Mirambell, pp. 243–298. Colección Científica 475, INAH, México.

Darras, Véronique 2006 Las relaciones entre Chupícuaro y el Centro de México durante el Preclásico reciente. Una crítica de las interpretaciones arqueológicas. *Journal de la Société des Amériques* 92–2: 69–110, Paris.

Darras, Véronique 2008 Estrategias para la producción de Navajas en la región de Zacapu y la Vertiente del Lerma (Michoacán, México) entre el Epiclásico y el Posclásico tardío. *Ancient Mesoamerica 19:* 243–264.

Darras, Véronique 2009 Peasant Artisans: Household Prismatic Blade Production in the Zacapu Region, Michoacan (Milpillas phase 1200–1450 A.D.). *Archaeological Papers of the American Anthropological Association,* Vol. 19(1): 92–114.

Darras, Véronique, and Brigitte Faugère 2007 Chupicuaro, entre el Occidente y el Altiplano central. Un balance de los conocimientos y las nuevas aportaciones. In *Dinámicas Culturales, entre el Occidente, el Centro-norte y la Cuenca de México, del Preclásico al Epiclásic,* edited by Brigitte Faugère, pp. 51–84. El Colegio de Michoacán, Zamora/CEMCA, México.

Darras, Véronique, Brigitte Faugère-Kalfon, Christophe Durlet, Catherine Liot, Javier Reveles, and Cybèle David 1999 Recherches récentes sur la culture Chupicuaro, Guanajuato. Journal de la Société des Américanistes, 85: 343–352.

Elam, J. Michael, Michael D. Glascock, and Hector Neff 1994 Obsidian Artifacts from Oaxaca, Mexico: Source identification and hydration dating. In *Archaeometry of Pre-Columbian Sites and Artifacts,* edited by Peter Meyers and David A. Scott, pp 135–159. The Getty Conservation Institute, Los Angeles.

Esparza, Rodrigo, and Carlos Ponce Ordaz 2005 La obsidiana en el contexto arqueológico de Los Guachimontones, Teuchitlan, Jalisco. In *El antiguo occidente de México. Nuevas perspectivas sobre el pasado prehispánico,* editado por Eduardo Williams, Phil C. Weigand, Lorenza Lopez Mestas y David Grove, pp. 145–170, El Colegio de Michoacan, Zamora.

Faugère-Kalfon, Brigitte 1989 Entre Nomades et Sédentaires : archéologie du versant méridional du Lerma au Michoacan, Mexique. Tesis de doctorado de la Universidad de Paris I.

Faugère-Kalfon, Brigitte 1996 *Entre Zacapu y Lerma: Culturas en una zona fronteriza*. Cuadernos de estudios Michoacanos 7, CEMCA, México.

Faugère, Brigitte In Press Sociedad y poder en el centro-norte de Mesoamérica (600-1200 d.C.): el caso del norte de Michoacán. In *Las Sociedades Complejas del Occidente de México en el mundo mesoamericano. Homenaje a Phil Weigand*, edited by Eduardo Williams, Lorenza López Mestas, and Rodrigo Esparza, El Colegio de Michoacán, Zamora.

Filini, Agapi 2004 Interacción cultural entre la Cuenca de Cuitzeo y Teotihuacan. In *Tradiciones arqueológicas,* edited by Efraín Cardenas García, pp. 307–328, Colegio de Michoacan/Gobierno de Michoacan, Zamora/Morelia.

Filini, Agapi, and Efrain Cardenas 2007 El Bajío, la Cuenca de México y el estado teotihuacano. Un estudio de relaciones y antagonismos. In *Dinámicas culturales entre el Occidente, el centro-norte y la Cuenca de México, del Preclásico al Epiclásico,* edited by Brigitte Faugère, pp. 137–156, Colegio de Michoacan/CEMCA.

Galván Villegas, Luís J. 1991 *Las tumbas de tiro del valle de Atemajac*, Jalisco. Colección Científica 239, INAH, México.

García Velázquez, Jorge, and Gianfranco Cassiano 1990 La producción de navajillas prismáticas en el Postclásico tardío: El caso de la plaza de la Banca Nacionalizada. In *Etnoarqueología: Primer Coloquio Bosch-Gimpera*, edited by Yoko Sugiura and Mari Carmen Serra, pp. 513–526. UNAM, Mexico City.

Gómez, Sergio, and Julie Gazzola 2007 Análisis de las relaciones entre Teotihuacan y el Occidente de México. In *Dinámicas culturales entre el Occidente, el centro-norte y la Cuenca de México, del Preclásico al Epiclásico,* editado por Brigitte Faugère, pp. 113–136. Colegio de Michoacán, Zamora/CEMCA, México.

Grove, David 1974 *San Pablo, Nexpa, and the early Formative Archaeology of Morelos, Mexico.* Vanderbilt University-Publications in Anthropology 12, Vanderbilt University, Nashville.

Grove, David 1987 (editor) *Ancient Chalcatzingo.* University of Texas Press, Austin.

Healan, Dan M. 1997 Prehispanic Quarrying in the Ucareo-Zinapecuaro Obsidian Source Area. *Ancient Mesoamérica* 8: 77–110.

Healan, Dan M. 2002 Producer Versus Consumer: Prismatic Core-Blade Technology at Epiclassic/Early Postclassic Tula and Ucareo. In *Pathways to Prismatic Blades. A study in Mesoamerican Obsidian Core-Blade Technology*, edited by Kenneth G. Hirth and Bradford Andrews, Monograph 45, pp. 27–36, The Cotsen Institute of Archaeology, University of California, Los Angeles.

Healan, Dan M. 2003 From the Quarry Pit to the Trash Pit: Comparative Core-Blade Technology at Tula, Hidalgo, and the Ucareo Obsidian Source Region. In *Mesoamerican Lithic Technology. Experimentation and Interpretation*, pp. 153–169, University of Utah Press, Salt Lake City.

Healan, Dan M. 2004 Extracción prehispánica de obsidiana en el área de Ucareo-Zinapecuaro, Michoacán. In *Bienes estratégicos del Antiguo Occidente de México*, edited by Eduardo Williams, pp. 77–112, El Colegio de Michoacán, Zamora.

Healan, Dan M. 2005 Nuevos datos acerca del desarrollo de la tecnología de núcleos prismáticos en la fuente de obsidiana, Ucareo, Michoacán. In *El antiguo occidente de México. Nuevas perspectivas sobre el pasado prehispánico,* edited by Eduardo Williams, Phil C. Weigand, Lorenza Lopez Mestas, and David Grove, pp. 171–184. El Colegio de Michoacán, Zamora.

Healan, Dan M. 2009 Ground Platform Preparation and the "Banalization" of the Prismatic Blade in Western Mesoamerica. *Ancient Mesoamérica 20*: 103–111.

Hernandez, Cristina 2000 A History of Prehispanic Ceramics, interactions, and Frontier development in the Ucareo-Zinapecuaro Obsidian Source Area, Michoacan, México. Unpublished Ph.D. dissertation, Tulane University, New Orleans.

Hernandez, Cristina 2006 La cerámica del periodo clásico en el noreste de Michoacán. In *La producción alfarera en el México antiguo II,* edited by Beatriz L. Merino Carrion y Angel Garcia Cook, pp. 313–334, Colección científica 495, INAH, México.

Hirth, Kenneth G. 1995 The investigation of obsidian craft production at Xochicalco, Morelos. *Ancient Mesoamerica 6*: 251–258.

Hirth, Kenneth G. 2002 Provisioning constraints and the production of obsidian prismatic blades at Xochicalco, Mexico. In *Pathways to Prismatic Blades: A study In Mesoamerican obsidian core-blade technology*, edited by Kenneth G. Hirth and Bradford Andrews, pp. 85–94. The Cotsen Institute of Archaeology, University of California, Los Angeles.

Hirth, Kenneth G. 2006 *Obsidian Craft Production in Ancient Central Mexico. Archaeological Research at Xochicalco.* University of Utah Press, Salt Lake City.

Hirth, Kenneth G. (editor) 2003 *Mesoamerican Lithic Technology. Experimentation and Interpretation*. University of Utah Press, Salt Lake City.

Hirth , Kenneth G., and Bradford Andrews (editors) 2002 *Pathways to Prismatic Blades: A study In Mesoamerican obsidian core-blade technology*. Monograph 45. The Cotsen Institute of Archaeology, University of California, Los Angeles.

Hirth, Kenneth G., and Jeffrey J. Flenniken 2002 Core-blade Technology in Mesoamerican Prehistory. In Pathways to Prismatic Blades. A study in Mesoamerican Obsidian Core-Blade Technology, edited by Kenneth G. Hirth and Bradford Andrews, pp. 121–130, Monograph 45, The Cotsen Institute of Archaeology, University of California, Los Angeles.

Kelley, Charles J. 2000 The Aztatlan Mercantile System: Mobile Traders and the Northwest-ward Expansion of Mesoamerica Civilization. In *Greater Mesoamerica. Cultural Dynamics of West and Northwest of Mexico*, edited by Michael Foster and Shirley Gorenstein, pp. 137–154, Utah University Press.

Kelly, Isabel 1980 *Ceramic Sequence in Colima: Capacha, an early phase.* Anthropological Papers of the University of Arizona 37. The University of Arizona Press, Tucson.

Kirchhoff, Paul 1943 *Mesoamérica.* Acta Mesoamericana, vol. 1, México.

Liot Catherine, Susana Ramirez, Javier Reveles, and Otto Schondube 2006 *Transformaciones socioculturales y tecnológicas en el sitio de La Peña, Cuenca de Sayula, Jalisco.* Universidad de Guadalajara/INAH, Guadalajara.

Liot Catherine, Susana Ramirez, Javier Reveles, and Carmen Melgarejo 2007 Producción, distribución y relaciones interregionales en la cuenca de Sayula del 500–1100 d.C. In *Dinámicas Culturales, entre el Occidente, el Centro-norte y la Cuenca de México, del Preclásico al Epiclásico*, edited by Brigitte Faugère, pp. 165–200, El Colegio de Michoacán, Zamora / CEMCA, México.

Lodeho, Laure 2007 *L'économie de l'obsidienne à Chupicuaro, Mexique (600 av. J.-C. – 250 apr. J.-C.). Etude typo-technologique de la collection du site TR6.* Master 2 en Archéologie préhistorique et protohistorique, Université de Paris I.

Lopez-Mestas, Lorenza 2007 La ideología. Un punto de acercamiento para el estudio de la interacción entre le Occidente de México y Mesoamérica. In *Dinámicas Culturales, entre el Occidente, el Centro-norte y la Cuenca de México, del Preclásico al Epiclásico*, edited by Brigitte Faugère, pp. 37–50, El Colegio de Michoacán, Zamora /CEMCA, México.

Macias Gotilla, Angelina 1990 *Huandacareo: Lugar de juicios, tribunal.* Colección Científica 222, INAH, México.

MacNeish, Richard S., Antoinette Nelken Terner, and Irmgard Weitlaner- Johnson 1967 *The Prehistory of the Tehuacán Valley, vol 2. The Non Ceramics Artifacts.* University of Texas Press, Austin.

Manzanilla, Linda 2003 El proceso de abandono de Teotihuacan y su reocupación por grupos epiclásicos. *Trace 43:* 70–76, CEMCA, México.

Manzanilla Lopez, Rubén 1984 *Loma de Santa Maria I, Morelia, Michoacán. Un sitio del periodo Clásico mesoamericano.* Tesis de licenciatura, ENAH, México.

Michelet, Dominique 1992 El Centro-Norte de Michoacán: características generales de su estudio arqueológico regional. In El proyecto Michoacán 1983-1987, edited by Dominique Michelet, pp. 12–54, Cuadernos de Estudios Michoacanos 4, CEMCA, México.

Michelet, Dominique 1998 Topografía y prospección sistemática de los grandes asentamientos del malpaís de Zacapu: claves para un acercamiento a las realidades sociopolíticas. In *Génesis, Culturas y Espacios en Michoacán,* edited by Véronique Darras, pp. 47–60, CEMCA, México.

Michelet, Dominique, Grégory Pereira, and Gérald Migeon 2005 La llegada de los uacúsechas a la región de Zacapu, Michoacán: datos arqueológicos y discusión. In *Reacomodos demográficos del Clásico al Posclásico en el centro de México*, edited by Linda Manzanilla, pp. 137–154, UNAM, México.

Migeon, Gérald 1990 Archéologie en pays tarasque. Structure de l'habitat et ethnohistoire des habitations tarasques de la région de Zacapu au Postclassique Récent. Tesis de doctorado de la Universidad de Paris I.

Migeon, Gérald 1998 El poblamiento del malpaís de Zacapu y de sus alrededores, del Clásico al Posclásico. In *Génesis, Culturas y Espacios en Michoacán,* edited by Dominique Michelet, pp. 35–46, CEMCA, México.

Migeon, Gérald, and Grégory Pereira 2007 La secuencia ocupacional y cerámica del cerro Barajas, Guanajuato, y sus relaciones con el centro, el occidente y el norte de México. In *Dinámicas culturales entre el Occidente, el Centro-norte y la cuenca de México, del Preclásico al Epiclásico,* edited by Brigitte Faugère, pp. 201–230. El Colegio de Michoacán, Zamora.

Mountjoy, Joseph 1990 El desarrollo de la cultura Aztatlan visto desde su frontera suroeste. In *Mesoamérica y norte de México. Siglo IX-XII,* Vol. 2, edited by F. Soti Miranda pp. 541–564. Museo Nacional de Antropologia, INAH, México.

Mountjoy, Joseph 2000 Prehispanic Cultural Development along the Southern Coast of West Mexico. In *Greater Mesoamerica. The Archaeology of West and Northwest Mexico,* edited by Michael S. Foster, and Shirley Gorenstein, pp. 81–106. University of UTAH Press, Salt Lake City.

Mountjoy, Joseph 2004 La cultura indigena en la costa de Jalisco, El municipio de Puerto Vallarta. In *Arqueologia del Occidente de México*, edited by Beatriz Braniff Cornejo, pp. 339–370. Universidad de Colima / INAH, México.

Narez, Jésus 1990 *Materiales arqueológicos de Tlapacoya, México.* Colección científica 204, INAH, México.

Niederberger, Christine 1976 *Zohapilco. Cinco milenios de ocupación humana en un sitio lacustre de la cuenca de México.* Colección Científica 30, INAH, México.

Niederberger, Christine 1986 Excavación de un área de habitación doméstica en la capital "olmeca" de Tlacozotitlan. Reporte preliminar. In *Arqueologia e Etnohistoria del Estado de Guerrero.* Primer Coloquio de Arqueología y Etnohistoria del Estado de Guerrero, edited by Roberto Cervantes-Delgado, pp. 83–103, INAH/Gobierno del estado de México.

Niederberger, Christine 1987 *Paléopaysages et Archéologie pré-urbaine du bassin de Mexico.* Collection Etudes Mésoaméricaines I-II, CEMCA, Mexico.

Oliveros, Arturo 2004 *Hacedores de tumbas en El Opeño, Jacona, Michoacán.* El Colegio de Michoacan/, Michoacán.

Parry, William J. 1994 Prismatic Blade Technology in North America. In *The Organization of North American Prehistoric Chipped Stone Tool Technologies,* edited by Philip J. Carr, pp. 87–98. Archaeological Series 7, International Monographs in Prehistory, Ann Arbor.

Parry, William J. 2001 Production an Exchange of obsidian tools in late Aztec city-states. *Ancient Mesoamerica* 12: 101–111.

Parry, William J. 2002 Aztec Blade Production Strategies in the Eastern Basin of Mexico. In *Pathways to Prismatic Blades. A study in Mesoamerican Obsidian Core-Blade Technology,* edited by Kenneth G. Hirth and Bradford Andrews, pp. 37–46, Monograph 45. The Cotsen Institute of Archaeology, University of California, Los Angeles.

Pereira, Grégory 1999 *Potrero de Guadalupe. Anthropologie funéraire d'une communauté pré-tarasque du nord du Michoacan, Mexique.* BAR International Series 816.

Pereira Grégory, Gérald Migeon, and Dominique Michelet 2005 Transformaciones demográficas y culturales en el centro-norte de México en vísperas del Posclásico: los sitios del Cerro Barajas (Suroeste de Guanajuato) In *Reacomodos demográficos del Clásico al Posclásico en el centro de México,* edited by Linda Manzanilla, pp. 123–136, UNAM/INAH, México.

Pires-Ferreira, Jane W. 1975 *Formative Mesoamerican Exchange networks with special reference to the valley of Oaxaca.* Memoirs of the museum of Anthropology of Michigan,Vol 3, No. 7, Prehistory and human ecology of the valley of Oaxaca, edited by Kent V. Flannery, Michigan.

Pires-Ferreira, Jane W. 1976 Obsidian Exchange in Formative Mesoamerica. In *The Early Mesoamerican Village*, edited by Kent V. Flannery, pp. 292–306. Academic Press, New York.

Pollard Perlstein, Helen 1993 *Tariacuri's Legacy. The Prehispanic Tarascan State.* University of Oklahoma Press, Norman.

Pollard Perlstein, Helen 2003 Development of Tarascan Core: The Lake Patzcuaro Basin. In *The Postclassic Mesoamerican World*, edited by Kent V. Flannery, pp. 227–237. University of Utah Press, Salt Lake City.

Pollard Perlstein, Helen, and Thomas A. Vogel 1994 Implicaciones políticas y económicas del intercambio de obsidiana des Estado tarasco. In *Arqueología del Occidente de México*, edited by Eduardo Williams and Roberto Novela, pp. 158–182. El Colegio de Michoacán, Zamora.

Porter Noé, Muriel 1956 Excavations at Chupícuaro, Guanajuato, Mexico. *Transactions of the American Philosophical Society* 46, pp. 515–637, Philadelphia.

Relación de Michoacán 1977 [1541] Relación de las ceremonias y ritos y población y gobierno de la provincia de Michoacán. Reproducción facsimilar del Ms IV de El Escorial, reedición con un estudio preliminar de J. Corona Nuñez. Balsal Editores, Morelia.

Reveles, Javier 2005 La lítica de la Cuenca de Sayula. In *Arqueología de la Cuenca de Sayula,* edited by Fransisco Valdez, Otto Schöndube, and Jean Pierre Emphoux, pp. 349–368, UDG-IRD, Guadalajara.

Saint Charles, Juan Carlos 1996 El reflejo del poder Teotihuacano en el sur de Guanajuato y Querétaro. In *Tiempo y Territorio en Arqueología. El Centro-Norte de México*, edited by Ana Maria Crespo and Carlos Viramontes, pp. 143–169, INAH, México.

Santley, Robert 1984 Obsidian Exchange, economic Stratification, and the evolution of complex society in the Basin of Mexico, In *Trade and Exchange Society in Early Mesoamérica*, edited by Kenneth G. Hirth, pp. 43–86. University of New Mexico, Albuquerque.

Santley, Robert S., Janet M. Kerley and, Ronald R. Kneebone 1986 Obsidian Working, Long-distance Exchange, and the Politico-Economic Organization of Early States in Central Mexico. In *Economic Aspects of Prehispanic Highland Mexico*, edited by Barry L. Isaac, pp. 101–132. Research in Economic Anthropology, Supplement 2. JAI Press, Greenwich.

Spence, Michael W. 1981 Obsidian Production and the State of Teotihuacan. *American Antiquity* 46: 769–788.

Spence, Michael W. 1987 The Scale and Structure of Obsidian Production in Teotihuacan. In *Teotihuacan: Nuevos Datos, Nuevas Síntesis, Nuevos Problemas*, edited by Emily McClung de Tapia and Evelyn Childs Rattray, pp. 429–450. INAH/UNAM, México

Spence, Michael W., Phil C. Weigand, and Dolores Soto de Arechavaleta 2002 Production and Distribution of Obsidian Artifacts in Western Jalisco. In *Pathways to Prismatic Blades: a study in Mesoamerican Core-Blade Technology,* edited by Kenneth G. Hirth and Bradford Andrews, pp. 61–80, Monograph 45. The Cotsen Institute of Archaeology, University of California, Los Angeles.

Soto de Arachevaleta, M. Dolores 1982 *Análisis de la tecnología de producción del taller de obsidiana de Guachimonton, Teuchitlan, Jalisco*, México. Tesis de licenciatura, ENAH, México.

Soto de Arachevaleta, M. Dolores 2005 Teuchitlan: un sitio con especialización en el trabajo. La manufactura de herramientas de obsidiana. In *Reflexiones sobre la industria lítica,* edited by Leticia Gonzalez Arratia and Lorena Mirambell, pp. 135–180, INAH, México.

Valdez Fransisco, Otto Schondube, and Jean Pierre Emphoux 2005 *Arqueología de la Cuenca de Sayula.* Universidad de Guadalajara/IRD.

Weigand, Phil C. 1990 The Teuchitlan tradition of western Mesoamérica. In *La época clásica: Nuevos hallazgos, nuevas ideas*, edited by Amalia Cardos de Méndez, pp. 25–54. INAH, México.

Weigand, Phil C. 1993 *Evolución de una civilización prehispánica. Arqueología de Jalisco, Nayarit y Zacatecas.* El Colegio de Michoacán, Zamora.

Weigand, Phil C. 1996 Teuchitlan Tradition of the Occidente of Mesoamerica. *Ancient Mesoamerica* 7: 91–101.

Weigand, Phil C. 2000 Evolution and Decline of a Core of Civilization: The Teuchitlan tradition and the Archaeology of Jalisco. In *Greater Mesoamerica. The Archaeology of West and Northwest Mexico,* edited by Michael S. Foster and Shirley Gorenstein, pp. 43–58. University of Utah Press, Salt Lake City.

Weigand, Phil C., and Michael W. Spence 1982 The obsidian Mining Complex at La Joya, Jalisco. *Anthropology VI* (1–2): 175–188. Stony Brook, Nueva York.

Part III
Recent Advances in Experimentation

Chapter 18
New Experimental Observations for the Characterization of Pressure Blade Production Techniques

Jacques Pelegrin

18.1 Introduction

We owe to D. Crabtree, through his successful replication of obsidian pressure blades – even if it was not the genuine 'Mexica technique' reconstructed by J. Clark (1982, and Chap. 3 of this volume) – a first explanation of the stigmata of the pressure technique. These early experiments of mesoamerican blade production by pressure from D. Crabtree (1968) allowed J. Tixier to first recognize this particular technique applied to flint in the Old World, specifically in the Upper Capsian from Algeria in the late 1960s (Tixier 1984). Since then, this technique has been widely recognized in the Old World. The main steps of this recognition took place in Europe as part of research on the 'Chasséen méridional' (a large middle Neolithic culture primarily defined in the South of France; Binder 1984), the Neolithic of Greece (Perlès 1984, 2001, 2004), and the Maglemosian and Kongemosian Mesolithic in Denmark (Callahan 1985; Chap. 9 by Sørensen, this volume). Meanwhile, M.-L. Inizan detected this technique in several cultures from the Near and Middle East (Inizan 1991, and Chap. 2 of this volume).

Considering different allusions to this technique, or recent confirmations of it within other contexts, there is still much to discover and study in order to complete the picture. Beyond the economical significance of the pressure technique in each of the techno-complexes where it was put into use, such a picture should form the basis for a historical reconstruction of its origin(s) and development, its significance, and above all, its diffusion. With this aim in mind, experimental replication reference collections remain a key for the identification of archaeological production techniques.

J. Pelegrin (✉)
Laboratoire "Préhistoire et Technologie", CNRS et Université Paris Ouest Nanterre, MAE, Nanterre, France
e-mail: jacques.pelegrin@mae.u-paris10.fr

P.M. Desrosiers (ed.), *The Emergence of Pressure Blade Making: From Origin to Modern Experimentation*, DOI 10.1007/978-1-4614-2003-3_18,

18.2 Background to This Research

I started my experimental investigations of the pressure technique in 1981, after Jean-Paul Thevenot showed me in 1980 a few very regular bladelets from his ongoing excavations at Chassey-le-Camp and asked me 'How could these have been made?' During 1980, I put a lot of effort into reproducing them by indirect percussion, but the friendly critique of D. Binder led me to admit 'that was not exactly it'. That is the reason I turned to testing pressure techniques, all the while ignoring Tixier's early experiment with a flint core, bearing in mind Bordes' opinion that flint was much harder than obsidian when it came to detaching products larger than microblades (1969).

I also felt that it was time to solve the practical aspects of Crabtree's technique (the copper point, the core fixation device with a strong steel screw, the core prepared by modern sawing). That led me to test different holding devices, one of them quite simple, using a somewhat different and more ergonomic standing position and a flexible crutch, and to test different ways of preparing pressure cores by percussion. These adaptations, which allowed me to produce a series of small flint blades up to 14 cm long and 18 mm wide with an antler point, were demonstrated during the *3rd meeting of lithic technology* held at Meudon in October 1982 (Pelegrin 1984a, b, c).

Aware that there existed much 'smaller' versions of pressure, for instance the small bullet cores from the Near East, and that the standing technique I used could not be invented from scratch, I tried, from 1984 to 1987, various different ways to reproduce microblades and bladelets (Pelegrin 1988).[1] I also started an early and successful test of lever pressure in 1983 at the Archéodrome for the detachment of large pressure blades, which proved to be easy but for an unstable holding device. I was, however, conscious that one could not recognize lever pressure without a practical exploration of its 'competing' technique, that being indirect percussion.

This second program, started in 1986 and 1987 at the Lejre Centre (Denmark) together with Bo Madsen, had the purpose of being a preliminary exploration of the many parameters of indirect percussion (length, curvature and stiffness of the punch, holding position of the core, length and curvature of the blade-products, etc.). This 'blade' experiment was continued over a period from 1988 to 1991 with the indirect percussion technique (carefully exploring and demonstrating the interest of an elastic support for the core, testing of copper-tipped punches, etc.), until it appeared that two of my archaeological reference cases could not be successfully reproduced using a punch (not so much regarding morphology and dimensions of the blades, but for discrete technical stigmata such as ripples on the bulb, cracks on the blade platform, terminations, etc.). That led me to explore once again the lever pressure technique, mainly in 1992 and 1993, with further tests in relation to specific archaeological cases (1997). In 1993, thanks to an invitation to the Western United States by friends and colleagues (J. Flenniken, P. Geib, G. Titmus, M. Warburton and

[1] In this chapter, 'microblade' refers to tiny bladelets generally less than 8 mm wide, such as those described in the Far East and Arctic regions. 'Bladelet' refers to small blades generally 8 to 12 mm wide. 'Blade', which is the general term for elongated flakes detached in parallel series, is expected here to be over 12 mm wide.

J. Woods, to whom I am very grateful), I was given access to obsidian from California and Oregon, and could mail 70 kg of prepared cores to France for further knapping tests. However, throughout these years, the main driving force of this research has been the progressive discovery of the diversity of the archaeological record.

18.3 Several 'Modes' of Pressure

The main principle of this step-by-step progression comes from a simple fact that is acknowledged by knappers (since Crabtree) with pressure experience, 'the wider the blade, the greater the amount of pressure that is required' (Crabtree 1968, p. 468, see also Titmus and Clark 2003, p. 84). Since the pressure force depends on the physico-mechanical conditions of its application, the maximal width of the products constitutes a good indicator of the pressure mode employed.

From a methodological point of view, after the collection has been classified according to eventual types of products and sequences, the distribution of the product width, or more specifically, of the width and thickness (cloud of points diagram), should reveal a rather concentrated cluster indicative of a technical mode. For example, a scatter of width values between 10 and 18 mm and a few points up to, but not more than, 20–22 mm for a series of flint blades and fragments identified as pressure detached according to their morphological features (regularity, almost straight profile, often with a thin section, small platform) would be indicative of a standing pressure technique (mode 4 infra). If some blades are wider, one should reconsider the pressure diagnosis and consider a percussion technique.

In many contexts, indirect percussion is known to exist alongside pressure for the production of larger blades. This can be from specific cores, or from the same cores before they are further reduced by pressure. Indirect percussion can also be limited to the refined pre-shaping of pressure core preforms, including a series of axial punch blades, so as to create regular ridges on the flaking surface that will help start the pressure production. Indirect percussion blades can also help in repairing or rejuvenating the flaking surface, or even be used instead of pressure for a simple blade detachment that happens to be too difficult to produce by pressure (possibly after a mis-preparation of the platform). Some of these punch blades can therefore be as regular as pressure blades, yet somewhat wider. If a consistent fraction of the blades appear wider than the envisioned limit (for a given pressure mode), one should reconsider the hypothesis of a very careful indirect percussion technique. A difficult case, as I have seen in a Cardial collection from Portugal, is that of the reduction of cores by indirect percussion for the first step, followed by a pressure technique for the second step, with heat treatment of the core in between.

In this study, I postulate that, if anything, the main trend of innovation regarding pressure blade production techniques was to make them wider (which is a necessary condition to eventually make them longer), which assumes an improvement in the force applied and/or the stability of the core. We start with flint and follow with obsidian.

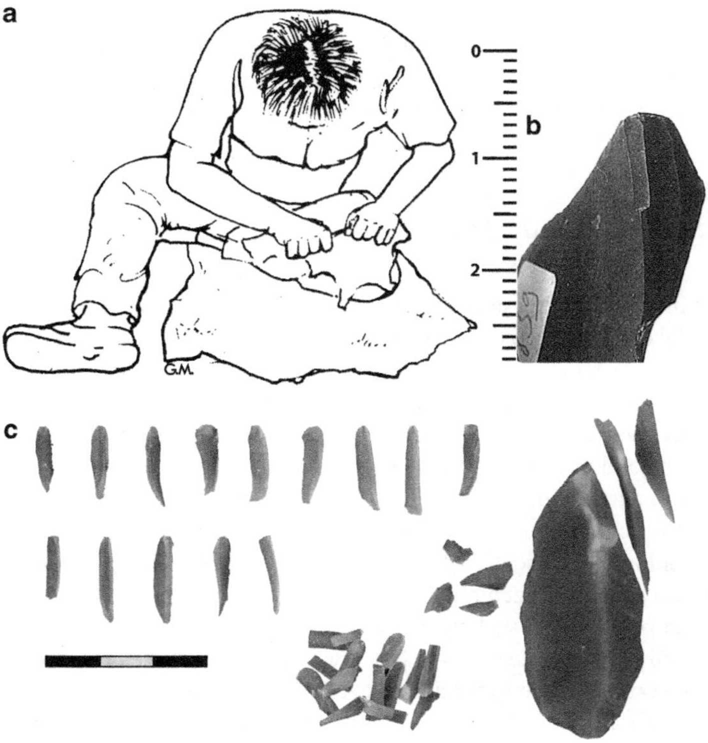

Fig. 18.1 Experimental model proposed for mode 1a: pressure microblade production using a hand-held antler baguette and holding the core directly in the other hand

18.3.1 Mode 1

By mode 1, we define the simplest way of applying pressure on a small core: using a hand-held pressure tool such as a 'baguette' or selected tine of antler (for a discussion about the quality of other materials such as bone, ivory, etc.; see Crabtree 1967). The core is directly held in the left hand (in this chapter, the right versus left indications apply to a right-handed worker), without any specific holding device except for a piece of leather to protect the palm and the fingers (Fig. 18.1). This seems to be the case for the detachment of tiny microblades from the evolved Aurignacian 'burins busqués', (Fig. 18.1 b) an archaeological 'burin busqué' from Laussel, Musée du Grand-Pressigny) which J.-G. Bordes has identified as microblade cores for 'lamelles Caminade' (Bordes and Lenoble 2002), and that I recognize, at least for some of them, as produced by pressure (Pelegrin and Bordes, in prep.; documentation in Michel 2010). These 'burins busqués' seem to be ordinarily rejected when reduced to 4.5 cm in length, which corresponds exactly to the minimum size for holding them (see Pataud level 7 in Bricker, 1995). The width of the microblades I could detach using mode 1 is about 5 mm, which is the same width attained by Callahan (1985) in similar conditions (with a maximum of up to 7–8 mm and 4 cm long).

In the Far East, some Horoka and Yubetsu cores (about 15,000 B.P., if not earlier) seem to provide the earliest evidence of microblades produced using this mode of pressure (Chap. 14 by Gomez-Coutouly, this volume).

18.3.2 Mode 1b

Still using a simple hand pressure tool, a first complementary tool can help to hold a flint core in the left hand. Indeed, if the core is not long enough – such as the so-called handle or keeled cores from Denmark – to be grasped transversally with several fingers, it seems impossible to squeeze the core with enough strength so as to stabilize it. For such slender cores that cannot be grasped firmly in the left hand, another device must therefore be used to help immobilize the core. As I have proposed elsewhere (Pelegrin 1988; see also Wilke, 1996), it can be a small, grooved piece of wood, bone or antler on which the core can be placed face down towards the palm and thus held more firmly (Fig. 18.2). The groove keeps the flaking surface free of contact and thus prevents the microblade from terminating in a hinge or step against the palm. A little piece of fur placed in the groove also reduces breakage of the microblades.

Using such a simple tool, microblades up to 8 mm wide can be produced from flint (Fig. 18.3). This device was effective for the reproduction of the tiny 'Rocher-de-la-Caille' microblade pressure cores from the middle Magdalenian in central France (Alix et al. 1995). E. Callahan (1985) proposed that having a composite clamp acting as a handle to hold elongated cores, and by using a short antler tine as a pressure flaker, he could detach flint microblades up to 3 cm long in a series with the longest reaching 4.5 cm long and 7–8 mm wide. The 'pocket' holding device from Tabarev (1997) reproduces the same principle as that from Callahan and does not seem to provide a better solution to the holding of slender cores.

18.3.3 Mode 2

A second mode can be proposed as an improvement of mode 1, replacing the hand pressure flaker with a shoulder crutch to produce a greater force (Fig. 18.4), as the strong muscles of the torso and shoulder will act to produce the pressure force instead of those of the right wrist (see Crabtree 1967, p 68). With such a tool, the width of flint microblades can reach 10 mm (Fig. 18.5). As Crabtree (1967) described it, such a shoulder crutch should be 30–40 cm long, adapted to the knapper's size and working position (sitting low or cross-legged on the ground).

When using such a tool with full force, a holding device for the core is certainly necessary for holding a flint core in the left hand. A larger, grooved piece was used here, but a hand-held clamp such as in Callahan's model is possible for large enough cores, or alternatively some other device may be developed.

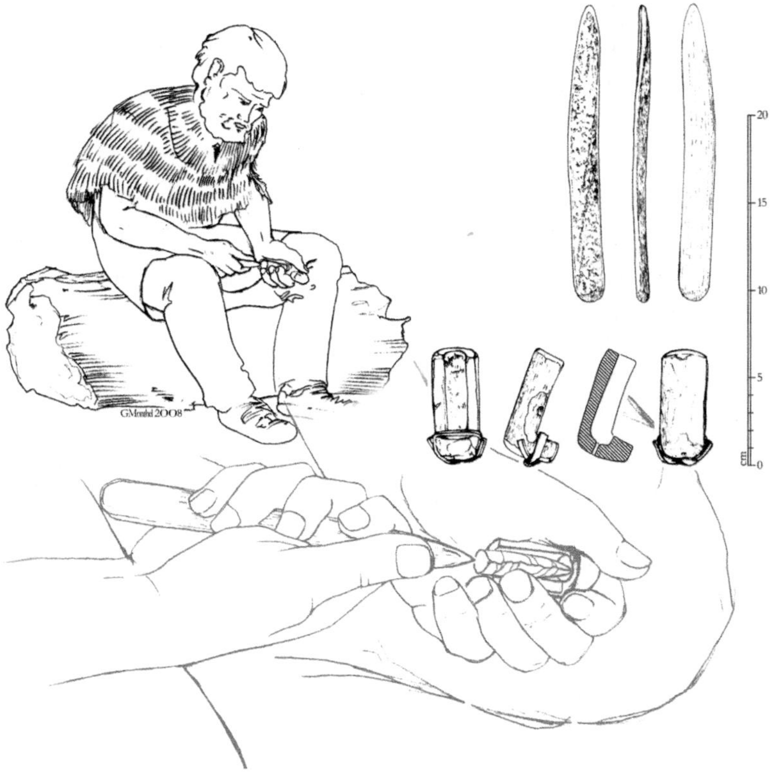

Fig. 18.2 Experimental model proposed for mode 1b: pressure microblade production using a hand-held baguette and holding the core with a grooved device

18.3.4 Mode 3

I postulate that the next step in microblade production consists of switching to the ground from the previous shoulder mode 2 (Fig. 18.6). It should be noted that the tools involved require little adaptation from those used in mode 2, and the low sitting position of the knapper can be maintained. The new key element lies in the fact that the core is placed on the ground so that the knapper can use a part of his or her own weight to build up the pressure force, with three positive consequences.

The first positive consequence is that the force is delivered in a straight axis that can be better controlled by the eye and exerted exactly in line with the blade to be removed, which is visualized while setting the core. This explains a 'jump' in the regularity of blades produced due to excellent control of the flaking direction, whereas many of the microblades detached by modes 1–2 are somewhat twisted or

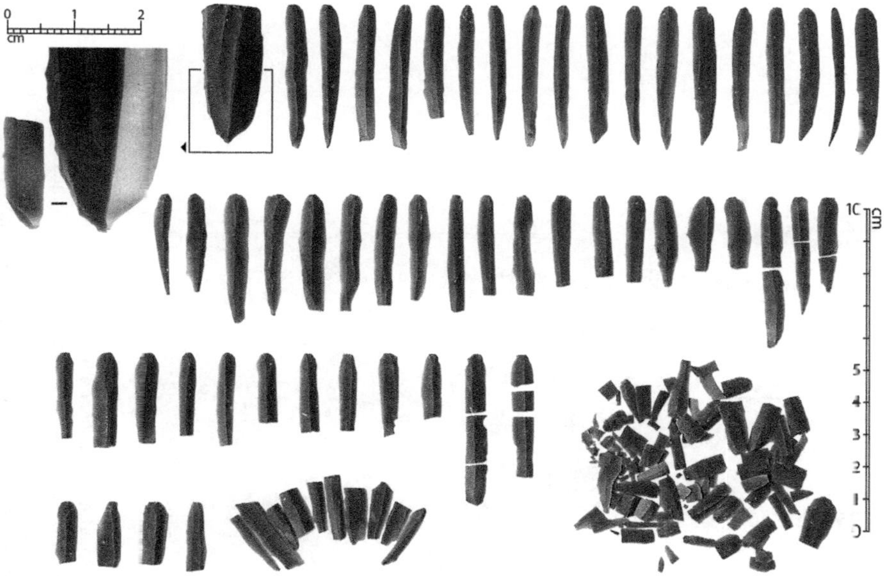

Fig. 18.3 A flint microcore (planar orthogonal platform created by a thick flake removal on a small nodule, shaped by axial removals + a dorsal crest) and its microblades pressure flaked using a hand-held baguette and core held in a grooved piece (mode 1b). The microblades are about 4 cm long and 5–8 mm wide. The detailed view of the core shows the curling distal end of the last microblade scar, an obvious stigmata of distal contact: the distal end of the core was pressed against the firm material of the grooved piece (hard wood, could be bone or antler as well) so that the fracture front started to plunge towards the too pointed distal tip of the core, but, feeling the compression, it 'curls up' 1 or 2 mm before the tip of the core. The distal fragment of the last microblade, broken during its detachment due to this distal contact, shows a complementary aspect of this 'curling' (described also by Titmus and Clark 2003, p. 97)

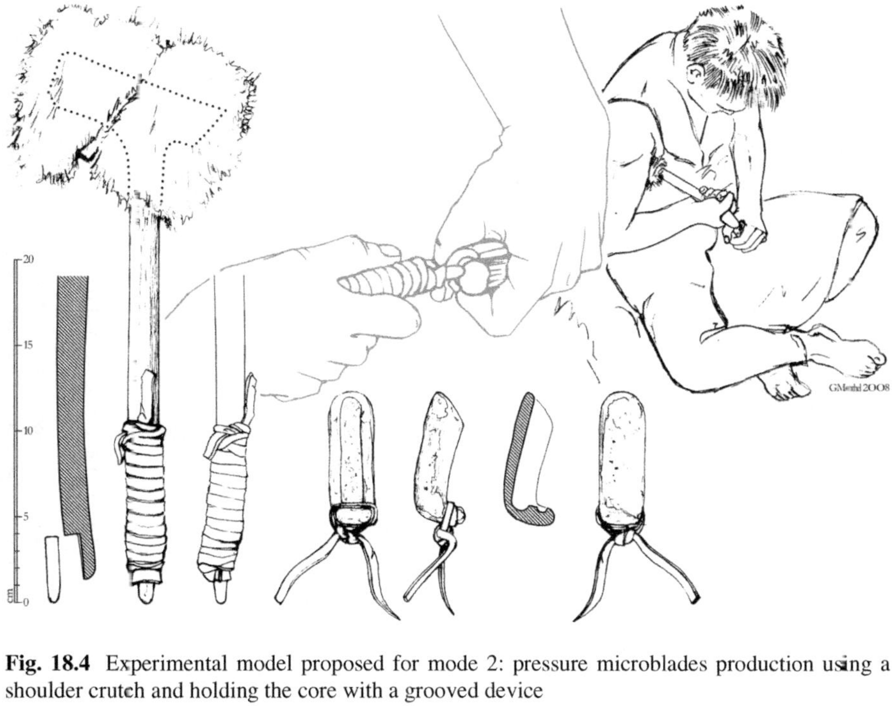

Fig. 18.4 Experimental model proposed for mode 2: pressure microblades production using a shoulder crutch and holding the core with a grooved device

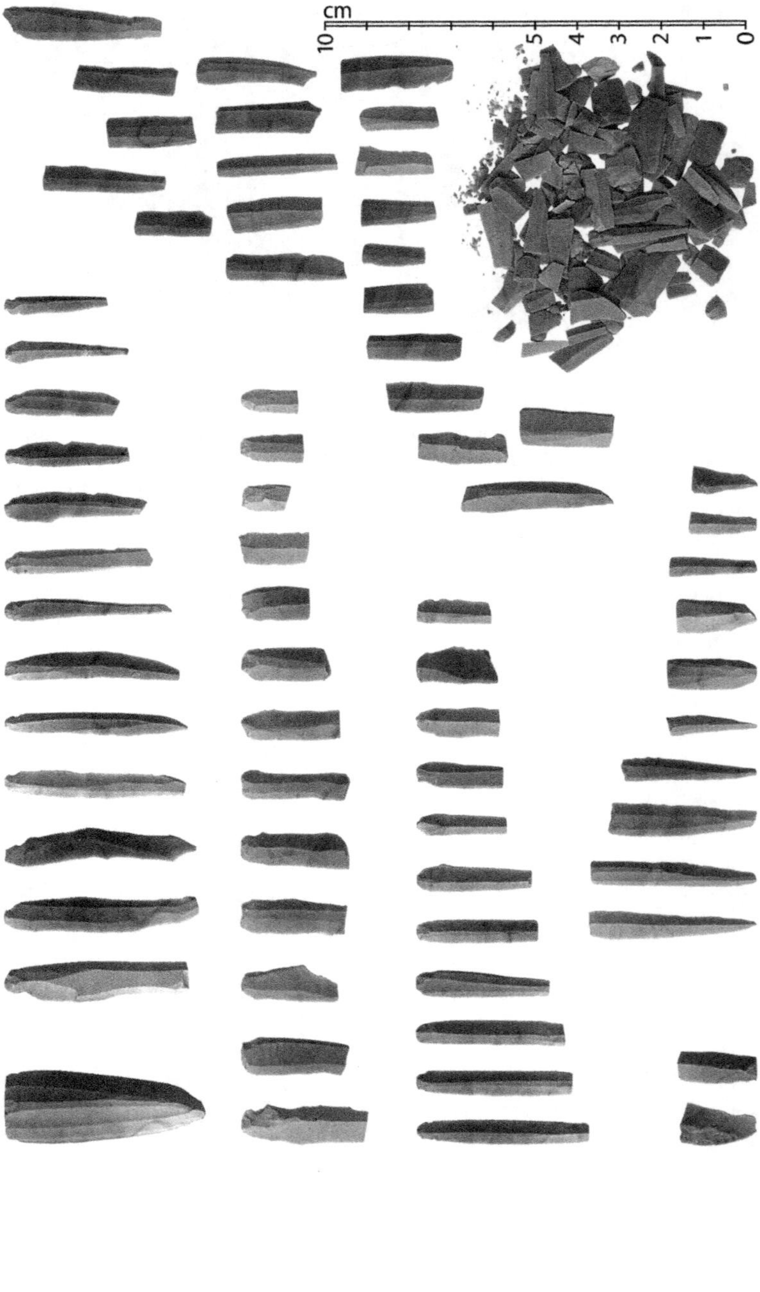

Fig. 18.5 A flint microcore (flat or planar orthogonal platform created by removing a thick flake on a small nodule, shaped by axial removals) and its microblades pressure flaked using a shoulder crutch and core held in a grooved piece (mode 2). The microblades can be up to 10 mm wide. The pointed tip of the nearly exhausted conical core (not far from a bullet shape) is discretely splintered (crushed), which is also a stigmata of distal contact. Most of the microblades reached the distal end of the core and therefore the material of the holding device, which explains the high rate of breakage. This could be improved

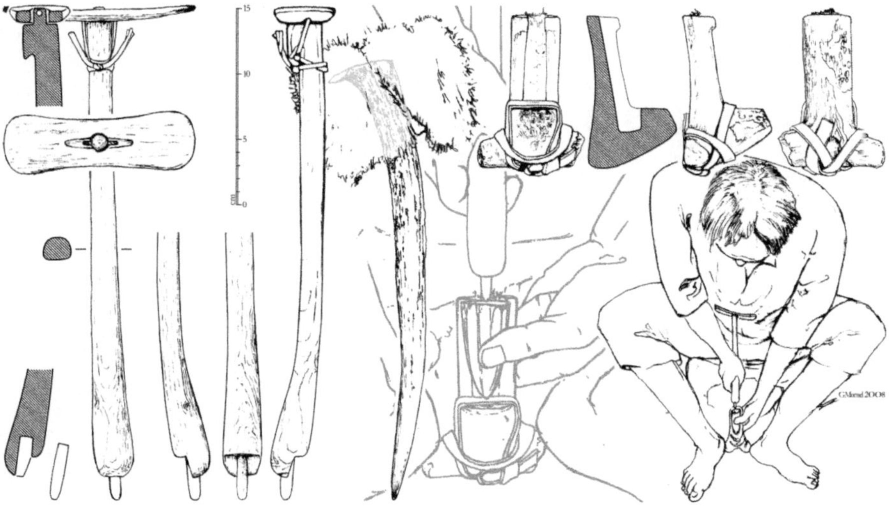

Fig. 18.6 Experimental model proposed for mode 3: pressure bladelets production using a short crutch in a sitting position, the core being held with a grooved device against the ground

skewed. The second positive consequence is the stability of the core, which, while using an adequate hold, does not move or tip at all. The third improvement is the significant ease of executing mode 3 compared to the more laborious modes 1 and 2. Once the position is properly adjusted, the knapper just needs to bend forward to push the short crutch placed at his belt straight to the front of the core platform and then initiate the fracture by an extra outward push with the hand(s) that hold the crutch. Indeed, some practice and coordination will permit the knapper to use the free left hand for maintaining the core in place during the setting of the pressure tool and the bladelet detachment.

Once again, a simple grooved device with a forked base kept in place by the heels offers good stability and practicality for different core shapes and sizes, and it can be adjusted by adding a small piece of wood or thick leather at the distal end of the core as it gets shorter. This device simply needs to be resting with the correct obliquity against a stone that is just high enough. In addition, one can easily imagine an improvement for sedentary knappers. One can select a superficial root, or stick a notched piece of wood in the ground in front of a small hole in which the core can be placed and adjusted with a set of small pieces of wood to secure its distal contact.

The regularity, ease and comfort of mode 3 accounts for the detachment of flint bladelets up to 12 mm wide and about 8 cm long (Fig. 18.7; in this case up to 11.5 mm wide and 7 cm long due to constraints imposed by the core). For extra force, if necessary, the knapper can lean forward on his heels to use more of his body weight. This can be considered as mode 3' and leads to the next mode.

Callahan's clamp might be adapted to mode 3 (in sitting position, the clamp being pressed on the ground with the left hand and/or a foot) but not to mode 4

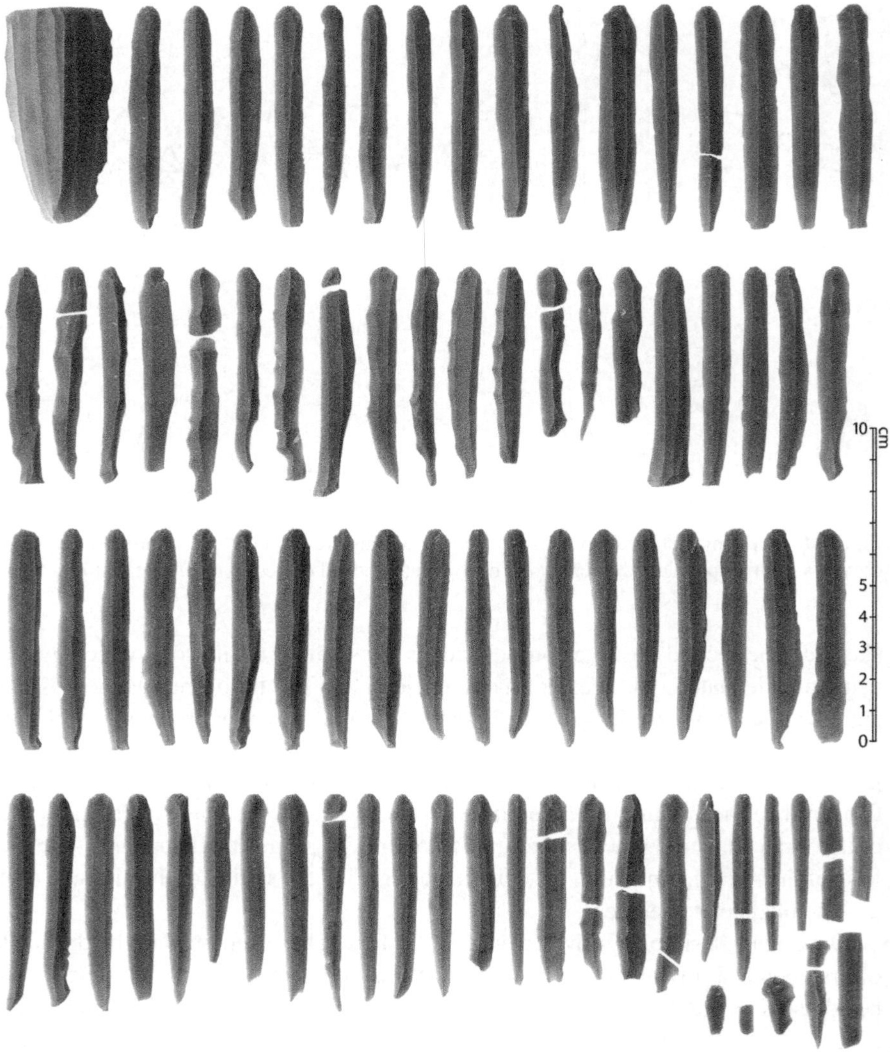

Fig. 18.7 A flint core (flat or planar orthogonal platform created by removing a thick flake on a small nodule, shaped by axial removals and two dorsal crests) and its bladelets pressure flaked using a short (abdominal) crutch in a sitting position, core held in a grooved device against the ground (mode 3). In this case, the distal contact of the removals was carefully avoided, explaining the low rate of breakage. The morphometric study of this experimental collection has been published by M. Gallet (1998)

(standing), as the left hand will not be able to stabilize the clamp. By any mode, the clamp – or any other device in which the core is squeezed vertically between its platform and distal end – is not well adapted to cores that finish with a bullet core shape, that is with a very small platform offering little grip to the clamp. Moreover, it becomes necessary to reposition the core quite often as its width or thickness is decreasing, which is impractical. Avoiding distal contact of the bladelet that is detached is possible if a notch has been carved in the lower jaw of the clamp, but the core will have to be repositioned in the clamp for each bladelet detachment that can be expected to reach the distal end of the core.

Conversely, it can be suggested that keeled core types which are always abandoned seemingly early (i.e. with a thickness greater than several cm) were held by their thickness and/or length, for example in a clamp (up to mode 3), in the hand (mode 1a or 2), in some hollow handle (for instance, I saw in Japan a few carinated 'burins' on long blades with the opposite end regularly shaped into a long ovate that can be suspected to have been forced into a hollow piece of bone or antler) or possibly pressed by the left hand and/or foot on the ground in a concavity or on some socket. To my knowledge however, there are no cores of such a shape that produced flint bladelets wider than 10–12 mm, if not 8 mm, which means that they were not reduced using modes greater than 3, and possibly just modes 1 and 2.

Consequently, I presume, but for new experiments that might prove otherwise, the flint microblade or bladelet cores that are exhausted and have a slender shape ('bullet core' conical or sub-conical, and those with an acute platform edge) were necessarily held in some kind of grooved or forked device, and not held in the hand or in some clamp device.

According to F. Brunet (Chap. 12, this volume), bullet cores appeared in Central Asia (Kazakhstan, Uzbekistan) possibly during the tenth or ninth millennium, and certainly during the eighth millennium cal B.C. In the Near East, they are common in the Pre-Pottery Neolithic and Early Neolithic from Anatolia (Cayönü circa 8500–8250 B.C.; Binder 2007) to Pakistan, from about the mid-ninth to mid-seventh millennium (most present in Iran; Wilke 1996; Chap. 2 by Inizan, this volume).

18.3.5 Mode 4

In the next step, mode 4 defines the potential use of the entire body weight as the knapper uses a longer crutch from a standing position (Fig. 18.8). The high force delivered requires an effective holding device for the core. After many attempts and adaptations with different clamps or 'squeezing vices' (Pelegrin 1984b), I am convinced that they constitute at best a mediocre, and possibly wrong, solution because any kind of binding used will still allow the core to move a little, judging in part from the grating sound that is produced (unless there is a metal screw somewhere that can be screwed very strongly). At any rate the holding device should certainly be rigid.

While trying to press off very long blades in flint using a forked device that bends somewhat during the push, it seemed very difficult to drive the blades all the way to the distal end of the core, and most of them came out rippled and broken. Switching to a rigid grooved device, with all other elements remaining the same, culminated in better results with complete blades running the full length of the core. It seems that

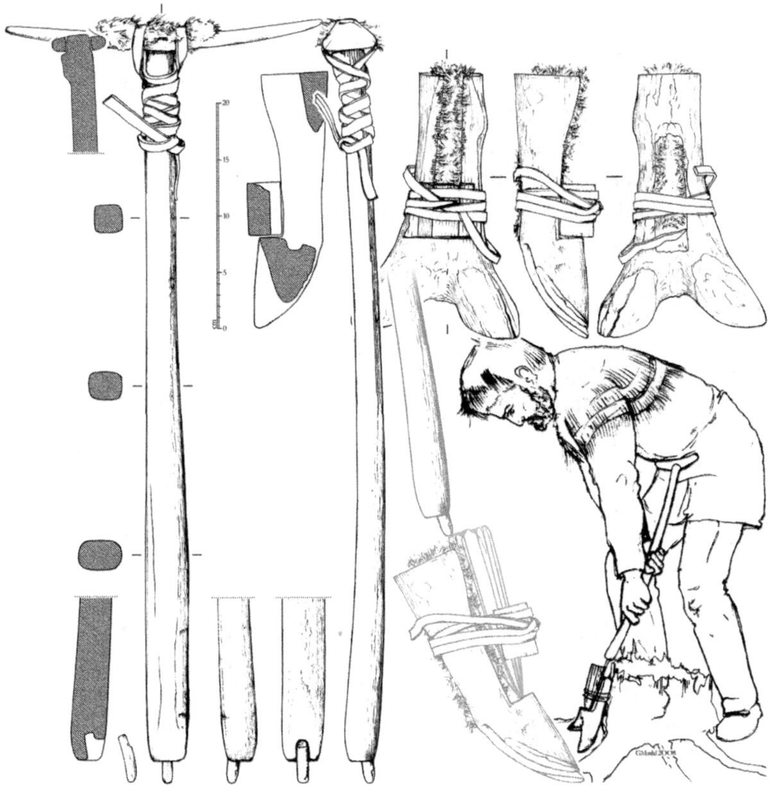

Fig. 18.8 Experimental model proposed for mode 4: pressure blade production using a long crutch in a standing position, the core being held with a grooved device against the ground

a non-rigid device absorbs, and thus reduces, the bending force of the crutch as the core appears to shake or vibrate. Curiously, this phenomenon occurs with long flint blades but not so much with shorter blades, nor obsidian blades, and it is possible that its manifestation depends on the duration of the detachment (if we assume that a 20 cm long blade takes twice the time to detach as a 10 cm long blade).

'Stuck' devices, like sticking a core into a hollow trunk or log, can be effective for some types of cores but they imply an adaptation of the core morphology (although the core decreases in size during knapping). They also need to be heavy, rendering them barely transportable and therefore inadequate for non-sedentary people.

Using the full weight of the body, flint blades up to 20–21 mm wide (for an antler tip, add 1 or 2 mm with a copper tip) can be detached with a well-adapted long crutch. The main shaft is made of boxwood or another strong wood, with a slight curve to produce a discrete but clear bending elastic effect, while the Mexican crutch used with obsidian should be more rigid (Titmus and Clark 2003). In my experience, more curved crutches are better adapted to the detachment of curved blades,

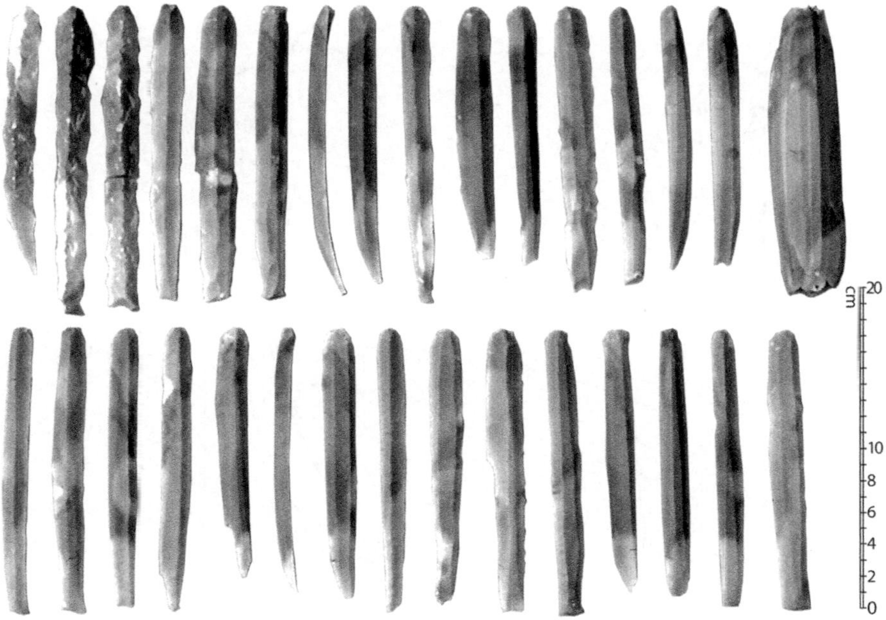

Fig. 18.9 A flint core (facetted platform, shaped by three axial crests) and 30 of its first 65 blades (the core was further reduced to a bullet shape 16 cm long and 17×22 mm wide, after producing 94 blades), using a long crutch in a standing position (mode 4). The widest blade (No. 10 from top left on this view) reaches 22 mm (a copper point was used for this test, reproducing the classic Harappan blade pressure production)

but producing curved blades was apparently never intentional. The length of pressure blades – up to 20 cm – depends mainly on the stability of the core holding device and on the quality of the pressure tool (the rigidity of the holding device allows for a full expression of the 'spring' of the bending crutch), assuming that the core allows for it in length, shape regularity and material homogeneity (Fig. 18.9).

18.3.6 *Mode 5*

Developing extra force by pressure – much more than the body force – implies the use of a lever that can multiply human strength by 10 or 15 times. Possibly used during the Paleolithic period to move mammoth carcasses or remove large stones out of a rock shelter, the lever principle was undoubtedly well understood by early Neolithic people who erected standing megaliths weighing several hundred kilograms in the Near East and who also constructed very large houses with strong poles in Europe. Although I could not measure it precisely, I estimate that about

Fig. 18.10 Experimental model proposed for mode 5: pressure blade production using a lever to act on a wood or antler pressure stick, the core being held in a single piece of wood

300 kg of pressure is needed to detach Neolithic or Chalcolithic pressure blades in flint that reach 3–4 cm in width and more than 30 cm long (the longest flint blade known comes from grave n°1 of the Varna Chalcolithic cemetery in Bulgaria; it is 43.4 cm long and 3.2 cm wide: Manolakakis 1996, 2005, 2006; Pelegrin 2006).

A strict immobilization of the core is required to adequately perform this technique, and preliminary attempts involved setting the core in the ground (with intermediary pieces allowing space for the blade to detach) and using a lever fixed through a socket in a tree or against a stone. This gave irregular results because the core tended to sink in the ground. I concluded that the whole device should be made from one single piece of wood, which could be a tree trunk about 20 cm wide, in which both the socket of the lever and the core would be fixed. Again, the groove principle can work here (with a double frontal support and a bottom rest), carved through the wood so that the detached blade 'flies' through it (Fig. 18.10).

In theory there is no dimensional limit for blades detached by lever pressure. While I was able to detach blades up to almost 6 cm wide and 40 cm long in flint, it is really the size and homogeneity of the raw material that determines the limit (Fig. 18.11). In the diagram presented in Fig. 18.12, we summarize the width of blades produced using these different pressure modes.

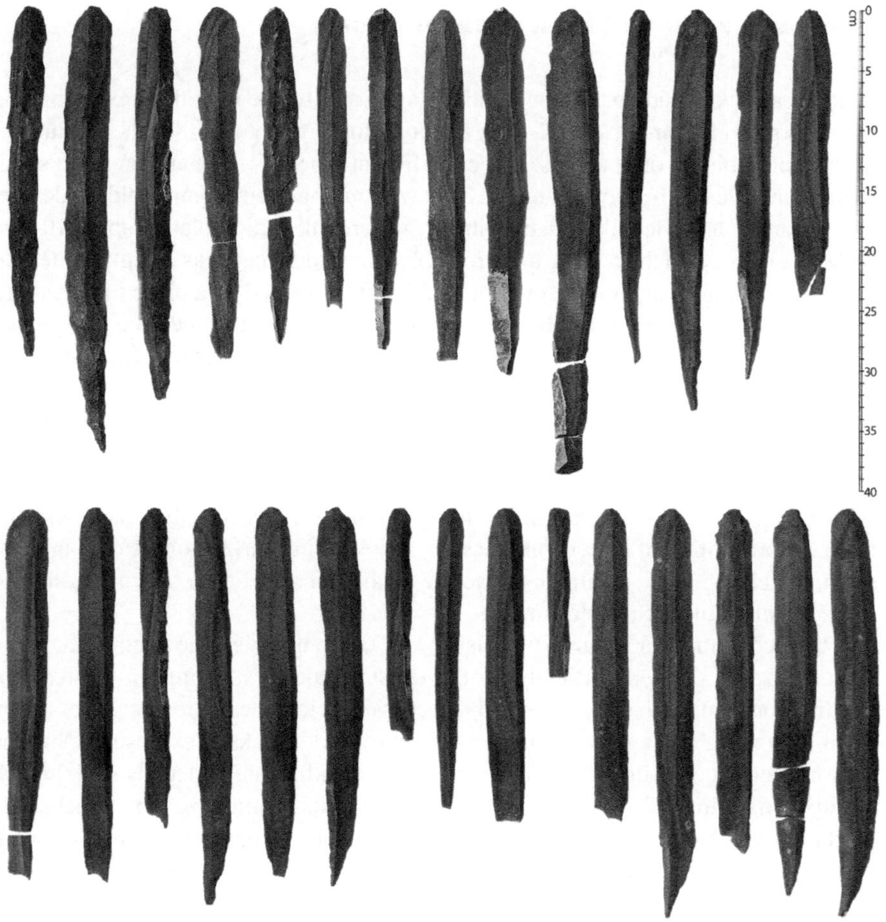

Fig. 18.11 A series of 29 blades detached by lever pressure (mode 5) from a core shaped by three crests (facetted platform, different modes of detachment preparation, use of a copper point)

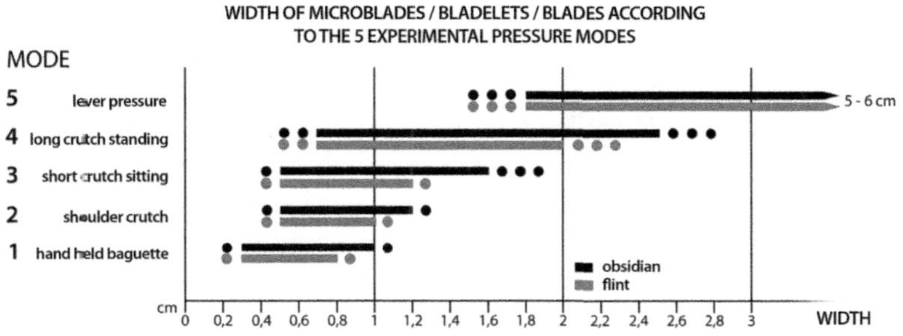

Fig. 18.12 Width range of pressure products for the five experimental modes for flint and obsidian (experimental data)

18.4 Obsidian and Heat-Treated Flint

With regards to obsidian, which is more fragile and brittle than flint, experimental attempts provide some results that are different to those obtained for the production of flint microblades or bladelets. Indeed, different types of obsidian cores of a small to medium size can be reduced by pressure without the need for any holding device. For instance, Jeff Flenniken demonstrated that regular microblades up to 10 mm wide could be detached from a narrow obsidian core that was simply squeezed between the fingers and the palm of the left hand (protected by a piece of soft skin), with the flaking surface of the core being turned face up and free of contact (Flenniken, 2003). This would be considered as a mode 1 pressure technique.

Ohnuma (1993) presented two different pressure techniques to detach obsidian microblades (and also some in siliceous shale from Yamagata in northern Honshu). The first is done using a short, hand-held tine and a vice made from a forked branch to hold a small core in the left palm. Ohnuma writes that the vice prevents the core from moving due to the flexibility of the palm, but we consider the real advantage to be the prevention of direct compression of the flaking surface of the core into the palm, which regularly results in a fracture of the microblade, or at worst a step or hinge termination of the detachment.

The second manner is performed using a 40 cm long antler tine controlled by the right hand and pressed vertically by the chest/shoulder (standing on his feet but bending forward) with the platform of a necessarily elongated core being pressed by the left hand on and against an organized set of rock blocks. As a result, Ohnuma demonstrated the production of microblades in obsidian and siliceous shale which are up to 6.5 mm wide and 4.5 cm long, but without mentioning which technique produced which results. Ohnuma's first technique can be considered consistent with mode 1b (his forked vice being equivalent to our grooved vice Pelegrin 1988). However his second technique, the archaeological plausibility of which requires more discussion (are such rocks available everywhere, at every site where such microblades were produced?), is difficult to classify.

The morphology of elongated cores allows them to be stabilized, with the knapper sitting in a low position (10–15 cm above the level of the core) while squeezing and pressing the core down onto the ground with the feet (with the help of the left hand if necessary), while the right hand controls the pressure stick pushed by the abdominal muscles (Fig. 18.13a). This is how I was able to reduce a large Yubetsu obsidian core into about 100 microblades up to 7 cm long and 15 mm wide (Fig. 18.13b, c), using a 35 cm long, slightly curved antler tine (Hokkaïdo red-deer) as a pressure tool placed at the belt. In this case, the technique resembles our flint mode 3, with the critical difference being that no complementary holding tool is needed. However, there should be a minimal size limit for the core to be immobilized in this way, a limit that remains to be further investigated. Fortunately, I was able to detect the stigmata of this 'between the feet' technique: the contact of one of the heels on the side of the detachment surface provokes a slackening of the fracture front so that the edge of the microblade, as well as the remaining ridge on the core, have overlapping lancets (Fig. 18.13d).

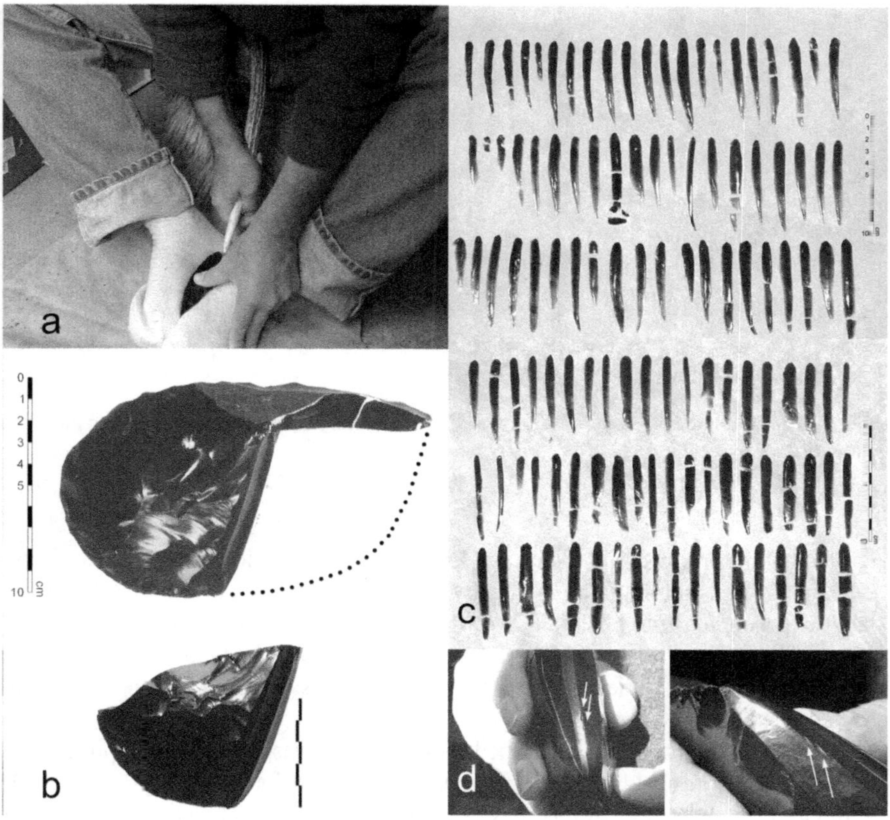

Fig. 18.13 Experimental model proposed for the detachment of pressure bladelets from an obsidian 'Yubetsu' core. (**a**) Holding the core between the feet with the help of the left hand and using an antler tine. (**b**) The experimental core with (*top*) 'crested blade' removed to create the platform and showing the volume reduced. (**c**) View of the bladelets detached. (**d**) Two views of overlapping 'lancets' produced by a contact between the bladelet detached and one of the ridges

Using mode 3 as described for flint (short crutch pressed with the abdomen in a sitting position, core set in a grooved piece facing the knapper), it is possible to detach obsidian bladelets up to 18 mm wide and 12–15 cm long. It is worth noting that in Kaletepe in Central Anatolia, the excavations by D. Binder and N. Balkan-Atli (2001) provided 'unipolar bladelet cores' with technical stigmata clearly indicative of the use of a similar device for holding the core to the one we developed for mode 3, that is evidence of distal contact and bilateral upper frontal contact for the core, with an 18 mm straight distance in between, which is exactly the maximum width that we experienced with obsidian in mode 3.

Mode 4 (a long crutch placed under the belt in a standing position using most of the body weight, with a core set in the ground) can produce obsidian blades up to 26–28 mm wide and 20, to even 30, cm long. That also seems to be the limit of both

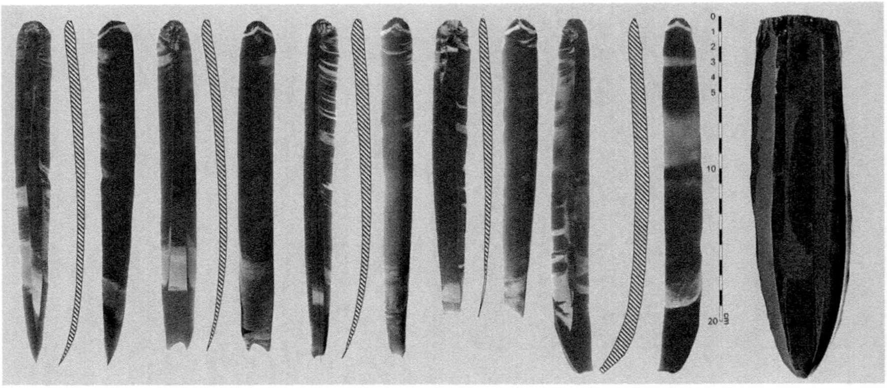

Fig. 18.14 Obsidian blades and core produced by experimental lever pressure (mode 5) using an antler tip

the technique proposed by Crabtree (1968) and the 'Mexica technique' rediscovered by J. Clark and put into practice by G. Titmus (Titmus and Clark 2003). These latter experiments with obsidian show that using the same pressure mode, obsidian allows for the detachment of blade(let)s about 30–40% wider than flint (not far from the P. Kelterborn experience of >50% with glass as compared to flint under controlled conditions, Kelterborn 2003).

In recent attempts, an antler tipped pressure stick was used in the application of lever pressure on obsidian, and as expected it proved its feasibility (Fig. 18.14). A mode 5 for the production of large obsidian pressure blades is presently recognized in Anatolia starting during the PPNB (Chap. 5 by Altinbilek et al., this volume), in the Neolithic and Chalcolithic of Armenia (Chap. 6 by Chabot and Pelegrin, this volume) and, I suspect, in the Early Classic Maya period (e.g. at the Kaminaljuyu site in Guatemala, Hirth 2003). A doctoral student from Japan (Oba Kobayashi) recently showed me some photos of a huge obsidian blade core collected in Northeast China that was obviously reduced in the form of very large blades (about 40 cm long and 4 cm wide) by lever pressure.

The heat treatment of flint 'pre-cores' prior to their bladelet production by pressure is presently demonstrated in different places around the Mediterranean Sea as early as the Middle, if not the Early, Neolithic period, and even earlier in some places in the Near and Middle East (Binder 1984; Pelegrin 1994; Inizan et al. 1975–76; Inizan and Tixier 2001; Chap. 7 by Binder et al., this volume). Some recent success in the heat-treatment of medium-sized flint cores (ongoing research with D. Binder and V. Léa) allowed me to test the increase in bladelet width using the pressure technique outlined for mode 3. Blades up to 16 mm wide were produced, as compared to the production of bladelets up to 12 mm wide using the same flint that was heat-treated. As measured by P. Kelterborn (2003) in laboratory conditions, heat-treatment seems to provide an increase of 20–30% to the width of flake products given the use of the same pressure mode.

18.5 Different Types of Platform Morphology and Preparation

Keeping different archaeological cases in mind, there are different types of core platform preparation used for pressure blade(let) production, which in turn lead to different types of blade platform (the portion of the platform that is detached with the blade and is still visible at the proximal end of it). They can be considered as more or less complex and thus can be presented with some order.

We start with those adapted to an organic pressure tool point, the most basic and widely used being the orthogonal, and thin blade platform. With this method, the pressure tool is placed just behind the edge of the core platform (a planar surface which forms an angle of about 90° with the detachment surface) after this edge was trimmed towards the flaking surface (reduction of the overhang, eventual lateral isolation +/− smoothing). The organic pressure tool point is offered a reasonable contact on a flat surface, spreading across the platform according to the force necessary for fracture initiation: from less than 1 mm in thickness for microblades, to 1–2 mm for mode 3 (Fig. 18.15) and 2–3 mm for mode 4 (Fig. 18.16) and even 5–7 mm for mode 5 (the respective thickness can be somewhat reduced on obsidian).

Depending on the minute preparation of the platform edge, the blade platform appears more or less elliptic, with a width larger than its thickness, but it is not really punctiform or linear. There is usually no visible crack on the blade platform because the organic point is too soft to create a circular crack, and because the pressure spreads on the whole of the blade platform. A discrete and regular lip can be seen and felt behind the platform because the fracture initiates from the tearing out of the outward component once the full compression of the inward component is produced (a pressure detachment movement involves a vertical equal inward component and a tear out equal outward component, according to Crabtree 1968).

Fig. 18.15 View of the flat platform of a flint core and refitted bladelets detached by sitting pressure technique using a short crutch (mode 3, antler tip)

Fig. 18.16 View of the flat platform of a flint core and refitted blades detached by the standing pressure technique with a long crutch (mode 4, antler tip). Note that the gap between the blade platforms, due to overhang removal and platform isolation, looks slightly exaggerated because of photographic distortion

This orthogonal thin modality seems to have existed everywhere that pressure blade(let) production occurred, both flint and obsidian. One should expect the blade platform size to remain relatively small, as the larger (wider and thicker) the blade platform, the more force necessary to initiate the fracture. As the pressure force is limited given the mode (except for mode 5), the knapper will avoid tearing out excessive blade platforms through unnecessary effort. An exception to this rule is given by the pressure detachment in Mesoamerica (Post-Classic, central Mexico, Titmus and Clark 2003, p. 91), of bladelets with a wide and thick platform (with unreduced overhang) from cores with a platform which has been pecked and ground, which clearly facilitates the fracture initiation using an organic point (hard wood, see Chap. 3 by Clark, this volume).

Another modality of platform preparation is that of faceting, where the platform of the core is not created as a flat surface but is corrected more or less frequently (for each blade or after a series of blades) so as to offer a small, flat surface or a little bump on the pressure tip. The platform of the blades can thus be somewhat thicker and quite variable, from a flat facet to a flat/convex/dihedral (non acute) facetted platform (Fig. 18.17). Such is the case, for example in the Upper Capsian (Tixier 1976, 1984) and in the Castelnovian (Chap. 7 by Binder et al., this volume).

A more specific platform morphology is that forming an acute edge angle with the flaking surface, as it has been already identified in the Recent Chassean (Binder 1984, 1991; Léa 2004a, b) and in Anatolia (Binder 2007 and Chap. 7 of this volume), seemingly adapted to a mode 3 (or 2?) pressure technique and using an antler point (practically, an antler point withstands the moderate force delivered without damage). In this case, the platform preparation for the next blade to be removed is very easy. From an acute edge, it can be done very precisely using a handheld pressure tool like in the Recent Chassean (Binder 1984, p. 83). Within this group can be mentioned the very recent diagnosis of the pressure production of small- and medium-sized pressure blades in Northern Finland (Rankama and Kankaanpää 2008).

Fig. 18.17 View of the facetted platform of a flint core and refitted blades detached by standing pressure technique (mode 4, antler tip). There is only one blade platform (indicated by the *arrow*) with a short crack indicative of a rather wide contact, 3–4 mm in diameter

A peculiar platform preparation observed in Poland and Ukraine for the detachment of lever pressure blades (mode 5) leads to thick convex facetted or dihedral blade platforms (Pelegrin in press). At first glance this could be evocative of the use of a copper pressure point, but the typical cracks created by a copper point are absent on the archaeological blades, and experiments demonstrate that a well-prepared deer antler pressure stick (tine or trimmed base) can withstand the pressure on such facetted or dihedral platforms. As another diagnostic feature, there is a tear out lip just behind the impact point or zone.

The use of a copper tip on the pressure tool certainly facilitates the initiation of a blade (let) fracture compared to an antler tip. That is because (1) copper is somewhat harder than antler which eases the fracture initiation and (2) copper can be shaped and used as a tiny end tip offering a very small impact point on the core platform or platform edge, while a similar antler or ivory point will split or crush under similar pressure (Inizan and Pelegrin 2002). A copper point also allows the detachment by pressure of minute faceting flakes on the core platform that procure additional facilitation of fracture initiation. The common consequence of these different elements explains how while using the same mode of pressure on flint, the use of a copper point instead of an antler point allows roughly for an extra 10% in blade(let) width (in Fig. 18.12, an extra point for the maximal width of mode 4 flint blades was recently added considering the experimental results of José Heredia, using a copper point on facetted platforms). Associated with the use of a metallic pressure tool point (more or less pure copper from native nuggets or some deliberate or non-deliberate alloy), one can describe different core platform preparation and related blade platform morphologies.

A pointed tip of copper (or bronze) is incisive enough to create a deep cone and thus facilitating fracture initiation. It is possible to detach a series of flint pressure

Fig. 18.18 View of the flat platform of a flint core and refitted blades detached by standing pressure technique (mode 4, copper tip). Most of the blade platforms show a clear circular crack indicative of a small contact 1.5–2 mm in diameter. Light brown semi-translucent flint. The dark traces on this and following pictures are those of copper

Fig. 18.19 View of the proximal end of five flint blades detached by lever pressure using a copper-tipped tool. The orthogonal core platform was prepared with reduced overhang removal creating rather thick blade platforms. Each of them presents a partial or complete circular crack indicative of a small contact, 2–3 mm wide. Grey, semi-translucent flint

blade(let)s using modes 1b–4 without any preparation from a flat orthogonal platform (Fig. 18.18) as seen in Pakistan near the Rohri (Pelegrin 1994, p. 592) dating possibly to the Indus period (Briois et al. 2005). The thick platforms of these blades, along with preserved overhangs, shows a clear circular crack of 1–2 mm in diameter which renders the diagnosis clear (indirect percussion using a copper-tipped punch can give a similar aspect with a larger diameter circular crack, but our experimental tests of the technique did not match the regularity of the blades produced by a pressure technique). Using mode 5 (lever pressure), the contact diameter reaches 2–3 mm

(Fig. 18.19). A very early mention of the probable use of a copper pressure tool point is given by D. Binder (2007, p. 237) in the Early PPNB from Cayönü in Anatolia.

The orthogonal rather thin abraded platform modality can also be used with a copper point, the pointed tip of which is placed just behind the edge of the platform after it has been abraded (overhang reduction) and smoothed to prevent edge crushing while the pressure is built up. The use of copper can be distinguished from that of antler because on most of the blade(lets) a punctiform initiation of the fracture will be visible at the back of the blade platform, the back line of which sometimes shows a half circle with a small diameter (about 2 mm). A similar observation has been made with obsidian (Pelegrin in Astruc et al. 2007).

Elliptical and small blade platforms detached from an acute edge core platform can be observed in different large blade productions (mode 5, possibly mode 4 for smaller blades) around the western Mediterranean (southeast of France with identified traces of copper, cf. Renault 1998, 2006; Renault et al. in prep.; Sardinia cf. Costa and Pelegrin 2004; Italy cf. Guilbeau 2010; Algeria cf. Pelegrin in press.). The platform of the core can be flat and generally inclined or it can be facetted, with each blade detachment being prepared by the removal of a small flake, the bulbar scar of which forms a localized acute edge angle. After lateral isolation and some smoothing by gentle grinding, the copper tip of the pressure tool is directly set on the acute edge with the resulting blade having round or ogival shoulders.

The fracture initiation usually starts right at the contact with the copper point, giving a very tiny blade platform (a few mm wide and about 1–2 mm thick for mode 4, 2–3 mm for mode 5). Using a somewhat translucent flint, a frontal light reveals diverse aspects of cracks that are a good indication of the use of a metallic material (Fig. 18.20). However, these cracks remain much less frequent (or invisible?) with opaque flint (Fig. 18.21). Note that the fracture initiation may start behind the actual contact, thus giving a somewhat larger blade platform with an obvious lip (Fig. 18.21: 5; an antler point would give a similar aspect), and that a few blade platforms can be discreetly split or splintered (Figs. 18.20: 3 and 18.21: 3). This is a crafty way of preparing the blade detachment, because such a tiny blade platform facilitates the fracture initiation and the bulb remains small or moderately prominent, which eases the preparation and detachment of further blades.

A very specific platform preparation seemingly adapted to a copper-tipped pressure tool is dihedral acute, as seen in South Iberia for flint cores of very different sizes corresponding to modes 1b–5 (Pelegrin 2006; Morgado et al. 2008). Two small flakes are detached (best done with a copper-tipped pressure or punch tool) on the platform so that their scars form an acute arris precisely in the axis of the blade to be detached. The pressure point is placed on this arris, a few millimeters back from the edge. The arris forms an angle of about 90° with the detachment surface. Experimentally, the copper point is hard enough to initiate the fracture most of the time at the very contact on the arris so that there is no lip just under the arris (an organic point would create a clear lip).

The same preparation, more or less systematic, can be observed in different 'Canaanean' productions (mode 5) in the Levant or Near East (Chabot 2002; Chap. 6 by Chabot and Pelegrin, this volume), and in Pakistan (Pelegrin 1994) where copper

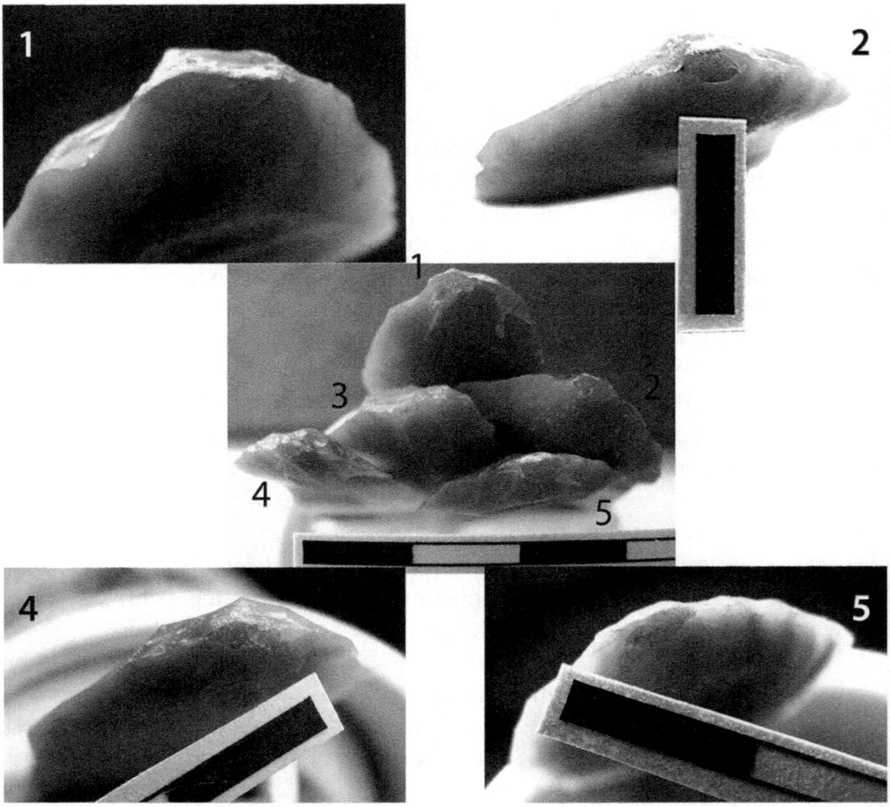

Fig. 18.20 View of the proximal end of five flint blades detached by lever pressure using a copper-tipped tool. The somewhat acute edge platform (75–85°) was prepared with repeated overhang reduction and lateral isolation. Each blade platform, except that of blade No. *3* which splintered during the detachment, presents a crack indicative of the reduced contact from a rather hard material. Blade No. *1*: three posterior cracks. Blade No. *2* and *4*: lateral cracks. Blade No. *5*: a complete front crack. Brown, semi-translucent flint

traces were identified on such a blade detached by mode 4 (Méry et al. 2007). A similar preparation, from convex faceting to dihedral, appears on obsidian (mode 3), with the recent Neolithic in Greece (Perlès 2004) probably associated with the use of a copper point. A series of experimental blades detached using mode 4 shows that about half of the blades present a crack on the platform (Figs. 18.22, 18.23).

Another specific preparation adapted to the use of a copper point on obsidian is done by isolating, using careful bilateral abrasion, a tiny beak (protruding 1 mm over 2 mm wide) at the edge of the core platform, and then smoothing it completely round by gentle grinding with a fine-grained sandstone. The copper point is then set precisely on this round 'nipple' and detaches a blade/bladelet with a minute, smooth blade platform (seen on pressure blades detached with mode 3 and 4 within

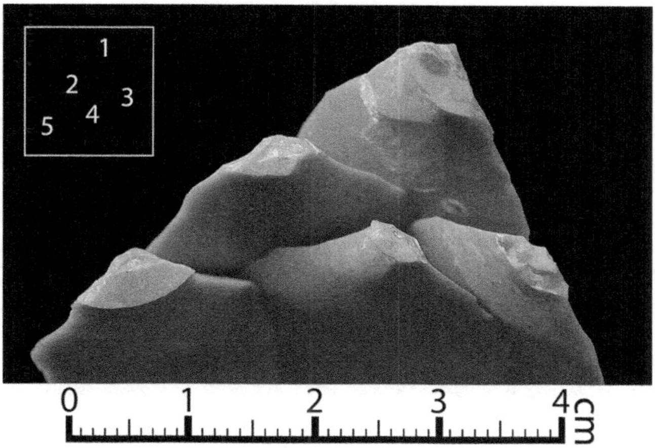

Fig. 18.21 View of the proximal end of five flint blades detached by lever pressure using a copper-tipped tool. The acute edge platform was prepared with overhang removal, isolation and smoothing. The first and last blades, with a large lip, show that the fracture initiation occurred well behind the contact area Blade No. *2* bears a transversal crack on its platform. The platform from blade No. *3* is splintered. Blade No. *4* has a very small platform (2 mm thick and 4 mm wide) with an irregular back-line, indicative of a hard tip. It is possible that a crack exists at the limit of the dark spot (copper trace of the contact) on blade platform of No. *1*, but it is not visible on this rather opaque and dry flint

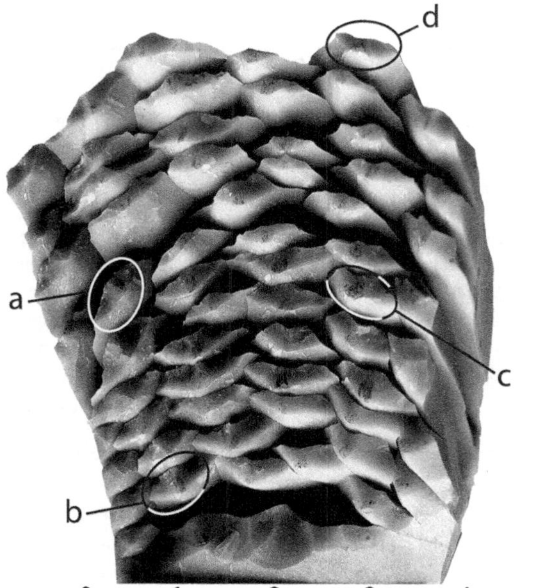

Fig. 18.22 View of the facetted platform of a flint core and refitted blades detached by standing pressure technique (mode 4) using a copper-tipped tool, following a convex facetted or dihedral preparation. In this barely translucent flint, about half of the blade platforms bear a crack, as can be seen on the photo: (**a**) back crack (located just behind the pressure point), (**b**) front crack (located in front of the pressure point), (**c**) lateral crack (occurred at one side of the pressure point, (**d**) '^' shape indicates bilateral crack, regarding their position to the pressure point

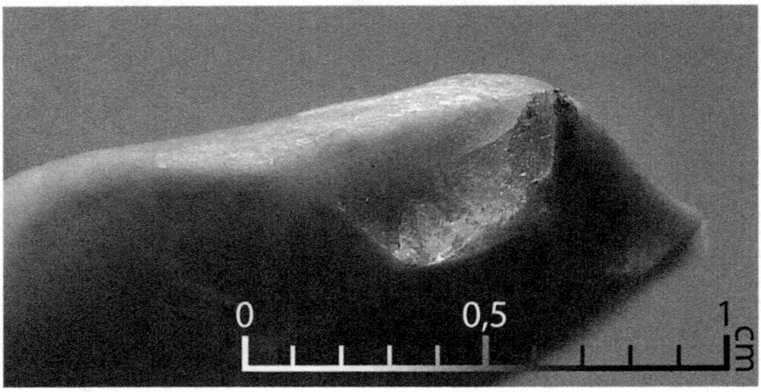

Fig. 18.23 View of a flint blade with a dihedral platform detached by mode 4, showing one of the possible forms of cracks due to the use of a copper point: '^' shape indicates bilateral crack starting at the back of the contact point. Semi-translucent flint

Table 18.1 Different platform preparations or aspects of pressure blades/bladelets/microblades related to different modes of pressure (see Fig. 18.24) on flint and obsidian using an organic or copper point

	Organic point		Copper point	
Platform prep/ aspect	Flint	Obsidian	Flint	Obsidian
Plain orthogonal thin +/– isolated	M 1–5 Europe M 1–3 or 4? Asia	M 1–3? Near-East M4-5 Europe MesoAm	M 4 and 5 Bulgaria Kar.V-VI	M 4 and 5 Anatolia, Armenia
Plain orthogonal thick	? (not adequate?)	M 1? to 5 MesoAm. on pecked platform	M 3 and 4 Pakistan	? (not adequate)
Facetted rather thick	M 3–4 Europe Castelnovian	?	M 5 Near-East Canaanean	? (not adequate)
Acute edge +/– rounded (ground)	M 4 + ? Finland M3 Anatolia, Chassean	?	M 5 West. Europe, Algeria	M 3–4 Neol. Anatolia
Convex facetted to dihedral acute	M 6 Ukrainia-Poland	? (not adequate?)	M 2–5 Andalousia M 5 Canaanean	M 3–4 Greece rec Neo-Chalc

different Anatolian archaeological contexts). A careful examination under light shows that most of the blade platforms bear stigmata of a very punctiform contact indicative of a hard material (an antler point would spread and produce a larger contact, and therefore a larger blade platform with a clear lip).

Table 18.1 summarizes some of the main archaeological cases that are referenced in the text above and in different chapters of this volume. It is obviously not exhaustive, but gives a general appraisal of the state of knowledge on the matter.

Fig. 18.24 Summary of the five experimental pressure modes (mode 1 with no holding device, mode 1b with a holding device)

18.6 Shaping of Pressure Cores: A Brief Overview

There are different methods of core shaping for pressure blade/bladelet production, the use of which is more or less dependant on the available raw material. Regarding the larger modes 4 and 5, there are only two ways to shape a core. In the first case, the raw material offers a large lenticular, or tabular, shape about 5–10 cm thick, with a rather thin cortex on both regular faces. Ideally, the shaping of the core can be limited to the creation of a platform and to the regularization of the flaking surface by 'opening' axial removals or by transversal, unifacial or bifacial flakes (cresting). In the second case, the raw material is nodular or irregular and a total, or nearly complete, core shaping must be completed. After an initial roughing out by hard hammer stone percussion, an elongated volume is built up using three or even four axial crests, giving the volume a triangular or quadrangular cross section. The blade production can thus start from one or two of these crests, by the detachment of crested blade(s).

Shaping out medium-sized cores (from 10 to15cm high) from nodules can be efficiently done by first creating the platform by removing a well-placed thick cortical flake, then using this platform for the detachment of axial shaping blades by indirect percussion, combined, if necessary, with transversal flakes (e.g. one partial temporary frontal crest and one or two back-crests). For smaller cores, a similar method can be applied to a short or long, but thick, flake or chunk, using the ventral face as a platform and shaped by axial flakes into a more or less elongated volume, for example the 'carinated scraper' type, similar to the Danish 'keeled-core' (Chap. 9 by Sørensen, this volume) or to the Japanese 'Horoka' method (Chap. 11 by Takakura, this volume). Another family of cores for small bladelets or microblades is the 'burin-like core'. In this case a platform is created by abrupt retouch and/or a burin spall at the corner of a flake or at the end of a blade and the bladelets/microblades will be detached on the side of the blank, the thickness of which is represented by the width of the flaking surface possibly further corrected by a notch or repeated abrupt retouch.

The well-known Yubetsu method (Inizan et al. 1999; Chap. 11 by Takakura, this volume) starts with the shaping of an asymmetrical biface from which a crested blade is detached, the negative scar of it serving as a platform for the detachment of microblades/bladelets. One can consider as simplified variants of this method the partial bifacial shaping of a large flake or thin slab, but treating the platform by transversal truncation should be considered as a significant variant (Chap. 14 by Gómez-Coutouly, this volume). Indeed, one has to consider the different factors of the raw material (dimensions, shape and quality) that can determine or influence the shaping method options observed within an archaeological assemblage prior to making any 'cultural' interpretation.

18.7 Different Sequences of Blade Production

After a careful examination of the 'diacritical schema' (order of the previous blade scars) on the cores and blades, D. Binder (1984; Binder and Gassin 1988) was the first to notice a clear difference between the Early and Recent Chassean. In the latter, the 2-1-2′ code (on the dorsal side of bladelets with three facets, the two lateral scars are chronologically later than the central scar) is clearly dominant, indicating that the reduction sequence was organized in a systematic order either by unidirectional (Binder and Gassin 1988) or by convergent, divergent or 'inserted' series (Pelegrin in Astruc et al. 2007). These observations, the code 2/1/2′ versus 1/2/3 or 3/2/1 from the blade(let)s and the eventual systematic order of the reduction sequence visible on the core(s), are indeed relevant to the characterization of a pressure production.

18.8 Development and Transmission of the Pressure Techniques

In the last part of this chapter, I discuss the archaeological development of pressure techniques (from the smallest to the largest), and examine the practical conditions of their diffusion or transmission. Let us first recall, as M.L. Inizan has already done (1991, 2002, Chap. 2 of this volume), the importance of pressure techniques in understanding cultural questions. It employs specific categories of knowledge. The first refers to its practicability: someone who would have never seen or heard that stone can be detached by pressure (but for edge regularization or steep retouch) has very little chance to discover it by himself. Another specific knowledge category relates to the specific tool(s) involved in pressure technique. Barring the detachment of tiny microblades from an elongated core simply grasped in the left hand (mode 1), a flint core has to be held with a specific tool, the minimal form of which being a grooved piece such as the one that I proposed in 1988, or a hollow shaft for the 'burin on blade' or 'carinated scraper' microblade core types.

From that base (mode 1b in flint) the particularity of the tools involved in performing pressure technique increases together with the size (mainly width) of the products. Mode 2 requires a shoulder crutch and a set of grooved pieces or another device for squeezing and holding the core. Mode 3 sitting requires a similar crutch to mode 2, but a larger grooved piece or a notched piece and a distal support set in the ground. Mode 4 standing necessitates a longer crutch and a holding device (without the help of the left hand) possibly similar to that of mode 3 but adapted to larger cores. Mode 5 – pressing with a lever – presupposes knowledge of the lever principle itself and that of rather sophisticated implements (Fig. 18.24).

Consequently, we can propose a few postulates about the development of pressure techniques. It is highly improbable that an inexperienced community could have invented from scratch an advanced mode of pressure like mode 3 or higher. It is, however, conceivable that an inventor of mode 1 can very quickly develop modes 1b and 2, so that the very first step (mode 1) may in fact be archaeologically invisible. It is highly improbable, however, that a group will directly invent mode 3, and even more improbable mode 4, not to mention lever pressure (mode 5).

In addition, modes 4 and 5 presuppose the mastery of other techniques to achieve the pre-shaping of the core, as well as access to raw material with adequate size and homogeneity. If a mode 3, 4 or 5 is recognized within a lithic industry, one should first look for the same or preceding mode within the same geographic or cultural space. If none of these are present, a source has to be detected within another cultural complex more or less adjacent geographically, taking into consideration the whole of the typo-technological characters of each that may account for their 'proximity' or distinction. Once a candidate is detected or suspected, the modalities of transmission or diffusion can be considered.

Other assumptions regarding the question of transmission or diffusion emerge from our experimental reproductions and experiences. To those assumptions, we can add anecdotal exchanges between modern flintknappers and experiences during practical teaching events. These observations help us to envision the practical conditions and circumstances that might be sufficient or necessary for the transmission of a given mode of pressure technique.

I can easily believe that a Paleolithic hunter and producer of his own microblades by percussion, who heard that it was possible to detach a microblade by pressing instead of striking on a core, would be tempted to try it due to curiosity and emulation, and that he could succeed in detaching a few microblades by pressure with mode 1 from a core already started by percussion. Considering the thinness and regularity of the microblade product that percussion cannot easily produce, our hunter may thus train in this new 'way of doing', and on the basis of this acquired knowledge, he would potentially master it rather quickly so as to be able to perform it efficiently and thus adopt it, all the while raising the interest of his fellow knappers.

The acquisition of this new way of doing would be even easier if that man could watch, be it only once, an expert in action, perhaps on the occasion of a meeting of our hunter with another initiated group or of a visit of the expert to the hunter's group. Through such a meeting, the transmission of the new technique does not even imply any description or explanation that would require a linguistic understanding, because the new 'way of doing' is essentially reducible to a knowledge that is visible and understandable in a few seconds (such as seeing someone using a bone needle).

The 'new way' can also be reported by miming it, and can thus be transmitted more quickly than by repeated meetings 'down the line' within related groups. It can even 'jump' over unrelated or scarcely related groups that would have little mutual linguistic understanding. In the context of groups and communities of hunter-gatherers that are already producers of microblades by percussion (assuming that the raw material(s) used would also be usable for pressure microblades), this elementary or basic mode 1 is susceptible to spreading quite quickly. Moreover, the greater productivity of the pressure technique mode would certainly stimulate the adoption of the 'new way' if suitable raw material is scarce or not available everywhere in the frequented territory.

In a context of ubiquitous raw material, permanent access to suitable raw materials allows for a more frequent and consumptive production by percussion, while in the first case of scarce or unevenly distributed raw material, the hunter-gatherer groups are used to managing raw material in a curated way, that is preparing cores in advance and transporting them for a later sequential reduction. In the opposite scenario, if the non-initiated groups are not microblade users, the transmission of the pressure mode of production can only accompany, and must rely upon, the adoption of those innovations that require microblades, such as the principle of fixing narrow lithic elements onto spearheads as a way to increase their haemorrhagic effect and the fabrication of adhesives. It would therefore mean much more to show and to explain, and might imply a mutual linguistic understanding.

The transmission of the modes 1b and 2 to inexperienced groups can be considered somewhat less easy because these modes require the fabrication of specific tool(s) (grooved piece +/− shoulder crutch), and some know-how for the use of them. Therefore a simple oral account has less chance of success in transmitting the technique. I presume that direct social contact, like meeting or visiting between experienced and non-initiated or inexperienced pressure knappers, would be necessary for the transmission of modes 1b and 2.

The transmission of modes 3, and above all 4 and 5, to a person ignorant of pressure technique would certainly require a demonstration including comments and explanations regarding the necessary morphology and regularity of the core, the platform preparation and setting of the tool including all critical details, the dynamic of the movement and the modalities of repair after accidents. All of these different conditions imply discrete knowledge and invisible know-how. The adoption of the 'new technique' by an inexperienced knapper would also imply, besides the access to adequate raw material, a respectable effort in the core shaping, the difficulty of which grows exponentially with the increasing dimensions of the blades being produced.

Conversely, the transmission of a mode X to a person who already practices the mode X-1 becomes relatively easy as, from one mode to the next, there are only one or two additional concepts to be mastered along with the improvement of some prior details. We have already assumed that it would be relatively 'easy' for a knapper to switch by himself from mode 1 to mode 2, that is, to replace his hand pressure tool by a shoulder crutch (possibly through an 'arm crutch' –or Ishi stick- step). Giving the knapper a short hands-on demonstration, or even simple indications, would certainly make it easier.

Switching from mode 2 to 3 is not that difficult either; the new concept is to steady the core on the ground, with the platform facing the knapper, through an easy adaptation of a larger but similar holding device (grooved piece or possibly another device). The very same shoulder crutch becomes useable by pressing it with the belly just under the belt, the knapper sitting low on some stone or piece of wood (a more refined adaptation of this is to dig out a little hole in the soil and/or use stones to stabilize the holding device). From mode 3 to 4, the adaptation consists in lengthening the crutch so as to employ it in a standing position, the holding device being again somewhat larger (a notch carefully carved on a superficial tree root can do as well, Pelegrin 2003).

So, from 1 to 2, 2 to 3 and 3 to 4, we assume that a brief demonstration, and possibly the narrative of an attentive observation, may well allow for a transmission of the innovation. That innovation should be of interest because each mode progressively gives the possibility of producing somewhat larger products with greater ease. However, the shaping of larger cores requires a qualitative improvement relating to the length of the expected products and the available raw material. From mode 4 to 5, which is from standing pressure to lever pressure, the innovations regarding the lever device are also rather understandable for an observer who is already an experienced knapper of pressure blades in a standing position.

If this knapper could attend a session of lever pressure blade production for a few hours, he or she would probably be able to reproduce blades on their own, especially if they were motivated (imagine what a master stroke in front of fellow knappers and for those who receive the products, staring with astonishment at such regular and huge blades). The report of an attentive observer, who could also mime the actions and draw some outlines on the ground, could easily motivate a knapper experienced in the standing pressure mode to try using a lever with success. I consider it highly improbable, however, that several inexperienced intermediaries might have transmitted, by word of mouth, sufficiently complete and precise information to be useful to a final receiver trying to replicate the pressure blade technique.

18.9 Conclusion: From Technical to Cultural Interest

We hope to have demonstrated in this chapter that pressure blade production involves much more than just one sole technique. Indeed, this peculiar mode of blade(let) production (mode in the sense of Newcomer 1975, p. 97) covers several 'techniques', or methods in the American sense: the particular mode of force application (each of them with their own range of products and dimensional limits), the way of holding or immobilizing the core, and two distinguishable types of material for the tip of the pressure tool (organic and metal). Furthermore, all of these techniques are visible on the archaeological material in the form of the platform preparation and detachment stigmata. The method of core shaping and the reduction sequence represent additional features.

Pressure blade production, from micro- to macroblades, constitutes a rich chapter of lithic technological evolution and opens up a wide field of cultural interpretation for archaeologists. In this chapter we chose not to consider the techno-economical and sociological aspects of pressure production, which are of great interest in Neolithic and later contexts as examples of specialized productions and exchange/diffusion mechanisms.

One of the more useful outcomes of this research is our postulate that the advanced 'modes' of pressure have very little chance (mode 3), or no chance at all (mode 4 and 5), to be invented de novo without the traditional knowledge base of less advanced modes and can hardly be transmitted by a casual contact from an experienced knapper to a knapper with no prior experience of pressure blade manufacture. As an example, consider the sudden appearance of pressure bladelets indicative of mode 3 in different regions of the western and southern Mediterranean basin during the seventh millennium (Chap. 4 by Rahmani and Lubell, this volume, Chap. 7 by Binder et al., this volume). This is indeed strongly evocative of the migration of some, possibly very few, experienced knappers of eastern Mediterranean origin.

The historical development of pressure techniques is not only interesting per se, regarding the amazing capacity of technical invention by prehistoric people, it is also relevant for the detection and interpretation of demographic events and social relations that made innovations spread over continents.

Acknowledgements I am grateful to all my colleagues from France and abroad who openly exchanged archaeological information with me. The photos of lithic pieces from Figs. 18.3, 18.5 and 18.7 are by V. Marcon (UMR 9993), Fig. 18.14 is by G. Monthel (UMR 7055), the rest are by the author. The pencil drawings from Figs. 18.1, 18.2, 18.4, 18.6, 18.8, 18.10 and 18.24 are the product of the talents of G. Monthel and are based on experimental documentation from J. Pelegrin. G. Monthel also prepared the layout of the figures.

I address a friendly thought to my friend Peter Kelterborn, who has shared many moments of these technical investigations over the past 30 years.

Last but not least, I want to express my sincere thanks to Pierre M. Desrosiers, who, after helping Noura Rahmani for the organization of the Lisbon meeting, did a remarkable work for the conception and scientific editing of this book. I am also grateful to Adrian Burke who corrected my approximate English and helped me, as Pierre did, to clarify some points.

References

Alix, Philippe, Jacques Pelegrin, and Huguette Deloge 1995 Un débitage original de lamelles par pression au Magdalénien du Rocher-de-la-Caille (Loire, France). *Paléo*, No.7: 187–99.

Astruc, Laurence, Bernard Gratuze, Jacques Pelegrin, and Peter Akkermans 2007 From production to use: a parcel of obsidian bladelets at Sabi Abyad II. In *Systèmes techniques et communautés du Néolithique précéramique. Technical systems and Near-Eastern PPN communities. (5ème colloque international sur les industries lithiques du Néolithique précéramique)*, edited by Laurence Astruc, Didier Binder and François Briois, pp 327–341. Editions APDCA, Antibes.

Binder, Didier 1984 Systèmes de débitage laminaire par pression: exemples chasséens provençaux. In *Préhistoire de la Pierre Taillée II, économie du débitage laminaire* (Table ronde de technologie lithique 3; Meudon-Bellevue octobre 1982), edited by Jacques Tixier, pp. 71–94. CREP, Paris.

Binder, Didier 1991 Facteurs de variabilité des outillages chasséens dans le sud-est de la France. In *Identité du Chasséen : actes du Colloque international de Nemours, 1989. (Mémoires du Musée de Préhistoire d'Ile-de-France 4),* edited by Alain Beeching, Didier Binder, Jean-Claude Blanchet and Claude Constantin, pp. 261–272. APRAIF, Nemours.

Binder, Didier 2007 PPN Pressure Technology: Views from Anatolia. In *Systèmes techniques et communautés du Néolithique précéramique. Technical systems and Near-Eastern PPN communities. (5ème colloque international sur les industries lithiques du Néolithique précéramique)*, edited by Laurence Astruc, Didier Binder and François Briois, pp. 235–243. Editions APDCA, Antibes.

Binder, Didier, and Bernard Gassin 1988 Le débitage laminaire chasséen après chauffe: technologie et traces d'utilisation. In *Industries lithiques, tracéologie et technologie*, edited by Sylvie Beyriès, pp. 93–125. British Archaeological Reports International Series, No. 411, Archeopress John & Erica Hedges Ltd, Oxford.

Binder, Didier, and Nur Balkan-Atli 2001 Obsidian exploitation and blade technology at Kömürcü-Kaletepe (Cappadocia, Turkey). In *Beyond tools: redefining the PPN lithic assemblages of the Levant, SENEPSE 9*, edited by Isabella Caneva, Christina Lemorini, Daniella Zampetti and Paolo Biagi, pp. 1–16. Ed. ex oriente, Berlin.

Bordes, François 1969 Reflexions on typology and techniques in the Palaeolithic. *Arctic Anthropology*, Vol. VI, 1: 1–29.

Bordes, Jean-Guillaume, and Arnaud Lenoble 2002 La « lamelle Caminade » : un nouvel outil lithique aurignacien ? *Bulletin de la Société Préhistorique Française,* 99(4): 735–749.

Bricker, Harvey M. (editor) 1995 *Le Paléolithique supérieur de l'abri Pataud (Dordogne): Les fouilles de H.-L. Movius jr. Suivi d'un inventaire analytique des sites aurignaciens et périgordiens de Dordogne*, (Document d'Archéologie Française, No. 50). Maison des Sciences de l'Homme, Paris.

Briois, François, Fabio Negrino, Jacques Pelegrin, and Elizabetha Starnini 2005 Flint exploitation and blade production during the Harappan period (Bronze Age): testing the evidence from the Rohri Hills flint mines (Sindh-Pakistan) throughout an experimental approach. In *Stone Age – Mining Age (actes du VIIIth International Flint Symposium Bochum 13–17 Sept. 1999). Der Anschnitt*, Beiheft 19: 185–191.

Callahan, Errett 1985 Experiments with Danish mesolithic microblade technology. *Journal of Danish Archaeology* 4: 23–39.

Chabot, Jacques 2002 *Tell'Atij, Tell Guded : Industrie lithique; Analyse technologique et fonctionnelle. Avec une annexe de Jacques Pelegrin (CNRS, France) sur la reconnaissance du débitage par pression au levier.* (Cahiers d'archéologie du CELAT, No. 13, Série archéométrie, No. 3). Université Laval, Québec.

Clark, John E. 1982 Manufacture of Mesoamerican Prismatic Blades: An Alternative Technique. *American Antiquity,* 47: 355–376.

Costa, Laurent, and Jacques Pelegrin 2004 Une production de grandes lames par pression à la fin du Néolithique dans le nord de la Sardaigne (Contraguda, Perfugas). *Bull. de la Société Préhistorique Française* 101(4): 867–873.

Crabtree, Donald E. 1967 Notes on experiment in flintknapping: 4. Tools used for making flaked stone artifacts. *Tebiwa* 10(1): 60–73.

Crabtree, Donald E. 1968 Meso-American polyhedral cores and prismatic blades. *American Antiquity* 33(4): 446–478.

Flenniken, Jeffrey J., and Kenneth G. Hirth 2003 Handheld prismatic blade manufacture in Mesoamerica. In *Experimentation and Interpretation in Mesoamerican Lithic Technology*, edited by Kenneth G. Hirth, pp. 98–107. The University of Utah Press, Salt Lake City.

Gallet, Michel 1998 *Pour une technologie des débitages laminaires préhistoriques.* (Dossier de documentation archéologique du CRA, No. 19). CNRS Editions, Paris.

Guilbeau, Denis 2010 *Les grandes lames et les lames par pression au levier du Néolithique et de l'Enéolithique en Italie.* Thèse de Doctorat de l'Université Paris Ouest, Ethnologie préhistorique, Vol. 3. MAE, Nanterre.

Hirth, Kenneth G. 2003 The Kaminaljuyu Production Sequence for Obsidian Prismatic Blades: Technological Characteristics and Research Questions. In *Experimentation and Interpretation in Mesoamerican Lithic Technology*, edited by Kenneth G. Hirth, pp. 170–181. The University of Utah Press, Salt Lake City.

Inizan, Marie-Louise 1991 Le débitage par pression: des choix culturels. In *25 ans d'études technologiques en Préhistoire. Bilan et perspectives. Actes des XIe Rencontres Internationales d'Archéologie et d'Histoire d'Antibes 18–20 Octobre 1990*, pp 367–377. Editions APDCA, Juan-les-Pins.

Inizan, Marie-Louise 2002 Tailler des roches par pression: émergence d'une technique et étapes de sa diffusion dans le monde. In *Matériaux, productions, circulation, du Néolithique à l'âge du Bronze*, edited by J. Guilaine, pp. 27–40. Errance, Paris.

Inizan, Marie-Louise, Hélène Roche, and Jacques Tixier 1975–76 Avantages d'un traitement thermique pour la taille des roches siliceuses. *Quaternaria*, XIX: 1–18.

Inizan, Marie-Louise, Michèle Reduron-Ballinger, Hélène Roche, and Jacques Tixier 1999 *Technology and Terminology of Knapped Stone, followed by a multilingual vocabulary Arabic, English, French, German, Greek, Italian, Portuguese, Spanish. (Préhistoire de la Pierre Taillée 5).* CREP, Nanterre (on line on the ARKEOTEK web site).

Inizan, Marie-Louise, and Jacques Tixier 2001 L'émergence des arts du feu: le traitement thermique des roches siliceuses. *Paléorient* 26(2): 23–36.

Inizan, Marie-Louise, and Jacques Pelegrin 2002 Débitage par pression et expérimentation : une question de méthodologie. *Paléorient* 28(2): 105–108.

Kelterborn, Peter 2003 Measurable flintknapping. In *Experimentation and Interpretation in Mesoamerican Lithic Technology*, edited by Kenneth G. Hirth, pp 120–131. The University of Utah Press, Salt Lake City.

Léa, Vanessa 2004a *Les industries lithiques du Chasséen en Languedoc oriental : caractérisation par l'analyse technologique.* British Archaeological Reports International Series, No. 1232, Archeopress John & Erica Hedges Ltd, Oxford.

Léa, Vanessa 2004b. Centre de production et diffusion des silex bédouliens au Chasséen. *Gallia-Préhistoire* 46: 231–250. (CNRS Editions, Paris).

Manolakakis, Laurence 1996 Production lithique et émergence de la hiérarchie sociale: l'industrie lithique de l'Enéolithique en Bulgarie (première moitié du IVème millénaire non cal.). *Bull. de la Société Préhistorique Française* 93(1): 119–123.

Manolakakis, Laurence 2005 *Les industries lithiques énéolithiques de Bulgarie*. Coll. Internationale Archäologie, Bd 88. Leidorf, Rahden/Westf.

Manolakakis, Laurence 2006 Les très longues lames de Varna, quelle fonction? In *La fin de l'Age de Pierre en Europe du Sud, actes de la table-ronde de l'EHESS Carcassonne 5–6 Septembre 2003*, edited by Jean Vaquer and François Briois, pp. 5–23. Editions des Archives d'Ecologie Préhistorique, Toulouse.

Méry, Sophie, Patricia Anderson, Marie-Louise Inizan, Monique Lechevallier, and Jacques Pelegrin 2007 A pottery workshop with flint tools on blades knapped with copper at Nausharo (Indus civilisation, ca. 2500 BC). *Journal of Archaeological Science* 34: 1098–1116.

Michel, Alexandre 2010 *L'Aurignacien récent (post-ancien) dans le sud-ouest de la France: variabilité des productions lithiques*. Thèse de Doctorat de l'Université de Bordeaux I, Préhistoire et Géologie du Quaternaire, 599 p.

Morgado, Antonio, Jacques Pelegrin, Gabriel Martinez-Fernandez, and José Afonso Marrero 2008 La production de grandes lames dans la Péninsule ibérique (IVe – IIIe mil. av.JC. cal.). In *Les industries lithiques taillées des 4e et 3e millénaires en Europe occidentale*, edited by Marie-Hélène Dias-Meirinho, Vanessa Léa, Karim Gernigon, Pierrick Fouéré, François Briois and Maxence Bailly, pp. 309–330. British Archaeological Reports International Series, No. 1884, Archeopress John & Erica Hedges Ltd, Oxford.

Newcomer, Mark 1975 "Punch technique" and Upper Paleolithic blades. In *Lithic Technology: making and using stone tools*, edited by Earl Swanson, pp. 97–102. Mouton, La Haye.

Ohnuma, Katsuhiko 1993 Experimental studies in the determination of manners of micro-blade detachment. *AL-RAFIDAN* XIV: 153–181 (The Institute for cultural studies of Ancient Iraq, Kokushikan University, Tokyo).

Pelegrin, Jacques 1984a Approche technologique expérimentale de la mise en forme de nucléus pour le débitage par pression. In *Préhistoire de la Pierre Taillée II, économie du débitage laminaire (Table ronde de technologie lithique 3; Meudon-Bellevue oct. 1982)* pp. 93–104. CREP, Paris.

Pelegrin, Jacques 1984b Systèmes expérimentaux d'immobilisation du nucléus pour le débitage par pression. In *Préhistoire de la Pierre Taillée II, économie du débitage* laminaire *(Table ronde de technologie lithique 3; Meudon-Bellevue oct. 1982)* pp. 105–116. CREP, Paris.

Pelegrin, Jacques 1984c. Débitage par pression sur silex : nouvelles expérimentations. In *Préhistoire de la Pierre Taillée II, économie du débitage laminaire (Table ronde de technologie lithique 3; Meudon-Bellevue oct. 1982)* pp. 117–127. CREP, Paris.

Pelegrin, Jacques 1988 Débitage expérimental par pression; "du plus petit au plus grand". In *Technologie préhistorique* (Notes et Monographies Techniques du CRA, No. 25), edited by Jacques Tixier, pp. 37–53. Editions du CNRS, Paris.

Pelegrin, Jacques 1994 Lithic technology in Harappan times. In *South Asian Archaeology 1993*, edited by Asko Parpolla and Petteri Koskikallio, Vol. 2, pp. 585–598. Annales Academiae Scientiarum Fennicae B 271, Helsinki.

Pelegrin, Jacques 2003 Blade making techniques from the Old World: insights and applications to Mesoamerican obsidian lithic technology. In *Experimentation and Interpretation in Mesoamerican Lithic Technology*, edited by Kenneth G. Hirth, pp. 55–71. The University of Utah Press, Salt Lake City.

Pelegrin, Jacques 2006 Long blade technology in the Old World: an experimental approach and some archaeological results. In *Skilled production and social reproduction – aspects on traditional stone-tool technology*, edited by Jan Appel and Kjel Knutsson, pp. 37–68. Societas Archeologica Upsaliensis, Upsalla.

Pelegrin, Jacques in press. Grandes lames de l'Europe néolithique et alentour. In *Sur quatre centres de production de grandes lames en Europe néolithique (Actes de la table ronde internationale de Tours, sept. 2007)*, edited by Jean-Claude Marquet and Christian Verjux, Supplément à la *Revue Archéologique du Centre*.

Perlès, Catherine 1984 Débitage laminaire de l'obsidienne dans le Néolithique de Franchthi (Grèce): techniques et place dans l'économie de l'industrie lithique. In *Préhistoire de la pierre taillée 2. Economie du débitage laminaire: technologie et expérimentation* (IIIe table ronde de technologie lithique, Meudon-Bellevue, Oct. 1982), edited by Jacques Tixier, pp. 129–137. CREP, Paris.

Perlès, Catherine 2001 *The early Neolithic in Greece; the first farming communities in Europe.* (collection Cambridge World Archaeology). Cambridge University Press, Cambridge.

Perlès, Catherine 2004 *Les industries lithiques taillées de Franchthi (Argolide, Grèce): tome III ; Du Néolithique ancien au Néolithique final.* (collection Excavations at Franchthi cave, Greece). Indiana University Press, Bloomington & Indianapolis.

Rankama, Tuija, and Jarmo Kankanpää 2008 Eastern arrivals in post-glacial Lapland: the Sujala site 10,000 cal BP. *Antiquity* 82(318): 884–899.

Renault, Stephane 1998 Economie de la matière première. L'exemple de la production, au Néolithique final en Provence, des grandes lames en silex zoné oligocène du bassin de Forcalquier (Alpes-de-Haute-Provence). In *Production et identité culturelle. Actualité de la recherche. Actes de la deuxième session des Rencontres méridionales de Préhistoire récente (Arles, 1996)*, edited by André D'Anna and Didier Binder, pp. 145–161. Editions APDCA, Antibes.

Renault, Stephane 2006 La production des grandes lames au Néolithique final en Provence: matériaux exploités, multiplicité des productions, aspects technologiques et chrono-culturels. In *La fin de l'Age de Pierre en Europe du Sud. actes de la table-ronde de l'EHESS Carcassonne 5–6 Septembre 2003*, edited by Jean Vaquer and François Briois, pp. 139–164. Editions des Archives d'Ecologie Préhistorique, Toulouse.

Renault, Stephane, Gérard Poupeau, and Stephan Dubernet in press. Reconnaissance de l'emploi d'un compresseur de cuivre pour la production de lames de la fin du Néolithique: le dépôt de lames de Sainte-Cécile-les-Vignes (Vaucluse).

Tabarev, Andrei V. 1997 Paleolithic wedge-shaped microcores and experiments with pocket devices. *Lithic Technology* 22(2): 139–149.

Titmus, Gene L., and John E. Clark 2003 Mexica Blade technique with wooden tools: recent experimental insights. In *Experimentation and Interpretation in Mesoamerican Lithic Technology*, edited by Kenneth G. Hirth, pp. 72–97. The University of Utah Press, Salt Lake City.

Tixier, Jacques 1976 L'industrie lithique capsienne de l'Aïn Dokkara Région de Tébessa, Algérie. *Lybica* XXIV: 21–53.

Tixier, Jacques 1984 Le débitage par pression. In *Préhistoire de la Pierre Taillée II, économie du débitage laminaire (Table ronde de technologie lithique 3; Meudon-Bellevue Oct. 1982),* pp. 57–70. CREP, Paris.

Wilke, Philip J. 1996 Bullet-Shaped Microblade Cores of the Near-Eastern Neolithic: Experimental Replicative Studies. In *Neolithic Chipped Stone Industries of the Fertile Crescent, and Their Contemporaries in Adjacent Regions, SENEPSE 9*, edited by Stephan K. Kozlowski and Hans Georg Gebel, pp. 289–310. ex oriente, Berlin.

Chapter 19
Measurable Flintknapping for Long Pressure Blades

Peter Kelterborn

19.1 Introduction

This chapter is a report on the ongoing developments in the lithic research method referred to as 'measurable flintknapping' and the design progress of my double lever detachment machine. These two subjects are explored by applying theoretical considerations in conjunction with concrete issues encountered by lithic analysts and replicative flintknappers. These issues are the prevention of longitudinal blade torsion, the unknown correlation between the blade length and width and the fracture propagation force, the growing importance of tuning and the correct follow-through[1] when the blade length increases. The latter point was made painfully clear to me by seemingly unexplainable breaks already occurring within the detachment phase during the earliest experiments with the larger double lever machine.

Tasks of this kind are, by their very nature, open ended and can best be presented as a condensed typical progress report. A particular feature of this chapter is the fact that all problems described and solutions provided are closely interrelated, and therefore, every new insight is relevant in more than one area. This did not make it easier to decide on the sequence of dealing with these subjects.

Furthermore, this essay reflects the considerable effort required in order to bridge the gap in the educational background, terminology and specific interests of the readers, which include archaeologists, lithic analysts and advanced flintknappers. Only the combined efforts of these three circles can lead to acceptable solutions to the problem of understanding highly complex stone artefacts. Wherever practicable, new terms or necessary definitions are integrated in the text, in the captions of the figures or in the footnotes.

[1] Follow-through refers to the skill of experienced flintknappers to keep their pressure tool in permanent contact with the platform of the emerging blade.

P. Kelterborn, dipl. Ing. ETHZ (✉)
Rainstrasse 372, 8706 Meilen, Switzerland
e-mail: pkelterborn@bluewin.ch

P.M. Desrosiers (ed.), *The Emergence of Pressure Blade Making: From Origin to Modern Experimentation*, DOI 10.1007/978-1-4614-2003-3_19,

19.2 The Extended Possibilities of Measurable Flintknapping

At the conference on Mesoamerican Lithic Technology held at the Pennsylvania State University in May 2000, measurable flintknapping was introduced to a small and specialized group of archaeologists and lithic analysts (Kelterborn 2002, 2003; Wilke 2007). Measurable flintknapping was characterized as a new approach to experimental percussion and pressure flaking research. An important part of the presentation dealt with feasibility studies for the first double lever pressure machine and the drop-weight percussion stick. I intended to prove that these two mechanical devices produce the same attributes as seen on both ancient artefacts and modern replicas and, therefore, can be put to use in lithic research with great confidence. What follows here will only briefly recapitulate the basics of measurable flintknapping, focusing instead on the extended range of possibilities of this approach and illustrating the improvements brought about by building a larger double lever detachment machine specially designed for the production of long pressure blades.

In order to ensure absolute repeatability and precision of the experiments, measurable flintknapping is inseparably linked to mechanical detachment machines, regardless of whether such devices were used in prehistoric times. It entails more than just quantifying the forces, as I had originally anticipated, but also includes measuring distances, angles, curvatures as well as stored energies and follow-through characteristics. Some of these parameters may be measured directly, while others can be interpolated with sufficient accuracy by using templates; energies and follow-through capacities have to be estimated based on the forces and masses involved and the deformations that occur under detachment loads.

The systematic use of fully engineered one- or two-component detachment machines and core fixation devices is another precondition for measurability, precision and working comfort. The tools should be ergonomic in design, require no excessive muscle power and allow a good view on the crucial observation and measurement areas as well as easy tuning. These features further increase the quality of the work because many experiments involve arduous repetitions and demanding measurement routines.

When using the traditional body tools from pressure blade experimentation,[2] little comfort and poor visibility had to be accepted. While there was never any doubt that the craftsmen perfectly knew what they were looking for and, based on their vast experience and great skills, would know instinctively when their detachment preparations were right, they were never able to see or measure exactly what they were doing.

For core shaping, maintenance and repair tasks, the introduction of lapidary equipment is efficient and saves both materials and time. However, this is only permissible under the strictest condition that it does not interfere with the actual purpose of the research project at hand. Depending on the geographical location of the

[2] Such as the widely known Crabtree crutch, the Pelegrin stick (Pelegrin 2003) and the Clark stick (Titmus and Clark 2003). The Clark stick is also called an itzcolotli.

researcher, the saving and cleverly exploiting the increasingly scarce good quality raw material is highly desirable and even necessary in the long run. The most frequently used machines are diamond saws and diamond grinding wheels.

Compared to the simple infrastructure used for my first measurable flintknapping experiments in the mid-1990s at the Lejre Research Station in Denmark, the workshops and support areas today consist of a diverse array of resources. These include a study area with the relevant literature and reports, good equipment for obsidian and glass photography with weather independent lighting, computing and printing equipment, as well as a large selection of different measuring instruments and templates, some of which I developed myself. Other equipment includes power tools for fabricating wood and metal devices plus a wet room with the various lapidary machines of the craft. Last but not least, a well-equipped ordinary flintknapping corner must be available to execute the different core shaping and core repair tasks and the diverse techniques of platform overhang removal and rim bevelling, to mention but the most frequently carried out tasks in manual flintknapping.

In order to handle the much increased volume of data, the short notes traditionally kept by replicative flintknappers had to be replaced by formalized laboratory journals and detailed records of operation procedures (as is customary in most industrial experimental and measurement activities). Additionally, this considerably lowers the individual rate of administrative working errors.

Finally, it is only fair to mention that this kind of a broadened scope of measurable flintknapping is not cheap and also takes time.

19.3 Development of More Versatile Double Lever Systems

Double lever systems, which can correctly reproduce the actions involved[3] in pressure blade production, need not necessarily consist of straight horizontal and vertical levers, as shown in the first publications (Kelterborn 2002, 2003) and in the top sketch in Fig. 19.1. A great number of variations have since been tested with the larger machine. These include the use of previously placed downward and outward preloads, changes in the stiffness and in the masses of the lever arms as well as modifications of the tool angle and platform tilt, as shown in the lower sketches of Fig. 19.1. Also shown (schematically) are the specific downward and outward force diagrams and a number of suitable core shapes. In the general context of measurable flintknapping, the designation of the 'downward' and 'outward' force vectors (D and O in Fig. 19.1) must be understood in relation to the momentary longitudinal core axis and not in relation to the earth's gravity.

[3] At first glance, these movements look like a single flowing gesture. However, closer inspection reveals that a downward component must first be applied to ensure a firm grip of the pressure bit on the platform. Only afterwards can more downward and outward force be added, which, together with an adjustment of the tool angle, initiates the detachment. I generally use round or rectangular copper bits, but rounded antler tips and tropical hard wood were also utilized.

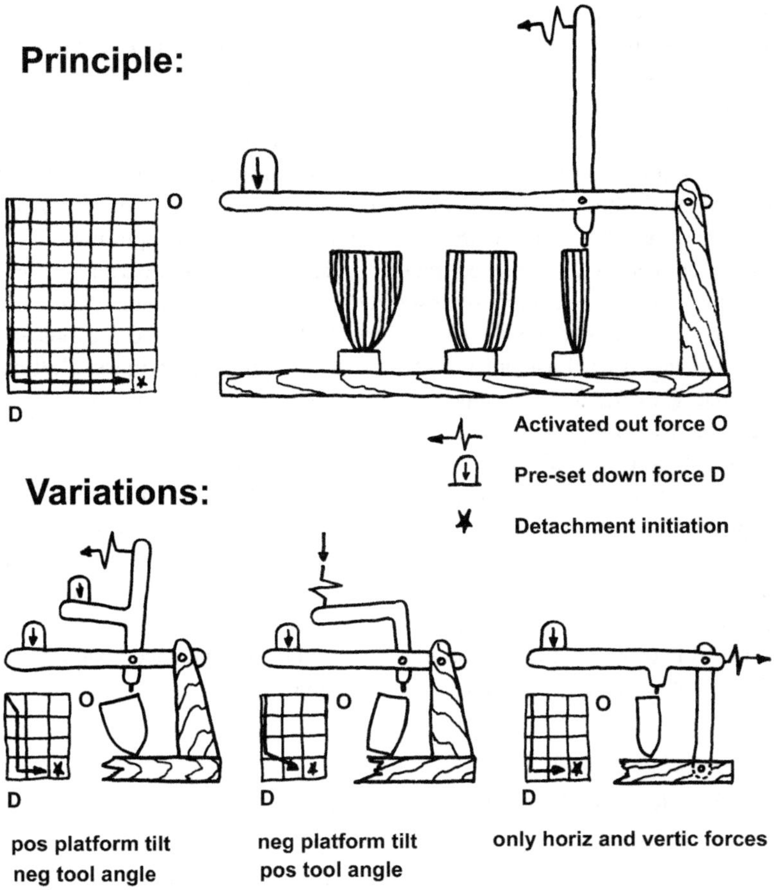

Fig. 19.1 Schematic sketches of the double lever pressure machine. *Above*: The basic principle. *Below*: Proven variations. Counterclockwise angles are defined as positive

The original small version and the current larger version of the double lever machine are based on the same principle: The approximately level lever arm rotates on an axis in a rigid frame that is firmly attached to a strong base plate. The approximately upright lever arm rotates on an axis within the horizontal lever arm, so that the pressure bit at the end can be moved upwards, downwards, forwards and backwards. In practice, the downward force must be determined in advance, while the outward force must be activated gradually and measured with a spring scale or an electronic force sensor. At the time of detachment initiation, the final downward and outward forces must be calculated from the maximum reading of the measuring instrument and the predetermined weights on both levers. The detached blade then falls after a few centimetres onto a soft piece of foam or fur. In my view of a structural engineer, such mechanical two-component machines, or 2C-machines, are transparent and predictable. They can be equipped with a great variety of instruments and

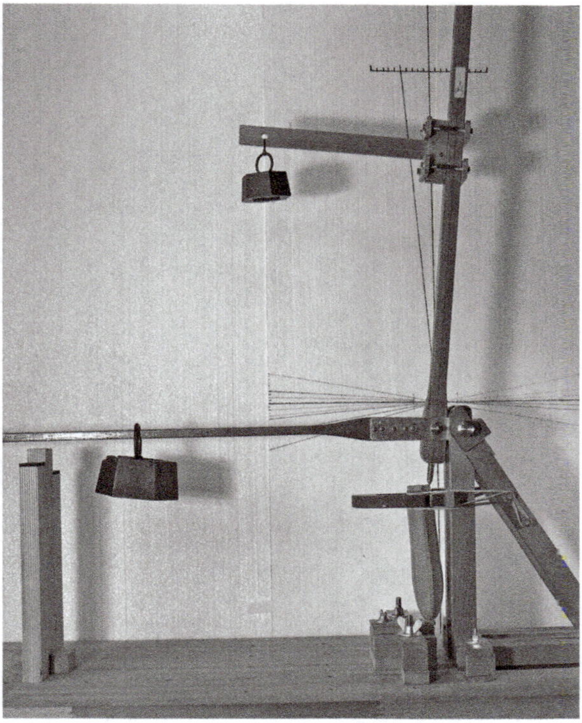

Fig. 19.2 Measurability, repeatability and comfort of the double lever pressure machine. The negative tool angle and positive platform tilt are measured by using the template in the background. Note the difference in stiffness of the exchangeable lever arms, the additional set-up weight fixed onto the upstanding lever arm and the pear shape of the 17 cm core (Photo by P. Kelterborn)

carry out exactly the same action as often as required without the fatigue that traditional flintknappers experience. This means that measurability, repeatability and comfort of experimentation are now made possible to a degree never known before, Fig. 19.2.

In the context of measurable flintknapping, the short term 'set-up' refers to the selection of all variable parameters of the detachment preparations before fracture initiation. This includes, for example, the platform tilt and the inclination of both lever arms, the stiffness of both lever arms and the weights fixed on both lever arms, the overall core orientation in three dimensions, the placement of the pressure point on the core platform, the platform overhang removal and rim bevelling. To ensure that nothing is forgotten, it is best to handle the set-up process in the style of a formal ritual or to follow an established checklist.

It is highly significant that many of the choices mentioned above do not leave behind any traces on the cores or blades. It is one of the strongest points of the method of measurable flintknapping that it specifically allows us to register and analyse parameters that were hitherto neglected or undiscovered. To illustrate this, the technique of applying slight dorsal finger pressure is presented in Fig. 19.3. For glancing percussion blows with hard stone hammers on handheld blanks, skilled and experienced modern flintknappers have been aware of this technique for a very long time. Another example frequently encountered is the detachment of elongated parallel pressure flakes on bifacial American projectile points. The method used

Fig. 19.3 Dorsal pressure as an example of set-up parameters which leave no trace in the archaeological record. *Left*: Light pressure with rubber bands. The white line on the core platform enables the core orientation in three dimensions with the help of templates behind the base plate. To reduce the fracture initiation force, the pressure point area is roughened by abrasion. *Right*: Finger pressure is applied in three locations to reduce the probability of breaks (Photos by P. Kelterborn)

involves holding and supporting the point in one hand on a soft pad of fur or on a folded piece of very fine leather while using the other hand to press off the flake from the core with an indenter. With 1C- and 2C-machines, this technique can now also be employed to make long pressure blades. Finger pressure can also be applied as an effective means of assistance when the existing tuning is not perfect or when small corrections of the core shape have to be made.

Of particular importance in order to understand the tuning process is the effect of the variable stiffness of the exchangeable lever arms while keeping all other set-up parameters the same as before. This has far-reaching consequences with regard to the two parameters stored energy in the system and the follow-through capacity of the pressure bit. The stored energy can be estimated by the work[4] that is performed by the applied detachment forces when they bend or compress the levers and the deformations of the core and holding devices just before detachment is initiated. By exercising the correct follow-through 'by feeling', traditional flintknappers have always maintained constant contact between pressure bit and blade platform until the blade has completely separated from the core.

[4] Work is proportional to the average force multiplied by the distance covered.

Geometrically this means that the pressure bit must cause an extremely quick, although small, downward and outward displacement of the blade platform in the right amount. The usable sources of this instant follow-through are, however, only those various stored energies that can be released instantly during the short duration of the fracture time. These energies stem primarily from the deformations due to the compression and bending of the levers, the muscle tensions of the flintknapper directly behind his body loading tool and an eventual upward follow-through from the elastic fixation below the core. With basic physical laws in mind only, the characteristics and possible consequences of changing the design and stiffness of the lever arms can be described as follows:

- Compression as such stores little energy, which however is available very quickly, similar to a hard spring.
- Bending stores much more energy which is not available as quickly, similar to a soft spring.
- Thick elastic lever arms store less energy from bending, but the energy is available more quickly than from thin lever arms under the same bending load.
- Thin elastic lever arms store more energy from bending, and the energy is not available as quickly as from thick lever arms under the same bending load.
- Stored energies that must first accelerate considerable masses before reaching the pressure bit, as is the case in compressed core fixation devices under weighty cores or heavy lever arms, might come too late to contribute much to the follow-through required immediately. Therefore, the stored energy below the core, although very interesting, is not expressly referred to in this chapter.

Large single lever devices were used successfully, and the findings became published in several articles by Jacques Pelegrin in the late 1980s (see comprehensive bibliography in Pelegrin 2003). When no outward force is applied, the 2C-machine works like a single lever arm device, provided that the second lever is light, not activated and free from preloads. However, by using the 2C option, the particularities of single lever devices can be measured conveniently and precisely and analysed in the usual manner. A fundamental difference between these two lever systems becomes immediately apparent: While 2C-machines and the body powered sticks designed by Crabtree (Pelegrin 2003; Titmus and Clark 2003) can reproduce all the different downward and outward sequences and combinations of forces and respective follow-throughs which can start a crack, 1C-machines can only produce the single critical force and inherent follow-through characteristics of the (straight or curved, thick or thin) pressure tool and tool angle involved.

19.4 The Controlling of Torsion

While quarrying for medium to long cores in ancient times, an understanding of longitudinal blade torsion or blade rotation was essential in order to efficiently and quickly shape, by percussion, the rough blocks into usable core preforms. In order to further improve the shape, torsion control was also required during the subsequent detachments of first and second series blades by percussion or pressure. The final step in producing functional blades for the end users was to keep torsion below an acceptable maximum during third series pressure blading. The amount of torsion that exists in each piece cannot be calculated with a mathematical formula but must be measured individually. With regard to the precision of such torsion measurements, it is permissible to use very simple devices and templates. In my personal view, a rotation up to approximately 15° in the length interval between the bulbar and distal zone of a core or blade is acceptable.

However, not only my own blades, but to my surprise, a considerable number of the cores and blades in my database showed a noticeably higher value than 15°, indicating that torsion was not easy to control even in prehistoric times. The data on which these observations were based did not just consist of a single excavation assemblage, but of a wide selection of 25 ancient and modern pressure cores and blades less than 12 cm in length, 10 ancient and modern pressure cores and blades of 12–18 cm in length and 25 ancient and modern pressure cores and blades in the 18–24-cm range. The archaeological records consist of notes and photos I took in the Field Museum in Chicago, the National Museum of Anthropology in Mexico DF and of collections at the Pennsylvania State University. The modern material is predominantly made up of my own work and gifts from Gene Titmus and Jacques Pelegrin.

Just like in the prehistoric times, every modern researcher must expect to encounter the obstacle of longitudinal torsion very early in his work, when entering himself into the field of producing long pressure blades. In technical terms, torsion is the result of an oblique force on the blade platform, combined with a skew bending moment at the proximal or platform side. Analysing blade torsion from an engineering theory point of view would involve the application of higher mathematics. For the benefit of flintknapping practitioners and lithic analysts, a less time consuming and more convenient solution had to be found, even if this meant sacrificing precision to a certain extent. My database was again scrutinized in much greater detail. This second study of the dorsal morphology revealed five clear tendencies. When viewed from above, as seen in Fig. 19.4, these are:

- Blades rotate away from a lateral ridge loss, see Fig. 19.4a.
- Blades rotate towards a lateral ridge pickup, see Fig. 19.4b.
- Blades with a laterally curved axis rotate towards the inside of the curve, see Fig. 19.4c.
- Blades rotate towards distally widening flanges, see Fig. 19.4d.
- Blades rotate away from distally narrowing flanges, see Fig. 19.4e.

An additional benefit of this analysis is the insight that good torsion experiments require cores that range in the length between 12 and 18 cm.

In the hope of discovering an applicable set of rules to prevent or reduce torsion from occurring in practice, a large series of 120 new blades were produced and a

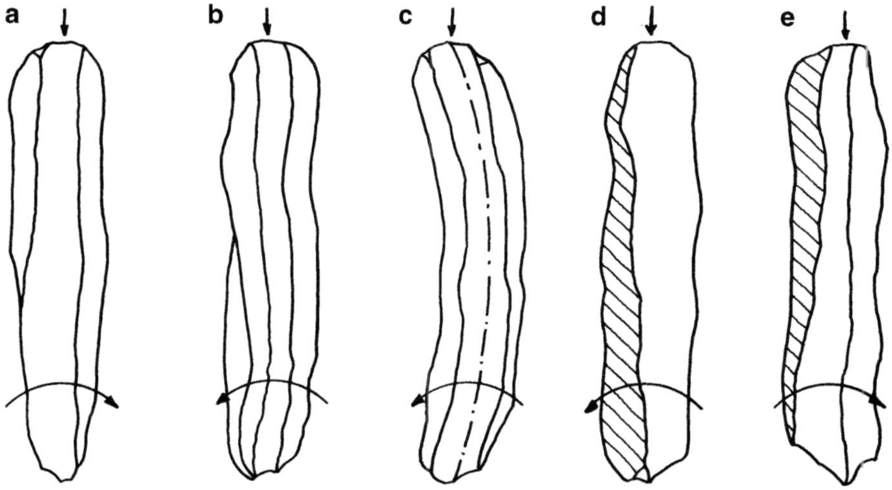

Fig. 19.4 Torsion indicators observed on the dorsal side of blades. (**a**) Lateral ridge loss on the *left*. (**b**) Lateral ridge pick-up on the *left*. (**c**) Curved axis of blade. (**d**) Distally widening flange on the *left*. (**e**) Distally narrowing flange on the *left*

number of completely new terms introduced.[5] These new terms are reference line, ridge loss, ridge pickup, online, offline, parallel and deviated, see Fig. 19.5. Using these expressions, the following principles regarding the torsion behaviour of prismatic blades can be formulated. These principles are NOT laws of physics, but statistically well-confirmed rules of experience:

- Overall, the fracture plane is governed by the direction of the outward force in relation to the geometry of the core cross section and the direction of the downward force in relation to the core ridges concerned. Because in practice the core cross section is not constant from the proximal beginning to the distal tip, the optimal interpretation requires some experience-based intuitive averaging.
- In principle, the proximal fracture plane is perpendicular to the direction of the outward force.
- For triangular blades[6] and trapezoidal blades with small upper widths (w smaller than 1/3 W), the distal fracture plane is perpendicular to the bisector of the distal core cross section at approximately 70% of the core length.

[5] For the readers who wish to find out more: The reason for going into such depth is the obligation to take into account the *similarity rule*, as it was postulated, justified and formulated in Kelterborn (2002: 45) and Kelterborn (2003: 129). For lack of space, the similarity rule is given here only in its most concentrated form: 'Qualitative and quantitative guidelines in flintknapping are only valid within the border conditions of similar morphologies and attributes shared by the groups of cores or detachments compared'.

[6] In this chapter, blades with triangular or trapezoidal cross section are simply referred to as triangular or trapezoidal blades.

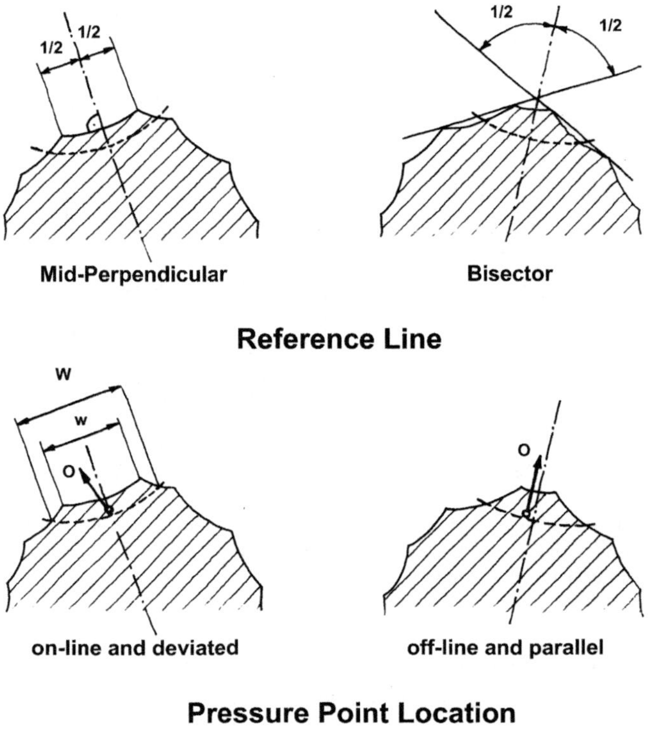

Fig. 19.5 Definition of the terms used in controlling torsion with the concept of the reference line. W represents the width of the intended (future) trapezoidal blade, w is its upper-width, which is equal to the distance between the two adjacent ridges of the core. The pressure point can be exactly on or off the reference line. The outward force vector O can be parallel to the reference line or deviate from it by a certain angle

- For trapezoidal blades with wide upper widths (w larger than 1/3 W), the distal fracture plane is at a right angle to the mid-perpendicular of w of the distal core cross section at approximately 70% of the core length.

These four conclusions were the starting point for what became the 'reference line concept'. The first step is to visualize the outline of the intended blade on the core. The second step is to classify this intended blade as a triangular blade (or a trapezoidal blade where w is smaller 1/3 W) or a trapezoidal blade (w is larger than 1/3 W). For triangular blades, the determining reference line is the bisector, while for wide trapezoidal blades, the determining reference line is the mid-perpendicular. Because cores are usually not exactly prismatic, the third step is to decide at which distance from the core platform the reference line shall be taken as governing or determining. In the fourth step, the pressure point must be placed exactly on or off the reference line and the force vector O can be selected parallel to or deviate with an angle from that line.

Based on these definitions, the five practical rules of thumb for a torsion-reduced or torsion-free detachment set-up are as follows:

1. The optimal location for the choice of the applicable reference line is the core cross section taken between the lower third and fourth of the core length.
2. For intended future triangular or trapezoidal blades with small upper widths, the applicable reference line for the outward force is the bisector.
3. For intended trapezoidal blades with wide upper widths, the applicable reference line for the outward force is the mid-perpendicular of the upper width w.
4. On the core platform, the pressure point of the pressure bit must be placed exactly above the proximal ridge of the intended triangular blade or exactly between the ridges of the intended trapezoidal blade.
5. After following rule 4, the downward force for an intended triangular blade should then point at the ridge in the distal zone of the core, or between the two ridges in the distal zone of the intended trapezoidal blade. In other words, the downward force should be directed in such a way that ridge pickups or ridge losses are avoided.

By following these five rules, it has been possible to practically eliminate, or dramatically reduce, torsion in my output of pressure blades.

Readers should note that when the intended blade is strongly asymmetrical, for example when the centre of gravity of the expected blade cross section is not located on one of the reference lines, one must choose between 'online and offline' and between 'parallel and deviating', as shown in the bottom sketches of Fig. 19.5. This delicate decision requires much experience. In general, the outward force vector must be moved closer to the centre of gravity of the asymmetrical future core cross section. If heavy core torsion remains from the preforming stage or first and second series blades, this must first be removed by detaching additional short triangular blades from the most protruding ridges with traditional flintknapping methods.

19.5 The Variable Fracture Propagation Force

In 2005, I switched my blading experiments from small prismatic cores below 10 cm in length to much longer ones in the 17-cm range. At the same time, the original small 2C-machine with its rigid and thick wooden lever arms was replaced by a similar, but much larger, machine with interchangeable carbon fibre–reinforced wooden lever arms (see Fig. 19.2). Although apparently nothing else had been altered, 16 of the first 27 blades (or 59%) inexplicably broke. It came as a surprise that this high breakage rate rapidly decreased when the stiffness and follow-through characteristics of the two lever arms were changed, while the amounts of downward and outward forces remained the same.

Once I was more accustomed to using the larger machine, a more acceptable breakage rate of less than 10% was gradually reached on a trial and error basis. A new surprise was to recognize that the last 32 breaks of this warm-up phase had occurred in only three typical locations: 4 proximal breaks had occurred in the

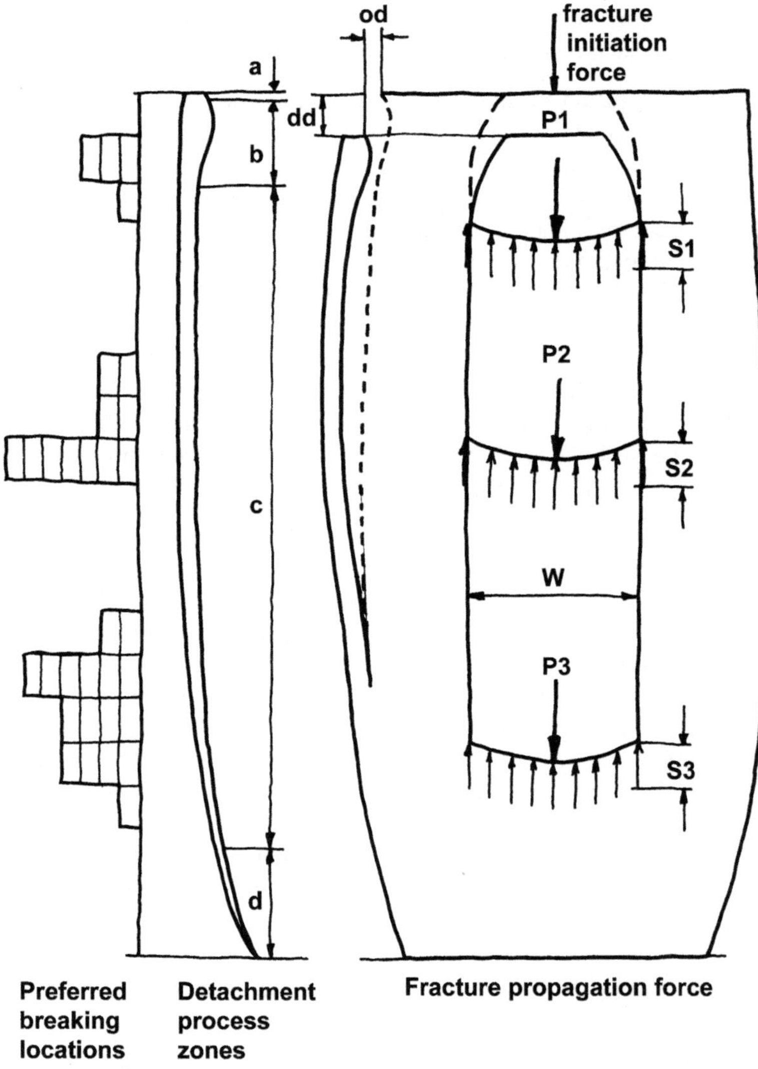

Fig. 19.6 From *left* to *right*: Frequency histogram of preferred breaking locations for a group of tests carried out in 2005. The four process zones of blade detachment. *Zone a*: crack initiation zone. *Zone b*: bulbar transition zone. *Zone c*: fracture propagation zone. *Zone d*: fracture termination zone. Distance *dd* downward displacement; *od* outward displacement of blade platform. Analysis of the fracture propagation force. P is equal to the sum of the ultimate shear strengths S along the fracture front. Provided that the core morphology and the material quality remain constant, these equations are valid: S1 = S2 = S3 and therefore P1 = P2 = P3 = CONSTANT

bulbar area, 11 breaks at approximately 40% of the blade length and 17 breaks at approximately 72% of the blade length (see the frequency histogram on the left side of Fig. 19.6). It must be borne in mind that this frequency histogram reflects only these above particular test series and not a general tendency to be expected in other circumstances.

The breakage rates outlined above proved to have been either the result of trying to detach blades that were too thin or of preparing and using a tool set-up with an unsuitable combination of downward and outward stored energy and follow-through in the two lever arms. The lessons are that every new 2C-machine requires fresh adjustments for each category of core length and that the most significant set-up parameters are the stored energy and flexibility of the lever arms, and each unsuitable set-up produces a number of 'most likely' or 'most frequent' breaking locations. Naturally, the probability of breaking increases with thinner blades and more brittle raw material (see Chap. 5 for further results).

With the newly adjusted larger 2C-machine, the question of fracture propagation could again be considered. The widespread opinion that longer blades require more force was not confirmed already during flintknapping sessions at the seminar on Mesoamerican lithic technology at Pennsylvania State University in May 2000. Subsequent investigations and measurements on prismatic blades have since clearly shown that the downward force required to detach longer blades is not directly correlated with blade length. The detachment of longer blades, however, takes more time and requires a longer lasting downward follow-through than that of shorter blades, so that the pressure bit can remain in permanent contact with the blade platform during the prolonged detachment time.

In order to elaborate further on this matter, let us look at the fracture front as it travels down from the platform to the tip, a few millimetres below the face of the core (see Fig. 19.6). To simplify the evaluations, the detachment process is divided into four zones since a different model is ideally suited to explain and describe the dynamics taking place in each zone of the blade. Furthermore, an intuitive, 'slow-motion view' will be applied to the detachment process, regardless of the fact that we are dealing with an extremely rapid process.

19.5.1 The Crack Initiation Zone A

A pressure blade can be initiated in two different ways, namely in the Hertzian mode (or shearing mode), as commonly seen in hard hammer percussion, or in the tension mode (sometimes incorrectly called bending mode), as known from the use of soft organic billets and punches (Fig. 19.6). Both fracture initiation modes are equivalent as far as this study is concerned, so that only the 'Hertzian mode' is assessed here. The basic relations were already described in 1886 with the mathematical formulas by Heinrich Hertz (see e.g. Bertouille 1989). In brief, the equations state that the crack starts directly around the contact area and develops in the shape of a complete cone. The crack initiates earlier when the ultimate material strength of the core is lower, when the contact diameter of the pressure bit is smaller and when an outward force is added to the downward load on the pressure bit. In the latter case, the crack will initially start on one side only. If the force vector is placed as an edge load close to the rim of a core, only a Hertzian half-cone will be observed on the bulb of the blade. This means that there exists no single fracture initiation force, but that this force may consist of any critical combination of a downward and an outward component. This theoretical deduction is confirmed by the actual experience that less downward force is required with 2C-machines when more outward force is applied. For pressure blades, the validity of the Hertzian formula is limited to the short crack initiation zone anyway because the whole theory was originally conceived for loads in the middle of a surface and not for edge loads like those applied to archaeological cores.

19.5.2 The Bulbar Transition Zone B

At the beginning of this zone, only the growing Hertzian half-cone exists, and the external fracture initiation force is in equilibrium with the internal reactions to the direct compression under the pressure bit plus the inclined tensions in both lateral wings (or ears) of the blade (Fig. 19.6). On the platform itself, the crack line has not yet reached the core rim. The situation is still static and reversible, which means that if the pressure bit is unloaded, the crack will immediately cease to expand. In order to make the crack advance further downwards, the flintknapper must continue to increase the crack initiation force until it finally reaches its highest value. Depending on the initial size and direction of the force vector, the flintknapper can do this by increasing just the outward component, or both the downward and outward components. Once the critical point is reached, the static fracture process changes suddenly and audibly to a dynamic fracture process, and the situation has become irreversible. From this point on, fracture dynamics has taken over, and the traditional laws of fracture mechanics are no longer all valid because time, speed, deformations, masses and inertias have entered into the picture. The growing bulb has reached its final size, and the blade is no longer partially suspended by its lateral wings and partially supported from below, as it had been in the first part of this zone. The crack line on the platform has reached, or even passed, the core rim, and the blade is about to attain its ultimate width.

19.5.3 The Fracture Propagation Zone C

This is by far the longest zone on each pressure blade, and this is where relevant new observations can be made (Fig. 19.6). Fine ripples and waves frequently occur across the ventral side of the blade or the dorsal side of the core. Striations (or hackles) can be seen on that lateral blade margin with the sharper blade angle. It is in this zone c (Fig. 19.6) that most breaks occur during detachment, sometimes even without leaving a trace on the core. As indicated by many tests, these breaks occur in a limited number of typical locations, depending on the core and set of tools (see the frequency histogram on the left side of Fig. 19.6).

When viewed from above, the shape of the fracture front is a curved line that does not change during the whole length of the fracture propagation zone, provided that the topography of the core remains constant. Furthermore, the foremost point of the fracture front tends to lie below the thickest part of the blade, while below the smaller blade edge angle, the front remains slightly recessed. These observations have far-reaching consequences.

Firstly, in order for the fracture to continue moving, the downward force under the pressure bit just needs to remain constant because it must never do more than overcome the constant shear resistance P along the curved fracture front line. This means that the fracture propagation force is independent of the blade length. On the other hand, the fracture propagation force P is directly proportional to the blade width because the sum of the local ultimate shear strengths S is proportional to the width W.

Secondly, distally narrowing blades require gradually less fracture propagation force as the fracture front travels further downwards because the blade width W also gradually decreases. This means that the fracture speed increases for distally narrowing blades. On the other hand, distally widening blades need more and more fracture propagation force that is not always provided by an extra amount of downward follow-through. The consequence of not providing sufficient follow-through can be an occurrence of mid-core and distal hinge or step fracture terminations.

Thirdly, in order to remain in permanent contact with the blade platform, the pressure bit must be displaced downwards instantaneously because stored energy and follow-through as such are not enough. Only this can compensate for the minute reduction in length due to the gradual shortening of the blade under pressure (see the distances dd and od in Fig. 19.6). The source of this instant downward follow-through is only that part of the stored energy in the tools that can be instantly released during the short duration of the fracture. This energy stems from the compressive and bending deformations of the levers or the muscle tension of the flintknapper directly behind the body loading tools.

19.5.4 The Fracture Termination Zone D

The blade leaves the core in this zone (Fig. 19.6). The longitudinal compression in the blade and the remaining bending deformations will generate a certain departure speed, a departure angle and an eventual departure rotation (tip over head). The details of the termination zone depend on the morphology of the core tip and the way the core is supported. Usually, this does not influence our study of the fracture propagation force so that the details of this zone, while manifold and interesting, are not dealt with in this chapter.

19.6 The Significance of Tuning

Tuning means creating for each core length and detachment device the correct interplay between the various tuning factors, such as the downward and outward components of the fracture initiation force, the masses, inertias, elasticities and deformations under load of the horizontal and vertical lever arms and the characteristics of the core fixation system, to mention but the most important elements.

Perfectly adjusted or tuned tools have always existed, but their tuning was hidden in the tool design and the knapping behaviour typical of each cultural tradition, including the manner of holding the tool. Tuning is also practised by experienced flintknappers today; however, they would call it ‘warming up’ or ‘getting used to’ a new core length or a different design of the pressure instrument. Therefore, the tuning issue was never directly addressed. However, today’s 1C- and 2C-machines do not ‘warm up’ on their own, so that the subject of tuning must be re-addressed as a separate issue for each new machine and core length category.

Inspired by the experiences outlined in Chap. 4, it seemed an interesting prospect to enter into the subject of tuning in a more systematic and theoretical manner. Again taking an intuitive, slow-motion approach to the detachment sequence, let us begin with a steadily increasing downward force and a constant, smaller outward force acting close to the edge of a prismatic core platform. The emerging blade shall be called a ‘column’ while the two force components initiate a downward moving crack, as well as a downward and an outward moving displacement of the top of that small column. Subsequently, this column grows in length; it is held or guided at the top by the pressure bit and embedded at the bottom in the core material. Furthermore, this column is standing next to a wall, namely the core face. I have built a large-sized imitation of this blade detachment mechanism, consisting of a rigid wooden frame in which various wooden columns of different stiffness can be placed. Simple lateral fixation devices allow the column length to be adjusted, and the downward and outward displacements can be determined by means of screws at the top (see again the distances od and dd in Fig. 19.6). This solid and straightforward model offered great assistance in visualizing the range of probable geometric blade deformations (Fig. 19.7). In order to further simplify this model, the core face was assumed to be straight, leaving the important effects of the core face curvature for later considerations. I am well aware that the use of models is not popular in all research circles, but experimenting in the field of flintknapping without explicit or implicit models is the equivalent of stabbing in the dark.

After having gained familiarity with the different manipulations and obtained some hands-on experiences with the device, systematic explorations of different downward and outward settings resulted in three clarifying observations[7]:

1. A pressure blade is different from the classical case of a column under buckling conditions, but rather behaves like a column under an oblique and inclined load plus a bending moment at the top. It is compressed bending, not buckling!
2. When the outward displacement (see the distance between o and d in Fig. 19.6) remains small (due to a slow outward follow-through by the pressure bit) and the

[7] Imaginative readers will recognize that this device opens a window to ‘experimental mathematics’ in order to solve the complex differential equations for the deformation of slender columns under downward and outward forces for various border conditions. These border conditions include bending moments at the top and time-dependent increases in the column length as well as the consequences of a wall at the side of the column.

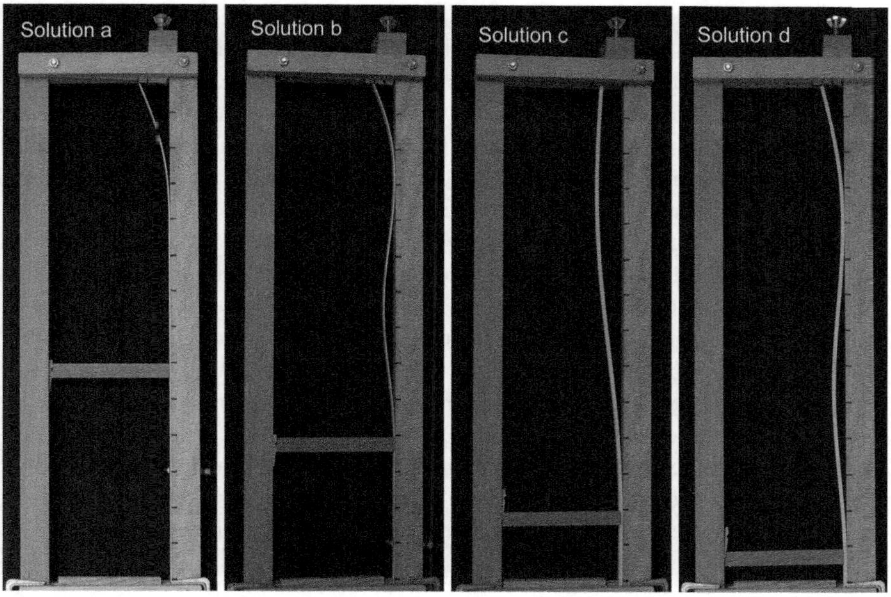

Fig. 19.7 Exaggerated visualization of possible geometries for the genesis of long blades with variations in the outward displacements of the blade platforms and differences in length and rigidity of the blades under development. Because breaks take time, occasional double breaks can occur, for example when proximal breaks are followed by mid-core breaks. *Solution a*: Short length with instant large out displacement results in preventing mid-core flexion, often leading to an immediate outward proximal bending break as step fracture or hinge termination. *Solution b*: Increased length with instant outward displacement results in mid-core flexion leading to a delayed outward mid-core bending break. *Solution c*: Increased length with medium to late outward displacement results in full length flexion leading to an immediate outward bending break. *Solution d*: Increased length with instant outward displacement results in mid-core flexion leading to a delayed outward mid-core bending break (All photos by P. Kelterborn)

column has reached a certain length, it can only bend outwards because the core face prevents an inward motion (Fig. 19.7c). In the context and terms of measurable flintknapping, I have termed this situation the 'immediate bending mode'.

3. Once the column has grown to the same (or even a smaller) length than in Fig. 19.7c, however with a comparatively much quicker outward displacement due to an instantaneous outward follow-through, the column will initially lean back onto the core, while the fracture front still continues to travel. Only under higher loads, or after a further increase of the column length, will the lower part of the blade once again begin to deflect to the outside, as in Fig. 19.7b, d. In the context and terms of measurable flintknapping, I have termed this situation the 'delayed bending mode'.

The purpose of learning the manual skills required to make long blades using traditional body tools was always to ensure that the blade detaches in the 'delayed bending mode'. The main variables were flexibility and follow-through of the chosen pressure instrument, the positioning of the body and hands which create the forces and bending deformations and the correct adjustment of the tool angle, until detachment initiation. The modern tuning of 1C- or 2C-machines requires balancing the ratios between the stored energies of the lever arms and their immediate availability amongst each other and the core lengths. The main variables are the flexibilities of the lever arms and the locations of the set-up weights.

In this early stage of research, and based on the model described above, the key insight is that the phenomenon of tuning exists, is very important and was obviously mastered in ancient times and by skilled flintknappers today when using their various sticks. The principal purpose of the downward follow-through was to maintain the fracture propagation force while the role of the outward follow-through was to fix or define the outward displacement of the blade platform in relation to the advancement of the fracture front, as visualized in Fig. 19.7. When using detachment machines, the newly recognized task is to design tools and select such set-up parameters that allow the correct follow-through to occur during the short fracture duration time and ensure detachment according to the delayed bending mode.

19.7 Summary and Perspectives

The motive for this study was the experience that unsolved problems create a serious barrier when trying to leave behind the blade lengths of 12–18 cm and venture into the 18–24-cm category. This chapter is a progress report on the research method called measurable flintknapping and the extended possibilities of the larger double lever machine. Compared with the methods used to date in established replicative flintknapping research, the following advantages stand out: Repeatability, measurability and working comfort. Relevant set-up parameters that produce no traces on the cores or blades can now also be recorded. The disadvantages are that detachment machines do not substitute considerable knowledge and experience in traditional flintknapping and cannot tune themselves but need to be tuned.

Three experimental projects were presented together with the results obtained so far:

1. The multifaceted question of longitudinal blade torsion was treated and a workable solution found in the reference line concept.
2. Contrary to my expectations, the fracture propagation force was found to be independent of the blade length, and the blade width appeared to be a major contributing factor. However, longer blades require longer lasting follow-through.
3. The complex task of tuning all the various set-up variables with regard to the double lever system could be reduced to just a few practical rules of thumb for choosing the correct degrees of stiffness and follow-through of the two lever arms. Because blades in the immediate bending mode will break earlier or under smaller loads than those in the delayed bending mode, the principal objective of

tool design and tuning is to ensure that the detachment process occurs in the delayed bending mode.

Many questions must remain unanswered for the time being. Depending on the problem at hand and whether an archaeologist, lithic analyst or a replicative flintknapper is asked for an opinion, priorities tend to shift. Amongst others, the following subjects are of interest to all three circles:

- The introduction of high-speed cameras into measurable flintknapping. It would be particularly interesting to improve on the simple models of the immediate and delayed bending modes currently in use.
- The detailed comparison between single and double lever machines.
- The development of improved techniques for the typical core maintenance tasks, such as maintaining, or even determining, the platform angle and core face curvature.

Acknowledgements I am deeply indebted to Tony Baker, John Clark, Ken Hirth, Bob Patten, Jacques Pelegrin and my family for their constant and loyal support during the step-by-step development of my research approach to measurable flintknapping. For the indispensible and constructive help with editing the final wording of this chapter, I am much obliged to Adrian Burke, John Clark, Pierre M. Desrosiers and Sandy Haemmerle.

References

Bertouille, H. 1989 *Theories Physiques et Mathematiques de la Taille des Outils Prehistoriques,* Cahiers du Quaternaire No. 15. Edition du Centre National de la Recherche Scientifique, Paris.

Kelterborn P. 2002 Measurable Flintknapping in *Experimentelle Archaeologie in Europa, Bilanz 2002*, edited by M. Fansa pp 35–49, Isensee Verlag, Oldenburg, BRD.

Kelterborn, P. 2003 Measurable Flintknapping. In *Mesoamerican Lithic Technology*, edited by Kenneth G. Hirth, pp. 120–131. University of Utah Press, Salt Lake City.

Pelegrin, J. 2003 Blade-Making Techniques from the Old World. In *Mesoamerican Lithic Technology*, edited by Kenneth G. Hirth, pp. 55–71. University of Utah Press, Salt Lake City.

Titmus, G. and J. Clark 2003 Mexica Blade Making with Wooden Tools. In *Mesoamerican Lithic Technology*, edited by Kenneth G. Hirth, pp. 72–97. University of Utah Press, Salt Lake City.

Wilke, P.J. 2007 Review of Mesoamerican Lithic Technology: Experimentation and Interpretation, edited by Kenneth Hirth. *Lithic Technology 32*: 219–232.

Index

P.M. Desrosiers (ed.), *The Emergence of Pressure Blade Making: From Origin to Modern Experimentation*, DOI 10.1007/978-1-4614-2003-3,

C

Q

R

GPSR Compliance

The European Union's (EU) General Product Safety Regulation (GPSR) is a set of rules that requires consumer products to be safe and our obligations to ensure this.

If you have any concerns about our products, you can contact us on ProductSafety@springernature.com

In case Publisher is established outside the EU, the EU authorized representative is:

Springer Nature Customer Service Center GmbH
Europaplatz 3
69115 Heidelberg, Germany

Batch number: 10837186

Printed by Printforce, the Netherlands